The World Muslim Population

The World Muslim Population
Spatial and Temporal Analyses

Houssain Kettani

Jenny Stanford
PUBLISHING

Published by

Jenny Stanford Publishing Pte. Ltd.
Level 34, Centennial Tower
3 Temasek Avenue
Singapore 039190

Email: editorial@jennystanford.com
Web: www.jennystanford.com

British Library Cataloguing-in-Publication Data
A catalogue record for this book is available from the British Library.

The World Muslim Population: Spatial and Temporal Analyses
Copyright © 2020 by Jenny Stanford Publishing Pte. Ltd.

All rights reserved. This book, or parts thereof, may not be reproduced in any form or by any means, electronic or mechanical, including photocopying, recording or any information storage and retrieval system now known or to be invented, without written permission from the publisher.

For photocopying of material in this volume, please pay a copying fee through the Copyright Clearance Center, Inc., 222 Rosewood Drive, Danvers, MA 01923, USA. In this case permission to photocopy is not required from the publisher.

ISBN 978-981-4800-31-0 (Hardcover)
ISBN 978-0-429-42253-9 (eBook)

In the Name of Allah, the Most Gracious, the Most Merciful

When the victory of Allah has come and the conquest (1)
And you see the people entering into the religion of
Allah in multitudes (2)
Then exalt [Him] with praise of your Lord and
ask forgiveness of Him.
Indeed, He is ever accepting of repentance (3)
Surat An-Nasr (The Victory, 110)

Contents

Preface xvii
Acknowledgments xix

1. **Introduction** 1
2. **Islam in Asia** 11
   - 2.1 Muslims in the Near East 14
     - 2.1.1 Armenia 16
     - 2.1.2 Azerbaijan 18
     - 2.1.3 Cyprus 23
     - 2.1.4 Georgia 28
     - 2.1.5 Iran 33
     - 2.1.6 Turkey 35
     - 2.1.7 Regional Summary and Conclusion 37
   - 2.2 Muslims in Central Asia 43
     - 2.2.1 Afghanistan 46
     - 2.2.2 Kazakhstan 47
     - 2.2.3 Kyrgyzstan 51
     - 2.2.4 Mongolia 54
     - 2.2.5 Tajikistan 55
     - 2.2.6 Turkmenistan 60
     - 2.2.7 Uzbekistan 64
     - 2.2.8 Regional Summary and Conclusion 67
   - 2.3 Muslims in Arabian Asia 74
     - 2.3.1 Bahrain 75
     - 2.3.2 Iraq 77
     - 2.3.3 Jordan 79
     - 2.3.4 Kuwait 82
     - 2.3.5 Lebanon 83
     - 2.3.6 Oman 85
     - 2.3.7 Palestine/Israel 86
     - 2.3.8 Qatar 90
     - 2.3.9 Saudi Arabia 92
     - 2.3.10 Syria 93

	2.3.11	United Arab Emirates	95
	2.3.12	Yemen	96
	2.3.13	Regional Summary and Conclusion	104
2.4		Muslims in Southeast Asia	105
	2.4.1	Brunei	107
	2.4.2	Cambodia	108
	2.4.3	Indonesia	110
	2.4.4	Laos	118
	2.4.5	Malaysia	119
	2.4.6	Myanmar (Burma)	123
	2.4.7	Philippines	127
	2.4.8	Singapore	136
	2.4.9	Thailand	140
	2.4.10	Timor-Leste (East Timor)	142
	2.4.11	Vietnam	143
	2.4.12	Regional Summary and Conclusion	145
2.5		Muslims in the Indian Subcontinent	152
	2.5.1	Bangladesh	154
	2.5.2	Bhutan	156
	2.5.3	India	157
	2.5.4	Maldives	161
	2.5.5	Nepal	169
	2.5.6	Pakistan	171
	2.5.7	Sri Lanka	174
	2.5.8	Regional Summary and Conclusion	179
2.6		Muslims in the Far East	186
	2.6.1	China	187
	2.6.2	Hong Kong	204
	2.6.3	Japan	206
	2.6.4	North Korea	209
	2.6.5	South Korea	210
	2.6.6	Macao	211
	2.6.7	Taiwan	214
	2.6.8	Regional Summary and Conclusion	215
2.7		Asia's Summary and Conclusion	228
3. Islam in Africa			**241**
3.1		Muslims in North Africa	244
	3.1.1	Algeria	246

	3.1.2	Chad	248
	3.1.3	Egypt	251
	3.1.4	Libya	253
	3.1.5	Mali	254
	3.1.6	Mauritania	256
	3.1.7	Morocco	257
	3.1.8	Niger	261
	3.1.9	Sudan	262
	3.1.10	Tunisia	263
	3.1.11	Regional Summary and Conclusion	265
3.2		Muslims in West Africa	272
	3.2.1	Benin	274
	3.2.2	Burkina Faso	276
	3.2.3	Cabo Verde	278
	3.2.4	Côte d'Ivoire (Ivory Coast)	280
	3.2.5	The Gambia	282
	3.2.6	Ghana	283
	3.2.7	Guinea	285
	3.2.8	Guinea-Bissau	287
	3.2.9	Liberia	289
	3.2.10	Senegal	291
	3.2.11	Sierra Leone	293
	3.2.12	Togo	295
	3.2.13	Regional Summary and Conclusion	297
3.3		Muslims in East Africa	304
	3.3.1	Comoros	306
	3.3.2	Djibouti	307
	3.3.3	Eritrea	309
	3.3.4	Ethiopia	310
	3.3.5	Kenya	312
	3.3.6	Madagascar	313
	3.3.7	Mauritius	316
	3.3.8	Mayotte	319
	3.3.9	Réunion	321
	3.3.10	Seychelles	322
	3.3.11	Somalia	323
	3.3.12	Tanzania	325
	3.3.13	Regional Summary and Conclusion	327
3.4		Muslims in Central Africa	334

		3.4.1	Burundi	336
		3.4.2	Cameroon	337
		3.4.3	Central African Republic	339
		3.4.4	Congo–Brazzaville	340
		3.4.5	Congo–Kinshasa	342
		3.4.6	Equatorial Guinea	343
		3.4.7	Gabon	344
		3.4.8	Nigeria	346
		3.4.9	Rwanda	347
		3.4.10	São Tomé and Príncipe	349
		3.4.11	South Sudan	350
		3.4.12	Uganda	351
		3.4.13	Regional Summary and Conclusion	353
	3.5	Muslims in Southern Africa		360
		3.5.1	Angola	361
		3.5.2	Botswana	363
		3.5.3	Eswatini (Swaziland)	364
		3.5.4	Lesotho	365
		3.5.5	Malawi	366
		3.5.6	Mozambique	369
		3.5.7	Namibia	371
		3.5.8	Saint Helena	372
		3.5.9	South Africa	376
		3.5.10	Zambia	379
		3.5.11	Zimbabwe	380
		3.5.12	Regional Summary and Conclusion	394
	3.6	Africa's Summary and Conclusion		394
4.	**Islam in Europe**			**407**
	4.1	Muslims in the Balkan Peninsula		410
		4.1.1	Albania	413
		4.1.2	Bosnia and Herzegovina	414
		4.1.3	Bulgaria	417
		4.1.4	Croatia	419
		4.1.5	Greece	422
		4.1.6	Kosovo	425
		4.1.7	Montenegro	427
		4.1.8	North Macedonia	429
		4.1.9	Romania	431

	4.1.10	Serbia	436
	4.1.11	Slovenia	441
	4.1.12	Regional Summary and Conclusion	443
4.2	Muslims in Eastern Europe		450
	4.2.1	Belarus	452
	4.2.2	Estonia	455
	4.2.3	Latvia	457
	4.2.4	Lithuania	460
	4.2.5	Moldova	463
	4.2.6	Russia	467
	4.2.7	Ukraine	477
	4.2.8	Regional Summary and Conclusion	483
4.3	Muslims in Western Europe		490
	4.3.1	Andorra	492
	4.3.2	Belgium	493
	4.3.3	France	496
	4.3.4	Gibraltar	500
	4.3.5	Luxembourg	501
	4.3.6	Monaco	504
	4.3.7	The Netherlands	506
	4.3.8	Portugal	510
	4.3.9	Spain	516
	4.3.10	Regional Summary and Conclusion	521
4.4	Muslims in Northern Europe		532
	4.4.1	Denmark	533
	4.4.2	Faroe Islands	537
	4.4.3	Finland	538
	4.4.4	Guernsey	541
	4.4.5	Iceland	542
	4.4.6	Ireland	544
	4.4.7	Isle of Man	546
	4.4.8	Jersey	547
	4.4.9	Norway	548
	4.4.10	Sweden	554
	4.4.11	The United Kingdom	557
	4.4.12	Regional Summary and Conclusion	560
4.5	Muslims in Central Europe		561
	4.5.1	Austria	569
	4.5.2	Czechia	572

		4.5.3	Germany	575
		4.5.4	Hungary	577
		4.5.5	Italy	580
		4.5.6	Liechtenstein	586
		4.5.7	Malta	587
		4.5.8	Poland	589
		4.5.9	San Marino	591
		4.5.10	Slovakia	593
		4.5.11	Switzerland	594
		4.5.12	Regional Summary and Conclusion	596
	4.6	Europe's Summary and Conclusion		597
5.	**Islam in the Americas**			**625**
	5.1	Muslims in Southern Caribbean Islands		627
		5.1.1	Aruba	629
		5.1.2	Barbados	630
		5.1.3	Caribbean Netherlands	632
		5.1.4	Curaçao	634
		5.1.5	Dominica	635
		5.1.6	Grenada	636
		5.1.7	Martinique	638
		5.1.8	Saint Lucia	639
		5.1.9	Saint Vincent and the Grenadines	641
		5.1.10	Trinidad and Tobago	643
		5.1.11	Regional Summary and Conclusion	645
	5.2	Muslims in North America		652
		5.2.1	Bermuda	653
		5.2.2	Canada	654
		5.2.3	Greenland	657
		5.2.4	Saint Pierre and Miquelon	661
		5.2.5	United States of America	662
		5.2.6	Regional Summary and Conclusion	672
	5.3	Muslims in Central Caribbean Islands		678
		5.3.1	Anguilla	679
		5.3.2	Antigua and Barbuda	681
		5.3.3	Guadeloupe	682
		5.3.4	Montserrat	684
		5.3.5	Saint Barthélémy	685
		5.3.6	Saint Kitts and Nevis	687

	5.3.7	Saint Martin	688
	5.3.8	Sint Maarten	690
	5.3.9	British Virgin Islands	691
	5.3.10	United States Virgin Islands	693
	5.3.11	Regional Summary and Conclusion	695
5.4	Muslims in South America		702
	5.4.1	Argentina	703
	5.4.2	Bolivia	704
	5.4.3	Brazil	706
	5.4.4	Chile	707
	5.4.5	Colombia	709
	5.4.6	Ecuador	711
	5.4.7	Falkland Islands (Islas Malvinas)	712
	5.4.8	French Guiana (Guyane)	714
	5.4.9	Guyana	716
	5.4.10	Paraguay	718
	5.4.11	Peru	719
	5.4.12	Suriname	720
	5.4.13	Uruguay	722
	5.4.14	Venezuela	724
	5.4.15	Regional Summary and Conclusions	725
5.5	Muslims in Northern Caribbean Islands		732
	5.5.1	The Bahamas	734
	5.5.2	Cayman Islands	735
	5.5.3	Cuba	736
	5.5.4	Dominican Republic	738
	5.5.5	Haiti	739
	5.5.6	Jamaica	740
	5.5.7	Puerto Rico	741
	5.5.8	Turks and Caicos	742
	5.5.9	Regional Summary and Conclusion	743
5.6	Muslims in Central America		750
	5.6.1	Belize	751
	5.6.2	Costa Rica	752
	5.6.3	El Salvador	753
	5.6.4	Guatemala	754
	5.6.5	Honduras	755
	5.6.6	Mexico	756
	5.6.7	Nicaragua	757

	5.6.8	Panama	758
	5.6.9	Regional Summary and Conclusion	770
5.7		The Americas' Summary and Conclusion	770

6. Islam in Oceania — 783

6.1	Muslims in Australasia			785
	6.1.1	Australia		786
		6.1.1.1	Antarctica	794
		6.1.1.2	Christmas Island	795
		6.1.1.3	Cocos (Keeling) Islands	796
		6.1.1.4	Norfolk Island	798
	6.1.2	New Zealand		800
	6.1.3	Regional Summary and Conclusion		807
6.2	Muslims in Melanesia			807
	6.2.1	Fiji		810
	6.2.2	New Caledonia		812
	6.2.3	Papua New Guinea		814
	6.2.4	Solomon Islands		816
	6.2.5	Vanuatu		817
	6.2.6	Regional Summary and Conclusion		819
6.3	Muslims in Micronesia			823
	6.3.1	Federated States of Micronesia		824
	6.3.2	Guam		825
	6.3.3	Kiribati		826
	6.3.4	Marshall Islands		828
	6.3.5	Nauru		829
	6.3.6	Northern Mariana Islands		830
	6.3.7	Palau		832
	6.3.8	Regional Summary and Conclusion		833
6.4	Muslims in Polynesia			840
	6.4.1	Cook Islands		840
	6.4.2	French Polynesia		842
	6.4.3	Niue		843
	6.4.4	Samoa		844
	6.4.5	American Samoa		846
	6.4.6	Tokelau		847
	6.4.7	Tonga		848
	6.4.8	Tuvalu		849

	6.4.9 Wallis and Futuna	851
	6.4.10 Regional Summary and Conclusion	858
6.5	Oceania's Summary and Conclusion	858

7. World Summary — **867**

Index — 885

Preface

The birth of Islam over fourteen centuries ago was a monumental event in human history with an everlasting effect on humanity. For centuries researchers contemplated on the growth and distribution of Muslims throughout the world. The purpose of this manuscript is to present a reliable estimate of the world Muslim population since the inception of Islam at the start of the seventh century to the end of the twenty-first century. In this book, the world is divided into five continents, each is divided into non-overlapping regions, and these in turn are divided into current countries. A centennial data estimate for each region and current country from 600 AD to 2100 AD (approximately 1 H to 1500 H) of the total population, and corresponding Muslim population and its percentage is provided. Furthermore, the same data in decennial order from 1790 to 2100 (or 1210 H to 1520 H) is provided for each region and country. These data are summarized to be a reference for other studies and discussions related to the Muslim population. The presented data show that the percentage of world Muslim population with respect to the total world population has increased steadily from 3% in 700 AD or 100 H to 7% in 800 AD or 200 H, to 11% in 900 AD or 300 H, to 13% in 1000 AD or 400 H, reaching 16% in 1700 AD or 1100 H. But it dropped to 13% in 1800 AD or 1200 H, to increase to 14% in 1900 AD or 1300 H. This percentage has been increasing by one percentage point per decade since 1950 AD or 1370 H, reaching 25% in 2020 AD or 1440 H. The rate of increase of the world Muslim population is expected to slow down, increasing their percentage to 30% by 2050 AD or 1470 H and 35% by 2100 AD or 1520 H.

Acknowledgments

I begin by saying, "Praise to Allah, who has guided us to this; and we would never have been guided if Allah had not guided us. Certainly, the messengers of our Lord had come with the truth" (Qur'an V.7:43).

The idea of this work stemmed from a conversation with my brother Dr. Hamza Kettani in March 2009 (Rabi I 1430 H) and motivated by my father's work and writings on Muslim minorities throughout the world. This led to five publications and a keynote speech at the International Conference on Social Sciences and Humanities, Singapore, October 2009, with the titles "Muslim Population in Asia," "Muslim Population in Africa," "Muslim Population in Europe," "Muslim Population in the Americas," and "Muslim Population in Oceania," followed by another publication titled "2010 World Muslim Population," at the 8th Hawaii International Conference on Arts and Humanities, Honolulu, Hawaii, January 2010. The work was then expanded into five publications at the *International Journal of Environmental Science and Development* (IJESD), Vol. 1, No. 2, June 2010, with the titles "Muslim Population in Asia: 1950–2020," "Muslim Population in Africa: 1950–2020," "Muslim Population in Europe: 1950–2020," "Muslim Population in the Americas: 1950–2020," and "Muslim Population in Oceania: 1950–2020." These papers received over 100 citations by researchers worldwide and multiple inquiries, which beckoned a more comprehensive work culminating in the first edition of this book in 1435 H: Kettani, H. (2014), *The World Muslim Population: History and Prospect*, Research Publishing Service, Singapore. In this book, we incorporate the latest census and survey data and estimate the total population from 1950 to 2100 based on the 2017 revision of the United Nations World Population Prospects (UNP).

I am in debt to my father Dr. Ali Kettani (1941–2001)/(1360–1422 H) for his upbringing and role model. His love and dedication to Muslims made an everlasting impression on his children and those who knew him. Without that I would not have been able to dedicate the time and effort to this work. I am also grateful to my mother,

Nezha Kettani, for her support to my father since their marriage in 1969/1389 H, and her upbringing of her four children: Hassna'a, Hassan, Hamza, and Houssain. I am grateful to her continuous support and motivation to me throughout my life and during my preparation of this work. Special thanks to my wife Mouna and to my children Ola, Ali, Salwa, and Abdullah for bringing joy to my life and their support, patience, and involvement throughout the years I spent on this project. I pray that they continue carrying this flame of serving Muslims and Islam.

This work is dedicated to those who sacrificed to protect and spread the message of Islam throughout the world and the centuries. Indeed, "And those who strive for Us, We will surely guide them to Our ways. And indeed, Allah is with the doers of good" (Qur'an V. 29:69).

Chapter 1

Introduction

The birth of Islam over fourteen centuries ago was a monumental event in human history with an everlasting effect on humanity. For centuries researchers contemplated on the growth and distribution of Muslims throughout the world [JAN, KET86, MAS, PEW]. A history of the spread of Islam until the fourteenth century was presented by Ismail Ibnu Kathir (704 H/1301 to 774 H/1373) in his book *Al-Bidaya wa Nihaya* (The Beginning and the End) [IIK], but it lacks a record of the numbers of Muslims. Nevertheless, we use such history to infer the percentage of Muslims in a particular country. However, the first global comprehensive work was done by Hubert Jansen (1854–1917) in 1897. He was followed by many others, including Louis Massignon (1883–1962) in 1923, Ali Kettani (1941–2001) in 1986 and Pew Research Center in 2009. All their work was concerned with the "current" number of Muslims worldwide or their increase within one to three decades. In contrast, the first comprehensive estimate of the world Muslim population since the inception of Islam around 600 to 2300 was introduced by Houssain Kettani in 2014 [KET14]. In the latter, the spread of the Muslim population, change in numbers, and distribution throughout the world and within current political boundaries was analyzed. The purpose of this book is to present a comprehensive and updated estimate of the world Muslim population since the inception of Islam at the start of the seventh century to the end of the twenty-first century, tracking its spread

---

The World Muslim Population: Spatial and Temporal Analyses
Houssain Kettani
Copyright © 2020 Jenny Stanford Publishing Pte. Ltd.
ISBN 978-981-4800-31-0 (Hardcover), 978-0-429-42253-9 (eBook)
www.jennystanford.com

and change in numbers and distribution throughout the world and within current political boundaries.

To begin, we need to distinguish a Muslim from non-Muslims as follows: A Muslim is a person who believes in one god and that prophet Mohammed is His messenger. The God or "Allah" in Arabic is the one who created us, all other creatures and the Universe. The distinction between god and The God is the fact that for someone who worships something, that thing is his god, but Allah is The God worthy of worship as He Is the one who created us and the Universe. In other words, a Muslim is someone who bears witness that "there is no god but Allah, and that Mohammed is His messenger" or in Arabic "la ilaha illa lah, mohammadun rasulu lah." Muslims believe that Mohammed is the final messenger of God in a chain of human Messengers to humanity that started with Adam and included Noah, Abraham, Moses and Jesus. Prophet Mohammed peace and blessings upon him (PBUH) was born in Mecca in Rabi I 53 BH (before Hijra) or May 570 AD. In Ramadan, 13 BH, or September 609 AD, at age forty in lunar years, the Prophet received the first revelation from God through the angel Gabriel while he was praying in Hira Cave at Thawr Mountain near Mecca. This revelation started with the word "read" from Sura 96 (al-Alaq), and by which the religion of Islam has started. Three years later, he was commanded by God to preach Islam and the number of Muslims was in the tens.

As the number of Muslims increased, the pagans of Mecca started persecuting them to abandon Islam. Accordingly, in Rajab 8 BH, or May 614 AD, sixteen Muslims migrated to Abyssinia (currently Ethiopia and Eritrea), where they were protected by its king, an-Najashi, who has also accepted Islam later. They were followed by 101 Muslims later in the same year. By Muharram 7 H, or May 628 AD, all those Muslims returned to Medina. The persecution of Muslims reached its extreme when Meccan pagans plotted to assassinate the Prophet (PBUH). Thus, in Rabi I 1 H, or September 622 AD, the Prophet entered the city of Medina, migrating from Mecca, which is referred to as Hijra, or the Migration. This was the start of the first Muslim state, and the Muslim (Hijri) Calendar, which is lunar and therefore about eleven days shorter than the Gregorian calendar which is solar. By now, the number of Muslims was in the hundreds.

In Ramadan 2 H, or March 624 AD, the Muslims won their first war with Meccan pagans at the Battle of Badr. After this victory, the number of Muslims was in the thousands. In 7 H and 8 H, or 628 AD and 629 AD, the Prophet sent messengers to nearby kings calling them to accept Islam. Those accepting Islam were Kings of Bahrain (currently Kuwait, Bahrain, Qatar, and East of Saudi Arabia), Oman (currently Oman and UAE), and Yemen. In Ramadan 8 H, or January 630 AD, the Muslim troops numbering 10,000, led by Prophet Mohammed (PBUH) conquered Mecca, and he pardoned its residents. By now, the number of Muslims was in the tens of thousands, or 0.01% of the world population, which is estimated at quarter of a billion.

Prophet Mohammed (PBUH) died in Rabi I 11 H, or June 632 AD, at age 62, at Medina, where he was buried. By then, the Arabian Peninsula was under Muslim control, and the number of Muslims was in the hundreds of thousands, or around 0.1% of the world population. The Prophet was followed by his first Caliph, Abu Bakr bnu Abi Quhafa, AKA Abdullah bnu Othman, who passed away in Jumada II 13 H or August 634 AD at age of 61, at Medina, where he was buried next to the Prophet (PBUH). During his two-year reign, Muslims expanded to West Iraq and Jordan. He then was followed by the second Caliph, Omar bnul Khattab, who passed away after an assassination attempt in Thul Hijja 23 H or November 644 AD at age 65, at Medina, where he was buried next to the Prophet (PBUH). During his ten-year reign, Muslims expanded to Syria, Lebanon, East Iraq, Iran, Palestine, Egypt, Libya, East Turkey, Azerbaijan, Armenia, Southeast Georgia, Afghanistan, Pakistan and South Turkmenistan. By now, the number of Muslims was in the millions or around 1% of the world population and continued to increase in numbers since then.

In this study, we track the change of the world Muslim population since the inception of Islam till 2100 AD or 1520 H. The global data show that the percentage of Muslims with respect to the world population has likely increased from 3.5% in 700 AD or 100 H to 7.2% in 800 AD or 200 H, to 10.5% in 900 AD or 300 H, to 13.1% in 1000 AD or 400 H and increased slowly to 16.1% by 1700. The Muslim population dropped to 13.0% in 1800 AD or 1200 H, to increase to 13.7% in 1900 AD or 1300 H. However, a remarkable trend happened after World War II, by which the rate of increase in each decade became over one percentage point, and the trend is expected to continue throughout this century. This caused the percentage of Muslims with respect to the world population to reach 25.1% in 2020/1440 H. Thus, Muslims increased from one out of eight of the world population in 1850/1270 H, to one out of seven in 1950/1370 H, to one out of six in 1970/1390 H, to one out of five in 1990/1410 H, to one out of four in 2020/1440 H, to one out of three by 2080/1500 H.

Official censuses inquiring on religious affiliation were considered in this study to be the most reliable. We exclude non-valid data such as missing, no response or refused to answer from the total of the population. Also, some censuses have a cut off at how old the respondent should be for religious data. Some start at five, twelve, sixteen or eighteen years old. However, not all censuses inquire on religious adherence, nor are held regularly. Thus, our second source of estimates is surveys that inquire on religion. These surveys interview a few numbers of people (hundreds to thousands) in a country. The data is then weighted to account for bias and underestimation, and excludes non-valid data such as missing, no response or refused to answer. Some common surveys are:

- Demographic and Health Surveys (DHS), prepared by ICF International since 1984. The survey typically includes 10,000 people aged 15 to 49, two-thirds of whom are female. We obtain the religion for men and women separately then take the average of the two.
- European Values Study (EVS), prepared by Faculty of Social and Behavioural Sciences at Tilburg University, The Netherlands in 1981, 1990, 1999 and 2008. The survey typically includes 1,000 people aged 18 and older.
- European Social Survey (ESS), prepared by the Centre for Comparative Social Surveys at City University London, The United Kingdom. The survey typically includes 1,000 people aged 15 and older and is conducted every two years since 2001.
- World Values Survey (WVS), prepared by WVS Association at the University of Aberdeen, The United Kingdom since 1981. The survey typically includes 1,000 people aged 18 and older.
- International Social Survey (ISS), prepared by the ISS Programme since 1985 and typically includes 1,000 people aged 18 and older.

- East Asia Barometer Survey (EAB), prepared by the Department of Political Science at the National Taiwan University, since 2001 and typically includes 1,500 people aged 18 and older.
- Asia Barometer Survey (ABM), prepared by the University of Niigata Prefecture, Tokyo, Japan, between 2003 and 2008 and typically includes 800 to 1,000 people.
- Asian Barometer Survey (ANM), prepared by the Center for East Asia Democratic Studies at the National University of Taiwan since 2001 and typically includes 800 to 1,200 people.
- Afro Barometer Survey (ABS), carried out through a partnership of the Institute for Justice and Reconciliation (IJR) in South Africa, the Ghana Center for Democratic Development (CDD-Ghana), Institute for Empirical Research in Political Economy (IREEP) in Benin, Institute for Development Studies (IDS) at the University of Nairobi, the Democracy in Africa Research Unit in the Centre for Social Science Research at the University of Cape Town and the Department of Political Science at Michigan State University. The survey is held since 2002 and typically includes 1,200 people aged 18 and older.

When no census or survey is available, we resort to other estimates from reliable sources applied to the total population obtained in the nearest census. Thus, in individual country data tables, the year and total population is based on census data with exception of countries where census is not taken for decades. In addition, the last column in these tables provides the reference on which the Muslim population estimate is based. Next to the reference, letters "c," "e," and "es" is added to indicate that the estimate is based on census where religious adherence is inquired, or ethnic census from which we can deduce religious affiliation, or just an estimate, respectively.

We divide each continent into various regions, and then sort countries that belong to each region in alphabetical order. These regions are sorted in terms of the percentage of Muslims from highest to lowest. Centennial data from 600 AD to 2100 AD (or approximately 1 H to 1500 H) and decennial data from 1790 AD to 2100 AD (or 1210 H to 1520 H) for each region and each current country in this region are presented in a table. History of the beginning of Islam in the corresponding country is discussed in a separate section. The data includes total population in thousands (P), the percentage of which is Muslim ($M\%$), the corresponding Muslim population in thousands (M), and the annual population growth rate (APGR, or $G\%$) of the total population in this region. The latter is calculated as $\ln(P_2/P_1)/(t_2 - t_1)*100$, where P_i is the population in year t_i and ln is the natural logarithm.

The total population estimate in each country since 1950 is based on the United Nations' World Population Prospects [UNP], while pre-1950 data is based on the work of Angus Maddison (1926–2010) in [MAD], which was based on the historical economies of the corresponding regions. Other estimates such as [AVA, OCE, PSH] and census data are used to fill in missing data from Maddison's work. Ancient census data were taken only in some major cities in Europe and the Middle East. Reliable censuses for extended geographical areas were only conducted in Europe starting in the eighteenth century. While global population data is only close to reliable after 1950. Nevertheless, most population models, including the ones adopted in this study, assume that global population has been increasing constantly for the last several millenniums, although acknowledging that regionally, populations did grow and decline following a cyclical path. Substantial population growth started after 1800 when world population reached one billion. The increase was due to improved health care, resulting in decrease in mortality rate and increase in life expectancy. Our estimate of the percentage of Muslims in each country prior to the nineteenth century is based on the history of the spread of Muslims and their ruling of the corresponding region.

Current area and map of each country is also presented in the country's corresponding section to illustrate its location, political boundaries, and neighboring countries. These maps are mostly obtained from the World Factbook [CIA]. Data for Asia, Africa, Europe, the Americas, and Oceania are considered in Chapters 2, 3, 4, 5, and 6, respectively. Chapter 7 presents a summary of the data for the whole world. Every attempt is sought to present reliable data, however, the statistics presented in this book, in the words of the

French demographer Jean-Baptiste Moheau (1745–1794) [MOH]: "These estimates based on likelihoods, constitute a first step to the truth, and the only proper way to criticize them, is to displace them by more accurate ones."

References

[ABS] Afrobarometer (2015). *Afrobarometer Online Data Analysis*. Retrieved from http://www.afrobarometer-online-analysis.com/

[ABM] Inoguchi, T. (2015). *AsiaBarometer Survey Data* [Computer Files]. Retrieved from http://www.asiabarometer.org/

[ANB] Center for East Asia Democratic Studies (2017). *Asian Barometer Survey Data* [Computer Files]. Retrieved from http://www.asianbarometer.org/

[AVA] Avakov, A. V. (2015). *Two Thousand Years of Economic Statistics, Years 1-2012: Population, GDP at PPP, and GDP Per Capita* (Vol. 2: by Country). New York, NY: Algora Publishing.

[CIA] Central Intelligence Agency (CIA) (2019). *The World Factbook*. Washington, DC: CIA. Retrieved from https://www.cia.gov/library/publications/the-world-factbook/

[DHS] ICF International (1984–2018). *The Demographic and Health Surveys (DHS) Program*. Rockville, MD: ICF International. Retrieved from http://dhsprogram.com/

[EAB] East Asia Barometer Network at National Taiwan University (2015). *East Asia Barometer Database*. Madrid: ASEP/JDS. Retrieved from http://www.jdsurvey.net/eab/

[ESS] Centre for Comparative Social Surveys at City University London [The United Kingdom] (2015). *European Social Survey Data*. Bergen: Norwegian Social Science Data Services (NSD). Retrieved from http://nesstar.ess.nsd.uib.no/. More info on ESS is available at http://www.europeansocialsurvey.org/

[EVS] European Values Study Foundation at Tilburg University [The Netherlands] (2015). *European Values Study, ZA4804: EVS 1981-2008 Longitudinal Data File*. Cologne: Leibniz Institute for the Social Sciences (GESIS). Retrieved from http://zacat.gesis.org/. More info on EVS is available at http://www.europeanvaluesstudy.eu/

[IIK] Ibnu-Kathir, I. (2003). *The Beginning and the End* [Arabic]. Riyadh: Dar Alam al Kutub.

[ISS] International Social Survey Programme (2015). *International Social Survey*. Cologne: Leibniz Institute for the Social Sciences (GESIS). Retrieved from http://zacat.gesis.org/. More info on ISS is available at http://www.issp.org/

[JAN] Jansen, H. (1897). *Verbreitung des Islâms: mit angabe der verschiedenen riten, sekten und religiösen bruderschaften in den verschiedenen ländern der erde 1890 bis 1897. Mit benutzung der neuesten angaben (zählungen, berechnungen, schätzungen und vermutungen). [German for Dissemination of Islam: with number of various rites, sects and religious brotherhoods in the various countries of the earth 1890 to 1897. With use of information on the latest disclosures (censuses, calculations, estimates and guesses)]*. Berlin: Selbstverlag des Verfassers.

[KET14] Kettani, H. (2014). *The World Muslim Population, History & Prospect*. Singapore: Research Publishing Services.

[KET86] Kettani, A. (1986). *Muslim Minorities in the World Today*. London: Mansell Publishing.

[MAD] Maddison, A. (2006). *The World Economy*. Paris: Organisation for Economic Co-operation and Development (OECD) Publishing.

[MAS23] Massignon, L. (1923). *Annuaire du Monde Musulman, Statistique, Historique, Social et Économique*. Paris: Ernest Leroux.

[MOH] Moheau, J-B. (1778). *Recherches et Considérations sur la population de la France.* (Ch. VI. Idée de la Population de la France). Paris: Chez Moutard, Imprimeur-Libraire de la Reine. The original quote in French is: "D'autres ont fait cette répartition un peu différemment: les uns et les autres peuvent soutenir la vérité de leurs conjectures. Nous ne sommes pas plus disposés à leur accorder une grande confiance, qu'en état de les contredire: nous ne disconviendrons pourtant pas que ces estimations fondées sur des vraisemblances, forment un premier pas vers la vérité, et que la seule manier juste de les critiquer, est d'en donner de plus exactes. Il en est comme des cartes anciennes des extrémités du globe où l'on n'avoitpoint encore pénetré elles ont été utiles, nécessaires même, jusqu'à ce que de nouvelles découvertes nous aient mis à portée de les rectifier."

[OCE] Caldwell, J., Missingham, B., & Marck, J. (2001). The Population of Oceania in the Second Millennium. *Proceedings of the International Union for the Scientific Study of Population's Seminar on the History of World Population in the Second Millennium*, Florence, Italy, pp. 1–34.

[PEW] Pew Research Center (PRC) (2009). *Mapping the Global Muslim Population, A Report on the Size and Distribution of the World's Muslim Population*. Washington, DC: PRC.

[PSH] Lahmeyer, J. (2006). Population Statistics: Historical Demography of All Countries, Their Divisions and Towns [Online Database]. Retrieved from www.populstat.info

[UNP] United Nations, Department of Economic and Social Affairs, Population Division (2019). *World Population Prospects 2019* (Online Edition). New York, NY: United Nations.

[WVS] World Values Survey Association at the University of Aberdeen [The United Kingdom] (2015). *World Values Survey Online Data Analysis.* Retrieved from http://www.worldvaluessurvey.org/WVSOnline.jsp

Chapter 2

Islam in Asia

Islam has started in Asia which explains why more than two-thirds of the world Muslim population resides in this continent. Islam started in Mecca in 609 AD, and then was established in Medina in 622 AD which marks the first Hijri year. Both cities are located in Hijaz, west of Saudi Arabia. By the death of Prophet Muhammad (PBUH) in 632 AD, Islam was ruling all of the Arabian Peninsula. By the death of his second Caliph Omar bnul Khattab in 644 AD, most of the Near East, parts of the Indian Subcontinent and Central Asia were under Muslim control. Islam continued spreading in Asia but at a much slower rate.

Significant portion of Central Asia and part of China was conquered during the reign of the sixth Umayyad Caliph al-Walid Ibnu Abdel Malik bnu Marwan who ruled from 705 to 715. Islam spread to the rest of Central Asia and Russia when the Mongols adopted Islam as their religion in the first half of the fourteenth century. Most of the Indian Subcontinent was under Muslim control by the end of the twelfth century. Islam spread in the rest of Asia with trade and preaching. It spread through much of the Malay Archipelago starting the twelfth century and by the sixteenth century it became the dominant religion. Islam only reached the Korean Peninsula and Japan during the twentieth century.

The World Muslim Population: Spatial and Temporal Analyses
Houssain Kettani
Copyright © 2020 Jenny Stanford Publishing Pte. Ltd.
ISBN 978-981-4800-31-0 (Hardcover), 978-0-429-42253-9 (eBook)
www.jennystanford.com

Thus, the Muslim population has likely increased from 7.6 million or 4.3% of the total Asian population in 700 AD, to 12.7 million or 7.0% in 800 AD, to 17 million or 9.2% in 900 AD, to 21 million or 11.3% in 1000 AD, to 25 million or 12.0% in 1100 AD, to 29 million or 12.8% in 1200 AD, to 35 million or 14.2% in 1300 AD, to 40 million or 15.1% in 1400 AD, to 47 million or 16.7% in 1500 AD, to 60 million or 15.9% in 1600 AD, to 70 million or 17.3% in 1700 AD, to 92 million or 13.9% in 1800 AD, to 162 million or 17.9% in 1900, to 0.93 billion or 24.8% in 2000, to 1.3 billion or 28.4% in 2020, and is projected to reach 1.8 billion or 33% by 2050 and 1.9 billion or 40% by 2100.

A plot of centennial estimates of the Muslim population and its percentage with respect to the total population in Asia from 600 to 2100 is provided in Fig. 2.0a. A zoom in of this plot, providing a plot of decennial estimates of the Muslim population and its percentage with respect to the total population in Asia from 1900 to 2100 is provided in Fig. 2.0b. This shows that the Muslim population in Asia will continue its sharp increase until 2080, at rate of over one percentage point per decade with respect to the total Asian population. It will peak slowly at 1.9 billion in 2080, and then it is estimated to decrease slightly in the remainder of this century. The Muslim population would have increased between 1920 and 2070 by ten folds, and double between 2000 and 2070. The percentage of Muslims in this continent will continue its substantial increase throughout this century, doubling between 1960 and 2100.

We divided Asia into six regions; the data for each is included in a separate section and are sorted in terms of the percentage of Muslims in descending order. These regions are the Near East (Section 2.1), Arabian Asia (Section 2.2), Central Asia (Section 2.3), Southeast Asia (Section 2.4), the Indian Subcontinent (Section 2.5), and the Far East (Section 2.6). The country of Russia was not included in Asia as most of its population lives in the European side of the country, although most of its territory is in Asia. Also, the European side of Turkey is included here. In Section 2.7, the total population in each of the six Asian regions and the corresponding percentage and number of

Muslims is presented centennially in Tables 2.7a and 2.7b from 600 to 2100, and decennially in Tables 2.7c to 2.7f from 1790 to 2100.

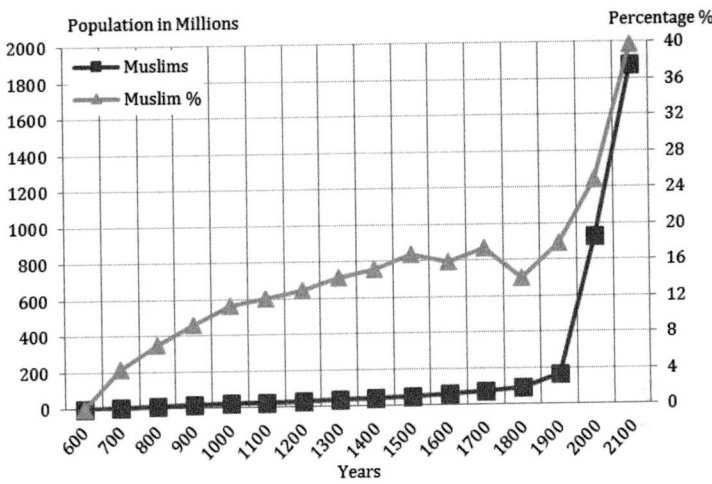

Figure 2.0a Plot of centennial estimates of the Muslim population and its percentage of the total population in Asia: 600–2100 AD (1–1500 H).

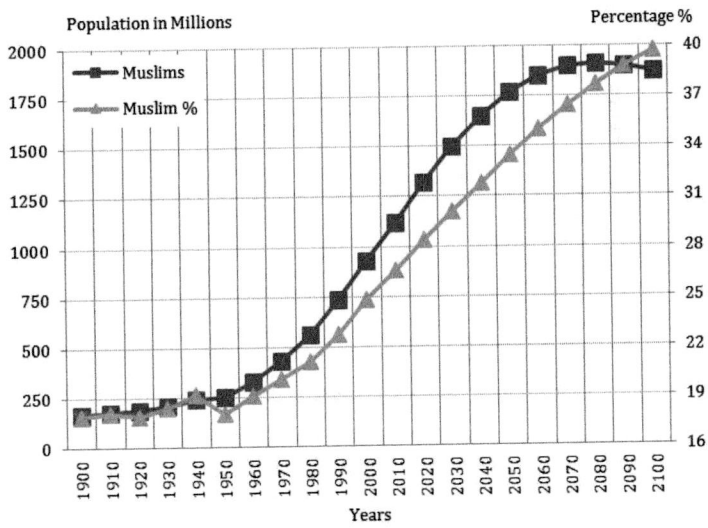

Figure 2.0b Plot of decennial estimates of the Muslim population and its percentage of the total population in Asia: 1900–2100 AD (1320–1520 H).

2.1 Muslims in the Near East

This region consists of six countries: Armenia, Azerbaijan, Cyprus, Georgia, Iran, and Turkey. Islam entered this region during the reign of the second Caliph Omar bnul Khattab from 17 H to 24 H, or 638 AD to 644 AD, when Muslims captured Iran, parts of the southern Caucasus, and southern and eastern part of Turkey. Muslims conquered the Island of Cyprus in 27 H or 647 AD under the reign of Caliph Muawiya bnu Abi Sufyan, the first ruler of the Umayyad Empire. They then continued advancing slowly northward, gaining more territory from the Byzantine Empire, until they conquered the European side of Istanbul in 1453, then Trabzon in 1461 AD during the reign of the Ottoman Sultan Mohammed II el-Fatih ben Murad II who ruled from 1451 AD to 1481 AD.

The Muslim population has likely increased from 1.3 million or 10.3% of the total population of this region in 700 AD, to 3.0 million or 24.5% in 800 AD, to 4.6 million or 38.3% in 900 AD, to 5.8 million or 48.2% in 1000 AD, to 6.9 million or 57.8% in 1100 AD, to 7.9 million or 66.8% in 1200 AD, to 8.4 million or 72.5% in 1300 AD, to 9.0 million or 78.1% in 1400 AD, to 9.5 million or 83.7% in 1500 AD, to 11.9 million or 83.6% in 1600 AD, to 12.4 million or 83.3% in 1700 AD, to 14.5 million or 82.1% in 1800 AD, to 23 million or 78.4% in 1900, to 137 million or 94.3% in 2000, to 179 million or 95.7% in 2020, and is projected to reach 212 million or 97% by 2050 and 194 million or 97% by 2100.

A plot of centennial estimates of the Muslim population and its percentage with respect to the total population in this region from 600 to 2100 is provided in Fig. 2.1a. A zoom in of this plot, providing a plot of decennial estimates of the Muslim population and its percentage with respect to the total population in this region from 1900 to 2100 is provided in Fig. 2.1b. This shows that the Muslim population in this region will continue its sharp increase, peaking at over 214 million in 2060, then start decreasing slowly afterwards to below 195 million by 2100. However, the percentage of Muslims in this region will continue its tiny increase throughout this century, reaching 97% by 2100. A spike in the percentage of Muslims occurred between 1920 and 1930, with an increase of about ten percentage points. This is due to the abolishment of the Ottoman Empire, and the subsequent departure of non-Muslims to the newly established non-Muslim states such as Armenia and Greece, and the expulsion of Muslims to Turkey.

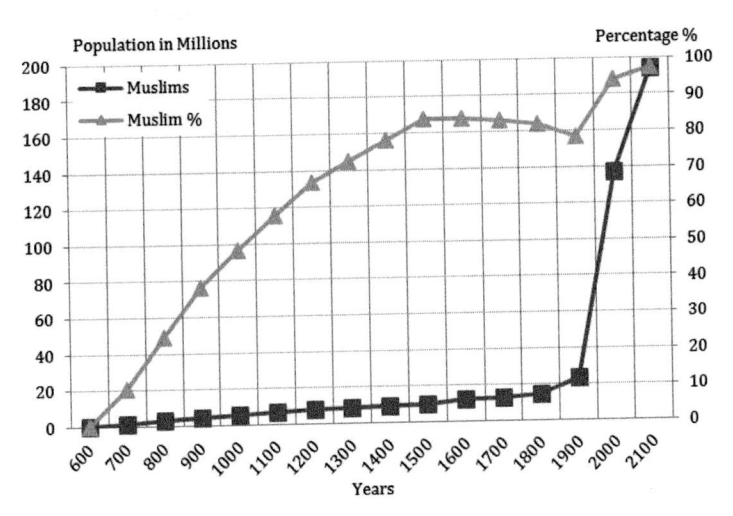

Figure 2.1a Plot of centennial estimates of the Muslim population and its percentage of the total population in the Near East: 600–2100 AD (1–1500 H).

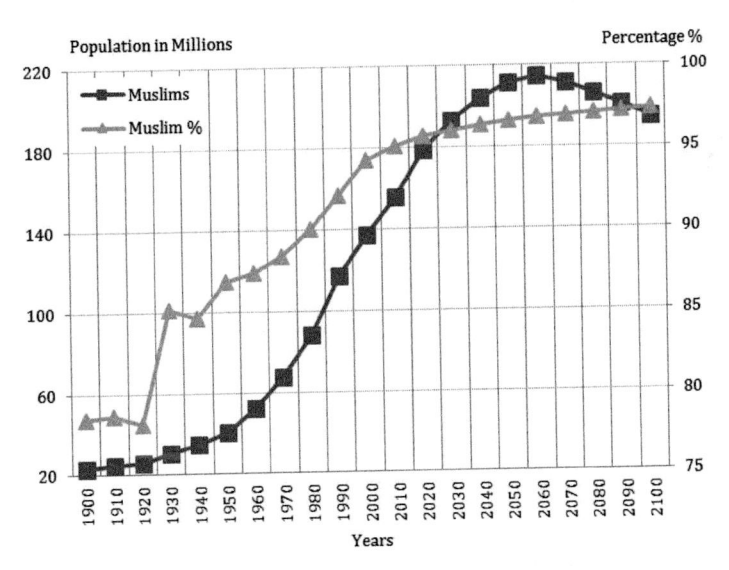

Figure 2.1b Plot of decennial estimates of the Muslim population and its percentage of the total population in the Near East: 1900–2100 AD (1320–1520 H).

The corresponding individual data for each country in this region is discussed below. In Section 2.1.7, the total population in each country in this region and the corresponding percentage and number of Muslims is presented centennially in Table 2.1a from 600 to 2100 and decennially in Tables 2.1b and 2.1c.

2.1.1 Armenia

The Republic of Armenia has an area of 29,743 sq. km and its map is presented in Fig. 2.1.1. It was taken by the Russians from the Persians and the Ottomans in 1813 and 1828, respectively. It gained its independence upon the dissolution of the Soviet Union in 1991. It was conquered by Muslims in 19 H or 640 AD during the time of Caliph Omar peacefully. The Muslim troops were led by the Prophet's companion Othman bnul Abil As.

The 1897 Russian Empire census indicated that Erivan Governorate had a Muslim population of 350,099 or 42.2% of the total population of 829,556. This governorate had an area of 26,094 sq. km covering most of current Armenia, Azerbaijan's Nakhichevan exclave with area 5,363 sq. km and 10.5% of the population at that time: 86,878; and Turkish Igdir province with area 3,587 sq. km and about 10.7% of the population at that time: 88,844; but excluded Zangezur County from Elisabetpol Governorate which is called now Syunik Province and is the land between Azerbaijan and its exclave. This county had 16.6% of the population (137,871) and area of 4,506 sq. km.

Censuses since 1926 collected ethnic affiliation demography of the population. As a result of war with Azerbaijan over Nagorno-Karabakh region, the only Muslim group left is the Kurds. Azerbaijani population was numerous until the breakup of the Soviet Union, but they were ethnically cleansed by ethnic Armin. There was also large Turkish, Persian and Karakalpak ethnicities that were recorded in the 1926 Census, but their number became negligible since then. The number of members of some historically numerous Muslim ethnicities since 1926 is provided in Table 2.1.1a.

Based on the ethnic census data, the Muslim population continued to decrease due to wars and ethnic cleansing of Turks and Azerbaijanis. Indeed, the Muslim population decreased in number and percentage from 350,000 or 42.2% in 1897, to 167,000 or 18.9% in 1926, to 131,000 or 10.2% in 1939, to 109,000 or 6.2% in 1959. It then increased to 149,000 or 6.0% in 1970, to 162,000 or 5.3% in 1979, but decreased to 86,000 or 2.6% in 1989. The 2011 census was the first to collect religious information and cross data with ethnicity. It recorded 812 or 0.03% Muslims and almost all Kurds and Yazidis were non-Muslim. The reason for the decrease of the Muslim population is the war with Azerbaijan over the Nagorno Karabagh region in the 1990s, which forced Azeris to leave Armenia to Azerbaijan.

As summarized in Table 2.1.1b and assuming that the percentage of Muslims will increase by 0.01 of a percentage point per decade, the number of Muslims is expected to remain around 2,000 or 0.1% throughout the second half of this century.

Figure 2.1.1 Map of the Republic of Armenia.

Table 2.1.1a Evolution of the largest ethnic Muslim populations in Armenia

	1926	1939	1959	1970	1979	1989	2001	2011
Azeri	76,870	130,896	107,748	148,189	160,841	84,860	29	
Tatar	27	324	577	581	761	367		
Kazakh		55	292	116	199	334		
Turk	78,386	18	19	10	28	13		
Karakalpak	6,311	0	8	4	5	43		
Persian	5,043	24	12	4	8	14		476
Total	166,637	131,317	108,656	148,904	161,842	85,631		

Table 2.1.1b Evolution of the Muslim population in Armenia

Year	Population	Muslims	%	Source
1897	829,556	350,099	42.20	[SU]c
1926	881,290	166,637	18.91	[SU]e
1939	1,282,338	131,317	10.24	[SU]e
1959	1,763,048	108,656	6.16	[SU]e
1970	2,491,873	148,904	5.98	[SU]e
1979	3,037,259	161,842	5.33	[SU]e
1989	3,304,776	85,631	2.59	[SU]e
2011	2,931,640	812	0.03	[AM]c
2050	2,816,000	2,000	0.07	es
2100	2,039,000	2,500	0.12	es

2.1.2 Azerbaijan

The Republic of Azerbaijan has an area of 86,600 sq. km, including Nakhichevan Autonomous Republic (5,363 sq. km), which is a landlocked exclave, and the breakaway republic of Nagorno Karabagh (11,458 sq. km), which is out of the Azerbaijani government control. Azerbaijan was occupied by the Russians in 1813 and gained its independence upon the dissolution of the Soviet Union in 1991. The Nagorno Karabakh Republic (NKR) or the Republic of Artsakh gained its de facto independence from Azerbaijan in 1992 but it is internationally unrecognized and is still claimed, but uncontrolled by Azerbaijan. A map of this Azerbaijan and NKR are presented in Figs. 2.1.2a and 2.1.2b.

Figure 2.1.2a Map of the Republic of Azerbaijan including Artsakh.

Azerbaijan was conquered by Muslims in 22 H or 643 AD during the time of Caliph Omar. The Muslim troops led by the Prophet's companion Bakeer bnu Abdellah al-Ashaj, followed by troops from the western side led by the Prophet companion Utba bnu Farqad Assulami, until they conquered Derbent city (Babel-Abwab) in southern Dagestan on the west coast of the Caspian Sea.

The 1897 Russian Empire census indicated that Muslims made up 72.1% of the total population of Baku and Elisabethpol Governorates that covered almost all of current Azerbaijan. Elisabethpol Governorate also included Zangezur County which is called now Syunik Province which belongs to Armenia and is the land between Azerbaijan and its exclave. This county had 15.7% (137,871) of the total population of the governorate and area of 4,506 sq. km. Elisabethpol Governorate also included the breakaway region of Nagorno Karabakh with an area of 11,458 sq. km and 31.4% (275,953) of the population of the governorate. The data excludes Azerbaijan's Nakhchivan Exclave with area 5,363 sq. km, and about 5.1% (86,878) of the population of both governorates at that time. Data of the 1897 census for the two governorates is summarized in Table 2.1.2a. For comparison, the table also contains an estimate based on the 1897 of the total population living in the current border of Azerbaijan including NKR and the current area of the country.

Censuses in 1886 and since 1926 collected ethnic affiliation demography of the population. In the 1886 Census, Azeris were recorded as Tatar. The largest Muslim ethnicities starting with the largest are Azerbaijanis, Lezgin, Talish, Avar, Turk, Tatar, Tat, Tsakhur, Kurd and Gryz (Kryz). The number of members of each of these ethnicities since 1926 is provided in Table 2.1.2b. As the name of the country suggests, about 95% of Muslims are Azeris.

Based on the ethnic census data, the Muslim population continued to decrease due to Russian policies of mass deportations and Armenian ethnic cleansing. Based on ethnic census data and as summarized in Table 2.1.2c, the Muslim population increased in number but decreased in percentage from 1.19 million or 72.6% in 1886, to 1.23 million or 72.1% in 1897 to 1.67 million or 72.0% in 1926, to 2.13 million or 66.4% in 1939. The decrease in percentage was due to Russian efforts in changing the demographic of this region and the persecution of Muslims. However, Muslims continued to increase both in number and percentage since 1959 from 2.65 million or 71.7% in 1959, to 4.0 million or 78.3% in 1970, to 4.97 million or 82.4% in 1979, to 6.12 million or 87.2% in 1989, to 7.62 million or 95.9% in 1999, to 8.63 million or 96.7% in 2009.

Thus, assuming the percentage of Muslims will increase by a quarter of a percentage point per decade, the Muslim population is expected to increase to eleven million or 97.8% by 2050, but decrease in number to nine million or 99.0% by 2100.

Table 2.1.2a 1897 Census data for territory covering current Azerbaijan and Artsakh but excluding Nakhichevan

Governorate	Population	Muslims	% Muslims	Area (km²)
Baku	826716	676243	81.80	39,096
Elisabethpol	878,415	552,822	62.93	43,307
Total	1,705,131	1,229,065	72.08	82,403
Azerbaijan	2,100,000			86,600

Table 2.1.2b Evolution of the largest ten Muslim ethnic populations in Azerbaijan and Artsakh

	1886	1926	1939	1959	1970	1979	1989	1999	2009
Azeri	NA	1,437,977	1,870,471	2,494,381	3,776,778	4,708,832	5,804,980	7,205,464	8,172,800
Lezgin	66,923	37,263	111,666	98,211	137,250	158,057	171,395	178,021	180,300
Talish	50,510	77,323	87,510	85	NA	NA	21,169	76,841	112,000
Avar	40,225	19,104	15,740	17,254	30,735	35,991	44,072	50,871	49,800
Turk	NA	95	600	202	8,491	7,926	17,705	43,454	38,000
Tatar	863,130	9,948	27,591	29,370	31,353	31,204	28,019	30,011	25,900
Tat	119,489	28,443	2,289	5,887	7,769	8,848	10,239	10,922	25,200
Tsakhur	NA	15,552	6,464	2,876	6,208	8,546	13,318	15,877	12,300
Kurd	32,541	41,193	6,005	1,487	5,488	5,676	12,226	13,075	6,100
Gryz	12,625	NA	NA	183	NA	NA	NA	NA	4,400
Total	1,185,443	1,666,898	2,128,336	2,649,936	4,004,072	4,965,080	6,123,123	7,624,536	8,626,800

Table 2.1.2c Evolution of the Muslim population in Azerbaijan (including Artsakh)

Year	Population	Muslims	%	Source
1886	1,524,385	1,185,443	72.55	[QQ]e
1897	1,705,131	1,229,065	72.08	[SU]c
1926	2,314,571	1,666,898	72.02	[SU]e
1939	3,205,150	2,128,336	66.40	[SU]e
1959	3,697,717	2,649,936	71.66	[SU]e
1970	5,117,081	4,004,072	78.25	[SU]e
1979	6,026,515	4,965,080	82.39	[AZ]e
1989	7,021,178	6,123,123	87.21	[AZ]e
1999	7,953,400	7,624,536	95.86	[AZ]e
2009	8,922,400	8,626,800	96.69	[AZ]e
2050	11,065,000	10,816,000	97.75	es
2100	9,192,000	9,100,000	99.00	es

Figure 2.1.2b Map of the Republic of Artsakh.

Due to Armenian migration to Nagorno-Karabakh, which was also encouraged by Russia, the Muslim population (all ethnic Azerbaijani) decreased in the territory of NKR from 79% in 1810 to 75% in 1871, to 14% in 1886 and then bounced to 59% in 1897. The decrease continued to 10% in 1926, as a result of Armenian-Azeri wars around WWI. Then to 9% in 1939 due to Russification policy implemented by Russia. However, Muslims picked up momentum, increasing to 14% in 1959, then 18% in 1970, and then 23% in 1979. As another Armenian-Azeri war started, the Muslim population decreased to 21% in 1989, and to almost none in 2005 and 2015, as a result of the independence of the territory, and the subsequent ethnic cleansing of Azeri Muslims. The data is summarized in Table 2.1.2d.

Table 2.1.2d Evolution of the Muslim population in Nagorno Karabakh

Year	Population	Muslims	%	Source
1810	12,000	9,500	79.17	[NKH]e
1871	117,000	87,800	75.04	[NKH]e
1886	121,216	17,038	14.06	[QQ]e
1897	275,953	164,098	59.47	[NKH]e
1926	125,162	12,746	10.18	[SU]e
1939	150,837	14,053	9.32	[QQ]e
1959	130,406	17,995	13.80	[QQ]e
1970	150,313	27,179	18.08	[QQ]e
1979	162,181	37,264	22.98	[QQ]e
1989	189,085	40,600	21.47	[QQ]e
2005	137,737	6	0.00	[QQ]e
2015	145,053	0	0.00	[QQ]c

2.1.3 Cyprus

Muslims conquered the Island of Cyprus in 27 H or 647 AD under the reign of Caliph Muawiya bnu Abi Sufyan, the first ruler of the Umayyad Empire. Muslims then lost control over it in 965 to the Byzantine Empire. It was recaptured by the Burji Dynasty in 1426 under al-Ashraf Sayfuddin Barsbay, but was lost by 1460 under al-Ashraf Sayfuddin Enal. The Island was recaptured by the Ottoman

Empire in 1570. In 1878, the Ottoman Empire under Sultan Abdul Hamid was forced to allow British troops in the Island in exchange to their help against the Russians. Eventually, the British Empire annexed the whole island in 1914. The Island then declared its independence from Britain in 1960, with two British enclaves in the south of the Island remaining till today: Akrotiri (123 sq. km) and Dhekelia (131 sq. km). From 1963 to 1974, Cypriot Greek Orthodox majority population who sought to combine the island with Greek carried ethnic cleansing against Muslims, which caused the Turkish military intervention in 1974, and the subsequent unilateral declaration of independence of the Turkish Republic of Northern Cyprus (North Cyprus) in 1983. A map of the Island of Cyprus is presented in Fig. 2.1.3.

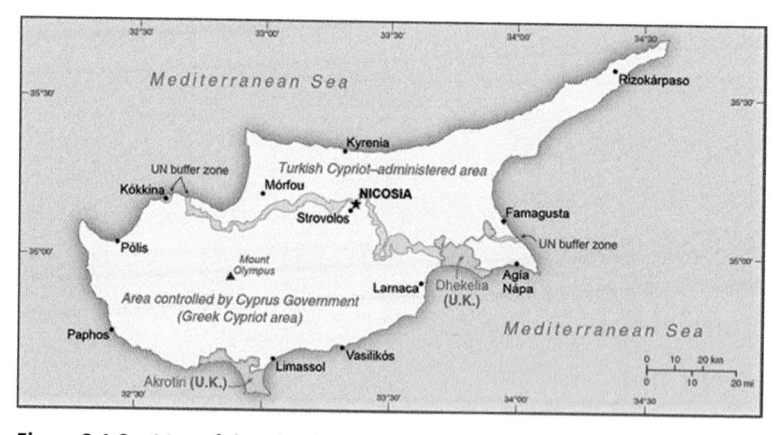

Figure 2.1.3 Map of the Island of Cyprus.

The Island of Cyprus has a total area of 9,505 sq. km. Currently, the South controls 5,896 sq. km, while the North controls 3,355 sq. km and the rest is under British control. Table 2.1.3a shows the change in Muslim population in the whole island before its division in 1974. Accordingly, estimates of the Muslim population changed from 20,000 or 19.1% in 1600, to 30,000 or 16.7% in 1670, to 28,000 or 14.9% in 1691, to 150,000 or 75.0% in 1745, to 47,000 or 56.0% in 1777 (census), to 60,000 or 75.0% in 1790, to 30,000 or 34.0% in 1831 (census), to 33,000 or 30.5% in 1841 (census), to 28,000 or 23.8% in 1847, to 44,000 or 26.7% in 1861, to 55,000 or 31.1% in 1872, to 45,000 or 31.5% in 1878, just before the British occupation.

Table 2.1.3a Evolution of the Muslim population in the Island of Cyprus

Year	Population	Muslims	%	Source
1600	105,000	20,000	19.05	[CYH]es
1670	180,000	30,000	16.67	[CYH]es
1691	188,000	28,000	14.89	[CYH]es
1745	200,000	150,000	75.00	[CYH]es
1777	84,000	47,000	55.95	[CYH]c
1790	80,000	60,000	75.00	[CYH]es
1831	88,166	29,966	33.99	[CYH]c
1841	108,300	33,000	30.47	[CYH]c
1847	117,700	28,000	23.79	[CYH]es
1861	165,000	44,000	26.67	[CYH]es
1872	176,750	55,000	31.12	[CYH]es
1878	143,000	45,000	31.47	[CYH]es
1881	186,173	45,458	24.42	[CY]c
1891	209,286	47,926	22.90	[CY]c
1901	237,022	51,309	21.65	[CY]c
1911	274,108	56,428	20.59	[CY]c
1921	310,715	61,339	19.74	[CY]c
1931	347,959	64,238	18.46	[CY]c
1946	450,114	80,548	17.90	[CY]c
1960	573,566	104,942	18.30	[CY]c
1973	631,778	120,200	19.03	[CY, CY06]e
1982	672,098	160,000	23.81	
1992	787,025	185,000	23.51	
2001	919,565	230,000	25.01	
2011	1,117,283	300,569	26.90	[CY11,WVS]c,s
2050	1,355,000	440,000	32.50	
2100	1,309,000	524,000	40.00	

British censuses show that the percentage of Muslims continued to decrease afterwards. This decrease was intentional by bringing Greek settlers and motivating Turks to migrate outside the Island, mainly to Australia, Britain, and Turkey. Accordingly, the Muslim population increased from 45,000 or 24.4% in 1881, to 48,000 or

22.9% in 1891, to 51,000 or 21.7% in 1901, to 56,000 or 20.6% in 1911, to 61,000 or 19.7% in 1921, to 64,000 or 18.5% in 1931, to 81,000 or 17.9% in 1946, to 105,000 or 18.3% in 1960, to 120,000 or 19.0% in 1973. The latter figure is obtained by combining the census data and official records of population exchange between the South and the North.

The decrease in Muslim Turkish population was not enough for Orthodox Greeks, which prompted the start of ethnic cleansing against Muslims from 1963 to 1974 that caused the Turkish military intervention in 1974 and the subsequent separation of the northern part of the Island in 1983. The figure of 1973 was obtained by using the following data: 120,200 Turkish Cypriots where in the whole island by mid-1974, and 32,213 migrated from the South to the North before 1979. Thus, the latter was assumed to be the Muslims in the south in 1973 census, while the remainder of the Turkish Cypriots was taken as Muslims in the North.

We can now obtain records of the Muslim population in either part of the Island from the detailed census records as follows: The south includes Larnaca, Limassol, Paphos, and half of Nicosia districts, while the north includes Kyrenia, Famagusta, and half of Nicosia districts. Tables 2.1.3b and 2.1.3c present a summary of the data for the south and the north, respectively. Thus, before WWII the Muslim population in the southern part decreased by one percentage point per decade. Indeed, the Muslim population changed from 27,000 or 25.3% in 1881, to 28,000 or 23.2% in 1891, to 30,000 or 22.3% in 1901, to 33,000 or 21.2% in 1911, to 36,000 or 20.3% in 1921, to 38,000 or 19.1% in 1931, to 47,000 or 18.4% in 1946. This population then increased to 61,000 or 18.7% in 1960, but dropped sharply to 32,000 or 9.0% in 1973, to 4,000 or 0.6% in 2001, then increased to 15,000 or 1.8% in 2011.

Combining data from census in both parts of the Island, we obtain the following estimate for the whole island as summarized in Table 2.1.3a. Accordingly, the Muslim population increased to 0.16 million or 23.8% in 1982, to 0.18 million or 23.5% in 1992, to 0.23 million or 25.0% in 2001, and 0.30 million or 26.9% in 2011. Thus, assuming that the Muslim population continues to increase by 1.5

percentage point per decade, the Muslim population is expected to increase to 0.4 million or 33% by 2050 and 0.5 million or 40% by 2100.

Table 2.1.3b Evolution of the Muslim population in (South) Cyprus

Year	Population	Muslims	%	Source
1881	106,594	27,006	25.34	[CY]c
1891	122,012	28,342	23.23	[CY]c
1901	136,062	30,281	22.26	[CY]c
1911	155,078	32,811	21.16	[CY]c
1921	178,463	36,181	20.27	[CY]c
1931	198,823	37,948	19.09	[CY]c
1946	254,484	46,921	18.44	[CY]c
1960	326,170	61,056	18.72	[CY]c
1973	358,985	32,213	8.97	[CY, CY06]e
2001	688,572	4,182	0.61	[UN]c
2011	831,026	15,279	1.84	[CY11]c

As shown in Table 2.1.3c, the Muslim population in the northern part of the Island of Cyprus decreased by one percentage point per decade before WWII. Indeed, the Muslim population changed from 18,000 or 23.2% in 1881, to 20,000 or 22.4% in 1891, to 21,000 or 20.8% in 1901, to 24,000 or 19.8% in 1911, to 25,000 or 19.0% in 1921, to 26,000 or 17.6% in 1931, to 34,000 or 17.2% in 1946. This population then increased to 44,000 or 17.7% and then increased sharply to 88,000 or 32.3% in 1973 and 144,000 or 98.3% in 1978, due to the exchange of population between the south and the north and migration of Turks to the northern part of the Island. The figures for 1978, and 1996, were obtained by subtracting 2,452 and 753 Greek Cypriots living in the North in 1977, and 1992, respectively, from the corresponding census. After independence censuses indicate that more than 99% of the population is Muslim. This population reached 0.20 million or 99.6% in 1996, and 0.26 million or 99.6% in 2006, to 0.29 million or 99.6% in 2011.

Table 2.1.3c Evolution of the Muslim population in North Cyprus

Year	Population	Muslims	%	Source
1881	79,579	18,452	23.19	[CY]c
1891	87,275	19,584	22.44	[CY]c
1901	100,961	21,028	20.83	[CY]c
1911	119,031	23,617	19.84	[CY]c
1921	132,253	25,158	19.02	[CY]c
1931	149,136	26,291	17.63	[CY]c
1946	195,631	33,627	17.19	[CY]c
1960	247,397	43,886	17.74	[CY]c
1973	272,793	87,987	32.25	[CY, CY06]e
1978	146,740	144,288	98.33	[CY, CY06]e
1996	200,587	199,834	99.62	[CY, CY06]e
2006	265,100	264,200	99.66	[WVS]s
2011	286,257	285,290	99.66	[WVS]s

2.1.4 Georgia

It has an area of 69,700 sq. km, including Tskhinvali Region or South Ossetia (3,900 sq. km), and the Autonomous Republics of Abkhazia (8,432 sq. km) and Ajaria (2,900 sq. km). Georgia was annexed by the Russians in 1801 and gained its independence upon the dissolution of the Soviet Union in 1991. A map of this country is presented in Fig. 2.1.4. Islam entered Georgia in 22 H/643 AD when Muslim troops during the reign of Caliph Omar bnul Khattab and under the leadership of Habib bnu Maslama occupied the capital Tbilisi.

The Republics of Abkhazia and South Ossetia gained their de facto independence from Georgia in 1992 but are internationally unrecognized and still are claimed, yet uncontrolled by Georgia. Abkhazia was raided by Muslims in 736 during the reign of the Umayyad Caliph Marwan II bnu Muhammad bnu Marwan. However, Islam only entered Abkhazia with the Ottoman conquest in 1570 during the reign of Sultan Selim II. The Ottomans lost this territory to the Russians in 1810. South Ossetia was conquered by Muslims in 735 during the reign of the Umayyad Caliph Marwan II bnu Muhammad bnu Marwan who ruled from 744 to 750.

![Map of Georgia]

Figure 2.1.4 Map of the Republic of Georgia with the breakaway regions of Abkhazia and South Ossetia.

The 1897 Russian Empire census indicated that there were 0.31 million Muslims or 14.5% of the total population of Kutaisi and Tiflis Governorates. The two governorates covered almost all of current Georgia, Abkhazia, South Ossetia, but also small parts of neighboring countries. Data of the 1897 census for the two governorates is summarized in Table 2.1.4a. For comparison, the table also contains an estimate based on the 1897 of the total population living in the current border of Georgia including Abkhazia and South Ossetia and the current area of the country.

Table 2.1.4a 1897 Census data for territory covering current Georgia, Abkhazia and South Ossetia

Governorate	Population	Muslims	% Muslims	Area (km^2)
Kutaisi	1,058,241	117,620	11.11	36,177
Tiflis	1,051,032	189,028	17.98	43,634
Total	2,109,273	306,648	14.54	79,810
Georgia	2,317,000			69,700

Censuses since 1926 collected ethnic affiliation demography of the population. The ten largest Muslim ethnicities starting with the largest are Azerbaijanis, Ossetians, Kist, Abkhaz, Kurd, Avar, Chechen, Turk, Persian and Lezgin. The number of members of each of these ethnicities since 1926 is provided in Table 2.1.4b.

Table 2.1.4b Evolution of the largest ten Muslim ethnic populations in Georgia including Abkhazia and South Ossetia (except in 2002 and 2014)

	1926	1939	1959	1970	1979	1989	2002	2014
Azeri	137,921	188,058	153,600	217,758	255,678	307,556	284,761	233,024
Ossetians	113,298	147,677	141,178	150,185	160,497	164,055	38,028	14,385
Kist							7,110	5,697
Abkhaz	56,847	57,805	62,878	79,449	85,285	95,853	3,527	4,551
Kurd	7,955	12,915	16,212	20,690	25,688	33,331	2,514	
Avar	1	114	585	450	3,680	4,230	1,996	
Chechen	66	2,538	105	232	158	609	1,271	
Turk	142,356	4,950	1,411	853	917	1,375	441	
Persian	2,220	1,150	73	64	91	123	46	
Lezgin	3,420	4,481	4,050	3,650	768	720	44	
Total	464,084	419,688	380,092	473,331	532,762	607,852	339,738	257,657

Based on the ethnic census data, the evolution of the Muslim population in the area consisting of Georgia, Abkhazia and South Ossetia is presented in Table 2.1.6c. Accordingly, the Muslim population decreased from 0.31 million or 14.5% in 1897, to 0.46 million or 17.3% in 1926, to 0.42 million or 11.9% in 1939, to 0.38 million or 9.4% in 1959, then increased to 0.47 million or 10.1% in 1970, to 0.53 million or 10.7% in 1979, to 0.61 million or 11.3% in 1989. The last two censuses did inquire on religious affiliation but did not include Abkhazia and South Ossetia. Accordingly, the Muslim population decreased in number but increase in percentage from 0.43 million of 9.9% in 2002 to 0.40 million or 10.9% in 2014.

Thus, assuming that the percentage of Muslim population continues to increase by half of a percentage point each decade, the Muslim population is expected to reach 0.4 million or 13% by 2050 and 0.4 million or 15% by 2100.

As for South Ossetia, we use ethnic censuses with the assumption that Muslim population makes up one third of the ethnic Ossetians and we arrive at the results presented in Table 2.1.4d. Accordingly, the Muslim population increased from 19,000 or 23.6% in 1886, to 20,000 or 23.0% in 1926, to 24,000 or 22.7% in 1939, decreased to 21,000 or 21.9% in 1959, and remained around 22,000 or 22% from 1970 to 1989, then 16,000 or 30% by 2015.

Table 2.1.4c Evolution of the Muslim population in Georgia, Abkhazia, and South Ossetia (excludes the latter two in 2002 and 2014)

Year	Population	Muslims	%	Source
1897	2,109,273	306,648	14.54	[SU]c
1926	2,677,233	464,084	17.33	[SU]e
1939	3,540,023	419,688	11.86	[SU]e
1959	4,044,045	380,092	9.40	[SU]e
1970	4,686,358	473,331	10.10	[SU]e
1979	4,993,182	532,762	10.67	[SU]e
1989	5,400,841	607,852	11.25	[SU]e
2002	4,371,523	433,784	9.92	[GE02]c
2014	3,669,918	398,677	10.86	[GE14]c
2050	3,394,000	440,000	12.50	es
2100	2,514,000	377,000	15.00	es

Table 2.1.4d Evolution of the Muslim population in South Ossetia

Year	Population	Muslims	%	Source
1886	81,762	19,262	23.56	[QQ]e
1926	87,375	20,117	23.02	[QQ]e
1939	106,118	24,089	22.70	[QQ]e
1959	96,807	21,233	21.93	[QQ]e
1970	99,421	22,024	22.15	[QQ]e
1979	97,988	21,692	22.14	[QQ]e
1989	98,527	21,744	22.07	[QQ]e
2015	53,532	16,060	30.00	[OS]e

As for Abkhazia, according to surveys held in 1997 and 2003, Muslims' percentage with respect to the total population decreased from 17% to 16%, respectively. Comparing this with the ethnic census of 2003, we can estimate that one third of the ethnic Abkhaz are Muslim. We can then apply this to previous ethnic censuses, with the assumption that Muslims make up one third of ethnic Abkhaz. Thus, based on census data and as presented in Table 2.1.4e, the Muslim population changed from 20,000 or 28.3% in 1886, to 23,000 or 21.4% in 1897, to 44,000 or 24.3% in 1913, to 20,000 or 10.1% in 1926, to 19,000 or 6.0% in 1939, to 20,000 or 5.0% in 1959, to 26,000 or 5.3% in 1970, to 28,000 or 5.7% in 1979, to 31,000 or 5.9% in 1989, to 32,000 or 14.7% in 2003, to 41,000 or 16.9% in 2011. According to a survey, the Muslim population changed to 25,000 or 17% in 1997, and 34,000 or 16% in 2003.

Table 2.1.4e Evolution of the Muslim population in Abkhazia

Year	Population	Muslims	%	Source
1886	68,773	19,655	28.58	[QQ]e
1897	106,179	22,740	21.42	[QQ]e
1913	181,947	44,169	24.28	[QQ]e
1926	201,016	20,374	10.14	[QQ]e
1939	311,885	18,733	6.01	[QQ]e
1959	404,738	20,398	5.04	[QQ]e
1970	486,959	25,760	5.29	[QQ]e
1979	486,082	27,700	5.70	[QQ]e
1989	525,061	31,089	5.92	[QQ]e
1997	145,986	24,820	17.00	[AB]s
2003	214,016	34,240	16.00	[AB]s
2011	240,705	40,725	16.92	[QQ]e

As for Ajaria, the 2002 census included religious data, from which about a third of ethnic Georgians were Muslim. We apply this fraction to previous censuses that collected ethnic data after 1926. Until then, Adjars or Muslim Georgians were listed as a separate ethnicity, until many were deported to Central Asia in late 1920s for their rebellion against anti-religious campaigns and collectivization by the new communist Russian regime. In 1886 and 1897 Adjaria was mostly covered by Batumi Oblast. The number of Ajars had increased from 38,094 in 1886, to 56,498 in 1897, to 90,314 in 1926. The number of Turks on the other hands, increased from 1,512 in 1886, to 3,199 in 1897, then decreased to 51 in 1926. Thus, based on census data and as presented in Table 2.1.4f, the Muslim population changed from 40,000 or 71.9% in 1886, to 60,000 or 67.5% in 1897, to 90,000 or 68.4% in 1926, then decreased by half to 43,000 or 21.3% in 1939, but has been increasing since then to 60,000 or 24.3% in 1959, to

79,000 or 25.5% in 1970, to 95,000 or 26.7% in 1979, to 108,000 or 27.6% in 1989, to 115,000 or 30.6% in 2002 and 133,000 or 40.8% in 2014.

Table 2.1.4f Evolution of the Muslim population in Ajaria

Year	Population	Muslims	%	Source
1886	55,088	39,606	71.90	[QQ]e
1897	88,444	59,697	67.50	[QQ]e
1926	131,957	90,314	68.44	[QQ]e
1939	200,106	42,514	21.25	[QQ]e
1959	245,286	59,546	24.28	[QQ]e
1970	309,768	78,976	25.50	[QQ]e
1979	354,224	94,624	26.71	[QQ]e
1989	392,432	108,271	27.59	[QQ]e
2002	376,016	115,161	30.63	[GE02]c
2014	325,812	132,852	40.78	[GE14]c

2.1.5 Iran

Muslims started conquering the current territory of Iran in 17 H or 638 AD from the West, both along the current border with Iraq, and by sea through the Hormuz Straight. The decisive battle for the collapse of the Persian Empire was the Battle of Nahawand, midway between Bagdad and Tehran. The battle occurred in 21 H or 642 AD and the Muslim troops were under the leadership of the Prophet's companion Ano'mano bnu Moqrin Al-Mozani. They then took the southern coast of the Caspian Sea in 22 H. In the same year, and towards the east, the Muslim troops under the leadership of Al-Ahnaf bnu Qais Attamimi, passed current Iran and reached Marw or Mary in southern Turkmenistan, Balakh near Mazar-i-Sharif in northern Afghanistan, and passed the Amu River, that separates Afghanistan from its neighboring countries to the north. By 23 H or 644 AD, Muslims conquered all current Iran, and the western half of current Pakistan up to the Indus River.

Currently, the Islamic Republic of Iran has an area of 1,648,195 sq. km and its map is presented in Fig. 2.1.5. Estimates of the Muslim population increased from 4.3 million or 98.3% in 1868, to 7.6 million or 98.8% in 1881, to 8.8 million or 97.8% in 1899, to 9.4 million or 98.4% in 1911. The first census was conducted in 1956 and every decade since then. Accordingly, as shown in Table 2.1.5, the Muslim population increased from 18.7 million or 98.7% in 1956, to 24.8 million or 98.8% in 1966, to 33.4 million or 99.1% in 1976, to 49.2 million or 99.6% in 1986, to 59.8 million or 99.7% in 1996, to 70.1 million or 99.7% in 2006, to 74.7 million or 99.7% in 2011, to 79.8 million or 99.7% in 2016.

Thus, assuming that the percentage of Muslims will increase by 0.02 of a percentage point per decade, the Muslim population is projected to reach 103 million or 99.8% by 2050 but decrease to 98 million or 99.9% by 2100.

Figure 2.1.5   Map of the Islamic Republic of Iran.

Table 2.1.5 Evolution of the Muslim population in Iran

Year	Population	Muslims	%	Source
1868	4,400,000	4,326,000	98.32	[SYB80]es
1881	7,653,600	7,560,000	98.78	[SYB90]es
1899	9,000,000	8,800,000	97.78	[SYB00]es
1911	9,500,000	9,350,000	98.42	[SYB15]es
1956	18,908,866	18,654,127	98.65	[UN63]c
1966	25,078,923	24,771,922	98.76	[UN71]c
1976	33,708,704	33,396,908	99.07	[UN83]c
1986	49,405,257	49,198,228	99.58	[ROD]c
1996	59,965,772	59,788,791	99.70	[UN]c
2006	70,290,465	70,097,741	99.73	[IR]c
2011	74,883,674	74,682,938	99.73	[IR]c
2016	79,801,698	79,598,054	99.74	[UN]c
2050	103,098,000	102,892,000	99.80	es
2100	98,588,000	98,489,000	99.90	es

2.1.6 Turkey

The Republic of Turkey was declared in 1923 as the successor state of the Ottoman Empire. It has a total area of 783,562 sq. km, covering the Anatolian Peninsula in Asia, and East Thrace (23,764 sq. km) in Europe. The map of Turkey is presented in Fig. 2.1.6.

The Muslim conquest of current Turkey was very slow as it was under the Byzantine Empire. This conquest was initiated by the capture of the southernmost city of Antakya in Shaban 15 H or September 636 AD during the reign of Caliph Omar bnul Khattab and the Muslim troops were under the leadership of the Prophet's companion Abu Obayda Aamer bnul Jarrah. Then Muslims conquered Malatya in 638, Diyarbakir in 639 and Van in 640. Then during the Umayyad Dynasty, Muslims captured Erzurum in 700, then Amorium, 200Km southwest of Ankara in 838 by the Abbassid Caliph al-Motassim.

Coming from the east, the Seljuks captured almost all of Anatolia after the battle of Manzikert with the Byzantines in 1071 north of Van Lake. Thus, Ankara fell in 1073, Izmir in 1076, and the Asian part of Istanbul in 1077. They captured Samsun in the north in 1200.

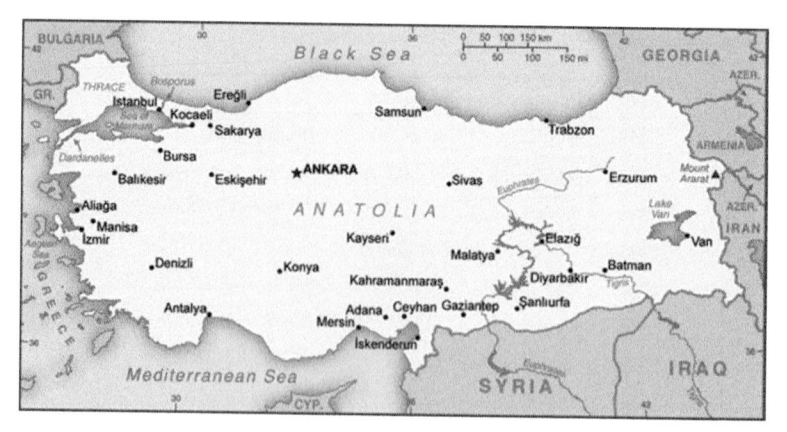

Figure 2.1.6 Map of the Republic of Turkey.

During the reign of Sultan Orhan I, the Ottomans conquered Bursa in 1326 and became their capital for a while. Edirne in East Thrace was captured by the Ottoman Sultan Murad I in 1365. Then the European side of Istanbul in 1453 during the reign of the Ottoman Sultan Mohammed II el-Fatih ben Murad II (ruled from 1451 to 1481). He gained his eternal nickname "el-Fatih" or the conqueror, due to this conquest. The conquest of current Turkey was culminated with the capture Trabzon in the north in 1461.

Based on census data and as shown in Table 2.1.6, the Muslim population increased from 11.0 million or 79.1% in 1897, to 12.0 million or 79.0% in 1906, to 13.3 million or 80.1% in 1914. The 1897 to 1914 estimates for current borders of Turkey were taken from [TR897], excluding the lands that were under Russian/Armenian occupation between 1878 and 1917. These were Kars, Artvin and Suramli and their data was taken from the 1897 Russian census, which gives the Muslims versus total population as 145,852 vs. 290,654 for Kars, 41,580 vs. 56,140 for Artvin, and 60,516 vs. 89,055 for Surmali. Religious data for Kars was taken from [SU], while that for Artivin and Surmali was inferred from ethnic data in [QQ]. Estimate of both populations for these areas was estimated to increase by 25% from 1897 to 1906, and 50% from 1897 to 1914. The Muslim and total population in 1906 for Edirne Vilayet (excluding parts outside current Turkey) was taken as the average of the data from 1897 and 1906.

Table 2.1.6 Evolution of the Muslim population in Turkey

Year	Population	Muslims	%	Source
1897	13,912,181	11,009,818	79.14	[TR897]c
1906	15,233,603	12,036,499	79.01	[TR897]c
1914	16,645,063	13,338,827	80.14	[TR897]c
1927	13,648,270	13,269,606	97.23	[SYB931]c
1935	16,158,018	15,838,673	98.02	[SYB950]c
1945	18,789,953	18,497,801	98.45	[UN56]c
1955	24,062,269	23,804,048	98.93	[TR965]c
1960	27,743,182	27,476,539	99.04	[TR965]c
1965	31,391,207	31,129,973	99.17	[TR965]c
2001	64,183,000	64,074,000	99.83	[EVS]s
2009	71,261,000	71,097,000	99.77	[EVS]s
2050	97,140,000	96,945,000	99.80	es
2100	86,170,000	85,998,000	99.80	es

The Muslim population decreased in number but increased in percentage from to 13.3 million or 97.1% in 1927 due WWI massacres and exchange of populations with neighboring countries. The Muslim population continued to increase since then to 18.5 million or 98.4% in 1945, to 23.8 million or 98.9% in 1955, to 27.5 million or 99.0% in 1960, to 31.1 million or 99.2% in 1965, which was the last census to collect data on religion. According to European Values Survey (EVS), the percentage of Muslims changed from 99.83% in 2001 to 99.77% in 2009.

Thus, assuming that the percentage of Muslims remains fixed at 99.8%, the Muslim population is expected to increase to 97 million by 2050 but decrease to 86 million by 2100.

2.1.7 Regional Summary and Conclusion

Islam entered the Near East within a decade of the death of Prophet Mohammed peace and blessings upon him. Islam spread widely, and the vast majority of this region remains Muslim. Accordingly, the Near East has the largest concentration of Muslims among the six regions spanning Asia, and any other region in the world. This is expected to remain so for the next three centuries.

The following tables present centennial data from 600 AD to 2100 AD (or approximately 1 H to 1500 H) in Table 2.1a and decennial data from 1790 AD to 2100 AD (or 1210 H to 1520 H) in Tables 2.1b and 2.1c for current countries in the Near East. The data includes total population in thousands (P), the percentage of which is Muslim ($M\%$), the corresponding Muslim population in thousands (M), and the annual population growth rate (APGR, or $G\%$) of the total population in this region.

Table 2.1a Centennial estimates of the Muslim population (×1000) in the Near East: 600–2100 AD (1–1500 H)

		600	700	800	900	1000	1100	1200	1300
Armenia	P	54	57	60	63	66	80	100	120
	M%	—	1.00	10.00	20.00	25.00	30.00	32.50	35.00
	M	—	1	6	13	17	24	33	42
Azerbaijan	P	139	145	151	157	163	210	260	310
	M%	—	10.00	20.00	30.00	40.00	50.00	60.00	70.00
	M	—	15	30	47	65	105	156	217
Cyprus	P	130	135	140	145	150	130	110	95
	M%	—	1.00	5.00	10.00	5.00	5.00	5.00	5.00
	M	—	1	7	15	8	7	6	5
Georgia	P	152	159	166	173	180	220	270	330
	M%	—	1.00	5.00	10.00	15.00	20.00	20.00	20.00
	M	—	2	8	17	27	44	54	66
Iran	P	4,300	4,350	4,400	4,450	4,500	4,400	4,300	4,200
	M%	—	20.00	50.00	70.00	80.00	90.00	99.00	99.00
	M	—	870	2,200	3,115	3,600	3,960	4,257	4,158
Turkey	P	7,400	7,300	7,200	7,100	7,000	6,900	6,750	6,600
	M%	—	5.00	10.00	20.00	30.00	40.00	50.00	60.00
	M	—	365	720	1,420	2,100	2,760	3,375	3,960
Total	P	12,175	12,146	12,117	12,088	12,059	11,940	11,790	11,655
	M%	—	10.32	24.52	38.27	48.23	57.78	66.84	72.48
	M	—	1,253	2,972	4,627	5,816	6,900	7,880	8,448
	G%		−0.002	−0.002	−0.002	−0.002	−0.010	−0.013	−0.012

Table 2.1a (*Continued*)

		1400	1500	1600	1700	1800	1900	2000	2100
Armenia	P	140	164	197	230	300	847	3,070	2,039
	M%	37.50	40.00	42.50	45.00	42.20	42.20	0.03	0.12
	M	53	66	84	104	127	357	1	2
Azerbaijan	P	360	408	498	571	760	2,100	8,123	9,192
	M%	80.00	90.00	85.00	80.00	77.77	72.08	95.86	99.00
	M	288	367	423	457	591	1,514	7,786	9,100
Cyprus	P	80	65	98	98	85	237	943	1,309
	M%	5.00	10.00	19.00	75.00	60.00	21.65	25.01	40.00
	M	4	7	19	74	51	51	236	524
Georgia	P	390	450	540	630	830	2,317	4,362	2,514
	M%	20.00	20.00	20.00	20.00	14.54	14.54	9.92	15.00
	M	78	90	108	126	121	337	433	377
Iran	P	4,100	4,000	5,000	5,000	6,100	10,140	65,623	98,588
	M%	99.00	99.00	99.00	99.00	98.78	97.78	99.70	99.73
	M	4,059	3,960	4,950	4,950	6,026	9,915	65,427	98,489
Turkey	P	6,450	6,300	7,900	8,400	9,600	13,948	63,240	86,170
	M%	70.00	80.00	80.00	80.00	79.14	79.01	99.83	99.80
	M	4,515	5,040	6,320	6,720	7,597	11,020	63,133	85,998
Total	P	11,520	11,387	14,233	14,929	17,675	29,589	145,361	199,812
	M%	78.09	83.69	83.63	83.26	82.11	78.39	94.26	97.34
	M	8,997	9,529	11,904	12,430	14,512	23,195	137,015	194,491
	G%	−0.012	−0.012	0.223	0.048	0.169	0.515	1.592	0.318

Table 2.1b Decennial estimates of the Muslim population (×1000) in the Near East: 1790–1940 AD (1210–1360 H)

		1790	1800	1810	1820	1830	1840	1850	1860
Armenia	P	290	300	310	319	370	420	490	550
	M%	42.20	42.20	42.20	42.20	42.20	42.20	42.20	42.20
	M	122	127	131	135	156	177	207	232
Azerbaijan	P	745	760	775	792	940	1,080	1,220	1,360
	M%	77.77	77.77	77.77	77.77	77.77	77.77	77.77	77.77
	M	579	591	603	616	731	840	949	1,058
Cyprus	P	80	85	90	95	100	110	130	150
	M%	75.00	60.00	50.00	40.00	33.99	30.47	23.79	26.67
	M	60	51	45	38	34	34	31	40
Georgia	P	810	830	850	874	1,040	1,200	1,360	1,520
	M%	14.54	14.54	14.54	14.54	14.54	14.54	14.54	14.54
	M	118	121	124	127	151	174	198	221
Iran	P	5,900	6,100	6,300	6,560	6,900	7,250	7,600	8,000
	M%	98.78	98.78	98.78	98.78	98.78	98.78	98.78	98.78
	M	5,828	6,026	6,223	6,480	6,816	7,162	7,507	7,902
Turkey	P	9400	9,600	9,800	10,074	10,200	10,350	10,500	10,650
	M%	79.14	79.14	79.14	79.14	79.14	79.14	79.14	79.14
	M	7,439	7,597	7,756	7,973	8,072	8,191	8,310	8,428
Total	P	17,225	17,675	18,125	18,714	19,550	20,410	21,300	22,230
	M%	82.13	82.11	82.10	82.12	81.64	81.22	80.76	80.44
	M	14,147	14,512	14,881	15,368	15,960	16,578	17,201	17,882
	G%		0.258	0.251	0.320	0.437	0.430	0.427	0.427

Table 2.1b (Continued)

		1870	1880	1890	1900	1910	1920	1930	1940
Armenia	P	603	674	753	847	952	1,052	1,170	1,282
	M%	42.20	42.20	42.20	42.20	42.20	18.91	18.91	10.24
	M	254	284	318	357	402	199	221	131
Azerbaijan	P	1,496	1,671	1,867	2,100	2,372	2,608	2,902	3,210
	M%	77.77	77.77	77.77	72.08	72.08	72.02	72.02	66.40
	M	1,163	1,300	1,452	1,514	1,710	1,878	2,090	2,131
Cyprus	P	170	186	209	237	274	311	348	461
	M%	31.12	24.42	22.90	21.65	20.59	19.74	18.46	17.90
	M	53	45	48	51	56	61	64	83
Georgia	P	1,650	1,844	2,060	2,317	2,601	2,877	3,201	3,542
	M%	14.54	14.54	14.54	14.54	14.54	17.33	17.33	11.86
	M	240	268	300	337	378	499	555	420
Iran	P	8,415	8,955	9,529	10,140	10,580	11,927	13,245	14,708
	M%	98.78	98.78	98.78	97.78	98.42	98.42	98.41	98.41
	M	8,312	8,846	9,413	9,915	10,413	11,739	13,034	14,474
Turkey	P	11,793	12,472	13,189	13,948	14,750	13,877	14,928	17,821
	M%	79.14	79.14	79.14	79.01	80.14	80.14	97.23	98.02
	M	9,333	9,870	10,438	11,020	11,821	11,121	14,514	17,468
Total	P	24,127	25,802	27,607	29,589	31,530	32,652	35,794	41,023
	M%	80.23	79.89	79.57	78.39	78.59	78.09	85.15	84.60
	M	19,356	20,614	21,968	23,195	24,780	25,497	30,479	34,707
	G%	0.819	0.671	0.676	0.693	0.635	0.350	0.919	1.364

Table 2.1c Decennial estimates of the Muslim population (×1000) in the Near East: 1950–2100 AD (1370–1520 H)

		1950	1960	1970	1980	1990	2000	2010	2020
Armenia	P	1,354	1,874	2,525	3,100	3,538	3,070	2,877	2,963
	M%	10.24	6.16	5.98	5.33	2.59	0.03	0.03	0.04
	M	139	115	151	165	92	1	1	1
Azerbaijan	P	2,928	3,895	5,180	6,151	7,243	8,123	9,032	10,139
	M%	71.66	71.66	78.25	82.39	87.21	95.86	96.69	97.00
	M	2,098	2,791	4,053	5,068	6,316	7,786	8,733	9,835
Cyprus	P	494	573	614	685	767	943	1,113	1,207
	M%	17.90	18.30	19.03	23.81	23.51	25.01	26.90	28.00
	M	88	105	117	163	180	236	299	338
Georgia	P	3,527	4,008	4,713	5,018	5,410	4,362	4,099	3,989
	M%	11.86	9.40	10.10	10.67	11.25	9.92	10.86	11.00
	M	418	377	476	535	609	433	445	439
Iran	P	17,119	21,907	28,514	38,650	56,366	65,623	73,763	83,993
	M%	98.41	98.65	98.76	99.07	99.58	99.70	99.73	99.74
	M	16,847	21,611	28,160	38,291	56,129	65,427	73,563	83,775
Turkey	P	21,408	27,472	34,876	43,976	53,922	63,240	72,327	84,339
	M%	98.45	99.04	99.17	99.17	99.83	99.83	99.77	99.80
	M	21,077	27,209	34,587	43,611	53,830	63,133	72,161	84,170
Total	P	46,830	59,729	76,422	97,580	127,246	145,361	163,211	186,631
	M%	86.84	87.41	88.38	90.01	92.07	94.26	95.09	95.67
	M	40,667	52,208	67,544	87,833	117,157	137,015	155,203	178,558
	G%	1.324	2.433	2.464	2.444	2.654	1.331	1.158	1.341

Table 2.1c (*Continued*)

		2030	2040	2050	2060	2070	2080	2090	2100
Armenia	P	2,967	2,905	2,816	2,689	2,525	2,351	2,191	2,039
	M%	0.05	0.06	0.07	0.08	0.09	0.10	0.11	0.12
	M	1	2	2	2	2	2	2	2
Azerbaijan	P	10,740	11,055	11,065	10,832	10,462	10,014	9,598	9,192
	M%	97.25	97.50	97.75	98.00	98.25	98.50	98.75	99.00
	M	10,444	10,779	10,816	10,615	10,279	9,863	9,478	9,100
Cyprus	P	1,275	1,324	1,355	1,362	1,354	1,337	1,319	1,309
	M%	29.50	31.00	32.50	34.00	35.50	37.00	38.50	40.00
	M	376	410	440	463	481	495	508	524
Georgia	P	3,853	3,689	3,517	3,321	3,115	2,915	2,721	2,514
	M%	11.50	12.00	12.50	13.00	13.50	14.00	14.50	15.00
	M	443	443	440	432	420	408	395	377
Iran	P	92,664	98,594	103,098	105,213	104,134	101,886	100,339	98,588
	M%	99.76	99.78	99.80	99.82	99.84	99.86	99.88	99.90
	M	92,441	98,377	102,892	105,024	103,967	101,744	100,219	98,489
Turkey	P	89,158	94,132	97,140	97,941	96,624	93,897	90,336	86,170
	M%	99.80	99.80	99.80	99.80	99.80	99.80	99.80	99.80
	M	88,979	93,943	96,945	97,745	96,431	93,709	90,155	85,998
Total	P	200,657	211,698	218,990	221,358	218,215	212,400	206,504	199,812
	M%	96.03	96.34	96.60	96.80	96.96	97.09	97.22	97.34
	M	192,686	203,954	211,535	214,281	211,581	206,221	200,757	194,491
	G%	0.725	0.536	0.339	0.108	−0.143	−0.270	−0.281	−0.329

2.2 Muslims in Central Asia

This region consists of seven countries: Afghanistan, Kazakhstan, Kyrgyzstan, Mongolia, Tajikistan, Turkmenistan, and Uzbekistan. Muslims entered this region coming from Persia in 22 H or 643 AD during the reign of Caliph Omar bnul Khattab. The Muslim troops under the leadership of Al-Ahnaf bnu Qais Attamimi, passed current Iran and reached Marw or Mary in southern Turkmenistan, Balakh near Mazar-i-Sharif in northern Afghanistan, and passed the Amu River, that separates Afghanistan from its neighboring countries to the north: Turkmenistan, Uzbekistan, and Tajikistan.

The Muslim conquest resumed during the reign of the sixth Umayyad Caliph al-Walid I bnu Abdel Malik bnu Marwan (ruled from 705 AD to 715 AD). The Muslim army was under the leadership of Qutaiba bnu Muslim al-Bahili (lived from 48 H or 668 AD to 96 H or 715 AD). Accordingly, he captured Bukhara of Uzbekistan in 90 H/709 AD, Samarkand of Uzbekistan in 93 H/712 AD, Fergana of Uzbekistan, Khudjand of Tajikistan, and Kabul; capital of Afghanistan in 94 H/ 713 AD, and Kashgar of the Chinese Xinjiang Uyghur Autonomous Region in 96 H/714 AD.

Thus, by early eighth century Afghanistan, Tajikistan, Turkmenistan, Uzbekistan and southern Kyrgyzstan were under Muslim control. Islam spread to the rest of the region when the Mongols adopted Islam as their religion in the first half of the fourteenth century. Islam did not spread further in current Mongolia as its rulers did not accept Islam, unlike the rest of this region and the Turkic parts of Russia and China.

Thus, the Muslim population has likely increased from 0.28 million or 9.6% of the total population of this region in 700 AD, to 1.0 million or 33.4% in 800 AD, to 1.6 million or 48.7% in 900 AD, to 2.1 million or 59.9% in 1000 AD, to 2.4 million or 62.2% in 1100 AD, to 2.6 million or 63.2% in 1200 AD, to 2.9 million or 64.2% in 1300 AD, to 3.4 million or 69.4% in 1400 AD, to 4.3 million or 80.9% in 1500 AD, to 5.1 million or 83.4% in 1600 AD, to 5.8 million or 85.2% in 1700 AD, to 7.8 million or 85.8% in 1800 AD, to 15 million or 90.6% in 1900, to 68 million or 87.0% in 2000, to 105 million or 90.4% in 2020, and is projected to reach 157 million or 93% by 2050 and 185 million or 95% by 2100.

A plot of centennial estimates of the Muslim population and its percentage with respect to the total population in this region from 600 to 2100 is provided in Fig. 2.2a. A zoom in of this plot, providing a plot of decennial estimates of the Muslim population and its percentage with respect to the total population in this region from 1900 to 2100 is provided in Fig. 2.2b. This shows that the Muslim population in this region was increasing slowly until 1960 and is increasing substantially afterwards towards the end of this century. The percentage of Muslims in this region on the other hand, dipped from 91% in 1900 to 84% in 1920, then 70% in 1960 due to Stalin's disastrous policies against Muslims in this region. This percentage bounced equally sharply afterwards reaching back to 90% in 2020 and is expected to continue to increase throughout this century, reaching 95% by 2100. The Muslim percentage increased by more than ten percentage points in the decade following the dissolution of the Soviet Union.

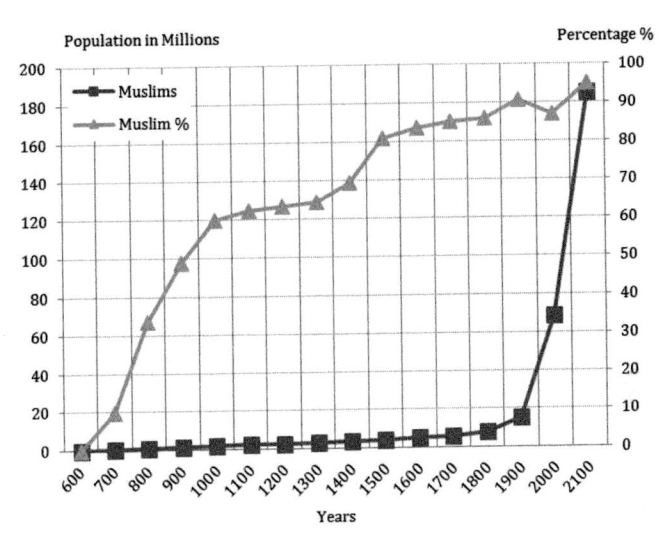

Figure 2.2a Plot of centennial estimates of the Muslim population and its percentage of the total population in Central Asia: 600–2100 AD (1–1500 H).

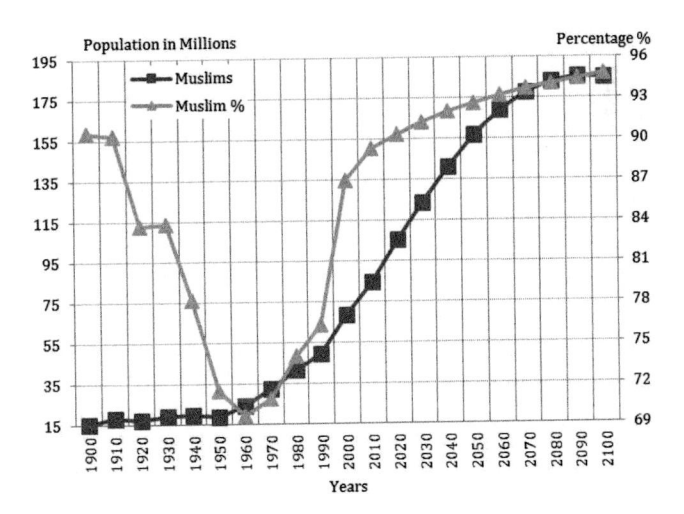

Figure 2.2b Plot of decennial estimates of the Muslim population and its percentage of the total population in Central Asia: 1900–2100 AD (1320–1520 H).

The corresponding individual data for each country in this region is discussed below. In Section 2.2.8, the total population in each country in this region and the corresponding percentage and number of Muslims is presented centennially in Table 2.2a from 600 to 2100 and decennially in Tables 2.2b and 2.2c from 1790 to 2100.

2.2.1 Afghanistan

The Muslim conquest of Afghanistan started by Al-Ahnaf bnu Qais Attamimi in 22 H/643 AD during the reign of Caliph Omar bnul Khattab. He conquered the areas near the Iranian and Turkmenistan border, from Herat to Mazar-i-Sharif. The conquest was completed by Qutaiba bnu Muslim al-Bahili in 94 H/713 AD during the reign of the sixth Umayyad Caliph al-Walid I bnu Abdel Malik bnu Marwan, when he captured Kabul. Currently, the Islamic Republic of Afghanistan has an area of 652,230 sq. km and a map of the country is presented in Fig. 2.2.1. It was occupied by the British in 1879 and gained its independence from the UK in 1919.

Figure 2.2.1 Map of the Islamic Republic of Afghanistan.

As shown in Table 2.2.1, estimates of the Muslims population increased from 5.8 million or 95% in 1908 to 6.4 million or 100% in 1922. According to the 1979 census, the Muslim population was 13.0 million or 99.8% of the total population. The same percentage (99.77%) is obtained by Asia Barometer Survey in 2005. Thus, assuming that this percentage remains constant, the Muslim population is expected to increase to 65 million by 2050 and 75 million by 2100.

Table 2.2.1 Evolution of the Muslim population in Afghanistan

Year	Population	Muslims	%	Source
1880	6,145,000	5,385,000	87.63	[JAN]es
1908	6,150,000	5,842,000	94.99	[AFH]es
1922	6,380,500	6,377,500	99.95	[MAS23]es
1979	13,051,358	13,020,810	99.77	[UN83]c
2005	24,400,000	24,344,000	99.77	[ABM]s
2050	64,683,000	64,534,000	99.77	es
2100	74,938,000	74,766,000	99.77	es

2.2.2 Kazakhstan

Islam started spreading into current Kazakhstan from its south in the eighth century. It was part of the Golden Horde Mongol Empire from the thirteenth to the fifteenth century. The whole country came under Muslim control when Sultan Mohammed Öz-Beg accepted Islam before taking the throne of the Golden Horde and ruled it from 1313 to 1341. He adopted Islam as the state's religion and continued the spread of Islam among the Turkic people. The leaders of the Golden Horde remained Muslims afterwards. By then the Empire controlled significant parts of current Russia (the region west of a line extending from the short Russian-Chinese border between Kazakhstan and Mongolia to the Arctic Ocean), Kazakhstan, Ukraine, Moldova, eastern Poland, Romania and Bulgaria.

The Republic of Kazakhstan has an area of 2,724,900 sq. km and its map is presented in Fig. 2.2.2. It was conquered by Russia in the eighteenth century and gained its independence upon the dissolution of the Soviet Union in 1991. The current territory of Kazakhstan was carved out by the Soviets from historic Turkistan in three stages. First, in 1920 the Kirghiz Autonomous Soviet Socialist Republic (ASSR) was created as part of the Russian Soviet Federative Socialist Republic (SFSR). Then in 1925, it was renamed Kazakh ASSR, and it included until 1930 the Karakalpak Autonomous Oblast (AO); a 160,000 sq. km area referred to now as Karakalpakstan and is the northwestern part of current Uzbekistan. Finally, in 1936 Kazakh ASSR was separated from Russian SFSR and elevated to Kazakh Soviet Socialist Republic (SSR), which in 1991 became the Republic of Kazakhstan.

The 1897 Russian Empire census indicated that Muslims made up 86.5% of the total population of five Oblasts that cover almost all of current Kazakhstan: Akmola, Semirechye (covers also eastern half of Kyrgyzstan), Semipalatinsk, Syr Darya (contains northern Uzbekistan), Turgay and Urals. Data of the 1897 census for these Oblasts is summarized in Table 2.2.2a. For comparison, the table also contains an estimate based on the 1897 of the total population living in the current border of Kazakhstan and the current area of the country.

Censuses since 1926 collected ethnic affiliation demography of the population, about fifty of whom are Muslim. The top fifteen Muslim ethnicities in Kazakhstan from largest in number are Kazakh, Uzbek, Uygur, Tatar, Turk, Azeri, Dungan, Kurd, Tajik, Chechen, Kyrgyz, Bashkir, Ingush, Lezghin and Romani (Gypsy). The number of members of each of these ethnicities since 1926 is provided in Table 2.2.2b. As the name of the country suggests, almost 90% of Muslims are Kazakh.

Based on the ethnic census data, the Muslim population continued to decrease due to Russian policies of mass deportations of ethnic Muslims and government planned migration of non-Muslim ethnicities. Indeed, the Muslim population decreased in number and percentage from 4.9 million or 86.5% in 1897, to 3.9 million or 62.5% in 1926 (excluding Karakalpakstan region, which is included in Uzbekistan), to 2.6 million or 43.0% in 1939, it then reached 3.5 million or 37.1% in 1959. The Muslim population continued to increase since then both in number and percentage to 5.0 million or 38.7% in 1970, to 6.3 million or 42.8% in 1979, to 7.7 million or 47.1% in 1989, to 9.2 million or 61.4% in 1999, to 11.2 million or 70.6% in 2009. Thus, the percentage of Muslims almost doubled in a half of a century; since 1959. A summary of the data is provided in Table 2.2.2c.

The surge in the Muslim percentage following the independence of Kazakhstan in 1991 was caused by the return of many ethnic Kazakhs from neighboring countries, and the departure of non-Muslim ethnicities such as Germans, Russians, and Ukrainians. The religion adherence question was included for the first time in the 2009 census. Using ethnic data for the same year give slightly higher number with about 110,000 extra persons. Assuming the percentage of Muslims will increase by two percentage points per decade, the Muslim population is expected to reach 19 million or 79% by 2050 and 25 million or 89% by 2100.

Figure 2.2.2 Map of the Republic of Kazakhstan.

Table 2.2.2a 1897 Census data for territory covering current Kazakhstan

Oblast	Population	Muslims	% Muslims	Area (km^2)
Akmola	682,608	438,983	64.31	594,685
Semirechye	987,863	890,270	90.12	394,403
Semipalatinsk	684,590	614,773	89.80	478,192
Syr Darya	1,478,398	1,425,313	96.41	504,667
Turgay	453,416	415,806	91.71	456,405
Urals	645,121	478,765	74.21	360,443
Total	4,931,996	4,263,910	86.45	2,788,796
Kazakhstan	3,944,000			2,724,900

Table 2.2.2b Evolution of the largest Muslim ethnic populations in Kazakhstan

	1926	1939	1959	1970	1979	1989	1999	2009
Kazakh	3,627,612	2,327,625	2,794,966	4,161,164	5,289,349	6,534,616	8,011,452	10,096,763
Uzbek	129,407	120,655	136,570	207,514	263,295	332,017	370,765	456,997
Uygur	11,631	35,409	59,840	120,784	147,943	185,301	210,377	224,713
Tatar	79,758	108,127	191,802	281,849	312,626	327,982	249,052	204,229
Turk	46	523	9,916	18,397	25,820	49,567	75,950	97,015
Azeri	20	12,996	38,362	56,166	73,345	90,083	78,325	85,292
Dungan	8,455	7,415	9,980	17,283	22,491	30,165	36,945	51,944
Kurd	—	2,387	6,109	12,299	17,692	25,425	32,764	38,325
Tajik	7,599	11,229	8,075	7,166	19,293	25,514	25,673	36,277
Chechen	3	2,639	130,232	34,492	38,256	49,507	31,802	31,431
Kyrgyz	10,200	5,033	6,810	9,474	9,352	14,112	10,925	23,274
Bashkir	470	3,450	8,742	21,134	32,499	41,847	23,225	NA
Ingush	3	322	47,867	18,356	18,337	19,914	16,900	NA
Lezghin	51	808	872	2,566	6,126	13,905	4,616	NA
Romani (Gypsy)	750	4,257	7,265	7,766	8,626	7,165	5,130	NA
Total	3,876,005	2,642,875	3,457,408	4,976,410	6,285,050	7,747,120	9,183,901	11,347,260

Table 2.2.2c Evolution of the Muslim population in Kazakhstan

Year	Population	Muslims	%	Source
1897	4,931,996	4,263,910	86.45	[SU]c
1926	6,198,465	3,876,005	62.53	[SU]e
1939	6,151,102	2,642,875	42.97	[SU]e
1959	9,309,847	3,457,408	37.14	[SU]e
1970	12,848,573	4,976,410	38.73	[SU]e
1979	14,684,283	6,285,050	42.80	[SU]e
1989	16,464,464	7,747,120	47.05	[SU]e
1999	14,953,126	9,183,901	61.42	[KZ]e
2009	15,928,587	11,239,176	70.56	[KZ]c
2050	24,024,000	18,979,000	79.00	es
2100	27,918,000	24,847,000	89.00	es

2.2.3   Kyrgyzstan

The Muslim conquest of Kyrgyzstan started by Qutaiba bnu Muslim al-Bahili during the reign of the sixth Umayyad Caliph al-Walid I bnu Abdel Malik bnu Marwan. In 94 H/713 AD he captured the southern part of current Kyrgyzstan. Islam spread to the rest of the country when the Mongols adopted Islam as their religion in the first half of the fourteenth century. The Kyrgyz Republic has an area of 199,951 sq. km and its map is presented in Fig. 2.2.3. It was conquered by Russia in the 1876 and gained its independence upon the dissolution of the Soviet Union in 1991. The current territory of Kyrgyzstan was carved out by the Soviets from historic Turkistan in three stages. First, in 1924 the Kara-Kyrgyz Autonomous Oblast (AO) was created as part of the Russian Soviet Federative Socialist Republic (SFSR). It was renamed in 1925 as Kyrgyz AO. Then in 1926, it was upgraded to Kyrgyz Autonomous Soviet Socialist Republic (ASSR), yet still part of the Russian SFSR. Finally, in 1936, it was separated from Russian SFSR and renamed Kyrgyz Soviet Socialist Republic (SSR), which in 1991 became the Republic of Kyrgyzstan.

The 1897 Russian Empire census indicated that Muslims made up 95.6% of the total population of two Oblasts that cover almost all of current Kyrgyzstan: Ferghana and Semirechye. Ferghana Oblast contained western half of Kyrgyzstan, eastern half of Tajikistan and eastern portion of Uzbekistan. Semirechye Oblast contained eastern half of Kyrgyzstan, while most of it in current Kazakhstan. Data of the 1897 census for these Oblasts is summarized in Table 2.2.3a. For comparison, the table also contains an estimate based on the 1897 of the total population living in the current border of Kyrgyzstan and the current area of the country.

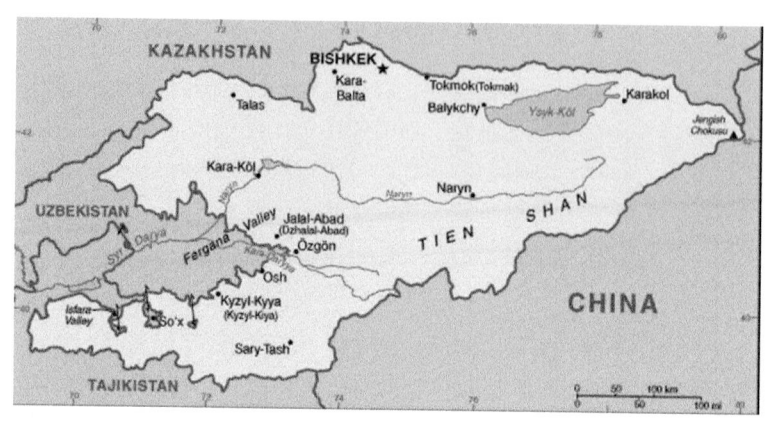

Figure 2.2.3   Map of the Kyrgyz Republic.

Table 2.2.3a   1897 Census data for territory covering current Kyrgyzstan

Oblast	Population	Muslims	% Muslims	Area (km²)
Ferghana	1,572,214	1,557,057	99.04	92,343
Semirechye	987,863	890,270	90.12	394,403
Total	2,560,077	2,447,327	95.60	486,747
Kyrgyzstan	980,000			199,951

Censuses since 1926 collected ethnic affiliation demography of the population, about fifty of whom are Muslim. The top seventeen Muslim ethnicities in Kyrgyzstan from largest in number are Kyrgyz, Uzbek, Dungan, Uighur (Uygur), Tatar, Kazakh, Tajik, Turk, Azerbaijani, Kurd, Balkar, Lezgin, Dargin, Chechen, Karachai, Turkmen and Bashkir. The number of members of each of these ethnicities since 1897 is provided in Table 2.2.3b. As the name of the country suggests, more than 80% of Muslims are Kyrgyz.

Table 2.2.3b Evolution of the largest Muslim ethnic populations in Kyrgyzstan

	1926	1939	1959	1970	1979	1989	1999	2009
Kyrgyz	661,171	754,323	836,831	1,284,773	1,687,382	2,229,663	3,128,147	3,804,800
Uzbek	109,776	151,551	218,640	332,638	426,194	550,096	664,950	419,600
Dungan	6,004	5,921	11,088	19,837	26,661	36,928	51,766	58,100
Uighur	73	9,412	13,757	24,872	29,817	36,779	46,944	48,500
Tatar	4,902	20,017	56,266	68,827	71,744	70,068	45,438	31,400
Kazakh	NA	23,925	20,067	21,998	27,442	37,318	42,657	33,200
Tajik	2,667	10,670	15,221	21,927	23,209	33,518	42,636	46,100
Turk	3,631	33	542	3,076	5,160	21,294	33,327	39,400
Azeri	NA	7,724	10,428	12,536	17,207	15,775	14,014	17,300
Kurd	NA	1,490	4,783	7,974	9,544	14,262	11,620	13,200
Balkar	NA	12	2,234	1,973	1,688	2,131	1,512	1,300
Lezgin	24	888	1,165	1,599	1,896	2,493	2,657	2,600
Dargin	NA	638	NA	1,419	1,890	2,479	2,704	2400
Chechen	1	7	25,208	3,391	2,654	2,873	2,612	1,900
Karachai	NA	9	4,575	2,631	2,458	2,509	2,200	1700
Turkmen	3	178	235	352	607	899	430	1,400
Bashkir	36	870	2,595	3,250	3,741	4,026	2,044	1,100
Total	788,288	987,668	1,223,635	1,813,073	2,339,294	3,063,111	4,095,658	4,524,000

Based on ethnic census data, the Muslim population increased in number but decreased in percentage from 0.79 million or 79.4% in 1926, to 0.99 million or 67.7% in 1939, to 1.22 million or 59.2% in 1959, to 1.81 million or 61.8% in 1970. The Muslim population continued increasing since then in both number and percentage, to 2.34 million or 66.4% in 1979, to 3.06 million or 71.9% in 1989, to 4.10 million or 84.9% in 1999, to 4.52 million or 84.4% in 2009. A summary of the data is provided in Table 2.2.3c. Thus, assuming that the Muslim population will continue to increase by half of a percentage point per decade, the Muslim population is expected to reach eight million or 86.5% by 2050, and ten million or 89% by 2100.

Table 2.2.3c Evolution of the Muslim population in Kyrgyzstan

Year	Population	Muslims	%	Source
1897	2,560,077	2,447,327	95.60	[SU]c
1926	993,004	788,288	79.38	[SU]e
1939	1,458,213	987,668	67.73	[SU]e
1959	2,065,837	1,223,635	59.23	[SU]e
1970	2,932,805	1,813,073	61.82	[SU]e
1979	3,522,832	2,339,294	66.40	[SU]e
1989	4,257,755	3,063,111	71.94	[SU]e
1999	4,822,900	4,095,658	84.92	[KG]e
2009	5,362,800	4,524,000	84.36	[KG]e
2050	9,126,000	7,894,000	86.50	es
2100	10,985,000	9,777,000	89.00	es

2.2.4 Mongolia

Islam entered this country when the Mongols adopted Islam as their religion in the first half of the fourteenth century. Unlike its neighbors to the east, Islam did not spread here as much since its rulers did not accept Islam. Currently, Mongolia has an area of 1,564,116 sq. km and its map is presented in Fig. 2.2.4. It was conquered by China in late seventeenth century and gained its independence in 1921.

The Muslim population was estimated in 1920 at 20,000 or 3.1% of the total population. Census data since 1956 included ethnic information from which we can infer religion with the assumption that all and only ethnic Kazakhs are Muslim. Thus, as summarized in Table 2.2.4, the Muslim population increased from 37,000 or 4.3% in 1956, to 48,000 or 4.7% in 1963, to 63,000 or 5.3% in 1969, to 84,000 or 5.5% in 1979, to 121,000 or 6.1% in 1989, then decreased to 103,000 or 4.4% in 2000, to 58,000 or 3.0% in 2010. The decrease is due to the migration of ethnic Kazakhs back to Kazakhstan after its independence in 1991. The 2010 census was the first census to collect information on religious adherence, which was collected from population aged 15 and over only. This population was 1,905,969, while the total population was 2,754,685. In this last census, the Kazakh population numbered 101,526 or 3.86% of the total population (all ages) and 114,506 of 2,970,355 or 3.85% in 2015. Thus, assuming that the percentage of Muslims remains constant at 3.0%, the Muslim population is expected to reach 0.14 million by 2050, and 0.16 million by 2100.

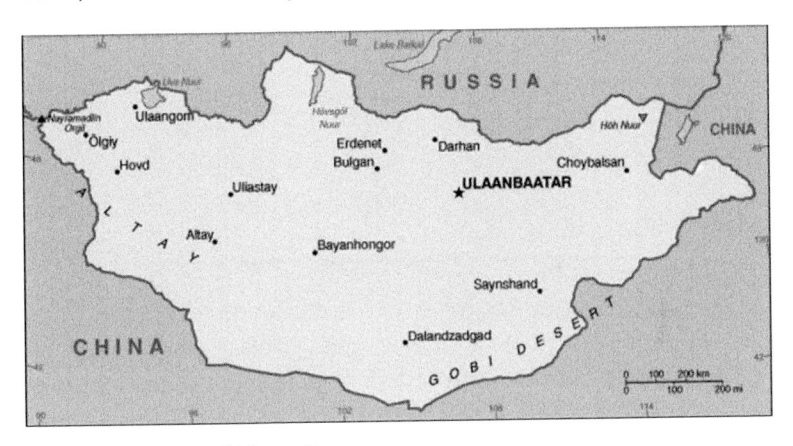

Figure 2.2.4   Map of Mongolia.

2.2.5   Tajikistan

The Muslim conquest of Tajikistan started by Qutaiba bnu Muslim al-Bahili during the reign of the sixth Umayyad Caliph al-Walid I bnu Abdel Malik bnu Marwan. In 96 H/714 AD he captured Khudjand; north of Tajikistan and capital of Sughd Province. Currently, the Republic of Tajikistan has an area of 143,100 sq. km and its map is presented in Fig. 2.2.5. It was conquered by Russia between 1860s and 1870s and gained its independence upon the dissolution of the Soviet Union in 1991. The current territory of Tajikistan was carved out by the Soviets from historic Turkistan in two stages. First, in 1924 the Tajik Autonomous Soviet Socialist Republic (ASSR) was created as part of Uzbek Soviet Socialist Republic (SSR). Then in 1929, Tajik ASSR was upgraded to Tajik SSR with the addition of the Sughd Province from Uzbek SSR, which has an area of 25,400 sq. km and is the northwest part of current Tajikistan.

Table 2.2.4 Evolution of the Muslim population in Mongolia

Year	Population	Muslims	%	Source
1920	645,000	20,000	3.10	[MAS23]es
1956	845,481	36,729	4.34	[MN00]e
1963	1,017,162	47,735	4.69	[MN00]e
1969	1,188,271	62,812	5.29	[MN00]e
1979	1,538,980	84,305	5.48	[MN00]e
1989	1,987,274	120,506	6.06	[MN00]e
2000	2,365,269	102,983	4.35	[MN00]e
2010	1,905,969	57,702	3.03	[MN10]c
2050	4,449,000	135,000	3.03	es
2100	5,387,000	163,000	3.03	es

The 1897 Russian Empire census indicated that Muslims made up 99% of the total population of Ferghana Oblast, which contained eastern half of Tajikistan, western half of Kyrgyzstan, and eastern portion of Uzbekistan. The rest of Tajikistan belonged to the Bukhara Emirate, which was a Russian dependent state, and census was not carried, but the total population was estimated at 2.5 million and all Muslim [SYB00]. Data of the 1897 census and estimate for these two regions is summarized in Table 2.2.5a. For comparison, the table also contains an estimate based on the 1897 of the total population living in the current border of Tajikistan and the current area of the country. Censuses since 1926 collected ethnic affiliation demography of the population. The number of members of each Muslim ethnicity since 1897 that was ever larger than one hundred is provided in Table 2.2.5b. As the name of the country suggests, the number of Tajiks with respect to the total population increased from

62% in 1989, to 80% in 2000 to 84% in 2010. Up to 1989, the Uzbek nationality included Lakai, Kongrat Durmen, Katagan, Barlos, Yuzi, Mingi, Kesamiry and Semizy.

According to census data and as summarized in Table 2.2.5c., the Muslim population increased in number but decreased in percentage from 0.64 million or 99.0% in 1897, to 0.82 million or 98.8% in 1926, to 1.23 million or 89.3% in 1937, to 1.31 million or 88.4% in 1939, to 1.62 million or 81.7% in 1959. However, it continued increasing since then in both number and percentage to 2.43 million or 83.9% in 1970, to 3.28 million or 86.1% in 1979, to 4.56 million or 89.6% in 1989, to 6.05 million or 98.7% in 2000, to 7.52 million or 99.5% in 2010.

Thus, assuming the percentage of Muslims will remain constant, the Muslim population is expected to reach 16 million by 2050 and 25 million by 2100.

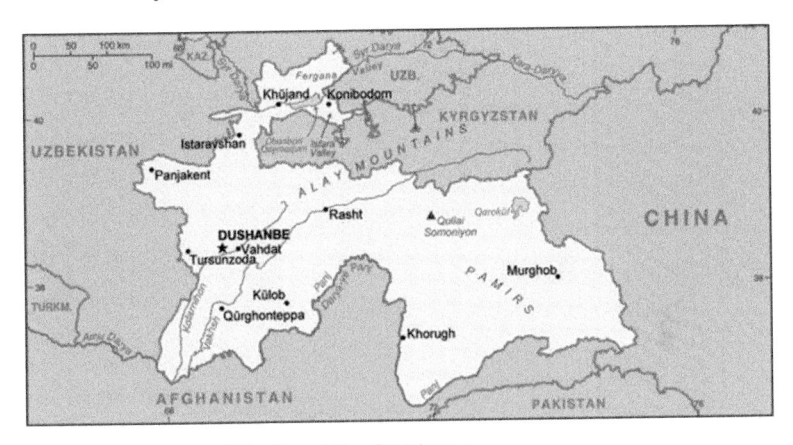

Figure 2.2.5 Map of the Republic of Tajikistan.

Table 2.2.5a 1897 Census data for territory covering current Tajikistan

Oblast	Population	Muslims	% Muslims	Area (km^2)
Bukhara	2,500,000	2,500,000	100	238,279
Ferghana	1,572,214	1,557,057	99.04	92,343
Total	4,072,214	4,057,057	99.63	330,622
Tajikistan	978,000			143,100

Table 2.2.5b Evolution of the top ten Muslim ethnic populations in Tajikistan

	1926	1937	1939	1959	1970	1979	1989	2000	2010
Tajik	618,954	840,600	883,966	1,051,164	1,629,920	2,237,048	3,172,420	4,898,382	6,373,834
Uzbek	175,627	332,300	353,478	454,433	665,662	873,199	1,197,841	936,703	926,344
Lakay								51,001	65,555
Kongrats								15,102	38,078
Durmeny								3,502	7,608
Katagany								4,888	7,601
Barlosy								3,743	5,271
Yuzy								1,053	3,798
Mingi								243	268
Kesamiry								13	156
Semizy								1	47
Kyrgyz	11,410	26,400	27,968	25,635	35,485	48,376	63,832	65,515	60,715
Turkmen	4,148	3,200	4,040	7,115	11,043	13,991	20,487	20,270	15,171
Tatar	950	16,600	18,296	56,893	70,803	79,529	79,442	19,077	6,530
Arab	3,260	2,300	2,290	1,297	248	176	276	14,450	4,184
Afghan	666	1,000	550	532	1,337	1,510	2,088	4,702	3,675
Gypsy	186		1,193	1,556	1,171	1,139	1,791	4,249	2,334

Turk	817,007		76	53	39	53	768	672	1,360
Kazakh	1,636	12,400	12,712	12,551	8,306	9,606	11,376	900	595
Persian					436	419	388	306	473
Azeri			6,064	1,182	1,553	2,153	3,556	798	371
Uighur			316	402	462	514	566	379	276
Bashkir	170		1,409	3,872	4,842	6,083	6,821	872	143
Chechen			12	67	54	80	128	47	20
Avar			5	39	68	120	191	19	13
Lezghin					258	245	307	51	13
Kabardian			78	133	124	135	244	11	8
Darghin			2		53	122	340	39	6
Kumyk					61	95	125	26	5
Circassian					95	87	86	14	5
Karakalpak			9	39	61	72	163	34	4
Lak			50	484	861	1,194	1,398	147	2
Total	817,007	1,234,800	1,312,514	1,617,447	2,432,942	3,275,946	4,564,634	6,047,209	7,524,463

Table 2.2.5c Evolution of the Muslim population in Tajikistan

Year	Population	Muslims	%	Source
1897	4,072,214	4,057,057	99.63	[SU]c
1926	827,167	817,007	98.77	[TJ00]e
1937	1,383,500	1,234,800	89.25	[TJ00]e
1939	1,484,440	1,312,514	88.42	[TJ00]e
1959	1,979,897	1,617,447	81.69	[TJ00]e
1970	2,899,602	2,432,942	83.91	[TJ00]e
1979	3,806,220	3,275,946	86.07	[TJ00]e
1989	5,092,603	4,564,634	89.63	[TJ10]e
2000	6,127,493	6,047,209	98.69	[TJ10]e
2010	7,564,502	7,524,463	99.47	[TJ10]e
2050	16,208,000	16,122,000	99.47	es
2100	25,328,000	25,194,000	99.47	es

2.2.6   Turkmenistan

The Muslim conquest of Turkmenistan started by Al-Ahnaf bnu Qais Attamimi in 22 H/643 AD during the reign of Caliph Omar bnul Khattab. Passing from current Iran, he conquered Marw or Merv, and currently Mary; pronounced Mur-ree. Current Turkmenistan has an area of 488,100 sq. km and its map is presented in Fig. 2.4.6. It was conquered by Russia between 1865 and 1885 and gained its independence upon the dissolution of the Soviet Union in 1991. The current territory of Tajikistan was carved out by the Soviets from historic Turkistan in 1924 as Turkmen Soviet Socialist Republic (SSR).

The 1897 Russian Empire census indicated that Muslims made up 88.1% of the total population of Trans-Caspian Oblast, which covered most of current Turkmenistan, and its northern part laid in Kazakhstan and Uzbekistan. The rest of Turkmenistan belonged to the Khiva Emirate, which was a Russian dependent state, and census was not carried, but the total population was estimated at 0.7 million and all Muslim [SYB00]. The northern part of Khiva laid in current Uzbekistan. Data of the 1897 census and estimate for these two regions is summarized in Table 2.2.6a. For comparison, the table also contains an estimate based on the 1897 of the total population living in the current border of Turkmenistan and the current area of the country.

Figure 2.2.6 Map of Turkmenistan.

Table 2.2.6a 1897 Census data for territory covering current Turkmenistan

Oblast	Population	Muslims	% Muslims	Area (km^2)
Khiva	700,000	700,000	100	57,809
Trans-Caspian	382,487	336,826	88.06	554,871
Total	1,082,487	1,036,826	95.78	612,680
Turkmenistan	835,000			1,167,551

Censuses since 1926 collected ethnic affiliation demography of the population, about fifty of whom are Muslim. The top sixteen Muslim ethnicities in Turkmenistan from largest in number are Turkmen, Uzbek, Kazakh, Tatar, Azeri, Balochi, Lezgin, Persian, Bashkir, Kurd, Tajik, Karakalpak, Lak, Dargin, Uighur and Afghan. The number of members of each of these ethnicities since 1926 is provided in Table 2.2.6b. As the name of the country suggests, almost 80% of Muslims are Turkmen.

Table 2.2.6b Evolution of the largest Muslim ethnic populations in Turkmenistan

	1926	1939	1959	1970	1979	1989	1995	2012
Turkmen	719,792	741,488	923,724	1,416,700	1,891,695	2,536,606	3,401,936	4,066,959
Uzbek	104,971	107,451	125,231	179,498	233,730	317,333	407,109	275,565
Kazakh	9,471	61,397	69,522	68,519	79,539	87,802	86,987	19,004
Tatar	4,769	19,517	29,946	36,457	40,432	39,277	36,355	
Azeri	4,229	7,442	12,868	16,775	23,548	33,365	36,586	49,852
Balochi	9,974	5,396	7,626	12,374	18,584	28,280	36,428	
Lezgin	206	539	1,831	4,243	7,562	10,425	9,553	
Persian	7,153	8,254	4,132	5,068	4,827	7,637	8,600	
Bashkir	426	957	1,894	2,607	3,914	4,678	3820	
Kurd	2,308	1,954	2,263	2,933	3,521	4,387	6,097	
Tajik	566	1,082	870	1,271	1,255	3,149	3,103	
Karakalpak	1,537	3,555	2,548	2,542	2,690	3,062	3,531	
Lak	NA	NA	1,120	1,590	1,939	2,441	2,821	
Dargin	NA	8	NA	1,599	1,091	1,626	1,550	
Ujghur	NA	822	836	1,111	1,205	1,308	1,204	
Afghan	3,947	493	578	879	984	1,256	1,808	
Total	869,349	960,355	1,184,989	1,754,166	2,316,516	3,082,632	4,047,488	

According to the 1897 census, which inquired on religious adherence of the population, based on ethnic census data and as shown in Table 2.2.6c, the Muslim population increased in number but decreased in percentage from 0.87 million or 86.9% in 1926, to 0.96 million or 76.7% in 1939. The decrease in percentage is due to the Russification policy of the Soviet Union. The Muslim population then continued to increase in both percentage and number to 1.18 million or 78.2% in 1959, to 1.75 million or 81.3% in 1970, to 2.3 million or 83.8% in 1979, to 3.1 million or 87.5% in 1989, to 4.0 million or 91.2% in 1995, to 4.5 million or 93.4% in 2012. According to survey data, the percentage of Muslims was 93.3% in 2000 [DHS] and 92.75% in 2005 [ABM].

The ethnic Russians decreased from 9.5% in 1989, to 6.7% in 1995, to 5.1% in 2012. The Ukrainians and Armenians decreased from 1.0% and 0.9% in 1989 to 0.5% and 0.8% in 1995, to 0.2% and 0.4% in 2012, respectively.

Thus, assuming that the percentage of the Muslim population continues to increase by a half of a percentage point per decade, the Muslim population is expected to be around eight million in the second half of this century, reaching 96% by 2050 and 98% by 2100.

Table 2.2.6c Evolution of the Muslim population in Turkmenistan

Year	Population	Muslims	%	Source
1897	1,082,487	1,036,826	95.78	[SU]c
1926	1,000,914	869,349	86.86	[SU]e
1939	1,251,883	960,355	76.71	[SU]e
1959	1,516,375	1,184,989	78.15	[SU]e
1970	2,158,880	1,754,166	81.25	[SU]e
1979	2,764,748	2,316,516	83.78	[SU]e
1989	3,522,717	3,082,632	87.51	[SU]e
1995	4,437,570	4,047,488	91.21	[TM95]e
2000	4,501,000	4,199,000	93.29	[DHS]s
2012	4,751,120	4,450,000	93.66	[TM12]e
2050	7,949,000	7,592,000	95.50	es
2100	8,421,000	8,253,000	98.00	es

2.2.7 Uzbekistan

The Muslim conquest of Uzbekistan started by Qutaiba bnu Muslim al-Bahili in 90 H/709 AD during the reign of the sixth Umayyad Caliph al-Walid I bnu Abdel Malik bnu Marwan. During this year he captured Bukhara in 90 H/709 AD, Samarkand in 93 H/712 AD and in 94 H/713 AD Fergana and Tashkent (formerly Chach or as-Shash until the tenth century). Currently, the Republic of Uzbekistan has an area of 447,400 sq. km and its map is presented in Fig. 2.2.7. It was conquered by Russia in the late nineteenth century and gained its independence upon the dissolution of the Soviet Union in 1991. The current territory of Tajikistan was carved out by the Soviets from historic Turkistan in two stages. First, in 1924 the Uzbek Soviet Socialist Republic (SSR) was created, which included current Uzbekistan and Tajikistan, but excluded Karakalpakstan; which is a 160,000 sq. km area northwestern part of current Uzbekistan. Then in 1929, Tajik ASSR was upgraded to Tajik SSR with the addition of the Sughd Province from Uzbek SSR, which has an area of 25,400 sq. km and is the northwest part of current Tajikistan. Karakalpak Autonomous Oblast (AO) was created in 1925 and was part of the Kazakh ASSR until 1930, then part of the Russian SFSR until 1932, after which it was elevated to Karakalpak ASSR, and then joined Uzbek SSR in 1936.

Figure 2.2.7 Map of Republic of Uzbekistan.

The 1897 Russian Empire census indicated that Muslims made up 97.5% of the total population of Samarkand Oblast, whose area belonged to current Uzbekistan. The census also showed that Muslims made up 99% of the total population of Ferghana Oblast, which contained eastern portion of Uzbekistan, eastern half of Tajikistan, and western half of Kyrgyzstan. The rest of Uzbekistan belonged to the Bukhara Emirate, which was a Russian dependent state, and census was not carried, but the total population was estimated at 2.5 million and all Muslim [SYB00]. The Northern half of Bukhara Emirate belonged to Uzbekistan, while the southern half belonged to Tajikistan. Data of the 1897 census and estimate for these three regions is summarized in Table 2.2.7a. For comparison, the table also contains an estimate based on the 1897 of the total population living in the current border of Turkmenistan and the current area of the country.

Table 2.2.7a 1897 Census data for territory covering current Uzbekistan

Oblast	Population	Muslims	% Muslims	Area (km²)
Bukhara	2,500,000	2,500,000	100	238,279
Ferghana	1,572,214	1,557,057	99.04	92,343
Samarkand	860,021	838,861	97.54	68,964
Total	4,932,235	4,895,918	99.26	161,307
Uzbekistan	4,202,000			447,400

Censuses since 1926 collected ethnic affiliation demography of the population, about fifty of whom are Muslim. The top seventeen Muslim ethnicities in Uzbekistan from largest in number are Uzbek, Tajik, Kazakh, Tatar, Karakalpak, Kyrgyz, Turkmen, Turk, Azeri, Uighur, Bashkir, Persian, Gypsy (Roma), Lezgin, Lak, Arab and Dargin. The number of members of each of these ethnicities since 1926 is provided in Table 2.2.7b. As the name of the country suggests, almost 80% of Muslims are Uzbek.

Based on ethnic census data as shown in Table 2.2.7b, the Muslim population changed from 4.93 million or 99.3% in 1897, to 4.39 million or 92.4% in 1926 (including Karakalpak AO), to 6.0 million or 95.6% in 1939, to 6.56 million or 81.0% in 1959.

The Muslim population then continued increasing in both number and percentage to 9.90 million or 82.8% in 1970, to 13.17 million or 85.6% in 1979, to 17.52 million or 88.5% in 1989. According to survey data, the percentage of Muslims was 94.5% in 1996, then 96.3% in 2002, and 95.1% in 2011.

Table 2.2.7b Evolution of the top seventeen Muslim ethnic populations in Uzbekistan

	1926	1939	1959	1970	1979	1989
Uzbek	3,467,226	4,804,096	5,038,273	7,733,541	10,569,007	14,142,475
Tajik	350,670	317,560	311,375	457,356	594,627	933,560
Kazakh	191,126	305,416	335,267	549,312	620,136	808,227
Tatar	28,335	147,157	444,810	577,757	648,764	656,601
Karakalpak	142,688	181,420	168,274	230,273	297,788	411,878
Kyrgyz	79,610	89,044	92,725	110,864	142,182	174,907
Turkmen	31,492	46,543	54,804	71,066	92,285	121,578
Turk	371	474	21,269	46,398	48,726	106,302
Azeris	20,764	3,645	40,511	40,431	59,779	44,410
Uighur	36,349	50,638	19,377	24,039	29,104	35,762
Bashkir	624	7,516	13,500	21,069	25,879	34,771
Persian	9,185	18,181	8,883	16,316	20,026	24,779
Gypsy	3,710	5,487	7,860	11,371	12,581	16,397
Lezgin	329	746	716	1,598	2,049	3,071
Lak	NA	NA	1,072	1,771	2,248	2,807
Arab	27,977	18,939	5,407	3,425	2,039	2,805
Dargin	NA	28	NA	1,346	1,385	2,030
Total	4,390,456	5,996,890	6,564,123	9,897,933	13,168,605	17,522,360

Thus, assuming that the percentage of the Muslim population continues to increase by a half of a percentage point per decade, the Muslim population is expected to reach 42 million in the second half of this century, making up 97% by 2050 and 99.5% by 2100. The data is summarized in Table 2.2.7c.

Table 2.2.7c Evolution of the Muslim population in Uzbekistan

Year	Population	Muslims	%	Source
1897	4,932,235	4,895,918	99.26	[SU]c
1926	4,750,175	4,390,456	92.43	[SU]e
1939	6,271,269	5,996,890	95.62	[SU]e
1959	8,105,704	6,564,123	80.98	[SU]e
1970	11,959,582	9,897,933	82.76	[SU]e
1979	15,389,307	13,168,605	85.57	[SU]e
1989	19,810,077	17,522,360	88.45	[SU]e
1996	23,087,000	21,840,000	94.60	[DHS]s
2002	25,089,000	24,160,000	96.30	[DHS]s
2011	28,158,000	26,770.000	95.07	[WVS]s
2050	42,942,000	41,654,000	97.00	es
2100	42,271,000	42,059,000	99.50	es

2.2.8 Regional Summary and Conclusion

Islam entered Central Asia after a decade from the death of Prophet Mohammed peace and blessings upon him. Islam spread fast the following century and by the start of the second millennium the majority of this region was Muslim. Huge efforts were exerted by the Russians in the last two centuries to exterminate Islam from this region, which reduced the Muslim population by twenty percentage points. However, with the collapse of the Soviet Union, the percentage of Muslims returned to that prior to the Russian occupation. This made Central Asia have the second largest concentration of Muslims among the six regions spanning Asia. This is expected to remain so for the next three centuries.

The following tables present centennial data from 600 AD to 2100 AD (or approximately 1 H to 1500 H) in Table 2.2a and decennial data from 1790 AD to 2100 AD (or 1210 H to 1520 H) in Tables 2.2b and 2.2c for current countries in Central Asia. The data includes total population in thousands (P), the percentage of which is Muslim ($M\%$), the corresponding Muslim population in thousands (M), and the annual population growth rate (APGR, or $G\%$) of the total population in this region.

Table 2.2a Centennial estimates of the Muslim population (×1000) in Central Asia: 600–2100 AD (1–1500 H)

		600	700	800	900	1000	1100	1200	1300
Afghanistan	P	850	900	950	1,000	1,078	1,200	1,400	1,600
	M%	—	30.00	70.00	80.00	85.00	85.00	85.00	85.00
	M	—	270	665	800	916	1,020	1,190	1,360
Kazakhstan	P	650	680	710	740	773	850	900	950
	M%	—	—	1.00	5.00	5.00	5.00	5.00	5.00
	M	—	—	7	37	39	43	45	48
Kyrgyzstan	P	160	168	176	184	192	210	225	240
	M%	—	—	5.00	10.00	10.00	10.00	20.00	30.00
	M	—	—	9	18	19	21	45	72
Mongolia	P	160	170	180	190	203	240	270	300
	M%	—	—	—	—	—	—	—	1.00
	M	—	—	—	—	—	—	—	3
Tajikistan	P	164	171	178	185	192	210	220	235
	M%	—	—	30.00	60.00	90.00	99.00	99.00	99.00
	M	—	—	53	111	173	208	218	233
Turkmenistan	P	136	143	150	157	164	180	190	200
	M%	—	10.00	50.00	90.00	99.00	99.00	99.00	99.00
	M	—	14	75	141	162	178	188	198
Uzbekistan	P	700	730	760	790	823	890	950	1,010
	M%	—	—	30.00	60.00	90.00	99.00	99.00	99.00
	M	—	—	228	474	741	881	941	1,000
Total	P	2,820	2,962	3,104	3,246	3,425	3,780	4,155	4,535
	M%	—	9.60	33.42	48.73	59.85	62.19	63.21	64.23
	M	—	284	1,037	1,582	2,050	2,351	2,626	2,913
	G%		0.049	0.047	0.045	0.054	0.099	0.095	0.088

Table 2.2a (Continued)

		1400	1500	1600	1700	1800	1900	2000	2100
Afghanistan	P	1,800	1,920	2,227	2,589	3,170	5,219	20,780	74,938
	M%	85.00	85.00	85.00	85.00	87.63	94.99	99.77	99.77
	M	1,530	1,632	1,893	2,201	2,778	4,958	20,732	74,766
Kazakhstan	P	1,000	1,082	1,236	1,391	1,960	3,944	14,923	27,918
	M%	20.00	70.00	80.00	90.00	86.45	86.45	61.42	89.00
	M	200	757	989	1,252	1,694	3,410	9,166	24,847
Kyrgyzstan	P	255	269	307	346	485	980	4,921	10,985
	M%	60.00	90.00	99.00	99.00	95.60	95.60	84.92	89.00
	M	153	242	304	343	464	937	4,179	9,777
Mongolia	P	330	362	420	489	600	707	2,397	5,387
	M%	2.00	3.00	4.00	5.00	3.10	3.10	4.35	3.03
	M	7	11	17	24	19	22	104	163
Tajikistan	P	250	268	307	245	480	978	6,216	25,328
	M%	99.00	99.00	99.00	100.00	99.63	99.63	98.69	99.47
	M	248	265	304	245	478	974	6,135	25,194
Turkmenistan	P	215	229	262	295	420	835	4,516	8,421
	M%	99.00	99.00	99.00	99.00	95.78	95.78	91.21	98.00
	M	213	227	259	292	402	800	4,119	8,253
Uzbekistan	P	1,080	1,153	1,318	1,482	2,000	4,202	24,770	42,271
	M%	99.00	99.00	99.00	99.00	99.26	99.26	96.30	99.50
	M	1,069	1,141	1,305	1,467	1,985	4,171	23,853	42,059
Total	P	4,930	5,283	6,077	6,837	9,115	16,865	78,523	195,248
	M%	69.35	80.94	83.44	85.18	85.80	90.55	86.97	94.78
	M	3,419	4,276	5,071	5,824	7,820	15,271	68,288	185,059
	G%	0.084	0.069	0.140	0.118	0.288	0.615	1.538	0.911

Table 2.2b Decennial estimates of the Muslim population (×1000) in Central Asia: 1790–1940 AD (1210–1360 H)

		1790	1800	1810	1820	1830	1840	1850	1860
Afghanistan	P	3,115	3,170	3,225	3,280	3,400	3,600	3,800	4,000
	M%	87.63	87.63	87.63	87.63	87.63	87.63	87.63	87.63
	M	2,730	2,778	2,826	2,874	2,979	3,155	3,330	3,505
Kazakhstan	P	1,900	1,960	2,020	2,081	2,220	2,360	2,500	2,650
	M%	86.45	86.45	86.45	86.45	86.45	86.45	86.45	86.45
	M	1,643	1,694	1,746	1,799	1,919	2,040	2,161	2,291
Kyrgyzstan	P	470	485	500	517	550	590	630	670
	M%	95.60	95.60	95.60	95.60	95.60	95.60	95.60	95.60
	M	449	464	478	494	526	564	602	641
Mongolia	P	590	600	610	619	630	640	650	660
	M%	3.10	3.10	3.10	3.10	3.10	3.10	3.10	3.10
	M	18	19	19	19	20	20	20	20
Tajikistan	P	460	480	500	516	550	590	630	670
	M%	99.63	99.63	99.63	99.63	99.63	99.63	99.63	99.63
	M	458	478	498	514	548	588	628	668
Turkmenistan	P	410	420	430	441	470	500	530	560
	M%	95.78	95.78	95.78	95.78	95.78	95.78	95.78	95.78
	M	393	402	412	422	450	479	508	536
Uzbekistan	P	1,900	2,000	2,100	2,217	2,350	2,500	2,650	2,800
	M%	99.26	99.26	99.26	99.26	99.26	99.26	99.26	99.26
	M	1,886	1,985	2,084	2,201	2,333	2,482	2,630	2,779
Total	P	8,845	9,115	9,385	9,671	10,170	10,780	11,390	12,010
	M%	85.66	85.80	85.92	86.07	86.28	86.52	86.74	86.93
	M	7,577	7,820	8,064	8,324	8,775	9,327	9,879	10,440
	G%		0.301	0.292	0.300	0.503	0.583	0.550	0.530

Table 2.2b (Continued)

		1870	1880	1890	1900	1910	1920	1930	1940
Afghanistan	P	4,207	4,520	4,857	5,219	5,730	6,125	6,673	7,494
	M%	87.63	87.63	87.63	94.99	94.99	99.95	99.95	99.95
	M	3,687	3,961	4,256	4,958	5,443	6,122	6,670	7,490
Kazakhstan	P	2,809	3,138	3,506	3,944	5,982	4,898	5,449	6,146
	M%	86.45	86.45	86.45	86.45	86.45	62.53	62.53	42.97
	M	2,428	2,713	3,031	3,410	5,171	3,063	3,407	2,641
Kyrgyzstan	P	698	780	871	980	1,100	1,217	1,354	1,484
	M%	95.60	95.60	95.60	95.60	95.60	79.38	79.38	67.73
	M	667	746	833	937	1,052	966	1,075	1,005
Mongolia	P	668	681	694	707	725	729	734	740
	M%	3.10	3.10	3.10	3.10	3.10	3.10	3.10	3.10
	M	21	21	22	22	22	23	23	23
Tajikistan	P	697	778	870	978	1,034	1,215	1,352	1,481
	M%	99.63	99.63	99.63	99.63	99.63	98.75	98.77	88.42
	M	694	775	867	974	1,030	1,200	1,335	1,310
Turkmenistan	P	595	665	742	835	1,081	1,037	1,154	1,265
	M%	95.78	95.78	95.78	95.78	95.78	86.86	86.86	76.71
	M	570	637	711	800	1,035	901	1,002	970
Uzbekistan	P	2,993	3,344	3,735	4,202	4,334	5,219	5,806	6,363
	M%	99.26	99.26	99.26	99.26	99.26	92.43	92.43	95.62
	M	2,971	3,319	3,707	4,171	4,302	4,824	5,366	6,084
Total	P	12,667	13,906	15,275	16,865	19,986	20,440	22,522	24,973
	M%	87.14	87.53	87.90	90.55	90.34	83.65	83.82	78.18
	M	11,038	12,172	13,426	15,271	18,055	17,098	18,879	19,523
	G%	0.533	0.933	0.939	0.990	1.698	0.225	0.970	1.033

Table 2.2c Decennial estimates of the Muslim population (×1000) in Central Asia: 1950–2100 AD (1370–1520 H)

		1950	1960	1970	1980	1990	2000	2010	2020
Afghanistan	P	7,752	8,997	11,174	13,357	12,412	20,780	29,186	38,928
	M%	99.77	99.77	99.77	99.77	99.77	99.77	99.77	99.77
	M	7,734	8,976	11,148	13,326	12,384	20,732	29,118	38,839
Kazakhstan	P	6,703	9,935	13,036	14,796	16,384	14,923	16,252	18,777
	M%	37.14	37.14	38.73	42.80	47.05	61.42	70.56	73.00
	M	2,489	3,690	5,049	6,333	7,709	9,166	11,468	13,707
Kyrgyzstan	P	1,740	2,170	2,970	3,611	4,373	4,921	5,422	6,524
	M%	59.23	59.23	61.82	66.40	71.94	84.92	84.36	85.00
	M	1,031	1,285	1,836	2,398	3,146	4,179	4,574	5,546
Mongolia	P	780	956	1,279	1,690	2,184	2,397	2,720	3,278
	M%	4.34	4.69	5.29	5.48	6.06	4.35	3.03	3.03
	M	34	45	68	93	132	104	82	99
Tajikistan	P	1,532	2,087	2,930	3,905	5,284	6,216	7,527	9,538
	M%	81.69	81.69	83.91	86.07	89.63	98.69	99.47	99.47
	M	1,251	1,705	2,459	3,361	4,736	6,135	7,487	9,487
Turkmenistan	P	1,211	1,603	2,195	2,877	3,684	4,516	5,087	6,031
	M%	78.15	78.15	81.25	83.78	87.51	91.21	93.66	94.00
	M	946	1,253	1,784	2,410	3,224	4,119	4,765	5,669
Uzbekistan	P	6,264	8,526	12,080	15,899	20,398	24,770	28,516	33,469
	M%	80.98	80.98	82.76	85.57	88.45	96.30	95.07	95.50
	M	5,073	6,905	9,998	13,605	18,042	23,853	27,110	31,963
Total	P	25,982	34,274	45,664	56,134	64,719	78,523	94,710	116,546
	M%	71.43	69.61	70.82	73.97	76.29	86.97	89.33	90.36
	M	18,558	23,859	32,340	41,525	49,373	68,288	84,605	105,310
	G%	0.396	2.770	2.869	2.064	1.423	1.933	1.874	2.075

Table 2.2c *(Continued)*

		2030	2040	2050	2060	2070	2080	2090	2100
Afghanistan	P	48,094	56,912	64,683	70,845	74,975	76,870	76,693	74,938
	M%	99.77	99.77	99.77	99.77	99.77	99.77	99.77	99.77
	M	47,983	56,781	64,534	70,682	74,803	76,693	76,517	74,766
Kazakhstan	P	20,639	22,370	24,024	25,243	26,223	27,042	27,650	27,918
	M%	75.00	77.00	79.00	81.00	83.00	85.00	87.00	89.00
	M	15,479	17,225	18,979	20,447	21,765	22,986	24,055	24,847
Kyrgyzstan	P	7,446	8,307	9,126	9,775	10,272	10,657	10,904	10,985
	M%	85.50	86.00	86.50	87.00	87.50	88.00	88.50	89.00
	M	6,366	7,144	7,894	8,505	8,988	9,378	9,650	9,777
Mongolia	P	3,716	4,089	4,449	4,738	4,947	5,133	5,296	5,387
	M%	3.03	3.03	3.03	3.03	3.03	3.03	3.03	3.03
	M	113	124	135	144	150	156	160	163
Tajikistan	P	11,557	13,846	16,208	18,437	20,633	22,494	24,110	25,328
	M%	99.47	99.47	99.47	99.47	99.47	99.47	99.47	99.47
	M	11,496	13,772	16,122	18,339	20,523	22,374	23,982	25,194
Turkmenistan	P	6,782	7,409	7,949	8,277	8,460	8,553	8,539	8,421
	M%	94.50	95.00	95.50	96.00	96.50	97.00	97.50	98.00
	M	6,409	7,038	7,592	7,946	8,164	8,297	8,326	8,253
Uzbekistan	P	37,418	40,608	42,942	44,083	44,403	44,039	43,314	42,271
	M%	96.00	96.50	97.00	97.50	98.00	98.50	99.00	99.50
	M	35,922	39,187	41,654	42,981	43,515	43,378	42,881	42,059
Total	P	135,653	153,542	169,382	181,398	189,912	194,788	196,506	195,248
	M%	91.24	92.01	92.64	93.19	93.68	94.08	94.44	94.78
	M	123,768	141,272	156,910	169,043	177,907	183,262	185,572	185,059
	G%	1.518	1.239	0.982	0.685	0.459	0.254	0.088	−0.064

2.3 Muslims in Arabian Asia

This region consists of twelve countries: Bahrain, Iraq, Jordan, Kuwait, Lebanon, Oman, Palestine/Israel, Qatar, Saudi Arabia, Syria, United Arab Emirates, and Yemen. Islam has started in this region in 609 and spread to the rest of the Arabian Peninsula and Jordan by 629 during the time of Prophet Muhammad peace and blessings be upon him (PBUH). Islam then spread to the rest of this region during the reign of Caliph Omar bnul Khattab, when Muslims advanced north capturing the southern and eastern part of Turkey by 19 H/640 AD.

The Muslim population has likely increased from 5.5 million or 62.2% of the total population of this region in 700 AD, to 6.6 million or 73.7% in 800 AD, to 7.4 million or 81.7% in 900 AD, to 8.1 million or 88.2% in 1000 AD, to 8.0 million or 90.3% in 1100 AD, to 7.8 million or 92.6% in 1200 AD, to 7.5 million or 93.1% in 1300 AD, to 7.2 million or 93.4% in 1400 AD, to 6.9 million or 93.7% in 1500 AD, to 7.1 million or 93.7% in 1600 AD, to 6.7 million or 92.8% in 1700 AD, to 7.5 million or 90.8% in 1800 AD, to 11 million or 91.5% in 1900, to 94 million or 89.3% in 2000, to 158 million or 89.5% in 2020, and is projected to reach 243 million or 91% by 2050 and 291 million or 92% by 2100.

A plot of centennial estimates of the Muslim population and its percentage with respect to the total population in this region from 600 to 2100 is provided in Fig. 2.3a. A zoom in of this plot, providing a plot of decennial estimates of the Muslim population and its percentage with respect to the total population in this region from 1900 to 2100 is provided in Fig. 2.3b. This shows that the Muslim population increased slowly until 1960, then it increased substantially afterwards and towards the end of this century. The percentage of Muslims in this region was around 91% until 1910, then dipped to 85% in 1960, then reached 90% in 1990 and is expected to exceed 91% by the end of this century.

The corresponding individual data for each country in this region is discussed below. In Section 2.3.13, the total population in each country in this region and the corresponding percentage and number of Muslims is presented centennially in Table 2.3a from 600 to 2100 and decennially in Tables 2.3b and 2.3c from 1790 to 2100.

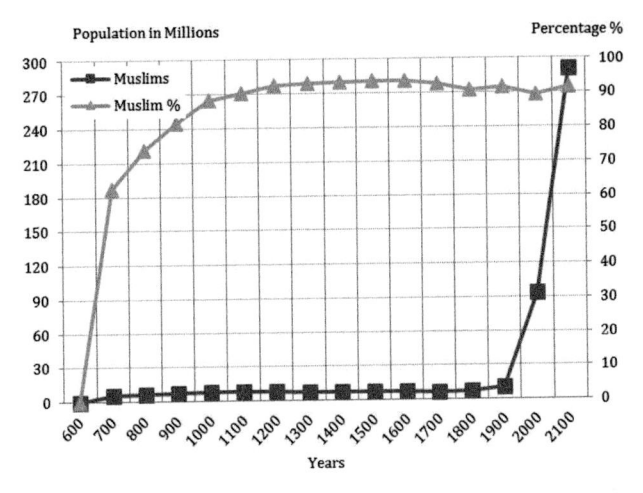

Figure 2.3a Plot of centennial estimates of the Muslim population and its percentage of the total population in Arabian Asia: 600–2100 AD (1–1500 H).

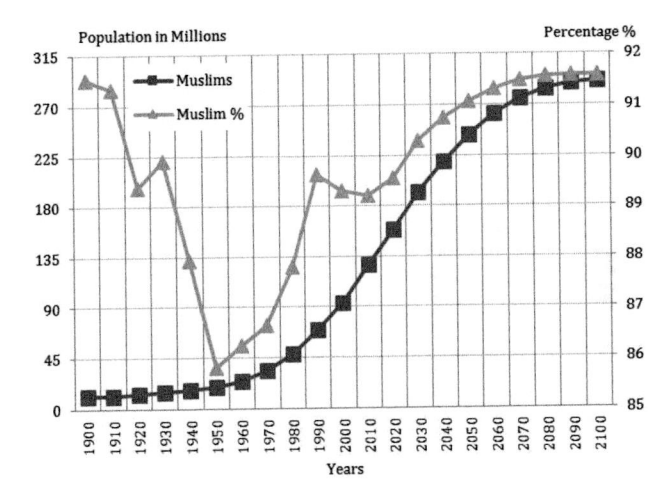

Figure 2.3b Plot of decennial estimates of the Muslim population and its percentage of the total population in Arabian Asia: 1900–2100 AD (1320–1520 H).

2.3.1 Bahrain

The Kingdom of Bahrain was occupied by the British in late nineteenth century and gained its independence from the UK in

1971. It is an Island nation with an area of 760 sq. km consisting of the main island of Bahrain (558 sq. km) and 32 nearby much smaller islands, the largest of which are Hawar (50 sq. km), Umm an Na'san (19 sq. km), Muharraq (18 sq. km) and Sitrah (10 sq. km). The surface area of this country is gradually expanding through reclamation of land from the nearby shallow sea and the construction of dozens of man-made small islands connected by a series of bridges to the main island. A map of the country is presented in Fig. 2.3.1.

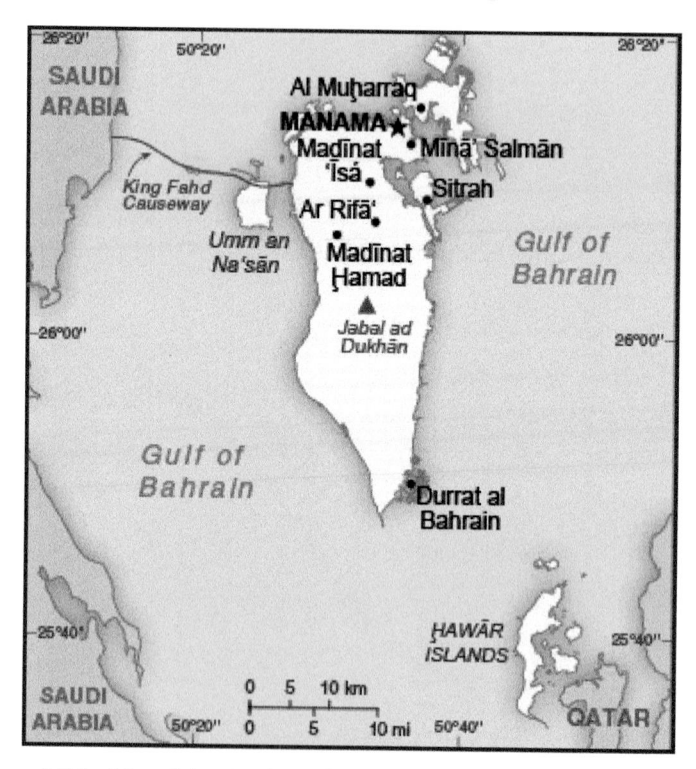

Figure 2.3.1 Map of the Kingdom of Bahrain.

In 628, Prophet Muhammad sent his companion Al-Ala bnul Hadrami as an envoy to Munther bnu Sawa Attamimi, the ruler of the historical region of Bahrain, which included current Kuwait, Bahrain, Qatar, and East of Saudi Arabia. The Prophet invited this ruler to Islam, which he accepted along with his population and Muslims remained the majority of the population of the whole region since then. The entire population was estimated to be Muslim by the turn of the twentieth century. The first census was conducted in 1941 and subsequent censuses were held each decade and they did include religious information. As shown in Table 2.3.1, these censuses show a constant decrease in the percentage of the Muslim population from 98% in 1941, to 70% in 2010. This is due to the discovery of oil and the subsequent import of foreign workers, some of which are non-Muslim. Indeed, the census data show that the Muslim population increased from 0.09 million or 98.1% in 1941, to 0.11 million or 96.1% in 1950, to 0.14 million or 94.8% in 1959, to 0.17 million or 95.3% in 1965, to 0.21 million or 95.7% in 1971, to 0.30 million or 85.0% in 1981, to 0.42 million or 81.8% in 1991, to 0.53 million or 81.2% in 2001, to 0.87 million or 70.2% in 2010. Assuming that this percentage remains constant, the Muslim population is expected to increase to 1.6 million throughout the second half of this century.

Table 2.3.1 Evolution of the Muslim population in Bahrain

Year	Population	Muslims	%	Source
1875	37,000	37,000	100	[SYB90]es
1910	90,000	90,000	100	[SYB10]es
1922	100,000	100,000	100	[MAS23]es
1941	89,970	88,298	98.14	[ROD]c
1950	109,650	105,401	96.12	[UN56]c
1959	143,135	135,720	94.82	[UN63]c
1965	182,203	173,594	95.28	[UN71]c
1971	216,078	206,708	95.66	[UN71]c
1981	350,798	298,140	84.99	[UN81]c
1991	508,037	415,427	81.77	[UN04]c
2001	650,604	528,393	81.22	[UN]c
2010	1,234,571	866,888	70.22	[BH]c
2050	2,316,000	1,626,000	70.22	es
2100	2,252,000	1,581,000	70.22	es

2.3.2 Iraq

The Republic of Iraq has an area of 438,317 sq. km and its map is presented in Fig. 2.3.2. It was taken by the British from the Ottomans in 1920 and gained its independence from the UK in 1932. The west part was conquered by Muslims during the reign of Caliph Abu Bakr in 12 H or 633 AD. The Muslim troops were under the leadership of the Prophet's companion Khalid bnul Walid al Makhzomi. In this year they conquered Basra, Kofa, and the Anbar province. Their fight was with Persians who were in control of Iraq. The Muslim attacks continued under Caliph Omar. The decisive battle in the conquest of Iraq was the battle of Qadissiya, in the middle of Iraq, south of Bagdad, in 14 H or 635 AD. Muslim troops were led by the Prophet's companion Saad bnu Abi Waqqas. The conquest of Iraq was culminated in 16 H or 637 AD when Mosul, in north Iraq, was conquered.

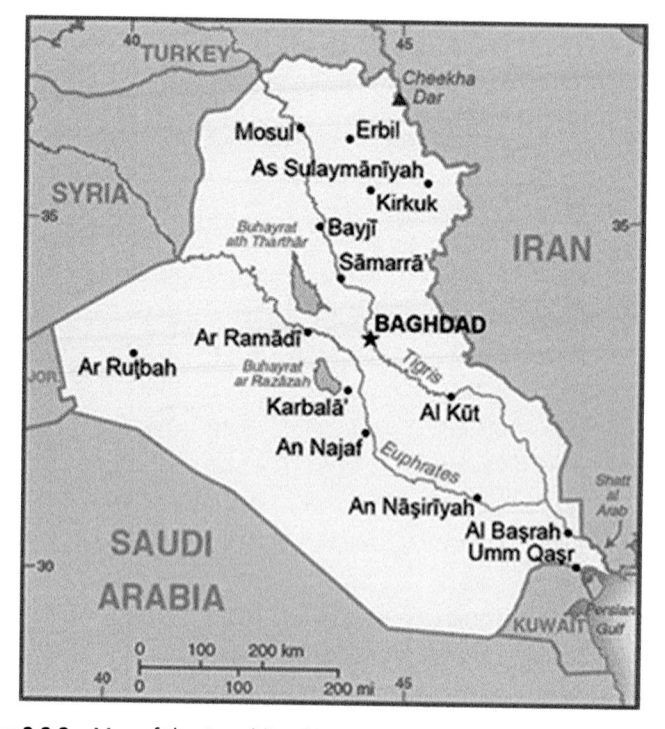

Figure 2.3.2   Map of the Republic of Iraq.

As shown in Table 2.3.2 and based on census data, the Muslim population increased from 0.42 million or 94.8% in 1580, to 0.79 million or 93.0% in 1881, to 1.35 million or 91.7% in 1894, to 1.38 million or 91.7% in 1914, to 2.6 million or 92.7% in 1920, to 3.1 million or 93.1% in 1935, to 4.5 million or 93.6% in 1947, to 6.1

million or 95.6% in 1957, to 7.7 million or 96.0% in 1965, to 11.5 million or 96.7% in 1977.

According to the World Values Survey (WVS), the percentage of Muslims was 99.15% in 2006 and 99.00% in 2013. Thus, assuming that the percentage of Muslims remains at 99%, the Muslim population is projected to increase to seventy million by 2050 and 107 million by 2100.

Table 2.3.2 Evolution of the Muslim population in Iraq

Year	Population	Muslims	%	Source
1580	441,485	418,295	94.75	[EGH]c
1881	845,404	786,386	93.02	[EGH]c
1894	1,471,065	1,349,555	91.74	[EGH]c
1914	1,504,306	1,380,050	91.74	[EGH]c
1920	2,849,282	2,640,701	92.68	[SYB20]c
1935	3,370,111	3,136,632	93.07	[SYB50]c
1947	4,816,185	4,508,779	93.62	[UN56]c
1957	6,337,354	6,057,493	95.58	[UN63]c
1965	8,032,520	7,711,712	96.01	[UN71]c
1977	11,862,620	11,474,293	96.73	[ROD]c
2006	27,717,000	27,482,000	99.15	[WVS]s
2013	34,107,000	33,766,000	99.00	[WVS]s
2050	70,940,000	70,231,000	99.00	es
2100	107,711,000	106,634,000	99.00	es

2.3.3 Jordan

The Hashemite Kingdom of Jordan has an area of 89,342 sq. km and its map is presented in Fig. 2.3.3. It was taken by the British from the Ottomans in 1920, gained its independence from the UK in 1946, and changed its name from Transjordan in 1950.

Prophet Mohammed (PBUH) has sent his companion Harith bnu Umair al Azdi to the leader of Bosra (currently ruins south of Syria near the borders with Jordan) as an envoy inviting him and his people to Islam. So, the governor of al-Balqa (now in Jordan) Shurahbil bnu Amr al-Ghassani, who was representative of the Caesar in this region, captured him and killed him in the city of Mota (in south Jordan) before reaching his destination. This was the only messenger of the Prophet that was killed and was interpreted as a declaration of war.

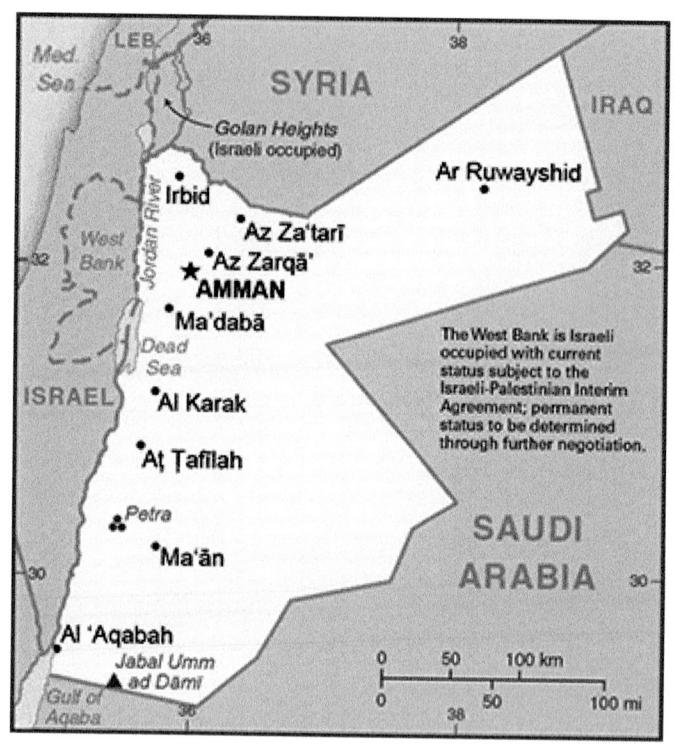

Figure 2.3.3 Map of the Hashemite Kingdom of Jordan.

The Prophet summoned 3,000 of his companions to answer the Ghassanids action. They in turn met the Muslim troops by 200,000 soldiers collected from neighboring regions, half of which was supplied by Heraclius, the Byzantine leader. The troops met in the city of Mota in 8 H or 629 AD, and the Battle of Mota started. Muslims lost a dozen people including three consecutive leaders of the Muslim troops, then Khalid bnul Walid took over and managed to retreat back to Medina. The casualties in the other side were in the thousands.

The second attempt was the Troops of Usama that the Prophet prepared and consisted of 700 of his companions led by Usama bnu Zaid bnu Haritha. The Prophet died before the troops continue their journey to now Jordan and Palestine (the lands of Balqa and Daroom). When the first Caliph took over, he went with the wish of the Prophet, and the troops went and came back in 11 H, or 632 AD. The conquest of Jordan was completed by the decisive battle of Yarmook in north Jordan by the Syrian border. The Muslim troops were led by the Prophet's companion Khalid bnul Walid al-Makhzomi in 13 H or 634 AD. During this battle, Caliph Abu Bakr passed away and Caliph Omar took over.

According to the 1914 Ottoman census, the Muslim population in this area was 0.12 million or 78.4% of the total population. Estimates of the Muslim population increased from 0.23 million or 88.5% in 1929, to 0.41 million or 91.1% in 1949. The first post-independence census was conducted in 1961. The census data indicate that the Muslim population increased from 1.6 million or 93.6% in 1961, to 2.0 million or 95.5% in 1979.

According to Demographic Health Survey (DHS), the percentage of Muslims was 97.10% in 1997. Later World Values Survey (WVS) put the percentage at 95.26% in 2001, then 97.67% in 2007 and 96.92% in 2014.

Thus, assuming that the percentage of Muslims remains at 98%, the Muslim population will increase to around thirteen million throughout the second half of this century. The data is summarized in Table 2.3.3.

Table 2.3.3 Evolution of the Muslim population in Jordan

Year	Population	Muslims	%	Source
1914	153,593	120,490	78.45	[TR14]c
1929	260,000	230,000	88.46	[SYB31]es
1949	450,000	410,000	91.11	[SYB51]es
1961	1,706,226	1,596,745	93.58	[UN63]c
1979	2,132,997	2,036,407	95.47	[UN88]c
1997	4,545,000	4,413,000	97.10	[DHS]s
2001	4,850,000	4,620,000	95.26	[WVS]s
2007	5,759,000	5,625,000	97.67	[WVS]s
2014	7,416,000	7,187,500	96.92	[WVS]s
2050	12,932,000	12,674,000	98.00	es
2100	13,644,000	13,371,000	98.00	es

2.3.4 Kuwait

The State of Kuwait's area is 17,818 sq. km and its map is presented in Fig. 2.3.4. It was occupied by the British in 1899 and gained its independence from the UK in 1961. In 1922, the entire population was Muslim, which decreased to 98.8% in 1949. The first census in this country was conducted in 1957, however, after the 1980 census information on religious affiliation were not published. The published data show a constant decrease in the percentage of the Muslim population. This is due to the discovery of oil and the subsequent import of foreign workers, some of which are non-Muslim. As shown in Table 2.3.4, the available census data show an increase in number but decrease in percentage of the Muslim population from 0.19 million or 94.4% in 1957, to 0.44 million or 94.1% in 1965, to 0.94 million or 95.0% in 1975, to 1.24 million or 91.5% in 1980 and 3.44 million or 74.4% in 2018. Thus, assuming that this percentage remains constant, the Muslim population is expected to increase to four million by 2050 and five million by 2100.

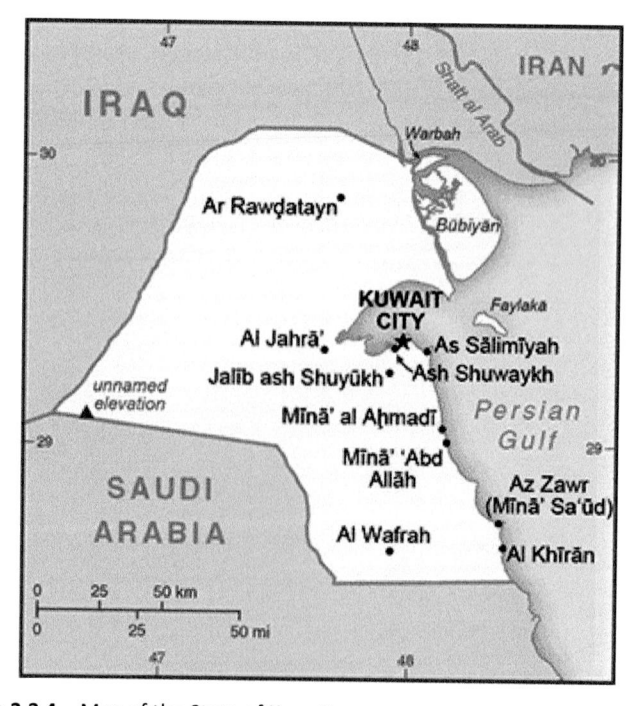

Figure 2.3.4 Map of the State of Kuwait.

Table 2.3.4 Evolution of the Muslim population in Kuwait

Year	Population	Muslims	%	Source
1922	40,000	40,000	100	[MAS23]es
1949	100,000	98,800	98.80	[MAS55]es
1957	206,473	194,894	94.39	[UN63]c
1965	439,687	439,687	94.08	[UN71]c
1975	944,781	944,781	94.97	[UN83]c
1980	1,242,708	1,242,708	91.51	[UN88]c
2018	4,621,638	3,437,061	74.37	[KW]c
2050	5,393,000	4,011,000	74.37	es
2100	6,189,000	4,603,000	74.37	es

2.3.5 Lebanon

The Muslim presence in Lebanon started in 13 H or 634 AD, during the reign of Caliph Omar, after opening Damascus. The Muslim troops were led by the Prophet's companion Khalid bnul Walid. Currently, the Republic of Lebanon has an area of 10,400 sq. km and its map is presented in Fig. 2.3.5. It was taken by the French from the Ottomans in 1920 and gained its independence from France in 1943. According to Ottoman census, the Muslim population of the Tripoli/Beirut province increased from 0.20 million or 76.4% in 1580 to 0.54 million or 59.1% in 1881. This province, however, included the costal side of current Syria and northern Palestine. A census in 1860 taken by the French Army found a Muslim population of 0.13 million or 27.0% of the total population.

As shown in Table 2.1.5 and based on census data, the Muslim population increased from 0.13 million or 27.0% in 1860, to 0.29 million or 66.5% in 1914, but decreased to 0.23 million or 37.7% in 1922, then bounced to 0.33 million or 42.0% in 1932, to 0.42 million or 40.5% in 1943, to 0.54 million or 38.4% in 1956. The relatively high percentage of Muslims in 1914 is also due to the inclusion of Druze who constituted from 6.8% to 7.2% of the total population in the first half of the twentieth century. During the second half of the latter, the percentage of Muslims increased by about two percentage points per decade. This is due to larger Muslim fertility rate, and the exodus of the non-Muslim population to Europe and the Americas where they had better opportunities. According to the World Values Survey (WVS), the percentage of Muslims was 51.83% in 2013.

Thus, assuming that this trend continues, the Muslim population will be around four million throughout the second half of this century.

Figure 2.3.5 Map of the Republic of Lebanon.

Table 2.3.5 Evolution of the Muslim population in Lebanon

Year	Population	Muslims	%	Source
1580	255,445	195,070	76.36	[EGH]c
1860	487,600	131,685	27.01	[LBH]c
1881	908,630	537,388	59.14	[EGH]c
1914	436,129	289,827	66.45	[TR14]c
1922	609,070	229,733	37.72	[LBH]c
1932	785,543	330,133	42.03	[LBH]c
1943	1,046,428	423,292	40.45	[LBH]c
1956	1,407,900	541,200	38.44	[EGH]c
2013	5,287,000	2,740,000	51.83	[WVS]s
2050	6,528,000	3,917,000	60.00	es
2100	5,707,000	3,995,000	70.00	es

2.3.6 Oman

The Sultanate of Oman has an area of 309,500 sq. km and its map is presented in Fig. 2.3.6. In 629, Prophet Muhammad sent his companion Amru bnul Aass as an envoy to the two brother rulers of historical Oman, which included current Oman and UAE. The Prophet invited the two brothers, Jayfar and Abd Bnul Julandi, to accept Islam. The rulers of Oman did accept Islam along with their population and Muslims remained the majority of the population of the whole region since then.

The entire population in 1922 and 1950 was Muslim. The first census was conducted in 1993, according to which Muslims numbered 1.77 million and made up 87.7% of the total population. The results are summarized in Table 2.3.6. Assuming that this percentage remains constant, the Muslim population is expected to be around six million throughout the second half of this century.

Figure 2.3.6 Map of the Sultanate of Oman.

Table 2.3.6 Evolution of the Muslim population in Oman

Year	Population	Muslims	%	Source
1922	500,000	500,000	100	[MAS23]es
1950	500,000	500,000	100	[MAS55]es
1993	2,018,074	1,769,851	87.70	[PEW]c
2050	6,915,000	6,065,000	87.70	es
2100	7,268,000	6,374,000	87.70	es

2.3.7 Palestine/Israel

Muslims conquered Palestine in 15 H or 636 AD, during the time of Caliph Omar, who entered Jerusalem without war. Currently, the total area of this region is 28,292 sq. km, distributed as follows: Israel 22,072 sq. km, West Bank 5,860 sq. km, and Gaza Strip 360 sq. km. A map of these territories is presented in Fig. 2.3.7. It was taken by the British from the Ottomans in 1922, then Israel was declared in 1948, Jordan administered the West Bank and Egypt administered Gaza Strip. The Strip and the Bank were occupied by Israel in 1967.

Based on census and Ottoman population registry data, the Muslim population increased in number but decreased in percentage from 0.30 million or 88.2% in 1850, to 0.33 million or 88.1 % in 1860, to 0.39 million or 87.6% in 1877, to 0.40 million or 87.4% in 1880, to 0.42 million or 87.0% in 1885, to 0.45 million or 86.4% in 1890, to 0.47 million or 85.6% in 1895, to 0.50 million or 85.1% in 1900, to 0.53 million or 84.6% in 1905, to 0.57 million or 84.0% in 1910, to 0.60 million or 83.4% in 1914, to 0.61 million or 81.7% in 1918, 0.64 million or 77.5% in 1922, to 0.70 million or 74.5% in 1926 (estimate), to 0.78 million or 73.5% in 1931. Later estimates show that the trend continued to 0.88 million or 63.3% in 1936, to 1.00 million or 63.2% in 1941, to 1.18 million or 60.5% in 1946. After the state of Israel was declared, the statistics for Israel included those living in 1948 occupied lands and Israeli citizens who live in colonies in the West Bank and Gaza Strip. The Palestinian statistics on the other hand, covers Palestinians who live in Gaza Strip and the West Bank and excludes those who live in Jerusalem. The statistics also exclude Israeli citizens who live in this region as they are covered under Israel.

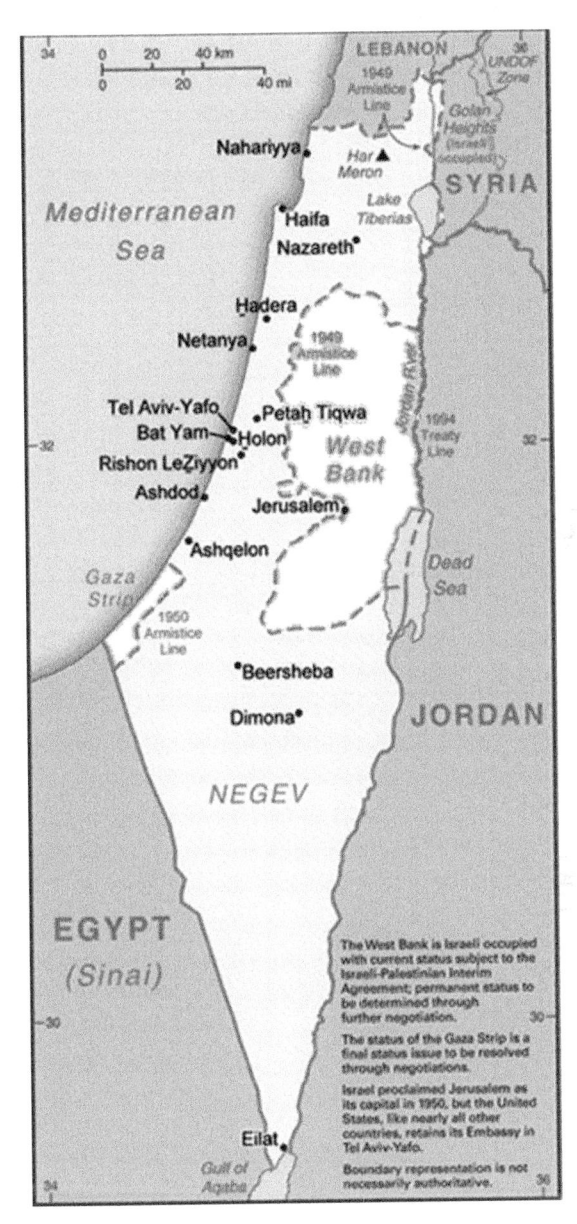

Figure 2.3.7 Map of Palestine/Israel.

Based on census data as shown in Table 2.3.7a, the Muslim population in Israel increased from 0.11 million or 9.5% in 1949, to 0.17 million or 7.8% in 1961, to 0.36 million or 11.2% in 1972, to 0.54 million or 13.2% in 1983, to 0.81 million or 14.0% in 1995, to 1.25 million or 16.9% in 2008 and 1.6 million or 17.9% in 2017. Assuming the Muslim population continues to increase by one percentage point each decade, then it is expected to reach three million or 21% by 2050 and five million or 26% by 2100.

Table 2.3.7a Evolution of the Muslim population in Israel: occupied lands of 1948, West Jerusalem and Jewish population in the rest of Historical Palestine

Year	Population	Muslims	%	Source
1949	1,173,900	111,500	9.50	[IL]c
1961	2,234,200	174,900	7.83	[IL]c
1972	3,225,000	360,600	11.18	[IL]c
1983	4,118,600	542,200	13.16	[IL]c
1995	5,812,300	811,200	13.96	[IL]c
2008	7,419,100	1,254,100	16.90	[IL]c
2017	8,713,300	1,561,700	17.92	[IL]s
2050	12,720,000	2,671,000	21.00	es
2100	18,128,000	4,713,000	26.00	es

Regarding Palestinians, the Muslim population in Gaza Strip increased from 0.35 million or 99.0% in 1967, to 1.0 million or 99.8% in 1997, to 1.4 million or 99.9% in 2007, to 1.9 million or 99.9% in 2017. On the other hand, the Muslim population in the West Bank increased from 1.56 million or 97.6% in 1997, to 2.2 million or 98.1% in 2007, to 2.7 million or 98.3% in 2017. The data is summarized in Tables 2.3.7b and 2.3.7c, for the Strip and the Bank, respectively. Assuming the percentage of Muslims will continue to increase by a fifth of a percentage point both territories combined, the Muslim population is expected to reach 5.3 million in 2020, then 9.7 million by 2050, and 15.1 million by 2100.

Table 2.3.7b Evolution of the Muslim population in Gaza Strip

Year	Population	Muslims	%	Source
1967	356,269	352,532	98.95	[UN71]c
1997	1,000,517	998,828	99.83	[PS97]c
2007	1,380,405	1,378,962	99.90	[PS07]c
2017	1,872,891	1,871,493	99.93	[PS17]c

Table 2.3.7c Evolution of the Muslim population in the West Bank

Year	Population	Muslims	%	Source
1997	1,596,697	1,558,031	97.58	[PS97]c
2007	2,229,595	2,187,495	98.11	[PS07]c
2017	2,791,026	2,744,190	98.32	[PS17]c

Table 2.3.7d Evolution of the Muslim population in Historical Palestine

Year	Population	Muslims	%	Source
1850	340,000	300,000	88.24	[MCC]c
1860	369,000	325,000	88.08	[MCC]c
1877	440,850	386,320	87.63	[MCC]c
1880	456,929	399,334	87.40	[MCC]c
1885	485,530	422,280	86.97	[MCC]c
1890	516,131	445,728	86.36	[MCC]c
1895	548,854	469,750	85.59	[MCC]c
1900	586,581	499,110	85.09	[MCC]c
1905	628,190	531,236	84.57	[MCC]c
1910	673,259	565,601	84.01	[MCC]c
1914	722,143	602,377	83.42	[MCC]c
1918	748,128	611,098	81.68	[MCC]c
1922	823,684	638,407	77.51	[MCC]c
1926	945,438	703,838	74.45	[MCC]es
1931	1,054,189	775,181	73.53	[MCC]c
1936	1,388,852	879,496	63.33	[MCC]es
1941	1,639,757	1,004,989	63.18	[MCC]es
1946	1,942,349	1,175,196	60.50	[MCC]es
1950	2,189,897	1,041,648	47.57	
1960	3,130,000	1,219,000	38.96	
1970	3,940,000	1,429,000	36.28	
1980	5,212,000	1,982,000	38.03	
1990	6,550,000	2,654,000	40.52	
2000	9,170,000	4,004,000	43.67	
2010	11,402,000	5,248,000	46.03	
2020	13,757,000	6,607,000	48.03	
2050	21,536,000	11,452,000	53.17	
2100	30,396,000	16,980,000	55.86	

All in all and as summarized in Table 2.3.7d, the percentage of Muslims in Historical Palestine continued to decrease to 47.6% or 1.0 million in 1950, to 39.0% or 1.2 million in 1960, to 36.3% or 1.4 million in 1970. However, Muslim percentage started to increase constantly since the Israeli occupation of the rest of Palestine. Accordingly, the percentage of Muslims increased to 38.0% or 2.0 million in 1980, to 40.5% or 2.7 million in 1990, to 43.7% or 4.0 million in 2000, to 46.0% or 5.2 million in 2010. With the previous assumptions, the Muslim population is expected to reach twelve million or 53% by 2050 and seventeen million or 56% by 2100.

2.3.8 Qatar

The State of Qatar's area is 11,586 sq. km and its map is presented in Fig. 2.3.8. It was taken by the British from the Ottoman Empire in 1913 and gained its independence from the UK in 1971. The entire population was estimated to be Muslim before 1970, increasing from 26,000 in 1922, to 111,000 in 1970, when the first census was conducted. As shown in Table 2.3.8, previous estimates show that the Muslims increased in number but decreased in percentage from 0.35 million or 95% in 1986, to 0.42 million or 80.5% in 1997, to 0.58 million or 78.5% in 2004. The reason of this decrease is the discovery of oil and the subsequent import of foreign workers, some of which are non-Muslim. Assuming the percentage of Muslims remains fixed, the Muslim population is expected to be around three million throughout the second half of this century.

Table 2.3.8 Evolution of the Muslim population in Qatar

Year	Population	Muslims	%	Source
1922	26,000	26,000	100	[MAS23]es
1949	25,000	25,000	100	[MAS55]es
1970	111,133	111,133	100	[CIA80]es
1986	369,079	350,625	95.00	[KET86]es
1997	521,997	420,367	80.53	[QA]c
2004	744,029	576,391	77.47	[UN]c
2050	3,851,000	2,983,000	77.47	es
2100	4,162,000	3,225,000	77.47	es

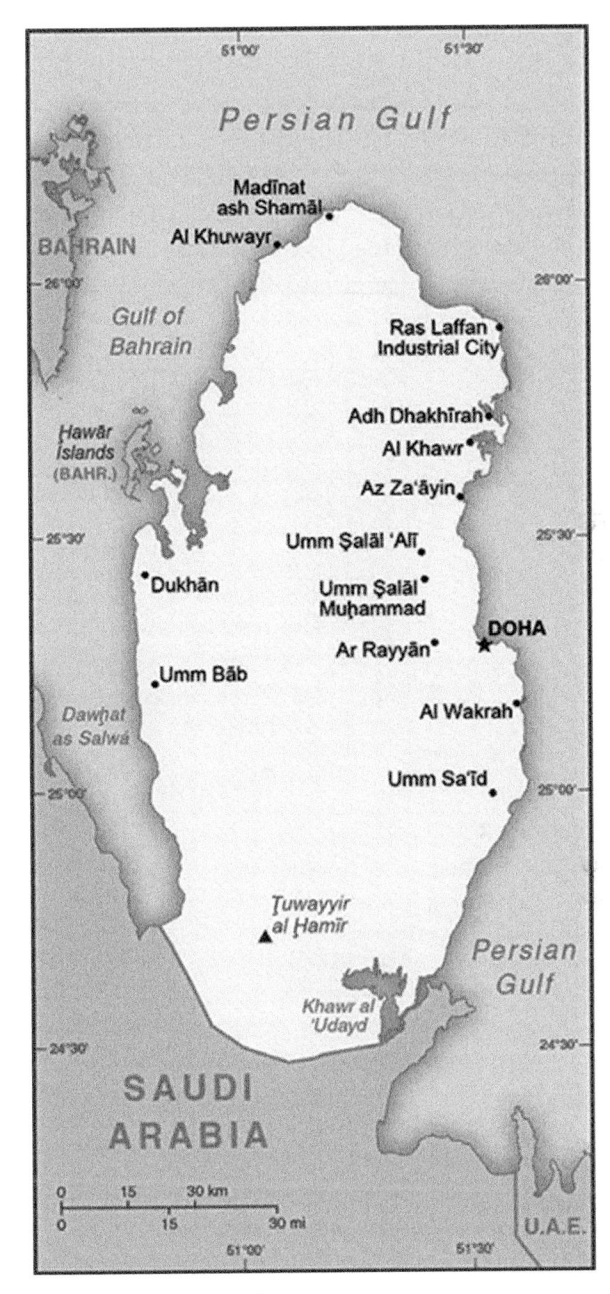

Figure 2.3.8 Map of the State of Qatar.

2.3.9 Saudi Arabia

Islam started in Mecca, West of Saudi Arabia in 609 AD. After the Hijra in 622 AD, the first Muslim state was established in Medina, 330 km north of Mecca. After the opening of Mecca in 630A D, all current Saudi Arabia became under Muslim control, and Muslims remained the majority of the population since then. Currently, the Kingdom of Saudi Arabia has an area of 2,149,690 sq. km and its map is presented in Fig. 2.3.9.

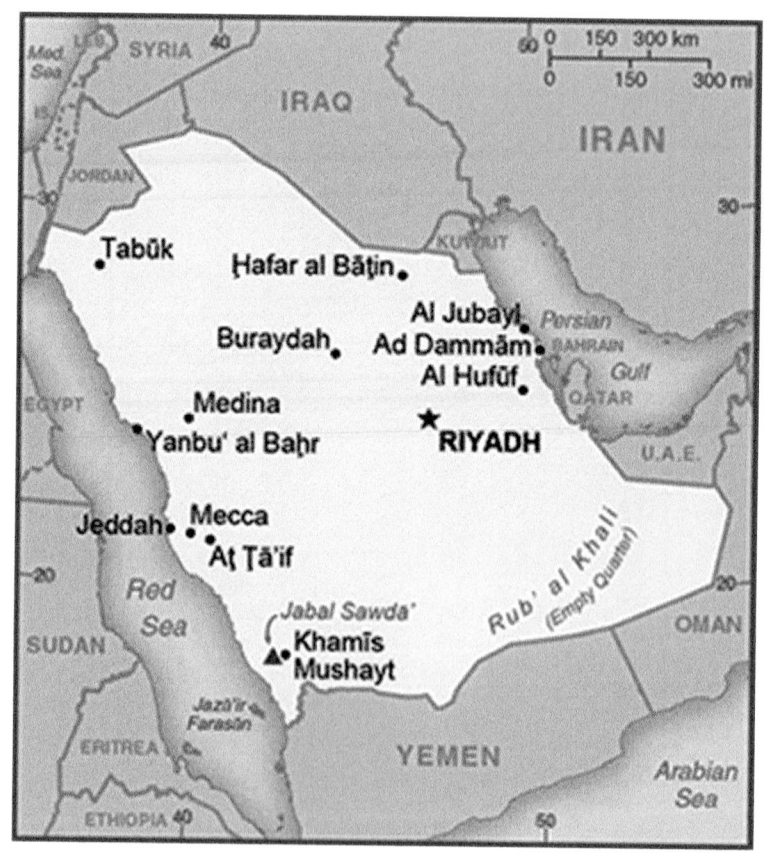

Figure 2.3.9 Map of the Kingdom of Saudi Arabia.

The first complete census was held in 1974. However, censuses in this country do not include a religion questionnaire, as the government claims that the entire population is Muslim. While this is true for Saudi Nationals, as is the case for the other countries in this region, the case is not true for the foreign population, who represent a third (32.4%) of the total population as of 2014. Oil exploration started in this region in 1920s and the following decade oil discoveries occurred in the western province. More non-Muslim workers started flocking to the country and region as a result. Around 1950, they were estimated at 2,000 Americans and 3,000 Italians in the city of Dammam, making the percentage of Muslims 99.8%. According to the World Values Survey (WVS), the percentage of Muslims was 97.20% in 2003. Thus, assuming this percentage remains constant, the Muslim population is expected to reach 43 million by 2050 and 41 million by 2100. A summary is presented in Table 2.3.9.

Table 2.3.9   Evolution of the Muslim population in Saudi Arabia

Year	Population	Muslims	%	Source
1922	2,228,000	2,228,000	100	[MAS23]es
1950	3,121,000	3,116,000	99.84	[MAS55]es
2003	23,358,000	22,704,000	97.20	[WVS]s
2050	44,562,000	43,315,000	97.20	es
2100	42,231,000	41,049,000	97.20	es

2.3.10   Syria

The Muslim presence in Syria started with the conquest of Damascus in 13 H or 634 AD, during the reign of Caliph Omar. The Muslim troops were led by the Prophet's companions Abu Obaida bnul Jarrah and Khalid bnul Walid. The conquest of current Syria was culminated with the opening of Aleppo and Antakya peacefully in 16 H or 637 AD.

Currently, the Syrian Arab Republic has an area of 185,180 sq. km and its map is shown in Fig. 2.3.10. It was taken by the French from the Ottomans in 1920 and gained its independence from France in 1946. The Muslim population of the Ottoman Provinces of Aleppo (half of which in current Turkey) and Damascus (includes northern Jordan) decreased from 1.10 million or 94.2% in 1580, to 1.03 million or 85.3% in 1881, then increased to 1.43 million or 87.8% in 1894. The 1914 Ottoman census showed that the Muslim population was 1.08 million or 86.4% of the total population. Based on census data as shown in Table 2.3.10, the Muslim population increased from 1.7 million or 86.8% in 1923 to 2.3 million or 84.4% in 1939, to 2.5 million or 81.8% in 1947, to 4.2 million or 92.1% in 1960, when last census data on religion was released. Thus, assuming this percentage remains constant, the Muslim population is expected to reach 31 million by 2050 and 33 million by 2100.

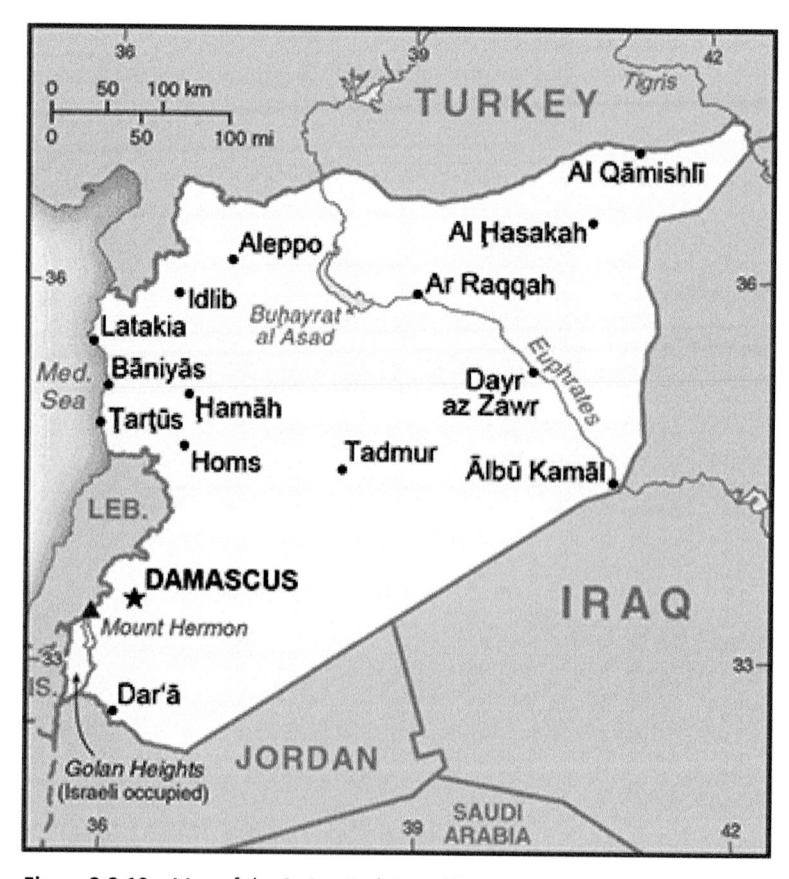

Figure 2.3.10   Map of the Syrian Arab Republic.

Table 2.3.10 Evolution of the Muslim population in Syria

Year	Population	Muslims	%	Source
1580	1,163,310	1,095,440	94.17	[EGH]c
1881	1,206,241	1,029,115	85.32	[EGH]c
1894	1,625,257	1,426,488	87.77	[EGH]c
1914	1,255,047	1,083,993	86.37	[TR14]c
1923	1,970,712	1,709,778	86.76	[MAS23]c
1939	2,770,669	2,337,633	84.37	[MAS55]c
1947	3,043,310	2,488,901	81.78	[MAS55]c
1960	4,565,121	4,204,476	92.10	[SYB70]c
2050	33,129,000	30,512,000	92.10	es
2100	36,103,000	33,251,000	92.10	es

2.3.11 United Arab Emirates

It is a union of seven Emirates that was formed in 1971: Abu Dhabi, Ajman, Dubai, Fujairah, Ras al Khaimah (joined in 1972), Sharjah, and Umm al Quwain. The total area of this union is 83,600 sq. km and its map is presented in Fig. 2.3.11. The first census was conducted in 1968, and subsequent censuses show a substantial decrease in the percentage of the Muslim population from 100% before 1947, to 0.17 million or 96.1% in 1968, to 3.12 million or 76% in 2005. Like other Gulf countries, this decrease is due to the economic boom caused by the discovery of oil, which resulted in the substantial import of foreign workers, many of whom were non-Muslim. Assuming the percentage of Muslims remains constant at 76%; the Muslim population is expected to reach eight million by 2050 and decrease to seven million by 2100. The data is summarized in Table 2.3.11.

Table 2.3.11 Evolution of the Muslim population in United Arab Emirates

Year	Population	Muslims	%	Source
1922	80,000	80,000	100	[MAS23]es
1947	95,000	95,000	100	[MAS55]es
1968	179,126	172,052	96.05	[UN71]c
1975	557,887	494,325	88.61	[UN83]c
2005	4,106,427	3,120,885	76.00	[DOS10]c
2050	10,425,000	7,923,000	76.00	es
2100	9,023,000	6,858,000	76.00	es

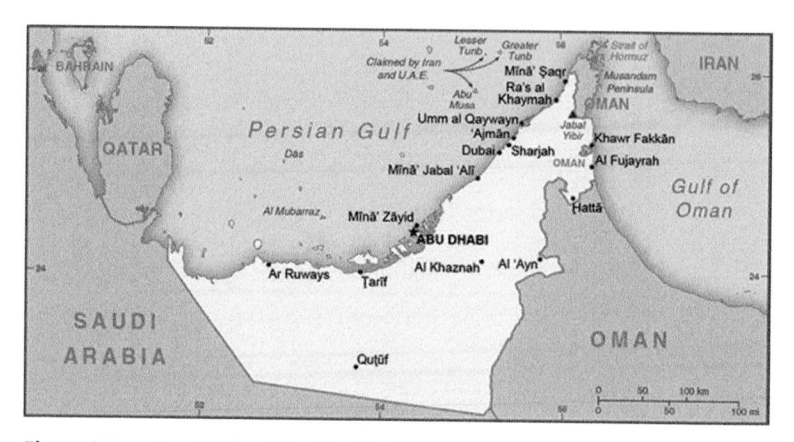

Figure 2.3.11 Map of the United Arab Emirates.

2.3.12 Yemen

In 628, Prophet Muhammad sent his companion Shuja bno Wahb as an envoy to the king of Persia Khosrau II inviting him to Islam. The Shahinshah (Persian for King of Kings) or Kisra (Arabic nickname for King of Persia) got furious, complaining that the Prophet started the letter with his name, "from Mohammed the messenger of God to Kisra, the leader of Persia," instead of starting with Kisra's name. Khorsau II then cut the letter into pieces, and sent to Batham, his proxy in Yemen which was under the Persian control, to send two strong men to bring the Prophet to Kisra. When they arrived to the Prophet, he informed them that Kisra was killed by his son Sheraweh, who is now the King of Persia. The Prophet also informed them to inform Batham of this, and to invite him to Islam. This was in Jumada I 7 H, or September 628 AD. The two men went back and informed Batham, who waited for the news from Persia, and received a letter from Sheraweh, informing him with what the Prophet has already told him. So, he knew that Mohammed (PBUH) must be God's Prophet, and accepted Islam together with his people, and Muslims remained the Majority of the population since then. Currently, the Republic of Yemen has an area of 527,968 sq. km and its map is presented in Fig. 2.3.12.

A 1922 estimate shows 93% of the population as Muslim. As a result of the decrease of the number of Jews as they left for Palestine in preparation for the establishment of Israel, the percentage of Muslims increased to 95% in 1929, then 98% in 1949. The first census covering all Yemen was conducted in 1973 and 1975 in the southern and northern parts, respectively. The former found 1,590,275 inhabitants, while the latter found 5,237,893. Both parts were united in 1990. Censuses in this country do not inquire on religious affiliation; however, the entire population is estimated to be Muslim according to the 2013 World Values Survey (WVS). A summary of the data is provided in Table 2.3.12. Assuming the entire population remains Muslim, the Muslim population is expected to increase to 48 million by 2050 and 53 million by 2100.

Table 2.3.12 Evolution of the Muslim population in Yemen

Year	Population	Muslims	%	Source
1922	1,420,000	1,320,000	92.96	[MAS23]es
1929	3,510,000	3,350,000	95.44	[MAS55]es
1949	4,680,000	4,600,000	98.29	[MAS55]es
1986	11,619,439	11,619,439	100	[KET86]es
2013	25,533,000	25,533,000	100	[WVS]s
2050	48,080,000	48,080,000	100	es
2100	53,171,000	53,171,000	100	es

Table 2.3a Centennial estimates of the Muslim population (×1000) in Arabian Asia: 600–2100 AD (1–1500 H)

		600	700	800	900	1000	1100	1200	1300
Bahrain	P	60	60	61	61	61	59	57	56
	M%	—	100.00	100.00	100.00	100.00	100.00	100.00	100.00
	M	—	60	61	61	61	59	57	56
Iraq	P	1,600	1,700	1,800	1,900	2,000	1,800	1,600	1,400
	M%	—	20.00	50.00	70.00	90.00	90.00	95.00	95.00
	M	—	340	900	1,330	1,800	1,620	1,520	1,330
Jordan	P	216	217	218	219	220	215	210	205
	M%	—	20.00	50.00	70.00	80.00	80.00	80.00	80.00
	M	—	43	109	153	176	172	168	164
Kuwait	P	79	80	80	80	80	78	76	74
	M%	—	100.00	100.00	100.00	100.00	100.00	100.00	100.00
	M	—	80	80	80	80	78	76	74
Lebanon	P	332	333	334	335	336	328	320	313
	M%	—	10.00	20.00	30.00	40.00	50.00	60.00	70.00
	M	—	33	67	101	134	164	192	219
Oman	P	305	306	307	308	309	305	300	295
	M%	—	100.00	100.00	100.00	100.00	100.00	100.00	100.00
	M	—	306	307	308	309	305	300	295
Palestine/Israel	P	182	183	184	185	186	182	178	174
	M%	—	10.00	20.00	30.00	40.00	50.00	60.00	70.00
	M	—	18	37	56	74	91	107	122
Qatar	P	13	14	14	14	14	14	13	13
	M%	—	100.00	100.00	100.00	100.00	100.00	100.00	100.00
	M	—	14	14	14	14	14	13	13
Saudi Arabia	P	2,006	2,012	2,018	2,024	2,032	2,000	1,950	1,900
	M%	—	100.00	100.00	100.00	100.00	100.00	100.00	100.00
	M	—	2,012	2,018	2,024	2,032	2,000	1,950	1,900
Syria	P	1,335	1,340	1,345	1,350	1,355	1,320	1,290	1,260
	M%	—	20.00	50.00	70.00	80.00	90.00	95.00	95.00
	M	—	268	673	945	1,084	1,188	1,226	1,197
UAE	P	39	40	40	40	40	39	38	37
	M%	—	100.00	100.00	100.00	100.00	100.00	100.00	100.00
	M	—	40	40	40	40	39	38	37
Yemen	P	2,500	2,505	2,510	2,515	2,520	2,480	2,420	2,360
	M%	—	90.00	90.00	90.00	90.00	90.00	90.00	90.00
	M	—	2,255	2,259	2,264	2,268	2,232	2,178	2,124
Total	P	8,667	8,790	8,911	9,031	9,153	8,820	8,452	8,087
	M%	—	62.22	73.66	81.66	88.20	90.27	92.57	93.12
	M	—	5,470	6,564	7,375	8,073	7,962	7,824	7,531
	G%		0.014	0.014	0.013	0.013	−0.037	−0.043	−0.044

Table 2.3a (*Continued*)

		1400	1500	1600	1700	1800	1900	2000	2100
Bahrain	P	55	54	54	54	61	74	665	2,252
	M%	100.00	100.00	100.00	100.00	100.00	100.00	81.22	70.22
	M	55	54	54	54	61	74	540	1,581
Iraq	P	1,200	1,000	1,250	1,000	1,073	2,244	23,498	107,711
	M%	95.00	95.00	95.00	95.00	93.02	91.74	99.15	99.00
	M	1,140	950	1,188	950	998	2,059	23,298	106,634
Jordan	P	200	196	191	186	211	321	5,122	13,644
	M%	80.00	80.00	80.00	80.00	78.45	78.45	95.26	98.00
	M	160	157	153	149	166	252	4,880	13,371
Kuwait	P	72	71	71	71	80	97	2,045	6,189
	M%	100.00	100.00	100.00	100.00	100.00	100.00	74.37	74.37
	M	72	71	71	71	80	97	1,521	4,603
Lebanon	P	306	299	292	284	324	591	3,843	5,707
	M%	75.00	80.00	76.00	70.00	27.01	66.45	50.00	70.00
	M	230	239	222	199	88	393	1,921	3,995
Oman	P	280	275	275	275	310	419	2,268	7,268
	M%	100.00	100.00	100.00	100.00	100.00	100.00	87.70	87.70
	M	280	275	275	275	310	419	1,989	6,374
Palestine/Israel	P	170	166	161	157	275	587	9,170	30,396
	M%	75.00	80.00	85.00	90.00	88.24	85.09	43.67	55.86
	M	128	133	137	141	243	499	4,004	16,980
Qatar	P	12	12	12	12	13	17	592	4,162
	M%	100.00	100.00	100.00	100.00	100.00	100.00	77.47	77.47
	M	12	12	12	12	13	17	459	3,225
Saudi Arabia	P	1,850	1,809	1,809	1,809	2,050	2,569	20,664	42,231
	M%	100.00	100.00	100.00	100.00	100.00	100.00	97.20	97.20
	M	1,850	1,809	1,809	1,809	2,050	2,569	20,085	41,049
Syria	P	1,230	1,206	1,175	1,145	1,305	1,859	16,411	36,103
	M%	95.00	95.00	95.00	90.00	85.32	86.37	92.10	92.10
	M	1,169	1,146	1,116	1,031	1,113	1,606	15,114	33,251
UAE	P	36	35	35	35	40	48	3,134	9,023
	M%	100.00	100.00	100.00	100.00	100.00	100.00	76.00	76.00
	M	36	35	35	35	40	48	2,382	6,858
Yemen	P	2,300	2,243	2,243	2,243	2,530	3,143	17,409	53,171
	M%	90.00	90.00	90.00	90.00	92.96	92.96	100.00	100.00
	M	2,070	2,019	2,019	2,019	2,352	2,922	17,409	53,171
Total	P	7,711	7,366	7,568	7,271	8,272	11,969	104,821	317,858
	M%	93.38	93.66	93.68	92.75	90.83	91.52	89.30	91.58
	M	7,201	6,899	7,090	6,744	7,513	10,954	93,602	291,092
	G%	−0.048	−0.046	0.027	−0.040	0.129	0.369	2.170	1.109

Table 2.3b Decennial estimates of the Muslim population (×1000) in Arabian Asia: 1790–1940 AD (1210–1360 H)

		1790	1800	1810	1820	1830	1840	1850	1860
Bahrain	P	60	61	62	63	63	63	63	63
	M%	100.00	100.00	100.00	100.00	100.00	100.00	100.00	100.00
	M	60	61	62	63	63	63	63	63
Iraq	P	1,065	1,073	1,085	1,093	1,200	1,300	1,400	1,500
	M%	93.02	93.02	93.02	93.02	93.02	93.02	93.02	93.02
	M	991	998	1,009	1,017	1,116	1,209	1,302	1,395
Jordan	P	207	211	214	217	230	240	250	260
	M%	78.45	78.45	78.45	78.45	78.45	78.45	78.45	78.45
	M	162	166	168	170	180	188	196	204
Kuwait	P	79	80	81	82	82	82	82	82
	M%	100.00	100.00	100.00	100.00	100.00	100.00	100.00	100.00
	M	79	80	81	82	82	82	82	82
Lebanon	P	320	324	328	332	360	385	410	440
	M%	27.01	27.01	27.01	27.01	27.01	27.01	27.01	27.01
	M	86	88	89	90	97	104	111	119
Oman	P	306	310	314	318	330	340	350	360
	M%	100.00	100.00	100.00	100.00	100.00	100.00	100.00	100.00
	M	306	310	314	318	330	340	350	360
Palestine/Israel	P	270	275	280	290	300	320	340	369
	M%	88.24	88.24	88.24	88.24	88.24	88.24	88.24	88.08
	M	238	243	247	256	265	282	300	325
Qatar	P	13	13	13	14	14	14	14	14
	M%	100.00	100.00	100.00	100.00	100.00	100.00	100.00	100.00
	M	13	13	13	14	14	14	14	14
Saudi Arabia	P	2,030	2,050	2,070	2,091	2,150	2,200	2,250	2,300
	M%	100.00	100.00	100.00	100.00	100.00	100.00	100.00	100.00
	M	2,030	2,050	2,070	2,091	2,150	2,200	2,250	2,300
Syria	P	1,290	1,305	1,320	1,337	1,350	1,400	1,450	1,500
	M%	85.32	85.32	85.32	85.32	85.32	85.32	85.32	85.32
	M	1,101	1,113	1,126	1,141	1,152	1,194	1,237	1,280
UAE	P	40	40	41	41	41	41	41	41
	M%	100.00	100.00	100.00	100.00	100.00	100.00	100.00	100.00
	M	40	40	41	41	41	41	41	41
Yemen	P	2,500	2,530	2,560	2,593	2,650	2,700	2,750	2,800
	M%	92.96	92.96	92.96	92.96	92.96	92.96	92.96	92.96
	M	2,324	2,352	2,380	2,410	2,463	2,510	2,556	2,603
Total	P	8,180	8,272	8,368	8,471	8,770	9,085	9,400	9,729
	M%	90.84	90.83	90.82	90.81	90.69	90.57	90.45	90.31
	M	7,430	7,513	7,600	7,693	7,954	8,228	8,503	8,786
	G%		0.112	0.115	0.122	0.347	0.353	0.341	0.344

Table 2.3b (Continued)

		1870	1880	1890	1900	1910	1920	1930	1940
Bahrain	P	63	66	70	74	79	84	91	99
	M%	100.00	100.00	100.00	100.00	100.00	100.00	100.00	98.14
	M	63	66	70	74	79	84	91	97
Iraq	P	1,580	1,776	1,997	2,244	2,448	2,991	3,559	4,235
	M%	93.02	93.02	91.74	91.74	91.74	92.68	93.07	93.62
	M	1,470	1,652	1,832	2,059	2,246	2,772	3,312	3,965
Jordan	P	266	283	301	321	348	369	397	428
	M%	78.45	78.45	78.45	78.45	78.45	78.45	88.46	88.46
	M	209	222	236	252	273	289	351	379
Kuwait	P	82	87	92	97	105	111	120	130
	M%	100.00	100.00	100.00	100.00	100.00	100.00	100.00	100.00
	M	82	87	92	97	105	111	120	130
Lebanon	P	476	512	550	591	649	755	917	1,114
	M%	27.01	59.14	66.45	66.45	66.45	37.72	42.03	40.45
	M	129	303	365	393	431	285	385	451
Oman	P	367	384	401	419	444	446	449	452
	M%	100.00	100.00	100.00	100.00	100.00	100.00	100.00	100.00
	M	367	384	401	419	444	446	449	452
Palestine/ Israel	P	400	457	516	587	683	900	1,036	1,532
	M%	87.63	87.40	86.36	85.09	84.01	77.51	73.53	63.18
	M	351	399	446	499	574	698	762	968
Qatar	P	14	15	16	17	18	19	21	22
	M%	100.00	100.00	100.00	100.00	100.00	100.00	100.00	100.00
	M	14	15	16	17	18	19	21	22
Saudi Arabia	P	2,338	2,413	2,490	2,569	2,676	2,768	2,892	3,020
	M%	100.00	100.00	100.00	100.00	100.00	100.00	100.00	99.84
	M	2,338	2,413	2,490	2,569	2,676	2,768	2,892	3,015
Syria	P	1,582	1,669	1,762	1,859	1,994	2,217	2,542	2,913
	M%	85.32	85.32	87.77	86.37	86.37	86.76	86.76	84.37
	M	1,350	1,424	1,547	1,606	1,722	1,923	2,205	2,458
UAE	P	41	43	46	48	52	55	60	65
	M%	100.00	100.00	100.00	100.00	100.00	100.00	100.00	100.00
	M	41	43	46	48	52	55	60	65
Yemen	P	2,840	2,938	3,039	3,143	3,284	3,525	3,862	4,230
	M%	92.96	92.96	92.96	92.96	92.96	92.96	95.44	95.44
	M	2,640	2,731	2,825	2,922	3,053	3,277	3,686	4,037
Total	P	10,049	10,643	11,280	11,969	12,780	14,240	15,946	18,240
	M%	90.08	91.51	91.90	91.52	91.34	89.38	89.90	87.93
	M	9,052	9,739	10,366	10,954	11,673	12,727	14,335	16,038
	G%	0.324	0.574	0.581	0.592	0.656	1.082	1.131	1.344

Table 2.3c Decennial estimates of the Muslim population (×1000) in Arabian Asia: 1950–2100 AD (1370–1520 H)

		1950	1960	1970	1980	1990	2000	2010	2020
Bahrain	P	116	162	213	360	496	665	1,241	1,702
	M%	96.12	94.82	95.66	84.99	81.77	81.22	70.22	70.22
	M	111	154	203	306	406	540	871	1,195
Iraq	P	5,719	7,290	9,918	13,653	17,419	23,498	29,742	40,222
	M%	93.62	95.58	96.01	96.73	98.00	99.15	99.00	99.00
	M	5,354	6,968	9,522	13,207	17,071	23,298	29,445	39,820
Jordan	P	481	933	1,721	2,378	3,566	5,122	7,262	10,203
	M%	93.58	93.58	93.58	95.47	97.10	95.26	97.67	98.00
	M	450	873	1,611	2,270	3,462	4,880	7,092	9,999
Kuwait	P	153	269	744	1,369	2,095	2,045	2,992	4,271
	M%	94.39	94.08	94.97	91.51	91.51	74.37	74.37	74.37
	M	145	253	707	1,252	1,917	1,521	2,225	3,176
Lebanon	P	1,335	1,805	2,297	2,589	2,803	3,843	4,953	6,825
	M%	38.44	38.44	40.00	44.00	48.00	50.00	51.83	54.00
	M	513	694	919	1,139	1,345	1,921	2,567	3,686
Oman	P	456	552	724	1,154	1,812	2,268	3,041	5,107
	M%	100.00	100.00	100.00	87.70	87.70	87.70	87.70	87.70
	M	456	552	724	1,012	1,589	1,989	2,667	4,479
Palestine/Israel	P	2,190	3,130	3,940	5,212	6,550	9,170	11,402	13,757
	M%	47.57	38.96	36.28	38.03	40.52	43.67	46.03	48.03
	M	1,042	1,219	1,429	1,982	2,654	4,004	5,248	6,607
Qatar	P	25	47	110	224	476	592	1,856	2,881
	M%	100.00	100.00	100.00	95.00	80.50	77.47	77.47	77.47
	M	25	47	110	212	383	459	1,438	2,232
Saudi Arabia	P	3,121	4,087	5,836	9,691	16,234	20,664	27,421	34,814
	M%	99.84	99.84	97.20	97.20	97.20	97.20	97.20	97.20
	M	3,116	4,080	5,673	9,420	15,779	20,085	26,654	33,839
Syria	P	3,413	4,574	6,351	8,931	12,446	16,411	21,363	17,501
	M%	81.78	92.10	92.10	92.10	92.10	92.10	92.10	92.10
	M	2,791	4,212	5,849	8,225	11,463	15,114	19,675	16,118
UAE	P	70	92	235	1,020	1,828	3,134	8,550	9,890
	M%	100.00	100.00	96.05	88.61	88.61	76.00	76.00	76.00
	M	70	92	225	903	1,620	2,382	6,498	7,517
Yemen	P	4,661	5,315	6,193	7,942	11,710	17,409	23,155	29,826
	M%	98.29	98.29	100.00	100.00	100.00	100.00	100.00	100.00
	M	4,582	5,224	6,193	7,942	11,710	17,409	23,155	29,826
Total	P	21,741	28,256	38,282	54,522	77,436	104,821	142,978	176,999
	M%	85.81	86.24	86.63	87.80	89.62	89.30	89.20	89.54
	M	18,655	24,369	33,166	47,872	69,401	93,602	127,535	158,493
	G%	1.756	2.621	3.037	3.536	3.508	3.028	3.104	2.135

Table 2.3c (Continued)

		2030	2040	2050	2060	2070	2080	2090	2100
Bahrain	P	2,013	2,200	2,316	2,386	2,414	2,397	2,337	2,252
	M%	70.22	70.22	70.22	70.22	70.22	70.22	70.22	70.22
	M	1,414	1,545	1,626	1,676	1,695	1,683	1,641	1,581
Iraq	P	50,194	60,584	70,940	80,712	89,542	97,129	103,203	107,711
	M%	99.00	99.00	99.00	99.00	99.00	99.00	99.00	99.00
	M	49,692	59,978	70,231	79,905	88,646	96,158	102,170	106,634
Jordan	P	10,655	11,887	12,932	13,615	14,024	14,146	13,996	13,644
	M%	98.00	98.00	98.00	98.00	98.00	98.00	98.00	98.00
	M	10,442	11,649	12,674	13,343	13,743	13,863	13,716	13,371
Kuwait	P	4,747	5,153	5,393	5,470	5,607	5,856	6,057	6,189
	M%	74.37	74.37	74.37	74.37	74.37	74.37	74.37	74.37
	M	3,530	3,832	4,011	4,068	4,170	4,355	4,504	4,603
Lebanon	P	6,195	6,376	6,528	6,563	6,457	6,241	5,981	5,707
	M%	56.00	58.00	60.00	62.00	64.00	66.00	68.00	70.00
	M	3,469	3,698	3,917	4,069	4,133	4,119	4,067	3,995
Oman	P	5,936	6,437	6,915	7,298	7,463	7,467	7,389	7,268
	M%	87.70	87.70	87.70	87.70	87.70	87.70	87.70	87.70
	M	5,206	5,646	6,065	6,400	6,545	6,549	6,480	6,374
Palestine/Israel	P	16,321	18,932	21,536	23,948	26,062	27,877	29,344	30,396
	M%	50.16	51.87	53.17	54.21	55.01	55.45	55.69	55.86
	M	8,187	9,820	11,452	12,981	14,336	15,457	16,343	16,980
Qatar	P	3,327	3,629	3,851	4,007	4,092	4,130	4,141	4,162
	M%	77.47	77.47	77.47	77.47	77.47	77.47	77.47	77.47
	M	2,577	2,811	2,983	3,104	3,170	3,199	3,208	3,225
Saudi Arabia	P	39,322	42,473	44,562	45,349	44,966	44,214	43,362	42,231
	M%	97.20	97.20	97.20	97.20	97.20	97.20	97.20	97.20
	M	38,221	41,284	43,315	44,079	43,707	42,976	42,147	41,049
Syria	P	26,677	30,153	33,129	35,235	36,147	36,425	36,491	36,103
	M%	92.10	92.10	92.10	92.10	92.10	92.10	92.10	92.10
	M	24,569	27,771	30,512	32,451	33,291	33,547	33,609	33,251
UAE	P	10,661	10,648	10,425	10,034	9,584	9,266	9,109	9,023
	M%	76.00	76.00	76.00	76.00	76.00	76.00	76.00	76.00
	M	8,102	8,093	7,923	7,626	7,284	7,042	6,923	6,858
Yemen	P	36,407	42,670	48,080	52,063	54,341	55,077	54,613	53,171
	M%	100.00	100.00	100.00	100.00	100.00	100.00	100.00	100.00
	M	36,407	42,670	48,080	52,063	54,341	55,077	54,613	53,171
Total	P	212,456	241,142	266,610	286,679	300,698	310,225	316,023	317,858
	M%	90.29	90.73	91.07	91.31	91.47	91.55	91.58	91.58
	M	191,817	218,796	242,789	261,765	275,061	284,025	289,423	291,092
	G%	1.826	1.267	1.004	0.726	0.477	0.312	0.185	0.058

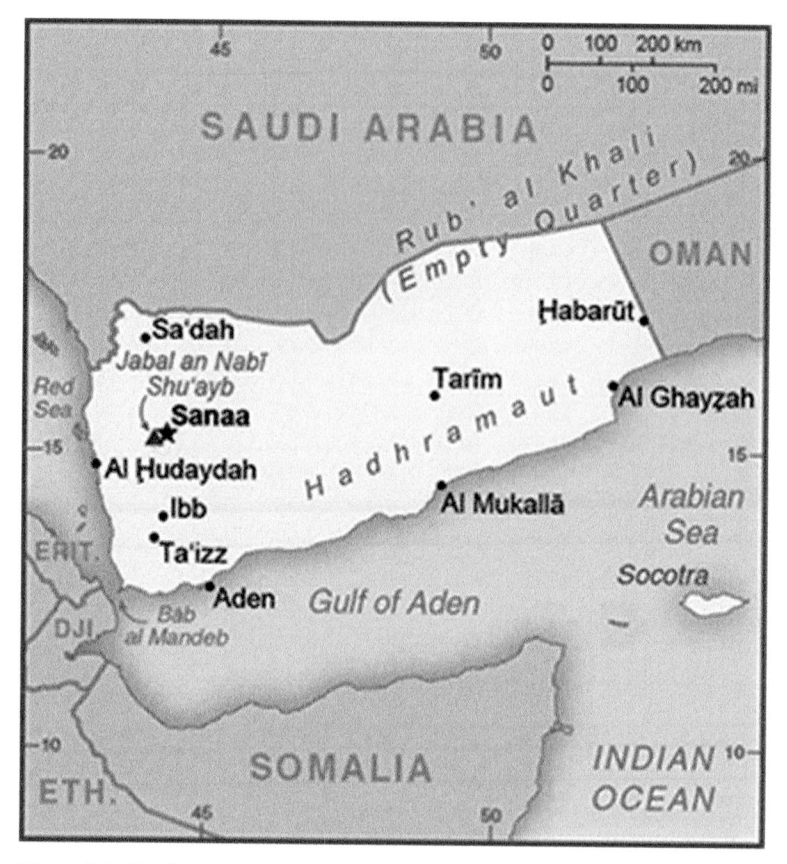

Figure 2.3.12 Map of the Republic of Yemen.

2.3.13 Regional Summary and Conclusion

Islam started in Arabian Asia and spread rapidly, making the vast majority of the population of this region Muslim. This is expected to remain so for the next three centuries. The following tables present centennial data from 600 AD to 2100 AD (or approximately 1 H to 1500 H) in Table 2.3a and decennial data from 1790 AD to 2100 AD (or 1210 H to 1520 H) in Tables 2.3b and 2.3c for current countries in Arabian Asia. The data includes total population in thousands (*P*), the percentage of which is Muslim ($M\%$), the corresponding Muslim population in thousands (M), and the annual population growth rate (APGR, or $G\%$) of the total population in this region.

2.4 Muslims in Southeast Asia

This region consists of eleven countries: Brunei, Cambodia, Indonesia, Laos, Malaysia, Myanmar (Burma), Philippines, Singapore, Thailand, Timor-Leste (East Timor), and Vietnam. Islam entered this region through Arab traders including prophet's companions as early as 54 H/674 AD in the Indonesian Island of Sumatra. However, masse conversion to Islam did not start until the eleventh century. The twelfth century saw the conversion of local kings and the establishment of Muslim Sultanates. This included Kedah in the Malay Peninsula in 1136, Ternate on the Maluku Islands east of Indonesia in 1257, Pasai on Sumatra in 1267, in addition to other subsequent Muslim sultanates. By the sixteenth century Islam became the dominant religion in this region and most of its territory became under Muslim control.

Thus, the Muslim population has likely increased from 46,000 or 0.5% of the total population of this region in 700 AD, to 0.10 million or 0.9% in 800 AD, to 0.27 million or 2.2% in 900 AD, to 0.56 million or 4.4% in 1000 AD, to 1.32 million or 9.4% in 1100 AD, to 2.34 million or 14.8% in 1200 AD, to 5.3 million or 30.4% in 1300 AD, to 7.0 million or 36.5% in 1400 AD, to 8.9 million or 42.6% in 1500 AD, to 10.9 million or 46.4% in 1600 AD, to 11.6 million or 42.9% in 1700 AD, to 15.3 million or 43.6% in 1800 AD, to 39 million or 41.8% in 1900, to 211 million or 40.1% in 2000, to 276 million or 41.2% in 2020, and is projected to reach 338 million or 43% by 2050 and 334 million or 45% by 2100.

106 | *Islam in Asia*

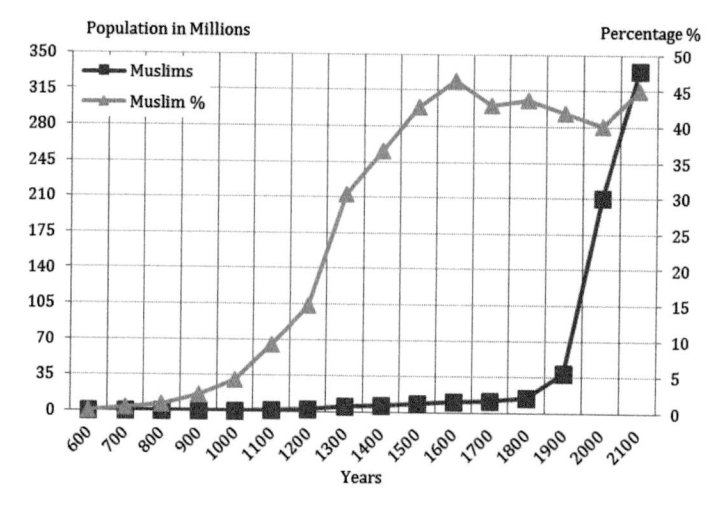

Figure 2.4a Plot of centennial estimates of the Muslim population and its percentage of the total population in Southeast Asia: 600–2100 AD (1–1500 H).

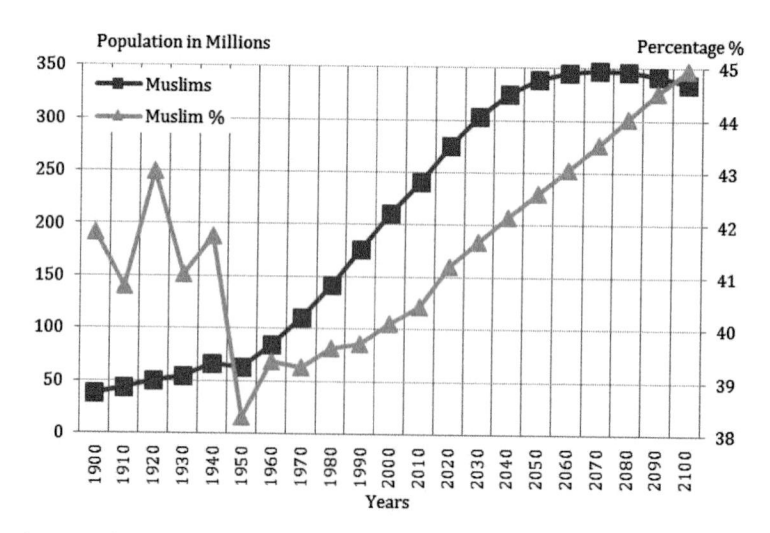

Figure 2.4b Plot of decennial estimates of the Muslim population and its percentage of the total population in Southeast Asia: 1900–2100 AD (1320–1520 H).

A plot of centennial estimates of the Muslim population and its percentage with respect to the total population in this region from 600 to 2100 is provided in Fig. 2.4a. A zoom in of this plot, providing a plot of decennial estimates of the Muslim population and its percentage with respect to the total population in this region from 1900 to 2100 is provided in Fig. 2.4b. This shows that the Muslim population in this region was increasing slowly until 1950 and is increasing substantially afterwards and remaining around 340 million throughout the second half of this century. The percentage of Muslims in this region on the other hand, peaked at 47% in 1870, then decreased to 37% in 1950, but is expected to increase steadily, reaching 45% by the end of this century.

The corresponding individual data for each country in this region is discussed below. In Section 2.4.12, the total population in each country in this region and the corresponding percentage and number of Muslims is presented centennially in Table 2.4a from 600 to 2100 and decennially in Tables 2.4b and 2.4c from 1790 to 2100.

2.4.1 Brunei

The Sultanate was established in the fifteenth century by Muslim Sultans. Currently, Brunei Darussalam consists of two close parts separated by Malaysia with total area of 5,765 sq. km and its map is presented in Fig. 2.4.1. It was occupied by the British in 1888 and gained its independence from the UK in 1984.

A 1921 estimate of the Muslim population was 12,000 or 47% of the total population. Based on census data and as shown in Table 2.4.1, the Muslim population increased in number but decreased in percentage from 27,000 or 67.1% in 1947, to 51,000 or 60.2% in 1960, then continued to increase to 85,000 or 62.2% in 1971, to 122,000 or 63.4% in 1981, to 175,000 or 67.2% in 1991, to 250,000 or 75.1% in 2001, to 310,000 or 78.8% in 2011 and 337,000 or 80.9% in 2016. Thus, assuming that the percentage of the Muslim population increases by two percentage points per decade, the Muslim population is expected to be around 0.4 million throughout the second half of this century, reaching 87% of the total population by 2050 and 97% by 2100.

Figure 2.4.1 Map of Sultanate of Brunei.

Table 2.4.1 Evolution of the Muslim population in Brunei

Year	Population	Muslims	%	Source
1921	25,434	12,000	47.18	[MAS23]es
1947	40,657	27,266	67.06	[UN56]c
1960	83,877	50,516	60.23	[UN63]c
1971	136,256	84,700	62.16	[UN73]c
1981	192,832	122,269	63.41	[BN11]c
1991	260,482	174,973	67.17	[BN11]c
2001	332,844	249,822	75.06	[BN11]c
2011	393,372	309,963	78.80	[BN11]c
2016	417,256	337,391	80.86	[BN16]c
2050	492,000	428,000	87.00	es
2100	390,000	378,000	97.00	es

2.4.2 Cambodia

The Kingdom of Cambodia has an area of 181,035 sq. km and its map is presented in Fig. 2.4.2. It was occupied by France in 1863 and became part of French Indochina in 1887. It gained its independence from France in 1953. Between 1975 and 1979, the country was under the rule of the Communist Khmer Rouge who executed at least 1.5 million people, a considerable percentage of which were Muslim.

Islam entered these lands around 674 through Arab traders during the time of Caliph Othman bnu Affan. However, influence by the Malay, the ethnic Cham embraced Islam in masse between the eleventh and thirteenth centuries. As shown in Table 2.4.2, estimate of the Muslim population based on ethnic Malay and Cham increased from 59,000 or 2.4% in 1921, to 68,000 or 2.4% in 1931, to 73,000 or 2.4% in 1936. Census data on religion was introduced in 1998, which indicate that the Muslim population increased in number but decreased in percentage from 245,000 or 2.2% in 1998, to 257,000 or 1.9% in 2008. Thus, assuming that this percentage remains constant, the Muslim population is expected to remain just over 0.4 million throughout the second half of this century.

Figure 2.4.2 Map of the Kingdom of Cambodia.

Table 2.4.2 Evolution of the Muslim population in Cambodia

Year	Population	Muslims	%	Source
1921	2,403,000	58,680	2.44	[KH921]e
1931	2,806,000	68,000	2.42	[KH931]e
1936	3,046,000	73,469	2.41	[KH936]e
1998	11,413,880	245,056	2.15	[KH]c
2008	13,395,682	257,022	1.92	[KH]c
2050	21,861,000	420,000	1.92	es
2100	21,355,000	410,000	1.92	es

2.4.3 Indonesia

The Republic of Indonesia was conquered by the Dutch in the seventeenth century, was referred to as Dutch East Indies, and gained its independence from the Netherlands in 1949. It has an area of 1,904,569 sq. km consisting of 17,508 islands, about a third (6,000) of which are inhabited. The most populous island is Java, with area 132,187 sq. km, where the capital Jakarta is located and more than half of the population. The second most populous island is Sumatra, with area 473,481 sq. km, and over a fifth of the population. Sumatra is also the sixth largest island in the world. The third most populous island is Sulawesi (174,600 sq. km) with 7% of the population. These three islands include over 86% of the total population. A map of Indonesia is presented in Fig. 2.4.3a.

The Indonesian archipelago includes the second and third largest islands in the world: New Guinea (785,753 sq. km) and Borneo (748,168 sq. km). The former is split with Papua New Guinea and includes less than 2% of the Indonesian population. Almost a third of Borneo is shared with Malaysia and Brunei and includes less than 6% of the Indonesian population. The Indonesian portion of Borneo is called Kalimantan.

The largest of the remaining islands is Timor (30,777 sq. km), which is split in half between Indonesia and East Timor. Then Halmahera (17,780 sq. km), Seram (17,100 sq. km), Sumbawa (15,448 sq. km), Flores (13,540 sq. km), Bangka (11,910 sq. km), Pulau Yos Sudarso (11,742 sq. km), Sumba (11,153 sq. km), Buru (9,505 sq. km), Bali (5,633 sq. km), Nias (5,121 sq. km), Lombok (4,725 sq. km), Belitung (4,800 sq. km), Madura (4,250 sq. km) with

2% of the population, Butung (4,200 sq. km), and Siberut (4,030 sq. km).

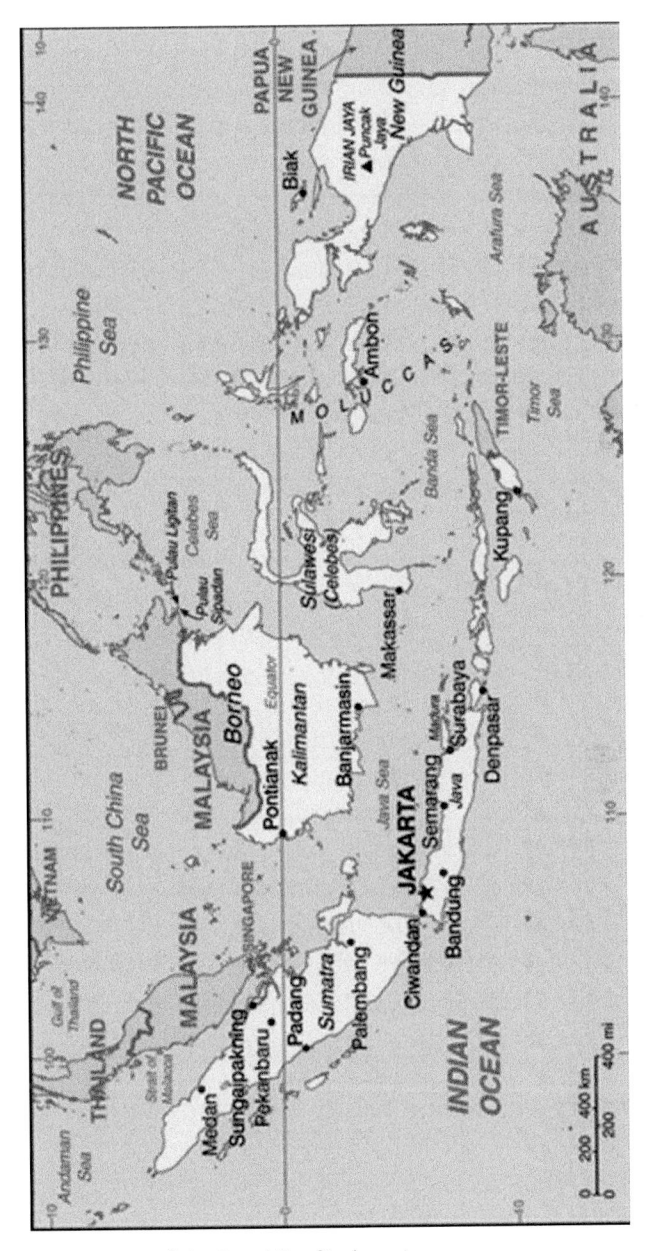

Figure 2.4.3a Map of the Republic of Indonesia.

Islam entered this archipelago through Arab traders including prophet's companions as early as 54 H/674 AD in the Island of Sumatra. However, masse conversion to Islam did not start until the eleventh century. The thirteenth century saw the establishment of Muslim Sultanates including Ternate in 1257. By the sixteenth century Islam became the dominant religion in this region and most of its territory became under Muslim control. Currently, Indonesia is the most populous majority Muslim country and has the largest Muslim population than any other country.

As shown in Table 2.4.3a, estimates of the Muslim population increased from 30 million or 85.9% in 1894, to 34 million or 81.6% in 1905, to 42 million or 85.9% in 1920, to 50 million or 83.2% in 1930. Post-independence censuses show an increase in number but decreased in percentage from 104 million or 87.5% in 1971, to 129 million or 87.1% in 1980, to 143 million or 86.9% in 1985, then continued to increase in both since then to 156 million or 87.2% in 1990, to 178 million or 88.2% in 2000, to 200 million or 88.6% in 2005, but dropped in percentage to 207 million or 87.5% in 2010. Assuming that the percentage of the Muslim population will remain at 88%, then the Muslim population is expected to reach 291 million by 2050 and 282 million by 2100.

Table 2.4.3b presents the evolution of the Muslim population since 1894 per island region. The data for 1920 and before is based on a mixture of census and guess as the regions were not thoroughly explored by the Dutch. This is particularly the case with Papua, which explains the huge difference in numbers. Data in 1971 and onward is based on census. A choropleth map illustrating the presence of Muslims per province is presented in Fig. 2.4.3b. The geographic regions presented in the table from west to east are as follows:

- Sumatra: It includes the Island of Sumatra and its surrounding islands, consisting of the following provinces: Aceh, North Sumatra, West Sumatra, Riau, Jambi, South Sumatra, Bengkulu, Lampung, Bangka-Belitung Islands (part of South Sumatra Province until 2000) and Riau Islands (part of Riau Province until 2004). Percentage of Muslims here continued to decrease from until 1971 and started increasing afterwards.

Accordingly, the number of Muslims increased from 3.3 million or 87.9% in 1894, to 3.5 million or 88.3% in 1905, to 5.3 million or 86.9% in 1920, to 7.0 million or 84.2% in 1930, to 17.4 million or 83.7% in 1971, to 23.8 million or 85.1% in 1980, to 31.4 million or 86.1% in 1990, to 37.4 million or 86.3% in 2000, to 44.1 million or 87.1% in 2010.

- Java: It includes the Island of Java and its surrounding islands, consisting of the following provinces: Special Capital Region of Jakarta, West Java, Central Java, Special Region of Yogyakarta, East Java and Banten (part of West Java Province until 2000). The number of Muslims here increased from 23.8 million or 95.0% in 1894, to 27.8 million or 92.3% in 1905, to 34.4 million or 98.3% in 1920, to 40.6 million or 97.4% in 1930, to 73.1 million or 96.1% in 1971, to 87.6 million or 96.0% in 1980, to 103 million or 95.6% in 1990, to 116 million or 95.9% in 2000, to 131 million or 95.6% in 2010.
- Bali: consists mainly of Bali Island, which is just east of Java. The number of Muslims here changed from 6,700 or 1.0% in 1894, to 3,000 or 0.4% in 1905, to 25,000 or 2.4% in 1920 and 1930, to 0.11 million or 5.1% in 1971, to 0.13 million or 5.2% in 1980, to 0.22 million or 8.0% in 1990, to 0.32 million or 10.3% in 2000, to 0.52 million or 13.4% in 2010.
- West Tenggara Barat: It is a province consisting mainly of the two large islands to the west of Bali: Lombok and Sumbawa. The number of Muslims here changed from 0.72 million or 93.3% in 1894, to 0.37 million or 91.4% in 1905, to 0.63 million or 93.9% in 1920, to 0.67 million or 94.2% in 1930, to 2.1 million or 95.8% in 1971, to 2.6 million or 96.5% in 1980, to 3.2 million or 96.0% in 1990, to 3.7 million or 92.3% in 2000, to 4.3 million or 96.5% in 2010.
- East Tenggara Barat: It is a province consisting of about 566 islands but is dominated by the three main islands of Flores, Sumba, and West Timor, the western half of the island of Timor, which is shared with East Timor. Other islands include Adonara, Alor, Komodo, Lembata, Menipo, Raijua, Rincah, Rote Island (the southernmost island

in Indonesia), Savu, Semau and Solor. The number of Muslims here increased from 21,000 or 2.9% in 1894, to 35,000 or 4.7% in 1905, to 60,000 or 6.2% in 1920, to 60,000 or 5.4% in 1930, to 0.19 million or 8.4% in 1971, to 0.22 million or 8.1% in 1980, to 0.32 million or 9.8% in 1990, to 0.34 million or 8.5% in 2000, to 0.42 million or 9.1% in 2010.

- Kalimantan: Consists of the Indonesian part of the Island of Borneo, consisting of the following provinces: West Kalimantan, Central Kalimantan, South Kalimantan and East Kalimantan. The number of Muslims here changed from 0.34 million or 30.0% in 1894, to 0.97 million or 73.2% in 1905, to 0.40 million or 26.3% in 1920, to 0.54 million or 24.9% in 1930, to 3.4 million or 65.6% in 1971, to 4.9 million or 73.5% in 1980, to 6.9 million or 75.8% in 1990, to 7.8 million or 68.6% in 2000, to 10.8 million or 78.2% in 2010.

- Sulawesi: Consists of Sulawesi Island, formerly known as Celebes, consisting of the following provinces: North Sulawesi, Central Sulawesi, South Sulawesi, Southeast Sulawesi, Gorontalo (part of North Sulawesi Province until 2000) and West Sulawesi. The number of Muslims here changed from 1.38 million or 72.3% in 1894, to 1.14 million or 43.0% in 1905, to 1.00 million or 32.5% in 1920, to 1.00 million or 23.6% in 1930, to 6.7 million or 78.6% in 1971, to 8.2 million or 79.1% in 1980, to 10.0 million or 79.9% in 1990, to 11.6 million or 77.8% in 2000, to 14.1 million or 80.9% in 2010.

- Maluku: Consists of the Maluku Islands between New Guinea and Sulawesi. It used to comprise one province of Maluku until 1999, when the northern part was declared a second province under the name of North Maluku. The number of Muslims here increased from 55,000 or 20.9% in 1894, to 0.17 million or 44.7% in 1905, to 0.29 million or 69.8% in 1920, to 0.30 million or 33.4% in 1930, to 0.54 million or 49.9% in 1971, to 0.78 million or 55.1% in 1980, to 1.06 million or 57.0% in 1990, to 1.14 million or 57.0% in 2000, to 1.55 million or 60.2% in 2010.

- Papua: It includes the Indonesian portion of New Guinea and surrounding islands and consists of two provinces: Papua and

West Papua (part of the former until 2003). The number of Muslims here changed from 200 or 0.1% in 1894, to 108,000 or 7.4% in 1905, to 8,000 or 4.1% in 1920, to 8,000 or 4.0% in 1930, to 33,000 or 3.6% in 1971, to 0.13 million or 11.3% in 1980, to 0.33 million or 20.1% in 1990, to 0.41 million or 18.5% in 2000, to 0.74 million or 20.7% in 2010.

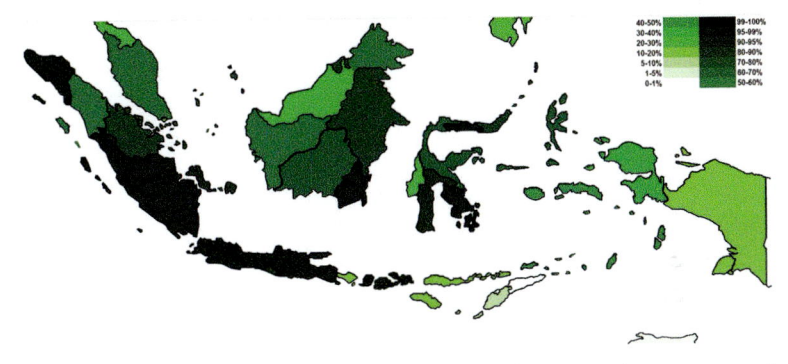

Figure 2.4.3b A choropleth map of Indonesia illustrating the presence of Muslims per province.

Table 2.4.3a Evolution of the Muslim population in Indonesia

Year	Population	Muslims	%	Source
1894	34,511,221	29,645,616	85.90	[JAN]es
1905	41,740,150	34,061,037	81.60	[IDH]es
1920	48,972,535	42,073,992	85.91	[MAS23]es
1930	60,364,198	50,221,981	83.20	[MAS55]es
1971	118,367,850	103,579,496	87.51	[UN73]c
1980	147,490,298	128,464,050	87.10	[ID90]c
1985	165,012,000	143,430,000	86.92	[ID05]s
1990	179,378,946	156,436,380	87.21	[ID05]c
2000	201,241,999	177,528,772	88.22	[ID05]c
2005	226,255,000	200,420,000	88.58	[ID05]s
2010	236,744,626	207,176,176	87.51	[ID10]c
2050	330,905,000	291,196,000	88.00	es
2100	320,782,000	282,289,000	88.00	es

Table 2.4.3b Evolution of the Muslim population in Indonesia per island region since 1894. *P*: Total population, *M*: Muslim population, *M%*: Percentage of Muslim population

		1894	1905	1920	1930	1971	1980	1990	2000	2010
Sumatra	P	3,765,734	3,949,773	6,070,490	8,306,414	20,808,148	28,016,160	36,506,703	43,309,707	50,630,931
	M	3,308,277	3,489,174	5,277,190	6,995,562	17,405,096	23,833,498	31,445,109	37,370,213	44,111,873
	M%	87.85	88.34	86.93	84.22	83.65	85.07	86.14	86.29	87.12
Java	P	25,067,551	30,090,008	34,984,171	41,713,268	76,086,327	91,269,528	107,581,306	121,352,608	136,610,590
	M	23,814,173	27,781,671	34,377,802	40,623,419	73,102,760	87,599,660	102,820,173	116,341,892	130,651,037
	M%	95.00	92.33	98.27	97.39	96.08	95.98	95.57	95.87	95.64
Bali	P	670,000	733,535	1,054,934	1,027,000	2,120,322	2,469,930	2,777,811	3,151,162	3,890,757
	M	6,700	3,000	25,000	25,000	108,414	128,436	222,225	323,853	520,244
	M%	1.00	0.41	2.37	2.43	5.11	5.20	8.00	10.28	13.37
West Nusa Tenggara	P	773,600	400,000	671,200	711,200	2,203,465	2,724,664	3,369,649	4,009,261	4,500,212
	M	721,510	365,418	630,000	670,000	2,110,054	2,629,301	3,234,863	3,699,018	4,341,284
	M%	93.27	91.35	93.86	94.21	95.76	96.50	96.00	92.26	96.47

Region		Col1	Col2	Col3	Col4	Col5	Col6	Col7	Col8	Col9
East Nusa Tenggara	P	740,000	742,000	970,708	1,107,376	2,295,287	2,737,166	3,268,644	3,952,279	4,683,827
	M	21,350	34,650	60,000	60,000	192,401	221,710	320,327	335,005	423,925
	M%	2.89	4.67	6.18	5.42	8.38	8.10	9.80	8.48	9.05
Kalimantan	P	1,118,931	1,321,496	1,533,798	2,168,661	5,154,774	6,723,086	9,099,874	11,331,558	13,787,831
	M	335,700	967,440	404,000	540,000	3,382,388	4,938,179	6,900,674	7,770,860	10,786,584
	M%	30.00	73.21	26.34	24.90	65.62	73.45	75.83	68.58	78.23
Sulawesi	P	1,914,000	2,650,000	3,075,108	4,231,919	8,526,901	10,409,533	12,520,711	14,946,488	17,371,782
	M	1,383,000	1,140,000	1,000,000	1,000,000	6,701,900	8,233,031	9,999,405	11,626,140	14,051,853
	M%	72.26	43.02	32.52	23.63	78.60	79.09	79.86	77.79	80.89
Maluku	P	261,405	383,338	418,276	898,360	1,089,565	1,411,006	1,857,790	1,990,598	2,571,593
	M	54,706	171,444	292,000	300,000	543,400	777,464	1,058,940	1,135,097	1,547,240
	M%	20.93	44.72	69.81	33.39	49.87	55.10	57.00	57.02	60.17
Papua	P	200,000	1,470,000	193,850	200,000	923,440	1,173,875	1,648,708	2,220,934	3,593,803
	M	200	108,240	8,000	8,000	33,083	132,879	331,229	410,231	742,122
	M%	0.10	7.36	4.13	4.00	3.58	11.32	20.09	18.47	20.65

2.4.4 Laos

Lao People's Democratic Republic has a total area of 236,800 sq. km and its map is presented in Fig. 2.4.4. It was occupied by France in 1893 and gained its independence from France in 1954. Islam entered here in the 1970s as Muslim Chams fled the Khmer massacres in Cambodia. By 1985, Muslims numbered 1,000 or 0.03% of the total population. Recent census data indicate that the Muslim population decreased from 1,133 or 0.02% in 1995, to 1,044 or 0.02% in 2005, then increased to 1,605 or 0.03% in 2015. Thus, assuming that the percentage of Muslims will increase by 0.01 of a percentage point per decade, the Muslim population is expected to remain less than 10,000 throughout this century and staying around 0.1% of the total population in the second half of this century. The data is summarized in Table 2.4.4.

Figure 2.4.4 Map of Lao People's Democratic Republic.

Table 2.4.4 Evolution of the Muslim population in Laos

Year	Population	Muslims	%	Source
1911	631,800	—	0.00	[VNH]es
1921	818,800	—	0.00	[MAS23]es
1952	1,763,761	—	0.00	[MAS55]es
1985	3,570,000	1,000	0.03	[KET86]es
1995	4,564,565	1,133	0.02	[LA95]c
2005	5,581,785	1,044	0.02	[LA05]c
2015	6,378,216	1,605	0.03	[LA15]c
2050	9,480,000	7,000	0.07	es
2100	8,424,000	10,000	0.12	es

2.4.5 Malaysia

It has an area of 330,803 sq. km, consisting of thirteen states and three federal territories. Eleven states are located in the West or Peninsular Malaysia with two federal territories, and two other states, Sabah (73,631 sq. km) and Sarawak (124,450 sq. km), in the East, on the north of the Island of Borneo and the Federal Territory of the Labuan Island (91 sq. km). The western states are Johor (19,210 sq. km), Kedah (9,500 sq. km), Kelantan (15,099 sq. km), Melaka (1,664 sq. km), Negeri Sembilan (6,686 sq. km), Pahang (36,137 sq. km), Perak (21,035 sq. km), Perlis (821 sq. km), Penang (1,048 sq. km), Selangor (8,153 sq. km), Terengganu (13,035 sq. km), and the two federal territories of Kuala Lumpur (243 sq. km) which is the largest city and Putrajaya (49 sq. km) which is the capital or administrative center. A map of Malaysia is presented in Fig. 2.4.5.

Melaka or Malacca was occupied by the Portuguese in 1511, who lost it to the Dutch in 1641, who in turn lost it to the British in 1824. The British also took Penang from the Kedah Sultanate in 1786. Then Labuan in 1826, Sarawak in 1841, taken from the Brunei Sultanate, Negeri Sembilan in 1873, Selangor in 1874, Perak in 1876, and Sabah in 1882, taken from the Sulu Sultanate, Pahang in 1887, Kedah, Kelantan, Perlis and Terengganu in 1909, and Johor in 1914. Malaysia gained its independence from the UK in 1957 (West) and 1963 (East).

Figure 2.4.5 Map of Malaysia.

In 1136, the ninth Hindu Rajah of Kedah Kingdom, Phra Ong Mahawangsa renounced Hinduism and converted to Islam, which was introduced by Muslims from neighboring Aceh, Sumatra. He also changed his name to Sultan Mudzafar Shah, and thereby starting the Kedah Sultanate which still exists as part of the Malay Federation. He ruled the northern region of Malay Peninsula from 1136 to 1179. In the thirteenth century, the Sultanate of Brunei was established in the island of Borneo. Other Muslim sultanates were established in the lands occupied by current Malaysia, all accelerated the spread of Islam.

In Sabah, the first census was conducted in 1911, then every ten years till 1951, then in 1960 and every ten years then after. Religious data was collected starting in 1951. Thus, the estimate on the size of the Muslim population in Sabah before 1951 is based on the results of the 1951 census. The data for Sabah and Labuan Island is summarized in Table 2.4.5a.

In Sarawak, the first census was conducted in 1939, then in 1947, then 1960, then every ten years then after. Religious data was collected starting in 1947. Thus, the estimate on the size of the Muslim population in Sarawak before 1947 is based on the results of the 1947 census. The data for Sarawak is summarized in Table 2.4.5b.

Table 2.4.5a Evolution of the Muslim population in Sabah; including Labuan

Year	Population	Muslims	%	Source
1891	190,853	38,170	20.00	[JAN]es
1911	208,183	60,165	28.90	[GB911]c
1921	263,252	80,000	30.39	es
1931	277,476	90,000	32.44	es
1951	334,141	115,126	34.45	[UN56]c
1960	454,421	172,324	37.92	[UN63]c
1970	651,304	260,945	40.07	[UN79]c
1980	950,556	487,627	51.30	[UN88]c
1991	1,788,926	1,101,740	61.59	[MY00]c
2000	2,539,117	1,618,858	63.76	[MY00]c
2010	3,293,650	2,162,218	65.65	[MY10]c

Table 2.4.5b Evolution of the Muslim population in Sarawak

Year	Population	Muslims	%	Source
1891	300,000	30,000	10.00	[JAN]es
1911	350,000	80,000	22.86	[BNH]es
1921	400,000	90,000	22.50	es
1931	450,000	100,000	22.22	es
1939	490,585	110,000	22.42	es
1947	546,385	134,318	24.58	[UN56]c
1960	744,529	174,123	23.39	[UN63]c
1970	887,292	229,590	25.88	[UN79]c
1980	1,233,103	324,575	26.32	[UN88]c
1991	1,642,771	471,451	28.70	[MY00]c
2000	2,009,893	637,496	31.72	[MY00]c
2010	2,471,140	796,239	32.22	[MY10]c

The first census covering all Peninsular Malaysia was conducted in 1911, then in 1921, 1931, 1947, 1957, 1970, then every ten years then after. Religious data was collected starting in 1970. Previous censuses included ethnic data, from which we can deduce religious data with the assumption that Malays are Muslim. The data for Peninsula Malaysia is summarized in Table 2.4.5c, where the number of Muslims from 1911 to 1957 is the recorded number of ethnic Malay. After independence, the constitution of Malaysia restricted Malay ethnic affiliation to Muslims. A new ethnic term was introduced since 1970, dividing the citizen population to indigenous people or Bumiputera which includes Malay, and others. In Peninsular Malaysia, the 2010 census showed that

- 100% of the 13,409,409 Malay population are Muslim;
- 25.35% of other Bumiputera are Muslim, or 82,743 out of 326,343;
- 0.51% of Chinese are Muslim, or 27,898 out of 5,509,302;
- 3.88% of Indians are Muslim, or 73,451out of 1,892,322;
- 44.89% of the other citizens are Muslim, or 58,450 out of 130,205;
- 58.80% of non-citizens are Muslim, or 765,386 out of 1,301,764.

By Malaysian law, a non-Muslim cannot be Malay.

Table 2.4.5c Evolution of the Muslim population in Peninsular Malaysia

Year	Population	Muslims	%	Source
1891	946,315	709,736	75.00	[JAN]es
1911	2,070,025	1,109,345	53.59	[GB11]c
1921	2,907,000	1,568,588	53.96	[MYH]e
1931	3,788,000	1,863,872	49.21	[MYH]e
1947	4,908,000	2,427,853	49.47	[MYH]e
1957	6,379,000	3,125,474	49.00	[MYH]e
1970	8,780,728	4,673,670	53.23	[UN79]c
1980	10,886,713	6,106,105	56.09	[UN88]c
1991	14,131,723	8,684,150	61.45	[MY00]c
2000	17,649,266	11,241,674	63.69	[MY00]c
2010	22,569,345	14,417,337	63.88	[MY10]c

Thus, by aggregating the above data, we can deduce the change in the size of the Muslim population in Malaysia as summarized in Table 2.4.5d. Accordingly, the Muslim population increased in number but decreased in percentage from 0.78 million or 54.1% in 1891, to 1.5 million or 50.7% in 1911, to 1.7 million or 48.4%

in 1921, to 2.1 million or 45.5% in 1931. The Muslim population continued to increase since then both in number and percentage to 2.7 million or 46.3% in 1947, to 3.5 million or 46.4% in 1960, to 5.2 million or 50.0% in 1970, to 6.9 million or 52.9% in 1980, to 10.3 million or 58.7% in 1991, to 14.0 million or 60.6% in 2000, to 17.4 million or 61.9% in 2010. Thus, assuming that the Muslim population continues to increase by one percentage point per decade, the Muslim population is expected to reach 27 million or 66% by 2050 and 28 million or 71% by 2100.

Table 2.4.5d Evolution of the Muslim population in Malaysia

Year	Population	Muslims	%	Source
1891	1,437,168	777,906	54.13	
1911	2,628,208	1,249,510	47.54	[GB911]c
1921	3,570,252	1,738,588	48.70	
1931	4,515,476	2,053,872	45.50	
1947	5,788,526	2,677,297	46.25	
1960	7,577,950	3,471,921	46.43	
1970	10,319,324	5,164,205	50.04	[UN73]c
1980	13,070,372	6,918,307	52.93	[UN88]c
1991	17,473,846	10,257,341	58.70	[UN04]c
2000	23,194,675	14,049,379	60.57	[UN]c
2010	28,062,370	17,375,794	61.92	[MY10]c
2050	40,550,000	26,763,000	66.00	es
2100	40,078,000	28,455,000	71.00	es

2.4.6 Myanmar (Burma)

The Republic of the Union of Myanmar has a total area of 676,578 sq. km and its map is presented in Fig. 2.4.6a. It was occupied by the British between 1824 and 1886 and gained its independence from the UK in 1948. It changed its name from Burma in 1989.

Islam entered here as early as the seventh century through Muslim Arab traders on northern coast and more to the south later. A Muslim state was established in Arakan state (later renamed Rakhine) when the Sultan of Muslim Bengal Naseerud Deen Mahmud Shah (ruled from 1442 to 1459) helped King Sulayman Naramithla establish a Muslim Mtauku state. This state lasted until 1784 and extended south covering most of the Burmese coastline during the reign of Sultan Salim Shah Razagri who ruled from 1593 to 1612. It extended as far south as Moulmein, which is later renamed Mawlamyaing.

Based on census data, the Muslim population changed from 0.10 million or 3.6% in 1871, to 0.17 million or 4.5% in 1881, to 0.25 million or 3.1% in 1891, to 0.34 million or 3.2% in 1901, to 0.42 million or 3.5% in 1911, to 0.50 million or 3.8% in 1921, to 0.59 million or 4.0% in 1931.

The religion data from the 1941 was destroyed in WWII. A post-independence multi-stage census was conducted in 1953–1954 but was not completed due to unrest in the country. This census counted almost all the urban population (2,940,704) in 1953 but covered only 15% of the rural area, counting 2,676,719 in 1954. Religious data was based on 20% sample and found that Muslims constituted 7.97% of the urban population (234,512) and 3.20% of the rural population. Thus, out of a total population the census estimated 738,365 or 3.95% were Muslim. Post-independence complete censuses recorded a continuous increase in the Muslim population to 1.08 million or 3.9% in 1973, to 1.31 million or 3.8% in 1983 and 2.24 million or 4.3% in 2014. The latter census showed that almost half of the Muslims reside in the State of Arakan (Rakhine) on the border with Bangladesh, where they constitute half of the population. A choropleth map illustrating the presence of Muslims per state and region is presented in Fig. 2.4.6b.

An increase of the Muslim population during the British occupation was due to the fact that Burma was part of British India which made Muslims from current India and Bangladesh to settle in Burma. After the independence, and especially after the military coup of 1962, Muslims are under constant oppression.

Assuming that the percentage of the Muslim population continues to increase by a tenth of a percentage point per decade, the Muslim population is expected to remain just less than three million throughout the second half of this century. The data is summarized in Table 2.4.6.

Figure 2.4.6a Map of the Republic of the Union of Myanmar.

Figure 2.4.6b A choropleth map of Myanmar illustrating the presence of Muslims per state and region.

Table 2.4.6 Evolution of the Muslim population in Myanmar

Year	Population	Muslims	%	Source
1871	2,747,148	99,846	3.63	[INH71]c
1881	3,736,771	168,881	4.52	[INH81]c
1891	8,223,071	253,031	3.08	[INH91]c
1901	10,490,621	339,446	3.24	[INH01]c
1911	12,115,217	420,777	3.47	[INH11]c
1921	13,169,099	500,592	3.80	[INH21]c
1931	14,647,470	584,839	3.99	[INH31]c
1953	18,686,110	738,365	3.95	[MM53,UN56]c
1973	28,084,513	1,082,318	3.86	[MM73]c
1983	34,124,908	1,308,524	3.83	[MM83]c
2014	51,486,253	2,237,495	4.35	[MM14]c
2050	62,253,000	2,926,000	4.70	es
2100	55,299,000	2,876,000	5.20	es

2.4.7 Philippines

The Republic of the Philippines was occupied by Spain in 1565 and ceded to the United States of America in 1898 following the Spanish-American War. It gained its independence from the US in 1946. It has a total area of 340,575 sq. km, consisting of 7,107 islands of which only 3,144 are named. The largest is Luzon (104,688 sq. km), with more than half of the population and were the capital Manila and the largest city: Quezon City, are located. The second largest is Mindanao (94,630 sq. km) where over a fifth of the population and two-third of Filipino Muslims live. The largest of the remaining islands are Samar (13,080 sq. km) 2% of the population, Negros (12,710 sq. km) 5%, Palawan (11,785 sq. km) 1%, Panay (11,515 sq. km) 4%, Mindoro (9,735 sq. km) 1%, Leyte (7,214 sq. km) 2%, Cebu (4,422 sq. km) 3%, Bohol (3,865 sq. km) 1%, Masbate (3,269 sq. km) 1%, Catanduanes (1,492 sq. km) 0.3%, Romblon (1,356 sq. km), Basilan (1,282 sq. km) 0.3%, Tawi-Tawi (1,087 sq. km) 0.4%, Marinduque (953 sq. km) 0.3% and Jolo (894 sq. km) 1%. A map of the Philippines is presented in Fig. 2.4.7.

Islam entered these Islands in 1380 through Karimul Makhdum, Arab trader coming from Malacca in current Malaysia. Then more

Muslims from current Indonesia and Malaysia helped spread Islam in the Islands. Eventually three Muslim Sultanates covering the whole archipelago were erected:

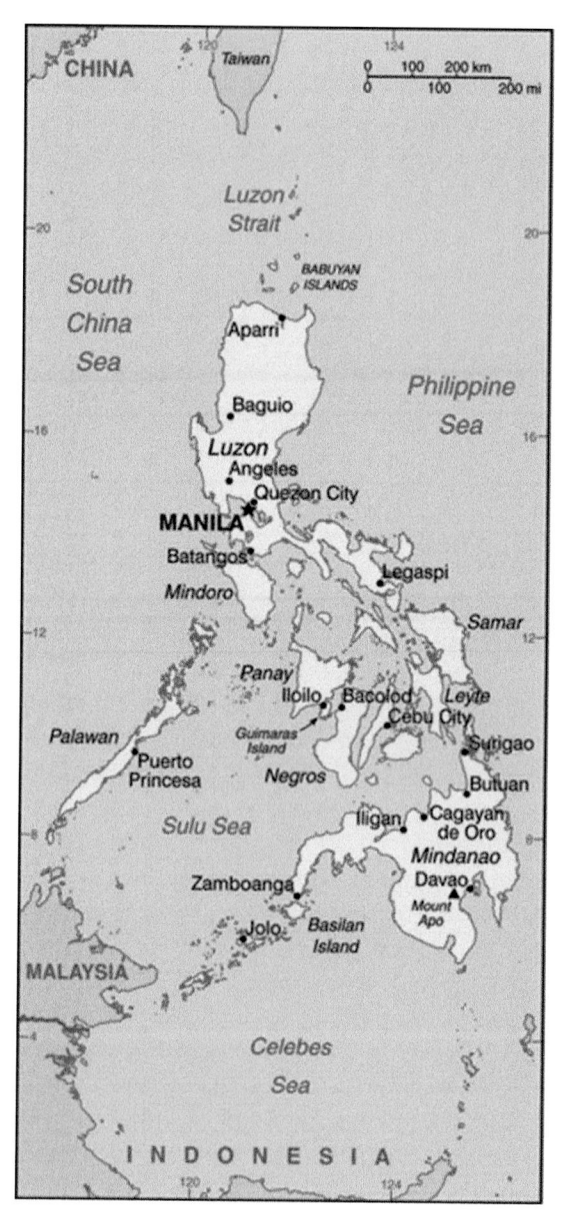

Figure 2.4.7 Map of the Republic of the Philippines.

- Sulu Sultanate: was established in 1457 by Sayyid Abu Bakr Abirin, an Arab religious scholar from Johor in current Malaysia, and lasted until 1917. At its peak, it controlled the Malaysian state of Sabah, the Sulu Archipelago, Palawan, Negros, Panay, Mindoro and Ilocos region in northwest Luzon.
- Maguindanao Sultanate: was established in 1520 by Shariff Mohammed Kabungsuwan from Johor and lasted until 1898. At its peak, it controlled the entire Island of Mindanao.
- Manilad or Maynila Sultanate: was established in the sixteenth century on the Island of Luzon but ended by the Spaniards in 1571. The name Manilad was converted to the name Manilla upon the Spanish occupation, which is the current capital of the Philippines.

Based on census data and as shown in Table 2.4.7a, the Muslim population increased from 0.28 million or 3.6% in 1903, to 0.44 million or 4.2% in 1918, to 0.68 million or 4.2% in 1939, to 0.79 million or 4.1% in 1948, to 1.32 million or 4.9% in 1960, to 1.58 million or 4.3% in 1970, to 2.77 million or 4.6% in 1990, to 3.86 million or 5.1% in 2000, to 5.13 million or 5.6% in 2010 and 6.06 million or 6.0% in 2015.

Assuming that the percentage of the Muslim population will continue to increase by a half of a percentage point each decade, the Muslim population is expected to reach eleven million or 7.5% by 2050 and fifteen million or 10.0% by 2100.

Table 2.4.7a Evolution of the Muslim population in the Philippines

Year	Population	Muslims	%	Source
1903	7,635,426	277,547	3.64	[PH903]c
1918	10,445,081	443,037	4.24	[PH918]c
1939	16,000,303	676,813	4.23	[PH939]c
1948	19,211,345	791,817	4.12	[PH948]c
1960	27,087,685	1,317,475	4.86	[PH60]c
1970	36,684,486	1,584,963	4.32	[PH70]c
1990	60,362,109	2,769,643	4.59	[UN04]c
2000	75,980,838	3,862,409	5.08	[UN]c
2010	92,094,656	5,126,831	5.57	[PH10]c
2015	100,913,714	6,064,744	6.01	[PH15]c
2050	144,488,000	10,837,000	7.50	es
2100	146,327,000	14,633,000	10.00	es

Table 2.4.7b Evolution of the Muslim population in the Philippines per island/archipelago since 1903. *P*: Total population, *M*: Muslim population, *M%*: Percentage of Muslim population, *MR%*: Muslim Ratio in percentage

		1903	1918	1939	1948	1960	1970	1990	2000	2010	2015
Tawi-Tawi	P	14,618	38,709	38,910	49,258	67,269	97,230	205,658	294,246	337,375	359,391
	M	14,500	38,265	35,071	47,743	64,582	92,915	198,883	280,235	326,402	348,406
	M%	99.19	98.85	90.13	96.92	96.01	95.56	96.71	95.24	96.75	96.94
Mapun &Turtles	P	2,020	6,090	6,851	9,273	11,325	12,966	21,668	25,611	27,937	31,324
	M	2,000	6,019	6,768	9,140	11,036	12,467	20,487	24,234	27,463	29,179
	M%	99.01	98.83	98.79	98.57	97.45	96.15	94.55	94.62	98.30	93.15
Jolo	P	90,589	127,977	201,343	182,295	248,304	315,421	463,756	614,340	718,142	824,731
	M	89,000	124,326	188,694	170,000	235,308	296,602	452,871	602,153	706,229	816,273
	M%	98.25	97.15	93.72	93.26	94.77	94.03	97.65	98.02	98.34	98.97
Basilan	P	11,331	23,089	57,561	110,297	155,712	143,829	236,473	331,553	390,736	459,366
	M	10,000	19,087	45,000	77,000	113,148	89,127	169,055	255,795	311,004	377,203
	M%	88.25	82.67	78.18	69.81	72.66	61.97	71.49	77.15	79.59	82.11

Muslims in Southeast Asia

Balabac	P	2,000	2,109	2,760	3,355	4,591	7,912	20,056	25,257	35,758	40,142
	M	1,800	1,854	2,147	2,800	3,547	5,360	14,985	20,326	29,884	34,558
	M%	90.00	87.91	77.79	83.46	77.26	67.75	74.72	80.48	83.57	86.09
Mindanao	P	447,532	848,304	1,820,314	2,468,328	4,786,375	7,250,433	13,138,405	16,570,252	20,114,265	22,218,390
	M	158,530	230,067	390,206	470,591	876,763	1,072,186	1,849,930	2,537,553	3,466,596	4,074,667
	M%	35.42	27.12	21.44	19.07	18.32	14.79	14.08	15.31	17.23	18.34
Palawan	P	15,789	36,630	50,812	64,024	111,375	175,282	404,246	611,638	806,211	907,639
	M	1,700	3,596	4,120	5,814	9,229	11,688	21,761	31,330	47,808	61,099
	M%	10.77	9.82	8.11	9.08	8.29	6.67	5.38	5.12	5.93	6.73
Cuyo	P	12,773.00	17,876.00	21,358.00	22,445.00	24,728.00	26,577.00	31,374.00	39,564.00	45,718.00	47,101
	M	—	—	—	—	—	—	2	70	80	117
	M%	0.00	0.00	0.00	0.00	0.00	0.00	0.01	0.18	0.17	0.25
Calamianes	P	5,134.00	12w,438.00	18,682.00	16,445.00	21,975.00	26,864.00	50,070.00	72,818.00	98,022.00	109,656
	M	—	81	128	—	—	21	18	103	181	488
	M%	0.00	0.65	0.69	0.00	0.00	0.08	0.04	0.14	0.18	0.45

Table 2.4.7b (Continued)

		1903	1918	1939	1948	1960	1970	1990	2000	2010	2015
Camiguin	P	30,754	37,839	40,803	69,599	44,717	59,913	64,164	74,072	83,676	88,478
	M	—	4	18	—	—	1	29	183	205	228
	M%	0.00	0.01	0.04	0.00	0.00	0.00	0.05	0.25	0.24	0.26
Siquijor	P	46,023	56,774	59,502	57,258	59,555	62,976	73,872	81,486	90,791	95,933
	M	10	16	—	50	100	189	15	27	51	115
	M%	0.02	0.03	0.00	0.09	0.17	0.30	0.02	0.03	0.06	0.12
Negros	P	509,766	669,160	1,156,357	1,482,219	1,930,084	2,219,022	3,173,937	3,681,700	4,187,878	4,413,690
	M	400	437	349	159	147	1,092	1,619	2,627	3,560	5,276
	M%	0.08	0.07	0.03	0.01	0.01	0.05	0.05	0.07	0.09	0.12
Cebu	P	653,727	855,065	1,067,471	1,123,107	1,332,847	1,634,182	2,636,536	3,334,804	4,156,408	4,631,549
	M	800	1,054	182	17	75	198	1,800	3,417	6,855	9,885
	M%	0.12	0.12	0.02	0.00	0.01	0.01	0.07	0.10	0.16	0.21

		Col1	Col2	Col3	Col4	Col5	Col6	Col7	Col8	Col9	Col10
Bohol	P	269,223	358,387	491,549	553,407	592,194	683,297	946,813	1,135,064	1,252,776	1,313,363
	M	—	—	24	17	5	140	689	919	1,615	2,167
	M%	0.00	0.00	0.00	0.00	0.00	0.02	0.07	0.08	0.13	0.16
Siargao	P	10,875.00	15,769.00	26,983.00	36,957.00	47,701.00	57,433.00	82,113.00	93,354.00	110,653.00	116,587
	M	—	—	7	—	—	—	5	59	137	181
	M%	0.00	0.00	0.03	0.00	0.00	0.00	0.01	0.06	0.12	0.16
Dinagats	P	5,243	8,382	16,156	17,317	22,761	32,227	98,671	106,592	126,699	127,148
	M	—	—	12	—	—	—	12	38	24	54
	M%	0.00	0.00	0.07	0.00	0.00	0.00	0.01	0.04	0.02	0.04
Leyte	P	388,922	597,950	861,019	1,006,891	1,172,972	1,362,051	1,686,743	1,944,265	2,181,250	2,388,443
	M	—	—	226	243	78	221	712	1,691	2,084	3,791
	M%	0.00	0.00	0.03	0.02	0.01	0.02	0.04	0.09	0.10	0.16
Guimaras	P	21,467	27,170	38,545	40,697	57,560	73,014	117,827	140,814	162,726	174,610
	M	—	3	—	—	—	—	18	36	23	87
	M%	0.00	0.01	0.00	0.00	0.00	0.00	0.02	0.03	0.01	0.05

Table 2.4.7b (Continued)

		1903	1918	1939	1948	1960	1970	1990	2000	2010	2015
Panay	P	813,290	923,443	1,309,513	1,451,062	1,688,422	2,043,530	3,012,560	3,491,028	4,023,422	3,854,207
	M	240	364	151	1,128	375	70	1,242	2,102	3,581	3,947
	M%	0.03	0.04	0.01	0.08	0.02	0.00	0.04	0.06	0.09	0.10
Samar	P	266,237	379,575	546,163	757,212	867,994	1,020,258	1,243,379	1,512,668	1,747,229	1,879,937
	M	360	521	86	17	9	156	450	848	1,168	1,464
	M%	0.14	0.14	0.02	0.00	0.00	0.02	0.04	0.06	0.07	0.08
Biliran	P	21,661	35,751	54,367	67,661	87,285	90,499	117,905	139,650	161,250	171,612
	M	—	—	—	—	—	—	31	71	262	427
	M%	0.00	0.00	0.00	0.00	0.00	0.00	0.03	0.05	0.16	0.25
Masabate	P	43,675	67,513	182,407	211,113	335,971	492,908	598,615	703,909	833,638	892,392
	M	50	81	89	381	49	177	267	736	813	1,215
	M%	0.11	0.12	0.05	0.18	0.01	0.04	0.04	0.10	0.10	0.14
Romblon	P	52,848	64,610	99,367	108,817	131,658	167,082	227,099	263,519	283,482	292,781
	M	40	53	30	26	—	22	37	98	102	180
	M%	0.08	0.08	0.03	0.02	0.00	0.01	0.02	0.04	0.04	0.06

Muslims in Southeast Asia | 135

Mindoro	P	39,582	72,431	131,483	167,705	313,314	472,396	823,262	1,055,732	1,235,152	1,331,473
	M	40	84	79	2,092	148	105	117	638	1,587	2,585
	M%	0.10	0.12	0.06	1.25	0.05	0.02	0.01	0.06	0.13	0.19
Luzon	P	3,829,285	5,006,883	7,460,327	9,070,204	12,832,026	17,994,488	30,598,207	39,224,630	48,377,576	53,283,017
	M	1,737	17,085	4,453	4,649	2,872	2,415	35,339	97,249	188,876	289,790
	M%	0.05	0.34	0.06	0.05	0.02	0.01	0.12	0.25	0.39	0.54
Marinduque	P	51,674	56,868	81,635	85,828	114,586	144,109	185,281	216,930	227,582	234,495
	M	70	83	32	—	—	—	67	230	106	181
	M%	0.14	0.15	0.04	0.00	0.00	0.00	0.04	0.11	0.05	0.08
Catanduanes	P	39,410	63,530	98,528	112,121	156,329	162,302	186,756	214,916	245,574	260,954
	M	70	116	31	—	4	—	18	101	132	252
	M%	0.18	0.18	0.03	0.00	0.00	0.00	0.01	0.05	0.05	0.10
Babuyanes & Batanes	P	8,946	8,214	12,236	14,206	14,895	18,180	32,243	30,705	32,730	33,948
	M	—	—	—	—	—	—	—	—	3	7
	M%	0.00	0.00	0.00	0.00	0.00	0.00	0.00	0.00	0.01	0.02

Over 90% of Filipino Muslims live in the southwest of the Philippines, mainly in eastern half of Mindanao Island and the Sulu Archipelago, while less than 0.5% of the rest of the Philippines is Muslim. Indeed, Table 2.4.7b shows the distribution and evolution of the Muslim population in the Philippines per island since 1903. The percentage of Muslims is increasing in virtually every island. Well over 90% of the population in the Sulu Archipelago is Muslim. Close to a fifth of Mindanao's population is Muslim, although the vast majority of them are in the east part of the island, where they constitute a third of the population. Nearly a tenth of Palawan's population is Muslim. However, the percentage of Muslims in the rest of the islands is well below 1%.

2.4.8 Singapore

The Republic of Singapore was occupied by the British in 1819 who separated it from the Johor Sultanate. It then gained its independence from the UK by joining the Malaysian Federation, and then opted out in 1965. It has a total area 697 sq. km consisting of 63 islands. By far the largest is the Main Island or Pulau Ujong (536 sq. km) where most of the population lives. The largest of the remaining islands are Jurong (32 sq. km), Tekong (24 sq. km), Ubin (10 sq. km) and Sentosa (5 sq. km). A map of Singapore is presented in Fig. 2.4.8.

Censuses inquiring on ethnic affiliation were conducted here since 1819. Religious adherence question was included for all ages of the population in the censuses of 1849, 1911, 1921 and 1931. Censuses since 1980 collected religious affiliation from resident population aged 15 and over. The ethnic distribution of the population was split into Malays, who are all Muslim, Indians, whose fifth to third are Muslim, Chinese, 0.2% are Muslim, and others, whose tenth to third are Muslim. The religious census data also distributed the religion among ethnicities, making inference of religious adherence from ethnic affiliation more reliable. These data are provided in Table 2.4.8a. These tables show the total ethnic affiliates (T) for each ethnicity and the number of Muslims (M) who belong to that ethnicity, and the percentage (%) of each ethnicity that is Muslim.

The data in 1990 onward are from the census results, as well as the data for Malay in 1980 and for Indians in 1980 and 1931. The data for 1849 and 1911, 1921, 1931 and 1980 are estimated based on the recorded total Muslim population and the percentage of Muslims in the closest census. The data in 1990 is based on the population ten years and over.

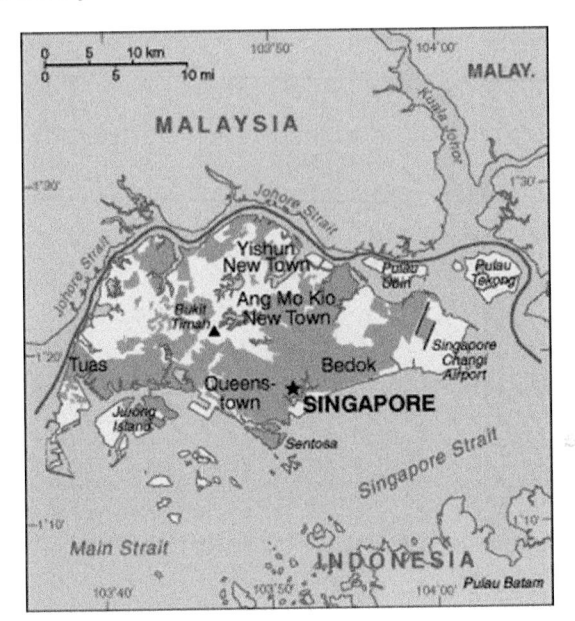

Figure 2.4.8 Map of the Republic of Singapore.

Table 2.4.8a Evolution of the Muslim populations per ethnicity in Singapore since 1819. T: Total, M: Muslims, M%: Percentage of Muslims

		1819	1824	1828	1830	1832	1836	1840	1849	1860
Malay	T	120	6,431	6,943	7,640	9,296	12,538	13,200	17,039	16,183
	M	120	6,431	6,943	7,640	9,296	12,538	13,200	17,039	16,183
	M%	100	100	100	100	100	100	100	100	100.00
Indian	T	—	756	1,389	1,913	1,943	2,932	3,362	6,284	12,996
	M	—	529	972	1,339	1,360	2,052	2,353	4,399	6,498
	M%	—	70.00	70.00	70.00	70.00	70.00	70.00	70.00	50.00
Chinese	P	30	3,317	6,210	6,555	7,762	13,749	17,695	27,988	50,021
	M	0	7	12	13	16	27	35	56	100
	M%	0.20	0.20	0.20	0.20	0.20	0.20	0.20	0.20	0.20
Other	P	—	179	235	526	609	765	1,097	1,580	2,534
	M	—	59	78	174	201	252	362	521	785
	M%	—	33.0	33.0	33.0	33.0	33.0	33.0	33.0	31.0
Total	P	150	10,683	14,885	16,634	19,715	29,984	35,389	52,891	81,734
	M	120	7,026	8,005	9,166	10,873	14,870	15,951	22,015	23,567
	M%	80.00	65.77	53.78	55.10	55.15	49.59	45.07	41.62	28.83

Table 2.4.8a *(Continued)*

		1871	1881	1891	1901	1911	1921	1931	1947
Malay	T	26,141	33,012	35,956	35,988	41,806	53,595	65,014	113,803
	M	26,141	33,012	35,956	35,988	41,806	53,595	65,014	113,803
	M%	100	100	100	100	100	100	100	100
Indian	T	10,313	12,086	16,009	17,047	27,755	32,314	50,811	71,927
	M	5,157	5,439	6,404	6,819	8,327	10,017	14,380	18,701
	M%	50.00	45.00	40.00	40.00	30.00	31.00	28.3	26.0
Chinese	P	54,572	86,766	121,906	164,041	219,577	315,151	418,640	729,473
	M	109	174	244	328	439	630	837	1,459
	M%	0.20	0.20	0.20	0.20	0.20	0.20	0.20	0.20
Other	P	3,790	5,858	7,727	9,768	14,183	17,298	23,380	22,941
	M	1,099	1,582	1,932	2,247	2,978	5,362	6,546	5,735
	M%	29.0	27.0	25.0	23.0	21.0	31.0	28.0	25.0
Total	P	94,816	137,722	181,602	226,842	303,321	418,358	557,745	938,144
	M	32,506	40,206	44,535	45,382	53,550	69,605	86,777	139,698
	M%	34.28	29.19	24.52	20.01	17.65	16.64	15.56	14.89

Table 2.4.8a *(Continued)*

		1957	1970	1980	1990	2000	2010
Malay	T	197,059	311,379	351,508	299,965	315,198	386,968
	M	197,059	311,379	350,102	298,765	313,780	382,017
	M%	100	100	99.6	99.6	99.55	98.72
Indian	T	129,510	145,169	154,632	158,385	179,187	265,223
	M	31,082	31,937	34,174	42,764	45,927	57,546
	M%	24.0	22.0	22.1	27.0	25.63	21.70
Chinese	P	1,090,596	1,579,866	1,856,237	1,795,825	1,969,357	2,349,505
	M	2,181	3,160	3,712	3,592	5,063	8,332
	M%	0.20	0.20	0.20	0.2	0.26	0.35
Other	P	28,764	38,093	51,568	22,559	30,888	104,053
	M	5,753	5,714	7,735	5,567	6,891	9,540
	M%	20.0	15.0	15.0	24.7	22.31	9.17
Total	P	1,445,929	2,074,507	2,413,945	2,276,734	2,494,630	3,105,749
	M	236,075	352,190	395,723	350,687	371,661	457,435
	M%	16.33	16.98	16.39	15.40	14.90	14.73

Table 2.4.8b Evolution of the Muslim population in Singapore

Year	Population	Muslims	%	Source
1819	150	120	80.00	[SG1]e
1824	10,683	7,026	65.77	[SG2]e
1828	14,885	8,005	53.78	[SG1]e
1830	16,634	9,166	55.10	[SG2]e
1832	19,715	10,873	55.15	[SG1]e
1836	29,984	14,870	49.59	[SG2]e
1840	35,389	15,951	45.07	[SG1]e
1849	52,891	22,007	41.61	[SG3]c
1860	81,734	23,567	28.83	[SG1]e
1871	94,816	32,506	34.28	[SG2]e
1881	137,722	40,206	29.19	[SG2]e
1891	181,602	44,535	24.52	[SG2]e
1901	226,842	45,382	20.01	[SG2]e
1911	303,321	53,595	17.67	[SG3]c
1921	418,358	69,604	16.64	[SG3]c
1931	557,745	86,827	15.57	[SG3]c
1947	938,144	139,698	14.89	[SG2]e
1957	1,445,929	236,075	16.33	[SG2]e
1970	2,074,507	352,190	16.98	[SG2]e
1980	1,981,962	323,867	16.34	[UN88]c
1990	2,253,900	346,200	15.36	[UN04]c
2000	2,494,630	371,660	14.90	[UN]c
2010	3,105,748	457,435	14.73	[SG10]c
2050	6,408,000	944,000	14.73	es
2100	5,733,000	845,000	14.73	es

The Muslim population is increasing steadily since 1819; however, the percentage of Muslims is in constant decrease as a result of non-Muslim migration to Singapore. Accordingly and as summarized in Table 2.4.8b, the Muslim population increased in number but decreased in percentage from 120 or 80.0% in 1819, to 7,000 or 65.8% in 1824, to 8,000 or 53.8% in 1928, to 9,200 or 55.1% in 1830, to 10,900 or 55.2% in 1932, to 14,900 or 49.6% in

1836, to 16,000 or 45.1% in 1840, to 22,000 or 41.6% in 1849, to 23,600 or 28.8% in 1860, to 32,500 or 34.3% in 1871, to 40,200 or 29.2% in 1881, to 44,500 or 24.5% in 1891, to 45,400 or 20.0% in 1901, to 53,600 or 17.7% in 1911, to 69,600 or 16.6% in 1921, to 86,800 or 15.6% in 1931, to 0.14 million or 14.9% in 1947, to 0.24 million or 16.3% in 1957, to 0.35 million or 17.0% in 1970, to 0.32 million or 16.3% in 1980, to 0.35 million or 15.4% in 1990, to 0.37 million or 14.9% in 2000, to 0.46 million or 14.7% in 2010. Thus, assuming that the percentage of Muslims remains constant; the Muslim population is expected to remain less than one million throughout this century.

2.4.9 Thailand

The Kingdom of Thailand has an area of 513,120 sq. km and its map is presented in Fig. 2.4.9. It avoided European colonization and changed its name from Siam in 1939. Islam entered here through the Malay Peninsula, where it still largely Muslim. About two-thirds of Thailand's Muslims are Malay who lives in Thailand's part of the Malay Peninsula.

In 1921 the Muslim population was estimated at 0.30 million or 3.3% of the total population. Based on census data and as shown in Table 2.4.9, the Muslim population increased from 0.50 million or 4.3% in 1929, to 0.63 million or 4.3% in 1937, to 0.67 million or 3.8% in 1947, to 1.03 million or 3.9% in 1960, to 1.33 million or 3.9% in 1970, to 1.72 million or 3.8% in 1980, to 2.25 million or 4.1% in 1990, to 2.78 million or 4.6% in 2000, to 3.26 million or 4.9% in 2010. The 2000 census showed that five-sixths (83%) of the Thai Muslim population live in the Southern Region, which is the Thai's part of the Malay Peninsula. This percentage dropped to 78% in 2010. The Muslim population makes up nearly a third (29%) of the total population of the Southern Region according to the 2010 and 2000 censuses.

Thus, assuming that the percentage of Muslims continues to increase by half of a percentage point per decade, the Muslim population is expected to remain around 4.5 million throughout the second half of this century, reaching 7% of the total population by 2050 and 9.5% by 2100.

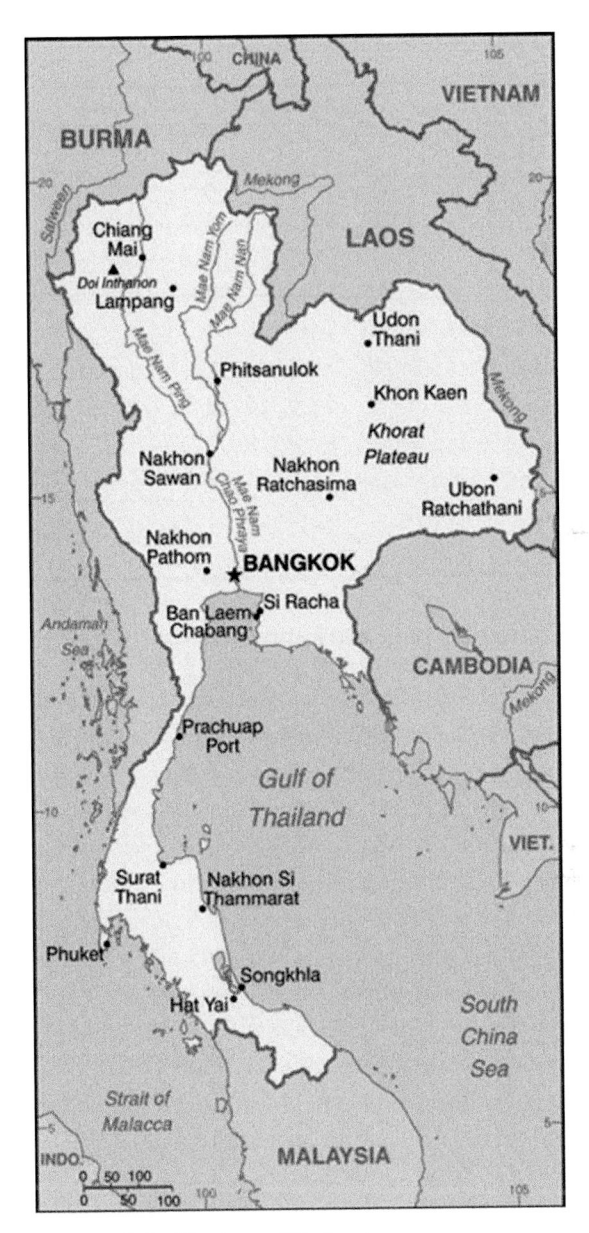

Figure 2.4.9 Map of the Kingdom of Thailand.

Table 2.4.9 Evolution of the Muslim population in Thailand

Year	Population	Muslims	%	Source
1919	9,207,355	300,000	3.26	[MAS23]es
1929	11,506,199	498,311	4.33	[TH929]c
1937	14,464,105	626,907	4.33	[TH937]c
1947	17,442,689	670,404	3.84	[UN56]c
1960	26,253,162	1,025,569	3.91	[UN63]c
1970	34,350,676	1,325,587	3.86	[UN73]c
1980	44,583,688	1,714,689	3.85	[UN83]c
1990	54,548,530	2,252,427	4.13	[TH]c
2000	60,694,241	2,777,542	4.56	[TH]c
2010	65,977,840	3,259,340	4.94	[UN]c
2050	65,940,000	4,616,000	7.00	es
2100	46,016,000	4,371,000	9.50	es

2.4.10 Timor-Leste (East Timor)

The Democratic Republic of Timor-Leste was occupied by the Portuguese in 1769, and then annexed by Indonesia in 1976, until it gained its independence in 2002. It has a total area of 14,874 sq. km and consists of the eastern half of the Island of Timor which it shares with Indonesia, and the Islands of Atauro (105 sq. km) and Jaco (10 sq. km). A map of Timor-Leste is presented in Fig. 2.4.10.

Islam entered here through current Indonesia. Estimates of the Muslim population were none in 1896 and 1930. Census data show that the number of Muslims changed from 900 or 0.15% in 1970, to 2,800 or 0.5% in 1980, to 39,600 or 5.3% due to the Indonesian annexation. The number decreased to 3,000 or 0.3% in 2004 as a result of the violence between Muslims and Christians and the subsequent independence. The Muslim population then increased to 3,600 or 0.3% in 2010 but decreased to 2,800 or 0.2% in 2015. Most of the Muslim population (61%) lives in the capital Dili. Thus, assuming that the percentage of the Muslim population will increase by 0.02 of a percentage point per decade, the Muslim population is expected to reach 8,000 or 0.4% by 2050 and 12,000 or 0.5% by 2100. A summary of the data is provided in Table 2.4.10.

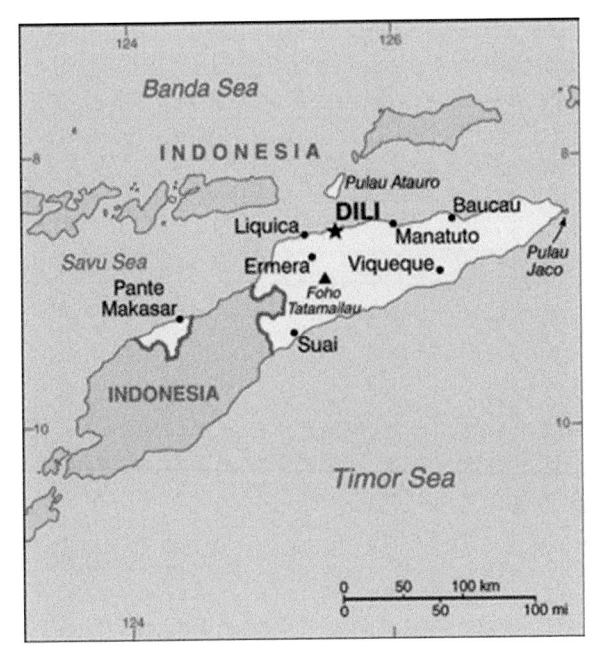

Figure 2.4.10 Map of the Democratic Republic of Timor-Leste.

Table 2.4.10 Evolution of the Muslim population in East Timor

Year	Population	Muslims	%	Source
1896	300,000	—	0.00	[JAN]es
1930	380,000	—	0.00	[MAS55]es
1970	609,477	900	0.15	[TL70]c
1980	555,350	2,777	0.50	[TL80]c
1990	747,557	39,620	5.30	[TL90]c
2004	924,642	2,970	0.32	[TL04]c
2010	1,053,971	3,623	0.34	[TL10]c
2015	1,179,654	2,824	0.24	[TL15]c
2050	2,019,000	8,000	0.42	es
2100	2,373,000	12,000	0.52	es

2.4.11 Vietnam

The Socialist Republic of Vietnam has an area of 331,210 sq. km and its map is presented in Fig. 2.4.11. It was occupied by France between 1858 and 1884 and became part of French Indochina in 1887. The French were expelled in 1954, but the country was split between communist north and American backed south. This led to decades of war between Soviet and Chinese backed North, and the American backed south, culminated by victory of the north and unification of the country in 1975.

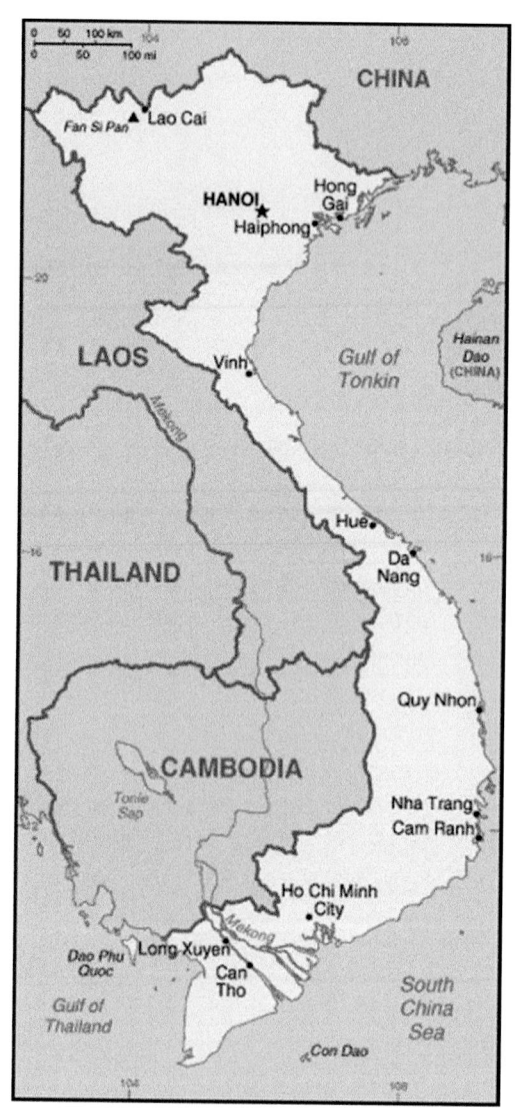

Figure 2.4.11 Map of the Socialist Republic of Vietnam.

About two-thirds of Muslims here are ethnic Cham, whose ancestors embraced Islam between eleventh and thirteenth centuries. As shown in Table 2.4.11, estimate of the Muslim population based on ethnic Malay and Cham increased from 21,000 or 0.1% in 1921, to 23,000 or 0.1% in 1931, to 31,000 or 0.2% in 1936. The 1999 census included for the first time a question about religious affiliation. The census data show that almost all the Muslim population lives in the southernmost provinces of Southeast and Mekong River Delta, with over 80% living in the former. Based on census data, the Muslim population increased from 63,000 or 0.1% in 1999 to 75,000 or 0.1% in 2009.

Thus, assuming that the percentage of Muslims will continue to increase by 0.01 of a percentage point per decade, the Muslim population is expected to reach 0.1 million or 0.1% by 2050 and 0.2 million or 0.2% by 2100. The data is summarized in Table 2.4.11.

Table 2.4.11 Evolution of the Muslim population in Vietnam

Year	Population	Muslims	%	Source
1921	15,584,000	21,210	0.14	[KH921]e
1931	17,702,000	23,000	0.12	[KH931]e
1936	18,972,000	31,000	0.16	[KH936]e
1999	76,289,515	63,147	0.08	[UN]c
2009	85,846,997	75,268	0.09	[VN09]c
2050	109,605,000	142,000	0.13	es
2100	97,437,000	175,000	0.18	es

2.4.12 Regional Summary and Conclusion

Islam spread widely in the western part of Southeast Asia as of the eleventh century, but this spread was slowed down by the European occupation and subsequent Christianization. Accordingly, the majority of the western half of this region is Muslim and well over a third of the population of Southeast Asia is Muslim. This situation is expected to remain so for the next three centuries. The following tables present centennial data from 600 AD to 2100 AD (or approximately 1 H to 1500 H) in Table 2.4a and decennial data from 1790 AD to 2100 AD (or 1210 H to 1520 H) in Tables 2.4b and 2.4c for current countries in Southeast Asia. The data includes total

Table 2.4a Centennial estimates of the Muslim population (×1000) in Southeast Asia: 600–2100 AD (1–1500 H)

		600	700	800	900	1000	1100	1200	1300
Brunei	P	1	2	3	4	5	5	5	6
	M%	—	—	—	—	—	—	—	1.00
	M	—	—	—	—	—	—	—	—
Cambodia	P	570	600	630	660	687	800	900	1,000
	M%	—	0.10	0.25	0.50	0.75	1.00	1.25	1.50
	M	—	1	2	3	5	8	11	15
Indonesia	P	4,000	4,300	4,600	4,900	5,200	6,300	7,400	8,500
	M%	—	1.00	2.00	5.00	10.00	20.00	30.00	60.00
	M	—	43	92	245	520	1,260	2,220	5,100
Laos	P	126	133	140	147	154	175	200	225
	M%	—	—	—	—	—	—	—	—
	M	—	—	—	—	—	—	—	—
Malaysia	P	75	80	85	90	94	110	125	140
	M%	—	1.00	2.00	3.00	4.00	5.00	20.00	40.00
	M	—	1	2	3	4	6	25	56
Myanmar	P	950	1,000	1,050	1,100	1,152	1,300	1,500	1,700
	M%	—	0.10	0.50	1.00	2.00	3.00	4.00	5.00
	M	—	1	5	11	23	39	60	85
Philippines	P	229	242	255	268	281	300	350	400
	M%	—	—	—	—	—	—	—	—
	M	—	—	—	—	—	—	—	—
Singapore	P	—	—	—	—	—	—	—	—
	M%	—	1.00	2.00	3.00	4.00	5.00	20.00	40.00
	M	—	—	—	—	—	—	—	—
Thailand	P	920	970	1,020	1,070	1,122	1,300	1,450	1,600
	M%	—	0.10	0.20	0.30	0.40	0.50	1.00	1.50
	M	—	1	2	3	4	7	15	24
Timor-Leste	P	12	14	16	18	20	22	25	30
	M%	—	—	—	—	—	—	—	—
	M	—	—	—	—	—	—	—	—
Vietnam	P	2,700	3,000	3,300	3,600	3,835	3,835	3,835	3,835
	M%	—	—	—	—	—	0.10	0.20	0.30
	M	—	—	—	—	—	4	8	12
Total	P	9,583	10,341	11,099	11,857	12,550	14,147	15,790	17,436
	M%	—	0.45	0.92	2.24	4.43	9.35	14.81	30.35
	M	—	46	103	265	556	1,323	2,339	5,292
	G%		0.076	0.071	0.066	0.057	0.120	0.110	0.099

Table 2.4a (*Continued*)

		1400	1500	1600	1700	1800	1900	2000	2100
Brunei	P	7	8	9	10	11	18	333	390
	M%	5.00	10.00	30.00	40.00	47.18	47.18	75.06	97.00
	M	—	1	3	4	5	8	250	378
Cambodia	P	1,100	1,224	1,419	1,650	2,000	2,828	12,155	21,355
	M%	1.60	1.70	1.80	1.90	2.00	2.00	2.15	1.92
	M	18	21	26	31	40	65	261	410
Indonesia	P	9,600	10,700	11,700	13,100	17,100	45,100	211,514	320,782
	M%	70.00	80.00	90.00	85.00	85.90	81.60	88.22	88.00
	M	6,720	8,560	10,530	11,135	14,689	36,802	186,597	282,289
Laos	P	250	275	319	371	454	1,154	5,324	8,424
	M%	—	—	—	—	—	—	0.02	0.12
	M	—	—	—	—	—	—	1	10
Malaysia	P	155	168	195	227	275	2,232	23,194	40,078
	M%	60.00	60.00	60.00	60.00	54.13	50.73	60.57	71.00
	M	93	101	117	136	149	1,132	14,049	28,455
Myanmar	P	1,900	2,052	2,380	2,768	2,900	10,491	46,720	55,299
	M%	5.00	5.00	5.00	5.00	4.52	3.24	4.00	5.20
	M	95	103	119	138	131	340	1,869	2,876
Philippines	P	450	500	791	1,250	1,561	7,635	77,992	146,327
	M%	1.00	2.00	3.00	4.00	3.64	3.64	5.08	10.00
	M	5	10	24	50	57	278	3,962	14,633
Singapore	P	—	—	—	—	—	227	4,029	5,733
	M%	60.00	80.00	90.00	90.00	80.00	20.01	14.90	14.73
	M	—	—	—	—	—	45	600	845
Thailand	P	1,800	2,000	2,236	2,500	4,300	7,320	62,953	46,016
	M%	2.00	2.50	3.00	3.50	4.33	4.33	4.56	9.50
	M	36	50	67	88	186	317	2,871	4,371
Timor-Leste	P	35	40	45	50	55	165	884	2,373
	M%	—	—	—	—	—	—	0.32	0.52
	M	—	—	—	—	—	—	3	12
Vietnam	P	3,835	3,835	4,447	5,171	6,350	16,091	79,910	97,437
	M%	0.40	0.50	0.60	0.70	0.10	0.10	0.08	0.18
	M	15	19	27	36	6	16	64	175
Total	P	19,132	20,802	23,541	27,097	35,006	93,261	525,008	744,215
	M%	36.49	42.61	46.35	42.88	43.60	41.82	40.10	44.94
	M	6,982	8,865	10,912	11,619	15,263	39,004	210,527	334,454
	G%	0.093	0.084	0.124	0.141	0.256	0.980	1.728	0.349

Table 2.4b Decennial estimates of the Muslim population (×1000) in Southeast Asia: 1790–1940 AD (1210–1360 H)

		1790	1800	1810	1820	1830	1840	1850	1860
Brunei	P	10	11	11	11	12	12	12	13
	M%	47.18	47.18	47.18	47.18	47.18	47.18	47.18	47.18
	M	5	5	5	5	6	6	6	6
Cambodia	P	1,950	2,000	2,050	2,090	2,150	2,200	2,250	2,300
	M%	2.00	2.00	2.00	2.00	2.00	2.00	2.00	2.00
	M	39	40	41	42	43	44	45	46
Indonesia	P	16,600	17,100	17,500	17,927	19,473	21,153	22,977	25,779
	M%	85.90	85.90	85.90	85.90	85.90	85.90	85.90	85.90
	M	14,259	14,689	15,033	15,399	16,727	18,170	19,737	22,144
Laos	P	446	454	462	470	530	590	650	700
	M%	—	—	—	—	—	—	—	—
	M	—	—	—	—	—	—	—	—
Malaysia	P	270	275	280	287	300	400	530	700
	M%	54.13	54.13	54.13	54.13	54.13	54.13	54.13	54.13
	M	146	149	152	155	162	217	287	379
Myanmar	P	2,850	2,900	2,950	3,506	3,650	3,800	3,932	4,100
	M%	4.52	4.52	4.52	4.52	4.52	4.52	4.52	4.52
	M	129	131	133	158	165	172	178	185
Philippines	P	1,400	1,561	1,933	2,106	2,593	3,096	3,857	4,290
	M%	3.64	3.64	3.64	3.64	3.64	3.64	3.64	3.64
	M	51	57	70	77	94	113	140	156
Singapore	P	—	—	—	—	17	30	53	75
	M%	80.00	80.00	80.00	80.00	55.10	45.07	41.61	28.83
	M	—	—	—	—	9	14	22	22
Thailand	P	4,100	4,300	4,500	4,665	4,800	5,000	5,230	5,500
	M%	4.33	4.33	4.33	4.33	4.33	4.33	4.33	4.33
	M	178	186	195	202	208	217	226	238
Timor-Leste	P	50	55	60	70	80	90	100	110
	M%	—	—	—	—	—	—	—	—
	M	—	—	—	—	—	—	—	—
Vietnam	P	6,200	6,350	6,450	6,551	7,300	8,100	8,900	9,700
	M%	0.10	0.10	0.10	0.10	0.10	0.10	0.10	0.10
	M	6	6	6	7	7	8	9	10
Total	P	33,876	35,006	36,196	37,683	40,905	44,471	48,491	53,267
	M%	43.73	43.60	43.20	42.58	42.59	42.63	42.59	43.53
	M	14,813	15,263	15,635	16,045	17,422	18,959	20,650	23,186
	G%		0.328	0.334	0.403	0.820	0.836	0.865	0.939

Table 2.4b (*Continued*)

		1870	1880	1890	1900	1910	1920	1930	1940
Brunei	P	13	14	16	18	30	25	30	39
	M%	47.18	47.18	47.18	47.18	47.18	47.18	47.18	67.06
	M	6	7	8	8	14	12	14	26
Cambodia	P	2,340	2,340	2,655	2,828	3,070	3,296	3,612	3,958
	M%	2.00	2.10	2.20	2.30	2.40	2.44	2.42	2.41
	M	47	49	58	65	74	80	87	95
Indonesia	P	32,743	36,203	40,532	45,100	50,034	54,993	60,727	74,376
	M%	85.90	85.90	85.90	81.60	81.60	85.91	83.20	83.20
	M	28,126	31,098	34,817	36,802	40,828	47,244	50,525	61,881
Laos	P	755	870	1,002	1,154	1,387	1,412	1,445	1,478
	M%	—	—	—	—	—	—	—	—
	M	—	—	—	—	—	—	—	—
Malaysia	P	800	1,126	1,585	2,232	2,950	3,570	4,413	5,434
	M%	54.13	54.13	54.13	50.73	47.54	48.70	48.42	46.25
	M	433	610	858	1,132	1,402	1,739	2,137	2,513
Myanmar	P	4,245	5,638	7,489	10,491	12,115	13,096	14,515	16,824
	M%	4.52	4.52	3.08	3.24	3.47	3.80	3.99	3.99
	M	192	255	231	340	420	498	579	671
Philippines	P	4,712	5,726	6,476	7,635	8,861	10,445	13,194	16,585
	M%	3.64	3.64	3.64	3.64	3.64	4.24	4.23	4.23
	M	172	208	236	278	323	443	558	702
Singapore	P	95	138	182	227	303	418	446	529
	M%	34.28	29.19	24.52	20.01	17.67	16.64	15.57	14.89
	M	33	40	45	45	54	70	69	79
Thailand	P	5,775	6,206	6,195	7,320	8,266	9,207	11,506	15,513
	M%	4.33	4.33	4.33	4.33	4.33	4.33	4.33	4.33
	M	250	269	268	317	358	399	498	672
Timor-Leste	P	120	140	150	165	265	406	448	461
	M%	—	—	—	—	—	—	—	—
	M	—	—	—	—	—	—	—	—
Vietnam	P	10,528	12,127	13,120	16,091	19,339	20,652	22,472	24,452
	M%	0.10	0.10	0.10	0.10	0.10	0.14	0.12	0.16
	M	11	12	13	16	19	29	27	39
Total	P	62,126	70,528	79,401	93,261	106,621	117,521	132,808	159,648
	M%	47.11	46.15	46.01	41.82	40.79	42.98	41.03	41.77
	M	29,269	32,548	36,533	39,004	43,492	50,513	54,495	66,678
	G%	1.538	1.268	1.185	1.609	1.339	0.973	1.223	1.841

Table 2.4c Decennial estimates of the Muslim population (×1000) in Southeast Asia: 1950–2100 AD (1370–1520 H)

		1950	1960	1970	1980	1990	2000	2010	2020
Brunei	P	48	82	130	194	259	333	389	437
	M%	67.06	60.23	62.16	63.41	67.17	75.06	78.80	80.89
	M	32	49	81	123	174	250	306	354
Cambodia	P	4,433	5,722	6,997	6,694	8,976	12,155	14,312	16,719
	M%	2.40	2.35	2.30	2.25	2.20	2.15	1.92	1.92
	M	106	134	161	151	197	261	275	321
Indonesia	P	69,543	87,751	114,793	147,448	181,413	211,514	241,834	273,524
	M%	83.20	87.51	87.51	87.10	87.21	88.22	87.51	88.00
	M	57,860	76,791	100,456	128,427	158,211	186,597	211,629	240,701
Laos	P	1,683	2,121	2,688	3,258	4,258	5,324	6,249	7,276
	M%	—	—	—	0.02	0.02	0.02	0.03	0.04
	M	—	—	—	1	1	1	2	3
Malaysia	P	6,110	8,156	10,804	13,798	18,030	23,194	28,208	32,366
	M%	46.25	46.43	50.04	52.93	58.70	60.57	61.92	63.00
	M	2,826	3,787	5,406	7,303	10,584	14,049	17,466	20,391
Myanmar	P	17,780	21,737	27,269	34,224	41,335	46,720	50,601	54,410
	M%	3.95	3.90	3.86	3.83	3.90	4.00	4.35	4.40
	M	702	848	1,053	1,311	1,612	1,869	2,201	2,394
Philippines	P	18,580	26,270	35,804	47,358	61,895	77,992	93,967	109,581
	M%	4.12	4.86	4.32	4.57	4.59	5.08	5.57	6.00
	M	766	1,277	1,547	2,164	2,841	3,962	5,234	6,575
Singapore	P	1,022	1,633	2,072	2,412	3,013	4,029	5,131	5,850
	M%	14.89	16.33	16.98	16.34	15.36	14.90	14.73	14.73
	M	152	267	352	394	463	600	756	862
Thailand	P	20,710	27,397	36,885	47,374	56,558	62,953	67,195	69,800
	M%	3.84	3.91	3.85	3.83	4.13	4.56	4.94	5.50
	M	795	1,071	1,420	1,814	2,336	2,871	3,319	3,839
Timor-Leste	P	415	475	572	600	738	884	1,094	1,318
	M%	—	—	0.15	0.50	5.30	0.32	0.34	0.36
	M	—	—	1	3	39	3	4	5
Vietnam	P	24,810	32,670	43,405	54,282	67,989	79,910	87,968	97,339
	M%	0.16	0.14	0.12	0.10	0.08	0.08	0.09	0.10
	M	40	46	52	54	54	64	79	97
Total	P	165,134	214,014	281,418	357,642	444,464	525,008	596,947	668,620
	M%	38.32	39.38	39.28	39.63	39.71	40.10	40.42	41.21
	M	63,279	84,270	110,528	141,745	176,511	210,527	241,272	275,541
	G%	0.338	2.593	2.738	2.397	2.173	1.665	1.284	1.134

Table 2.4c (*Continued*)

		2030	2040	2050	2060	2070	2080	2090	2100
Brunei	P	471	489	492	482	461	437	413	390
	M%	83.00	85.00	87.00	89.00	91.00	93.00	95.00	97.00
	M	391	415	428	429	419	406	393	378
Cambodia	P	18,781	20,527	21,861	22,640	22,889	22,649	22,116	21,355
	M%	1.92	1.92	1.92	1.92	1.92	1.92	1.92	1.92
	M	361	394	420	435	439	435	425	410
Indonesia	P	299,198	318,638	330,905	336,444	337,225	334,740	329,171	320,782
	M%	88.00	88.00	88.00	88.00	88.00	88.00	88.00	88.00
	M	263,295	280,401	291,196	296,071	296,758	294,571	289,671	282,289
Laos	P	8,226	8,972	9,480	9,706	9,661	9,387	8,957	8,424
	M%	0.05	0.06	0.07	0.08	0.09	0.10	0.11	0.12
	M	4	5	7	8	9	9	10	10
Malaysia	P	36,095	38,755	40,550	41,732	42,062	41,625	40,887	40,078
	M%	64.00	65.00	66.00	67.00	68.00	69.00	70.00	71.00
	M	23,101	25,190	26,763	27,960	28,602	28,721	28,621	28,455
Myanmar	P	58,478	61,202	62,253	62,157	61,238	59,568	57,451	55,299
	M%	4.50	4.60	4.70	4.80	4.90	5.00	5.10	5.20
	M	2,632	2,815	2,926	2,984	3,001	2,978	2,930	2,876
Philippines	P	123,698	135,619	144,488	150,220	153,039	153,010	150,497	146,327
	M%	6.50	7.00	7.50	8.00	8.50	9.00	9.50	10.00
	M	8,040	9,493	10,837	12,018	13,008	13,771	14,297	14,633
Singapore	P	6,262	6,445	6,408	6,265	6,089	5,915	5,789	5,733
	M%	14.73	14.73	14.73	14.73	14.73	14.73	14.73	14.73
	M	922	949	944	923	897	871	853	845
Thailand	P	70,346	69,008	65,940	61,692	57,258	53,224	49,459	46,016
	M%	6.00	6.50	7.00	7.50	8.00	8.50	9.00	9.50
	M	4,221	4,486	4,616	4,627	4,581	4,524	4,451	4,371
Timor-Leste	P	1,574	1,809	2,019	2,203	2,330	2,395	2,408	2,373
	M%	0.38	0.40	0.42	0.44	0.46	0.48	0.50	0.52
	M	6	7	8	10	11	11	12	12
Vietnam	P	104,164	107,795	109,605	109,363	107,062	103,916	100,709	97,437
	M%	0.11	0.12	0.13	0.14	0.15	0.16	0.17	0.18
	M	115	129	142	153	161	166	171	175
Total	P	727,294	769,258	794,002	802,904	799,313	786,866	767,857	744,215
	M%	41.67	42.16	42.61	43.05	43.52	44.03	44.52	44.94
	M	303,087	324,287	338,287	345,616	347,885	346,466	341,833	334,454
	G%	0.841	0.561	0.317	0.111	−0.045	−0.157	−0.245	−0.313

population in thousands (P), the percentage of which is Muslim ($M\%$), the corresponding Muslim population in thousands (M), and the annual population growth rate (APGR, or $G\%$) of the total population in this region.

2.5 Muslims in the Indian Subcontinent

This region consists of seven countries: Bangladesh, Bhutan, India, Maldives, Nepal, Pakistan, and Sri Lanka. Islam first entered this region through Arab traders who were doing business with communities on the eastern Indian shore and Sri Lanka. Muslim Armies entered Baluchistan in 644 during the time if Caliph Omar, and Sindh in 711 during the time of Umayyad Caliph al-Walid I. The Muslim armies continued their advance under different dynasties, until they conquer current Bangladesh in 1192. Then in the sixteenth century, Muslim troops under the Moghul Empire moved deeper south in India, occupying much of its territory.

Thus, the Muslim population has likely increased from 0.54 million or 0.7% of the total population of this region in 700 AD, to 1.76 million or 2.4% in 800 AD, to 2.58 million or 3.4% in 900 AD, to 3.73 million or 4.9% in 1000 AD, to 5.27 million or 6.3% in 1100 AD, to 7.1 million or 7.8% in 1200 AD, to 9.3 million or 9.4% in 1300 AD, to 12 million or 11.2% in 1400 AD, to 16 million or 14.0% in 1500 AD, to 22 million or 15.7% in 1600 AD, to 30 million or 17.8% in 1700 AD, to 39 million or 19.1% in 1800 AD, to 64 million or 21.8% in 1900, to 0.40 billion or 28.9% in 2000, to 0.57 billion or 31.6% in 2020, and is projected to reach 0.79 billion or 36% by 2050 and 0.84 billion or 41% by 2100.

A plot of centennial estimates of the Muslim population and its percentage with respect to the total population in this region from 600 to 2100 is provided in Fig. 2.5a. A zoom in of this plot, providing a plot of decennial estimates of the Muslim population and its percentage with respect to the total population in this region from 1900 to 2100 is provided in Fig. 2.5b. This shows that the Muslim population in this region was increasing slowly until 1950, and is increasing substantially afterwards until 2060, peaking at 0.86 billion in 2080. The percentage of Muslims in this region on the other hand, had been generally increasing at between one and two percentage points per decade, and is expected to continue its increase towards the end of this century.

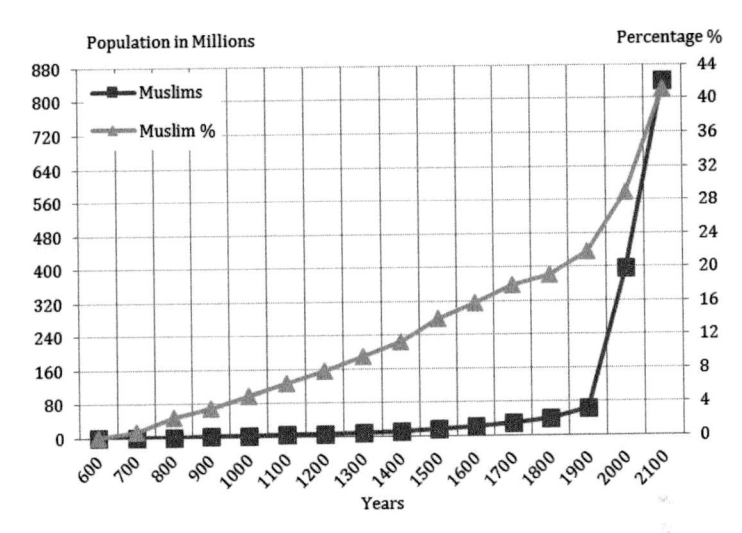

Figure 2.5a Plot of centennial estimates of the Muslim population and its percentage of the total population in the Indian Subcontinent: 600–2100 AD (1–1500 H).

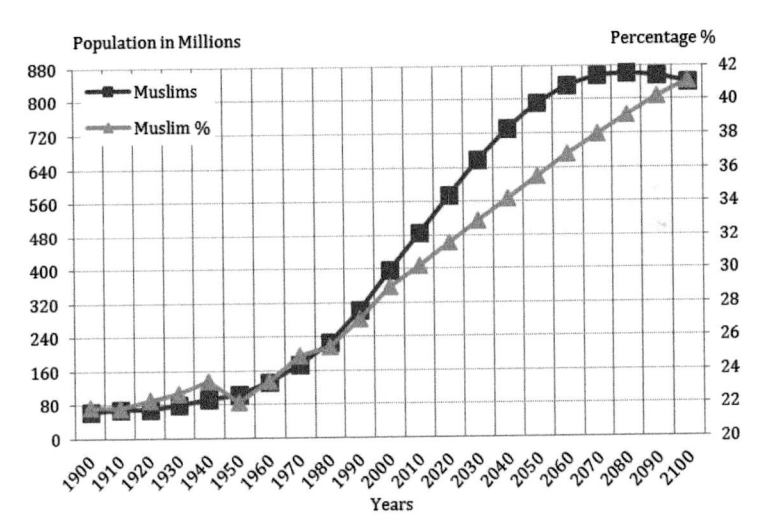

Figure 2.5b Plot of decennial estimates of the Muslim population and its percentage of the total population in the Indian Subcontinent: 1900–2100 AD (1320–1520 H).

The corresponding individual data for each country in this region is discussed below. In Section 2.5.8, the total population in each country in this region and the corresponding percentage and number of Muslims is presented centennially in Table 2.5a from 600 to 2100 and decennially in Tables 2.5b and 2.5c from 1790 to 2100.

2.5.1 Bangladesh

Islam entered current Bangladesh in 1192, when Muízzuddin Muhammad Ghori of the Ghorid Dynasty captured these lands together with north India. Currently, the People's Republic of Bangladesh has an area of 143,998 sq. km and its map is presented in Fig. 2.5.1a. It was occupied by the British in 1757, gained its independence from the UK as East Pakistan in 1947, and from Pakistan as Bangladesh in 1971.

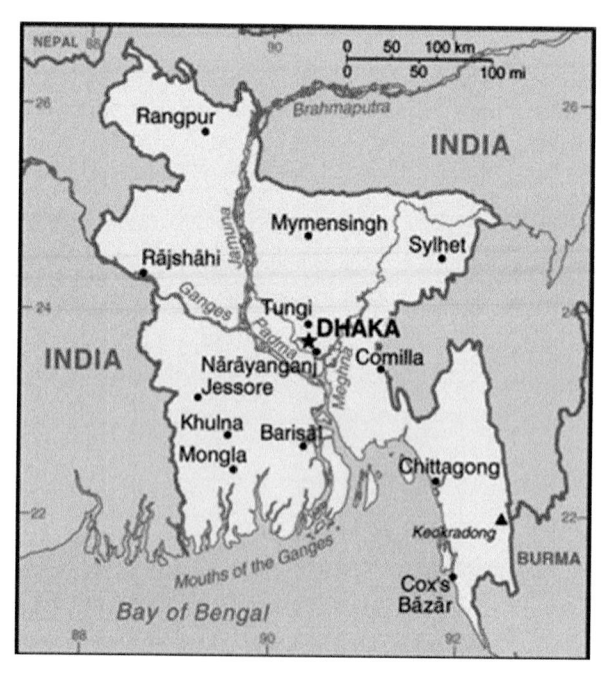

Figure 2.5.1a Map of the People's Republic of Bangladesh.

Based on census data and as shown in Table 2.5.1, the Muslim population increased from 14.7 million or 63.7% in 1871, to 16.3 million or 65.1% in 1881, to 17.7 million or 65.5% in 1891, to 19.1 million or 66.1% in 1901, to 21.2 million or 67.2% in 1911, to 22.6 million or 1921, to 24.7 million or 69.5% in 1931, to 29.5 million or 70.3% in 1941. Then after its independence from the British, Muslims increased to 34.5 million or 76.9% in 1951, and 40.9 million or 80.4% in 1961. The increase in percentage of Muslims is due to the forced population exchange between Pakistan and India, whereby Muslims migrated to East and West Pakistan, while Hindus moved to India. After independence from Pakistan, the Muslim population increased to 61 million or 85.4% in 1974. Again, the increase of Muslim percentage is due to the war of independence from Pakistan, during which many Hindus fled to India and did not return. The number of Muslims continued to increase to 75 million or 86.7% in 1981, to 94 million or 88.3% in 1991, to 111 million or 89.6% in 2001, to 135 million or 90.4% in 2011. A choropleth map illustrating the presence of Muslims per division is presented in Fig. 2.5.1b.

Table 2.5.1 Evolution of the Muslim population in Bangladesh

Year	Population	Muslims	%	Source
1871	23,000,000	14,665,373	63.76	[INH71]c
1881	25,086,000	16,323,588	65.07	[INH81]c
1891	27,103,000	17,743,760	65.47	[INH91]c
1901	28,928,000	19,112,730	66.07	[BD]c
1911	31,555,000	21,201,805	67.19	[BD]c
1921	33,255,000	22,646,655	68.10	[BD]c
1931	35,602,000	24,729,149	69.46	[BD]c
1941	41,997,000	29,507,092	70.26	[BD]c
1951	44,832,000	34,453,592	76.85	[BD]c
1961	50,840,613	40,890,138	80.43	[BD]c
1974	71,477,748	61,038,929	85.40	[BD11]c
1981	87,119,965	75,486,980	86.65	[BD11]c
1991	106,314,992	93,886,770	88.31	[BD11]c
2001	124,355,263	111,393,250	89.58	[BD11]c
2011	144,043,697	130,204,860	90.39	[BD11]c
2050	192,568,000	178,125,000	92.50	es
2100	151,393,000	143,823,000	95.00	es

Thus, assuming that the percentage of the Muslim population will continue to increase by a half of a percentage point per decade, the Muslim population is expected to reach 178 million or 92.5% by 2050 and 144 million or 95% by 2100.

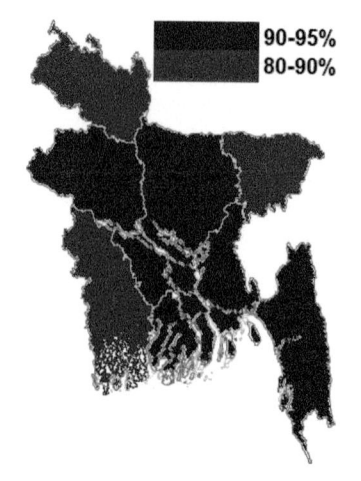

Figure 2.5.1b A choropleth map of Bangladesh illustrating the presence of Muslims per division.

2.5.2 Bhutan

The Kingdom of Bhutan has an area of 38,394 sq. km and its map is presented in Fig. 2.5.2. It did not come under Muslim control but was influenced by the spread of Islam in India. While the British had some influence and occupied some Bhutanese land from 1865 to 1947, they never occupied the whole country. As shown in Table 2.5.2, estimates for the Muslim population increased from none in 1891, to 400 or 0.1% in 1908 and according to ABM survey data, the percentage of Muslims was 0.12% in 2005. Thus, assuming that the percentage of Muslims will increase by 0.02 of a percentage point per decade, the Muslim population is expected to remain around 2,000 throughout the second half of this century.

Table 2.5.2 Evolution of the Muslim population in Bhutan

Year	Population	Muslims	%	Source
1891	200,000	—	0.00	[JAN]es
1908	400,000	400	0.10	[AFH]es
2005	672,425	800	0.12	[ABM]s
2050	905,000	2,000	0.20	es
2100	686,000	2,000	0.30	es

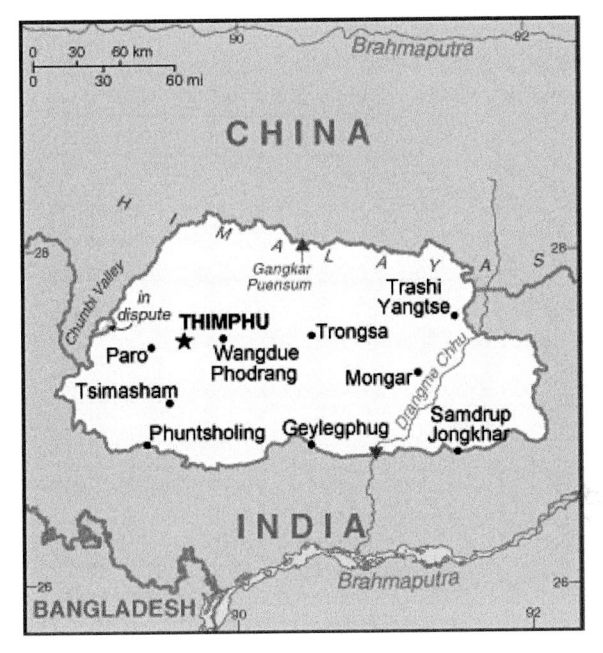

Figure 2.5.2 Map of the Kingdom of Bhutan.

2.5.3   India

Islam started entering current India as early as the seventh century along the entire western Indian shore through Arab merchants. However, the land occupied by current India did not come under Muslim control until the eleventh century. This occurred during the reign of Mahmud Sebük Tigin al Ghaznawi, the fourth ruler of the Ghaznavid Dynasty who ruled from 998 to 1030. During his reign, his dynasty extended to the northeastern part of current India. He captured Thanesar in 1011, Uttar Pradesh all the way to Varanasi in 1017, Ajmer and Gujarat in 1024.

The conquest of current India continued During the Ghorid Dynasty. It was established by the brothers Ghiyasuddin (ruled 1173 to 1202) and Muízzuddin Muhammad Ghori (ruled 1202 to 1206). They captured Gujarat in 1178, then Ajmer, Delhi, Utter Pradesh, Bihar, Bengal and Assam in 1192, capturing all of north India. The rest of India, with exception of its southernmost, was captured during the second half of the sixteenth century by the Moghul Muslim Empire. With the fall of this empire, some of these lands were captured by the Maratha Hindu Empire, while others remained under independent Muslim states until the arrival of the British.

The Republic of India has an area of 3,287,263 sq. km and is the seventh largest country in the world. It is also the second most populous after China, with over a sixth of the world population. A map of India is presented in Fig. 2.5.3a. European occupation of Indian lands started in the sixteenth century and by 1856, most of India was under British control. India gained its independence from the UK in 1947. Based on census data and as shown in Table 2.5.3a, the Muslim population in British India increased from 40.9 million or 21.5% in 1871, to 50.1 million or 19.7% in 1881, to 57.3 million or 20.0% in 1891, to 62.4 million or 21.2% in 1901, to 66.6 million or 21.3% in 1911, to 68.7 million or 21.7% in 1921, to 77.7 million or 22.0% in 1931, to 92.1 million or 24.0% in 1941, just before the partition of British India.

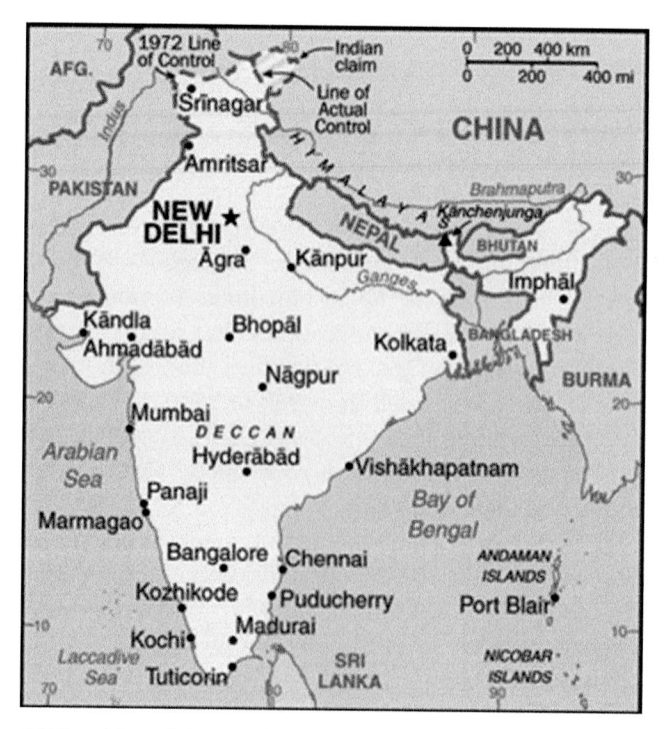

Figure 2.5.3a   Map of the Republic of India.

Table 2.5.3a Evolution of the Muslim population in British India

Year	Population	Muslims	%	Source
1871	190,563,048	40,882,537	21.45	[INH71]c
1881	253,891,821	50,121,585	19.74	[INH81]c
1891	287,223,431	57,321,164	19.96	[INH91]c
1901	294,361,056	62,458,077	21.22	[INH01]c
1911	313,517,840	66,647,299	21.26	[INH11]c
1921	316,128,721	68,735,233	21.74	[INH21]c
1931	352,837,778	77,677,545	22.02	[INH31]c
1941	383,643,745	92,058,096	24.00	[INH41]c

To get the data for the area currently controlled by India from 1871 to 1891, we deduct the Muslim population of Bangladesh, Burma, and Pakistan from British India. The total population from 1901 to 1941 was obtained from the corresponding census bureau while the Muslim population in 1901 till 1941 was obtained from [INH]. Accordingly, as summarized in Table 2.5.3b, the Muslim population in India increased from 14.4 million or 9.6% in 1871, to 20.3 million or 9.8% in 1881, to 24.2 million or 10.4% in 1891, to 29.1 million or 12.2% 1901, to 30.3 million or 12.0% in 1911, to 30.7 million or 12.2% in 1921, to 35.8 million or 12.8% in 1931, to 42.6 million or 13.4% in 1941, just before the partition. After independence censuses, show that the Muslim population decreased to 35.4 million or 9.9% in 1951 due to forced population exchange with Pakistan and related massacres. The Muslim population continued increasing since then, to 46.9 million or 10.7% in 1961, to 61.4 million or 11.2% in 1971, to 75.6 million or 11.4% in 1981, to 102 million or 12.1% in 1991, to 138 million or 13.4% in 2001, to 172 million or 14.3% in 2011.

Thus, assuming that the percentage of Muslims will continue to increase by 0.75 of a percentage point per decade, the Muslim population is expected to remain around 0.3 billion in the second half of this century, reaching 17% by 2050 and 21% by 2100.

Table 2.5.3c presents the distribution of total population (P) and Muslims (M) with their percentage per Indian state and territory, respectively, since 1941. The data show that the increase of Muslims happens in every state, albeit in various degrees. Underlined data in these tables are estimates based on census data. For example, census data shows that the Indian Punjab had 5.33 million Muslims in 1941 but dropped to 0.29 million in 1951 due to the population exchange between Pakistan and India that was based on religious lines. The Indian Punjab later was reduced in size, part went to Himachal Pradesh, small territories of Delhi and Chandigarh, and the rest where Most Muslims were located, was split between the new Punjab state and Haryana. In 2000, the states of Chhattisgarh, Uttarakhand (formerly, Uttaranchal), and Jharkhand, were carved from Madhya Pradesh, Uttar Pradesh, Bihar, respectively.

A choropleth map of India illustrating the presence of Muslims per province is presented in Fig. 2.5.3b, where white is less than 1% of the provincial population and the darker the region, the more is the percentage of Muslims.

Table 2.5.3b Evolution of the Muslim population in India

Year	Population	Muslims	%	Source
1871	149,850,034	14,366,000	9.59	[INH71]c
1881	207,911,967	20,339,879	9.78	[INH81]c
1891	233,842,379	24,203,843	10.35	[INH91]c
1901	238,396,327	29,102,000	12.21	[INH01]c
1911	252,093,390	30,269,000	12.01	[INH11]c
1921	251,321,213	30,739,000	12.23	[INH21]c
1931	278,977,238	35,818,000	12.84	[INH31]c
1941	318,660,580	42,645,000	13.38	[INH41]c
1951	357,248,383	35,414,279	9.91	[UN56]c
1961	438,661,708	46,939,592	10.70	[UN63]c
1971	547,913,726	61,418,269	11.21	[IN71]c
1981	665,227,632	75,571,514	11.36	[UN88]c
1991	838,168,419	101,596,057	12.12	[UN04]c
2001	1,027,882,740	138,188,240	13.44	[UN]c
2011	1,207,987,674	172,245,158	14.26	[IN11]c
2050	1,639,176,000	282,758,000	17.25	es
2100	1,450,421,000	304,588,000	21.00	es

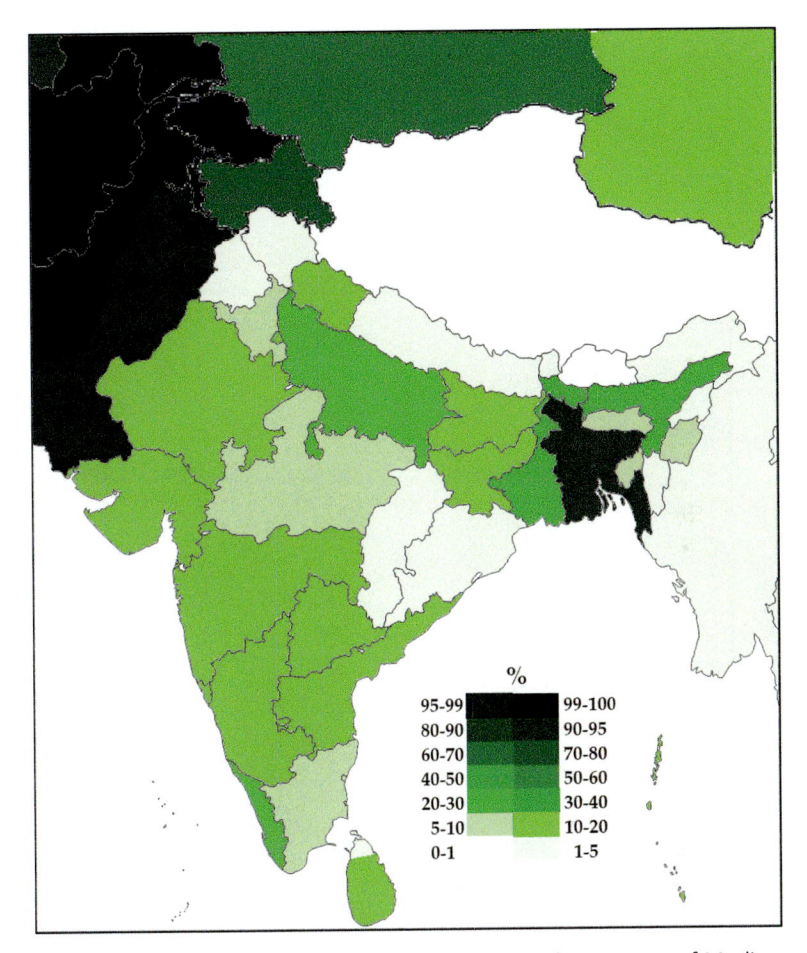

Figure 2.5.3b A choropleth map of India illustrating the presence of Muslims per state.

2.5.4 Maldives

In 1153, the king of these islands accepted Islam after his encounter with a Moroccan visitor named Abu al Barakat Yusuf al Barbari (Berber), who remained in the Islands and buried in the capital Malé. His name was Dhovemi Kalaminja and ruled the islands from 1141 to 1176. He changed his name to Sultan Muhammad bnu Abdullah when he became Muslim. His population followed suit, and the Islands were predominantly Muslim since then.

Table 2.5.3c Evolution of Muslim population in Indian states and territories since 1941

		1941	1951	1961	1971	1981	1991	2001	2011
Jammu Kashmir	P	2,947,000	3,254,000	3,561,000	4,617,000	5,987,000	7,719,000	10,143,700	12,541,302
	M	2,134,000	2,277,800	2,432,000	3,040,000	3,843,000	5,055,945	6,793,240	8,567,485
	%	72.41	70.00	68.30	65.84	64.19	65.50	66.97	68.31
Himachal Pradesh	P	2,263,000	2,386,000	2,812,000	3,460,000	4,280,818	5,170,877	6,077,900	6,864,602
	M	30,570	15,200	37,980	50,330	69,613	89,134	119,512	149,881
	%	1.35	0.64	1.35	1.45	1.63	1.72	1.97	2.18
Punjab	P	9,600,000	9,161,000	11,135,000	13,551,000	16,788,915	20,281,969	24,358,999	27,743,338
	M	3,023,100	30,900	89,050	114,400	168,094	239,401	382,045	535,489
	%	31.49	0.34	0.80	0.84	1.00	1.18	1.57	1.93
Haryana	P	5,273,000	5,674,000	7,591,000	10,037,000	12,922,618	16,463,648	21,144,564	25,351,462
	M	2,000,000	160,000	290,400	405,700	523,536	763,775	1,222,916	1,781,342
	%	37.93	2.82	3.83	4.04	4.05	4.64	5.78	7.03
Rajasthan	P	13,864,000	15,971,000	20,156,000	25,766,000	34,261,862	44,005,990	56,507,188	68,548,437
	M		991,200	1,315,000	1,778,000	2,492,145	3,525,339	4,788,227	6,215,377
	%		6.21	6.52	6.90	7.27	8.01	8.47	9.07

Uttar Pradesh	P	56,347,000	63,216,000	73,746,000	88,341,000	110,862,013	132,061,653	166,197,921	199,812,341
	M	8,692,000	9,029,000	10,788,000	13,677,000	17,756,735	23,404,386	30,740,158	38,483,967
	%	15.43	14.28	14.63	15.48	16.02	17.72	18.50	19.26
Uttarakhand	P						7,050,634	8,489,349	10,086,292
	M						705,298	1,012,141	1,406,825
	%						10.00	11.92	13.95
Bihar	P	26,306,000	29,089,000	34,850,000	42,127,000	52,302,665	64,530,554	82,998,509	104,099,452
	M	4,030,000	3,590,000	4,698,000	6,121,000	7,893,000	10,128,120	13,722,048	17,557,809
	%	15.32	12.34	13.48	14.53	15.09	15.70	16.53	16.87
Jharkhand	P	8,868,000	9,697,000	11,606,000	14,226,000	17,612,069	21,843,911	26,945,829	32,988,134
	M	689,000	783,000	1,088,000	1,473,000	1,982,000	2,659,865	3,731,308	4,793,994
	%	7.77	8.07	9.37	10.35	11.25	12.18	13.85	14.53
Sikkim	P	121,500	137,700	162,200	209,800	316,385	406,457	540,851	610,577
	M	83	124	1,207	335	3,241	3,849	7,693	9,867
	%	0.07	0.09	0.74	0.16	1.02	0.95	1.42	1.62
Arunachal Pradesh	P			336,600	467,500	631,839	864,558	1,097,968	1,383,727
	M			1,008	842	5,073	11,922	20,675	27,045
	%			0.30	0.18	0.80	1.38	1.88	1.95

Table 2.5.3c (Continued)

		1941	1951	1961	1971	1981	1991	2001	2011
Nagaland	P	189,600	213,000	369,200	516,400	774,930	1,209,546	1,990,036	1,978,502
	M	531	520	891	2,966	11,810	20,642	35,005	48,963
	%	0.28	0.24	0.24	0.57	1.52	1.71	1.76	2.47
Manipur	P	512,100	577,600	780,000	1,073,000	1,420,953	1,837,149	2,166,788	2,855,794
	M	29,560	37,200	48,590	70,970	99,330	133,535	190,939	239,836
	%	5.77	6.44	6.23	6.61	6.99	7.27	8.81	8.40
Mizoram	P	152,800	196,200	266,100	332,400	493,757	689,756	888,573	1,097,206
	M	101	131	203	1,882	2,205	4,538	10,099	14,832
	%	0.07	0.07	0.08	0.57	0.45	0.66	1.14	1.35
Tripura	P	513,000	639,000	1,142,000	1,556,000	2,053,058	2,757,205	3,199,203	3,673,917
	M	123,600	137,000	230,000	103,962	138,529	196,495	254,442	316,042
	%	24.09	21.44	20.14	6.68	6.75	7.13	7.95	8.60
Meghalaya	P	555,800	605,700	769,400	1,012,000	1,335,819	1,774,778	2,318,822	2,966,889
	M	14,060	13,950	23,020	26,350	41,430	61,462	99,169	130,399
	%	2.53	2.30	2.99	2.60	3.10	3.46	4.28	4.40
Assam	P	6,695,000	8,028,856	10,837,329	14,625,152	18,041,248	22,414,322	26,655,528	31,205,576
	M	1,683,000	1,982,000	2,742,000	3,592,000	4,690,724	6,373,204	8,240,611	10,679,345
	%	25.14	24.69	25.30	24.56	26.00	28.43	30.92	34.22

State									
West Bengal	P	23,230,000	26,300,000	34,926,279	44,312,011	54,580,647	68,077,965	80,176,197	91,276,115
	M	6,848,000	5,118,000	6,985,287	9,064,338	11,743,209	16,075,836	20,240,543	24,654,825
	%	29.48	19.46	20.00	20.46	21.52	23.61	25.25	27.01
Odisha	P	13,768,000	14,646,000	17,549,000	21,945,000	26,370,271	31,659,736	36,804,660	41,974,218
	M	165,700	176,300	215,300	326,500	422,266	577,775	761,985	911,670
	%	1.20	1.20	1.23	1.49	1.60	1.82	2.07	2.17
Madhya Pradesh	P	23,991,000	26,072,000	32,372,000	41,654,000	52,178,844	48,566,242	60,348,023	72,626,809
	M		1,050,000	1,318,000	1,816,000	2,501,919	2,983,127	3,841,449	4,774,695
	%		4.03	4.07	4.36	4.79	6.14	6.37	6.57
Chhattisgarh	P						17,614,928	20,833,803	25,545,198
	M						299,673	409,615	514,998
	%						1.70	1.97	2.02
Gujarat	P	13,701,000	16,262,000	20,633,000	26,697,000	34,085,799	41,309,582	50,671,017	60,439,692
	M		1,451,000	1,745,000	2,249,000	2,907,744	3,606,920	4,592,854	5,846,761
	%		8.92	8.46	8.42	8.53	8.73	9.06	9.67
Maharashtra	P	26,833,000	32,003,000	39,554,000	50,412,000	62,784,171	78,937,187	96,878,627	112,374,333
	M		2,436,000	3,034,000	4,233,000	5,805,785	7,628,755	10,270,485	12,971,152
	%		7.61	7.67	8.40	9.25	9.66	10.60	11.54

Table 2.5.3c (Continued)

		1941	1951	1961	1971	1981	1991	2001	2011
Andhra Pradesh	P	27,289,000	31,115,000	35,983,000	43,503,000	53,549,763	66,508,008	76,210,007	84,580,777
	M	2,137,000	2,418,000	2,715,000	3,520,000	4,533,700	5,923,954	6,986,856	8,082,412
	%	7.83	7.77	7.55	8.09	8.47	8.91	9.17	9.56
Karnataka	P	16,255,000	19,402,000	23,587,000	29,299,000	37,135,714	44,977,201	52,850,562	61,095,297
	M	1,592,990	1,950,000	2,328,000	3,113,000	4,104,616	5,234,023	6,463,127	7,893,065
	%	9.80	10.05	9.87	10.62	11.05	11.64	12.23	12.92
Goa	P	540,900	547,400	590,000	795,100	1,008,000	1,169,793	1,347,668	1,458,545
	M	8,222	8,813	11,150	26,480	41,320	61,455	92,210	121,564
	%	1.52	1.61	1.89	3.33	4.10	5.25	6.84	8.33
Kerala	P	11,032,000	13,549,000	16,904,000	21,347,000	25,453,680	29,098,518	31,841,374	33,406,061
	M	1,884,000	2,375,000	3,028,000	4,163,000	5,409,687	6,788,364	7,863,842	8,873,472
	%	17.08	17.53	17.91	19.50	21.25	23.33	24.70	26.56
Tamil Nadu	P	26,268,000	30,119,000	33,687,000	41,199,000	48,408,007	55,858,946	62,405,679	72,147,030
	M	1,443,000	1,560,000	2,104,000	2,519,947	3,052,717	3,470,647	4,229,479
	%	4.79	4.63	5.11	5.21	5.47	5.56	5.86
Chandigarh	P	22,570	24,260	119,900	257,300	451,610	642,015	900,635	1,055,450
	M	500	700	1,467	3,720	9,115	17,477	35,548	51,447
	%	2.22	2.89	1.22	1.45	2.02	2.72	3.95	4.87

Region									
Delhi	P	917,900	1,744,000	2,659,000	4,066,000	6,220,406	9,420,644	13,850,507	16,787,941
	M	304,900	99,500	155,500	263,000	481,802	889,641	1,623,520	2,158,684
	%	33.22	5.71	5.85	6.47	7.75	9.44	11.72	12.86
Daman & Diu	P	42,810	48,610	36,670	62,650	78,980	101,586	158,204	243,247
	M	5,107	5,194	3,013	5,770	7,144	9,048	12,281	19,277
	%	11.93	10.69	8.22	9.21	9.05	8.91	7.76	7.92
Dadra & Nagar Haveli	P	40,440	41,530	57,960	74,170	103,700	138,477	220,490	343,709
	M	175	159	443	740	1,932	3,341	6,524	12,922
	%	0.43	0.38	0.76	1.00	1.86	2.41	2.96	3.76
Lakshadweep	P	18,360	21,040	24,110	31,810	40,250	51,707	60,650	64,473
	M	18,280	21,020	23,790	30,020	38,170	48,765	57,903	62,268
	%	99.56	99.90	98.67	94.37	94.83	94.31	95.47	96.58
Puducherry	P	285,000	317,300	369,100	471,700	604,471	807,785	974,345	1,247,953
	M			23,470	29,140	36,663	52,867	59,358	75,556
	%			6.36	6.18	6.07	6.54	6.09	6.05
Andaman & Nicobar	P	33,770	30,970	63,550	115,100	188,741	280,661	356,152	380,581
	M	8,005	4,783	7,398	11,660	16,190	21,354	29,265	32,413
	%	23.70	15.44	11.64	10.13	8.58	7.61	8.22	8.52

The Republic of Maldives was occupied by the British in 1887 and gained its independence from the UK in 1965. It has an area of 298 sq. km, consists of 1,192 coral islands grouped into 26 atolls. Of these islands, 193 are inhabited while 91 have been developed as tourist resorts. The largest of the islands are Huvadu (39 sq. km) but only a tenth of the population, Haddunmathi, Milandunmadulu (26 sq. km each), Thiladunmathi (25 sq. km), Addu (16 sq. km) and South Maalosmadulu (15 sq. km). However, a third of the population lives in the Island of Malé (11 sq. km), where the capital Malé is located. A map of the Maldives is presented in Fig. 2.5.4 and by the end of this century; the whole country is expected to be under sea level.

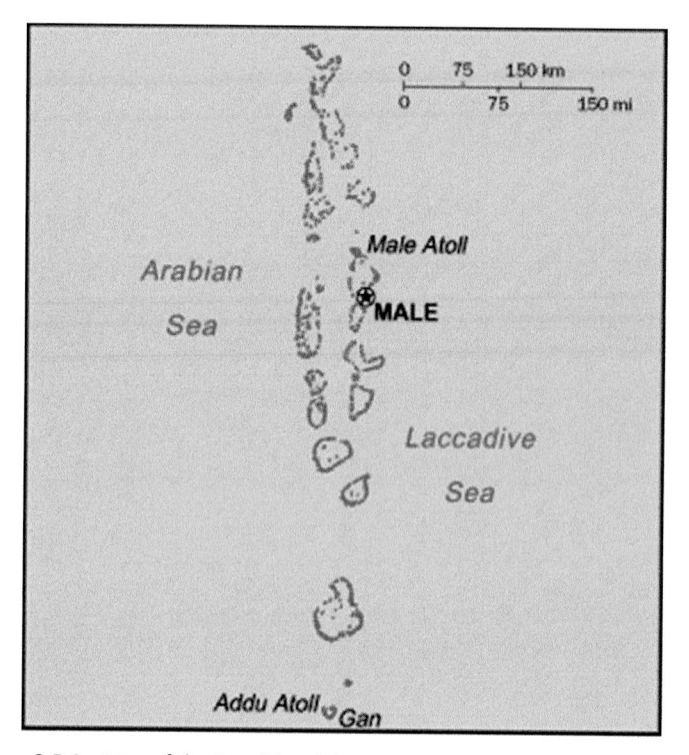

Figure 2.5.4   Map of the Republic of the Maldives.

The first census was conducted in 1911 while the first census to collect religious data was in 2014. The total population quadrupled since 1960 as shown in Table 2.5.4. The data did not include foreign population that was only included in the census of 2014, which recorded 0.38 million or 94.2% of the total population as Muslim. The entire indigenous population is Muslims, while the non-Muslim population comprises the third of the foreign workers.

Thus, assuming that the percentage of Muslims remains constant, the Muslim population is expected to remain around half million throughout the second half of this century.

Table 2.5.4 Evolution of the Muslim population in the Maldives

Year	Population	Muslims	%	Source
1895	30,000	30,000	100	[JAN]es
1911	72,237	72,237	100	[MV]c
1921	70,413	70,413	100	[MV]c
1931	79,281	79,281	100	[MV]c
1946	82,068	82,068	100	[MV]c
1960	92,247	92,247	100	[MV]c
1970	114,469	114,469	100	[MV]c
1977	142,832	142,832	100	[MV]c
1990	213,215	213,215	100	[MV]c
2000	270,101	270,101	100	[MV]c
2006	298,968	298,968	100	[MV]c
2014	400,843	377,651	94.21	[UN]c
2050	586,000	552,000	94.21	es
2100	490,000	462,000	94.21	es

2.5.5 Nepal

The Federal Democratic Republic of Nepal has an area of 147,181 sq. km and its map is presented in Fig. 2.5.5a. It did not come under Muslim control but was influenced by the spread of Islam in India. Thus, Islam entered here in the fifteenth century with Muslim settlers from Kashmir, and the first mosque was built in 1524.

Early estimates showed that the number of Muslims increased from 1,000 or 0.3% in 1891, to 3,500 or 0.06% of the total population.

Based on census data as shown in Table 2.5.5, the Muslim population increased dramatically since then. Accordingly, Muslims increased from 0.21 million or 2.5% in 1954, to 0.28 million or 3.0% in 1961, to 0.35 million or 3.0% in 1971, to 0.40 million or 2.7% in 1981, to 0.65 million or 3.5% in 1991, to 0.95 million or 4.2% in 2001, to 1.16 million or 4.4% in 2011. A choropleth map illustrating the presence of Muslims per district is presented in Fig. 2.5.5b.

Thus, assuming that the percentage of Muslims will continue to increase by a fifth of a percentage point per decade, the Muslim population is expected to reach 1.8 million or 5.2% by 2050 and 1.5 million or 6.2% by 2100.

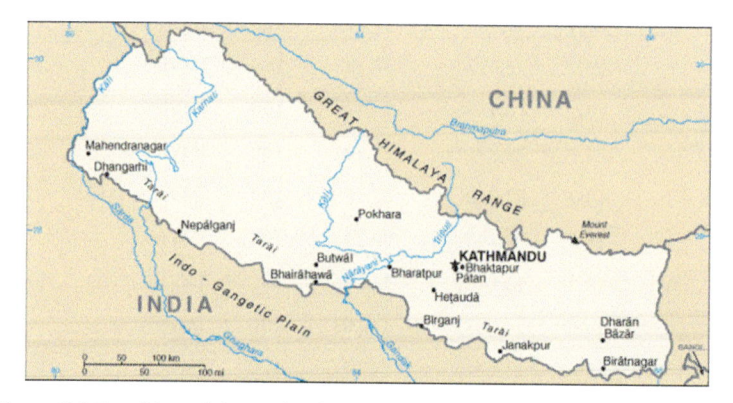

Figure 2.5.5a Map of the Federal Democratic Republic of Nepal.

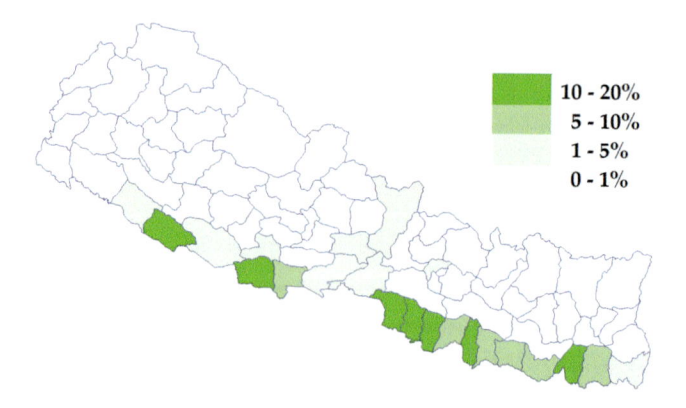

Figure 2.5.5b A choropleth map of Nepal illustrating the presence of Muslims per district.

Table 2.5.5 Evolution of the Muslim population in Nepal

Year	Population	Muslims	%	Source
1891	3,000,000	1,000	0.03	[JAN]es
1911	5,639,100	3,500	0.06	[AFH]es
1954	8,235,079	208,899	2.54	[NP]c
1961	9,407,280	280,597	2.98	[NP]c
1971	11,555,983	351,186	3.04	[NP]c
1981	15,022,839	399,197	2.66	[NP]c
1991	18,472,959	653,218	3.54	[NP]c
2001	22,736,934	954,023	4.20	[NP]c
2011	26,432,923	1,162,370	4.40	[NP11]c
2050	35,324,000	1,837,000	5.20	es
2100	23,708,000	1,470,000	6.20	es

2.5.6 Pakistan

The Muslim conquest of lands occupied by current Pakistan started in 644 AD during the reign of Caliph Omar bnul Khattab and completed in 711 AD during the reign of the sixth Umayyad Caliph al Walid I bnu Abdel Malik bnu Marwan. Indeed, in 23 H or 644 AD Caliph Omar sent troops under the leadership of Suhail bnu Adiy al Khazraji who conquered the eastern half of current Pakistani up to the Indus River. The conquest continued in 92 H or 711 AD under the leadership of Muhammad bnul Qassim Athaqafi by capturing the Sind and parts of the Pakistani Punjab. The Muslim conquest of current Pakistan was completed in 1021 by the conquest of Lahore during the reign of Mahmud ben Sebük Tegin al Ghaznawi. He was the fourth ruler of the Ghaznavid Dynasty who ruled from 998 to 1030. Currently, the Islamic Republic of Pakistan has an area of 796,095 sq. km and its map is presented in Fig. 2.5.6. It was occupied by the British in 1849 and gained its independence from the UK in 1947.

Based on census data, we obtain the following estimates: For 1871 and 1881, the total population is derived by two-third, one-sixth and half of the population in Punjab and Bombay provinces (excluding Princely States) and Kashmir State, respectively. The Muslim population is obtained by all and two-thirds of the Muslim population in Punjab and Bombay provinces, respectively, and half the population of Kashmir State. For 1891, we used total population and Muslims of Sind province instead of Bombay. The total population from 1901 to 1961 was obtained from [PK61]. The Muslim population in 1901 till 1941 was obtained from [INH]. The population of the Frontier Regions of 1,622,000 recorded in 1911 was added to the total population of 16,576,000 recorded in 1901 which excluded the Frontier Regions. All these were assumed Muslim and added to the number of Muslims in 1901.

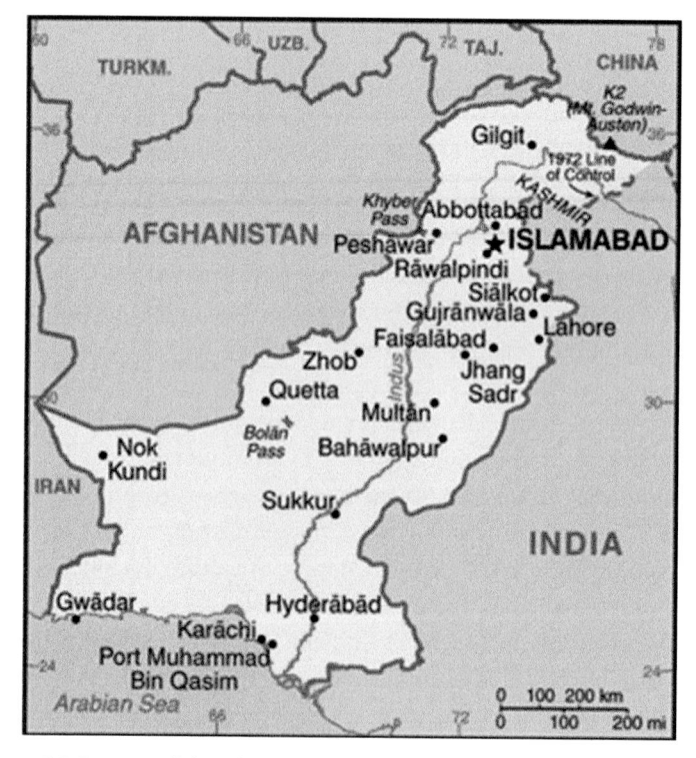

Figure 2.5.6 Map of the Islamic Republic of Pakistan.

Thus, as shown in Table 2.5.6, before independence the Muslim population increased from 11.8 million or 78.5% in 1871, to 13.3 million or 77.5% in 1881, to 15.1 million or 83.8% in 1891, to 18.2 million or 85.3% in 1901, to 16.4 million or 84.4% in 1911, to 17.6 million or 83.5% in 1921, to 18.8 million or 79.7% in 1931, to 22.3 million or 78.8% in 1941. After independence, the Muslim population jumped to 32.7 million or 97.1% in 1951 due to forced population exchange with India. The Muslim population continued to increase to 41.7 million or 97.2% in 1961, to 63.3 million or 96.9% in 1972, to 81.5 million or 96.7% in 1981, to 127.4 million or 96.3% in 1998. Thus, assuming that the percentage of Muslims remains constant, the Muslim population is expected to reach 0.3 billion by 2050 and 0.4 billion by 2100.

Table 2.5.6 Evolution of the Muslim population in Pakistan

Year	Population	Muslims	%	Source
1871	14,965,866	11,751,318	78.52	[INH71]c
1881	17,157,083	13,289,237	77.46	[INH81]c
1891	18,054,981	15,120,530	83.75	[INH91]c
1901	18,198,000	15,526,000	85.32	[INH]c
1911	19,382,000	16,364,000	84.43	[INH]c
1921	21,109,000	17,620,000	83.47	[INH]c
1931	23,542,000	18,757,000	79.68	[INH]c
1941	28,282,000	22,293,000	78.82	[INH]c
1951	33,730,000	32,732,000	97.12	[UN56,BD]c
1961	42,880,378	41,666,153	97.17	[PK61]c
1972	65,309,000	63,281,776	96.90	[SYB80]c
1981	84,253,644	81,450,057	96.67	[UN88]c
1998	132,352,279	127,433,409	96.28	[PK]c
2050	338,013,000	325,439,000	96.28	es
2100	403,103,000	388,107,000	96.28	es

2.5.7 Sri Lanka

The Democratic Socialist Republic of Sri Lanka is an island nation with total area of 65,610 sq. km and its map is presented in Fig. 2.5.7a. Islam entered it in the seventh century through Arab traders from the southern Arabian Peninsula. Then the number of Muslims increased as Islam spread in India during the thirteenth century and onward. The Island was occupied by the Portuguese in 1502, which lost it to the Dutch in 1658, which in turn lost it to the British between 1796 and 1815. The Island gained its independence from the UK in 1948. It changed its name from Ceylon in 1972. The Dutch settled Muslims from Java here, and their descendants remain in Sri Lanka to this day.

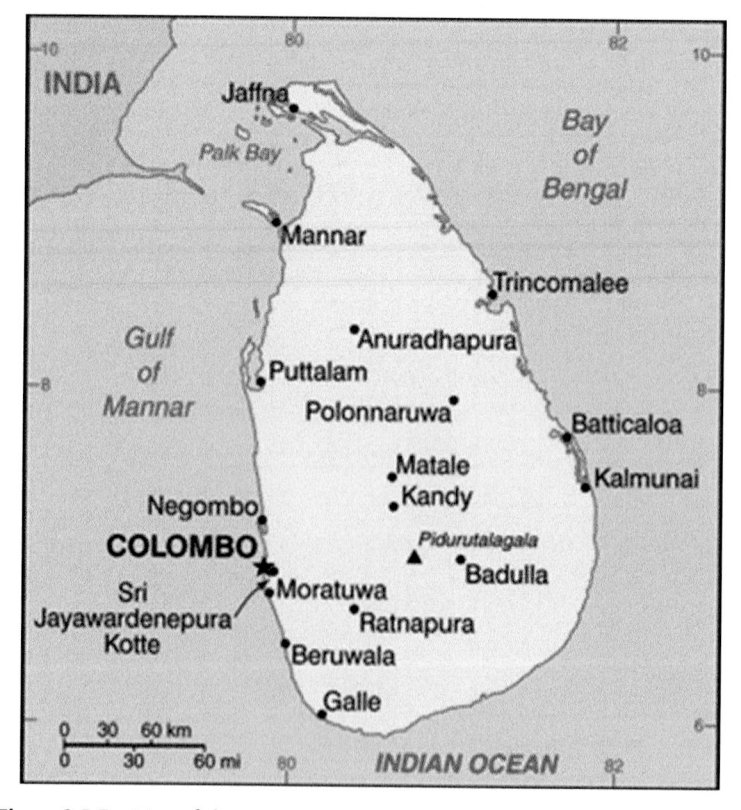

Figure 2.5.7 Map of the Democratic Socialist Republic of Sri Lanka.

Sri Lanka is divided into nine provinces; each consists of several districts. The distribution of Muslims per province since 1946 is provided in Table 2.5.7a which shows that there was not a significant change in the concentration and distribution of Muslims in the past century and a half. Almost a third of the Muslim population lives in Eastern Province where they make over a third of the population. Over a quarter of the Muslim population resides in Western Province, where the capital Colombo is located, and they make almost a tenth of the total population of the province. Over a quarter of the Muslim population is split in Central and North Western provinces, where the make over a tenth of the population in each province. A choropleth map illustrating the presence of Muslims per district is presented in Fig. 2.5.7b.

The Northern Province consists of five districts: Jaffna, Kilinochchi, Mannar, Mullaitivu and Vavuniya, while the Eastern Province consists of three districts: Ampara, Batticaloa and Trincomalee. In 2001, all these districts except Ampara were under control of Tamil Tigers who led a separation movement between 1983 and 2009 to establish an ethnic based country named Tamil Eelam. In 1990, Tamil Tigers expelled Muslims from most of the Northern Province except Vavuniya District which was mostly under Sri Lankan government control. However, some of those expelled Muslims returned in mid-1990s. The total population in 2001 in the districts outside government control was officially estimated based on the 2012 census, while Muslim population is estimated by assuming growth same to population growth in each district where the data is missing.

Based on census data and as shown in Table 2.5.7b, the Muslim population increased from 0.17 million or 7.2% in 1871, to 0.20 million or 7.1% in 1881, to 0.21 million or 7.1% in 1891, to 0.25 million or 6.9% in 1901, to 0.28 million or 6.9% in 1911, to 0.30 million or 6.7% in 1921, to 0.35 million or 6.7% in 1931, to 0.44 million or 6.6% in 1946, to 0.54 million or 6.7% in 1953, to 0.72 million or 6.8% in 1963, to 0.91 million or 7.2% in 1971, to 1.12 million or 7.6% in 1981, to 1.43 million or 8.5% in 2001, to 1.97 million or 9.7% in 2012.

Table 2.5.7a Evolution of the Muslim population in Sri Lanka per province since 1871. P: Total population, M: Muslim population, M%: Percentage of Muslim population, MR%: Muslim Ratio in percentage.

Province		1871	1881	1891	1901	1911	1921	1946
Central	P	365,703	473,669	474,487	622,832	672,258	717,739	1,135,290
	M%	6.96	7.01	6.55	5.88	6.14	6.17	6.26
	M	25,442	33,197	31,059	36,604	41,281	44,257	71,016
	MR%	14.83	16.96	14.65	14.87	14.55	14.63	16.27
Eastern	P	117,353	127,555	148,444	173,602	183,698	192,821	279,112
	M%	33.61	34.17	35.01	36.58	38.72	39.46	39.32
	M	39,441	43,583	51,969	63,504	71,126	76,094	109,758
	MR%	22.99	22.26	24.51	25.80	25.08	25.16	25.14
North Central	P	58,972	66,146	75,333	79,110	85,961	96,525	139,534
	M%	8.05	11.33	11.79	11.43	11.69	12.10	10.97
	M	4,745	7,494	8,884	9,043	10,047	11,675	15,301
	MR%	2.77	3.83	4.19	3.67	3.54	3.86	3.50
North Western	P	276,013	293,327	320,070	353,626	434,116	492,181	667,889
	M%	7.72	7.86	7.31	7.22	6.83	6.71	5.99
	M	21,306	23,056	23,385	25,543	29,641	33,025	40,010
	MR%	12.42	11.78	11.03	10.38	10.45	10.92	9.16
Northern	P	281,640	302,500	319,296	340,936	369,966	374,829	479,572
	M%	3.46	3.48	3.79	3.54	3.51	3.53	3.86
	M	9,732	10,518	12,098	12,081	12,998	13,231	18,514
	MR%	5.67	5.37	5.71	4.91	4.58	4.38	4.24
Sabara-gamuwa	P	197,564	225,829	258,626	321,755	408,521	471,814	745,382
	M%	3.00	3.15	3.30	3.20	3.24	3.29	2.92
	M	5,922	7,103	8,547	10,281	13,244	15,537	21,797
	MR%	3.45	3.63	4.03	4.18	4.67	5.14	4.99
Southern	P	398,648	433,520	489,799	566,736	628,817	671,234	961,418
	M%	4.30	4.14	3.71	3.56	3.52	3.30	2.85
	M	17,122	17,928	18,168	20,169	22,110	22,164	27,446
	MR%	9.98	9.16	8.57	8.19	7.80	7.33	6.29
Uva	P	129,000	165,692	159,201	186,674	216,692	233,864	372,238
	M%	4.11	4.23	4.42	3.89	3.62	3.56	3.52
	M	5,303	7,009	7,044	7,263	7,840	8,321	13,085
	MR%	3.09	3.58	3.32	2.95	2.76	2.75	3.00
Western	P	575,721	671,500	762,533	920,683	1,106,321	1,246,847	1,876,904
	M%	7.39	7.12	6.67	6.69	6.81	6.26	6.37
	M	42,529	47,806	50,841	61,630	75,344	78,102	119,629
	MR%	24.79	24.42	23.98	25.04	26.56	25.83	27.40

Table 2.5.7a (*Continued*)

		1953	1963	1971	1981	2001	2012
Central	P	1,366,596	1,697,018	1,953,044	2,009,248	2,423,966	2,558,716
	M%	6.31	6.78	6.96	7.82	9.60	10.26
	M	86,214	115,051	135,903	157,108	232,669	262,525
	MR%	15.92	15.89	14.94	14.01	16.20	13.34
Eastern	P	355,231	546,454	717,571	975,251	1,419,602	1,551,381
	M%	38.13	34.28	34.67	32.54	30.21	37.12
	M	135,456	187,332	248,808	317,354	428,896	575,936
	MR%	25.01	25.87	27.34	28.29	29.87	29.28
North Central	P	229,174	393,759	552,423	849,492	1,104,677	1,259,567
	M%	8.59	7.81	7.04	7.07	8.15	8.08
	M	19,691	30,757	38,877	60,089	90,022	101,813
	MR%	3.64	4.25	4.27	5.36	6.27	5.18
North Western	P	854,915	1,155,207	1,404,063	1,704,334	2,169,892	2,370,075
	M%	6.20	6.11	6.30	6.72	10.73	11.39
	M	53,037	70,595	88,429	114,463	232,866	269,977
	MR%	9.79	9.75	9.72	10.20	16.22	13.72
Northern	P	571,214	741,351	877,768	1,109,404	1,040,963	1,058,762
	M%	4.16	4.26	4.31	4.92	5.83	3.22
	M	23,786	31,591	37,855	54,534	60,700	34,040
	MR%	4.39	4.36	4.16	4.86	4.23	1.73
Sabara-gamuwa	P	890,838	1,124,543	1,316,096	1,482,031	1,801,331	1,918,880
	M%	3.04	2.98	3.07	3.46	4.08	4.44
	M	27,089	33,552	40,369	51,248	73,576	85,106
	MR%	5.00	4.63	4.44	4.57	5.12	4.33
Southern	P	1,130,014	1,430,740	1,661,870	1,882,661	2,278,271	2,464,732
	M%	2.68	2.68	2.72	2.78	3.10	3.24
	M	30,240	38,376	45,149	52,379	70,657	79,964
	MR%	5.58	5.30	4.96	4.67	4.92	4.06
Uva	P	468,835	654,105	808,425	914,522	1,177,358	1,259,900
	M%	3.65	3.53	3.60	3.82	4.21	4.51
	M	17,103	23,091	29,083	34,901	49,530	56,874
	MR%	3.16	3.19	3.20	3.11	3.45	2.89
Western	P	2,231,820	2,838,877	3,402,134	3,919,807	5,381,197	5,821,710
	M%	6.66	6.82	6.98	7.13	8.20	8.61
	M	148,566	193,698	237,587	279,639	441,397	500,992
	MR%	27.44	26.75	26.11	24.93	30.74	25.47

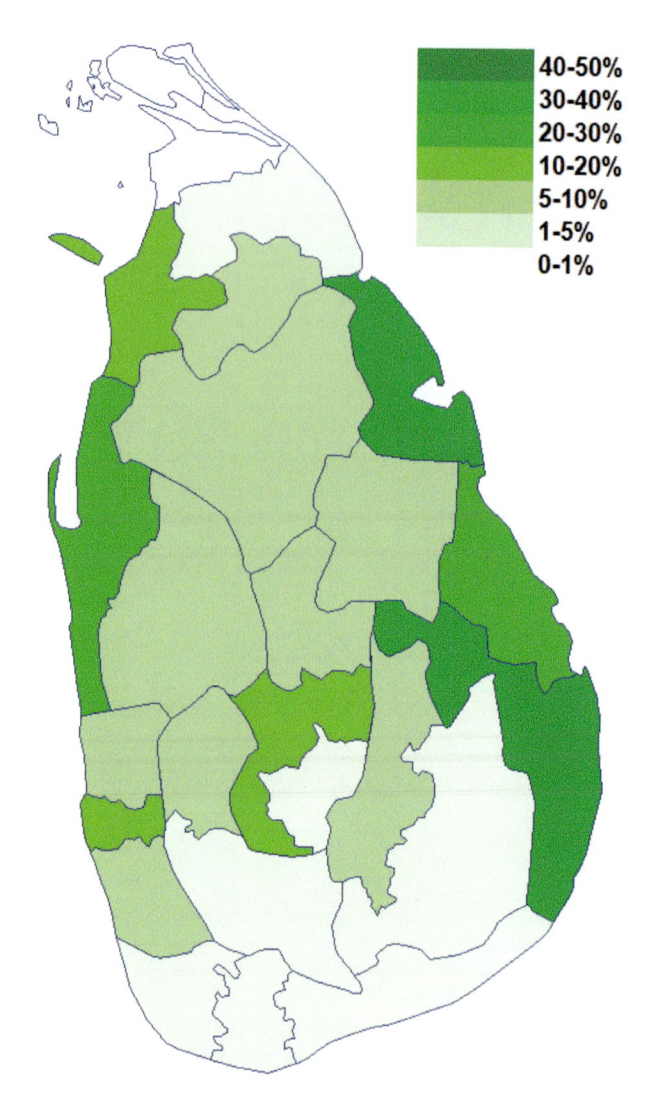

Figure 2.5.7b A choropleth map of Sri Lanka illustrating the presence of Muslims per district.

Thus, assuming that the percentage of Muslims will continue to increase by a half of a percentage point per decade, the Muslim population is expected to reach 2.5 million or 11.5% by 2050 and 2.1 million or 14.0% by 2100.

Table 2.5.7b Evolution of the Muslim population in Sri Lanka

Year	Population	Muslims	%	Source
1871	2,400,380	171,542	7.15	[LKH]c
1881	2,759,738	195,775	7.09	[LKH]c
1891	3,007,789	211,995	7.05	[LKH]c
1901	3,565,954	246,118	6.90	[LKH]c
1911	4,106,350	283,631	6.91	[LKH]c
1921	4,498,605	302,532	6.73	[LKH]c
1931	5,306,900	354,200	6.67	[LKH]c
1946	6,657,339	436,556	6.56	[LKH]c
1953	8,097,895	541,506	6.69	[LKH]c
1963	10,582,064	724,043	6.84	[LKH]c
1971	12,711,143	909,941	7.16	[LKH]c
1981	14,846,750	1,121,717	7.56	[UN83]c
2001	16,929,689	1,435,896	8.48	[UN]c
2012	20,263,723	1,967,227	9.71	[LK]c
2050	21,814,000	2,509,000	11.50	es
2100	15,275,000	2,139,000	14.00	es

2.5.8 Regional Summary and Conclusion

The Indian Subcontinent started entering into Muslim control after a decade from the death of Prophet Mohammed peace and blessings upon him and by the eleventh century almost all of it was under Muslim control until the British arrival in mid-nineteenth century. Almost a third of the total population is Muslim and is expected to continue to increase for the next three centuries. The following tables present centennial data from 600 AD to 2100 AD (or approximately 1 H to 1500 H) in Table 2.5a and decennial data from 1790 AD to 2100 AD (or 1210 H to 1520 H) in Tables 2.5b and 2.5c for current countries in the Indian Subcontinent. The data includes total population in thousands (P), the percentage of which is Muslim ($M\%$), the corresponding Muslim population in thousands (M), and the annual population growth rate (APGR, or $G\%$) of the total population in this region.

Table 2.5a Centennial estimates of the Muslim population (×1000) in the Indian Subcontinent: 600–2100 AD (1–1500 H)

		600	700	800	900	1000	1100	1200	1300
Bangladesh	P	7,000	7,050	7,100	7,150	7,177	7,900	8,600	9,300
	M%	—	—	—	—	—	—	1.00	10.00
	M	—	—	—	—	—	—	86	930
Bhutan	P	2	3	4	5	6	7	8	9
	M%	—	—	—	—	—	—	0.01	0.01
	M	—	—	—	—	—	—	—	—
India	P	59,300	60,200	61,100	62,000	62,924	68,800	74,700	80,600
	M%	—	0.10	0.50	1.00	2.00	3.00	4.00	5.00
	M	—	60	306	620	1,258	2,064	2,988	4,030
Maldives	P	2	3	4	5	6	7	8	9
	M%	—	1.00	2.00	3.00	4.00	5.00	50.00	100.00
	M	—	—	—	—	—	—	4	9
Nepal	P	1,050	1,100	1,150	1,200	1,275	1,500	1,700	1,900
	M%	—	—	—	—	—	—	0.01	0.02
	M	—	—	—	—	—	—	—	—
Pakistan	P	4,700	4,750	4,800	4,850	4,899	5,300	5,700	6,100
	M%	—	10.00	30.00	40.00	50.00	60.00	70.00	70.00
	M	—	475	1,440	1,940	2,450	3,180	3,990	4,270
Sri Lanka	P	460	485	510	535	561	600	650	700
	M%	—	1.00	2.00	3.00	4.00	5.00	6.00	7.00
	M	—	5	10	16	22	30	39	49
Total	P	72,514	73,591	74,668	75,745	76,848	84,114	91,366	98,618
	M%	—	0.73	2.35	3.40	4.85	6.27	7.78	9.42
	M	—	540	1,756	2,576	3,731	5,274	7,107	9,288
	G%		0.015	0.015	0.014	0.014	0.090	0.083	0.076

Table 2.5a (*Continued*)

		1400	1500	1600	1700	1800	1900	2000	2100
Bangladesh	P	10,000	10,526	12,919	15,789	19,200	29,164	127,658	151,393
	M%	20.00	40.00	50.00	60.00	65.07	66.07	89.58	95.00
	M	2,000	4,210	6,460	9,473	12,493	19,269	114,356	143,823
Bhutan	P	10	11	12	13	14	250	591	686
	M%	0.01	0.01	0.01	0.01	0.01	0.10	0.12	0.30
	M	—	—	—	—	—	—	1	2
India	P	86,500	92,289	113,263	138,434	167,600	238,396	1,056,576	1,450,421
	M%	6.00	7.00	8.00	9.00	9.59	12.21	13.44	21.00
	M	5,190	6,460	9,061	12,459	16,073	29,108	142,004	304,588
Maldives	P	10	11	12	13	14	50	279	490
	M%	100.00	100.00	100.00	100.00	100.00	100.00	100.00	94.21
	M	10	11	12	13	14	50	279	462
Nepal	P	2,100	2,272	2,635	3,064	3,760	5,283	23,941	23,708
	M%	0.03	0.04	0.05	0.06	0.06	0.06	4.20	6.20
	M	1	1	1	2	2	3	1,006	1,470
Pakistan	P	6,500	7,185	8,818	10,777	13,200	18,544	142,344	403,103
	M%	70.00	70.00	70.00	75.00	78.52	85.32	96.28	96.28
	M	4,550	5,030	6,173	8,083	10,365	15,822	137,048	388,107
Sri Lanka	P	750	800	850	900	850	3,566	18,778	15,275
	M%	8.00	9.00	8.00	7.00	7.15	6.90	8.48	14.00
	M	60	72	68	63	61	246	1,592	2,139
Total	P	105,870	113,094	138,509	168,990	204,638	295,253	1,370,166	2,045,076
	M%	11.16	13.96	15.72	17.81	19.06	21.84	28.92	41.10
	M	11,811	15,784	21,774	30,093	39,008	64,498	396,286	840,591
	G%	0.071	0.066	0.203	0.199	0.191	0.367	1.535	0.401

Table 2.5b Decennial estimates of the Muslim population (×1000) in the Indian Subcontinent: 1790–1940 AD (1210–1360 H)

		1790	1800	1810	1820	1830	1840	1850	1860
Bangladesh	P	18,800	19,200	19,600	20,000	20,900	21,800	22,700	23,600
	M%	65.07	65.07	65.07	65.07	65.07	65.07	65.07	65.07
	M	12,233	12,493	12,754	13,014	13,600	14,185	14,771	15,357
Bhutan	P	13	14	15	16	17	18	19	20
	M%	0.01	0.01	0.01	0.01	0.01	0.01	0.01	0.01
	M	—	—	—	—	—	—	—	—
India	P	163,200	167,600	172,000	175,349	182,677	190,205	198,100	205,149
	M%	9.59	9.59	9.59	9.59	9.59	9.59	9.59	9.59
	M	15,651	16,073	16,495	16,816	17,519	18,241	18,998	19,674
Maldives	P	13	14	15	16	17	18	19	20
	M%	100.00	100.00	100.00	100.00	100.00	100.00	100.00	100.00
	M	13	14	15	16	17	18	19	20
Nepal	P	3,700	3,760	3,820	3,881	4,000	4,200	4,352	4,500
	M%	0.06	0.06	0.06	0.06	0.06	0.06	0.06	0.06
	M	2	2	2	2	2	3	3	3
Pakistan	P	13,000	13,200	13,400	13,651	14,000	14,500	15,000	15,500
	M%	78.52	78.52	78.52	78.52	78.52	78.52	78.52	78.52
	M	10,208	10,365	10,522	10,719	10,993	11,385	11,778	12,171
Sri Lanka	P	800	850	900	950	1,009	1,230	1,500	1,919
	M%	7.15	7.15	7.15	7.15	7.15	7.15	7.15	7.15
	M	57	61	64	68	72	88	107	137
Total	P	199,526	204,638	209,750	213,863	222,620	231,971	241,690	250,708
	M%	19.13	19.06	19.00	19.00	18.96	18.93	18.90	18.89
	M	38,164	39,008	39,852	40,635	42,203	43,920	45,676	47,361
	G%		0.253	0.247	0.194	0.401	0.411	0.410	0.366

Table 2.5b (Continued)

		1870	1880	1890	1900	1910	1920	1930	1940
Bangladesh	P	24,721	25,536	28,210	29,164	31,555	32,161	35,602	39,934
	M%	65.07	65.07	65.47	66.07	67.19	68.10	69.46	70.26
	M	16,086	16,616	18,469	19,269	21,202	21,902	24,729	28,058
Bhutan	P	30	61	124	250	300	350	400	470
	M%	0.01	0.01	0.01	0.10	0.10	0.10	0.10	0.10
	M	—	—	—	—	—	—	—	—
India	P	212,189	215,172	233,338	238,396	252,093	251,321	278,977	318,661
	M%	9.59	9.78	10.35	12.21	12.01	12.23	12.84	13.38
	M	20,349	21,044	24,150	29,108	30,276	30,737	35,821	42,637
Maldives	P	25	35	45	50	72	70	79	81
	M%	100.00	100.00	100.00	100.00	100.00	100.00	100.00	100.00
	M	25	35	45	50	72	70	79	81
Nepal	P	4,698	4,939	5,192	5,283	5,639	6,113	6,783	7,525
	M%	0.06	0.06	0.06	0.06	0.06	0.06	0.06	2.54
	M	3	3	3	3	3	4	4	191
Pakistan	P	16,090	16,492	18,055	18,544	19,382	20,477	22,812	28,282
	M%	78.52	77.46	83.75	85.32	84.43	83.47	79.68	78.82
	M	12,634	12,775	15,121	15,822	16,364	17,092	18,177	22,292
Sri Lanka	P	2,400	2,760	3,008	3,566	4,106	4,499	5,307	6,134
	M%	7.15	7.09	7.05	6.90	6.91	6.73	6.73	6.56
	M	172	196	212	246	284	303	357	402
Total	P	260,154	264,995	287,972	295,253	313,148	314,991	349,960	401,087
	M%	18.94	19.12	20.14	21.84	21.78	22.26	22.62	23.35
	M	49,268	50,668	58,001	64,498	68,202	70,108	79,167	93,661
	G%	0.370	0.184	0.832	0.250	0.588	0.059	1.053	1.364

Table 2.5c Decennial estimates of the Muslim population (×1000) in the Indian Subcontinent: 1950–2100 AD (1370–1520 H)

		1950	1960	1970	1980	1990	2000	2010	2020
Bangladesh	P	37,895	48,014	64,232	79,639	103,172	127,658	147,575	164,689
	M%	76.79	80.43	85.40	86.65	88.31	89.58	90.40	91.00
	M	29,099	38,617	54,855	69,008	91,111	114,356	133,408	149,867
Bhutan	P	177	223	297	407	531	591	686	772
	M%	0.11	0.11	0.11	0.11	0.11	0.12	0.12	0.14
	M	—	—	—	—	1	1	1	1
India	P	376,325	450,548	555,190	698,953	873,278	1,056,576	1,234,281	1,380,004
	M%	9.91	10.70	11.21	11.36	12.12	13.44	14.26	15.00
	M	37,294	48,209	62,237	79,401	105,841	142,004	176,008	207,001
Maldives	P	74	90	116	158	223	279	366	541
	M%	100.00	100.00	100.00	100.00	100.00	100.00	94.21	94.21
	M	74	90	116	158	223	279	345	509
Nepal	P	8,483	10,105	12,075	15,016	18,905	23,941	27,013	29,137
	M%	2.54	2.98	3.04	2.66	3.54	4.20	4.40	4.60
	M	215	301	367	399	669	1,006	1,189	1,340
Pakistan	P	37,542	44,989	58,142	78,054	107,648	142,344	179,425	220,892
	M%	97.12	97.17	96.90	96.67	96.28	96.28	96.28	96.28
	M	36,461	43,716	56,340	75,455	103,643	137,048	172,750	212,675
Sri Lanka	P	7,971	9,874	12,486	15,036	17,326	18,778	20,262	21,413
	M%	6.69	6.84	7.16	7.56	8.48	8.48	9.71	10.00
	M	533	675	894	1,137	1,469	1,592	1,967	2,141
Total	P	468,467	563,843	702,538	887,264	1,121,083	1,370,166	1,609,607	1,817,448
	M%	22.13	23.34	24.88	25.42	27.02	28.92	30.17	31.56
	M	103,677	131,608	174,808	225,559	302,958	396,286	485,668	573,535
	G%	1.553	1.853	2.199	2.334	2.339	2.006	1.611	1.214

Table 2.5c (Continued)

		2030	2040	2050	2060	2070	2080	2090	2100
Bangladesh	P	178,994	188,417	192,568	191,443	185,611	176,199	164,351	151,393
	M%	91.50	92.00	92.50	93.00	93.50	94.00	94.50	95.00
	M	163,779	173,343	178,125	178,042	173,546	165,627	155,312	143,823
Bhutan	P	843	885	905	901	870	816	752	686
	M%	0.16	0.18	0.20	0.22	0.24	0.26	0.28	0.30
	M	1	2	2	2	2	2	2	2
India	P	1,503,642	1,592,692	1,639,176	1,651,590	1,630,706	1,582,193	1,519,176	1,450,421
	M%	15.75	16.50	17.25	18.00	18.75	19.50	20.25	21.00
	M	236,824	262,794	282,758	297,286	305,757	308,528	307,633	304,588
Maldives	P	519	556	586	593	578	548	515	490
	M%	94.21	94.21	94.21	94.21	94.21	94.21	94.21	94.21
	M	489	524	552	559	545	516	485	462
Nepal	P	33,390	34,889	35,324	34,738	33,045	30,374	27,075	23,708
	M%	4.80	5.00	5.20	5.40	5.60	5.80	6.00	6.20
	M	1,603	1,744	1,837	1,876	1,851	1,762	1,624	1,470
Pakistan	P	262,959	302,129	338,013	366,792	387,094	399,523	404,491	403,103
	M%	96.28	96.28	96.28	96.28	96.28	96.28	96.28	96.28
	M	253,177	290,890	325,439	353,148	372,694	384,661	389,444	388,107
Sri Lanka	P	22,023	22,186	21,814	20,972	19,842	18,501	16,989	15,275
	M%	10.50	11.00	11.50	12.00	12.50	13.00	13.50	14.00
	M	2,312	2,440	2,509	2,517	2,480	2,405	2,294	2,139
Total	P	2,002,370	2,141,754	2,228,386	2,267,029	2,257,745	2,208,155	2,133,349	2,045,076
	M%	32.87	34.17	35.51	36.76	37.95	39.11	40.16	41.10
	M	658,186	731,738	791,222	833,429	856,875	863,501	856,794	840,591
	G%	0.969	0.673	0.397	0.172	−0.041	−0.222	−0.345	−0.423

2.6   Muslims in the Far East

This region consists of seven countries and territories: China, Hong Kong, Japan, North and South Korea, Macao, and Taiwan. Islam entered China in the seventh century through messengers of the third Caliph Othman bnu Affan, then through Umayyad Muslim armies in the eighth century. However, Islam did not reach the rest of this region with non-negligible numbers until the twentieth century.

Thus, the Muslim population has likely increased from 6,000 or 0.01% of the total population of this region in 700 AD, to 0.30 million or 0.4% in 800 AD, to 0.59 million or 0.8% in 900 AD, to 0.89 million or 1.2% in 1000 AD, to 1.09 million or 1.3% in 1100 AD, to 1.31 million or 1.4% in 1200 AD, to 1.55 million or 1.5% in 1300 AD, to 1.81 million or 1.6% in 1400 AD, to 2.1 million or 1.6% in 1500 AD, to 3.4 million or 1.8% in 1600 AD, to 3.0 million or 1.7% in 1700 AD, to 8.0 million or 2.1% in 1800 AD, to 9.4 million or 2.0% in 1900, to 21 million or 1.4% in 2000, to 26 million or 1.6% in 2020, and is projected to reach 30 million or 1.8% by 2050 and 28 million or 2.3% by 2100.

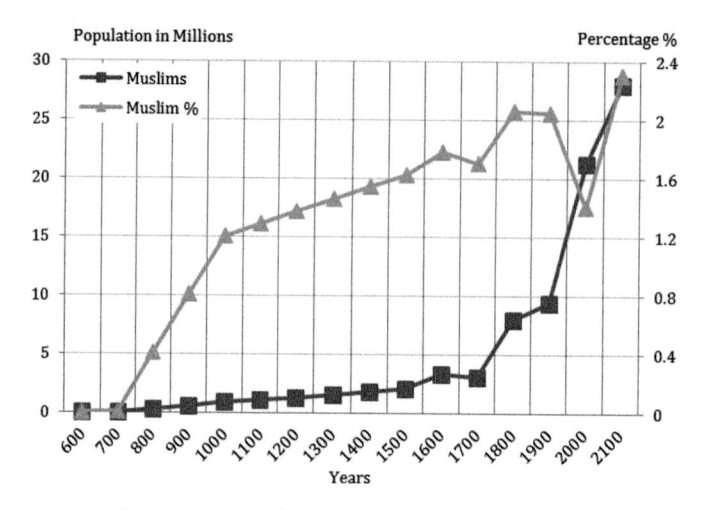

Figure 2.6a Plot of centennial estimates of the Muslim population and its percentage of the total population in the Far East: 600–2100 AD (1–1500 H).

A plot of centennial estimates of the Muslim population and its percentage with respect to the total population in this region from 600 to 2100 is provided in Fig. 2.6a. A zoom in of this plot, providing a plot of decennial estimates of the Muslim population and its percentage with respect to the total population in this region from 1900 to 2100 is provided in Fig. 2.6b. This shows that the Muslim population in this region is increasing slowly since 1950 and will remain almost steady at just below thirty million for the remainder of this century.

The corresponding individual data for each country in this region is discussed below. In Section 2.6.8, the total population in each country in this region and the corresponding percentage and number of Muslims is presented centennially in Table 2.6a from 600 to 2100 and decennially in Tables 2.6b and 2.6c from 1790 to 2100.

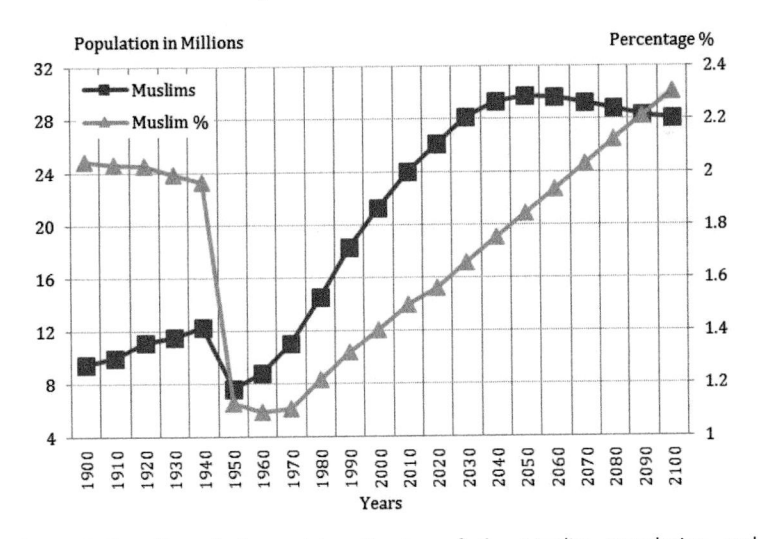

Figure 2.6b Plot of decennial estimates of the Muslim population and its percentage of the total population in the Far East: 1900–2100 AD (1320–1520 H).

2.6.1 China

In 31 H or 651 AD, the third Muslim Caliph Othman bnu Affan sent a delegation of fifteen people to Yung Way, the Emperor of Tang Dynasty. They arrived by sea, landed in Canton. They then visited the Emperor in Shang-An (currently, Sian or Xi'an), they were well received and built the first Mosque there. This was the first mosque built in China and it exists to this day.

In 96 H or 714 AD, Muslim armies reached the border of China for the first time and entered Kashgar, which is now in the northwest of China, in the Xinjiang Uyghur Autonomous Region. This occurred during the time of the sixth Umayyad Caliph al-Walid I bnu Abdel Malik bnu Marwan (ruled from 705 AD to 715 AD). Accordingly, the governor of Iraq, al-Hajjaj bnu Yusuf al-Thaqafi sent an army under the leadership of Qutaiba bnu Muslim al-Bahili (lived from 48 H or 668A D to 96 H or 715 AD). The Chinese Emperor then agreed to pay tribute to the Muslims as a sign of allegiance to the Muslim State. Islam then spread in China through commerce and Muslim state support to the Chinese state whenever requested during the Umayyad and Abbasid Dynasties. With the collapse of the Ming Dynasty in 1644, Muslim remained oppressed until the Republican Revolution in 1911. Oppression returned and continued since 1948 with the Communists takeover.

Xinjiang Uyghur Autonomous Region or East Turkestan as known historically, remained under Muslim control from 715 AD until 1073 AD, when non-Muslim Turkish tribes gained control over it, followed by Genghis Khan Mongol armies in 1221. After the Islamization of the Mongols, this region became an independent Muslim state until it was occupied by Chinese Manchu rulers in 1644. It gained brief independence under Muslim revolts in 1872–76, 1931–34 and 1940–44.

Currently, the People's Republic of China has an area of 9,640,011 sq. km, excluding the Special Administrative Regions or Hong Kong (1,104 sq. km) and Macao (28 sq. km) and their populations. It is the third largest country in the world after Russia and Canada, with similar size to the United States of America. It is also the most populous country in the world; with one fifth of the world population. A map of China is presented in Fig. 2.6.1a.

M. Broomhall estimated the Muslim population in 1850 at 9,641,000 or 2.3% of the total population. Then in 1910, he puts the estimate between five and ten million, or up to 2.4% of the total population. It is also reported that a 1936 census puts the Muslim population at 10.5%. However, there is no evidence that such census took place.

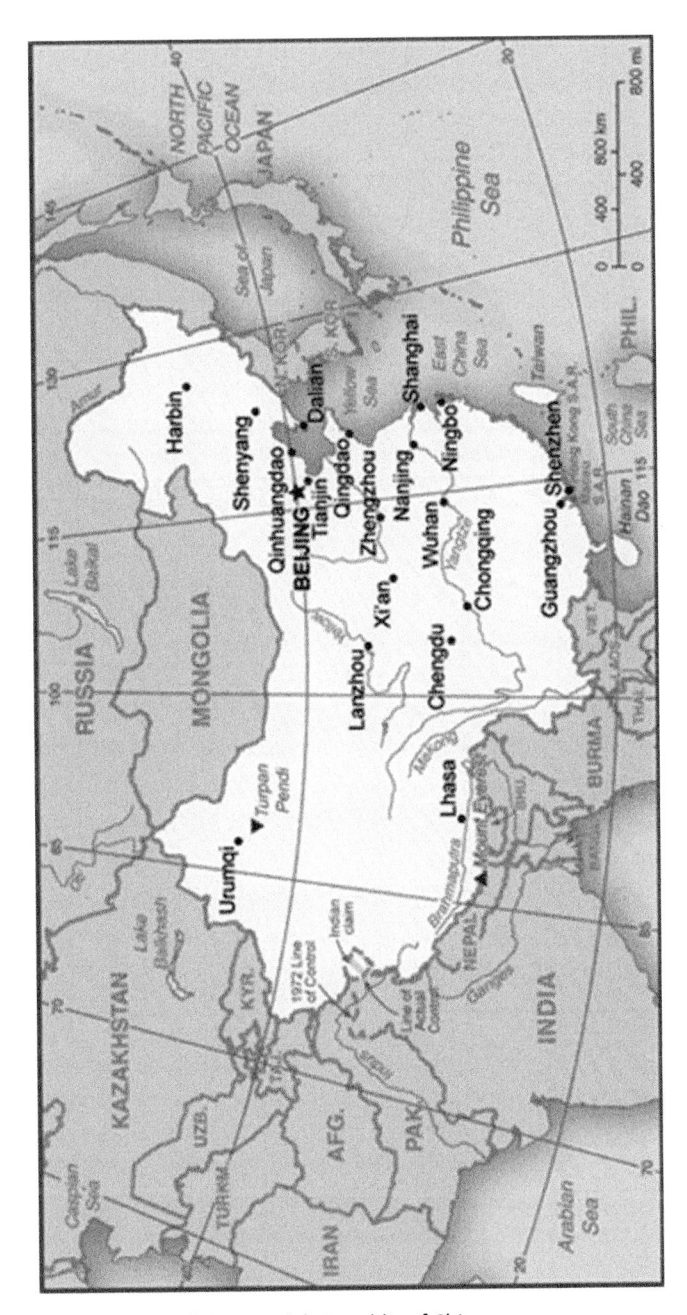

Figure 2.6.1a Map of the People's Republic of China.

The first comprehensive census was taken in 1953, and almost every decade since then. These censuses recorded ethnic affiliation of the population, where ten of the ethnic minorities are Muslim. These are Bonan (Bao'An), Dongxiang, Hui, Kazakh, Kyrgyz, Salar, Tajik, Tatar, Uygur (Uighur) and Uzbek. Almost half of Muslims in China are Hui, while over 40% are Uygur. More than 90% of the population of China belongs to the non-Muslim Han ethnicity. Table 2.6.1a presents the census count of each Muslim ethnicity since 1953 and their percentage of all Muslims in China.

Table 2.6.1a Evolution of the Muslim ethnic populations in China: 1953–2010

	1953	1964	1982	1990	2000	2010
Hui	3,559,350	4,473,147	7,228,398	8,612,001	9,816,805	10,586,087
Uygur	3,640,125	3,996,311	5,963,491	7,207,024	8,399,393	10,069,346
Kazakh	509,375	491,637	907,546	1,110,758	1,250,458	1,462,588
Dongxiang	155,761	147,443	279,397	373,669	513,805	621,500
Kyrgyz	70,944	70,151	113,386	143,537	160,823	186,708
Salar	30,658	34,644	69,135	82,398	104,503	130,607
Tajik	14,462	16,236	26,600	33,223	41,028	51,069
Bonan	4,957	5,125	9,017	11,683	16,505	20,074
Uzbek	13,626	7,717	12,213	14,763	12,370	10,569
Tatar	6,929	2,294	4,122	5,064	4,890	3,556
Total	8,006,187	9,244,705	14,613,305	17,594,120	20,320,580	23,142,104

Hence, based on ethnic census data, the Muslim population in China increased from 8.0 million or 1.4% in 1953, to 9.2 million or 1.3% in 1964, to 14.6 million or 1.5% in 1982, to 17.6 million or 1.6% in 1990, to 20.3 million or 1.6% in 2000, to 23.1 million or 1.7% in 2010. Thus, assuming that the percentage of Muslims will increase by a tenth of a percentage point per decade, the Muslim population is expected to reach 29 million or 2.1% by 2050 and 28 million or 2.6% by 2100. A summary of this data is provided in Table 2.6.1b.

As shown in Tables 2.6.1c and 2.6.1d, more than half of Muslims live in the northeastern most province of Xinjiang Uyghur Autonomous Region or East Turkistan, where they make up almost 60% of the total population. According to official estimates, the percentage of Muslims decreased from 93.5% in 1941 to 91.3% in 1949. With the migration of ethnic Han Chinese (non-Muslims), the percentage of Muslims decreased significantly according to subsequent censuses; from 90% in 1953 to 58% in 2000 and 2010.

Table 2.6.1b Evolution of the Muslim population in China

Year	Population	Muslims	%	Source
1850	412,000,000	9,641,000	2.34	[KET86]es
1910	423,000,000	10,000,000	2.36	[SYB20]es
1953	582,603,417	8,006,187	1.37	[CN]e
1964	694,581,759	9,244,705	1.33	[CN]e
1982	1,008,175,288	14,613,305	1.45	[CN]e
1990	1,133,682,501	17,594,120	1.55	[CN]e
2000	1,242,612,226	20,320,580	1.64	[UNE]e
2010	1,332,810,869	23,142,104	1.74	[UNE]e
2050	1,402,405,000	29,451,000	2.10	es
2100	1,064,993,000	27,690,000	2.60	es

Table 2.6.1c Evolution of Muslim ethnic populations in Xinjiang province (East Turkistan): 1941–1964 [CNO]

	1941	1949	1953	1964
Uyghur	2,984,000	3,291,100	3,607,609	3,991,577
Kazakh	326,000	443,700	506,390	489,261
Hui	92,000	122,500	134,215	264,017
Kirghiz	65,000	66,100	70,900	69,600
Tajik	9,000	13,500	14,500	16,400
Uzbek	5,000	12,200	13,600	7,500
Tatar	6,900	5,900	6,900	2,200
Total Pop.	3,730,000	4,333,400	4,783,603	7,270,067
Muslims	3,487,900	3,955,000	4,354,114	4,840,455
Muslim%	93,51	91.27	91.02	66.58

Table 2.6.1d Evolution of Muslim ethnic populations in China per province: 1982–2010 [CNO]

		1982	1990	2000	2010
Xinjiang Uighur	Hui	567,689	682,912	839,837	983,015
	Uyghur	5,955,947	7,191,845	8,345,622	10,001,302
	Kazakh	903,337	1,106,271	1,245,023	1,418,278
	Dongxiang	40,346	56,690	55,841	61,613
	Kirghiz	112,366	141,840	158,775	180,472
	Salar	2,933	3,706	3,762	3,728
	Tajik	26,573	33,197	39,493	47,261
	Bonan	283	482	571	568
	Uzbek	12,188	14,715	12,096	10,114
	Tatar	4,078	4,921	4,501	3,242
	Total Pop.	13,081,538	15,156,883	18,459,511	21,815,815
	Muslims	7,625,740	9,236,579	10,705,521	12,709,593
	Muslim%	58.29	60.94	57.99	58.26
Nigxia Hui	Hui	1,235,182	1,525,336	1,862,474	2,173,820
	Uyghur	21	70	312	613
	Kazakh	4	11	69	190
	Dongxiang	129	2,598	2,168	1,261
	Kirghiz	—	1	5	12
	Salar	9	44	38	72
	Tajik	—	—	—	1
	Bonan	10	8	14	21
	Uzbek	—	—	2	3
	Tatar	—	—	2	2
	Total Pop.	3,895,576	4,655,445	5,486,393	6,301,350
	Muslims	1,235,355	1,528,068	1,865,084	2,175,995
	Muslim%	31.71	32.82	33.99	34.53
Qinghai	Hui	533,859	639,766	753,378	834,298
	Uyghur	136	122	431	209
	Kazakh	1,495	523	407	680
	Dongxiang	651	1,475	2,498	6,331
	Kirghiz	—	—	8	4
	Salar	60,981	76,818	87,043	107,089
	Tajik	—	1	15	3
	Bonan	377	602	635	904
	Uzbek	—	3	1	2
	Tatar	—	—	12	2
	Total Pop.	3,895,695	4,456,952	4,822,963	5,626,723
	Muslims	597,499	719,310	844,428	949,522
	Muslim%	15.34	16.14	17.51	16.88

		1982	1990	2000	2010
Gansu	Hui	957,170	1,095,668	1,184,930	1,258,641
	Uyghur	377	938	2,131	1,937
	Kazakh	2,366	3,146	2,963	4,444
	Dongxiang	237,879	311,902	451,622	546,255
	Kirghiz	4	13	13	48
	Salar	5,117	6,740	11,784	13,517
	Tajik	1	2	99	13
	Bonan	8,322	10,555	15,170	18,170
	Uzbek	—	7	18	15
	Tatar	—	1	186	18
	Total Pop.	19,569,191	22,371,085	25,124,282	25,575,263
	Muslims	1,211,236	1,428,972	1,668,916	1,843,058
	Muslim%	6.19	6.39	6.64	7.21
Yunnan	Hui	437,933	521,561	643,238	698,265
	Uyghur	13	38	1,161	1,282
	Kazakh	—	20	199	896
	Dongxiang	5	37	179	181
	Kirghiz	—	14	42	34
	Salar	6	12	55	265
	Tajik	—	1	17	7
	Bonan	—	—	7	49
	Uzbek	—	1	6	11
	Tatar	—	5	12	10
	Total Pop.	32,553,699	36,972,587	42,360,089	45,966,766
	Muslims	437,957	521,689	644,916	701,000
	Muslim%	1.35	1.41	1.52	1.53
Tianjin	Hui	143,585	161,229	172,357	177,734
	Uyghur	63	199	974	2,170
	Kazakh	1	2	34	476
	Dongxiang	—	2	51	121
	Kirghiz	1	3	13	41
	Salar	—	2	32	122
	Tajik	—	—	30	11
	Bonan	—	—	1	10
	Uzbek	—	—	7	19
	Tatar	—	—	12	4
	Total Pop.	7,764,137	8,785,427	9,848,731	12,938,693
	Muslims	143,650	161,437	173,511	180,708
	Muslim%	1.85	1.84	1.76	1.40

Table 2.6.1d (Continued)

		1982	1990	2000	2010
Beijing	Hui	185,228	207,050	235,837	249,223
	Uyghur	757	2,020	3,129	6,975
	Kazakh	217	346	400	1,602
	Dongxiang	27	21	101	484
	Kirghiz	43	27	45	189
	Salar	28	23	413	644
	Tajik	6	3	22	21
	Bonan	4	3	7	24
	Uzbek	7	12	23	51
	Tatar	14	23	29	23
	Total Pop.	9,230,663	10,819,414	13,569,194	19,612,368
	Muslims	186,331	209,528	240,006	259,236
	Muslim%	2.02	1.94	1.77	1.32
Henan	Hui	730,282	870,327	953,531	957,964
	Uyghur	737	1,833	4,623	3,035
	Kazakh	3	10	38	4,805
	Dongxiang	12	9	29	72
	Kirghiz	—	—	15	37
	Salar	1	16	92	279
	Tajik	—	—	73	23
	Bonan	—	3	2	4
	Uzbek	—	—	10	16
	Tatar	—	8	8	4
	Total Pop.	74,422,573	85,534,200	91,236,854	94,029,939
	Muslims	731,035	872,206	958,421	966,239
	Muslim%	0.98	1.02	1.05	1.03
Inner Mongolia	Hui	168,997	192,725	209,850	221,483
	Uyghur	95	166	1,259	658
	Kazakh	19	13	20	377
	Dongxiang	36	52	118	574
	Kirghiz	67	111	153	141
	Salar	2	1	26	53
	Tajik	—	—	29	6
	Bonan	4	5	7	90
	Uzbek	3	3	8	6
	Tatar	—	1	6	6
	Total Pop.	19,274,281	21,456,518	23,323,347	24,706,291
	Muslims	169,223	193,077	211,476	223,394
	Muslim%	0.88	0.90	0.91	0.90

		1982	1990	2000	2010
Hebei	Hui	420,380	493,899	542,639	570,170
	Uyghur	37	279	1,785	864
	Kazakh	1	4	29	320
	Dongxiang	—	1	37	78
	Kirghiz	—	3	15	25
	Salar	5	13	30	131
	Tajik	—	4	35	8
	Bonan	1	2	—	7
	Uzbek	—	1	2	7
	Tatar	—	17	6	17
	Total Pop.	53,005,507	61,082,755	66,684,419	71,854,210
	Muslims	420,424	494,223	544,578	571,627
	Muslim%	0.79	0.81	0.82	0.80
Tibet (Xizang)	Hui	1,772	2,956	9,031	12,630
	Uyghur	2	14	701	205
	Kazakh	5	2	8	2,143
	Dongxiang	—	54	111	757
	Kirghiz	—	—	—	2,678
	Salar	—	14	228	255
	Tajik	—	5	4	—
	Bonan	—	2	24	15
	Uzbek	—	—	1	4
	Tatar	—	2	—	—
	Total Pop.	1,863,623	2,196,029	2,616,329	3,002,165
	Muslims	1,779	3,049	10,108	18,687
	Muslim%	0.10	0.14	0.39	0.62
Loaoning	Hui	239,449	263,481	264,407	245,798
	Uyghur	96	390	2,407	1,917
	Kazakh	17	30	121	448
	Dongxiang	—	—	8	27
	Kirghiz	5	19	45	125
	Salar	—	—	45	88
	Tajik	—	—	46	10
	Bonan	—	—	—	2
	Uzbek	1	—	1	5
	Tatar	1	4	9	14
	Total Pop.	35,721,694	39,459,694	41,824,412	43,746,323
	Muslims	239,569	263,924	267,089	248,434
	Muslim%	0.67	0.67	0.64	0.57

Table 2.6.1d (Continued)

		1982	1990	2000	2010
Shandong	Hui	389,989	461,230	497,597	535,679
	Uyghur	44	238	2,386	4,635
	Kazakh	2	9	49	1,116
	Dongxiang	7	12	28	126
	Kirghiz	—	2	16	90
	Salar	1	5	21	609
	Tajik	—	—	81	32
	Bonan	—	3	3	7
	Uzbek	—	1	4	13
	Tatar	—	26	9	11
	Total Pop.	74,419,152	84,392,104	89,971,789	95,792,719
	Muslims	390,043	461,526	500,194	542,318
	Muslim%	0.52	0.55	0.56	0.57
Anhui	Hui	255,104	302,052	337,521	328,062
	Uyghur	42	130	1,733	710
	Kazakh	7	16	24	1,203
	Dongxiang	2	14	9	112
	Kirghiz	—	2	1	14
	Salar	2	8	26	78
	Tajik	—	—	98	3
	Bonan	—	1	1	15
	Uzbek	—	1	4	4
	Tatar	—	1	10	9
	Total Pop.	49,665,947	56,181,005	58,999,948	59,500,468
	Muslims	255,157	302,225	339,427	330,210
	Muslim%	0.51	0.54	0.58	0.55
Guizhou	Hui	98,452	127,118	168,734	184,788
	Uyghur	22	42	1,149	548
	Kazakh	—	—	226	2,093
	Dongxiang	291	580	412	958
	Kirghiz	—	—	17	2
	Salar	15	15	87	99
	Tajik	3	—	16	—
	Bonan	1	—	3	8
	Uzbek	—	—	1	—
	Tatar	—	—	2	3
	Total Pop.	28,552,942	32,391,051	35,247,695	34,748,556
	Muslims	98,784	127,755	170,647	188,499
	Muslim%	0.35	0.39	0.48	0.54

		1982	1990	2000	2010
Jilin	Hui	110,283	122,422	125,620	118,799
	Uyghur	24	264	1,500	1,127
	Kazakh	5	38	89	467
	Dongxiang	1	2	—	32
	Kirghiz	13	17	10	36
	Salar	—	7	9	50
	Tajik	—	—	3	2
	Bonan	5	4	8	6
	Uzbek	—	—	1	2
	Tatar	—	—	2	5
	Total Pop.	22,560,024	24,659,790	26,802,191	27,452,815
	Muslims	110,331	122,754	127,242	120,526
	Muslim%	0.49	0.50	0.47	0.44
Shaanxi	Hui	118,528	131,878	139,232	138,716
	Uyghur	84	583	1,187	1,570
	Kazakh	10	102	122	794
	Dongxiang	13	17	26	116
	Kirghiz	3	5	19	23
	Salar	9	21	22	156
	Tajik	1	2	7	3
	Bonan	—	1	1	4
	Uzbek	3	3	6	7
	Tatar	1	5	1	1
	Total Pop.	28,904,369	32,882,286	35,365,072	37,327,379
	Muslims	118,652	132,617	140,623	141,390
	Muslim%	0.41	0.40	0.40	0.38
Shanghai	Hui	44,329	50,392	57,514	78,163
	Uyghur	176	492	1,701	5,254
	Kazakh	13	62	62	639
	Dongxiang	—	1	20	362
	Kirghiz	2	3	5	58
	Salar	7	5	15	941
	Tajik	—	—	24	14
	Bonan	—	—	1	16
	Uzbek	—	4	5	14
	Tatar	—	6	5	12
	Total Pop.	11,859,700	13,341,852	16,407,734	23,019,196
	Muslims	44,527	50,965	59,352	85,473
	Muslim%	0.38	0.38	0.36	0.37

Table 2.6.1d (Continued)

		1982	1990	2000	2010
	Hui	31,188	92,464	109,880	115,978
	Uyghur	10	33	1,080	1,159
	Kazakh	—	2	6	1,492
	Dongxiang	1	8	15	339
	Kirghiz	—	—	7	50
	Salar	—	—	13	80
Fujian	Tajik	—	1	22	19
	Bonan	—	1	3	5
	Uzbek	—	3	8	9
	Tatar	—	—	—	17
	Total Pop.	25,872,917	30,048,275	34,097,947	36,894,217
	Muslims	31,199	92,512	111,034	119,148
	Muslim%	0.12	0.31	0.33	0.32
	Hui	127,068	139,357	124,003	101,749
	Uyghur	72	211	1,189	884
	Kazakh	9	10	48	165
	Dongxiang	9	5	64	19
	Kirghiz	874	1,436	1,473	1,431
	Salar	—	3	4	7
Heilongjiang	Tajik	—	1	73	9
	Bonan	5	5	7	3
	Uzbek	1	2	1	5
	Tatar	3	10	11	3
	Total Pop.	32,665,512	35,215,932	36,237,576	38,313,991
	Muslims	128,041	141,040	126,873	104,275
	Muslim%	0.39	0.40	0.35	0.27
	Hui	104,295	121,695	132,582	130,757
	Uyghur	54	361	2,213	4,367
	Kazakh	3	31	73	2,124
	Dongxiang	1	14	35	420
	Kirghiz	—	4	30	61
	Salar	3	10	51	353
Jiangsu	Tajik	2	1	39	17
	Bonan	—	—	3	32
	Uzbek	1	1	5	15
	Tatar	2	1	10	19
	Total Pop.	60,521,113	67,056,812	73,043,577	78,660,941
	Muslims	104,361	122,118	135,041	138,165
	Muslim%	0.17	0.18	0.18	0.18

		1982	1990	2000	2010
Shanxi	Hui	51,917	58,098	61,690	59,709
	Uyghur	19	66	1,084	670
	Kazakh	2	11	23	277
	Dongxiang	1	2	9	30
	Kirghiz	—	—	1	6
	Salar	1	1	8	43
	Tajik	—	—	94	1
	Bonan	—	—	2	—
	Uzbek	—	—	3	3
	Tatar	—	2	—	1
	Total Pop.	25,291,450	28,758,846	32,471,242	35,712,101
	Muslims	51,940	58,180	62,914	60,740
	Muslim%	0.21	0.20	0.19	0.17
Hunan	Hui	66,953	93,583	97,368	94,705
	Uyghur	4,450	5,794	7,939	6,716
	Kazakh	3	5	110	2,775
	Dongxiang	8	3	28	62
	Kirghiz	1	9	10	62
	Salar	3	7	272	79
	Tajik	5	—	105	3
	Bonan	—	2	6	13
	Uzbek	3	—	22	33
	Tatar	—	11	16	15
	Total Pop.	54,010,155	60,657,992	63,274,173	65,700,762
	Muslims	71,426	99,414	105,876	104,463
	Muslim%	0.13	0.16	0.17	0.16

Table 2.6.1d (*Continued*)

		1982	1990	2000	2010
Hainan	Hui		5,674	8,372	10,670
	Uyghur		10	354	393
	Kazakh		6	36	1,553
	Dongxiang		—	11	19
	Kirghiz		—	16	1
	Salar		2	49	21
	Tajik		2	38	—
	Bonan		—	—	—
	Uzbek		—	29	—
	Tatar		3	9	—
	Total Pop.		6,558,076	7,559,035	8,671,485
	Muslims		5,697	8,914	12,657
	Muslim%		0.09	0.12	0.15
Guangdong	Hui	10,891	8,886	25,307	45,073
	Uyghur	12	262	3,057	6,438
	Kazakh	1	7	46	4,602
	Dongxiang	1	16	61	403
	Kirghiz	—	6	15	577
	Salar	4	5	89	823
	Tajik	—	2	95	165
	Bonan	1	2	5	25
	Uzbek	4	2	42	73
	Tatar	—	6	10	55
	Total Pop.	59,299,620	62,829,741	85,225,007	104,320,459
	Muslims	10,914	9,194	28,727	58,234
	Muslim%	0.02	0.01	0.03	0.06

		1982	1990	2000	2010
Sichuan	Hui	91,092	108,285	109,960	104,544
	Uyghur	57	209	2,158	1,945
	Kazakh	25	64	94	1,882
	Dongxiang	14	30	60	191
	Kirghiz	5	11	14	74
	Salar	4	40	108	223
	Tajik	4	—	58	23
	Bonan	3	—	16	43
	Uzbek	—	3	3	17
	Tatar	—	3	7	15
	Total Pop.	99,713,246	107,218,310	82,348,296	80,417,528
	Muslims	91,204	108,645	112,478	108,957
	Muslim%	0.09	0.10	0.14	0.14
Chongqing	Hui			10,064	9,056
	Uyghur			1,194	1,162
	Kazakh			22	259
	Dongxiang			25	102
	Kirghiz			2	28
	Salar			36	71
	Tajik			177	2
	Bonan			3	9
	Uzbek			—	4
	Tatar			6	1
	Total Pop.			30,512,763	28,846,170
	Muslims			11,529	10,694
	Muslim%			0.04	0.04

Table 2.6.1d (Continued)

		1982	1990	2000	2010
Hubei	Hui	70,037	77,934	77,759	67,185
	Uyghur	68	277	1,457	2,577
	Kazakh	—	11	40	2,191
	Dongxiang	87	105	78	67
	Kirghiz	—	2	4	36
	Salar	1	3	58	135
	Tajik	—	—	36	9
	Bonan	—	—	3	6
	Uzbek	—	1	—	9
	Tatar	—	5	2	10
	Total Pop.	47,808,118	53,970,501	59,508,870	57,237,727
	Muslims	70,193	78,338	79,437	72,225
	Muslim%	0.15	0.15	0.13	0.13
Zhejiang	Hui	9,418	17,294	19,609	38,192
	Uyghur	11	65	785	5,377
	Kazakh	—	4	19	695
	Dongxiang	—	14	55	322
	Kirghiz	2	8	37	279
	Salar	—	23	45	424
	Tajik	—	—	7	3,368
	Bonan	—	1	—	17
	Uzbek	—	—	8	86
	Tatar	—	2	3	9
	Total Pop.	38,884,593	41,446,015	45,930,651	54,426,891
	Muslims	9,431	17,411	20,568	48,769
	Muslim%	0.02	0.04	0.04	0.09

		1982	1990	2000	2010
Guangxi Zhuang	Hui	19,374	27,174	32,512	32,319
	Uyghur	58	53	1,550	1,795
	Kazakh	1	—	49	1,884
	Dongxiang	2	4	98	17
	Kirghiz	—	1	9	28
	Salar	3	2	30	128
	Tajik	5	1	135	2
	Bonan	1	1	2	1
	Uzbek	2	—	47	13
	Tatar	23	1	4	24
	Total Pop.	36,421,421	42,244,884	43,854,538	46,023,761
	Muslims	19,469	27,237	34,436	36,211
	Muslim%	0.05	0.06	0.08	0.08
Jiangxi	Hui	7,954	9,555	9,972	8,902
	Uyghur	7	20	1,142	852
	Kazakh	—	2	9	1,718
	Dongxiang	—	1	8	49
	Kirghiz	—	—	8	46
	Salar	—	—	12	44
	Tajik	—	—	57	33
	Bonan	—	—	—	—
	Uzbek		—	6	9
	Tatar	—	—	—	4
	Total Pop.	33,185,471	37,710,177	40,397,598	44,567,797
	Muslims	7,961	9,578	11,214	11,657
	Muslim%	0.02	0.03	0.03	0.03

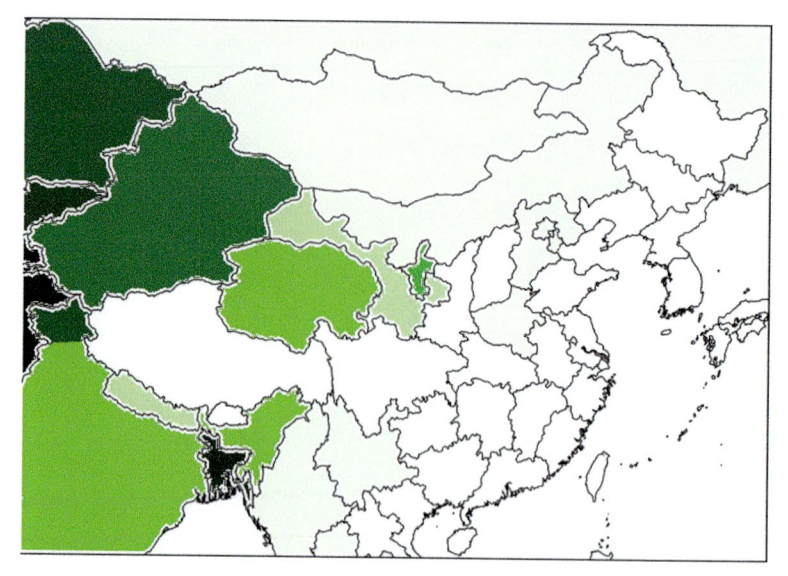

Figure 2.6.1b A choropleth map of China illustrating the presence of Muslims per province

A quarter of the Muslim population lives in the northeastern provinces of Ningxia, Gansu and Qinghai, where their share of the population is 35%, 17% and 7%, respectively. These percentages did not change much since the 1953 census. Table 2.6.1d shows the distribution of Muslim ethnic minority groups in each Chinese province according to census data since 1982. Hainan and Chongqing provinces were split from Guangdong and Sichuan provinces, respectively in 1988 and 1997. A choropleth map of China illustrating the presence of Muslims per province is presented in Fig. 2.6.1b, where white is less than 1% of the provincial population and the darker the region, the more is the percentage of Muslims.

2.6.2 Hong Kong

It is a Special Administrative Region of the People's Republic of China since 1997, when it was returned by the UK after occupying it since 1841. It has an area of 1,104 sq. km, comprising a peninsula from mainland China and over 260 islands. The main island was captured by the British in 1841, while the peninsula was ceded to the British in 1861. A map of Hong Kong is presented in Fig. 2.6.2. Islam arrived at Hong Kong in the mid nineteenth century with the arrival of the British East India Company which brought along the first group of Muslims into Hong Kong. With the outbreak of the Opium war, the Hui population of the Guangdong province headed for Hong Kong, thus the Muslim society was further developed. The 1871 census recorded the presence of 151 Muslims among the non-Chinese population, who were mainly Indians. The 1911 Census recorded the presence of 1,779 Muslims or 0.4% of the total population.

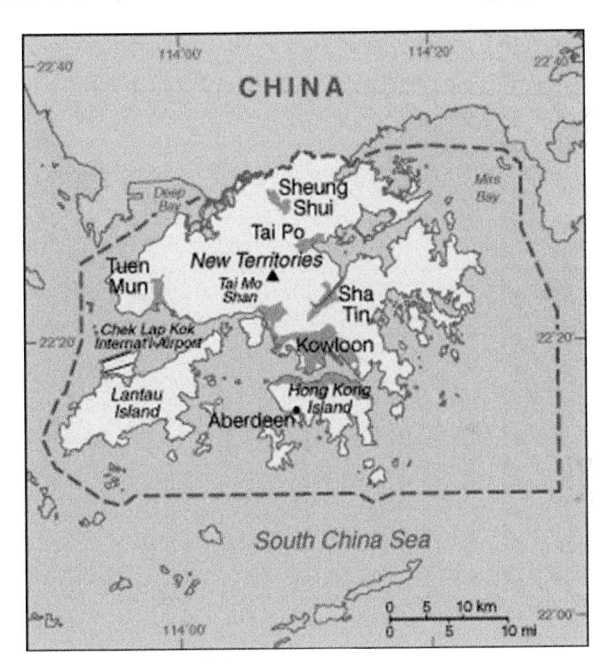

Figure 2.6.2 Map of Hong Kong SAR.

After WWII, the British brought workers from India and Indonesia, some of whom were Muslims. Towards the end of the twentieth century, the number of Muslims increased due to immigration from Muslim countries. According to East Asia Barometer Survey (EAB), the percentage of Muslims was 0.12% in 2001 while the World Values Survey (WVS) put the Muslim percentage at 0.08% in 2008 and 0.10% in 2013. Thus, assuming that the percentage of Muslims will increase by 0.02 of a percentage point per decade, the Muslim population is expected to reach 14,000 or 0.2% by 2050 and 21,000 or 0.3% by 2100. The data is summarized in Table 2.6.3.

Table 2.6.3 Evolution of the Muslim population in Hong Kong

Year	Population	Muslims	%	Source
1871	120,124	154	0.13	[GB871]c
1911	456,739	1,779	0.39	[GB911]c
2001	6,837,000	8,000	0.12	[EAB]s
2008	6,910,000	5,500	0.08	[WVS]s
2013	7,164,000	7,000	0.10	[WVS]s
2050	8,041,000	14,000	0.18	es
2100	7,647,000	21,000	0.28	es

2.6.3 Japan

It has a total area of 377,915 sq. km consisting of 6,852 islands. By far, the largest island is Honshu with area 227,963 sq. km, four-fifth of the population, and where the capital Tokyo is located. The largest of the remaining islands are Hokkaido (83,454 sq. km) with 5% of the population, Kyushu (35,640 sq. km) with 10% of the population, Shikoku (18,800 sq. km) with 3% of the population and Okinawa (1,201 sq. km) with 1% of the population. A map of Japan is presented in Fig. 2.6.3.

Islam entered this archipelago at the turn of the twentieth century when Japanese embraced Islam abroad and returned home. This includes Omar Yamaoka who founded the Islamic fraternity in Tokyo in 1911. The number increased when Muslim Tatar from Central Asia came as refugees to Japan after the 1917 Bolshevik Revolution. The 1940 Census recorded 499 foreign nationals from Muslim-majority countries: 239 Turkey, 56 Indonesia, 52 Iran, 46 Syria, 36 Afghanistan, thirty Egypt, sixteen Albania, twelve Iraq, eleven Yemen and one from Morocco. The Muslim population remained less than 0.01% of the total population until the end of 1980s. After WWII it consisted primarily of Indonesian nationals, whose number increased from 191 in 1947, to 257 in 1950, to 284 in 1955. Noticeable increase in the Muslim population started in the 1980s due to Japan's economic boom. The largest of Muslim nationalities are Indonesia, Bangladesh, Pakistan, Malaysia and Egypt. Evolution of members of these nationalities since 1960 is summarized in Table 2.6.3a.

Figure 2.6.3 Map of Japan.

Accordingly, the Muslim population increased from 600 in 1960, to 1,500 in 1965, to 1,700 in 1970, to 2,200 in 1975, to 3,100 in 1980, to 5,400 in 1985, but remained at 0.00% of the total population. The Muslim population then increased to 13,000 or 0.01% in 1990, to 23,000 or 0.02% in 1995, to 44,000 or 0.03% in 2000, to 54,000 or 0.04% in 2005, to 55,000 or 0.04% in 2010 and 70,000 or 0.06% in 2015. Thus, assuming that the Muslim population will continue to increase by 0.01 of a percentage point per decade, the Muslim population is expected to remain around 0.1% of the total population, remaining around 0.1 million throughout the second half of this century. The data is summarized in Table 2.6.3b.

Table 2.6.3a Evolution of the largest six Muslim foreign nationalities in Japan since 1960

	1960	1965	1970	1975	1980	1985
Indonesia	420	1,206	1,036	1,119	1,448	1,704
Pakistan	89	159	243	248	437	1,032
Bangladesh	0	0	0	108	260	684
Malaysia	71	320	451	718	744	1,761
Egypt	—	—	—	—	206	268
Total	580	1,505	1,730	2,193	3,095	5,449
	1990	**1995**	**2000**	**2005**	**2010**	**2015**
Indonesia	3,623	6,956	19,346	25,097	24,895	35,910
Pakistan	2,067	4,753	7,498	8,789	10,299	12,708
Bangladesh	2,109	4,935	7,176	11,015	10,175	10,835
Malaysia	4,683	5,354	8,386	7,910	8,364	8,738
Egypt	368	636	1,103	1,366	1,593	1,747
Total	12,850	22,634	43,509	54,177	55,326	69,938

Table 2.6.3b Evolution of the Muslim population in Japan

Year	Population	Muslims	%	Source
1900	43,847,000	0	0.00	[KET86]es
1920	55,963,053	100	0.00	[MAS23]es
1940	73,114,308	499	0.00	[JP40]e
1947	78,101,473	191	0.00	[JPH]e
1950	84,114,574	257	0.00	[JPH]e
1955	90,076,594	284	0.00	[JPH]e
1960	94,301,623	580	0.00	[JPH]e
1965	99,209,137	1,505	0.00	[JPH]e
1970	104,665,171	1,730	0.00	[JPH]e
1975	111,939,643	2,193	0.00	[JPH]e
1980	117,060,396	3,095	0.00	[JPH]e
1985	121,048,923	5,449	0.00	[JPH]e
1990	123,611,167	12,850	0.01	[JP10]e
1995	125,570,246	22,634	0.02	[JP10]e
2000	126,925,843	43,509	0.03	[JP10]e
2005	127,767,994	54,177	0.04	[JP10]e
2010	128,057,352	55,326	0.04	[JP10]e
2015	127,110,000	69,938	0.06	[JP15]e
2050	105,804,000	95,000	0.09	es
2100	74,959,000	105,000	0.14	es

2.6.4 North Korea

The Democratic People's Republic of Korea has a total area of 120,538 sq. km and its map is presented in Fig. 2.6.4. The Korean Peninsula was occupied by the Japanese in 1905. They were kicked out in 1945 after their WWII defeat. The Peninsula was divided with the Russians and Chinese supporting the North and Americans supporting the South. The two countries were declared in 1948. An unsuccessful war to unify the peninsula erupted between 1950 and 1953. The Muslim population increased from none in 1970 and before, to 500 in 1982, to 2,000 or 0.01% in 2008. Assuming that the percentage of Muslims will remain constant, the Muslim population is expected to remain less than 3,000 for the rest of this century. A summary of the data is provided in Table 2.6.4.

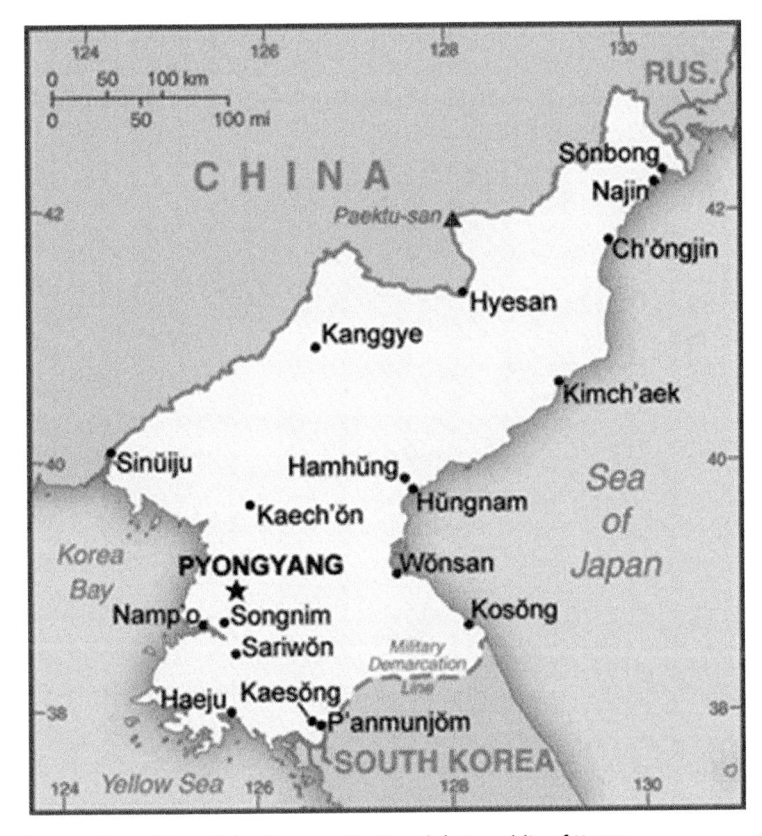

Figure 2.6.4 Map of the Democratic People's Republic of Korea.

Table 2.6.4 Evolution of the Muslim population in North Korea

Year	Population	Muslims	%	Source
1935	7,000,000	—	—	[MAS55]es
1970	13,900,000	—	—	[KET73]es
1982	19,000,000	500	0.00	[KET86]es
2008	23,349,859	2,000	0.01	[PEW]es
2050	26,562,000	3,000	0.01	es
2100	22,793,000	2,000	0.01	es

2.6.5 South Korea

The Republic of Korea has an area of 99,720 sq. km and its map is presented in Fig. 2.6.5. There were ninety Muslim refugees in Seoul in 1935. However, the seeds of Islam here started with Turkish troops who were stationed here following the Korean War (1950–54) as part of the United Nations peace keeping force. Thus, estimates for the Muslim population increased from 200 in 1955, to 3,000 or 0.01% in 1970, to 15,000 or 0.04% in 1980, to 35,000 or 0.1% in 1990.

Figure 2.6.5 Map of the Republic of Korea.

According to the World Values Survey (WVS), the percentage of Muslims was 0.21% in 1982, 0.08% in 1996, 0.17% in 2001, 0.08% in 2005 and 0.08% in 2010. Thus, assuming that the percentage of Muslims will continue to increase by 0.01 of a percentage point per decade, the Muslim population is expected to remain over 50,000 throughout the second half of this century. The data is summarized in Table 2.6.5.

Table 2.6.5 Evolution of the Muslim population in South Korea

Year	Population	Muslims	%	Source
1930	13,900,000	—	0.00	[MAS29]es
1935	17,000,000	90	0.00	[MAS55]es
1955	21,502,386	200	0.00	[KRH]es
1970	31,435,252	3,000	0.01	[KRH]es
1980	37,406,815	15,000	0.04	[KRH]es
1990	43,390,374	35,000	0.08	[KRH]es
1996	44,967,000	36,000	0.08	[WVS]s
2005	47,606,000	38,000	0.08	[WVS]s
2010	49,090,000	40,000	0.08	[WVS]s
2050	46,830,000	56,000	0.12	es
2100	29,542,000	50,000	0.17	es

2.6.6   Macao

It is a Special Administrative Region of the People's Republic of China since 1999, when it was returned by Portugal after occupying it since 1557. Originally it consisted of Macao Peninsula dropping from mainland China and two islands below it named Taipa and Coloane. However, Macao is constantly growing through land claimed from sea, so the islands of Taipa and Coloane are now connected by COTAI reclamation area and other man-made islands to the east have been developed. Thus, Macau has increased its sq. km size from ten in 1930, to fifteen in 1980 to thirty in 2017. The map of Macao in 1980 and 2017 are presented in Figs. 2.6.6a and 2.6.6b.

Muslims arrived here through trade before the Ming Dynasty in the fourteenth century, but there is no record on their numbers. The first and only mosque was erected in 1880 and the Islamic Association of Macao was founded in 1935, which is in close tie with Muslims in nearby Hong Kong. The number of Muslims remained low throughout the twentieth century and was recorded as 117 or 0.05% in 1970 census, which was the only one to collect information on Muslims. Based on nationality, mainly Malaysia and Indonesia, the number increased to around 500 or 0.1% by the end of the twentieth century. The number then increased with the immigration of more guest workers from Malaysia and Indonesia, which increased to 2,000 or 0.4% in 2005, then 5,000 or 0.9% in 2010, and 5,500 or 0.9% in 2015.

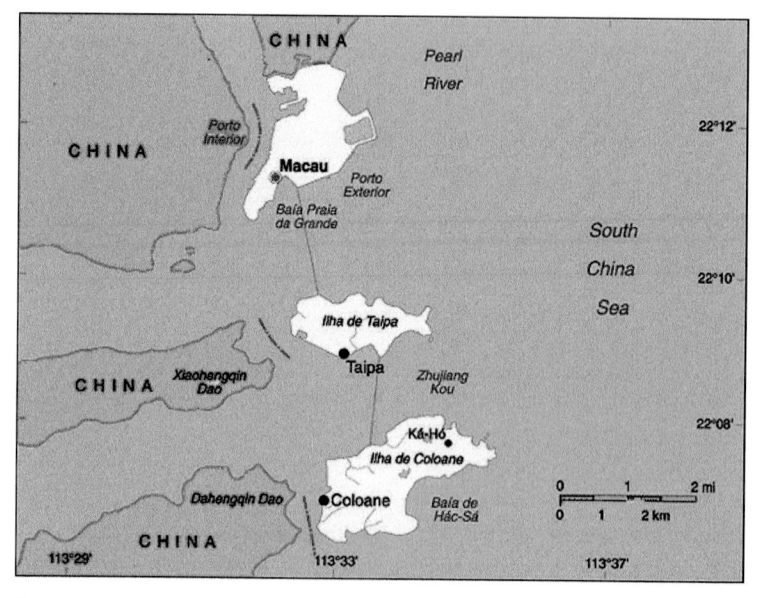

Figure 2.6.6a   Map of Macao SAR in 1980.

Thus, assuming that the percentage of Muslims increases by 0.05 of a percentage point per decade, the Muslim population is expected to remain around 10,000 throughout the second half of this century. The data is summarized in Table 2.6.6b.

Table 2.6.6 Evolution of the Muslim population in Macao

Year	Population	Muslims	%	Source
1970	248,636	117	0.05	[MO70]c
1985	408,500	354	0.09	[MO85]e
1990	339,510	293	0.09	[MO90]e
1995	424,430	548	0.13	[MO95]e
2000	437,900	477	0.11	[MO00]e
2005	484,300	2,155	0.44	[MO15]e
2010	540,600	4,961	0.92	[MO15]e
2015	646,800	5,503	0.85	[MO15]e
2050	838,000	8,000	1.00	es
2100	1,012,000	13,000	1.25	es

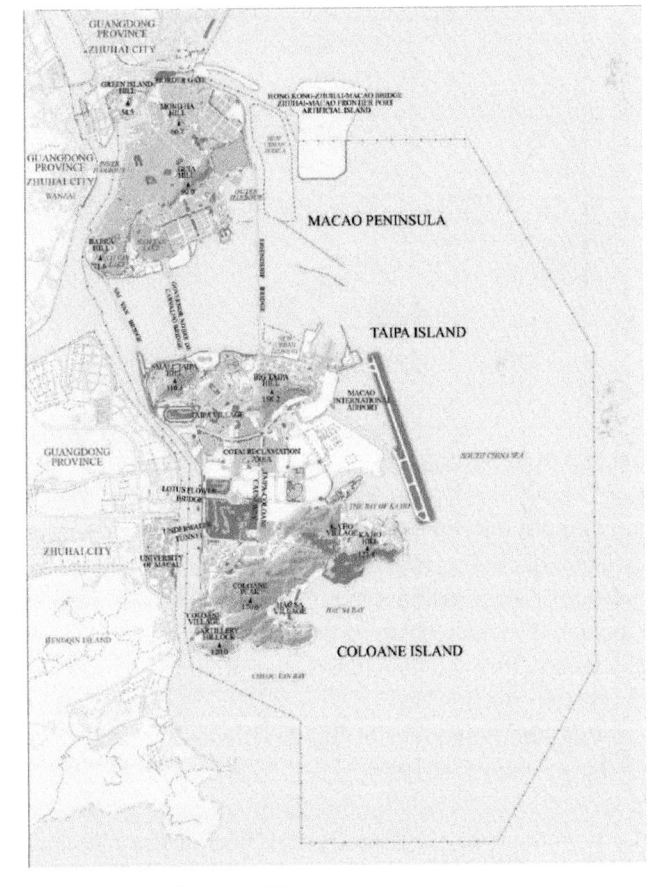

Figure 2.6.6b Map of Macao SAR.

2.6.7 Taiwan

Formally the Republic of China, it is an island nation with total area of 36,191 sq. km, consisting of the main island of Taiwan (35,980 sq. km) which also used to be called Formosa and hundreds of tiny largely uninhabited islands. A map of Taiwan is presented in Fig. 2.6.7. It was taken by Japan from Mainland China in 1895 and returned in 1945. After Communist takeover of Mainland China in 1949, Taiwan de facto separated from the Mainland.

Figure 2.6.7 Map of the Republic of China.

Islam entered here in 1661 when the army of the Chinese Ming Dynasty landed on the island to free it from Dutch occupation. Among those Chinese soldiers many were Muslims from Quanzhou in the Fukien Province. Over the years, the Muslim community gradually shrank through intermarriage, adoption of other customs, and persecution by the Japanese colonial rule. By the end of the Japanese rule, the remainder of the decedents of this first Muslim wave no longer embraced Islam.

The second wave of Muslim migration arrived in 1949 with the central government to the Island after fleeing the mainland due to the communist revolution there. This wave consisted of 20,000

Chinese Muslims who were mostly soldiers, civil servants or food service workers. According to Asian Barometer Survey data, the percentage of Muslims was 0.07% in 2001, 0.3% in 2006, 2010 and 2014. Thus, assuming that the percentage of Muslims will increase by 0.05 of a percentage point per decade, the Muslim population is expected to remain over 0.1 million in the second half of this century, reaching 0.5% by 2050 and 0.7% by the end of this century. The data is summarized in Table 2.6.7.

Table 2.6.7 Evolution of the Muslim population in Taiwan

Year	Population	Muslims	%	Source
1944	8,554,857	—	0.00	[TW01]es
1949	7,521,495	20,000	0.27	[TW01]es
2001	22,405,568	16,000	0.07	[ANB]s
2006	22,876,527	73,000	0.32	[ANB]s
2010	23,162,123	73,000	0.32	[ANB]s
2014	23,492,074	71,000	0.30	[ANB]s
2050	22,413,000	112,000	0.50	es
2100	16,259,000	122,000	0.75	es

2.6.8   Regional Summary and Conclusion

The Far East has the least concentration of Muslims among the six regions spanning Asia. Although Islam entered this region in the eighth century, most of this region was never under Muslim control. The Muslim population has been increasing slowly. It is currently almost 2% of the total population but is expected to remain less than 5% for the next three centuries. The following tables present centennial data from 600 AD to 2100 AD (or approximately 1 H to 1500 H) in Table 2.6a and decennial data from 1790 AD to 2100 AD (or 1210 H to 1520 H) in Tables 2.6b and 2.6c for current countries in the Far East. The data includes total population in thousands (P), the percentage of which is Muslim ($M\%$), the corresponding Muslim population in thousands (M), and the annual population growth rate (APGR, or $G\%$) of the total population in this region.

Table 2.6a Centennial estimates of the Muslim population (×1000) in the Far East: 600–2100 AD (1–1500 H)

		600	700	800	900	1000	1100	1200	1300
China	P	59,400	59,300	59,200	59,100	59,000	68,000	77,000	86,000
	M%	—	0.01	0.50	1.00	1.50	1.60	1.70	1.80
	M	—	6	296	591	885	1,088	1,309	1,548
Hong Kong	P	3	3	3	3	3	4	4	4
	M%	—	—	—	—	—	—	—	—
	M	—	—	—	—	—	—	—	—
Japan	P	5,700	6,150	6,600	7,050	7,500	9,000	10,500	12,000
	M%	—	—	—	—	—	—	—	—
	M	—	—	—	—	—	—	—	—
S. Korea	P	1,140	1,210	1,280	1,350	1,420	1,600	1,850	2,100
	M%	—	—	—	—	—	—	—	—
	M	—	—	—	—	—	—	—	—
N. Korea	P	3,900	4,300	4,700	5,100	5,470	5,500	5,550	5,600
	M%	—	—	—	—	—	—	—	—
	M	—	—	—	—	—	—	—	—
Macao	P	3	3	3	3	3	4	4	4
	M%	—	—	—	—	—	—	—	—
	M	—	—	—	—	—	—	—	—
Taiwan	P	90	95	100	105	112	120	140	160
	M%	—	—	—	—	—	—	—	—
	M	—	—	—	—	—	—	—	—
Total	P	70,236	71,061	71,886	72,711	73,508	84,228	95,048	105,868
	M%	—	0.01	0.41	0.81	1.20	1.29	1.38	1.46
	M	—	6	296	591	885	1,088	1,309	1,548
	G%		0.012	0.012	0.011	0.011	0.136	0.121	0.108

Muslims in the Far East | 217

Table 2.6a (Continued)

		1400	1500	1600	1700	1800	1900	2000	2100
China	P	95,000	103,000	160,000	138,000	340,000	400,000	1,290,551	1,064,993
	M%	1.90	2.00	2.10	2.20	2.34	2.36	1.64	2.60
	M	1,805	2,060	3,360	3,036	7,956	9,440	21,165	27,690
Hong Kong	P	5	5	8	7	16	306	6,606	7,647
	M%	—	—	—	—	0.01	0.39	0.12	0.28
	M	—	—	—	—	—	1	8	21
Japan	P	13,500	15,400	18,500	27,000	30,400	43,847	127,524	74,959
	M%	—	—	—	—	—	—	0.03	0.14
	M	—	—	—	—	—	—	38	105
S. Korea	P	2,350	2,530	3,124	3,858	4,250	4,777	22,929	22,793
	M%	—	—	—	—	—	—	0.01	0.01
	M	—	—	—	—	—	—	2	2
N. Korea	P	5,550	5,470	6,755	8,342	9,200	9,896	47,379	29,542
	M%	—	—	—	—	—	—	0.08	0.17
	M	—	—	—	—	—	—	38	50
Macao	P	5	5	8	7	16	64	428	1,012
	M%	—	—	—	—	0.05	0.05	0.11	1.25
	M	—	—	—	—	0	0	—	13
Taiwan	P	180	200	447	1,000	1,850	2,794	21,967	16,259
	M%	—	—	—	0.01	0.01	0.01	0.07	0.75
	M	—	—	—	—	—	—	15	122
Total	P	116,590	126,610	188,842	178,214	385,732	461,684	1,517,384	1,217,205
	M%	1.55	1.63	1.78	1.70	2.06	2.05	1.40	2.30
	M	1,805	2,060	3,360	3,036	7,956	9,442	21,267	28,003
	G%	0.096	0.082	0.400	-0.058	0.772	0.180	1.190	-0.220

Table 2.6b Decennial estimates of the Muslim population (×1000) in the Far East: 1790–1940 AD (1210–1360 H)

		1790	1800	1810	1820	1830	1840	1850	1860
China	P	320,000	340,000	360,000	381,000	409,000	412,000	412,000	377,000
	M%	2.34	2.34	2.34	2.34	2.34	2.34	2.34	2.34
	M	7,488	7,956	8,424	8,915	9,571	9,641	9,641	8,822
Hong Kong	P	14	16	18	20	24	28	33	63
	M%	0.01	0.01	0.01	0.01	0.01	0.01	0.10	0.10
	M	—	—	—	—	—	—	—	—
Japan	P	30,100	30,400	30,700	31,000	31,330	31,663	32,000	33,196
	M%	—	—	—	—	—	—	—	—
	M	—	—	—	—	—	—	—	—
S. Korea	P	4,200	4,250	4,300	4,345	4,370	4,400	4,440	4,480
	M%	—	—	—	—	—	—	—	—
	M	—	—	—	—	—	—	—	—
N. Korea	P	9,100	9,200	9,300	9,395	9,450	9,500	9,545	9,650
	M%	—	—	—	—	—	—	—	—
	M	—	—	—	—	—	—	—	—
Macao	P	14	16	18	20	22	25	30	51
	M%	0.04	0.04	0.04	0.04	0.04	0.04	0.04	0.04
	M	—	—	—	—	—	—	—	—
Taiwan	P	1,775	1,850	1,925	2,000	2,075	2,150	2,200	2,250
	M%	0.01	0.01	0.01	0.01	0.01	0.01	0.01	0.01
	M	—	—	—	—	—	—	—	—
Total	P	365,203	385,732	406,261	427,780	456,271	459,766	460,248	426,690
	M%	2.05	2.06	2.07	2.08	2.10	2.10	2.09	2.07
	M	7,488	7,956	8,424	8,916	9,571	9,641	9,641	8,822
	G%		0.547	0.519	0.516	0.645	0.076	0.010	−0.757

Table 2.6b (Continued)

		1870	1880	1890	1900	1910	1920	1930	1940
China	P	358,000	368,000	380,000	400,000	423,000	472,000	489,000	518,770
	M%	2.34	2.36	2.36	2.36	2.36	2.36	2.36	2.36
	M	8,377	8,685	8,968	9,440	9,983	11,139	11,540	12,243
Hong Kong	P	123	162	214	306	457	648	785	1,786
	M%	0.13	0.13	0.20	0.39	0.39	0.39	0.30	0.20
	M	—	—	—	1	2	3	2	4
Japan	P	34,437	36,649	39,902	43,847	49,589	55,818	64,203	72,967
	M%	—	—	—	—	0.00	0.00	0.00	0.00
	M	—	—	—	—	—	—	—	1
S. Korea	P	4,511	4,598	4,097	4,777	4,897	5,577	6,592	7,791
	M%	—	—	—	—	—	—	—	—
	M	—	—	—	—	—	—	—	—
N. Korea	P	9,753	9,800	9,848	9,896	10,096	11,804	13,900	15,627
	M%	—	—	—	—	—	—	—	0.00
	M	—	—	—	—	—	—	—	—
Macao	P	72	67	75	64	75	84	157	375
	M%	0.04	0.04	0.04	0.04	0.04	0.04	0.04	0.04
	M	—	—	—	—	—	—	—	0.150
Taiwan	P	2,345	2,421	2,500	2,794	3,276	3,721	4,563	5,872
	M%	0.01	0.01	0.01	0.01	0.01	0.01	0.01	0.01
	M	—	—	—	—	—	—	—	1
Total	P	409,241	421,697	436,636	461,684	491,389	549,652	579,200	623,188
	M%	2.05	2.06	2.05	2.05	2.03	2.03	1.99	1.97
	M	8,378	8,685	8,969	9,441	9,985	11,142	11,543	12,248
	G%	−0.418	0.300	0.348	0.558	0.624	1.120	0.524	0.732

Table 2.6c Decennial estimates of the Muslim population (×1000) in the Far East: 1950–2100 AD (1370–1520 H)

		1950	1960	1970	1980	1990	2000	2010	2020
China	P	554,419	660,408	827,601	1,000,089	1,176,884	1,290,551	1,368,811	1,439,324
	M%	1.37	1.33	1.33	1.45	1.55	1.64	1.74	1.80
	M	7,596	8,783	11,007	14,501	18,242	21,165	23,817	25,908
Hong Kong	P	1,974	3,046	3,849	4,869	5,728	6,606	6,966	7,497
	M%	0.12	0.12	0.12	0.12	0.12	0.12	0.10	0.12
	M	2	4	5	6	7	8	7	9
Japan	P	82,802	93,674	104,929	117,817	124,505	127,524	128,542	126,476
	M%	0.00	0.00	0.00	0.00	0.01	0.03	0.04	0.06
	M	—	1	2	4	15	38	51	76
S. Korea	P	10,549	11,424	14,410	17,472	20,293	22,929	24,549	25,779
	M%	—	—	—	0.00	0.01	0.01	0.01	0.01
	M	—	—	—	0.524	2	2	2	3
N. Korea	P	19,211	25,330	32,196	38,046	42,918	47,379	49,546	51,269
	M%	0.00	0.00	0.01	0.04	0.08	0.08	0.08	0.09
	M	—	—	3	15	34	38	40	46
Macao	P	196	168	246	238	344	428	538	649
	M%	0.05	0.05	0.05	0.05	0.09	0.11	0.92	0.85
	M	—	—	—	—	—	—	5	6
Taiwan	P	7,602	10,876	14,924	17,905	20,479	21,967	23,188	23,817
	M%	0.27	0.20	0.10	0.07	0.07	0.07	0.32	0.35
	M	21	22	15	13	14	15	74	83
Total	P	676,755	804,925	998,156	1,196,436	1,391,151	1,517,384	1,602,140	1,674,811
	M%	1.13	1.09	1.11	1.22	1.32	1.40	1.50	1.56
	M	7,619	8,810	11,032	14,539	18,315	21,267	23,997	26,130
	G%	0.825	1.734	2.152	1.812	1.508	0.869	0.544	0.444

Muslims in the Far East | 221

Table 2.6c (Continued)

		2030	2040	2050	2060	2070	2080	2090	2100
China	P	1,464,340	1,449,031	1,402,405	1,333,031	1,258,054	1,185,891	1,120,467	1,064,993
	M%	1.90	2.00	2.10	2.20	2.30	2.40	2.50	2.60
	M	27,822	28,981	29,451	29,327	28,935	28,461	28,012	27,690
Hong Kong	P	8,019	8,140	8,041	7,912	7,786	7,623	7,552	7,647
	M%	0.14	0.16	0.18	0.20	0.22	0.24	0.26	0.28
	M	11	13	14	16	17	18	20	21
Japan	P	120,758	113,356	105,804	98,326	90,472	83,925	79,047	74,959
	M%	0.07	0.08	0.09	0.10	0.11	0.12	0.13	0.14
	M	85	91	95	98	100	101	103	105
S. Korea	P	26,651	26,858	26,562	26,015	25,297	24,506	23,684	22,793
	M%	0.01	0.01	0.01	0.01	0.01	0.01	0.01	0.01
	M	3	3	3	3	3	2	2	2
N. Korea	P	51,152	49,784	46,830	42,702	38,546	34,934	31,918	29,542
	M%	0.10	0.11	0.12	0.13	0.14	0.15	0.16	0.17
	M	51	55	56	56	54	52	51	50
Macao	P	732	791	838	875	904	932	963	1,012
	M%	0.90	0.95	1.00	1.05	1.10	1.15	1.20	1.25
	M	7	8	8	9	10	11	12	13
Taiwan	P	24,011	23,593	22,413	20,967	19,513	18,082	16,962	16,259
	M%	0.40	0.45	0.50	0.55	0.60	0.65	0.70	0.75
	M	96	106	112	115	117	118	119	122
Total	P	1,695,663	1,671,554	1,612,893	1,529,828	1,440,572	1,355,894	1,280,593	1,217,205
	M%	1.66	1.75	1.84	1.94	2.03	2.12	2.21	2.30
	M	28,075	29,255	29,740	29,623	29,235	28,763	28,318	28,003
	G%	0.124	−0.143	−0.357	−0.529	−0.601	−0.606	−0.571	−0.508

Table 2.7a Centennial estimates of the Muslim population (×1000) in Asia: 600–2100 AD (1–1500 H)

		600	700	800	900	1000	1100	1200	1300
Near East	P	12,175	12,146	12,117	12,088	12,059	11,940	11,790	11,655
	M%	—	10.32	24.52	38.27	48.23	57.78	66.84	72.48
	M	—	1,253	2,972	4,627	5,816	6,900	7,880	8,448
Central Asia	P	2,820	2,962	3,104	3,246	3,425	3,780	4,155	4,535
	M%	—	9.60	33.42	48.73	59.85	62.19	63.21	64.23
	M	—	284	1,037	1,582	2,050	2,351	2,626	2,913
Arabian Asia	P	8,667	8,790	8,911	9,031	9,153	8,820	8,452	8,087
	M%	—	62.22	73.66	81.66	88.20	90.27	92.57	93.12
	M	—	5,470	6,564	7,375	8,073	7,962	7,824	7,531
Southeast Asia	P	9,583	10,341	11,099	11,857	12,550	14,147	15,790	17,436
	M%	—	0.45	0.92	2.24	4.43	9.35	14.81	30.35
	M	—	46	103	265	556	1,323	2,339	5,292
Indian Subcontinent	P	72,514	73,591	74,668	75,745	76,848	84,114	91,366	98,618
	M%	—	0.73	2.35	3.40	4.85	6.27	7.78	9.42
	M	—	540	1,756	2,576	3,731	5,274	7,107	9,288
Far East	P	70,236	71,061	71,886	72,711	73,508	84,228	95,048	105,868
	M%	—	0.01	0.41	0.81	1.20	1.29	1.38	1.46
	M	—	6	296	591	885	1,088	1,309	1,548
Total	P	175,995	178,891	181,785	184,678	187,543	207,029	226,601	246,199
	M%	—	4.25	7.00	9.21	11.26	12.03	12.84	14.22
	M	—	7,599	12,727	17,015	21,111	24,897	29,085	35,020
	G%	—	0.016	0.016	0.016	0.015	0.099	0.090	0.083

Table 2.7a (Continued)

		1400	1500	1600	1700	1800	1900	2000	2100
Near East	P	11,520	11,387	14,233	14,929	17,675	29,589	145,361	199,812
	M%	78.09	83.69	83.63	83.26	82.11	78.39	94.26	97.34
	M	8,997	9,529	11,904	12,430	14,512	23,195	137,015	194,491
Central Asia	P	4,930	5,283	6,077	6,837	9,115	16,865	78,523	195,248
	M%	69.35	80.94	83.44	85.18	85.80	90.55	86.97	94.78
	M	3,419	4,276	5,071	5,824	7,820	15,271	68,288	185,059
Arabian Asia	P	7,711	7,366	7,568	7,271	8,272	11,969	104,821	317,858
	M%	93.38	93.66	93.68	92.75	90.83	91.52	89.30	91.58
	M	7,201	6,899	7,090	6,744	7,513	10,954	93,602	291,092
Southeast Asia	P	19,132	20,802	23,541	27,097	35,006	93,261	525,008	744,215
	M%	36.49	42.61	46.35	42.88	43.60	41.82	40.10	44.94
	M	6,982	8,865	10,912	11,619	15,263	39,004	210,527	334,454
Indian Subcontinent	P	105,870	113,094	138,509	168,990	204,638	295,253	1,370,166	2,045,076
	M%	11.16	13.96	15.72	17.81	19.06	21.84	28.92	41.10
	M	11,811	15,784	21,774	30,093	39,008	64,498	396,286	840,591
Far East	P	116,590	126,610	188,842	178,214	385,732	461,684	1,517,384	1,217,205
	M%	1.55	1.63	1.78	1.70	2.06	2.05	1.40	2.30
	M	1,805	2,060	3,360	3,036	7,956	9,442	21,267	28,003
Total	P	265,753	284,542	378,770	403,338	660,438	908,620	3,741,263	4,719,416
	M%	15.13	16.66	15.87	17.29	13.94	17.87	24.78	39.70
	M	40,214	47,413	60,111	69,746	92,073	162,362	926,986	1,873,690
	G%	0.076	0.068	0.286	0.063	0.493	0.319	1.415	0.232

Table 2.7b Decennial estimates of the Muslim population (×1000) in Asia: 1790–1940 AD (1210–1360 H)

		1790	1800	1810	1820	1830	1840	1850	1860
Near East	P	17,225	17,675	18,125	18,714	19,550	20,410	21,300	22,230
	M%	82.13	82.11	82.10	82.12	81.64	81.22	80.76	80.44
	M	14,147	14,512	14,881	15,368	15,960	16,578	17,201	17,882
Central Asia	P	8,845	9,115	9,385	9,671	10,170	10,780	11,390	12,010
	M%	85.66	85.80	85.92	86.07	86.28	86.52	86.74	86.93
	M	7,577	7,820	8,064	8,324	8,775	9,327	9,879	10,440
Arabian Asia	P	8,180	8,272	8,368	8,471	8,770	9,085	9,400	9,729
	M%	90.84	90.83	90.82	90.81	90.69	90.57	90.45	90.31
	M	7,430	7,513	7,600	7,693	7,954	8,228	8,503	8,786
Southeast Asia	P	33,876	35,006	36,196	37,683	40,905	44,471	48,491	53,267
	M%	43.73	43.60	43.20	42.58	42.59	42.63	42.59	43.53
	M	14,813	15,263	15,635	16,045	17,422	18,959	20,650	23,186
Indian Subcontinent	P	199,526	204,638	209,750	213,863	222,620	231,971	241,690	250,708
	M%	19.13	19.06	19.00	19.00	18.96	18.93	18.90	18.89
	M	38,164	39,008	39,852	40,635	42,203	43,920	45,676	47,361
Far East	P	365,203	385,732	406,261	427,780	456,271	459,766	460,248	426,690
	M%	2.05	2.06	2.07	2.08	2.10	2.10	2.09	2.07
	M	7,488	7,956	8,424	8,916	9,571	9,641	9,641	8,822
Total	P	632,855	660,438	688,085	716,182	758,286	776,483	792,519	774,635
	M%	14.16	13.94	13.73	13.54	13.44	13.74	14.08	15.04
	M	89,619	92,073	94,456	96,981	101,885	106,653	111,550	116,477
	G%		0.427	0.410	0.400	0.571	0.237	0.204	−0.228

Table 2.7b (Continued)

		1870	1880	1890	1900	1910	1920	1930	1940
Near East	P	24,127	25,802	27,607	29,589	31,530	32,652	35,794	41,023
	M%	80.23	79.89	79.57	78.39	78.59	78.09	85.15	84.60
	M	19,356	20,614	21,968	23,195	24,780	25,497	30,479	34,707
Central Asia	P	12,667	13,906	15,275	16,865	19,986	20,440	22,522	24,973
	M%	87.14	87.53	87.90	90.55	90.34	83.65	83.82	78.18
	M	11,038	12,172	13,426	15,271	18,055	17,098	18,879	19,523
Arabian Asia	P	10,049	10,643	11,280	11,969	12,780	14,240	15,946	18,240
	M%	90.08	91.51	91.90	91.52	91.34	89.38	89.90	87.93
	M	9,052	9,739	10,366	10,954	11,673	12,727	14,335	16,038
Southeast Asia	P	62,126	70,528	79,401	93,261	106,621	117,521	132,808	159,648
	M%	47.11	46.15	46.01	41.82	40.79	42.98	41.03	41.77
	M	29,269	32,548	36,533	39,004	43,492	50,513	54,495	66,678
Indian Subcontinent	P	260,154	264,995	287,972	295,253	313,148	314,991	349,960	401,087
	M%	18.94	19.12	20.14	21.84	21.78	22.26	22.62	23.35
	M	49,268	50,668	58,001	64,498	68,202	70,108	79,167	93,661
Far East	P	409,241	421,697	436,636	461,684	491,389	549,652	579,200	623,188
	M%	2.05	2.06	2.05	2.05	2.03	2.03	1.99	1.97
	M	8,378	8,685	8,969	9,442	9,985	11,142	11,543	12,248
Total	P	778,363	807,571	858,171	908,620	975,453	1,049,497	1,136,230	1,268,159
	M%	16.23	16.65	17.39	17.87	18.06	17.83	18.39	19.15
	M	126,361	134,426	149,262	162,362	176,187	187,085	208,899	242,856
	G%	0.048	0.368	0.608	0.571	0.710	0.732	0.794	1.099

Table 2.7c Decennial estimates of the Muslim population (×1000) in Asia: 1950–2100 AD (1370–1520 H)

		1950	1960	1970	1980	1990	2000	2010	2020
Near East	P	46,830	59,729	76,422	97,580	127,246	145,361	163,211	186,631
	M%	86.84	87.41	88.38	90.01	92.07	94.26	95.09	95.67
	M	40,667	52,208	67,544	87,833	117,157	137,015	155,203	178,558
Central Asia	P	25,982	34,274	45,664	56,134	64,719	78,523	94,710	116,546
	M%	71.43	69.61	70.82	73.97	76.29	86.97	89.33	90.36
	M	18,558	23,859	32,340	41,525	49,373	68,288	84,605	105,310
Arabian Asia	P	21,741	28,256	38,282	54,522	77,436	104,821	142,978	176,999
	M%	85.81	86.24	86.63	87.80	89.62	89.30	89.20	89.54
	M	18,655	24,369	33,166	47,872	69,401	93,602	127,535	158,493
Southeast Asia	P	165,134	214,014	281,418	357,642	444,464	525,008	596,947	668,620
	M%	38.32	39.38	39.28	39.63	39.71	40.10	40.42	41.21
	M	63,279	84,270	110,528	141,745	176,511	210,527	241,272	275,541
Indian Subcontinent	P	468,467	563,843	702,538	887,264	1,121,083	1,370,166	1,609,607	1,817,448
	M%	22.13	23.34	24.88	25.42	27.02	28.92	30.17	31.56
	M	103,677	131,608	174,808	225,559	302,958	396,286	485,668	573,535
Far East	P	676,755	804,925	998,156	1,196,436	1,391,151	1,517,384	1,602,140	1,674,811
	M%	1.13	1.09	1.11	1.22	1.32	1.40	1.50	1.56
	M	7,619	8,810	11,032	14,539	18,315	21,267	23,997	26,130
Total	P	1,404,909	1,705,041	2,142,480	2,649,578	3,226,099	3,741,263	4,209,594	4,641,055
	M%	17.97	19.07	20.04	21.10	22.74	24.78	26.57	28.39
	M	252,456	325,124	429,418	559,073	733,714	926,986	1,118,280	1,317,568
	G%	1.024	1.936	2.284	2.124	1.969	1.481	1.179	0.976

Table 2.7c (Continued)

		2030	2040	2050	2060	2070	2080	2090	2100
Near East	P	200,657	211,698	218,990	221,358	218,215	212,400	206,504	199,812
	M%	96.03	96.34	96.60	96.80	96.96	97.09	97.22	97.34
	M	192,686	203,954	211,535	214,281	211,581	206,221	200,757	194,491
Central Asia	P	135,653	153,542	169,382	181,398	189,912	194,788	196,506	195,248
	M%	91.24	92.01	92.64	93.19	93.68	94.08	94.44	94.78
	M	123,768	141,272	156,910	169,043	177,907	183,262	185,572	185,059
Arabian Asia	P	212,456	241,142	266,610	286,679	300,698	310,225	316,023	317,858
	M%	90.29	90.73	91.07	91.31	91.47	91.55	91.58	91.58
	M	191,817	218,796	242,789	261,765	275,061	284,025	289,423	291,092
Southeast Asia	P	727,294	769,258	794,002	802,904	799,313	786,866	767,857	744,215
	M%	41.67	42.16	42.61	43.05	43.52	44.03	44.52	44.94
	M	303,087	324,287	338,287	345,616	347,885	346,466	341,833	334,454
Indian Subcontinent	P	2,002,370	2,141,754	2,228,386	2,267,029	2,257,745	2,208,155	2,133,349	2,045,076
	M%	32.87	34.17	35.51	36.76	37.95	39.11	40.16	41.10
	M	658,186	731,738	791,222	833,429	856,875	863,501	856,794	840,591
Far East	P	1,695,663	1,671,554	1,612,893	1,529,828	1,440,572	1,355,894	1,280,593	1,217,205
	M%	1.66	1.75	1.84	1.94	2.03	2.12	2.21	2.30
	M	28,075	29,255	29,740	29,623	29,235	28,763	28,318	28,003
Total	P	4,974,092	5,188,949	5,290,263	5,289,195	5,206,455	5,068,328	4,900,833	4,719,416
	M%	30.11	31.78	33.47	35.05	36.47	37.73	38.82	39.70
	M	1,497,619	1,649,302	1,770,483	1,853,757	1,898,545	1,912,239	1,902,696	1,873,690
	G%	0.693	0.423	0.193	−0.002	−0.158	−0.269	−0.336	−0.377

2.7 Asia's Summary and Conclusion

Islam started in Asia and the continent was always home to most Muslims. By now over two thirds of the world Muslim population lives in Asia. This percentage is decreasing as more Muslims spread to the rest of the world and is expected to reach less than a half by the end of this century. The ratio of Muslims in Asia is currently more than a quarter and is increasing to more than a third (39%) by the end of this century. The following tables present centennial data from 600 AD to 2100 AD (or approximately 1 H to 1500 H) in Table 2.7a and decennial data from 1790 AD to 2100 AD (or 1210 H to 1520 H) in Tables 2.7b and 2.7c for the six regions of Asia. The data includes total population in thousands (P), the percentage of which is Muslim (M%), the corresponding Muslim population in thousands (M), and the annual population growth rate (APGR, or G%) of the total population in Asia and each of its six regions.

References

[AB] Krylov, A. (2004, April 18). Secret of Abkhaz tolerance. *Caucasian Knot.* Retrieved from http://eng.kavkaz-uzel.ru/

[ABM] Inoguchi, T. (2015). *AsiaBarometer Survey Data* [Computer Files]. Retrieved from http://www.asiabarometer.org/

[AFH] Djinguiz, M. (1908). L'Islam dans l'Inde. *Revue du Monde Musulman.* 6(9), 85–118.

[AM] National Statistical Service of the Republic of Armenia (ARMSTAT) (2013). *Socio-Economic Situation in Armenia in January–March 2013* [Armenian] (5.2. 2011 October 12–21 Census conducted in the Republic of Armenia). Yerevan: ARMSTAT.

[ANB] Center for East Asia Democratic Studies (2017). *Asian Barometer Survey Data* [Computer Files]. Retrieved from http://www.asianbarometer.org/

[AZ] The State Statistical Committee of the Republic of Azerbaijan (2011). Population of Azerbaijan, Table 1.5: Population by Ethnic Groups. *Statistical Yearbook* [Online].

[BD11] Bangladesh Bureau of Statistics (2015). *Population and Housing Census 2011, National Report* (Vol. 1: Analytical Report). Dhaka: BBS.

[BD] Bangladesh Bureau of Statistics (2016). Table 2.12: Percentage Distribution and Variation of Major Communities by Religion, 1901–2011. *2014 Statistical Yearbook of Bangladesh.* Dhaka: BBS.

[BH] Central Informatics Organization of Bahrain (CIO) (2011). *Census Results 2010.* Manama: CIO.

[BN16] Department of Statistics, Department of Economic Planning and Development (DEPD) (2017). *Population and Housing Census Update Final Report 2016.* Bandar Seri Begawan: DEPD.

[BN11] Department of Statistics, Department of Economic Planning and Development (DEPD) (2013). *Brunei Darussalam Statistical Yearbook 2011.* Bandar Seri Begawan: DEPD.

[BNH] Djinguiz, M. (1908). L'Islam au Bornéo Britannique Septentrional. *Revue du Monde Musulman.* 5(6), 294–302.

[CIA80] Central Intelligence Agency (CIA) (1980). *The World Factbook.* Washington, DC: CIA.

[CNO] China Data Center, University of Michigan (2015). *China Data Online.* Online database retrieved from http://chinadataonline.org/

[CN] Mackerras, C. (2001). *The New Cambridge Handbook of Contemporary China.* Cambridge: Cambridge University Press.

[CY11] Statistical Service (2014). *Latest Figures: Results of the Census of Population, 2011.* Population–Country of Birth, Citizenship Category, Country of Citizenship, Language, Religion, 2011. Online data file retrieved from http://www.cystat.gov.cy/

[CY06] State Planning Organization (2007). *The Final Results of TRNC General Population and Housing Unit Census.* Nicosia: Turkish Republic of Northern Cyprus (TRNC) Prime Ministry.

[CY] Statistical Service (2010). *Demographic Report 2009* (Population Statistics, Series II, Report No. 47). Nicosia: The Printing Office of the Republic of Cyprus.

[CYH] Hatay, M. (2007). *Is the Turkish Cypriot Population Shrinking? An Overview of the Ethnic-Demography of Cyprus in the Light of the Preliminary Results of the 2006 Turkish Census.* Oslo: International Peace Research Institute (PRIO).

[DHS] ICF International (1984–2018). *The Demographic and Health Surveys (DHS) Program.* Rockville, MD: ICF International. Retrieved from http://dhsprogram.com/

[DOS] United States Department of State (DOS), Bureau of Democracy, Human Rights, and Labor (2002–2015). *Annual Report on International Religious Freedom.* Washington, DC: DOS. Retrieved from http://www.state.gov/j/drl/rls/irf/

[EAB] East Asia Barometer Network at National Taiwan University (2015), *East Asia Barometer Database.* Madrid: ASEP/JDS. Retrieved from http://www.jdsurvey.net/eab/

[EGH] Saleh, M. (2013). *On the Road to Heaven: Taxation, Conversions, and the Coptic-Muslim Socioeconomic Gap in Medieval Egypt*. Toulouse School of Economics Working Paper, No. 13-428.

[EVS] European Values Study Foundation at Tilburg University [The Netherlands] (2015). *European Values Study, ZA4804: EVS 1981-2008 Longitudinal Data File*. Cologne: Leibniz Institute for the Social Sciences (GESIS). Retrieved from http://zacat.gesis.org/. More info on EVS is available at http://www.europeanvaluesstudy.eu/

[GB911] His Majesty's Stationary Office (HMSO) (1917). *Census of England and Wales. 1911.* (General Report with Appendices). London: Darling & Son.

[GB871] Her Majesty's Stationary Office (HMSO) (1873). *Census of England and Wales for the Year 1871* (General Report. Vol. IV). London: George Edward Eyre and William Spottiswoode.

[GE14] National Statistics Office of Georgia (GEOSTAT) (2016). *2014 General Population Census–Main Results*. Tbilisi: GEOSTAT.

[GE02] State Department for Statistics of Georgia (SDSG) (2003). *Population Census 2002*. Tbilisi: SDSG.

[ID10] Statistics Indonesia (2011). *Citizenship, Racial, Religious and Everyday Linguistic of Indonesia People.* 2010 Population Census Results [Indonesian]. Jakarta: Statistics Indonesia.

[ID05] Cholil, S., Bagir, Z. A., Rahayu, M., & Asyhari, B. (2010). *Annual Report on Religious Life in Indonesia 2009*. Jakarta: Center for Religious & Cross-cultural Studies (CRCS).

[ID90] Ricklefs, M. C. (2008). *A History of Modern Indonesia since c. 1200* (4th ed.). Palo Alto, CA: Stanford University Press.

[IDH] Djinguiz, M. (1909). L'Islam dans les Possessions Hollandaises, Portugaises, et Americaines de l'Archipel Indien. *Revue du Monde Musulman*. 7(1-2), 104–120.

[IR] Statistical Centre of Iran (CSI) (2012). *Selected Findings of the 2011 National Population and Housing Census* [Persian]. Tehran: CSI.

[IN11] Census India (2015, August 25). *C-1 Population by religious data*. Office of the Registrar General & Census Commissioner, India. Online database retrieved from http://www.censusindia.gov.in/2011census/c-01.html

[IN71] Ministry of Home Affairs (1974). *The Population of India*. 1974 Country Monographs, CICRED (Comite International de Coordination de Recherches Nationales de Demographie) Series. Paris: CICRED.

Retrieved from http://www.cicred.org/Eng/Publications/content/3MonographNational/Index.html

[INH71] Waterfield, H. (1875). *Memorandum on the Census of British India 1871–72*. London: Eyre and Spttiswoode.

[INH81] Plowden, W. C. (1883). *Report on the Census of British India Taken on the 17th February 1881*. London: Eyre and Spttiswoode.

[INH91] Baines, J. A. (1893). *General Report on the Census of India 1891*. London: Her Majesty's Stationary Office.

[INH01] Risley, H. H., & Gait, E. A. (1903). *Report on the Census of India 1901*. Calcutta: Superintendent of Government Printing.

[INH11] Gait, E. A. (1913). *Census of India 1911*. Calcutta: Superintendent of Government Printing.

[INH21] Marten, J. T. (1923). *Census of India 1921*. Calcutta: Superintendent of Government Printing.

[INH31] Hutton, J. H. (1933). *Census of India 1931*. Delhi: Manager of Publications.

[INH41] Yeatts, M. W. M. (1943). *Census of India 1941*. Simla: Manager of Publications.

[INH] Joshi, A. P., Srinivas, M. D., & Bajaj, J. K. (2003*). Religious Demography of I*ndia. Chennai: Centre for Policy Studies.

[IL] Central Bureau of Statistics (CBS) (2018). Table 2.2 Population, by Religion. *Statistical Abstract of Israel 2018*. Jerusalem: CBS.

[JAN] Jansen, H. (1897). *Verbreitung des Islâms: mit angabe der verschiedenen riten, sekten und religiösen bruderschaften in den verschiedenen ländern der erde 1890 bis 1897. Mit benutzung der neuesten angaben (zählungen, berechnungen, schätzungen und vermutungen). [German for Dissemination of Islam: with number of various rites, sects and religious brotherhoods in the various countries of the earth 1890 to 1897. With use of information on the latest disclosures (censuses, calculations, estimates and guesses)]*. Berlin: Selbstverlag des Verfassers.

[JP15] Statistical Japan (2018). Table 2.10 Foreign National Residents by Nationality (2014–2016). *Japan Statistical Yearbook 2018*. Tokyo: Statistics Bureau. Retrieved from http://www.stat.go.jp/english/data/nenkan/

[JP10] Statistical Research and Training Institute (ed.) (2015). Table 2.14 Registered Foreigners by Nationality (1990–2013). *Japan Statistical Yearbook 2015*. Tokyo: Statistics Bureau. Retrieved from http://www.stat.go.jp/english/data/nenkan/

[JPH] Statistical Research and Training Institute (ed.) (2005). Table 2.14 Foreigners by Nationality and Age (Five-Year Groups) (1950–2005). *Historical Statistics of Japan*. Tokyo: Statistics Bureau. Retrieved from http://www.stat.go.jp/english/data/chouki/

[JP40] Bureau of Statistics (1961). Table 5: Race or Nationality by Sex, for All Japan (All Persons including Military Personnel). *1940 Population Census of Japan* (Vol. I: Total Population, Sex, Age, Marital Status and Race or Nationality). Tokyo: Bureau of Statistics.

[KET86] Kettani, A. (1986). *Muslim Minorities in the World Today.* London: Mansell Publishing.

[KET73] Kettani, A. (1973). *Muslims in Communist Block.* [Arabic] Mecca: Organization of Islamic Conference

[KG] National Statistical Committee of the Kyrgyz Republic (NSCKR) (2009). *Book I: Main Social and Demographic Characteristics of Population and Number of Housing Units.* Population and Housing Census of the Kyrgyz Republic of 2009. Bishkek: NSCKR.

[KH] National Institute of Statistics (NIS) (2009). *General Population Census of Cambodia 2008: National Report on Final Census Results*. Phnom Penh: NIS.

[KH936] Direction des Services Economiques, Service de la Statistique Générale (1939). *Annuaire statistique de l'Indochine* (Vol. 8: 1937–1938). Hanoi: Imprimerie d'Extrême-Orient.

[KH931] Inspection Générale des Mines et de l'Industrie – Service de la Statistique Générale de l'Indochine (1932). *Annuaire statistique de l'Indochine* (Vol. 3: 1930–1931). Hanoi: Imprimerie d'Extrême-Orient.

[KH921] Inspection Générale des Mines et de l'Industrie – Service de la Statistique Générale de l'Indochine (1931). *Annuaire statistique de l'Indochine* (Vol. 2: 1923–1929). Hanoi: Imprimerie d'Extrême-Orient.

[KRH] Baker, D. (2006). Islam Struggles for a Toehold in Korea. *Harvard Asia Quarterly*, X(1), 25–30.

[KW] The Public Authority for Civil Information (2018). *Statistics Service System Issue 55: Population* (Table 17: Total Population by Religion and Nationality Groups and gender). Retrieved from http://www.paci.gov.kw/

[KZ] The Agency of the Republic of Kazakhstan on Statistics (ARKS) (2011). *Results of the 2009 National Population Census of the Republic of Kazakhstan–Analytical Report*. Astana: ARKS.

[LA15] Lao Statistics Bureau (2016). *Results of Population and Housing Census 2015*. Vientiane Capital: LSB.

[LA05] National Statistics Center (2006). *Results from the Population and Housing Census 2005*. Vientiane Capital: NSC.

[LA95] National Statistics Center (1997). *Lao Census 1995–Country Report*. Vientiane Capital: NSC.

[LBH] Courbage, Y., & Fargues, P. (1974). *La Population du Liban*. 1974 Country Monographs, CICRED (Comite International de Coordination de Recherches Nationales de Demographie) Series. Beyrouth: Publications de l'Université Libanaise. Retrieved from http://www.cicred.org/Eng/Publications/content/3MonographNational/Index.html

[LK] Department of Census and Statistics (DCS) (2012). Population Atlas of Sri Lanka 2012. Series. Colombo: DCS.

[LKH] Department of Census and Statistics (DCS) (1974). *The Population of Sri Lanka*. 1974 Country Monographs, CICRED (Comite International de Coordination de Recherches Nationales de Demographie) Series. Colombo: DCS. Retrieved from http://www.cicred.org/Eng/Publications/content/3MonographNational/Index.html

[MCC] McCarthy, J. (1990). *The Population of Palestine: Population History and Statistics of the Late Ottoman Period and the Mandate*. Institute for Palestine Studies Series. New York, NY: Columbia University Press.

[MAS55] Massignon, L. (1955). *Annuaire du Monde Musulman, Statistique, Historique, Social et Économique* (4ème éd.). Paris: Presses Universitaires de France

[MAS29] Massignon, L. (1929). *Annuaire du Monde Musulman, Statistique, Historique, Social et Économique* (2ème éd.). Paris: Librairie Ernest Leroux.

[MAS23] Massignon, L. (1923). *Annuaire du Monde Musulman, Statistique, Historique, Social et Économique*. Paris: Ernest Leroux.

[MM14] Department of Population [Myanmar] (2016). *The 2014 Myanmar Population and Housing Census* (Census Report Volume 2-C: The Union Report: Religion). Nay Pyi Taw: Ministry of Labour, Immigration and Population.

[MM83] Immigration and Manpower Department (1986). *Burma 1983 Population Census*. Rangoon: IMD.

[MM73] Immigration and Manpower Department (1976). *Burma 1973 Population Census*. Rangoon: IMD.

[MM53] Central Statistical and Economics Department (1969). *Statistical Year Book 1969*. Rangoon: CSED.

[MN10] National Statistical Office of Mongolia (NSOM) (2011). *2010 Population and housing census of Mongolia* [in Mongolian], Ulaanbaatar: NSOM.

[MN00] National Statistical Office of Mongolia (NSOM) (2004). *2000 Population and housing census of Mongolia* [in Mongolian], Ulaanbaatar: NSOM.

[MO15] Government of Macao Special Administrative Region Statistics and Census Service (Direcção dos Serviços de Estatística e Censos [DSEC]) (2016). *Yearbook of Statistics 2015*. Macao: DSEC

[MO00] Government of Macao Special Administrative Region Statistics and Census Service (Direcção dos Serviços de Estatística e Censos [DSEC]) (2001). *Yearbook of Statistics 2000*. Macao: DSEC

[MO95] Government of Macao Statistics and Census Service (Direcção dos Serviços de Estatística e Censos [DSEC]) (1996). *Yearbook of Statistics 1995*. Macao: DSEC

[MO90] Government of Macao Statistics and Census Service (Direcção dos Serviços de Estatística e Censos [DSEC]) (1991). *Yearbook of Statistics 1990*. Macao: DSEC

[MO85] Government of Macao Statistics and Census Department (Estatistica) (1986). *Yearbook of Statistics 1985*. Macao: Imprensa Oficial de Macau.

[MO70] Repartição dos Serviços de Estatística (1981). *Macau Year Book of Statistics 1980*. Macau: Imprensa Nacional.

[MY10] Department of Statistics Malaysia (DSM) (2011). *Population Distribution and Basic Demographic Characteristics 2010*. Population and Housing Census of Malaysia, Putrajaya: DSM.

[MY00] Saw, S. H. (2007). *The Population of Malaysia*. Singapore: Institute of Southeast Asian Studies.

[MYH] Teik, K. B. (2005). Ethnic Structure, Inequality and Governance in the Public Sector, Malaysian Experience. Democracy, *Governance and Human Rights Programme*, Paper No. 20. Geneva: United Nations Research Institute for Social Development.

[MV] Ministry of Planning and National Development (MPND) (2008). *Analytical Report 2006: Population and Housing Census 2006*. Male: MPND, and email correspondence in February 2017 with Ms. Ihrisha Abdul Wahid from the National Bureau of Statistics of the Maldives.

[NKH] Baguirov, A. (2008). Nagorno-Karabakh: Basis and Reality of Soviet-Era Legal and Economic Claims Used to Justify the Armenia-Azerbaijan War. *Caucasian Review of International Affairs*, 2(1), 11–24.

[NP11] Central Bureau of Statistics (CBS) (2012). National Population and Housing Census 2011. Kathmandu: CBS.

[NP] Dahal, D.R. (2003). Social Composition of the Population: Caste/Ethnicity and Religion in Nepal. In CBS (ed.), *2001 Population Monograph of Nepal*. (pp. 87–135). Kathmandu: Central Bureau of Statistics.

[OS] DFWatch Staff (2016, August 13). Census shows only 7% Georgians living in South Ossetia. *Democracy & Freedom Watch*. Retrieved from http://dfwatch.net/

[PEW] Pew Research Center (PRC) (2009). *Mapping the Global Muslim Population, A Report on the Size and Distribution of the World's Muslim Population*. Washington, DC: PRC.

[PH15] Philippine Statistics Authority (PSA) (2017). *Results from the 2015 Census of Population*. Quezon City: PSA

[PH10] Philippine Statistics Authority (PSA) (2015). *Philippines in Figures 2015*. Quezon City: PSA

[PH70] National Census and Statistics Office (1974). *1970 Census of Population and Housing–Final Report*, Vol. I. Manila: NCSO.

[PH60] Bureau of the Census and Statistics (1963). *Census of the Philippines: 1960 Population and Housing* (Vol. I, Report by Province). Manila: NCSO.

[PH948] Philippine Bureau of the Census and Statistics (1954). *Census of the Philippine: 1948* (Vol. I, Part I and II and Vol. II, Part I and II, Report by Province for Census of Population). Manila: Bureau of Printing.

[PH939] Census Office of the Philippine Islands (1940). *Census of the Philippine: 1939* (Vol. I. Reports by Provinces for Census of Population, Part I–IV). Manila: Bureau of Printing.

[PH918] Census Office of the Philippine Islands (1920). *Census of the Philippine Islands taken under the direction of the Philippine Legislature in the Year 1918* (Vol. II: Population and Mortality). Manila: Bureau of Printing.

[PH903] United States Bureau of the Census (1905). *Census of the Philippine Islands taken under the direction of the Philippine Commission in the Year 1903* (Vol. II: Population). Washington, DC: Government Printing Office.

[PK] Population Census Organization (PCO) (1998). *Census Report of Pakistan*. Islamabad: PCO.

[PK61] Afzal, M. (1974). *The Population of Pakistan*. 1974 Country Monographs, CICRED (Comite International de Coordination de Recherches Nationales de Demographie) Series. Rawalpindi: Ferozsons Ltd. Retrieved from http://www.cicred.org/Eng/Publications/content/3MonographNational/Index.html

[PS17] Palestinian Central Bureau of Statistics (PCBS) (2018). *Preliminary Results of the Population, Housing and Establishments Census 2017*. Ramallah: PCBS.

[PS07] Palestinian Central Bureau of Statistics (PCBS) (2009). *Population, Housing and Establishment Census 2007 - Census Final Results, Population Report - West Bank*. Ramallah: PCBS.

[PS97] Palestinian Central Bureau of Statistics (PCBS) (1999). *Population, Housing and Establishment Census 1997 - Final Results, Population Report, Palestinian Territory*. Statistical Reports Series, Ramallah: PCBS.

[QA] The Planning Council of the Central Statistics Office (1999). 1997 General Population and Housing Census. Doha: Central Statistics Office [Arabic]

[QQ] Ethno Cacausus (2013). Caucasus Ethnographies: Detailed Data of the Census of Population and Ethnographic Maps. [Online Collection]. Retrieved from http://www.ethno-kavkaz.narod.ru/

[ROD] Radončić, E. (2013). *World Almanac of the demographic history of Muslims: Rich cartographic representation of the geographical distribution of Muslim populations* (M. Durmić, Trans.). Sarajevo: Edin Radončić.

[SG1] Aye, D. K. K. (2005). *Bazaar Malay: History, Grammar and Contact* (Unpublished doctoral dissertation). National University of Singapore.

[SG2] Saw, S. H. (2007). *The Population of Singapore* (2nd ed.). Singapore: Institute of Southeast Asian Studies.

[SG3] Kiong, T. C. (2008). Religious Trends and Issues in Singapore. In A.E. Lai (ed.), *Religious Diversity in Singapore*. Singapore: Institute of Southeast Asian Studies.

[SG10] Singapore Department of Statistics (SDS) (2011). *Census of Population 2010 Statistical Release 1: Demographic Characteristics, Education, Language and Religion*. Singapore: SDS.

[SYB80] Paxton, J. (ed.) (1980). *The Statesman's Year-Book*. London: MacMillan.

[SYB70] Paxton, J. (ed.) (1970). *The Statesman's Year-Book.* London: MacMillan.

[SYB50] Steinberg, S. H. (ed.) (1950). *The Statesman's Year-Book.* London: MacMillan.

[SYB30] Epstein, M. (ed.) (1930). *The Statesman's Year-Book.* London: MacMillan.

[SYB20] Scott-Keltie, J. & Epstein, M. (eds.) (1920). *The Statesman's Year-Book.* London: MacMillan.

[SYB15] Scott-Keltie, J. (ed.) (1915). *The Statesman's Year-Book.* London: MacMillan.

[SYB10] Scott-Keltie, J. (ed.) (1910). *The Statesman's Year-Book.* London: MacMillan.

[SYB00] Scott-Keltie, J. (ed.) (1900). *The Statesman's Year-Book.* London: MacMillan.

[SYB890] Scott-Keltie, J. (ed.) (1890). *The Statesman's Year-Book.* London: MacMillan.

[SYB880] Martin, F. (ed.) (1880). *The Statesman's Year-Book.* London: MacMillan.

[SU] Demoscope Analysis and Information System (2012). Census of the Russian Empire, the USSR, and the 15 New Independent States [Russian]. *Demoscope Weekly.* Retrieved from http://demoscope.ru/weekly/pril.php

[TH929] Bureau of General Statistics (BGS) (1933). *Statistical Year Book of the Kingdom of Siam* (No. 17, B.E. 2474–75 (1931–33)). Bangkok: Ministry of Economic Affairs.

[TH937] Central Service of Statistics (CSS) (1939). *Statistical Year Book Thailand*, (No. 20, B.E. 2480 (1937–38) & 2481 (1938–39)). Bangkok: CSS.

[TH] National Statistics Office (NSO) (2001). *The 2000 Population and Housing Census.* Bangkok: NSO.

[TJ12] Statistical Agency under the President of the Republic of Tajikistan (TAJSTAT) (2013). *Census of Population and Housing of Tajikistan in 2010* [Russian] (Vol. III: Ethnic Composition and Language Skills, Citizenship of the Republic of Tajikistan). Dushanbe: TAJSTAT.

[TJ10] Statistical Agency under the President of the Republic of Tajikistan (TAJSTAT) (2013, March 1). NEWS: *Information on the Number of Muslims in the Republic of Tajikistan.* Dushanbe: TAJSTAT. Retrived online from http://www.stat.tj/en/news/

[TJ00] Tula, M. (2005). The Census Population of Tajikistan in 2000: A National, Age, Gender, Family Composition and Educational [in Russian]. *Demoscope Weekly*. No. 191–192.

[TL15] Statistics Timore-Leste (STL) (2016). *Timor-Leste Population and Housing Census 2015* (Vol. 2: Population Distribution by Administrative Areas (Nationality, Citizenship and Religion)). Dili: STL.

[TL10] National Statistics Directorate (NSD) [Timor Leste], & United Nations Population Fund (UNFPA) (2011). *Population and Housing Census of Timor-Leste, 2010* (Vol. 2: Population Distribution by Administrative Areas). Dili: NSD & UNFPA.

[TL04] United Nations International Human Rights Instruments (HRI) (2007). *Core Document Forming Part of the Reports of States Parties: Timor-Leste*. HRI/CORE/TLS/2007. Geneva: HRI.

[TL90] Ministry of Health and National Statistics Office [Timor-Leste], University of Newcastle, [Australia], The Australian National University [Australia], & ACIL Australia Pty Ltd [Australia] (2004). *Timor-Leste 2003 Demographic and Health Survey*. Newcastle: University of Newcastle.

[TL80] Cribb, R. (2010). *Digital Atlas of Indonesian History*. Copenhagen: NIAS Press.

[TL70] Chrystello, J. C. (1992). *East Timor: The secrete file 1973–1975*. Sao Paulo: eBooksBrasil.

[TM12] Globe, P. (2015, February 10). Unpublished Census provides rare and unvarnished look at Turkmenistan. *Eurasia Daily Monitor*, 12 (26). Retrieved from http://www.jamestown.org/programs/edm/

[TM95] Asgabat.net (2016, March 10). "Итоги всеобщей переписи населения Туркменистана по национальному составу в 1995 году" [The results of the general census of the population of Turkmenistan on the ethnic composition in 1995]. Retrieved from asgabat.net

[TR965] Spitler, J. F., & Roof, M. K. (1982). *Detailed Statistics on the Urban and Rural Population of Turkey: 1950 to 2000*. Washington, DC: International Demographic Data Center

[TR914] Turkish General Staff Military History and Strategic Research and Inspection Directorate (ATASE ve Dent.) (2005). *Armenian Activities in the Archive Documents 1914–1918* (Vol. I). Ankara: Genelkurmay Basim Evi.

[TR897] Mutlu, S. (2003). Late Ottoman Population and Its Ethnic Distribution. *Turkish Journal of Population Studies*, 25, 3–38.

[TW01] Government Information Office (GIO) (2001). Religion: Islam. *The Republic of China Yearbook – Taiwan 2001*. Taipei: GIO.

[UN] United Nations Statistics Division, UNSD Demographic Statistics (2018). Population by Religion, Sex and Urban/Rural Residence. *UNdata* [Online Database]. Retrieved from http://data.un.org

[UNE] United Nations Statistics Division, UNSD Demographic Statistics (2017). Population by National and/or Ethnic Group, Sex and Urban/Rural Residence, *UNdata* [Online Database]. Retrieved from http://data.un.org

[UN04] United Nations, Department of International Economic and Social Affairs, Statistical Office (2006). Table 6 - Population by Religion, Sex and Urban/Rural Residence: Each Census, 1985–2004. *Demographic Yearbook Special Census Topics* (Vol. 2b: Ethnocultural Characteristics). New York, NY: United Nations.

[UN88] United Nations, Department of International Economic and Social Affairs, Statistical Office (1990). Table 29 - Population by Religion and Sex: Each Census, 1979–1988. *Demographic Yearbook 1988*. New York, NY: United Nations.

[UN83] United Nations, Department of International Economic and Social Affairs, Statistical Office (1985). Table 29 - Population by Religion and Sex: Each Census, 1974–1983. *Demographic Yearbook 1983*. New York, NY: United Nations.

[UN79] United Nations, Department of International Economic and Social Affairs, Statistical Office (1980). Table 29 - Population by Religion, Sex and Urban/Rural Residence: Each Census, 1970–1979. *Demographic Yearbook 1979*. New York, NY: United Nations.

[UN71] United Nations, Department of International Economic and Social Affairs, Statistical Office (1972). Table 17 - Population by Religion, Sex and Urban/Rural Residence: Each Census, 1962–1971. *Demographic Yearbook 1971*. New York, NY: United Nations.

[UN63] United Nations, Department of International Economic and Social Affairs, Statistical Office (1964). Table 11 - Population by Religion and Sex: Each Census, 1955–1963. *Demographic Yearbook 1963*. New York, NY: United Nations.

[UN56] United Nations, Department of International Economic and Social Affairs, Statistical Office (1957). Table 8 - Population by Religion and Sex: Each Census, 1945–1955. *Demographic Yearbook 1956*. New York, NY: United Nations.

[VN09] General Statistics Office of Vietnam (GSOV) (2010). The 2009 Vietnam Population and Housing Census: Completed Results. Hanoi: GSOV.

[VNH] Cabaton, A. (1906). Note sure l'Islam dans l'Indo-Chine Française. *Revue du Monde Musulman.* 1(1), 27–47.

[WVS] World Values Survey Association at the University of Aberdeen [The United Kingdom] (2015). *World Values Survey Online Data Analysis.* Retrieved from http://www.worldvaluessurvey.org/WVSOnline.jsp

Chapter 3

Islam in Africa

Africa was the second continent that Islam spread into, which explains why almost one-third of world Muslim population resides in this continent. Muslims crossed current Djibouti and Eritrea to seek refuge in current Ethiopia from Pagan Arabs. On the advice of Prophet Muhammad, in Rajab 8 BH, or May 614 AD, sixteen Muslims migrated to Abyssinia where they were protected by its king, an-Najashi, who has also accepted Islam later. They were followed by 101 Muslims later in the same year. By Muharram 7 H, or May 628 AD, all those Muslims returned to Medina, but locals who embraced Islam remained there.

In 20 H/641 AD during the reign of Caliph Omar bnul Khattab, Muslim troops took over current Egypt and conquered current Libya the following year. Muslims then expanded to current Tunisia in 27 H/647 AD during the reign of the third Muslim Caliph, Othman bnu Affan. The conquest of North Africa continued under the Umayyad Dynasty, taking Algeria by 61 H/680 AD, and Morocco the following year. From the latter Muslim troops crossed the Strait of Gibraltar to Europe in 711.

Islam then spread slowly in much of the continent through trade and preaching. By the ninth century Muslim Sultanates started being established in the Horn of Africa, and by the twelfth century

---

The World Muslim Population: Spatial and Temporal Analyses
Houssain Kettani
Copyright © 2020 Jenny Stanford Publishing Pte. Ltd.
ISBN 978-981-4800-31-0 (Hardcover), 978-0-429-42253-9 (eBook)
www.jennystanford.com

the Kilwa Muslim Dynasty spread as far south as Mozambique. Islam only crossed deeper to Malawi and Congo in the second half of the nineteenth century under Zanzibar Sultanate. Then the British brought labor force from India to their African colonies towards the end of the nineteenth and beginning of the twentieth centuries. Islam gained momentum during the tenth century in West and Central Africa with the start of Almoravids movement on Senegal River and as rulers and kings embraced Islam.

Thus, the Muslim population has likely increased from 1.2 million or 4.5% of the total African population in 700 AD, to 5.1 million or 17.2% in 800 AD, to 9.1 million or 29.7% in 900 AD, to 12 million or 37.5% in 1000 AD, to 13 million or 37.6% in 1100 AD, to 14 million or 36.4% in 1200 AD, to 14 million or 35.5% in 1300 AD, to 15 million or 35.1% in 1400 AD, to 16 million or 34.8% in 1500 AD, to 21 million or 37.7% in 1600 AD, to 23 million or 38.3% in 1700 AD, to 26 million or 37.1% in 1800 AD, to 46 million or 40.4% in 1900, to 0.36 billion or 44.1% in 2000, to 0.59 billion or 44.0% in 2020, and is projected to reach 1.1 billion or 44% by 2050 and 1.9 billion or 45% by 2100.

A plot of centennial estimates of the Muslim population and its percentage with respect to the total population in Africa from 600 to 2100 is provided in Fig. 3.0a. A zoom in of this plot, providing a plot of decennial estimates of the Muslim population and its percentage with respect to the total population in this continent from 1900 to 2100 is provided in Fig. 3.0b. This shows that the Muslim population in Africa was increasing slowly until 1950, but is increasing substantially since then, doubling every quarter of a century or so. However, the percentage of Muslims peaked to 44% in 1980, remains so through much of this century, reaching 45% by its end. This is due to higher fertility rate among non-Muslims in sub-Saharan Africa.

We divided Africa into five regions; the data for each is included in a separate section and are sorted in terms of the percentage of Muslims in descending order. These regions are North Africa (Section 3.1), West Africa (Section 3.2), East Africa (Section 3.3), Central Africa (Section 3.4), and Southern Africa (Section 3.5). In Section 3.6, the total population in each of the five African regions and the corresponding percentage and number of Muslims is presented centennially in Table 3.6a from 600 to 2100 and decennially in Tables 3.6b and 3.6c from 1790 to 2100.

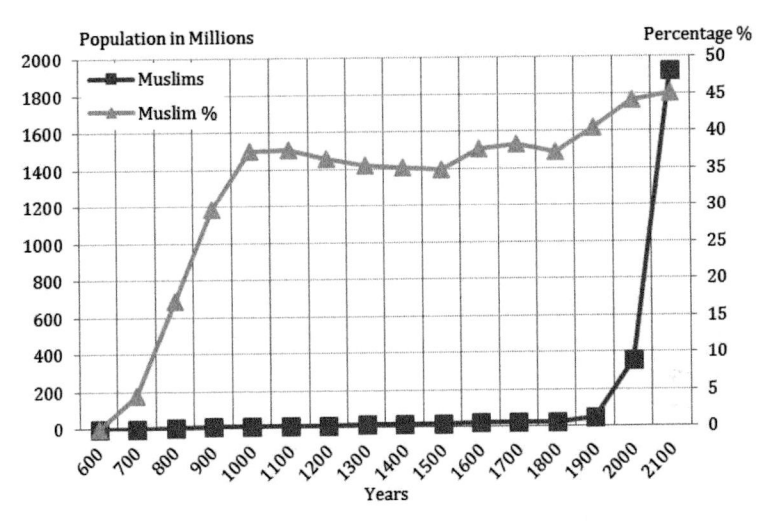

Figure 3.0a Plot of centennial estimates of the Muslim population and its percentage of the total population in Africa: 600–2100 AD (1–1500 H).

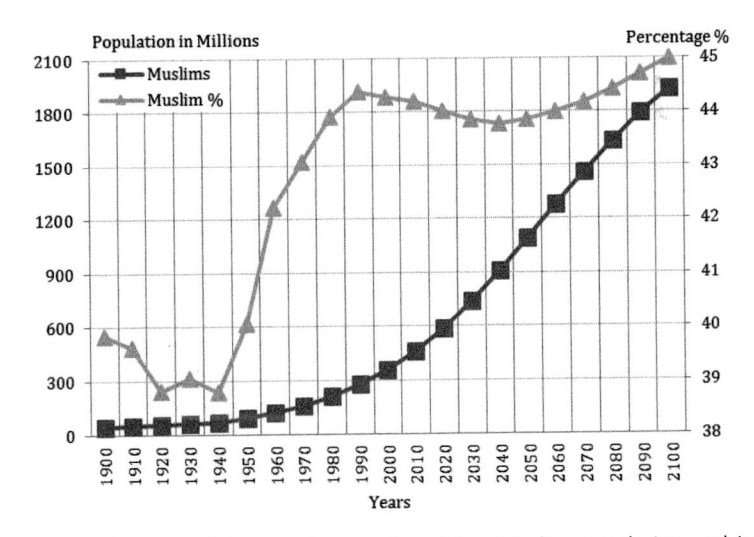

Figure 3.0b Plot of decennial estimates of the Muslim population and its percentage of the total population in Africa: 1900–2100 AD (1320–1520 H).

3.1 Muslims in North Africa

This region consists of ten countries: Algeria, Chad, Egypt, Libya, Mali, Mauritania, Morocco, Niger, Sudan, and Tunisia. Islam entered this region in 20 H/641 AD during the reign of Caliph Omar bnul Khattab. The Muslim troops were led by the Prophet companion Amr bnul Aass who conquered Egypt and Libya. Muslims then conquered Tunisia in 27 H/647 AD, Algeria between 51 H and 61 H, or 671 AD and 680 AD, Morocco between 62 H/681 AD and 89 H/708 AD. Islam spread in southern part of this region during the eleventh century, from the preaching efforts of Almoravids movement in the west and trade in the other parts. Islam was strengthened by its adoption of the rulers of Songhai Empire in current Mali and east Niger in 1010 and the rulers of the Kanem-Bornu Empire around Lake Chad in 1085.

Thus, the Muslim population has likely changed from 1.2 million or 8.8% of the total population of this region in 700 AD, to 4.9 million or 35.1% in 800 AD, to 8.8 million or 61.5% in 900 AD, to 11.5 million or 78.5% in 1000 AD, to 12 million or 83–84% in 1100 AD to 1500 AD, to 17 million or 85.4% in 1600 AD, to 14 million or 85.2% in 1700 AD, to 15 million or 86.3% in 1800 AD, to 28 million or 88.3% in 1900, to 0.19 billion or 95.3% in 2000, to 0.30 billion or 95.6% in 2020, and is projected to reach 0.50 billion or 96% by 2050 and 0.80 billion or 97% by 2100.

A plot of centennial estimates of the Muslim population and its percentage with respect to the total population in this region from 600 to 2100 is provided in Fig. 3.1a. A zoom in of this plot, providing a plot of decennial estimates of the Muslim population and its percentage with respect to the total population in this region from 1900 to 2100 is provided in Fig. 3.1b. This shows that the Muslim population in this region was increasing slowly until 1950 and is increasing substantially afterwards and throughout this century. The percentage of Muslims on the other hand, increased sharply after its dip in 1940 to less than 87%, reaching 95% in 1980 and 97% by the end of this century.

The corresponding individual data for each country in this region is discussed below. In Section 3.1.11, the total population in each country in this region and the corresponding percentage and number of Muslims is presented centennially in Table 3.1a from 600 to 2100 and decennially in Tables 3.1b and 3.1c from 1790 to 2100.

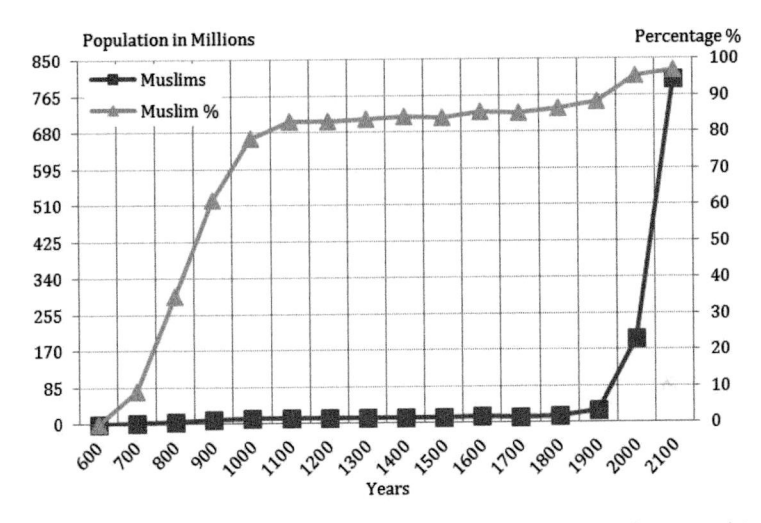

Figure 3.1a Plot of centennial estimates of the Muslim population and its percentage of the total population in North Africa: 600–2100 AD (1–1500 H).

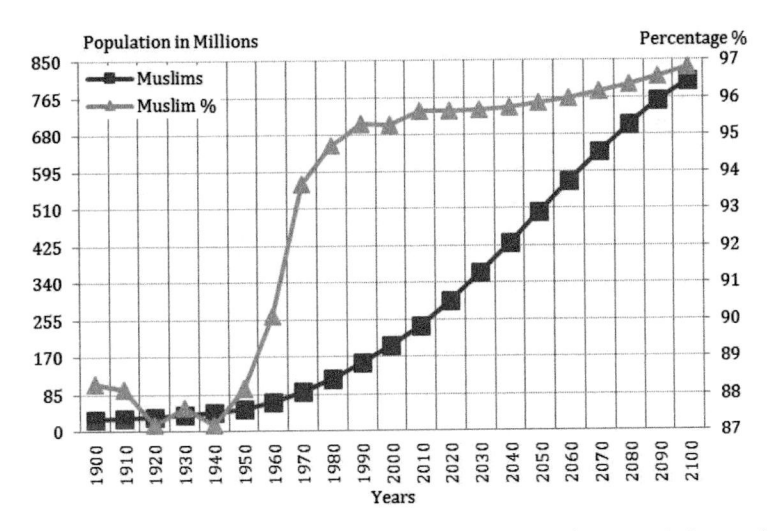

Figure 3.1b Plot of decennial estimates of the Muslim population and its percentage of the total population in North Africa: 1900–2100 AD (1320–1520 H).

3.1.1 Algeria

The People's Democratic Republic of Algeria has a total area of 2,381,741 sq. km, and is the tenth largest country in the world, and the largest in Africa. The map of Algeria is presented in Fig. 3.1.1. It was taken by France from the Ottoman Empire in 1830 and gained its independence from France in 1962. The French conquest expanded gradually by taking Algiers in 1830 and by 1844 the northern one-sixth or 390,000 sq. km was occupied when the French conquered Biskra. The area doubled when the French seized Ouargla in 1853. Another sixth was gained with the capture of El Menia in 1873. The rest or about two-third was seized by 1901 from Morocco, which included the areas of Touat, Gourara, Tidikelt and Hoggar.

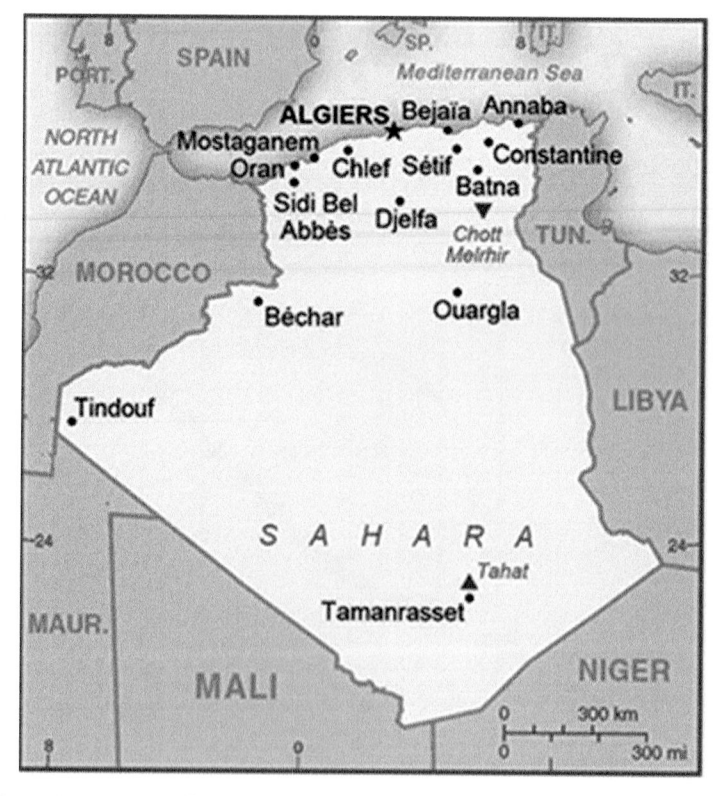

Figure 3.1.1 Map of the People's Republic of Algeria.

Islam entered Algeria between 51 H and 61 H, or 671 AD and 680 AD during the reign of the founder of the Umayyad Dynasty Muawiya bnu Abi Sufyan. The Muslim troops were under the leadership of Abul Mohajir Dinar.

Table 3.1.1 Evolution of the Muslim population in Algeria

Year	Population	Muslims	%	Source
1845	2,142,000	2,028,000	94.68	[DZH54,DZH66]c
1851	2,476,300	2,324,000	93.85	[DZH54,DZH66]c
1856	2,487,700	2,307,300	92.75	[DZH54]c
1861	2,953,700	2,732,900	92.53	[DZH54]c
1866	2,904,000	2,652,100	91.33	[DZH54]c
1872	2,404,700	2,125,100	88.37	[DZH54]c
1876	2,807,700	2,462,900	87.72	[DZH54]c
1881	3,254,900	2,842,500	87.33	[DZH54]c
1886	3,752,000	3,287,200	87.61	[DZH54]c
1891	4,108,000	3,577,100	87.08	[DZH54]c
1896	4,359,600	3,781,100	86.73	[DZH54]c
1901	4,723,000	4,089,200	86.58	[DZH54]c
1906	5,158,100	4,477,800	86.81	[DZH54]c
1911	5,492,600	4,740,500	86.31	[DZH54]c
1921	5,714,600	4,923,200	86.15	[DZH54]c
1926	5,984,100	5,150,800	86.08	[DZH54]c
1931	6,470,000	5,588,300	86.37	[DZH54]c
1936	7,147,200	6,201,100	86.76	[DZH54]c
1948	8,601,400	7,679,100	89.28	[DZH54]c
1954	9,433,400	8,449,300	89.57	[DZH54]c
1966	12,101,994	11,959,000	98.82	[DZH66]c
2014	38,934,000	38,740,000	99.50	[WVS]s
2050	60,923,000	60,619,000	99.50	es
2100	70,705,000	70,351,000	99.50	es

According to census data and as shown in Table 3.1.1, the Muslim population changed from 2.0 million or 95% in 1845, to 2.3 million or 94% in 1851, to 2.3 million or 93% in 1856, to 2.7 million or 93% in 1861, to 2.7 million or 91% in 1866, to 2.1 million or 88% in 1872, to 2.5 million or 88% in 1876, to 2.8 million or 87% in 1881, to 3.3 million or 88% in 1886, to 3.6 million or 87% in 1891, to 3.8 million 87% in 1896, to 4.1 million or 87% in 1901, to 4.5 million or 87% in 1906, to 4.7 million or 86% in 1911, to 4.9 million or 86% in 1921, to 5.2 million or 86% in 1926, to 5.6 million or 86% in 1931, to 6.2 million or 87% in 1936, to 7.7 million or 89% in 1948, to 8.4 million or 90% in 1954, to 12.0 million or 99% in 1966. The post-independence's increase in percentage is due to the migration of the Jewish population and European colonizers. According to the World Values Survey (WVS), the percentage of Muslims was 100.00% in 2002 and 99.50% in 2014. Thus, assuming that the percentage of Muslims remains fixed, the Muslim population is expected to reach sixty million by 2050 and seventy million by 2100.

3.1.2   Chad

The Republic of Chad has an area of 1,284,000 sq. km and its map is presented in Fig. 3.1.2a. It was occupied by France in 1900 and gained its independence from France in 1960. Islam entered here during the eleventh century with Muslim traders. By 1085, Hummay bnu Abdul Jalil founded the Sayfawa Muslim Dynasty of the Kanem-Bornu Empire and ruled until 1097. This helped spread of Islam in this region and the Dynasty lasted until 1389 but the rulers afterwards remained Muslim.

By 1921, the Muslim population was estimated at 0.92 million or 72.4% of the total population, then 0.97 million or 44.1% in 1948. Later census data indicated that the Muslim population increased from 1.04 million or 41.0% in 1964, to 2.10 million or 52.0% in 1975 (estimate), to 3.34 million or 53.9% in 1993, to 6.4 million or 58.4% in 2009. A choropleth map illustrating the presence of Muslims per administrative region is presented in Fig. 3.1.2b.

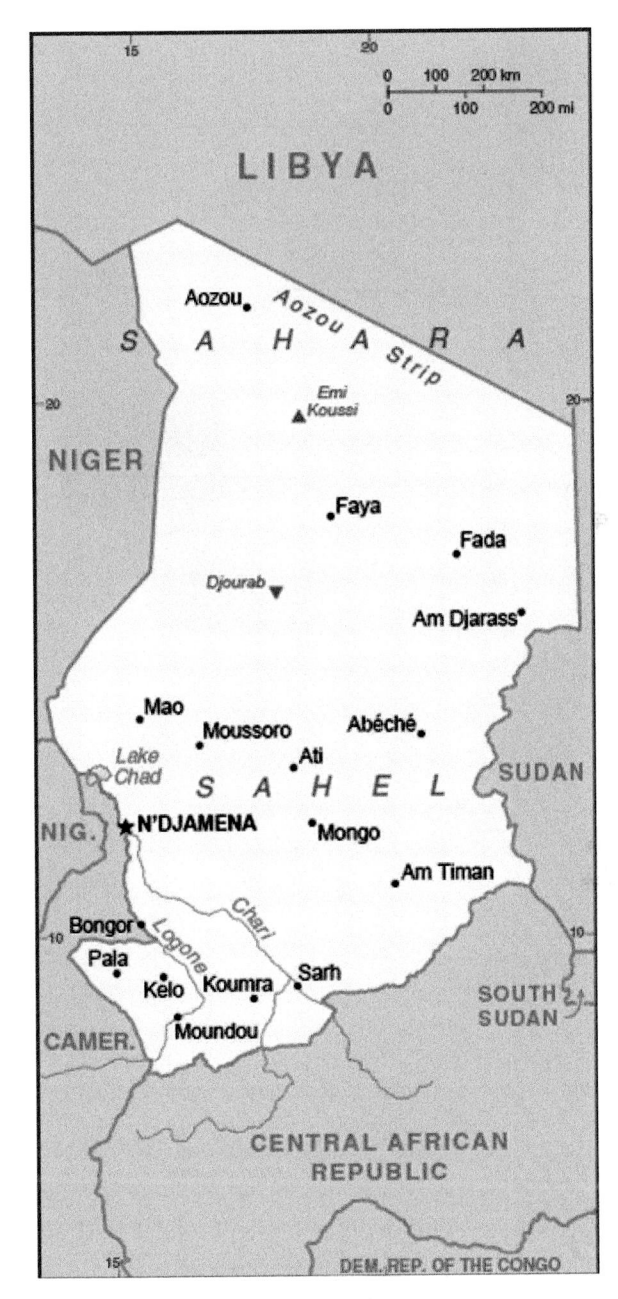

Figure 3.1.2a Map of the Republic of Chad.

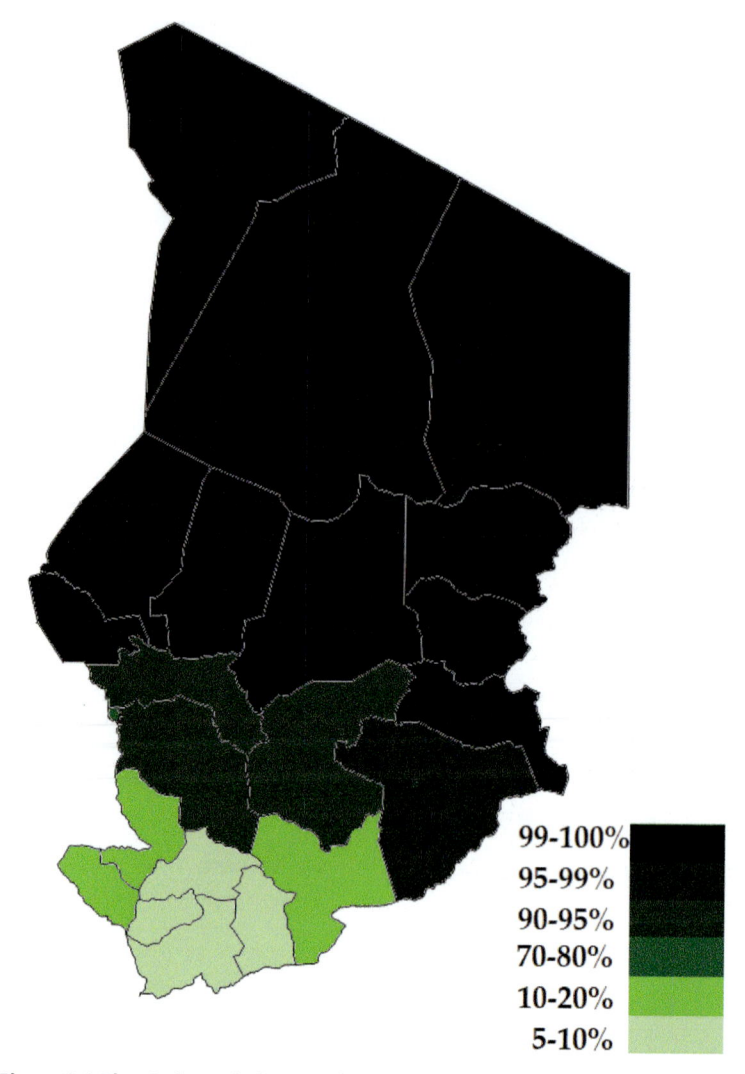

Figure 3.1.2b A choropleth map of Chad illustrating the presence of Muslims per administrative region.

Thus, assuming that the percentage of Muslims will continue to increase by two percentage points per decade, the Muslim population is expected to increase to 22 million or 66% by 2050 and 47 million or 76.0% by 2100. A summary of the data is provided in Table 3.1.2.

Table 3.1.2 Evolution of the Muslim population in Chad

Year	Population	Muslims	%	Source
1921	1,271,371	920,000	72.36	[MAS23]es
1948	2,200,000	970,000	44.09	[MAS55]es
1964	2,524,370	1,035,450	41.02	[UN71]c
1975	4,089,000	2,126,100	52.00	[SYB85]es
1993	6,193,538	3,335,869	53.86	[UN04]c
2009	10,941,682	6,392,040	58.42	[TD]c
2050	34,031,000	22,461,000	66.00	es
2100	61,850,000	47,005,000	76.00	es

3.1.3 Egypt

The Arab Republic of Egypt has a total area of 1,001,450 sq. km including the Sinai Peninsula (59,570 sq. km) which is in Asia. The map of Egypt is presented in Fig. 3.1.3. It was taken by the British from the Ottoman Empire in 1882 and gained its independence from the UK in 1922. It was conquered by Muslims in 20 H/641 AD during the reign of Caliph Omar bnul Khattab. The Muslim troops were led by the Prophet companion Amru bnul Aass and the territories remained under Muslim control since then.

Based on census data and as shown in Table 3.1.3, the Muslim population increased from 3.4 million or 92.7% in 1848, to 4.2 million or 92.5% in 1868, to 9.0 million or 92.2% in 1897, to 10.4 million or 91.8% in 1907, to 11.7 million or 91.4% in 1917, to 12.9 million or 91.0% in 1927, to 14.6 million or 91.4% in 1937, to 17.4 million or 91.7% in 1947, to 24.1 million or 92.8% in 1960, to 27.9 million or 92.9% in 1966, to 34.3 million or 93.7% in 1976, to 45.4 million or 94.1% in 1986, to 56 million or 94.4% in 1996, to 69 million or 94.7% in 2006. Thus, assuming that the percentage of Muslims will increase by a quarter of a percentage point per decade, the Muslim population is expected to reach 153 million or 95.8% by 2050 and 218 million or 97.0% by 2100.

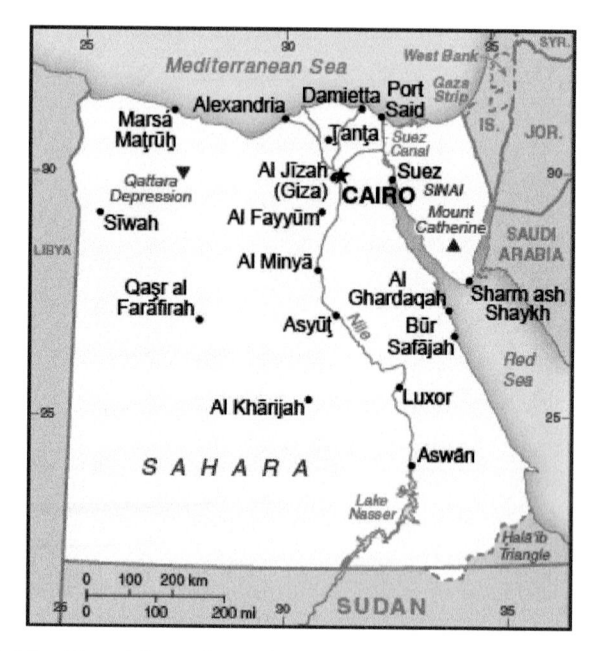

Figure 3.1.3 Map of the Arab Republic of Egypt.

Table 3.1.3 Evolution of the Muslim population in Egypt

Year	Population	Muslims	%	Source
1848	3,642,951	3,377,135	92.70	[EG868]c
1868	4,539,162	4,200,505	92.54	[EG868]c
1897	9,734,405	8,978,775	92.24	[SYB00]c
1907	11,287,359	10,366,526	91.84	[SYB10]c
1917	12,750,918	11,658,148	91.43	[SYB20]c
1927	14,213,364	12,929,260	90.97	[SYB31]c
1937	15,920,694	14,552,695	91.41	[SYB50]c
1947	18,966,767	17,397,946	91.73	[UN56]c
1960	25,984,101	24,105,450	92.77	[EG60]c
1966	30,075,858	27,925,659	92.85	[UN71]c
1976	36,626,204	34,334,328	93.74	[UN83]c
1986	48,205,049	45,368,453	94.12	[UN04]c
1996	59,312,914	55,969,068	94.36	[IPUMS]c
2006	72,798,031	68,936,896	94.70	[IPUMS]c
2050	159,957,000	153,159,000	95.75	es
2100	224,735,000	217,993,000	97.00	es

3.1.4 Libya

Libya has a total area of 1,759,540 sq. km and its map is presented in Fig. 3.1.4. It was taken by Italy from the Ottoman Empire in 1911 and passed to the United Nations in 1943 after the Italian defeat in WWII. It gained its independence in 1951. It used to consist of Cyrenaica (Eastern half with area 934,934 sq. km and a third of the population) and Tripolitania (Western half with area 824,606 and two thirds of the population). It was conquered by Muslims in 21 H/642 AD under the reign of Caliph Omar bnul Khattab. The Muslim troops were led by the Prophet companion Amr bnul Aass and the territories remained under Muslim control since then.

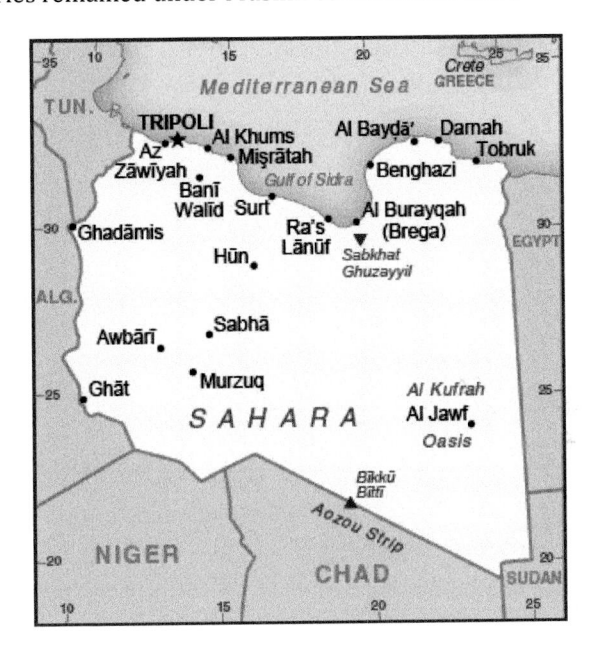

Figure 3.1.4 Map of Libya.

Based on census data and as shown in Table 3.1.4, the Muslim population continued to decrease in percentage due to the Italian occupation from half million or 95.8% in 1911, to 0.8 million or 94.7% in 1921, to 0.6 million or 90.5% in 1931, to 0.7 million or 88.7% in 1936, to 0.8 million or 85.9% in 1938. The Muslim population continued to increase since then with the Italian departure and subsequent independence to a million or 93.3% in 1948, to 1.5 million or 97.2% in 1964, to 2.1 million or 100% in 1973. According to the World values Survey (WVS), the percentage of Muslims was 97.81% in 2013. Thus, assuming that this percentage remains constant, the Muslim population is expected to increase to nine million by 2050 but will decrease to eight million by 2100.

Table 3.1.4 Evolution of the Muslim population in Libya

Year	Population	Muslims	%	Source
1911	523,176	501,000	95.76	[SYB20]c
1921	840,213	795,581	94.69	[MAS23]c
1931	692,016	626,417	90.52	[LY36]c
1936	814,573	722,479	88.69	[LY36]c
1938	888,401	763,179	85.90	[SYB50]c
1948	1,072,000	1,000,000	93.28	[SYB50]es
1964	1,564,369	1,520,517	97.20	[UN71]c
1973	2,052,372	2,051,731	99.97	[UN79]c
2013	6,266,000	6,129,000	97.81	[WVS]s
2050	8,525,000	8,012,000	97.81	es
2100	8,339,000	7,836,000	97.81	es

3.1.5 Mali

The Republic of Mali has an area of 1,240,192 sq. km and its map is presented in Fig. 3.1.5. It was occupied by France in 1892 and gained its independence from France in 1960. Its name was changed in 1958 from French Sudan to the Sudanese Republic and then after independence in 1960 to Mali. Islam entered here during the tenth century through trade. By 1010, Kusoy Muslim Dam, the ruler of the Za Dynasty of the Songhai Empire converted to Islam and ruled from 1008 to 1020. This Dynasty was based in the city of Gao, which is in the east of current Mali.

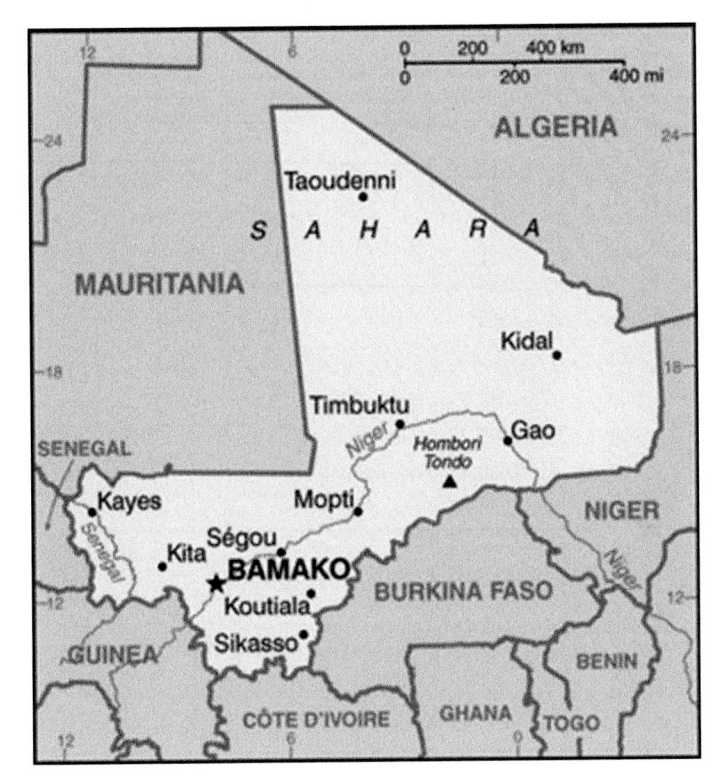

Figure 3.1.5 Map of the Republic of Mali.

Earlier estimates of the size of the Muslim population increased from 1.06 million or 42.9% in 1921, to 1.55 million or 43.1% in 1936, to 1.90 million or 50.2% in 1945, to 4.07 million or 76.5% in 1961. According to Demographic and Health Surveys (DHS), the percentage of Muslims was 76.51% in 1961, 93.37% in 1987, 90.78% in 1996, 93.15% in 2001, 92.58% in 2006 and 93.50% in 2013. However, the first census to record religious affiliation was in 2009, according to which the Muslim population reached 13.8 million or 95.1% of the total population. Thus, assuming that the percentage of Muslims will increase by a quarter of a percentage point per decade, the Muslim population is expected to reach 42 million or 96.3% by 2050 and eighty million or 97.5% by 2100. A summary of the data is provided in Table 3.1.5.

Table 3.1.5 Evolution of the Muslim population in Mali

Year	Population	Muslims	%	Source
1921	2,474,589	1,061,000	42.88	[MAS23]es
1936	3,600,000	1,550,000	43.06	[ML936]c
1945	3,793,000	1,903,000	50.17	[MAS55]es
1961	5,322,000	4,072,000	76.51	[ML61]s
1987	8,073,000	7,538,000	93.37	[DHS]s
1996	9,901,000	8,988,000	90.78	[DHS]s
2001	11,376,000	10,597,000	93.15	[DHS]s
2006	13,310,000	12,323,000	92.58	[DHS]s
2009	14,528,662	13,817,700	95.11	[ML09]c
2050	43,586,000	41,951,000	96.25	es
2100	80,383,000	78,373,000	97.50	es

3.1.6 Mauritania

The Islamic Republic of Mauritania has an area of 1,030,700 sq. km and its map is presented in Fig. 3.1.6. It was occupied by France in 1900 and gained its independence from France in 1960. Islam entered here in the eleventh century Abdullah bnu Yassin and Abu Bakr bnu Omar, the founders of the Almoravid Dynasty. As shown in Table 3.1.6, after independence, the entire population is and remains Muslim. Thus, assuming that the total population remains Muslim, the Muslim population is expected to increase to nine million by 2050 and seventeen million by 2100.

Table 3.1.6 Evolution of the Muslim population in Mauritania

Year	Population	Muslims	%	Source
1910	600,000	600,000	100	[SNH]es
1921	261,746	254,000	97.04	[MAS23]c
1948	518,000	494.100	95.39	[MAS55]es
1977	923,175	923,175	100	[KET86]es
2000	2,508,159	2,508,159	100	[DOS00]es
2013	3,537,368	3,537,368	100	[DOS15]es
2050	9,025,000	9,025,000	100	es
2100	17,065,000	17,065,000	100	es

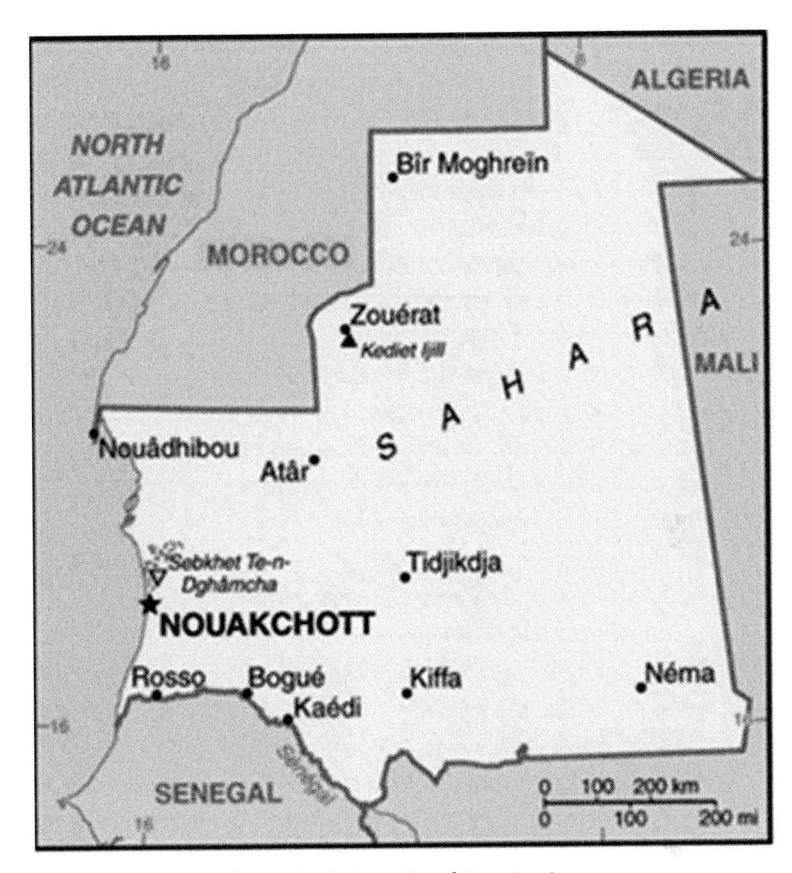

Figure 3.1.6 Map of the Islamic Republic of Mauritania.

3.1.7 Morocco

The Kingdom of Morocco has an area of 712,550 sq. km, including Western Sahara (266,000 sq. km), and its map is presented in Fig. 3.1.7a. Morocco was occupied by Spain and France, while Tangier (583 sq. km) was an international zone from 1912 to 1956. France occupied most of current Morocco proper or 398,521 sq. km from 1912 to 1956. Spain occupied the following:

- Northern coast with area 19,617 sq. km from 1912 to 1956, with the exception of some area that remains under

Spanish control, mainly the enclaves of Ceuta 18 sq. km and the uninhabited island of Perejil or Laila 0.15 sq. km (both occupied since 1415, first by Portugal then Spain since 1580), Melilla 12 sq. km (occupied since 1497), the Chafarinas Islands 0.5 sq. km, three islands that are three km off the coast of Morocco (occupied since 1847), Isla de Alborán 0.07 sq. km and fifty km off the coast of Morocco (occupied since 1540), Peñón de Vélez de la Gomera 0.02 sq. km (a little peninsula occupied since 1564), three little islands off the shore of Alhucima occupied by Spain since 1559: Peñón de Alhucemas 0.02 sq. km, Isla de Tierra 0.02 sq. km and Isla de Mar 0.01 sq. km. A total of 31 sq. km area that Morocco considers occupied territory and demands its return to Moroccan sovereignty. See Fig. 3.1.7b. for a map of this region.

- Ifni, an enclave around and including the city of Sidi Ifni with area 1,367 sq. km, about 150 km south of Agadir and was occupied from 1860 to 1969.
- Cabo Juby or Tarfaya Strip with 32,893 sq. km occupied from 1916 to 1958. It stretched from the Atlantic coast to the Algerian border and north of Western Sahara to Draa River.
- Spanish Sahara, later known as Western Sahara, which consisted of two zones: Sekia el Hamra to the north with area 82,000 sq. km and Rio de Oro in the south with area 184,000 sq. km or about two-thirds of the territory. Western Sahara was occupied by Spain in 1884, which left in 1975 but did not formally return it to Morocco, which annexed much of the territory. However, an area of 70,300 sq. km stretched along the Mauritanian border, comprising 26.4% of Western Sahara, remains under the self-proclaimed Sahrawi Arab Democratic Republic (SADR), but most of its population lives in the de facto capital of Tindouf, Algeria. SADR is recognized by about thirty countries.

Islam entered the northern parts in 62 H/681 AD during the time of the second Umayyad ruler Yazid bnu Muawiya (ruled 680–683). The Muslim troops were under the leadership of Oqba bnu Nafie. The rest of the upper part of Morocco was conquered in 89 H/708

AD during the reign of the sixth Umayyad ruler al-Walid bnu Abdel Malik bnu Marwan (ruled 705–715). The Muslim troops were under the leadership of Musa bnu Nusair. Islam entered the southern part (Western Sahara) during the eleventh century by Abdullah bnu Yassin and Abu Bakr bnu Omar, the founders of the Almoravid Dynasty.

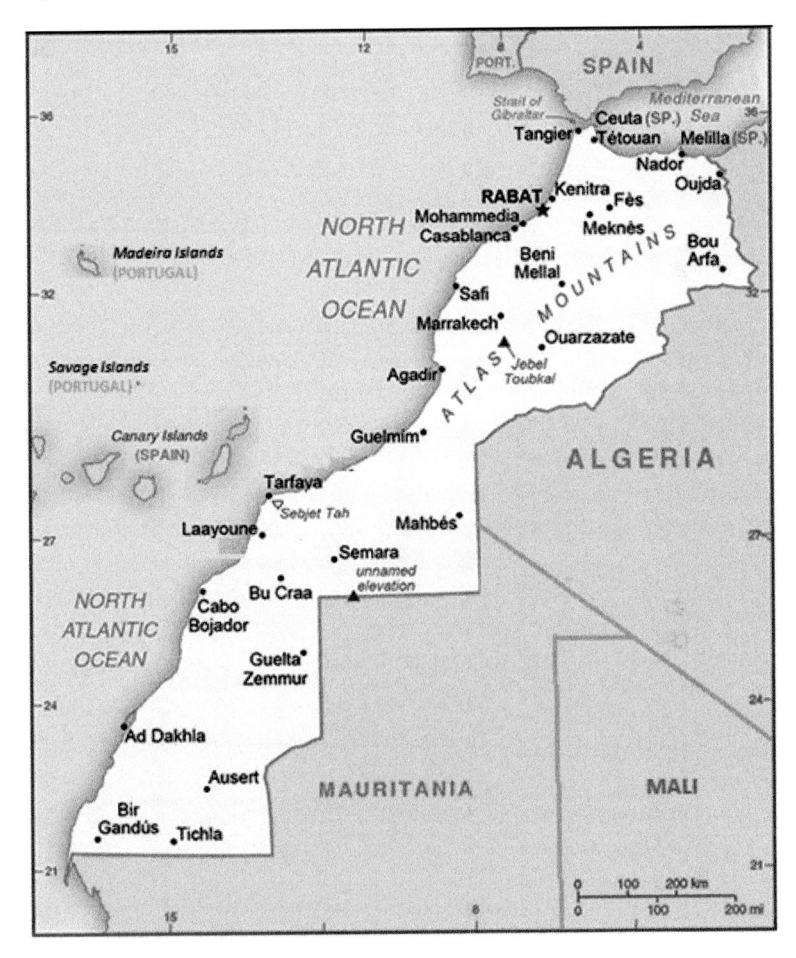

Figure 3.1.7a Map of the Kingdom of Morocco.

Figure 3.1.7b Spanish possessions in Northern Morocco.

The non-Muslim population was estimated in 1910 to be 150,000 Jews and 10,000 Christians, making the Muslim population 4.84 million or 96.8%. Based on census data and as shown in Table 3.1.7, the Muslim population increased from 5.1 million or 93.8% in 1926, to 6.6 million or 93.6% in 1935, to 8.6 million 91.9% in 1952, to 11.1 million or 95.2% in 1960, to 15.2 million or 99.1% in 1971. The post-independence increase in percentage is due to the migration of the Jewish population and European colonizers. According to Afro Barometer Survey (ABS), the percentage of Muslims was 99.58% in 2013. Thus, assuming that the percentage of Muslims will remain constant at 99.6%, the Muslim population is expected to reach 47 million by 2050 and 46 million by 2100.

Table 3.1.7 Evolution of the Muslim population in Morocco

Year	Population	Muslims	%	Source
1910	5,000,000	4,840,000	96.80	[SYB10]es
1926	5,404,860	5,067,743	93.76	[MAH]c
1935	7,040,000	6,590,000	93.61	[MAH]c
1952	9,342,000	8,585,000	91.90	[MAH]c
1960	11,626,470	11,070,841	95.22	[MAH]c
1971	15,379,259	15,236,231	99.07	[MAH]c
2013	34,002,000	33,860,000	99.58	[ABS]s
2050	47,150,000	46,951,000	99.58	es
2100	46,092,000	45,898,000	99.58	es

3.1.8 Niger

The Republic of Niger has an area of 1,267,000 sq. km and its map is presented in Fig. 3.1.8. It was occupied by France between 1900 and 1922 and gained its independence in 1960. Islam entered here during the tenth century through trade. By 1010, Kusoy Muslim Dam, the ruler of the Za Dynasty of the Songhai Empire converted to Islam. This Dynasty was based on Gao, east of current Mali. Starting 1085, the eastern part of current Niger was ruled by the Sayfawa Muslim Dynasty of the Kanem-Bornu Empire, which was centered in Kanem in west Chad.

By 1921, the Muslim population was estimated at 881,000 or 81.3% of the total population. Based on census data, the Muslim population increased from 1.5 million or 98.5% in 1960, to 7.2 million or 98.7% in 1988, to 11 million or 99.3% in 2001 and 17 million or 99.3% in 2012. According to ABS survey data, the percentage of Muslims was 98.50% in 2013 and 99.75% in 2015. Thus, assuming that the percentage of Muslims remains fixed at 99.3%, the Muslim population is expected to increase to 66 million by 2050 and 164 million by 2100. A summary of the data is provided in Table 3.1.8.

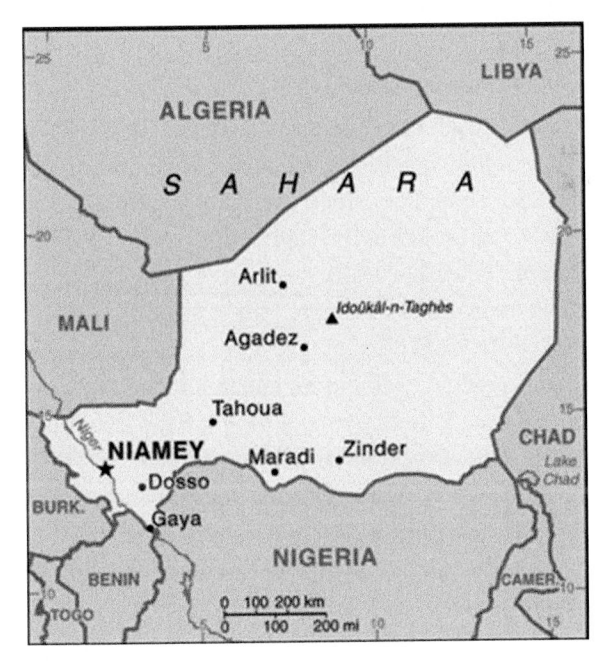

Figure 3.1.8 Map of the Republic of Niger.

Table 3.1.8 Evolution of the Muslim population in Niger

Year	Population	Muslims	%	Source
1910	1,073,800	600,000	55.88	[SNH]es
1921	1,084,043	881,000	81.27	[MAS23]es
1960	1,506,490	1,484,710	98.55	[UN64]c
1988	7,343,519	7,245,037	98.66	[ROD]c
2001	11,043,874	10,968,773	99.32	[NE]c
2012	17,096,099	16,978,889	99.31	[UN]c
2050	65,593,000	65,140,000	99.31	es
2100	164,947,000	163,809,000	99.31	es

3.1.9 Sudan

The Republic of the Sudan has an area of 1,886,068 sq. km and its map is presented in Fig. 3.1.9. It was occupied by the British in 1898 as Anglo-Egyptian Sudan and gained its independence from the UK in 1956, but its southern part seceded as South Sudan (619,745 sq. km) in 2011. Islam entered here through the immigration of Muslim Arabs from Egypt and Arabia between the seventh and ninth centuries. Eventually, the following cities came under Muslim control: Dongola (North) in 1350, Soba, the ruined capital of the Nubian kingdom of Alodia in 1504 (just south of the capital Khartoum), Darfur (East) in 1596, Kordofan (South) in 1600.

According to a 1923 estimate, the total population was 2.8 million all of which were Muslim. The 1993 census is the only census that inquired on religious affiliation and it indicated that the Muslim population was 20.5 million or 96.4% of the total population. According to ABS survey data, the percentage of Muslims was 99.25% in 2015. Thus, assuming that the percentage of Muslims remains constant at 99.25%, the Muslim population is expected to increase to 81 million by 2050 and 141 million by 2100. The data is summarized in Table 3.1.9.

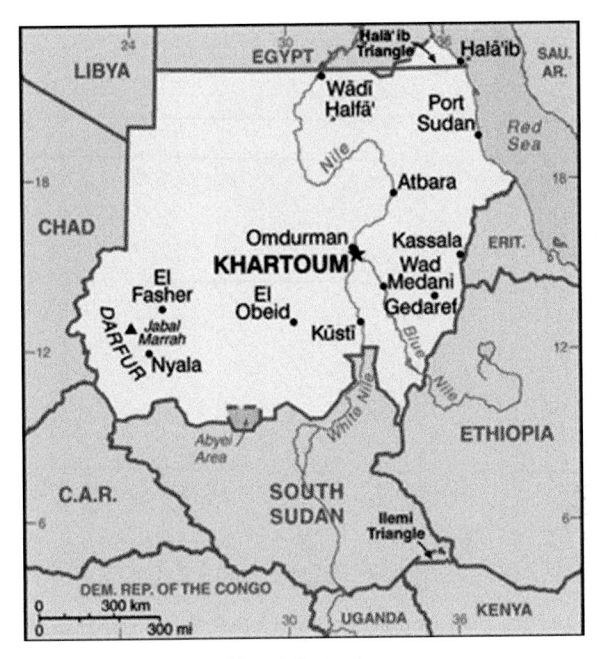

Figure 3.1.9 Map of the Republic of the Sudan.

Table 3.1.9 Evolution of the Muslim population in Sudan

Year	Population	Muslims	%	Source
1923	2.800,000	2,800,000	100	[MAS23]es
1993	21,266,641	20,495,781	96.38	[SD93]c
2015	40,235,000	39,933,238	99.25	[ABS]s
2050	81,193,000	80,584,000	99.25	es
2100	142,342,000	141,275,000	99.25	es

3.1.10 Tunisia

The Tunisian Republic has an area of 163,610 sq. km and its map is presented in Fig. 3.1.10. It was taken by France from the Ottoman Empire in 1881 and gained its independence from France in 1956. Islam entered here in 27 H/647 AD during the reign of the third Muslim Caliph, Othman bnu Affan. The Muslim troops were under the leadership of Abdullah bnu Saad bnu Abi Sarh.

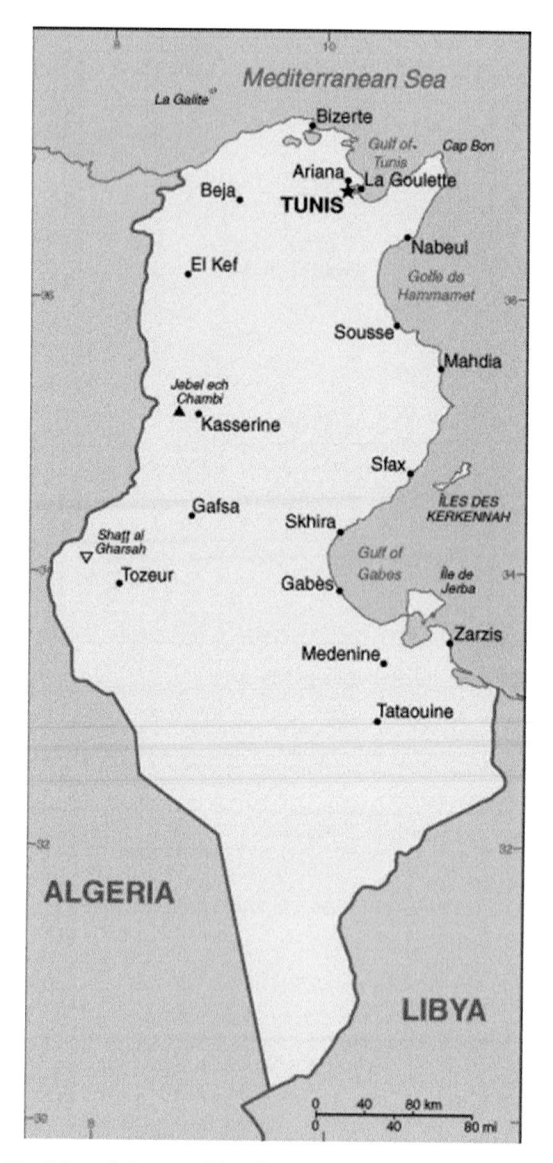

Figure 3.1.10 Map of the Republic of Tunisia.

Estimates of the Muslim population increased from 1.43 million or 95.3% in 1877, to 1.62 million or 95.3% in 1895. Based on census data and as shown in Table 3.1.10, the Muslim population increased from 1.7 million or 89.7% in 1911, to 1.9 million or 88.3% in 1921, to 1.9 million or 87.4% in 1926, to 2.4 million or 88.9% in 1931, to

2.3 million or 87.6% in 1936, to 2.8 million or 88.2% in 1946, to 3.4 million or 90.2% in 1956, to 4.5 million or 99.3% in 1966. The post-independence increase in percentage is due to the migration of the Jewish population and European colonizers. According to Afro Barometer Survey (ABS), the percentage of Muslims was 99.75% in 2015. Thus, assuming that the percentage of Muslims will remain fixed at 99.75%, then the Muslim population is expected to reach fourteen million by 2050 but decrease to thirteen million by 2100.

Table 3.1.10 Evolution of the Muslim population in Tunisia

Year	Population	Muslims	%	Source
1877	1,500,000	1,429,400	95.29	[SYB880]es
1895	1,700,000	1,619,350	95.26	[SYB900]es
1911	1,939,087	1,740,144	89.74	[TN911]c
1921	2,141,669	1,890,132	88.26	[TNH]c
1926	2,212,579	1,933,254	87.38	[TNH]c
1931	2,754,122	2,448,049	88.89	[TNH]c
1936	2,667,300	2,335,651	87.57	[TNH]c
1946	3,230,380	2,848,889	88.19	[TNH]c
1956	3,783,175	3,412,267	90.20	[TNH]c
1966	4,496,834	4,464,067	99.27	[TNH]c
2015	11,254,000	11,226,000	99.75	[ABS]s
2050	13,797,000	13,762,000	99.75	es
2100	12,972,000	12,940,000	99.75	es

3.1.11 Regional Summary and Conclusion

North Africa was the first part of Africa to come under Muslim control within a decade of the death of Prophet Mohammed peace and blessings upon him. Islam spread widely, and the vast majority of this region remains Muslim. Accordingly, North Africa has the largest concentration of Muslims among the five regions spanning Africa. This is expected to remain so for the next three centuries. The following tables present centennial data from 600 AD to 2100 AD (or approximately 1 H to 1500 H) in Table 3.1a and decennial data from 1790 AD to 2100 AD (or 1210 H to 1520 H) in Tables 3.1b and 3.1c for current countries in North Africa. The data includes total population in thousands (P), the percentage of which is Muslim (M%), the corresponding Muslim population in thousands (M), and the annual population growth rate (APGR, or G%) of the total population in this region.

Table 3.1a Centennial estimates of the Muslim population (×1000) in North Africa: 600–2100 AD (1–1500 H)

		600	700	800	900	1000	1100	1200	1300
Algeria	P	2,000	2,000	2,000	2,000	2,000	1,900	1,800	1,700
	M%	—	10.00	40.00	70.00	90.00	90.00	90.00	90.00
	M	—	200	800	1,400	1,800	1,710	1,620	1,530
Chad	P	160	170	180	190	200	220	240	260
	M%	—	1.00	10.00	20.00	30.00	40.00	50.00	60.00
	M	—	2	18	38	60	88	120	156
Egypt	P	4,800	4,850	4,900	4,950	5,000	4,800	4,600	4,400
	M%	—	10.00	40.00	70.00	90.00	90.00	90.00	90.00
	M	—	485	1,960	3,465	4,500	4,320	4,140	3,960
Libya	P	460	470	480	490	500	500	500	500
	M%	—	10.00	40.00	70.00	90.00	90.00	90.00	90.00
	M	—	47	192	343	450	450	450	450
Mali	P	800	850	900	950	1,000	1,100	1,200	1,300
	M%	—	0.10	1.00	5.00	10.00	15.00	20.00	25.00
	M	—	1	9	48	100	165	240	325
Mauritania	P	160	170	180	190	200	220	240	260
	M%	—	—	1.00	5.00	50.00	100.00	100.00	100.00
	M	—	—	2	10	100	220	240	260
Morocco	P	1,600	1,700	1,800	1,900	2,000	1,900	1,800	1,700
	M%	—	10.00	40.00	70.00	80.00	95.00	95.00	95.00
	M	—	170	720	1,330	1,600	1,805	1,710	1,615
Niger	P	480	510	540	570	600	640	680	720
	M%	—	1.00	1.00	5.00	10.00	20.00	30.00	40.00
	M	—	5	5	29	60	128	204	288
Sudan	P	1,820	1,890	1,960	2,030	2,100	2,250	2,400	2,550
	M%	—	10.00	40.00	70.00	90.00	100.00	100.00	100.00
	M	—	189	784	1,421	1,890	2,250	2,400	2,550
Tunisia	P	920	940	960	980	1,000	960	920	880
	M%	—	10.00	40.00	70.00	90.00	90.00	90.00	90.00
	M	—	94	384	686	900	864	828	792
Total	P	13,200	13,550	13,900	14,250	14,600	14,490	14,380	14,270
	M%	—	8.80	35.07	61.53	78.49	82.82	83.12	83.57
	M	—	1,193	4,874	8,769	11,460	12,000	11,952	11,926
	G%		0.026	0.026	0.025	0.024	−0.008	−0.008	−0.008

Table 3.1a (*Continued*)

		1400	1500	1600	1700	1800	1900	2000	2100
Algeria	P	1,600	1,500	2,250	1,750	2,300	4,675	31,042	70,705
	M%	90.00	90.00	90.00	90.00	88.37	86.58	99.50	99.50
	M	1,440	1,350	2,025	1,575	2,033	4,048	30,887	70,351
Chad	P	280	300	350	400	475	675	8,356	61,850
	M%	70.00	70.00	70.00	70.00	72.36	72.36	55.63	76.00
	M	196	210	245	280	344	488	4,648	47,006
Egypt	P	4,200	4,000	5,000	4,500	3,854	10,186	68,832	224,735
	M%	90.00	90.00	90.00	90.00	92.70	92.24	94.36	97.00
	M	3,780	3,600	4,500	4,050	3,573	9,396	64,949	217,993
Libya	P	500	500	500	500	533	747	5,358	8,012
	M%	90.00	90.00	90.00	90.00	95.76	95.76	99.97	97.81
	M	450	450	450	450	510	715	5,356	7,836
Mali	P	1,400	1,600	1,800	2,000	2,440	2,672	10,946	80,383
	M%	30.00	30.00	35.00	40.00	42.88	42.88	93.15	97.50
	M	420	480	630	800	1,046	1,146	10,197	78,373
Mauritania	P	280	300	350	400	440	502	2,630	17,065
	M%	100.00	100.00	100.00	100.00	100.00	100.00	100.00	100.00
	M	280	300	350	400	440	502	2,630	17,065
Morocco	P	1,600	1,500	2,250	1,750	2,300	4,675	29,108	46,092
	M%	95.00	95.00	95.00	95.00	96.80	96.80	99.58	99.58
	M	1,520	1,425	2,138	1,663	2,226	4,525	28,986	45,898
Niger	P	760	800	850	900	960	1,060	11,332	164,947
	M%	50.00	60.00	70.00	75.00	81.27	81.27	99.32	99.31
	M	380	480	595	675	780	861	11,255	163,809
Sudan	P	2,700	2,800	2,940	3,080	3,360	4,522	27,275	142,342
	M%	100.00	100.00	100.00	100.00	100.00	100.00	96.38	99.25
	M	2,700	2,800	2,940	3,080	3,360	4,522	26,288	141,275
Tunisia	P	840	800	1,000	800	825	1,900	9,708	12,972
	M%	90.00	90.00	90.00	90.00	95.29	89.74	99.75	99.75
	M	756	720	900	720	786	1,705	9,684	12,940
Total	P	14,160	14,100	17,290	16,080	17,487	31,614	204,587	829,102
	M%	84.19	83.79	85.44	85.15	86.34	88.28	95.26	96.80
	M	11,922	11,815	14,773	13,693	15,098	27,909	194,880	802,546
	G%	−0.008	−0.004	0.204	−0.073	0.084	0.592	1.867	1.399

Table 3.1b Decennial estimates of the Muslim population (×1000) in North Africa: 1790–1940 AD (1210–1360 H)

		1790	1800	1810	1820	1830	1840	1850	1860
Algeria	P	2,100	2,300	2,500	2,689	2,900	3,100	3,300	3,500
	M%	88.37	88.37	88.37	88.37	88.37	88.37	88.37	88.37
	M	1,856	2,033	2,209	2,376	2,563	2,739	2,916	3,093
Chad	P	470	475	480	485	490	495	500	505
	M%	72.36	72.36	72.36	72.36	72.36	72.36	72.36	72.36
	M	340	344	347	351	355	358	362	365
Egypt	P	3,575	3,854	4,115	4,194	4,662	4,304	4,752	5,000
	M%	92.70	92.70	92.70	92.70	92.70	92.70	92.70	92.54
	M	3,314	3,573	3,815	3,888	4,322	3,990	4,405	4,627
Libya	P	530	533	536	538	560	585	610	635
	M%	95.76	95.76	95.76	95.76	95.76	95.76	95.76	95.76
	M	508	510	513	515	536	560	584	608
Mali	P	2,430	2,440	2,450	2,460	2,470	2,480	2,490	2,500
	M%	42.88	42.88	42.88	42.88	42.88	42.88	42.88	42.88
	M	1,042	1,046	1,051	1,055	1,059	1,063	1,068	1,072
Mauritania	P	435	440	445	450	455	460	465	470
	M%	100.00	100.00	100.00	100.00	100.00	100.00	100.00	100.00
	M	435	440	445	450	455	460	465	470
Morocco	P	2,100	2,300	2,500	2,689	2,900	3,100	3,300	3,500
	M%	96.80	96.80	96.80	96.80	96.80	96.80	96.80	96.80
	M	2,033	2,226	2,420	2,603	2,807	3,001	3,194	3,388
Niger	P	950	960	970	980	990	1,000	1,010	1,020
	M%	81.27	81.27	81.27	81.27	81.27	81.27	81.27	81.27
	M	772	780	788	796	805	813	821	829
Sudan	P	3,320	3,360	3,400	3,437	3,500	3,600	3,750	3,900
	M%	100.00	100.00	100.00	100.00	100.00	100.00	100.00	100.00
	M	3,320	3,360	3,400	3,437	3,500	3,600	3,750	3,900
Tunisia	P	800	825	850	875	900	950	1,000	1,100
	M%	95.29	95.29	95.29	95.29	95.29	95.29	95.29	95.29
	M	762	786	810	834	858	905	953	1,048
Total	P	16,710	17,487	18,246	18,797	19,827	20,074	21,177	22,130
	M%	86.07	86.34	86.59	86.74	87.05	87.13	87.44	87.67
	M	14,382	15,098	15,798	16,305	17,259	17,490	18,518	19,401
	G%		0.455	0.425	0.298	0.533	0.124	0.535	0.440

Table 3.1b (*Continued*)

		1870	1880	1890	1900	1910	1920	1930	1940
Algeria	P	3,776	4,183	4,325	4,675	5,378	5,785	6,507	7,614
	M%	88.37	87.33	87.08	86.58	86.31	86.15	86.37	86.76
	M	3,337	3,653	3,766	4,048	4,642	4,984	5,620	6,606
Chad	P	510	513	588	675	888	1,271	1,536	2,021
	M%	72.36	72.36	72.36	72.36	72.36	72.36	72.36	44.09
	M	369	371	425	488	643	920	1,111	891
Egypt	P	5,252	6,817	8,777	10,186	11,740	13,222	14,767	16,887
	M%	92.54	92.54	92.24	92.24	91.84	91.43	90.97	91.41
	M	4,860	6,309	8,096	9,396	10,782	12,089	13,434	15,436
Libya	P	660	688	717	747	780	811	841	873
	M%	95.76	95.76	95.76	95.76	95.76	94.69	90.52	85.90
	M	632	659	687	715	747	768	761	750
Mali	P	2,510	2,520	2,529	2,672	2,824	2,984	3,153	3,331
	M%	42.88	42.88	42.88	42.88	42.88	42.88	50.00	60.00
	M	1,076	1,081	1,084	1,146	1,211	1,280	1,577	1,999
Mauritania	P	475	480	482	502	600	256	676	747
	M%	100.00	100.00	100.00	100.00	100.00	97.04	97.04	95.39
	M	475	480	482	502	600	248	656	713
Morocco	P	3,776	4,183	4,325	4,675	5,000	5,137	6,206	7,498
	M%	96.80	96.80	96.80	96.80	96.80	93.76	93.76	93.61
	M	3,655	4,049	4,187	4,525	4,840	4,816	5,819	7,019
Niger	P	1,030	1,040	1,050	1,060	1,074	1,084	1,567	1,944
	M%	81.27	81.27	81.27	81.27	81.27	81.27	81.27	81.27
	M	837	845	853	861	873	881	1,274	1,580
Sudan	P	4,080	4,222	4,369	4,522	4,728	4,843	4,995	5,151
	M%	100.00	100.00	100.00	100.00	100.00	100.00	100.00	100.00
	M	4,080	4,222	4,369	4,522	4,728	4,843	4,995	5,151
Tunisia	P	1,176	1,200	1,850	1,900	1,957	2,094	2,411	2,841
	M%	95.29	95.29	95.26	89.74	89.74	88.26	88.89	87.57
	M	1,121	1,143	1,762	1,705	1,756	1,848	2,143	2,488
Total	P	23,245	25,846	29,012	31,614	34,969	37,487	42,659	48,907
	M%	87.94	88.26	88.62	88.28	88.14	87.17	87.65	87.17
	M	20,442	22,812	25,712	27,909	30,821	32,677	37,389	42,632
	G%	0.492	1.061	1.155	0.859	1.009	0.696	1.292	1.367

Table 3.1c Decennial estimates of the Muslim population (×1000) in North Africa: 1950–2100 AD (1370–1520 H)

		1950	1960	1970	1980	1990	2000	2010	2020
Algeria	P	8,872	11,058	14,465	19,222	25,759	31,042	35,977	43,851
	M%	89.28	89.57	98.82	98.82	99.50	99.50	99.50	99.50
	M	7,921	9,905	14,294	18,995	25,630	30,887	35,798	43,632
Chad	P	2,502	3,002	3,644	4,514	5,963	8,356	11,952	16,426
	M%	44.09	41.02	52.00	52.00	53.86	55.63	58.42	60.00
	M	1,103	1,231	1,895	2,348	3,212	4,648	6,982	9,856
Egypt	P	20,452	26,633	34,514	43,309	56,134	68,832	82,761	102,334
	M%	91.73	92.77	92.85	93.74	94.12	94.36	94.70	95.00
	M	18,761	24,707	32,046	40,598	52,834	64,949	78,375	97,218
Libya	P	1,125	1,448	2,134	3,219	4,437	5,358	6,198	6,871
	M%	88.69	97.20	99.97	99.97	99.97	99.97	97.81	97.81
	M	997	1,408	2,133	3,219	4,435	5,356	6,062	6,721
Mali	P	4,708	5,264	5,949	7,090	8,450	10,946	15,049	20,251
	M%	70.00	76.51	80.00	90.00	93.37	93.15	95.11	95.50
	M	3,296	4,027	4,759	6,381	7,890	10,197	14,313	19,340
Mauritania	P	651	850	1,147	1,541	2,034	2,630	3,494	4,650
	M%	95.39	95.39	100.00	100.00	100.00	100.00	100.00	100.00
	M	621	811	1,147	1,541	2,034	2,630	3,494	4,650
Morocco	P	9,000	12,361	16,082	20,141	25,025	29,108	32,824	37,508
	M%	91.90	95.22	99.07	99.07	99.58	99.58	99.58	99.58
	M	8,271	11,770	15,932	19,954	24,920	28,986	32,686	37,350
Niger	P	2,560	3,389	4,511	5,989	8,027	11,332	16,464	24,207
	M%	98.55	98.55	98.55	98.66	98.66	99.32	99.31	99.31
	M	2,523	3,340	4,445	5,909	7,919	11,255	16,350	24,040
Sudan	P	5,734	7,544	10,282	14,507	20,148	27,275	34,545	43,849
	M%	96.38	96.38	96.38	96.38	96.38	96.38	99.25	99.25
	M	5,526	7,271	9,910	13,982	19,418	26,288	34,286	43,520
Tunisia	P	3,605	4,178	5,064	6,374	8,242	9,708	10,635	11,819
	M%	88.19	90.20	99.27	99.27	99.75	99.75	99.75	99.75
	M	3,180	3,769	5,027	6,328	8,222	9,684	10,609	11,789
Total	P	59,209	75,728	97,790	125,907	164,219	204,587	249,900	311,766
	M%	88.16	90.11	93.66	94.72	95.31	95.26	95.62	95.62
	M	52,199	68,240	91,588	119,253	156,514	194,880	238,955	298,114
	G%	1.912	2.461	2.557	2.527	2.657	2.198	2.001	2.212

Table 3.1c (Continued)

		2030	2040	2050	2060	2070	2080	2090	2100
Algeria	P	50,361	55,640	60,923	64,979	67,357	69,019	70,306	70,705
	M%	99.50	99.50	99.50	99.50	99.50	99.50	99.50	99.50
	M	50,109	55,362	60,619	64,654	67,021	68,674	69,954	70,351
Chad	P	21,690	27,643	34,031	40,563	46,923	52,782	57,824	61,850
	M%	62.00	64.00	66.00	68.00	70.00	72.00	74.00	76.00
	M	13,448	17,691	22,461	27,583	32,846	38,003	42,790	47,006
Egypt	P	120,832	140,350	159,957	177,538	193,263	206,565	217,250	224,735
	M%	95.25	95.50	95.75	96.00	96.25	96.50	96.75	97.00
	M	115,092	134,035	153,159	170,437	186,015	199,335	210,189	217,993
Libya	P	7,606	8,151	8,525	8,637	8,557	8,431	8,263	8,012
	M%	97.81	97.81	97.81	97.81	97.81	97.81	97.81	97.81
	M	7,440	7,972	8,339	8,448	8,370	8,246	8,082	7,836
Mali	P	26,957	34,939	43,586	52,430	61,045	68,868	75,399	80,383
	M%	95.75	96.00	96.25	96.50	96.75	97.00	97.25	97.50
	M	25,812	33,541	41,951	50,595	59,061	66,802	73,326	78,373
Mauritania	P	5,967	7,432	9,025	10,684	12,361	14,020	15,608	17,065
	M%	100.00	100.00	100.00	100.00	100.00	100.00	100.00	100.00
	M	5,967	7,432	9,025	10,684	12,361	14,020	15,608	17,065
Morocco	P	41,625	44,840	47,150	48,527	48,934	48,500	47,456	46,092
	M%	99.58	99.58	99.58	99.58	99.58	99.58	99.58	99.58
	M	41,450	44,652	46,951	48,323	48,729	48,296	47,257	45,898
Niger	P	34,846	48,746	65,593	84,726	105,284	126,183	146,411	164,947
	M%	99.31	99.31	99.31	99.31	99.31	99.31	99.31	99.31
	M	34,606	48,410	65,140	84,141	104,558	125,312	145,401	163,809
Sudan	P	55,254	67,919	81,193	94,749	108,222	120,912	132,318	142,342
	M%	99.25	99.25	99.25	99.25	99.25	99.25	99.25	99.25
	M	54,839	67,410	80,584	94,038	107,410	120,005	131,326	141,275
Tunisia	P	12,756	13,353	13,797	13,964	13,847	13,614	13,339	12,972
	M%	99.75	99.75	99.75	99.75	99.75	99.75	99.75	99.75
	M	12,724	13,320	13,762	13,929	13,812	13,580	13,305	12,940
Total	P	377,894	449,013	523,779	596,797	665,793	728,894	784,174	829,102
	M%	95.66	95.73	95.84	95.98	96.15	96.35	96.57	96.80
	M	361,486	429,824	501,991	572,833	640,182	702,274	757,238	802,546
	G%	1.924	1.724	1.540	1.305	1.094	0.905	0.731	0.557

3.2 Muslims in West Africa

This region consists of twelve countries: Benin, Burkina Faso, Cabo Verde, Côte d'Ivoire (Ivory Coast), Gambia, Ghana, Guinea, Guinea Bissau, Liberia, Senegal, Sierra Leone, and Togo. Islam entered this region towards the end of the tenth century through interaction with Muslim traders. It was strengthened by its adoption by the rulers of the Tekrur Kingdom in current Senegal in 1030 and the rulers of the Songhai Empire in current Mali in 1010 and the start of Almoravid movement by Abdullah bnu Yassin in 1040. Islam then continued to spread in this region through preaching and commerce.

Thus, the Muslim population remained negligible for the first millennium, and has likely reached 32,000 or 1.1% in 1100 AD, increasing to 68,000 or 2.0% in 1200 AD, to 0.11 million or 2.9% in 1300 AD, to 0.21 million or 5.1% in 1400 AD, to 0.32 million or 7.2% in 1500 AD, to 0.53 million or 9.4% in 1600 AD, to 1.1 million or 14.7% in 1700 AD, to 1.6 million or 20.1% in 1800 AD, to 3.1 million or 26.6% in 1900, to 41 million or 46.3% in 2000, to 74 million or 50.3% in 2020, and is projected to reach 0.15 billion or 54% by 2050 and 0.29 billion or 60% by 2100.

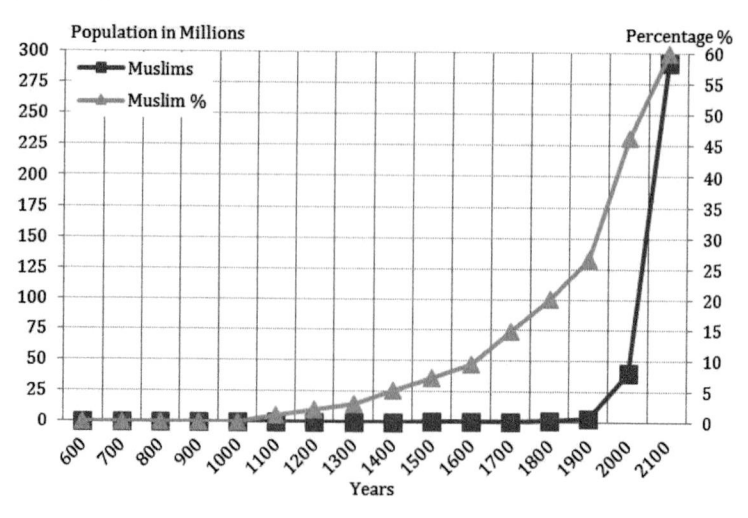

Figure 3.2a Plot of centennial estimates of the Muslim population and its percentage of the total population in West Africa: 600–2100 AD (1–1500 H).

A plot of centennial estimates of the Muslim population and its percentage with respect to the total population in this region from 600 to 2100 is provided in Fig. 3.2a. A zoom in of this plot, providing a plot of decennial estimates of the Muslim population and its percentage with respect to the total population in this region from 1900 to 2100 is provided in Fig. 3.2b. This shows that the Muslim population in this region was increasing slowly until 1970 and is increasing substantially afterwards and towards the end of this century. The percentage of Muslims on the other hand, more than doubled from 22% in 1860, to 46% in 1990, and is continuing its sharp increase at around one percentage point per decade for the rest of this century, reaching 60% by its end.

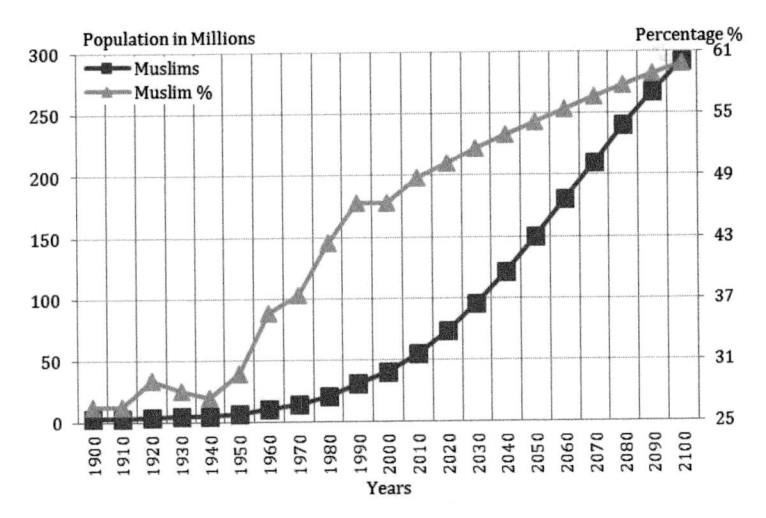

Figure 3.2b Plot of decennial estimates of the Muslim population and its percentage of the total population in West Africa: 1900–2100 AD (1320–1520 H).

The corresponding individual data for each country in this region is discussed below. In Section 3.2.13, the total population in each country in this region and the corresponding percentage and number of Muslims is presented centennially in Table 3.2a from 600 to 2100 and decennially in Tables 3.2b and 3.2c from 1790 to 2100.

3.2.1 Benin

The Republic of Benin has an area of 112,622 sq. km and its map is presented in Fig. 3.2.1a. It was occupied by the French in 1872

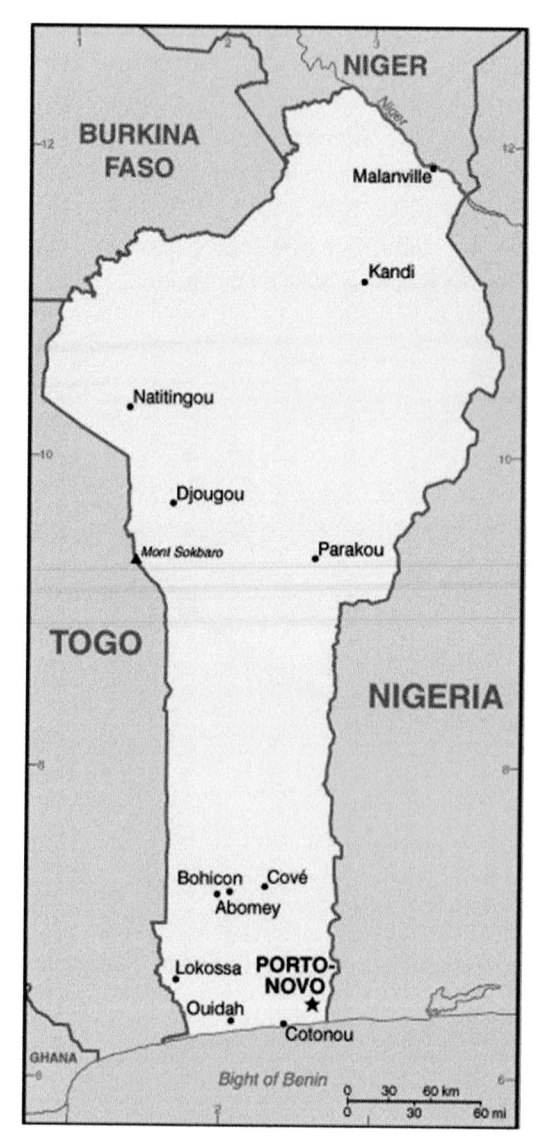

Figure 3.2.1a Map of the Republic of Benin.

and gained its independence from France in 1960 as the Republic of Dahomey but changed its name to Benin in 1975. Islam entered here through trade as early as the fourteenth century. Estimate of the Muslim population increased from 50,000 or 7.6% in 1910 to 70,000 or 8.3% in 1921, to 0.11 million or 7.0% in 1948. According to census data, the Muslim population increased to 0.15 million or 13.6% in 1961, to 1.01 million or 20.7% in 1992, to 1.65 million or 24.4% in 2002 and 2.77 million or 27.7% in 2013. A choropleth map illustrating the presence of Muslims per department is presented in Fig. 3.2.1b.

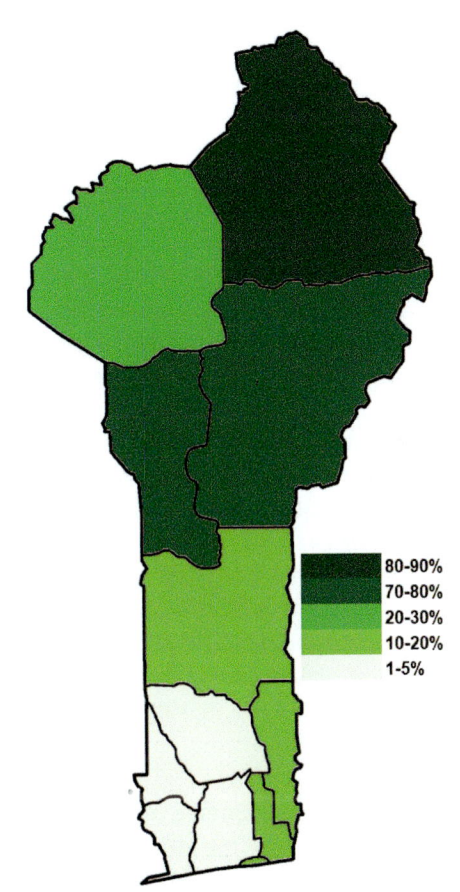

Figure 3.2.1b A choropleth map of Benin illustrating the presence of Muslims per department.

Thus, assuming that the percentage of Muslims will increase by two percentage points per decade, the Muslim population is expected to reach nine million or 36% by 2050 and 22 million or 46% by 2100. A summary of the data is provided in Table 3.2.1.

Table 3.2.1 Evolution of the Muslim population in Benin

Year	Population	Muslims	%	Source
1910	655,000	50,000	7.63	[SNH]es
1921	842,137	70,000	8.31	[MAS23]es
1948	1,505,000	105,000	6.98	[MAS55]es
1961	1,111,113	151,486	13.63	[BJ61]c
1992	4,882,950	1,011,193	20.71	[UN04]c
2002	6,769,914	1,652,953	24.42	[BJ02]c
2013	10,008,749	2,772,000	27.70	[BJ13]c
2050	24,280,000	8,741,000	36.00	es
2100	47,209,000	21,716,000	46.00	es

3.2.2 Burkina Faso

It has a total area of 274,200 sq. km and its map is presented in Fig. 3.2.1a. It was occupied by France in 1896. It gained independence from France in 1960 as Upper Volta but changed its name to Burkina Faso in 1984. Islam entered here with Muslim merchants coming from the northwest during the fifteenth century. However, by 1890 the Muslim population was estimated at 30,000 or 1.4% of the total population. The Muslim population then increased substantially to 250,000 or 10.2% in 1910, to 0.44 million or 14.9% in 1921, to 0.70 million or 22.8% in 1948, to 1.18 million or 27.5% in 1961 (demographic survey), then almost doubled to 2.03 million or 36% in 1975, then almost doubled again to 3.58 million or 45% in 1985. Later census data showed that the Muslim population increased to 5.8 million or 55.9% in 1996, to 8.5 million or 60.5% in 2006. A choropleth map illustrating the presence of Muslims per administrative region is presented in Fig. 3.2.2b.

Thus, assuming that the percentage of Muslims will increase by one percentage point per decade, the Muslim population is expected to reach 28 million or 65% by 2050 and 58 million or 70% by 2100. The data is summarized in Table 3.2.2.

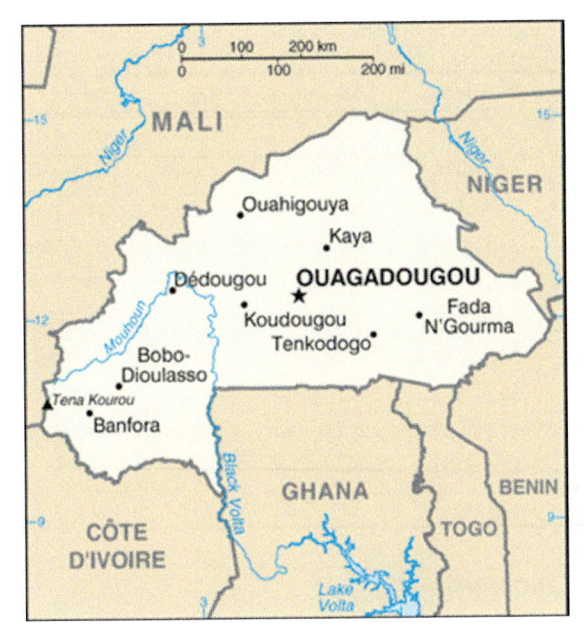

Figure 3.2.2a Map of Burkina Faso.

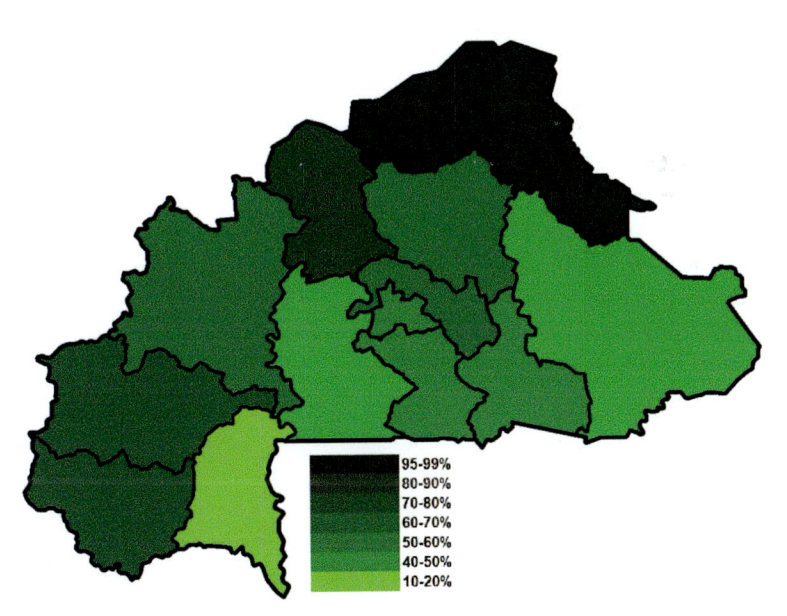

Figure 3.2.2b A choropleth map of Burkina Faso illustrating the presence of Muslims per administrative region.

Table 3.2.2 Evolution of the Muslim population in Burkina Faso

Year	Population	Muslims	%	Source
1890	2,135,000	30,000	1.41	[IML]es
1910	2,455,076	250,000	10.18	[SNH]es
1921	2,973,442	444,000	14.93	[MAS23]es
1948	3,070,000	700,000	22.80	[MAS55]es
1961	4,310,900	1,184,786	27.48	[BF61]]s
1975	5,638,203	2,030,000	36.00	[SYB85]es
1985	7,964,705	3,584,120	45.00	[SYB90]es
1996	10,312,609	5,764,748	55.90	[UN]c
2006	14,017,262	8,485,149	60.53	[BF06]c
2050	43,432,000	28,231,000	65.00	es
2100	83,194,000	58,236,000	70.00	es

3.2.3 Cabo Verde

The Republic of Cabo Verde is an island nation located in the Atlantic Ocean off the coast of Senegal and Mauritania and its map is presented in Fig. 3.2.3. It changed its name from Cape Verde in 2013. It was uninhabited when it was discovered by the Portuguese in 1456 who named it the Green Cape. It remained under their control until it gained its independence from Portugal in 1975. It has a total area of 4,033 sq. km, comprising ten islands and five islets, the largest of which are Santiago (991 sq. km) where more than half the population lives and the capital Praia is located, Santo Antão (779 sq. km) with a tenth of the total population, Boavista (620 sq. km), Fogo (476 sq. km) with 8% of the total population, São Nicolau (388 sq. km), Maio (269 sq. km), Sao Vicente (227 sq. km) with a sixth of the total population, Sal (216 sq. km), Brava (67 sq. km), Santa Luzia (35 sq. km) and Raso (7 sq. km).

Islam was brought to these islands in mid 1990s by immigrants from Guinea Bissau, Senegal and Nigeria. According to 1950 census, there were five people affiliated with a non-Christian religion. Estimates of the Muslim population increased from none in 1990, to 1,000 or 0.2% in 2000. The first census recording Muslim was in 2010, which recorded a total population of 491,683 of which

336,050 are fifteen years or older, of whom 6,008 or 1.79% are Muslim. The Muslim population continues to increase because of immigration from predominant Muslim countries and conversion of locals. The data is summarized in Table 3.2.3. Thus, assuming that the percentage of Muslims will continue to increase by a quarter of a percentage point per decade, the Muslim population is expected to reach 19,000 or 2.8% by 2050 and 24,000 or 4.0% by 2100.

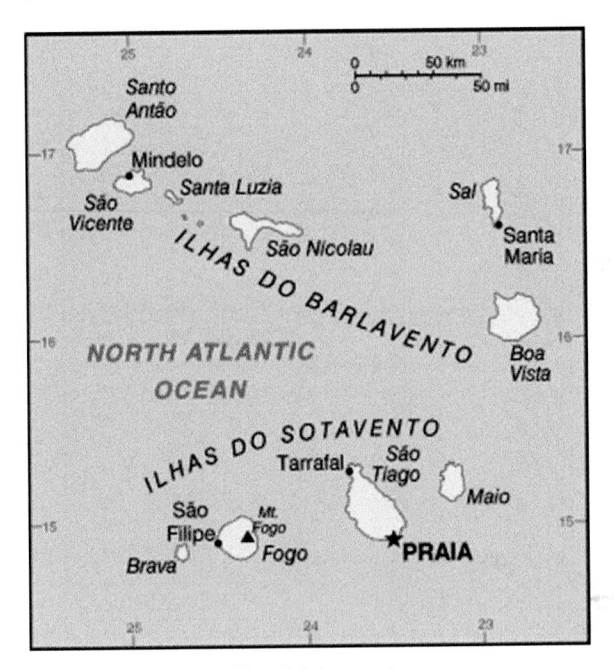

Figure 3.2.3 Map of the Republic of Cabo Verde.

Table 3.2.3 Evolution of the Muslim population in Cabo Verde

Year	Population	Muslims	%	Source
1950	148,331	<5	0.00	[UN56]c
1990	341,491	0	0.00	[CV90]es
2000	436,823	1,000	0.23	[PEW]es
2010	336,050	6,008	1.79	[CV10]c
2050	679,000	19,000	2.75	es
2100	604,000	24,000	4.00	es

3.2.4 Côte d'Ivoire (Ivory Coast)

The Republic of Côte d'Ivoire has an area of 322,463 sq. km and its map is presented in Fig. 3.2.4a. It was occupied by France in 1886 and gained its independence from France in 1960. Islam entered here through trade starting in the eighteenth century. The traders, who were mostly from Dyula, established Muslim communities in the northern districts which became centers of Islamization for the whole region.

Figure 3.2.4a Map of the Republic of Côte d'Ivoire.

Accordingly, estimates of the Muslim population decreased from 0.3 million or 12.5% in 1910, to 0.2 million or 9.9% in 1925, and continued to increase since then to 0.7 million or 21.7% in 1957, to 0.9 million or 23.5% in 1965. Post-independence censuses indicate an increase to 2.2 million or 33.5% in 1975, to 4.2 million or 38.7% in 1988, to 5.9 million or 38.7% in 1998 and 9.7 million or 42.9% in 2014. At the last census, 24.2% or 5.5 million of the total population were foreigners, of whom 72.7% or 4.0 million are Muslim (70.5% in 1998), and therefore, 33.7% of Ivoirian citizens were Muslim, which increased from 27.4% in 1998. A choropleth map illustrating the presence of Muslims per administrative region is presented in Fig. 3.2.4b.

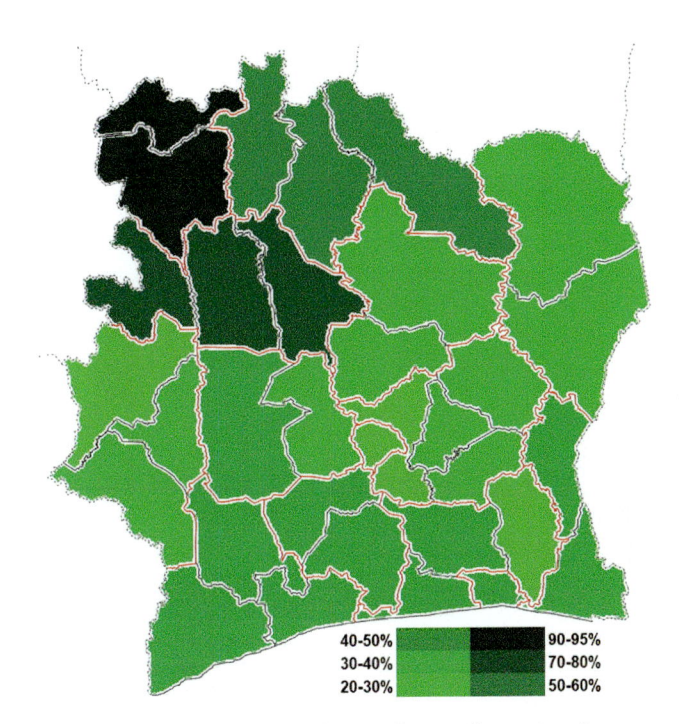

Figure 3.2.4b A choropleth map of Côte d'Ivoire illustrating the presence of Muslims per administrative region.

Thus, assuming that the percentage of Muslims will increase by one percentage point per decade, the Muslim population is expected to increase to 24 million or 47% by 2050 and fifty million or 52% by 2100. The data is summarized in Table 3.2.4.

Table 3.2.4 Evolution of the Muslim population in Côte d'Ivoire

Year	Population	Muslims	%	Source
1910	2,000,000	250,000	12.50	[SNH]es
1925	1,724,545	170,112	9.86	[MAS55]c
1957	3,048,000	661,416	21.70	[CIH]es
1965	3,840,000	902,400	23.50	[SYB70]es
1975	6,663,199	2,231,044	33.48	[CI98]c
1988	10,815,694	4,182,410	38.67	[CI98]c
1998	15,258,024	5,931,958	38.88	[CI98]c
2014	22,436,889	9,618,121	42.87	[UN]c
2050	51,264,000	24,094,000	47.00	es
2100	96,633,000	50,249,000	52.00	es

3.2.5 The Gambia

The Republic of the Gambia has an area of 11,295 sq. km and is the smallest country in the continent of Africa. Its map is presented in Fig. 3.2.5. It was occupied by the British in 1783 and gained its independence from the UK in 1965. Islam entered here after 1050 by Abdullah bnu Yassin and Abu Bakr bnu Omar, the founders of the Almoravid Dynasty with its later capital in Marrakech, current Morocco.

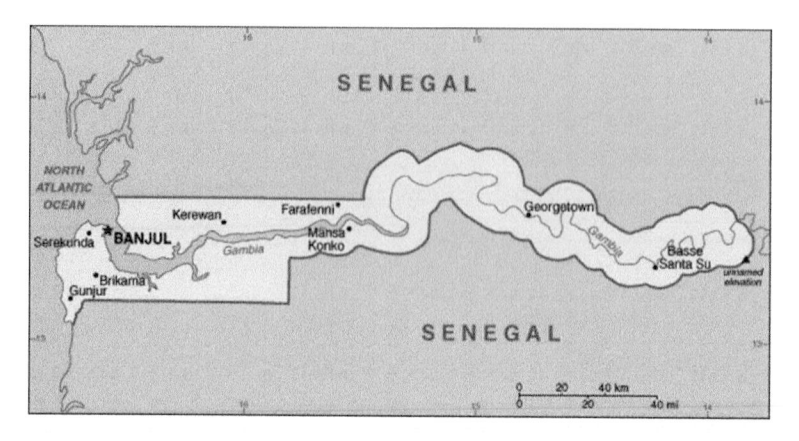

Figure 3.2.5 Map of the Republic of the Gambia.

Table 3.2.5 Evolution of the Muslim population in the Gambia

Year	Population	Muslims	%	Source
1871	14,190	5,717	40.29	[GB71]c
1911	146,101	121,113	82.90	[GB11]c
1963	315,486	268,163	85.00	[GM63]c
1983	683,834	656,192	95.96	[GM83]c
1993	1,029,376	986,009	95.79	[GM93]c
2003	1,357,190	1,298,165	95.65	[GM03]c
2013	1,856,211	1,782,859	96.05	[GM13]c
2050	4,882,000	4,689,000	96.05	es
2100	8,176,000	7,853,000	96.05	es

An 1871 census was held in the colony with an area of 54 sq. km, and recorded 5,717 Muslims, or 40.29% out of total population of 14,190. Later censuses showed that the Muslim population increased from 0.12 million or 82.9% in 1911, to 0.27 million or 85.0% in

1963, to 0.60 million or 95.1% in 1983, to 0.99 million or 95.8% in 1993, then to 1.30 million or 95.7% in 2003 and 1.78 million or 96.1% in 2013. Thus, assuming that the percentage of Muslims will remain fixed, the Muslim population is expected to increase to five million by 2050 and eight million by 2100. A summary of the data is provided in Table 3.2.5.

3.2.6 Ghana

The Republic of Ghana has an area of 238,533 sq. km and its map is presented in Fig. 3.2.6a. It was formed from the merger of the British colony of the Gold Coast (most of current Ghana) and the Togoland trust territory (the eastern part of current Ghana). Gold Coast was occupied by the British in 1874, while Togoland was taken in 1916 from Germans who occupied it since 1884 and split between the British and French. The latter became the nation of Togo after its independence. Ghana in 1957 became the first sub-Saharan country in colonial Africa to gain its independence.

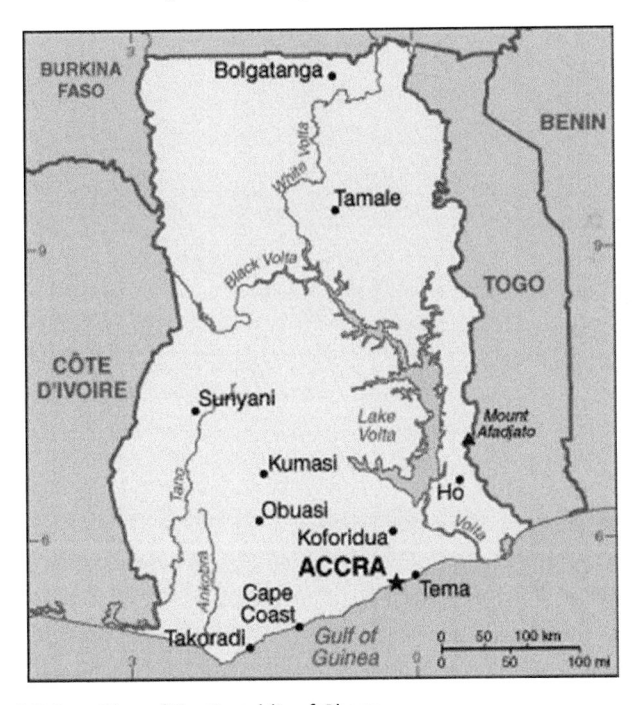

Figure 3.2.6a Map of the Republic of Ghana.

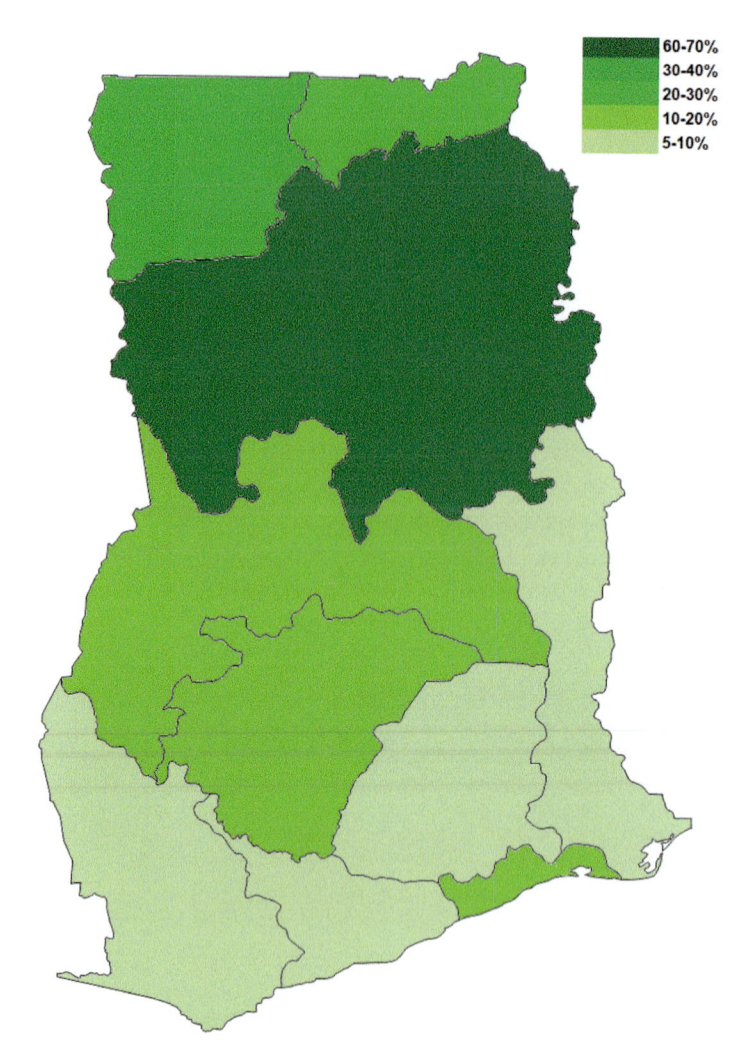

Figure 3.2.6b A choropleth map of Ghana illustrating the presence of Muslims per administrative region.

Islam entered here through trade as early as the fourteenth century. Estimates of the Muslim population increased from 50,000 or 3.5% in 1891, to 75,000 or 5.0% in 1911. Per census data, the number of Muslims increased dramatically from 0.05 million or 1.6% in 1931, to 0.18 million or 4.0% in 1948, to 0.81 million or 12.1% in 1960, to 1.85 million or 15.0% in 1984, to 3.00 million or 15.9% in 2000, to 4.35 million or 17.6% in 2010. A choropleth map illustrating the presence of Muslims per administrative region is presented in Fig. 3.2.6b.

Thus, assuming that the percentage of Muslims will continue to increase by a percentage point per decade, the Muslim population is expected to increase to eleven million or 22.0% by 2050 and 21 million or 27% by 2100. A summary of the data is provided in Table 3.2.6.

Table 3.2.6 Evolution of the Muslim population in Ghana

Year	Population	Muslims	%	Source
1891	1,473,882	50,000	3.46	[JAN]es
1911	1,508,000	75,000	4.97	[MAS23]es
1931	3,457,282	54,662	1.58	[ROD]c
1948	4,501,218	180,000	4.00	[ROD]c
1960	6,726,820	814,000	12.10	[ROD]c
1984	12,296,081	1,845,000	15.00	[ROD]c
2000	18,912,079	3,004,794	15.89	[UN]c
2010	24,658,823	4,345,723	17.62	[GH10]c
2050	52,016,000	11,444,000	22.00	es
2100	79,011,000	21,333,000	27.00	es

3.2.7 Guinea

The Republic of Guinea has an area of 245,857 sq. km and its map is presented in Fig. 3.2.7. It was occupied by France in 1898 and gained independence from France in 1958, when it changed its name from French Guinea to simply Guinea. Islam entered here in the fourteenth century through Mali Kingdom. It spread further starting the seventeenth century through Qadiri Sufi order (tariqa). Earlier estimates of the Muslim population increased from 1.6 million or 82.8% in 1911, to 1.8 million or 87.4% in 1921. Later census data showed an increase of the Muslim population to 4.2 million or 86.8% in 1983, then 6.2 million or 86.8% in 1996 and 9.4 million or 89.10% in 2014. A choropleth map illustrating the presence of Muslims per administrative region is presented in Fig. 3.2.7b.

Thus, assuming that the percentage of Muslims will continue to increase by one percentage point per decade, the Muslim population is expected to reach 24 million or 93% by 2050 and 44 million or 98% by 2100. A summary of the data is provided in Table 3.2.7.

Figure 3.2.7a Map of the Republic of Guinea.

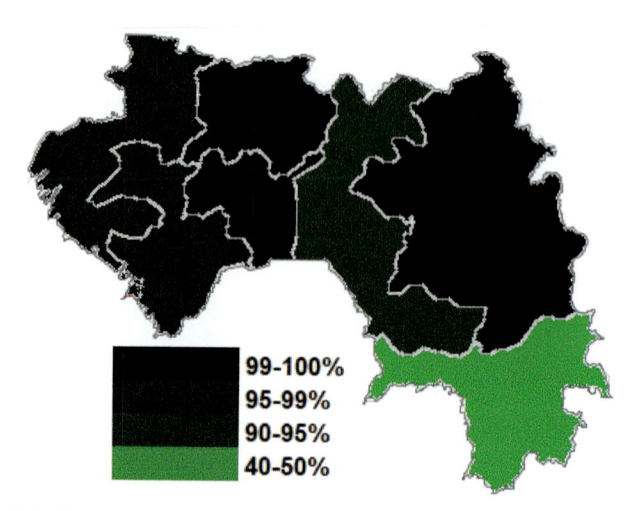

Figure 3.2.7b A choropleth map of Guinea illustrating the presence of Muslims per administrative region.

Table 3.2.7 Evolution of the Muslim population in Guinea

Year	Population	Muslims	%	Source
1911	1,876,000	1,553,000	82.78	[MAS23]es
1921	2,008,485	1,754,422	87.35	[GNH]es
1983	4,794,733	4,161,245	86.79	[GN83]c
1996	7,151,370	6,209,647	86.83	[GN96]c
2014	10,523,261	9,358,718	89.10	[GN14]c
2050	25,972,000	24,154,000	93.00	es
2100	45,257,000	44,352,000	98.00	es

3.2.8 Guinea-Bissau

The Republic of Guinea-Bissau has an area of 36,125 sq. km and its map is presented in Fig. 3.2.8. It was occupied by the Portuguese in 1446 and gained its independence from Portugal in 1974, when it changed its name from Portuguese Guinea to Guinea-Bissau, where Bissau is its capital. Islam entered here in the fourteenth century through Mali Kingdom which conquered these lands. The Muslim population was estimated in 1921 at 40,000 or 18.6% of the total population. Based on census data, the Muslim population continued to increase from 0.18 million or 35.6% in 1950, to 0.45 million or 45.9% in 1991, to 0.65 million or 53.6% in 2009. In the last census, 282,296 were excluded from the total as their religion was not collected. A choropleth map illustrating the presence of Muslims per administrative region is presented in Fig. 3.2.8b.

Thus, assuming that the percentage of Muslims will increase at two percentage points per decade, the Muslim population is expected to be over two million or 62% by 2050 and over four million or 72% by 2100. A summary of the data is provided in Table 3.2.8.

Table 3.2.8 Evolution of the Muslim population in Guinea-Bissau

Year	Population	Muslims	%	Source
1921	215,000	40,000	18.60	[MAS23]es
1950	508,970	181,284	35.62	[GP50]c
1991	979,203	448,933	45.85	[GP91]c
2009	1,213,509	650,402	53.60	[UN]c
2050	3,557,000	2,205,000	62.00	es
2100	5,706,000	4,109,000	72.00	es

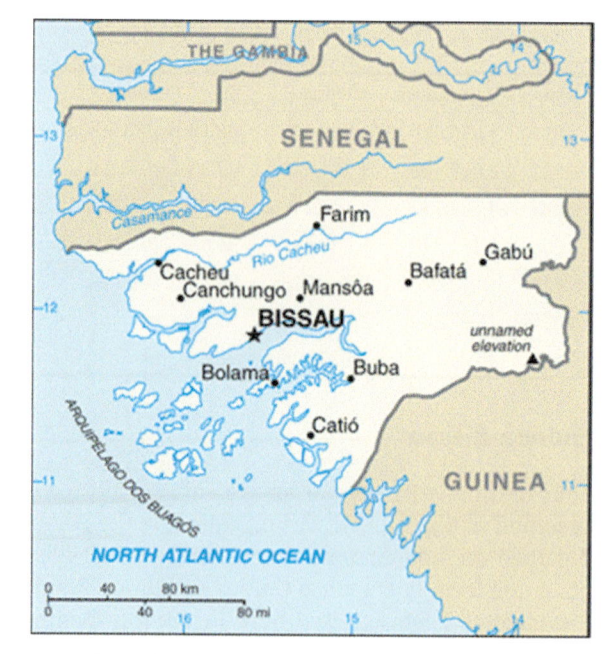

Figure 3.2.8a Map of the Republic of Guinea-Bissau.

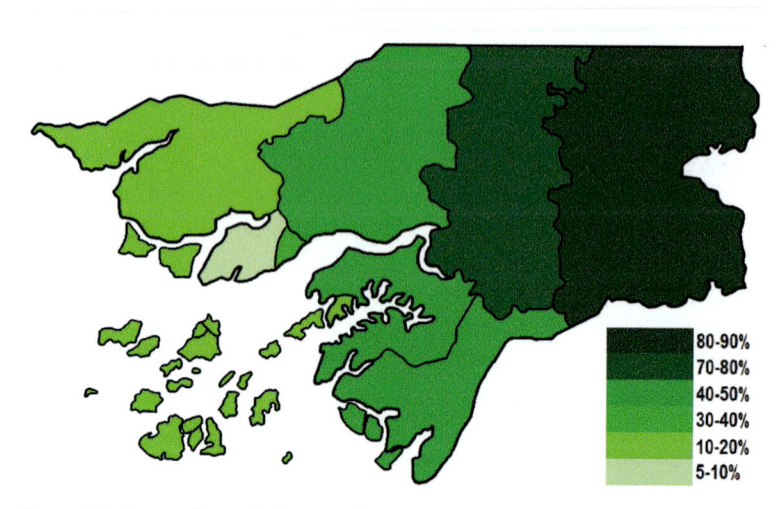

Figure 3.2.8b A choropleth map of Guinea-Bissau illustrating the presence of Muslims per administrative region.

3.2.9 Liberia

The Republic of Liberia has an area of 111,369 sq. km and its map is presented in Fig. 3.2.9a. It was established in 1847 by the Americans as a nation of freed black slaves from the USA. Islam entered here in the sixteenth century through Mali Kingdom. By 1921 the Muslim population was estimated at 200,000 or 11.8% of the total population. Muslim majority ethnic groups are Gola, Vai, Mandingo and Mende. Their total number increased from 0.11 million or 10.9% in 1962 to 0.18 million or 12.3% in 1974. These numbers are used here to estimate the number of Muslims based on ethnic affiliation. Later censuses indicate that the Muslim population increased from 0.29 million or 13.8% in 1984, to 0.42 million or 12.2% in 2008. The decrease in percentage may be due to the civil war that took place between 1989 and 1996, then between 1999 and 2003. A choropleth map illustrating the presence of Muslims per administrative region is presented in Fig. 3.2.9b.

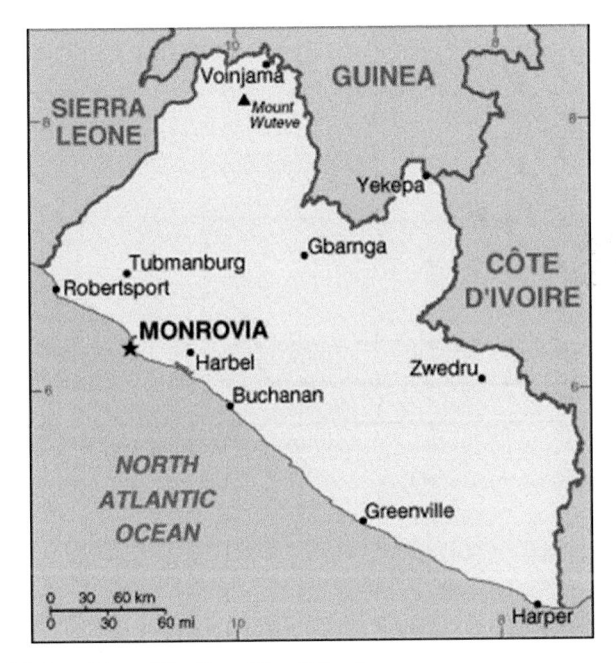

Figure 3.2.9a Map of the Republic of Liberia.

Thus, assuming that the percentage of Muslims will remain fixed, the Muslim population is expected to reach one million by 2050 and two million by 2100. A summary of the data is provided in Table 3.2.9.

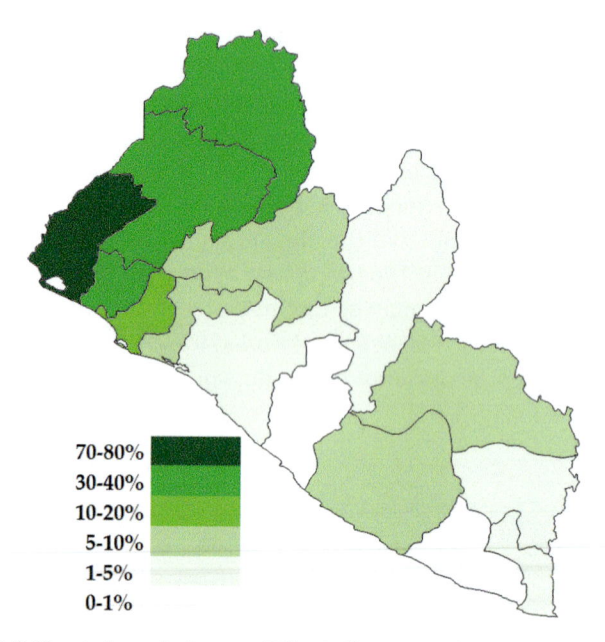

Figure 3.2.9b A choropleth map of Liberia illustrating the presence of Muslims per county.

Table 3.2.9 Evolution of the Muslim population in Liberia

Year	Population	Muslims	%	Source
1921	1,700,000	200,000	11.76	[MAS23]es
1962	1,016,443	110,917	10.91	[ROD]e
1974	1,503,368	184,415	12.27	[ROD]e
1984	2,101,628	290,828	13.84	[LR84]c
2008	3,476,608	424,685	12.22	[LR08]c
2050	9,340,000	1,141,000	12.22	es
2100	15,525,000	1,897,000	12.22	es

3.2.10 Senegal

The Republic of Senegal has an area of 196,722 sq. km and its map is presented in Fig. 3.2.10a. It was occupied by France in 1850s and gained its independence from France in 1960. Islam entered here in the 1030s when the War Jabi, King of Tekrur Kingdom embraced Islam. The Kingdom was situated in the north of current Senegal around the Senegal River. War Jabi died in 1040. Islam spread further around 1040 when Abdullah bnu Yassin and Yahya bnu Ibrahim, the founders of the Almoravid Dynasty, isolated themselves in an island at the end of the Senegal River, which separates current Mauritania from Senegal. They were followed by hundreds from which was the seed of Almoravid Dynasty. King Labi son of War Jabi continued the spread of Islam in this region. The Muslim population increased slowly as more local chiefs converted to Islam. The Muslim population was estimated at half million or 44.6% in 1910 [SNH], which increased to 0.9 million or 74.7% in 1921 and 1.3 million or 65.8% in 1948.

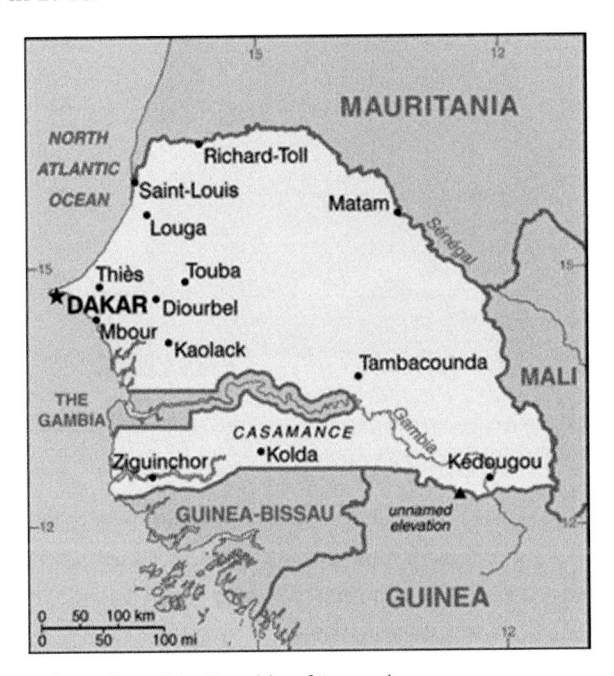

Figure 3.2.10a Map of the Republic of Senegal.

A census of Goree and Saint Louis recorded an increase of Muslims from 13,730 (83.1% of 16,521) in 1835 to 15,150 (84.4% of 17960) in 1838. A partial census in 1904 recorded 202,397 Muslims or 40.29% of a total population 502,389 distributed as follows:

- Saloum: 48,785 Muslims out of 82,197 (59.3%)
- East Baol: 23,533 Muslims out of 75,954 (31.0%)
- Wast Baol: 58,084 Muslims out of 78,981 (73.5%)
- Serrer and Lebou: 15,569 Muslims out of 53,636 (29.0%)
- Niani, Wouli, Tambakounda: 11,761 Muslims out of 28,034 (42.0%)

Post-independence census data show that the Muslim population increased to 2.8 million or 89.7% in 1961, then 6.4 million or 93.8% in 1988, to 9.5 million or 95.7% in 2002 and 12.8 million or 96.0% in 2013. A choropleth map illustrating the presence of Muslims per administrative region is presented in Fig. 3.2.10b.

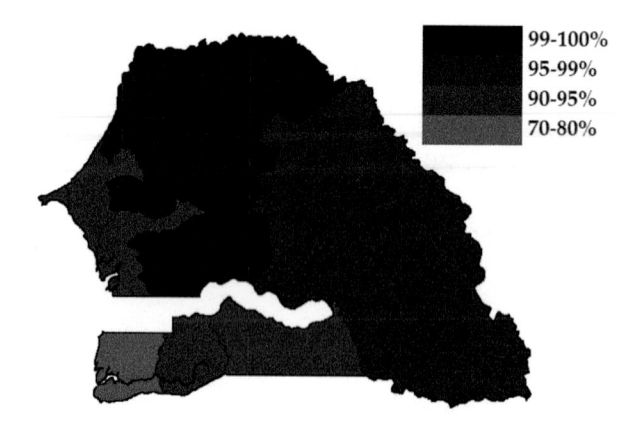

Figure 3.2.10b A choropleth map of Senegal illustrating the presence of Muslims per administrative region.

Thus, assuming that the percentage of Muslims will increase by a quarter of a percentage point per decade, the Muslim population is expected to reach 32 million or 97% by 2050 and 62 million or 98.3% by 2100. A summary of the data is provided in Table 3.2.10.

Table 3.2.10 Evolution of the Muslim population in Senegal

Year	Population	Muslims	%	Source
1904	502,389	202,397	44.29	[SN904]c
1910	1,121,150	500,000	44.60	[SNH]es
1921	1,225,523	915,000	74.66	[MAS23]es
1948	1,992,539	1,310,000	65.75	[MAS55]es
1961	3,109,840	2,789,320	89.69	[UN63]c
1988	6,773,417	6,353,464	93.80	[UN04]c
2002	9,926,673	9,496,038	95.66	[SN02]c
2013	13,281,722	12,750,000	96.00	[SN13]c
2050	33,187,000	32,191,000	97.00	es
2100	63,515,000	62,404,000	98.25	es

3.2.11 Sierra Leone

The Republic of Sierra Leone has an area of 71,740 sq. km and its map is presented in Fig. 3.2.11a. It was occupied by the British in 1787 and gained its independence from the UK in 1961. Islam entered here in the eighteenth century through preachers from neighboring Guinea. Estimates of the Muslim population increased from half a million or 32.1% in 1911, to 0.8 million or 35.0% in 1963, to 2.1 million or 60.0% in 1985. According to census data, the number of Muslims increased to 3.8 million or 76.7% in 2004 and 5.5 million or 77.0% in 2015. A choropleth map illustrating the presence of Muslims per district is presented in Fig. 3.2.11b.

Thus, assuming that the percentage of Muslims will increase by a half of a percentage point per decade, the Muslim population is expected to exceed ten million or 79% by 2050 and fourteen million or 82% by 2100. A summary of the data is provided in Table 3.2.11.

Table 3.2.11 Evolution of the Muslim population in Sierra Leone

Year	Population	Muslims	%	Source
1911	1,403,560	450,000	32.06	[MAS23]es
1963	2,180,355	763,124	35.00	[IML]es
1985	3,515,812	2,109,487	60.00	[IML]es
2004	4,930,461	3,780,473	76.68	[SL04]c
2015	7,092,113	5,460,927	77.00	[SL15]c
2050	12,945,000	10,226,000	79.00	es
2100	16,675,000	13,590,000	81.50	es

Figure 3.2.11a Map of the Republic of Sierra Leone.

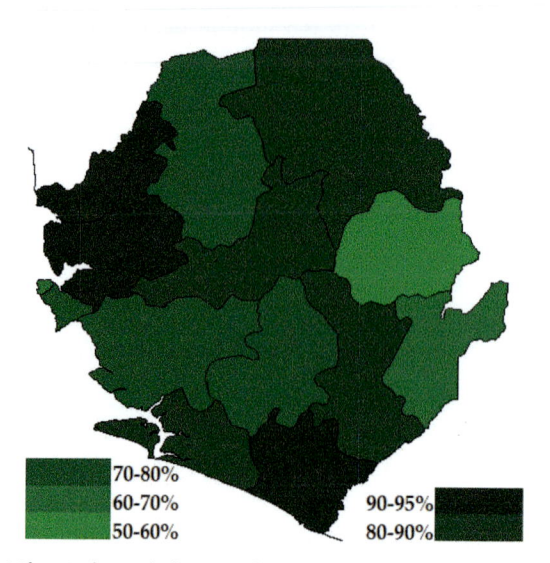

Figure 3.2.11b A choropleth map of Sierra Leone illustrating the presence of Muslims per district.

3.2.12 Togo

The Togolese Republic has an area of 56,785 sq. km and its map is presented in Fig. 3.2.12a. Togoland was taken in 1916 from Germans who occupied it since 1884 and split between the British and French. The latter became the nation of Togo after its independence from France in 1960. Islam entered here through trade as early as the fourteenth century. Estimate of the Muslim population was 30,000 or 3% of the total population in 1914 and 1951. According to census data, this number increased to 0.22 million or 11.6% in 1970, then 0.33 million or 12.2% in 1981 and 0.61 million or 15.7% in 2010 (age twelve and above). Islam is increasing rapidly in Togo, and there are reports of entire villages joining Islam as once. A choropleth map illustrating the presence of Muslims per prefecture is presented in Fig. 3.2.12b.

Thus, assuming that the percentage of Muslims will continue to increase by one percentage point per decade, then the Muslim population is expected to reach three million or 20% by 2050 and seven million or 25% by 2100. A summary of the data is provided in Table 3.2.12.

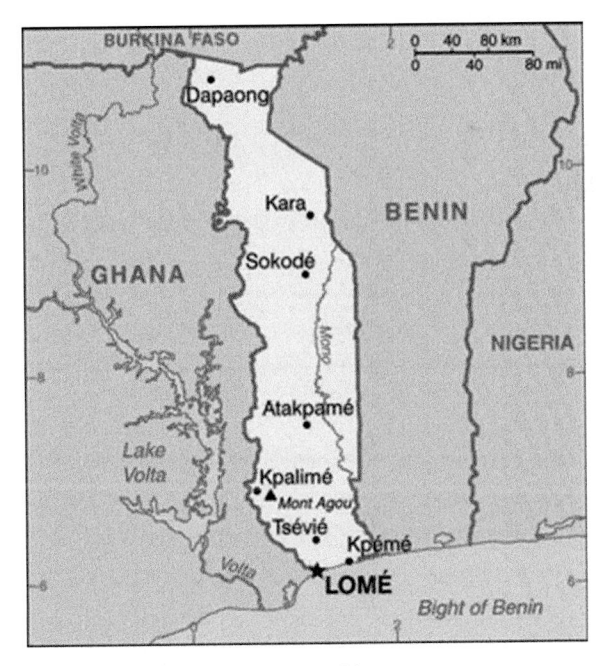

Figure 3.2.12a Map of the Togolese Republic.

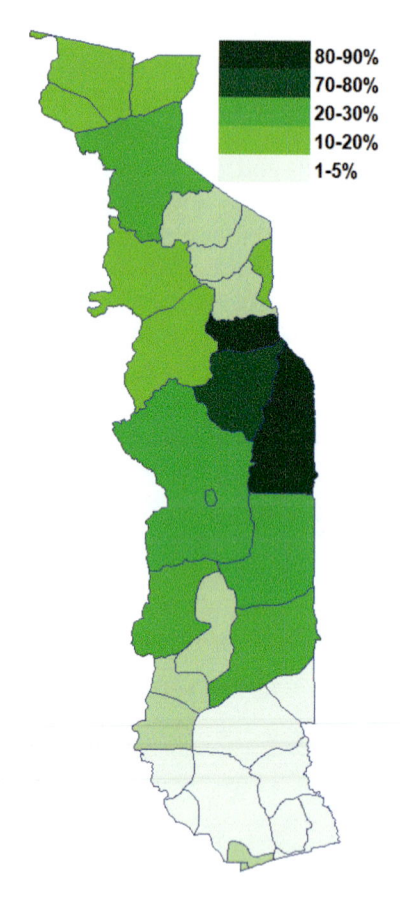

Figure 3.2.12b A choropleth map of Togo illustrating the presence of Muslims per prefecture.

Table 3.2.12 Evolution of the Muslim population in Togo

Year	Population	Muslims	%	Source
1914	1,032,000	30,000	2.91	[MAS23]es
1951	1,014,669	30,000	2.96	[MAS55]es
1970	1,949,470	226,186	11.60	[UN79]c
1981	2,698,090	328,094	12.16	[TG81]c
2010	3,886,658	609,218	15.67	[TG10]c
2050	15,415,000	3,083,000	20.00	es
2100	26,949,000	6,737,000	25.00	es

3.2.13   Regional Summary and Conclusion

Islam entered West Africa in the eleventh century and spread through Sufi orders. Islam spread rapidly and almost half of the population of this region is now Muslim. Accordingly, West Africa has the second largest concentration of Muslims among the five regions spanning Africa. The Muslim population is expected to continue to increase in number and percentage for the next three centuries. A choropleth map illustrating the presence of Muslims per administrative region in West Africa is presented in Fig. 3.2.13.

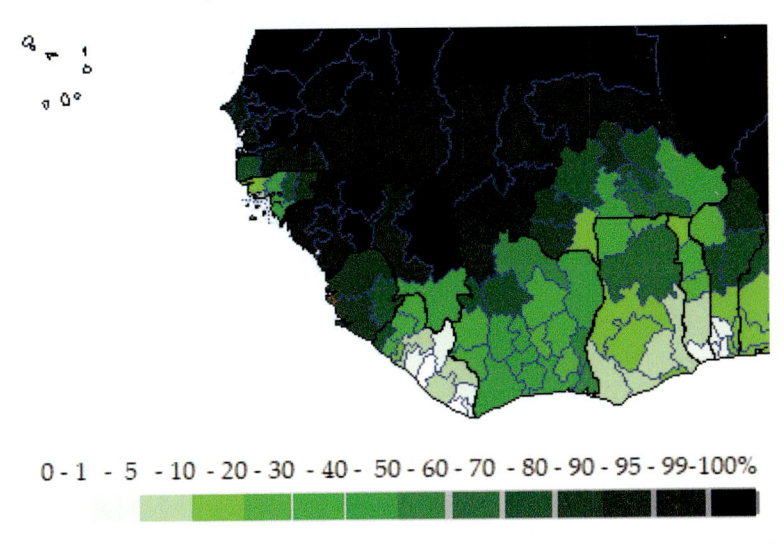

Figure 3.2.13   A choropleth map of West Africa illustrating the presence of Muslims per administrative regions.

The following tables present centennial data from 600 AD to 2100 AD (or approximately 1 H to 1500 H) in Table 3.2a and decennial data from 1790 AD to 2100 AD (or 1210 H to 1520 H) in Tables 3.2b and 3.2c for current countries in West Africa. The data includes total population in thousands (P), the percentage of which is Muslim (M%), the corresponding Muslim population in thousands (M), and the annual population growth rate (APGR, or G%) of the total population in this region.

Table 3.2a Centennial estimates of the Muslim population (×1000) in West Africa: 600–2100 AD (1–1500 H)

		600	700	800	900	1000	1100	1200	1300
Benin	P	80	85	90	95	100	110	120	130
	M%	—	—	—	—	—	—	—	—
	M	—	—	—	—	—	—	—	—
Burkina Faso	P	400	425	450	475	500	550	700	850
	M%	—	—	—	—	—	—	—	—
	M	—	—	—	—	—	—	—	—
Cabo Verde	P	—	—	—	—	—	—	—	—
	M%	—	—	—	—	—	—	—	—
	M	—	—	—	—	—	—	—	—
Côte d'Ivoire	P	400	425	450	470	490	510	530	550
	M%	—	—	—	—	—	—	—	—
	M	—	—	—	—	—	—	—	—
Gambia	P	0	0	0	0	0	1	1	1
	M%	—	—	—	—	—	10.00	20.00	30.00
	M	—	—	—	—	—	0	0	0
Ghana	P	350	375	400	430	463	500	550	600
	M%	—	—	—	—	—	—	—	—
	M	—	—	—	—	—	—	—	—
Guinea	P	240	260	280	300	315	350	385	420
	M%	—	—	—	—	—	—	—	—
	M	—	—	—	—	—	—	—	—
Guinea Bissau	P	53	57	61	65	69	76	84	92
	M%	—	—	—	—	—	—	—	—
	M	—	—	—	—	—	—	—	—
Liberia	P	160	170	180	190	200	210	220	230
	M%	—	—	—	—	—	—	—	—
	M	—	—	—	—	—	—	—	—
Senegal	P	240	255	270	285	300	320	340	360
	M%	—	—	—	—	—	10.00	20.00	30.00
	M	—	—	—	—	—	32	68	108
Sierra Leone	P	182	195	208	221	234	260	285	310
	M%	—	—	—	—	—	—	—	—
	M	—	—	—	—	—	—	—	—
Togo	P	96	102	108	114	120	130	140	150
	M%	—	—	—	—	—	—	—	—
	M	—	—	—	—	—	—	—	—
Total	P	2,201	2,349	2,497	2,645	2,791	3,017	3,355	3,693
	M%	—	—	—	—	—	1.06	2.03	2.93
	M	—	—	—	—	—	32	68	108
	G%		0.065	0.061	0.058	0.054	0.078	0.106	0.096

Table 3.2a (*Continued*)

		1400	1500	1600	1700	1800	1900	2000	2100
Benin	P	140	150	190	235	270	482	6,866	47,209
	M%	1.00	2.00	3.00	4.00	7.63	7.63	24.42	46.00
	M	1	3	6	9	21	37	1,677	21,716
Burkina Faso	P	1,000	1,100	1,400	1,700	1,940	2,222	11,608	83,194
	M%	—	0.10	0.50	1.00	1.41	10.18	55.90	70.00
	M	—	1	7	17	27	226	6,489	58,236
Cabo Verde	P	—	1	10	25	60	147	428	604
	M%	—	—	—	—	—	—	0.23	4.00
	M	—	—	—	—	—	—	1	24
Côte d'Ivoire	P	570	600	800	1,050	1,150	1,355	16,455	96,633
	M%	—	—	—	—	12.50	12.50	38.88	52.00
	M	—	—	—	—	144	169	6,398	50,249
Gambia	P	1	1	2	2	2	90	1,318	8,176
	M%	40.00	50.00	60.00	70.00	82.90	82.90	95.65	96.05
	M	0	1	1	1	2	75	1,260	7,853
Ghana	P	650	727	926	1,190	1,325	1,800	19,279	79,011
	M%	0.50	1.00	2.00	3.00	3.46	4.97	15.89	27.00
	M	3	7	19	36	46	89	3,063	21,333
Guinea	P	460	495	630	810	915	1,762	8,241	45,257
	M%	10.00	20.00	40.00	80.00	82.78	82.78	86.83	98.00
	M	46	99	252	648	757	1,459	7,155	44,352
Guinea Bissau	P	100	108	138	177	201	386	1,201	5,706
	M%	1.00	3.00	5.00	7.00	18.60	18.60	50.00	72.00
	M	1	3	7	12	37	72	601	4,109
Liberia	P	250	270	350	480	520	641	2,848	15,525
	M%	—	—	1.00	5.00	11.76	11.76	12.22	12.22
	M	—	—	4	24	61	75	348	1,897
Senegal	P	380	400	450	600	700	1,200	9,798	63,515
	M%	40.00	50.00	50.00	50.00	44.60	44.60	95.66	98.25
	M	152	200	225	300	312	535	9,373	62,404
Sierra Leone	P	335	367	468	601	665	1,024	4,585	16,675
	M%	—	—	—	—	32.06	32.06	76.68	81.50
	M	—	—	—	—	213	328	3,515	13,590
Togo	P	160	170	230	330	360	484	4,924	26,949
	M%	1.00	2.00	3.00	3.00	2.91	2.91	14.00	25.00
	M	2	3	7	10	10	14	689	6,737
Total	P	4,046	4,389	5,594	7,200	8,108	11,594	87,551	488,455
	M%	5.08	7.23	9.42	14.69	20.12	26.57	46.34	59.88
	M	206	318	527	1,058	1,631	3,080	40,569	292,500
	G%	0.091	0.081	0.243	0.252	0.119	0.358	2.022	1.719

Table 3.2b Decennial estimates of the Muslim population (×1000) in West Africa: 1790–1940 AD (1210–1360 H)

		1790	1800	1810	1820	1830	1840	1850	1860
Benin	P	260	270	280	290	300	310	320	330
	M%	7.63	7.63	7.63	7.63	7.63	7.63	7.63	7.63
	M	20	21	21	22	23	24	24	25
Burkina Faso	P	1,920	1,940	1,960	1,980	2,000	2,020	2,040	2,060
	M%	1.41	1.41	1.41	1.41	1.41	1.41	1.41	1.41
	M	27	27	28	28	28	28	29	29
Cabo Verde	P	55	60	65	70	75	80	87	84
	M%	—	—	—	—	—	—	—	—
	M	—	—	—	—	—	—	—	—
Côte d'Ivoire	P	1,130	1,150	1,170	1,190	1,210	1,230	1,250	1,270
	M%	12.50	12.50	12.50	12.50	12.50	12.50	12.50	12.50
	M	141	144	146	149	151	154	156	159
Gambia	P	2	2	3	3	4	5	6	7
	M%	82.90	82.90	82.90	82.90	82.90	82.90	82.90	82.90
	M	1	2	2	2	3	4	5	6
Ghana	P	1,300	1,325	1,350	1,374	1,400	1,440	1,480	1,520
	M%	3.46	3.46	3.46	3.46	3.46	3.46	3.46	3.46
	M	45	46	47	48	48	50	51	53
Guinea	P	905	915	925	935	1,000	1,060	1,140	1,240
	M%	82.78	82.78	82.78	82.78	82.78	82.78	82.78	82.78
	M	749	757	766	774	828	877	944	1,026
Guinea Bissau	P	199	201	203	205	220	240	260	280
	M%	18.60	18.60	18.60	18.60	18.60	18.60	18.60	18.60
	M	37	37	38	38	41	45	48	52
Liberia	P	510	520	530	540	550	560	570	580
	M%	11.76	11.76	11.76	11.76	11.76	11.76	11.76	11.76
	M	60	61	62	64	65	66	67	68
Senegal	P	650	700	750	800	850	900	950	1,000
	M%	44.60	44.60	44.60	44.60	44.60	44.60	44.60	44.60
	M	290	312	335	357	379	401	424	446
Sierra Leone	P	650	665	680	694	725	750	775	800
	M%	32.06	32.06	32.06	32.06	32.06	32.06	32.06	32.06
	M	208	213	218	222	232	240	248	256
Togo	P	350	360	370	380	390	400	410	420
	M%	2.91	2.91	2.91	2.91	2.91	2.91	2.91	2.91
	M	10	10	11	11	11	12	12	12
Total	P	7,931	8,108	8,286	8,461	8,724	8,995	9,287	9,591
	M%	20.04	20.12	20.19	20.27	20.75	21.13	21.63	22.24
	M	1,589	1,631	1,673	1,715	1,810	1,901	2,009	2,133
	G%		0.221	0.217	0.210	0.306	0.306	0.320	0.322

Table 3.2b (*Continued*)

		1870	1880	1890	1900	1910	1920	1930	1940
Benin	P	340	350	361	482	655	842	1,147	1,532
	M%	7.63	7.63	7.63	7.63	7.63	8.31	8.31	6.98
	M	26	27	28	37	50	70	95	107
Burkina Faso	P	2,080	2,100	2,135	2,222	2,455	2,973	2,995	3,308
	M%	1.41	1.41	1.41	10.18	10.18	14.93	14.93	22.80
	M	29	30	30	226	250	444	447	754
Cabo Verde	P	76	99	106	147	143	150	151	181
	M%	—	—	—	—	—	—	—	—
	M	—	—	—	—	—	—	—	—
Côte d'Ivoire	P	1,290	1,310	1,330	1,355	1,367	1,825	2,000	2,200
	M%	12.50	12.50	12.50	12.50	12.50	9.86	9.86	9.86
	M	161	164	166	169	171	180	197	217
Gambia	P	14	30	64	90	133	193	201	223
	M%	82.90	82.90	82.90	82.90	82.90	82.90	82.90	82.90
	M	12	25	53	75	110	160	167	185
Ghana	P	1,579	1,600	1,650	1,800	2,000	2,447	3,164	3,960
	M%	3.46	3.46	3.46	4.97	4.97	1.58	1.58	3.00
	M	55	55	57	89	99	39	50	119
Guinea	P	1,389	1,504	1,628	1,762	1,910	2,065	2,180	2,217
	M%	82.78	82.78	82.78	82.78	82.78	87.35	87.35	87.35
	M	1,150	1,245	1,348	1,459	1,581	1,804	1,904	1,937
Guinea Bissau	P	304	329	356	386	416	452	485	500
	M%	18.60	18.60	18.60	18.60	18.60	18.60	25.00	30.00
	M	57	61	66	72	77	84	121	150
Liberia	P	590	600	610	641	674	709	745	784
	M%	11.76	11.76	11.76	11.76	11.76	11.76	11.76	10.91
	M	69	71	72	75	79	83	88	86
Senegal	P	1,050	1,100	1,150	1,200	1,240	1,460	1,630	1,860
	M%	44.60	44.60	44.60	44.60	44.60	74.66	74.66	65.75
	M	468	491	513	535	553	1,090	1,217	1,223
Sierra Leone	P	850	900	950	1,024	1,400	1,541	1,768	1,859
	M%	32.06	32.06	32.06	32.06	32.06	32.06	32.06	32.06
	M	273	289	305	328	449	494	567	596
Togo	P	430	440	450	484	592	725	887	1,064
	M%	2.91	2.91	2.91	2.91	2.91	2.91	2.91	2.96
	M	13	13	13	14	17	21	26	31
Total	P	9,992	10,362	10,790	11,594	12,984	15,382	17,352	19,688
	M%	23.14	23.83	24.56	26.57	26.47	29.05	28.12	27.45
	M	2,312	2,469	2,650	3,080	3,437	4,469	4,879	5,404
	G%	0.410	0.364	0.405	0.718	1.133	1.695	1.205	1.263

Table 3.2c Decennial estimates of the Muslim population (×1000) in West Africa: 1950–2100 AD (1370–1520 H)

		1950	1960	1970	1980	1990	2000	2010	2020
Benin	P	2,255	2,432	2,912	3,717	4,978	6,866	9,199	12,123
	M%	6.98	13.63	16.00	18.00	20.71	24.42	27.70	30.00
	M	157	331	466	669	1,031	1,677	2,548	3,637
Burkina Faso	P	4,284	4,829	5,625	6,823	8,811	11,608	15,605	20,903
	M%	22.80	27.44	36.00	45.00	55.90	55.90	60.53	62.00
	M	977	1,325	2,025	3,070	4,925	6,489	9,446	12,960
Cabo Verde	P	178	202	269	284	338	428	493	556
	M%	—	—	—	—	—	0.23	1.79	2.00
	M	—	—	—	—	—	1	9	11
Côte d'Ivoire	P	2,630	3,504	5,102	8,034	11,925	16,455	20,533	26,378
	M%	21.70	21.70	23.50	33.48	38.67	38.88	42.87	44.00
	M	571	760	1,199	2,690	4,611	6,398	8,802	11,606
Gambia	P	305	365	464	637	956	1,318	1,793	2,417
	M%	85.00	85.00	85.00	95.96	95.79	95.65	96.05	96.05
	M	259	310	395	612	915	1,260	1,722	2,321
Ghana	P	5,036	6,635	8,735	11,056	14,773	19,279	24,780	31,073
	M%	4.00	12.10	14.00	15.00	15.89	15.89	17.62	19.00
	M	201	803	1,223	1,658	2,347	3,063	4,366	5,904
Guinea	P	3,013	3,494	4,155	4,871	6,352	8,241	10,192	13,133
	M%	86.79	86.79	86.79	86.79	86.77	86.83	89.10	90.00
	M	2,615	3,033	3,606	4,228	5,512	7,155	9,081	11,820
Guinea Bissau	P	535	616	705	782	975	1,201	1,523	1,968
	M%	35.62	38.00	40.50	43.00	45.85	50.00	53.60	56.00
	M	191	234	286	336	447	601	816	1,102
Liberia	P	930	1,119	1,401	1,853	2,076	2,848	3,891	5,058
	M%	10.91	10.91	12.27	13.84	13.84	12.22	12.22	12.22
	M	101	122	172	256	287	348	476	618
Senegal	P	2,487	3,207	4,258	5,583	7,526	9,798	12,678	16,744
	M%	65.75	89.69	89.69	93.80	93.80	95.66	96.00	96.25
	M	1,635	2,876	3,819	5,237	7,060	9,373	12,171	16,116
Sierra Leone	P	2,041	2,318	2,745	3,388	4,320	4,585	6,416	7,977
	M%	35.00	35.00	35.00	60.00	76.68	76.68	77.00	77.50
	M	714	811	961	2,033	3,312	3,515	4,940	6,182
Togo	P	1,395	1,581	2,116	2,721	3,774	4,924	6,422	8,279
	M%	2.96	11.60	11.60	12.16	13.00	14.00	15.67	17.00
	M	41	183	245	331	491	689	1,006	1,407
Total	P	25,091	30,300	38,486	49,749	66,805	87,551	113,524	146,608
	M%	29.75	35.61	37.40	42.45	46.31	46.34	48.79	50.26
	M	7,463	10,789	14,395	21,120	30,940	40,569	55,384	73,685
	G%	2.425	1.887	2.391	2.567	2.948	2.704	2.598	2.557

Table 3.2c (Continued)

		2030	2040	2050	2060	2070	2080	2090	2100
Benin	P	15,672	19,775	24,280	29,056	33,933	38,698	43,182	47,209
	M%	32.00	34.00	36.00	38.00	40.00	42.00	44.00	46.00
	M	5,015	6,724	8,741	11,041	13,573	16,253	19,000	21,716
Burkina Faso	P	27,404	35,051	43,432	52,152	60,921	69,276	76,791	83,194
	M%	63.00	64.00	65.00	66.00	67.00	68.00	69.00	70.00
	M	17,265	22,433	28,231	34,420	40,817	47,107	52,986	58,236
Cabo Verde	P	610	652	679	688	679	658	632	604
	M%	2.25	2.50	2.75	3.00	3.25	3.50	3.75	4.00
	M	14	16	19	21	22	23	24	24
Côte d'Ivoire	P	33,713	42,083	51,264	60,924	70,641	80,069	88,836	96,633
	M%	45.00	46.00	47.00	48.00	49.00	50.00	51.00	52.00
	M	15,171	19,358	24,094	29,243	34,614	40,035	45,306	50,249
Gambia	P	3,171	4,008	4,882	5,737	6,525	7,212	7,770	8,176
	M%	96.05	96.05	96.05	96.05	96.05	96.05	96.05	96.05
	M	3,045	3,850	4,689	5,510	6,267	6,927	7,463	7,853
Ghana	P	37,833	44,883	52,016	58,880	65,220	70,846	75,480	79,011
	M%	20.00	21.00	22.00	23.00	24.00	25.00	26.00	27.00
	M	7,567	9,425	11,444	13,542	15,653	17,712	19,625	21,333
Guinea	P	17,021	21,365	25,972	30,640	35,091	39,087	42,500	45,257
	M%	91.00	92.00	93.00	94.00	95.00	96.00	97.00	98.00
	M	15,489	19,655	24,154	28,802	33,337	37,524	41,225	44,352
Guinea Bissau	P	2,461	2,996	3,557	4,105	4,613	5,062	5,431	5,706
	M%	58.00	60.00	62.00	64.00	66.00	68.00	70.00	72.00
	M	1,428	1,798	2,205	2,627	3,045	3,442	3,802	4,109
Liberia	P	6,372	7,825	9,340	10,850	12,294	13,591	14,677	15,525
	M%	12.22	12.22	12.22	12.22	12.22	12.22	12.22	12.22
	M	779	956	1,141	1,326	1,502	1,661	1,794	1,897
Senegal	P	21,551	27,088	33,187	39,547	45,987	52,315	58,246	63,515
	M%	96.50	96.75	97.00	97.25	97.50	97.75	98.00	98.25
	M	20,797	26,208	32,191	38,460	44,838	51,138	57,081	62,404
Sierra Leone	P	9,649	11,339	12,945	14,346	15,454	16,205	16,599	16,675
	M%	78.00	78.50	79.00	79.50	80.00	80.50	81.00	81.50
	M	7,526	8,901	10,226	11,405	12,363	13,045	13,445	13,590
Togo	P	10,422	12,850	15,415	18,027	20,603	23,018	25,154	26,949
	M%	18.00	19.00	20.00	21.00	22.00	23.00	24.00	25.00
	M	1,876	2,441	3,083	3,786	4,533	5,294	6,037	6,737
Total	P	185,879	229,914	276,970	324,951	371,962	416,037	455,297	488,455
	M%	51.63	52.96	54.24	55.45	56.61	57.73	58.82	59.88
	M	95,970	121,765	150,219	180,182	210,563	240,161	267,786	292,500
	G%	2.373	2.126	1.862	1.598	1.351	1.120	0.902	0.703

3.3 Muslims in East Africa

This region consists of twelve countries and territories: Comoros, Djibouti, Eretria, Ethiopia, Kenya, Madagascar, Mauritius, Mayotte, Réunion, Seychelles, Somalia, and Tanzania. Islam entered here when in Rajab 8B H, or May 614 AD, sixteen persecuted Muslims fled to the Axumite port city of Zeila in present-day Somalia near the border with Djibouti, to seek protection from the Quraysh at the court of the Axumite Emperor, an-Najashi, in modern Ethiopia. They were followed by 101 Muslims later in the same year and preached Islam in these lands. After the Muslim conquest of Mecca Ramadan 8 H, or January 630 AD and the subsequent spread of Islam throughout the Arabian Peninsula, Islam continued to spread in this region through trade. Islam entered the islands further in the Indian Ocean in the nineteenth century with workers brought from India by the British.

Thus, through commerce and migration, Islam spread amongst the coastal lines of this region from Eritrea in the north to Tanzania in the south and deeper into Ethiopia. Eventually, several Muslim sultanates were established covering Djibouti and parts of its neighboring countries: Shewa (896–1285), followed by Ifat (1285–1415), then Adal (1415–1555), in addition to others. By 1806, the entire coast from Northern Mozambique to southern Somalia was under the control of the Sultanate of Oman, based in the south of the Arabian Peninsula.

Hence, the Muslim population has likely increased from 52,000 or 2.0% of the total population of this region in 700 AD, to 0.18 million or 6.4% in 800 AD, to 0.37 million or 11.8% in 900 AD, to 0.65 million or 19.0% in 1000 AD, to 0.84 million or 20.7% in 1100 AD, to 1.1 million or 22.8% in 1200 AD, to 1.3 million or 24.0% in 1300 AD, to 1.6 million or 25.6% in 1400 AD, to 1.9 million or 27.4% in 1500 AD, to 2.3 million or 28.3% in 1600 AD, to 2.8 million or 30.4% in 1700 AD, to 2.7 million or 24.0% in 1800 AD, to 6.1 million or 28.3% in 1900, to 47 million or 28.8% in 2000, to 85 million or 30.4% in 2000, to 86 million or 30.5% in 2020, and is projected to reach 0.17 billion or 33% by 2050 and 0.32 billion or 36% by 2100.

A plot of centennial estimates of the Muslim population and its percentage with respect to the total population in this region from 600 to 2100 is provided in Fig. 3.3a. A zoom in of this plot, providing a plot of decennial estimates of the Muslim population and its percentage with respect to the total population in this region from 1900 to 2100 is provided in Fig. 3.3b. This shows that the Muslim population in this region was increasing slowly until 1970 and is increasing substantially afterwards towards the end of this century. The percentage of Muslims on the other hand, increased sharply from 26.5% in 1870, peaking at almost 33% in 1910, but decreased equally fast to less than 28% in 1970, and is generally increasing since then and expected to be over 36% by the end of this century.

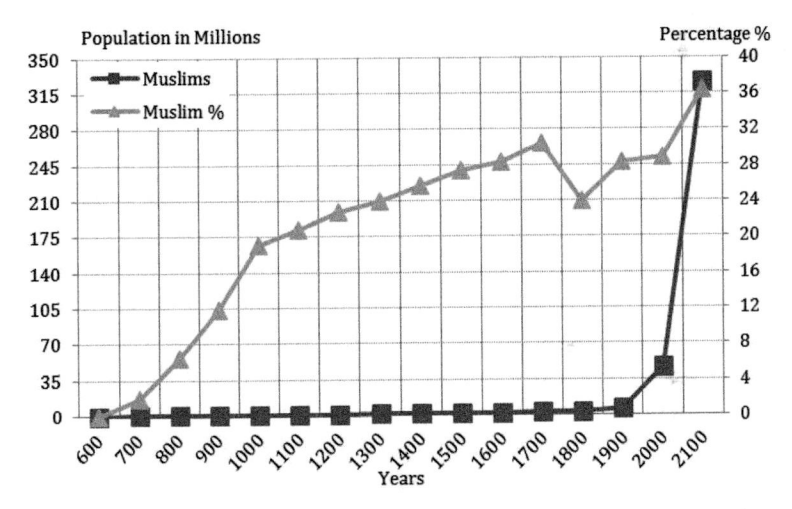

Figure 3.3a Plot of centennial estimates of the Muslim population and its percentage of the total population in East Africa: 600–2100 AD (1–1500 H).

The corresponding individual data for each country in this region is discussed below. In Section 3.3.13, the total population in each country in this region and the corresponding percentage and number of Muslims is presented centennially in Table 3.3a from 600 to 2100 and decennially in Tables 3.3b and 3.3c from 1790 to 2100.

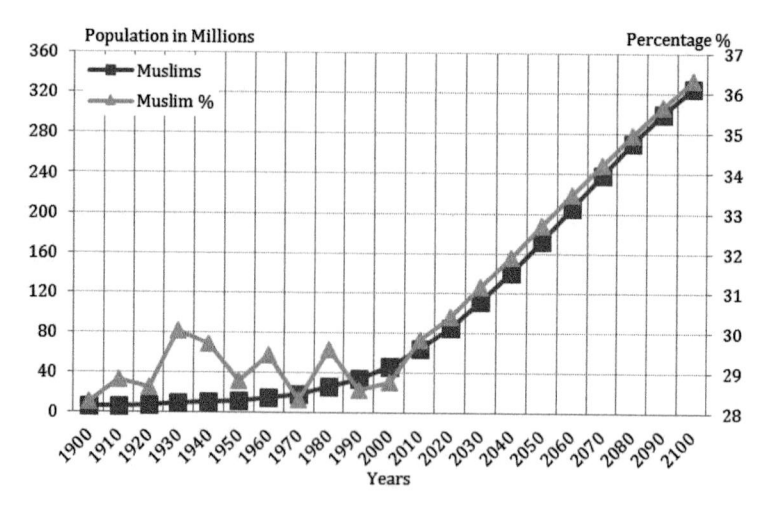

Figure 3.3b Plot of decennial estimates of the Muslim population and its percentage of the total population in East Africa: 1900–2100 AD (1320–1520 H).

3.3.1 Comoros

The Union of the Comoros has an area 2,235 sq. km consisting of three main islands: Njazidja (1,146 sq. km) with half of the population and where the capital Moroni is located, Nzwani (424 sq. km), and Mwali (290 sq. km), with only 6% of the population. A map of these islands is presented in Fig. 3.3.1. The French occupied the Islands in 1886 and gained their independence from France in 1975. Islam entered these islands early ninth century through Muslim traders from the Arabian Peninsula. By the eleventh century it became part of the Kilwa Muslim Sultanate. By 1497, and with the fall of the Kilwa Sultanate, each of the three main islands became an independent sultanate until they were all captured by the French. Almost all the population is Muslim: 100% up to 1966, and more than 99% after 1980. According to a 2012 DHS, 0.72 million or 99.2% of the total population is Muslim. Thus, assuming that this percentage remains fixed, the Muslim population is expected to reach 1.5 million by 2050 and 2.2 million by 2100. A summary of the data is provided in Table 3.3.1.

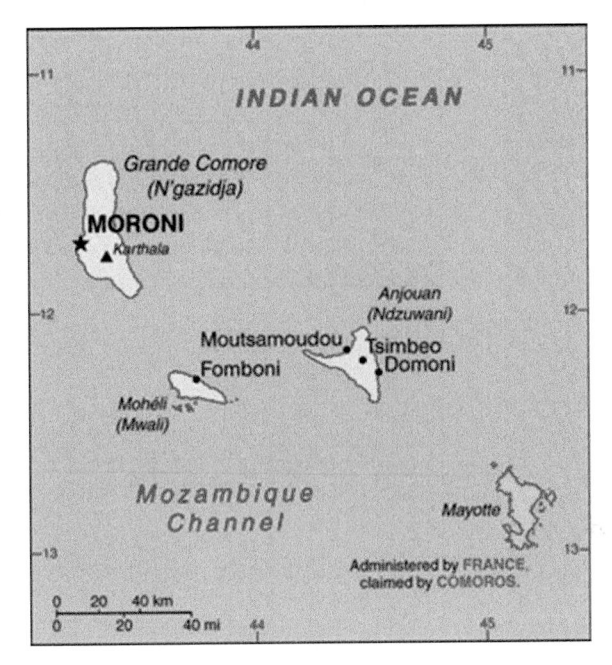

Figure 3.3.1 Map of the Union of the Comoros.

Table 3.3.1 Evolution of the Muslim population in the Comoros

Year	Population	Muslims	%	Source
1918	96,180	96,180	100	[MAS23]es
1948	123,754	123,754	100	[SYB50]es
1966	216,587	216,587	100	[SYB70]es
1980	335,150	331,800	99.00	[KET86]es
1991	446,817	442,349	99.00	[DOS00]es
2003	575,660	569,903	99.00	[DOS05]es
2012	734,000	731,700	99.69	[DHS]s
2050	1,472,000	1,467,000	99.69	es
2100	2,187,000	2,180,000	99.69	es

3.3.2 Djibouti

The Republic of Djibouti has an area of 23,200 sq. km and its map is presented in Fig. 3.3.2. It was occupied by France in 1897 and used to be called French Somaliland, then changed its name in 1958 to French Territory of the Afars and the Issas, and then became Djibouti upon independence in 1977. Almost the whole population is Muslim, with percentage fluctuating between 97% in 1946 to 100% in 1917 and 1960, to 99% in 2009. The non-Muslim population consists of foreigners. From 1960 to 2009 the total population increased by more than tenfold. A summary of the data is provided in Table 3.3.2. Thus, assuming that the percentage of the Muslim population remains at 99%, the Muslim population is expected to reach 1.3 million throughout the second half of this century.

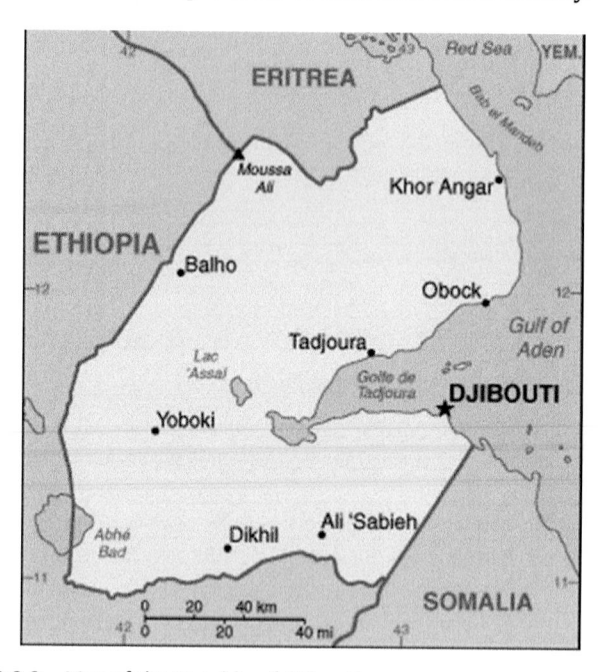

Figure 3.3.2 Map of the Republic of Djibouti.

Table 3.3.2 Evolution of the Muslim population in Djibouti

Year	Population	Muslims	%	Source
1885	200,000	200,000	100	[JAN]es
1917	210,000	210,000	100	[MAS23]es
1946	44,800	43,300	96.65	[SYB50]es
1960	81,200	81,200	100	[KET86]es
2009	818,159	809,977	99.00	[DOS10]es
2050	1,295,000	1,282,000	99.00	es
2100	1,332,000	1,318,000	99.00	es

3.3.3 Eritrea

The State of Eritrea has an area of 117,600 sq. km and its map is presented in Fig. 3.3.3a. It was taken by Italy from the Ottoman Empire in 1885. It was then taken from them by the British in 1941 and gained independence from the UK as part of Ethiopia in 1952. It gained its independence from Ethiopia in 1991. According to census data, the Muslim population increased from 0.15 million or 55.4% in 1905, to 0.31 million or 52.3% in 1931. According to Demographic and Health Surveys (DHS), the percentage of Muslims was 38.18% in 1995 and 39.59% in 2002. The population is predominantly Muslim in the eastern and western lowlands and mainly Christian in the highlands. A choropleth map illustrating the presence of Muslims per region is presented in Fig. 3.3.3b.

Thus, assuming that the percentage of Muslims remains fixed, the Muslim population is expected to exceed two million by 2050 and near four million by 2100. A summary of the data is provided in Table 3.3.3.

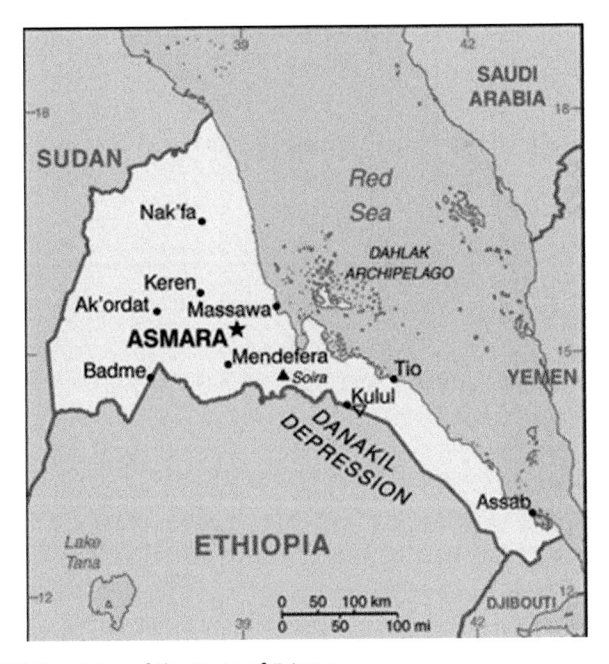

Figure 3.3.3a Map of the State of Eritrea.

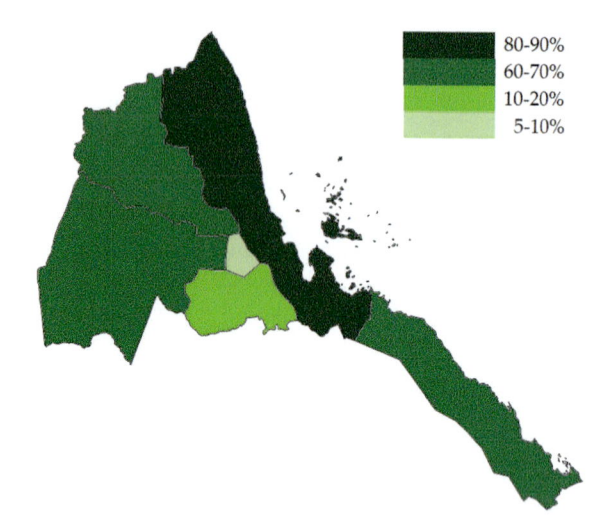

Figure 3.3.3b A choropleth map of Eritrea illustrating the presence of Muslims per region.

Table 3.3.3 Evolution of the Muslim population in Eritrea

Year	Population	Muslims	%	Source
1905	274,944	152,177	55.35	[EJB]c
1931	595,433	311,414	52.30	[ER31]c
1995	3,164,000	1,208,000	38.18	[DHS]s
2002	3,789,000	1,499,400	39.57	[DHS]s
2050	6,005,000	2,376,000	39.57	es
2100	9,062,000	3,586,000	39.57	es

3.3.4 Ethiopia

The Federal Democratic Republic of Ethiopia has an area of 1,104,300 sq. km and its map is presented in Fig. 3.3.4a. It was briefly occupied by the Italians between 1936 and 1941 and formerly known as Abyssinia. A 1924 estimate of the Muslim population puts it at 3.0 million or 37.5% of the total population [MAS23]. However, subsequent census data showed an increase from 12.6 million or 32.9% in 1984, to 17 million or 32.8% in 1994, to 25 million or 33.9% in 2007.

The 1984 census excluded the rural population of Tigray (2.2 million or 90%) and Eritrea (2.2 million or 85%). According to the

2007 Census, only 3% of the rural and 8% of the urban population of Tigray is Muslim. Thus, Muslim population from the 1984 Census could be estimated at 31% of the total population. A choropleth map illustrating the presence of Muslims per region is presented in Fig. 3.3.4b.

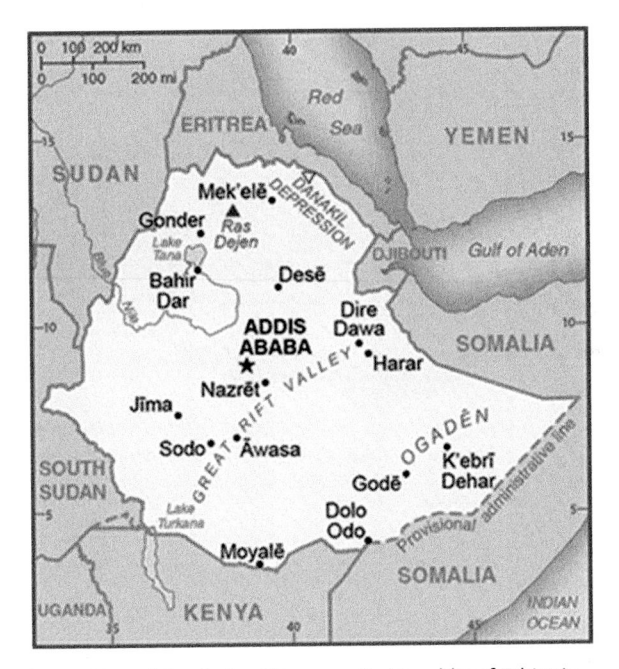

Figure 3.3.4a Map of the Federal Democratic Republic of Ethiopia.

Thus, assuming that the percentage of Muslims will continue to increase by half of a percentage point per decade, the Muslim population is expected to reach 74 million or 36.0% by 2050 and 113 million or 38.5% by 2100. A summary of the data is provided in Table 3.3.4.

Table 3.3.4 Evolution of the Muslim population in Ethiopia

Year	Population	Muslims	%	Source
1984	38,154,799	12,569,995	32.94	[ET84]c
1994	53,130,782	17,412,432	32.77	[UN04]c
2007	73,918,505	25,045,550	33.88	[ET07]c
2050	205,411,000	73,948,000	36.00	es
2100	294,393,000	113,341,000	38.50	es

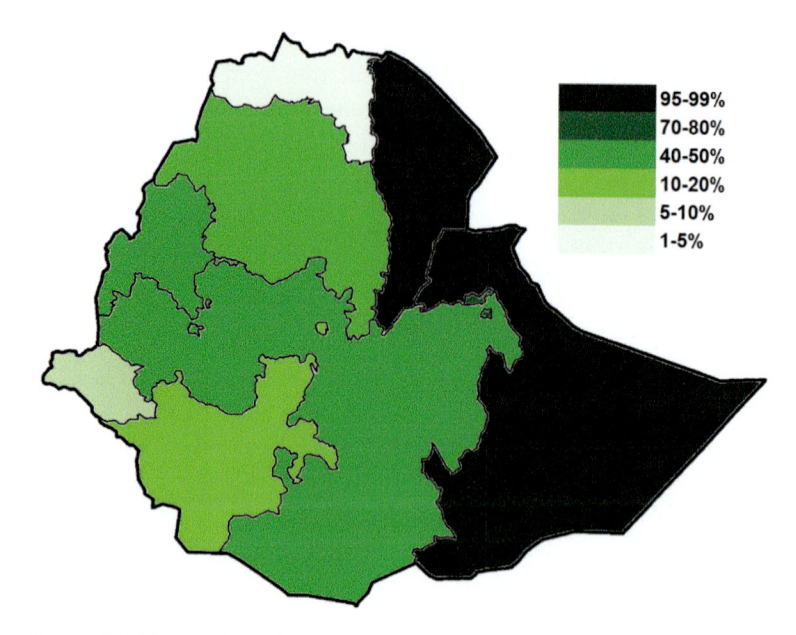

Figure 3.3.4b A choropleth map of Ethiopia illustrating the presence of Muslims per region.

3.3.5 Kenya

The Republic of Kenya has an area of 580,367 sq. km and its map is presented in Fig. 3.3.5a. It was occupied by the British in 1888 and gained its independence from the UK in 1963. According to census data, the Muslim population increased from 0.23 million or 4.3% in 1948, to 0.31 million or 3.8% in 1962, to 2.26 million or 8.0% in 1999 and 4.30 million or 11.2% in 2009. A choropleth map illustrating the presence of Muslims per county is presented in Fig. 3.3.5b.

Thus, assuming that the percentage of Muslims will continue to increase by one percentage point per decade, the Muslim population is expected to reach fourteen million or 15% by 2050 and 25 million or 20% by 2100. A summary of the data is provided in Table 3.3.5.

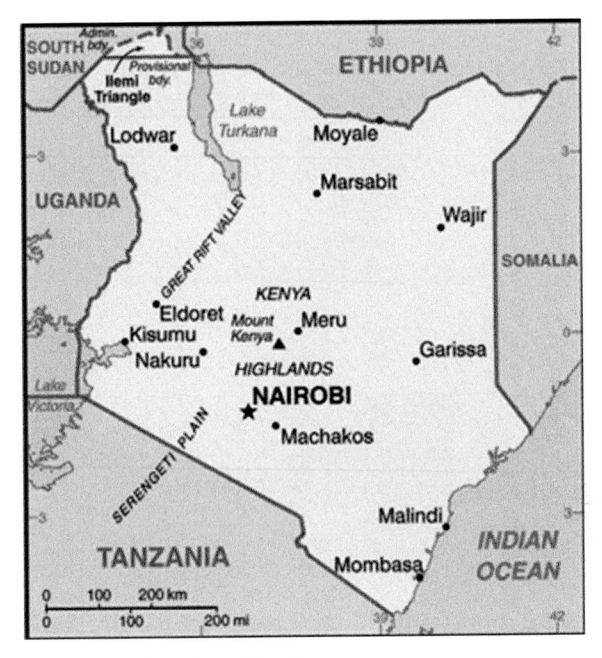

Figure 3.3.5a Map of the Republic of Kenya.

Table 3.3.5 Evolution of the Muslim population in Kenya

Year	Population	Muslims	%	Source
1948	5,373,078	231,040	4.30	[KE48]c
1962	8,048,000	309,100	3.84	[UN71]c
1999	28,159,915	2,258,523	8.02	[ROD]c
2009	38,412,088	4,304,798	11.21	[KE09]c
2050	91,575,000	13,736,000	15.00	es
2100	125,424,000	25,085,000	20.00	es

3.3.6 Madagascar

The Republic of Madagascar has an area of 587,041 sq. km occupying the fourth largest island in the world. A map of this island nation is presented in Fig. 3.3.6. It was occupied by the French in 1883 and gained its independence from France in 1960. Islam entered the northern part of this island during the ninth century through Muslim traders from the Arabian Peninsula. Islam also entered from the south in 1480 with Malay settlers. Estimates of the Muslim population increased from 70,000 or 2.0% in 1887, to 0.20 million or 4.6% in 1950, to 0.38 million or 5.0% in 1975. According to Demographic and Health Surveys (DHS), the percentage of Muslims was 0.94% in 1992, 0.71% in 1997 and 0.81% in 2009.

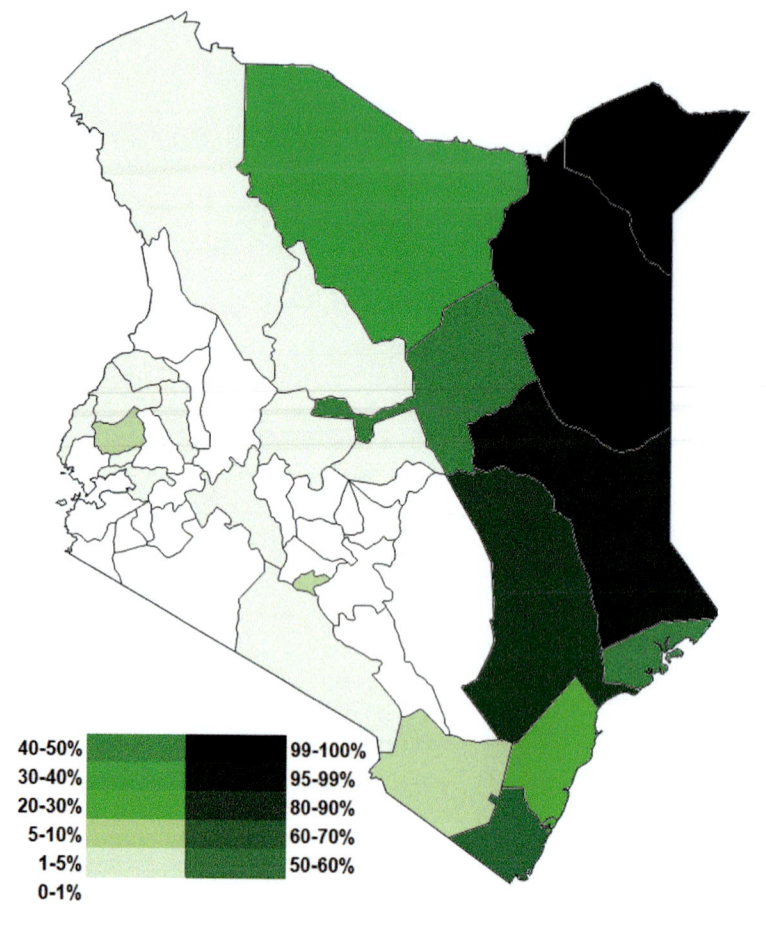

Figure 3.3.5b A choropleth map of Kenya illustrating the presence of Muslims per county.

Thus, assuming that the percentage of Muslims remains constant, the Muslim population is expected to reach 0.4 million by 2050 and 0.8 million by 2100. A summary of the data is provided in Table 3.3.6.

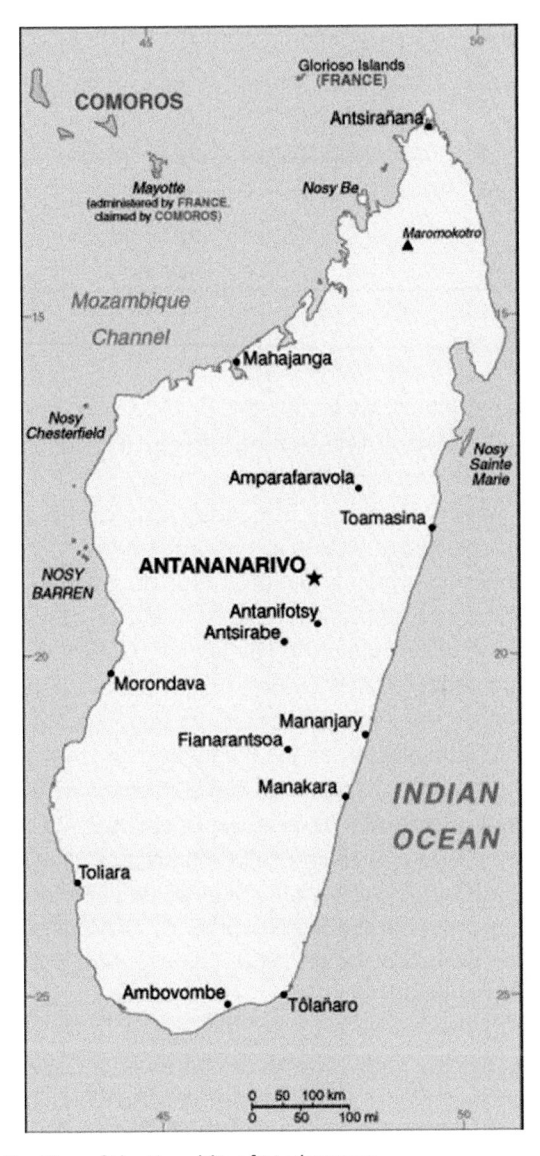

Figure 3.3.6 Map of the Republic of Madagascar.

Table 3.3.6 Evolution of the Muslim population in Madagascar

Year	Population	Muslims	%	Source
1887	3,520,000	70,000	1.99	[JAN]es
1950	4,350,700	200,000	4.60	[MAS55]es
1975	7,603,790	380,000	5.00	[SYB85]es
1992	12,264,000	115,000	0.94	[DHS]s
1997	14,329,000	102,000	0.71	[DHS]s
2009	20,496,000	166,000	0.81	[DHS]s
2050	54,048,000	438,000	0.81	es
2100	99,957,000	810,000	0.81	es

3.3.7 Mauritius

The Republic of Mauritius is an island nation with area 2,040 sq. km, consisting of Mauritius (1,865 sq. km) and its dependent islands of Rodrigues (104 sq. km), Agalega Archipelago (70 sq. km) and St. Brandon Rocks (1 sq. km). The latter also known as Cargados Carajos Shoals, consist of sixteen islands and islets 300 Km northeast of Mauritius that are mainly fishing outposts with no permanent residents. Agalega consists of two islands 1,000 km east of Mauritius with a population of around 300. The Island of Rodrigues has 3% of the population and is located 560 km east of the Island of Mauritius. A map of these islands is presented in Figs. 3.3.7a and 3.3.7b. Mauritius was sighted by the by the Portuguese in 1507 who claimed possession of it nearly the whole of the sixteenth century. The first who made any settlement in it were the Dutch, in 1598, who named it Mauritius, in honor of their Prince Maurice. It was abandoned by them in 1710, and afterwards taken possession by the French in 1715, which lost it to the British in 1810. The Islands gained its independence from the UK in 1968. Islam entered here in early eighteenth century with immigrant workers from India brought by the French. The first mosque was established in 1805.

After the British took over, they continued bringing workers from India as part of the British indentured workers system. The 1891 and 1901 censuses show that about 13% and 16% of all Indian population was Muslim, respectively. Thus, inferring religious adherence from ethnic affiliation from ethnic census data, we assume that 13% of Indians before 1871 were Muslims. This leads to the estimated increase in the Muslim population from 7,000 or 4.4% in 1846, to 10,000 or 5.5% in 1851, to 25,000 or 8.1% in 1861.

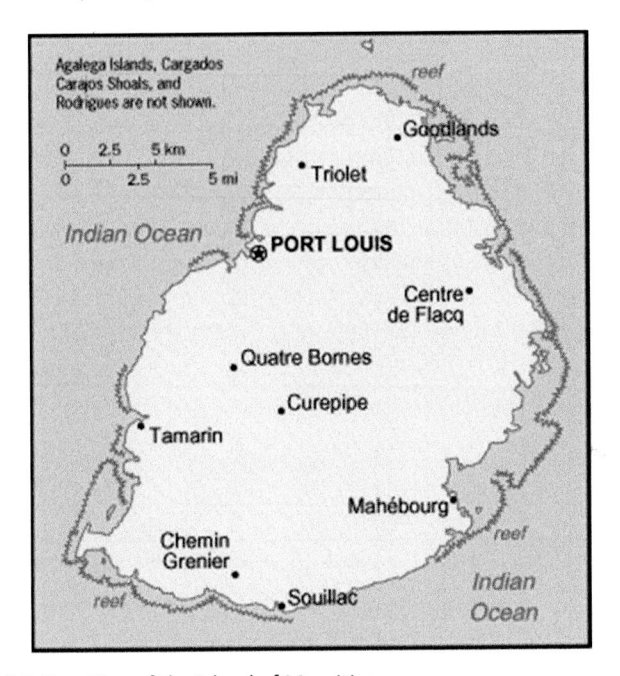

Figure 3.3.7a Map of the Island of Mauritius.

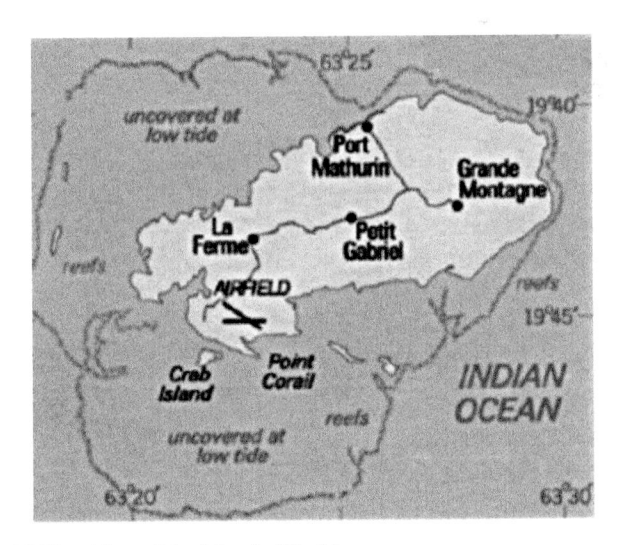

Figure 3.3.7b Map of the Island of Rodrigues.

According to census data, the Muslim population changed from 42,000 or 13.1% in 1871, to 35,000 or 9.8% in 1881, to 35,000 or 9.4% in 1891, to 41,000 or 10.9% in 1901, to 39,000 or 10.5% in 1911. The Muslim population continued to increase since then to 45,000 or 11.8% in 1921, to 51,000 or 12.6% in 1931, to 58,000 or 13.4% in 1944, to 77,000 or 15.2% in 1952, to 0.11 million or 16.2% in 1962, to 0.14 million or 16.6% in 1972, to 0.16 million or 15.0% in 1983, to 0.17 million or 16.3% in 1990, to 0.20 million or 16.7% in 2000, and 0.21 million or 17.3% in 2011. Thus, assuming that the percentage of Muslims will continue to increase by a half of a percentage point per decade, the Muslim population is expected to remain over 0.2 million throughout this century, reaching 19.5% of the total population by 2050 and 22% by 2100. A summary of the data is provided in Table 3.3.7a.

Table 3.3.7a Evolution of the Muslim population in the Republic of Mauritius

Year	Population	Muslims	%	Source
1846	158,462	7,000	4.42	[GB01]e
1851	181,318	10,000	5.52	[GB01]e
1861	310,743	25,000	8.05	[GB01]e
1871	317,150	41,575	13.11	[ROD]c
1881	361,305	35,316	9.77	[ROD]c
1891	371,655	34,763	9.35	[SYB00]c
1901	378,195	41,235	10.90	[GB01]c
1911	377,438	39,375	10.43	[GB11]c
1921	383,069	44,995	11.75	[ROD]c
1931	401,440	50,678	12.62	[ROD]c
1944	430,800	57,919	13.44	[MU944]c
1952	514,533	77,086	14.98	[UN56]c
1962	699,954	110,414	15.77	[UN63]c
1972	850,968	137,087	16.11	[UN73]c
1983	996,974	158,701	15.02	[MU83,MU83b]c
1990	1,055,792	172,047	16.30	[MU90]c
2000	1,178,848	196,240	16.65	[UN04]c
2011	1,236,817	213,969	17.30	[MU11]c
2050	1,186,000	231,000	19.50	es
2100	827,000	182,000	22.00	es

Almost all the Muslim population lives in the Island of Mauritius. On the Island of Rodrigues and per census data as shown in Table 3.3.7b, the Muslim population increased from none in 1871, to 34 or 1.1% in 1901, to 71 or 0.6% in 1944, to 72 or 0.5% in 1952, to 82 or 0.5% in 1962, to 90 or 0.4% in 1972, to 140 or 0.4% in 1983, to 184 or 0.5% in 1990, to 301 or 0.8% in 2000, and 386 or 1.0% in 2011.

Table 3.3.7b Evolution of the Muslim population on the Island of Rodrigues

Year	Population	Muslims	%	Source
1871	1,108	0	0.00	[GB71]e
1901	3,162	34	1.08	[GB01]e
1944	11,884	71	0.60	[MU944]c
1952	13,333	72	0.54	[UN56]c
1962	18,335	82	0.45	[UN63]c
1972	24,769	90	0.36	[UN73]c
1983	32,989	140	0.42	[MU83b]c
1990	34,204	184	0.54	[MU90]c
2000	35,779	301	0.84	[MU11]c
2011	40,434	386	0.95	[MU11]c

3.3.8 Mayotte

This is an Overseas Department of France that it occupied since 1843. It has a total area of 375 sq. km consisting of the main island of Mayotte or Mahoré (360 sq. km), Pamanzi (13 sq. km), and about a dozen much smaller islands surrounding the main island. Its map is presented in Fig. 3.3.8. Mayotte used to be the fourth and southernmost of the Comoros Islands group until it was separated by the French in 1975, when the Comoros gained their independence from France.

Like the other Comoros Islands, Islam entered these islands early ninth century through Muslim traders from the Arabian Peninsula. By the eleventh century it became part of the Kilwa Muslim Sultanate. By 1497, and with the fall of the Kilwa Sultanate, Mayotte became an independent sultanate until it was taken by the French. Before its separation from the Comoros Islands, all the population was estimated to be Muslim. However, the percentage of Muslims decreased to 98.4% by 2012. Thus, assuming that the percentage of Muslims will remain constant, the Muslim population is expected to reach half a million by 2050 and 0.7 million by 2100. A summary of the data is provided in Table 3.3.8.

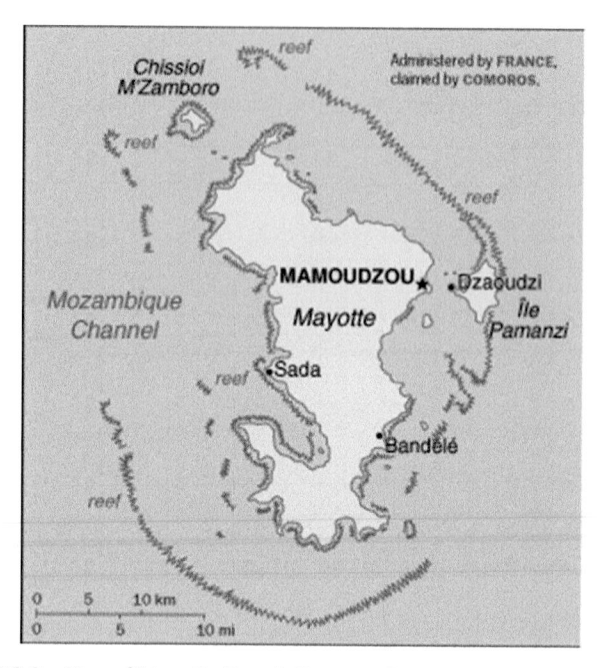

Figure 3.3.8 Map of Mayotte French Overseas Department.

Table 3.3.8 Evolution of the Muslim population in Mayotte

Year	Population	Muslims	%	Source
1918	13,425	13,425	100	[MAS23]es
1946	18,000	18,000	100	[SYB50]es
1966	32,607	32,607	100	[SYB70]es
2012	212,645	209,243	98.40	[PEW]es
2050	495,000	487,000	98.40	es
2100	746,000	734,000	98.40	es

3.3.9 Réunion

This is an Overseas Department of France that it occupied since 1638. It is a single island with total area of 2,512 sq. km and its map is presented in Fig. 3.3.9. It was known to Arabs and Africans centuries before its discovery by the Portuguese in 1507, when it was uninhabited. It was claimed by the French in 1642 and named Bourbon Island. However, it was only settled in 1665 with a population of twenty, then ninety in 1671, 269 in 1969, and 734 in 1704. It was named Réunion in 1793 to commemorate the union of revolutionaries from Marseille with the National Guard in Paris, which took place on August 10th, 1792. Islam came here in 1854 with the arrival of Muslim immigrant workers from Gujarat, India. Estimates of the Muslim population increased from none in 1849, to 204 or 0.1% in 1897 when they built their first mosque in the island, to 3,000 or 1.6% in 1926, to 5,000 or 2.3% in 1941, to 14,000 or 2.7% in 1982, to 30,000 or 4.3% in 1999, and to 50,000 or 6.1% in 2010. Thus, assuming that the percentage of Muslims will increase by one percentage point per decade, the Muslim population is expected to reach 0.10 million or 10% by 2050 and 0.14 million or 15% by 2100. A summary of the data is provided in Table 3.3.9.

Table 3.3.9 Evolution of the Muslim population in Réunion

Year	Population	Muslims	%	Source
1849	120,900	0	0	[REH]es
1897	173,192	204	0.12	[REH]es
1926	182,637	3,000	1.64	[MAS23]es
1941	220,955	5,000	2.26	[MAS55]es
1982	515,814	14,000	2.71	[KET86]es
1999	706,300	30,000	4.25	[PEW]es
2010	821,136	50,000	6.09	[RE]es
2050	1,010,000	101,000	10.00	es
2100	901,000	135,000	15.00	es

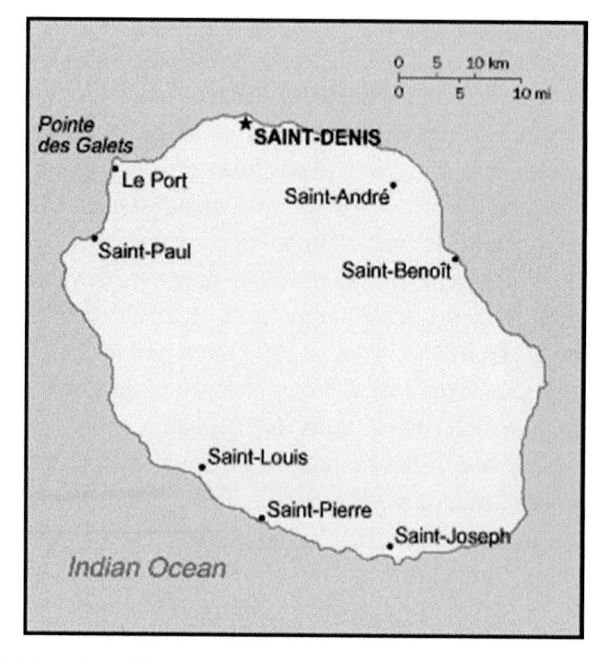

Figure 3.3.9 Map of Réunion French Overseas Department.

3.3.10 Seychelles

The Republic of Seychelles has an area of 455 sq. km consisting of 116 islands spread over 1.4 million sq. km of the Southwest Indian Ocean. The largest islands are Mahé (155 sq. km) with over 85% of the population and where the capital Victoria is located, Aldabra (129 sq. km), Praslin (38 sq. km), Silhouette (20 sq. km) and La Dique (10 sq. km). A map of these islands is presented in Fig. 3.3.10. The islands were occupied by the British in 1810 and gained their independence from the UK in 1976.

Islam entered here around mid-nineteenth century with labor workers brought by the British from India as part of the British indentured labor system. According to census data, the Muslim population increased from fifteen or 0.1% in 1871, to 68 or 0.4% in 1901, to 93 or 0.4% in 1911, to 132 or 0.3% in 1960, to 176 or 0.3% in 1971, to 328 or 0.5% in 1987, to 506 or 0.7% in 1994, to 866 or 1.1% in 2002, to 1,459 or 1.6% in 2010. Thus, assuming that the percentage of Muslims will increase by 0.4 of a percentage point per decade, the Muslim population is expected to reach 3,000 or 3.2% by 2050 and 5,000 or 5.2% by 2100. A summary of the data is provided in Table 3.3.10.

Table 3.3.10 Evolution of the Muslim population in Seychelles

Year	Population	Muslims	%	Source
1851	6,811	0	0.00	[KET86]es
1871	11,082	15	0.14	[GB71]c
1901	19,258	68	0.35	[GB01]c
1911	22,691	93	0.41	[GB11]c
1960	41,425	132	0.32	[UN56]c
1971	52,650	176	0.33	[UN63]c
1987	68,598	328	0.48	[UN73]c
1994	74,331	506	0.68	[UN88]c
2002	81,755	866	1.06	[UN]c
2010	90,945	1,459	1.60	[SC]c
2050	105,000	3,000	3.20	es
2100	88,000	5,000	5.20	es

3.3.11 Somalia

The Federal Republic of Somalia has a total area of 637,657 sq. km and its map is presented in Fig. 3.3.11. The current land occupied by Somalia used to consist of Italian Somaliland (southern two-thirds) and British Somaliland (northern third). The former was occupied by the Italians in 1888, while the latter was occupied by the British in 1888. Britain withdrew from British Somaliland in 1960 to allow its protectorate to join with Italian Somaliland and form the new nation of Somalia. The Italian census of 1931 recorded 1,018,412 Muslims, or 99.7% of the total population in the Italian occupied lands, which had two-thirds of the population. No other reliable statistics exist with regard to religious distribution of the population, but many sources assume that the entire population is Muslim. In 1991, the Somali government had collapsed, and the country entered a civil war. Concurrently, the lands of former British Somaliland, was declared the Republic of Somaliland. It remains independent but internationally unrecognized. Assuming that the percentage of Muslims will remain at 99.7%, then, as summarized in Table 3.3.11, the Muslim population will increase to sixteen million in 2020, then 35 million by 2050 and to 75 million by 2100.

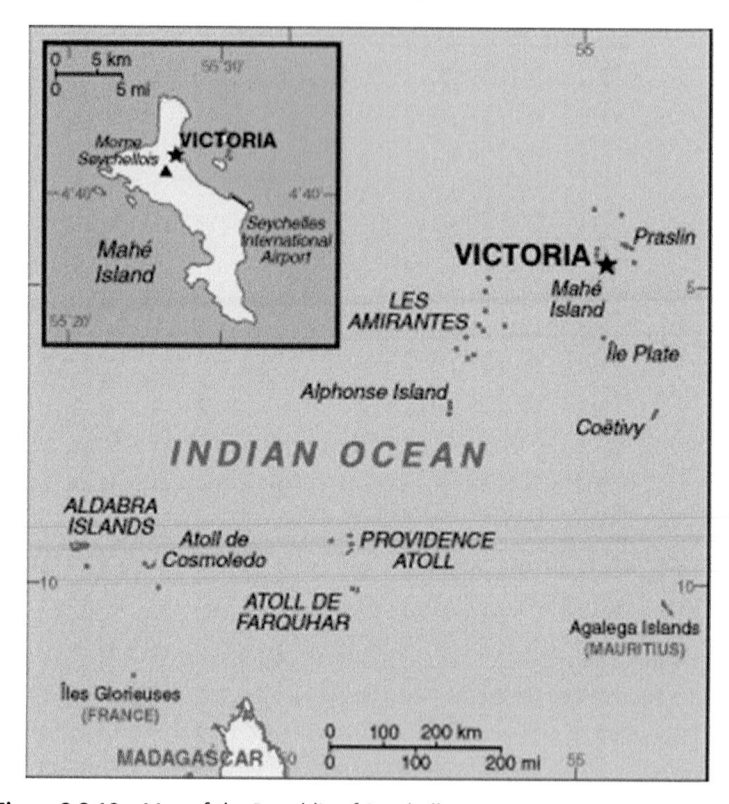

Figure 3.3.10   Map of the Republic of Seychelles.

Table 3.3.11   Evolution of the Muslim population in Somalia

Year	Population	Muslims	%	Source
1931	1,021,572	1,018,412	99.69	[ROD]c
2020	16,105,000	16,055,000	99.69	es
2050	34,922,000	34,814,000	99.69	es
2100	75,716,000	75,482,000	99.69	es

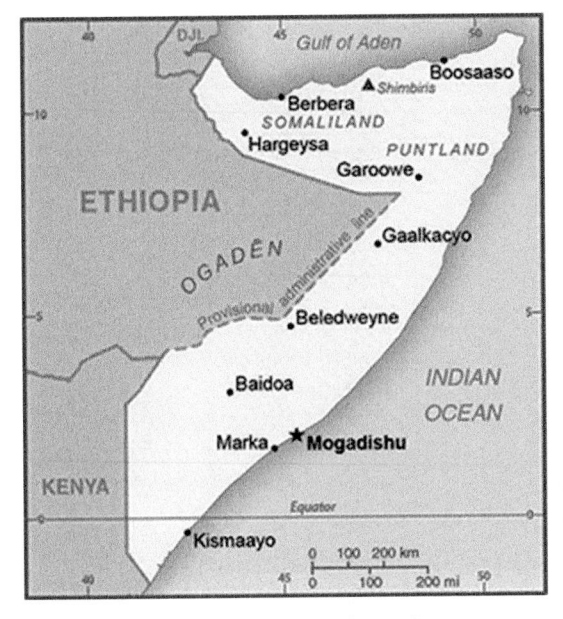

Figure 3.3.11 Map of the Federal Republic of Somalia.

3.3.12 Tanzania

The United Republic of Tanzania has an area of 947,300 sq. km, including the islands of Mafia (422 sq. km), Pemba (984 sq. km), and Unguja or Zanzibar (1,575 sq. km). The last two islands and small islands surrounding them comprise the semi-autonomous region of Zanzibar. A map of this country is presented in Fig. 3.3.12a.

The mainland was occupied by the Germans in 1884 but they lost it to the British between 1914 and 1919. The British also occupied the Islands of Zanzibar in 1919. The mainland was named Tanganyika by the British in 1920. Tanganyika and Zanzibar gained independence from the UK in 1961 and 1963, respectively. They united in 1964 to form Tanzania, where the name is formed using the first three letters of each of the two united countries.

Islam entered the eastern coast around 830 through Muslim traders from the Arabian Peninsula. Eventually, Ali ben Hassan established the Kilwa Sultanate, which lasted until it was conquered by the Portuguese in 1508. By the end of the seventeenth century, the Portuguese where driven out with the help of the Sultanate of Oman, which claimed these lands in 1698. In 1861, with the help of the British, this African side of the Sultanate of Oman was separated as Zanzibar Sultanate.

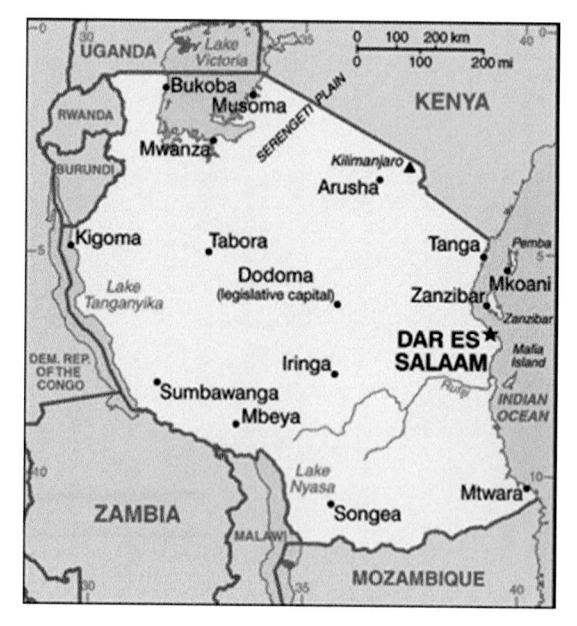

Figure 3.3.12a Map of the United Republic of Tanzania.

The Muslim population was estimated in 1921 at 0.6 million or 14.3% of the total population. According to census data, the Muslim population increased in number from 2.1 million or 27.5% in 1948, to 3.0 million or 33.2% in 1957, to 3.4 million or 28.6% in 1967.

Table 3.3.12 Evolution of the Muslim population in Tanzania

Year	Population	Muslims	%	Source
1921	4,199,462	599,462	14.28	[MAS23]es
1948	7,674,431	2,109,219	27.49	[TZH]c
1957	8,964,447	2,976,700	33.21	[TZH]c
1967	11,762,915	3,361,981	28.58	[TZ67]c
1994	29,030,000	8,874,500	30.57	[DHS]s
2004	37,935,000	11,475,000	30.25	[DHS]s
2015	53,470,000	17,220,000	31.79	[ABS]s
2050	129,387,000	43,345,000	33.50	es
2100	285,652,000	102,835,000	36.00	es

According to DHS survey data, the percentage of Muslims was 32.05% in 1992, 30.57% in 1994, 31.11% in 1996, 32.60% in 1999, 30.25% in 2004 and 30.13% in 2005. The ABS put it at 29.38% in 2003, 25.20% in 2006, 28.91% in 2009, 35.40% in 2012 and 31.79%

in 2015. A choropleth map illustrating the presence of Muslims per district is presented in Fig. 3.3.12b.

Thus, assuming that the percentage of Muslims will increase by half of a percentage point per decade, the Muslim population is expected to increase to 43 million or 33.5% by 2050 and 103 million or 36% by 2100. A summary of the data is provided in Table 3.3.12.

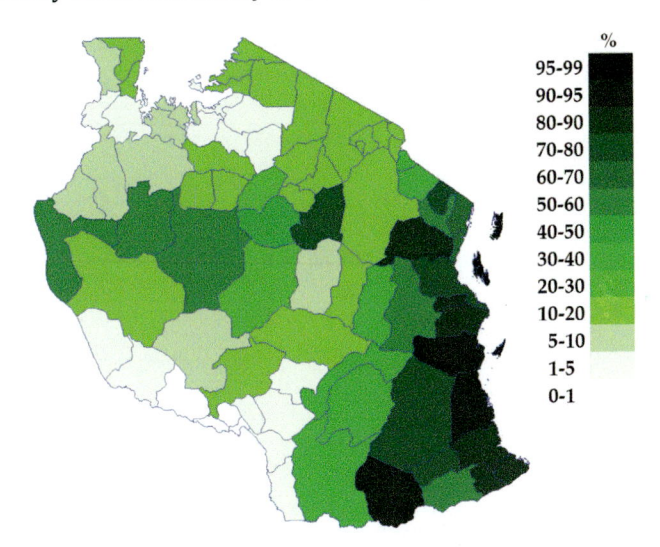

Figure 3.3.12b A choropleth map of Tanzania illustrating the presence of Muslims per district.

3.3.13 Regional Summary and Conclusion

Islam entered East Africa during the first decade of the start of Islam as many companions of the prophet Mohammed migrated to Abyssinia. Islam spread slowly, and by now almost a third of the population of this region is Muslim. The Muslim population is expected to continue its slow increase for the next three centuries. The following tables present centennial data from 600 AD to 2100 AD (or approximately 1 H to 1500 H) in Table 3.3a and decennial data from 1790 AD to 2100 AD (or 1210 H to 1520 H) in Tables 3.3b and 3.3c for current countries in East Africa. The data includes total population in thousands (P), the percentage of which is Muslim (M%), the corresponding Muslim population in thousands (M), and the annual population growth rate (APGR, or G%) of the total population in this region.

Table 3.3a Centennial estimates of the Muslim population (×1000) in East Africa: 600–2100 AD (1–1500 H)

		600	700	800	900	1000	1100	1200	1300
Comoros	P	0	0	0	0	0	1	2	3
	M%	—	—	—	50.00	100.00	100.00	100.00	100.00
	M	—	—	—	0	0	1	2	3
Djibouti	P	3	4	4	4	5	5	5	6
	M%	—	10.00	30.00	60.00	100.00	100.00	100.00	100.00
	M	—	0	1	2	5	5	5	6
Eritrea	P	53	56	59	62	65	80	95	110
	M%	—	10.00	20.00	30.00	40.00	50.00	60.00	60.00
	M	—	6	12	19	26	40	57	66
Ethiopia	P	755	800	845	890	935	1,100	1,300	1,500
	M%	—	1.00	5.00	10.00	15.00	18.00	20.00	22.00
	M	—	8	42	89	140	198	260	330
Kenya	P	320	365	410	455	500	600	700	800
	M%	—	1.00	5.00	5.00	10.00	10.00	15.00	15.00
	M	—	4	21	23	50	60	105	120
Madagascar	P	120	140	160	180	200	300	400	500
	M%	—	0.10	0.20	0.30	0.40	0.50	0.60	0.70
	M	—	0	0	1	1	2	2	4
Mauritius	P	—	—	—	—	—	—	—	—
	M%	—	—	—	—	—	—	—	—
	M	—	—	—	—	—	—	—	—
Mayotte	P	0	0	0	0	0	0	0	0
	M%	—	—	—	50.00	100.00	100.00	100.00	100.00
	M	—	—	—	0	0	0	0	0
Réunion	P	—	—	—	—	—	—	—	—
	M%	—	—	—	—	—	—	—	—
	M	—	—	—	—	—	—	—	—
Seychelles	P	0	0	0	0	0	0	0	0
	M%	—	—	—	—	—	—	—	—
	M	—	—	—	—	—	—	—	—
Somalia	P	320	340	360	380	400	475	550	625
	M%	—	10.00	30.00	60.00	100.00	100.00	100.00	100.00
	M	—	34	108	228	400	475	550	625
Tanzania	P	820	940	1,060	1,180	1,300	1,500	1,700	1,900
	M%	—	—	—	1.00	2.00	4.00	6.00	8.00
	M	—	—	—	12	26	60	102	152
Total	P	2,391	2,645	2,898	3,151	3,405	4,061	4,752	5,444
	M%	—	1.96	6.35	11.84	19.04	20.70	22.80	23.98
	M	—	52	184	373	648	841	1,084	1,306
	G%		0.101	0.091	0.084	0.078	0.176	0.157	0.136

Table 3.3a (*Continued*)

		1400	1500	1600	1700	1800	1900	2000	2100
Comoros	P	4	5	6	7	15	67	542	2,187
	M%	100.00	100.00	100.00	100.00	100.00	100.00	99.00	99.69
	M	4	5	6	7	15	67	537	2,180
Djibouti	P	7	8	9	10	11	36	718	1,332
	M%	100.00	100.00	100.00	100.00	100.00	100.00	99.00	99.00
	M	7	8	9	10	11	36	710	1,318
Eritrea	P	120	130	146	162	195	669	2,292	9,062
	M%	60.00	60.00	60.00	60.00	55.35	55.35	39.57	39.57
	M	72	78	88	97	108	370	907	3,586
Ethiopia	P	1,700	1,870	2,104	2,338	2,850	9,630	66,225	294,393
	M%	24.00	26.00	28.00	30.00	32.94	32.94	33.88	38.50
	M	408	486	589	701	939	3,172	22,437	113,341
Kenya	P	900	1,000	1,200	1,400	1,730	1,830	31,965	125,424
	M%	20.00	25.00	30.00	35.00	4.30	4.30	8.02	20.00
	M	180	250	360	490	74	79	2,564	25,085
Madagascar	P	600	700	800	1,000	1,550	2,969	15,767	99,957
	M%	0.80	0.90	1.00	1.50	1.99	1.99	0.71	0.81
	M	5	6	8	15	31	59	112	810
Mauritius	P	—	—	—	0	59	378	1,185	827
	M%	—	—	0.01	0.01	1.00	10.90	16.65	22.00
	M	—	—	—	0	1	41	197	182
Mayotte	P	0	1	1	1	3	12	150	746
	M%	100.00	100.00	100.00	100.00	100.00	100.00	98.40	98.40
	M	0	1	1	1	3	12	148	734
Réunion	P	—	—	—	1	70	173	737	901
	M%	—	—	—	—	—	0.12	4.25	15.00
	M	—	—	—	—	—	0	31	135
Seychelles	P	0	1	1	1	5	19	81	88
	M%	—	—	—	—	—	0.35	1.06	5.20
	M	—	—	—	—	—	0	1	5
Somalia	P	700	800	800	950	990	1,653	8,872	75,716
	M%	100.00	100.00	100.00	100.00	99.69	99.69	99.69	99.69
	M	700	800	800	950	987	1,648	8,845	75,482
Tanzania	P	2,200	2,600	3,000	3,400	3,870	3,968	33,499	285,652
	M%	10.00	12.00	14.00	16.00	14.28	14.28	30.25	36.00
	M	220	312	420	544	553	567	10,134	102,835
Total	P	6,231	7,115	8,067	9,270	11,348	21,404	162,033	896,285
	M%	25.61	27.36	28.27	30.37	23.98	28.27	28.77	36.34
	M	1,596	1,947	2,281	2,816	2,721	6,051	46,623	325,691
	G%	0.135	0.133	0.126	0.139	0.202	0.635	2.024	1.710

Table 3.3b Decennial estimates of the Muslim population (×1000) in East Africa: 1790–1940 AD (1210–1360 H)

		1790	1800	1810	1820	1830	1840	1850	1860
Comoros	P	10	15	20	25	30	35	40	45
	M%	100.00	100.00	100.00	100.00	100.00	100.00	100.00	100.00
	M	10	15	20	25	30	35	40	45
Djibouti	P	10	11	12	14	16	18	20	23
	M%	100.00	100.00	100.00	100.00	100.00	100.00	100.00	100.00
	M	10	11	12	14	16	18	20	23
Eritrea	P	190	195	200	205	230	260	300	350
	M%	55.35	55.35	55.35	55.35	55.35	55.35	55.35	55.35
	M	105	108	111	113	127	144	166	194
Ethiopia	P	2,800	2,850	2,900	2,949	3,500	4,100	4,700	5,300
	M%	32.94	32.94	32.94	32.94	32.94	32.94	32.94	32.94
	M	922	939	955	971	1,153	1,351	1,548	1,746
Kenya	P	1,720	1,730	1,740	1,750	1,760	1,770	1,780	1,790
	M%	4.30	4.30	4.30	4.30	4.30	4.30	4.30	4.30
	M	74	74	75	75	76	76	77	77
Madagascar	P	1,500	1,550	1,600	1,683	1,800	1,950	2,100	2,250
	M%	1.99	1.99	1.99	1.99	1.99	1.99	1.99	1.99
	M	30	31	32	33	36	39	42	45
Mauritius	P	40	59	78	98	90	89	181	312
	M%	0.50	1.00	2.00	3.00	4.00	4.42	5.52	8.05
	M	0	1	2	3	4	4	10	25
Mayotte	P	2	3	3	4	5	6	7	9
	M%	100.00	100.00	100.00	100.00	100.00	100.00	100.00	100.00
	M	2	3	3	4	5	6	7	9
Réunion	P	61	70	80	87	101	110	106	165
	M%	—	—	—	—	—	—	—	0.12
	M	—	—	—	—	—	—	—	0
Seychelles	P	4	5	6	7	7	7	7	7
	M%	—	—	—	—	—	—	—	0.10
	M	—	—	—	—	—	—	—	0
Somalia	P	985	990	995	1,000	1,070	1,140	1,210	1,280
	M%	99.69	99.69	99.69	99.69	99.69	99.69	99.69	99.69
	M	982	987	992	997	1,067	1,136	1,206	1,276
Tanzania	P	3,860	3,870	3,880	3,890	3,900	3,910	3,920	3,930
	M%	14.28	14.28	14.28	14.28	14.28	14.28	14.28	14.28
	M	551	553	554	555	557	558	560	561
Total	P	11,183	11,348	11,514	11,712	12,509	13,395	14,371	15,461
	M%	24.02	23.98	23.93	23.84	24.54	25.14	25.58	25.88
	M	2,687	2,721	2,755	2,792	3,070	3,367	3,675	4,001
	G%		0.147	0.145	0.171	0.658	0.685	0.703	0.731

Table 3.3b (*Continued*)

		1870	1880	1890	1900	1910	1920	1930	1940
Comoros	P	50	55	60	67	80	96	110	128
	M%	100.00	100.00	100.00	100.00	100.00	100.00	100.00	100.00
	M	50	55	60	67	80	96	110	128
Djibouti	P	26	29	32	36	40	45	49	54
	M%	100.00	100.00	100.00	100.00	100.00	100.00	100.00	100.00
	M	26	29	32	36	40	45	49	54
Eritrea	P	429	498	577	669	780	900	1,040	1,100
	M%	55.35	55.35	55.35	55.35	55.35	55.42	52.30	52.30
	M	237	276	319	370	432	499	544	575
Ethiopia	P	6,179	7,164	8,306	9,630	11,100	12,945	14,900	17,000
	M%	32.94	32.94	32.94	32.94	32.94	32.94	32.94	32.94
	M	2,035	2,360	2,736	3,172	3,656	4,264	4,908	5,600
Kenya	P	1,800	1,810	1,820	1,830	1,837	2,496	3,392	4,610
	M%	4.30	4.30	4.30	4.30	4.30	4.30	4.30	4.30
	M	77	78	78	79	79	107	146	198
Madagascar	P	2,400	2,576	2,766	2,969	3,200	3,422	3,648	3,888
	M%	1.99	1.99	1.99	1.99	1.99	1.99	1.99	4.60
	M	48	51	55	59	64	68	73	179
Mauritius	P	317	361	375	378	374	383	401	425
	M%	13.11	9.77	9.35	10.90	10.43	11.75	12.62	13.44
	M	42	35	35	41	39	45	51	57
Mayotte	P	10	12	9	12	10	13	15	18
	M%	100.00	100.00	100.00	100.00	100.00	100.00	100.00	100.00
	M	10	12	9	12	10	13	15	18
Réunion	P	183	172	166	173	174	173	198	221
	M%	0.12	0.12	0.12	0.12	0.12	1.64	2.26	2.26
	M	0	0	0	0	0	3	4	5
Seychelles	P	11	13	17	19	23	25	27	32
	M%	0.14	0.20	0.30	0.35	0.41	0.41	0.41	0.41
	M	0	0	0	0	0	0	0	0
Somalia	P	1,369	1,458	1,553	1,653	1,794	1,875	1,984	2,100
	M%	99.69	99.69	99.69	99.69	99.69	99.69	99.69	99.69
	M	1,365	1,453	1,548	1,648	1,788	1,869	1,978	2,093
Tanzania	P	3,940	3,950	3,960	3,968	4,063	4,199	5,299	5,535
	M%	14.28	14.28	14.28	14.28	14.28	14.28	27.49	27.49
	M	563	564	565	567	580	600	1,457	1,522
Total	P	16,714	18,099	19,640	21,404	23,474	26,573	31,064	35,111
	M%	26.64	27.15	27.69	28.27	28.83	28.64	30.05	29.70
	M	4,453	4,914	5,439	6,051	6,769	7,610	9,334	10,429
	G%	0.779	0.796	0.817	0.860	0.923	1.240	1.561	1.225

Table 3.3c Decennial estimates of the Muslim population (×1000) in East Africa: 1950–2100 AD (1370–1520 H)

		1950	1960	1970	1980	1990	2000	2010	2020
Comoros	P	159	191	230	308	412	542	690	870
	M%	99.00	99.00	99.00	99.00	99.00	99.00	99.69	99.69
	M	158	189	228	305	407	537	688	867
Djibouti	P	62	84	160	359	590	718	840	988
	M%	99.00	99.00	99.00	99.00	99.00	99.00	99.00	99.00
	M	61	83	158	355	584	710	832	978
Eritrea	P	822	1,008	1,311	1,733	2,259	2,292	3,170	3,546
	M%	52.30	52.30	52.30	38.18	38.18	39.57	39.57	39.57
	M	430	527	686	662	862	907	1,255	1,403
Ethiopia	P	18,128	22,151	28,415	35,142	47,888	66,225	87,640	114,964
	M%	32.94	32.94	32.94	32.94	32.77	33.88	33.88	34.50
	M	5,971	7,297	9,360	11,576	15,693	22,437	29,692	39,662
Kenya	P	6,077	8,120	11,301	16,417	23,725	31,965	42,031	53,771
	M%	4.30	3.84	5.00	6.00	7.00	8.02	11.21	12.00
	M	261	312	565	985	1,661	2,564	4,712	6,453
Madagascar	P	4,084	5,099	6,576	8,717	11,599	15,767	21,152	27,691
	M%	4.60	4.60	5.00	5.00	0.94	0.71	0.81	0.81
	M	188	235	329	436	109	112	171	224
Mauritius	P	493	660	826	966	1,056	1,185	1,248	1,272
	M%	14.98	15.77	16.11	15.02	16.30	16.65	17.30	18.00
	M	74	104	133	145	172	197	216	229
Mayotte	P	15	24	37	55	95	150	209	273
	M%	100.00	100.00	100.00	100.00	100.00	98.40	98.40	98.40
	M	15	24	37	55	95	148	205	268
Réunion	P	248	336	462	509	611	737	831	895
	M%	2.26	2.71	2.71	2.71	2.71	4.25	6.09	7.00
	M	6	9	13	14	17	31	51	63
Seychelles	P	36	42	52	66	71	81	91	98
	M%	0.32	0.32	0.33	0.48	0.68	1.06	1.60	2.00
	M	0	0	0	0	0	1	1	2
Somalia	P	2,264	2,756	3,445	6,281	7,225	8,872	12,044	15,893
	M%	99.69	99.69	99.69	99.69	99.69	99.69	99.69	99.69
	M	2,257	2,747	3,434	6,262	7,203	8,845	12,007	15,844
Tanzania	P	7,650	10,052	13,535	18,538	25,204	33,499	44,347	59,734
	M%	27.49	33.21	28.48	30.00	30.57	30.25	31.79	32.00
	M	2,103	3,338	3,855	5,561	7,705	10,134	14,098	19,115
Total	P	40,039	50,523	66,351	89,092	120,732	162,033	214,291	279,996
	M%	28.78	29.42	28.33	29.58	28.58	28.77	29.83	30.40
	M	11,525	14,865	18,797	26,356	34,508	46,623	63,927	85,109
	G%	1.313	2.326	2.725	2.947	3.039	2.942	2.795	2.674

Table 3.3c (*Continued*)

		2030	2040	2050	2060	2070	2080	2090	2100
Comoros	P	1,063	1,266	1,472	1,666	1,841	1,990	2,106	2,187
	M%	99.69	99.69	99.69	99.69	99.69	99.69	99.69	99.69
	M	1,060	1,262	1,467	1,661	1,835	1,983	2,100	2,180
Djibouti	P	1,117	1,217	1,295	1,346	1,364	1,363	1,353	1,332
	M%	99.00	99.00	99.00	99.00	99.00	99.00	99.00	99.00
	M	1,106	1,205	1,282	1,332	1,351	1,350	1,339	1,318
Eritrea	P	4,240	5,114	6,005	6,836	7,605	8,248	8,731	9,062
	M%	39.57	39.57	39.57	39.57	39.57	39.57	39.57	39.57
	M	1,678	2,023	2,376	2,705	3,009	3,264	3,455	3,586
Ethiopia	P	144,944	175,466	205,411	232,994	256,441	274,558	287,056	294,393
	M%	35.00	35.50	36.00	36.50	37.00	37.50	38.00	38.50
	M	50,731	62,290	73,948	85,043	94,883	102,959	109,081	113,341
Kenya	P	66,450	79,470	91,575	102,398	111,411	118,214	122,807	125,424
	M%	13.00	14.00	15.00	16.00	17.00	18.00	19.00	20.00
	M	8,638	11,126	13,736	16,384	18,940	21,279	23,333	25,085
Madagascar	P	35,622	44,471	54,048	64,059	74,035	83,598	92,343	99,957
	M%	0.81	0.81	0.81	0.81	0.81	0.81	0.81	0.81
	M	289	360	438	519	600	677	748	810
Mauritius	P	1,274	1,245	1,186	1,116	1,046	970	894	827
	M%	18.50	19.00	19.50	20.00	20.50	21.00	21.50	22.00
	M	236	237	231	223	214	204	192	182
Mayotte	P	343	419	495	564	626	679	719	746
	M%	98.40	98.40	98.40	98.40	98.40	98.40	98.40	98.40
	M	337	413	487	555	616	668	708	734
Réunion	P	955	996	1,010	1,003	988	970	942	901
	M%	8.00	9.00	10.00	11.00	12.00	13.00	14.00	15.00
	M	76	90	101	110	119	126	132	135
Seychelles	P	103	105	105	102	99	96	93	88
	M%	2.40	2.80	3.20	3.60	4.00	4.40	4.80	5.20
	M	2	3	3	4	4	4	4	5
Somalia	P	21,191	27,591	34,922	43,128	51,689	60,207	68,369	75,716
	M%	99.69	99.69	99.69	99.69	99.69	99.69	99.69	99.69
	M	21,125	27,505	34,814	42,995	51,529	60,021	68,157	75,482
Tanzania	P	79,163	102,587	129,387	158,869	190,389	222,833	254,983	285,652
	M%	32.50	33.00	33.50	34.00	34.50	35.00	35.50	36.00
	M	25,728	33,854	43,345	54,016	65,684	77,992	90,519	102,835
Total	P	356,464	439,946	526,911	614,083	697,535	773,727	840,396	896,285
	M%	31.14	31.91	32.69	33.47	34.23	34.96	35.67	36.34
	M	111,006	140,368	172,229	205,546	238,784	270,526	299,768	325,691
	G%	2.415	2.104	1.804	1.531	1.274	1.037	0.827	0.644

3.4 Muslims in Central Africa

This region consists of twelve countries: Burundi, Cameroon, Central African Republic, Congo-Brazzaville, Congo-Kinshasa, Equatorial Guinea, Gabon, Nigeria Rwanda, São Tomé & Principe, South Sudan, and Uganda. Islam entered this region towards the end of the eleventh century through interaction with Muslim traders and was strengthened through its adoption by the rulers of the Kanem-Bornu Empire around Lake Chad in 1085. Islam also advanced slowly from the east coast during the nineteenth century. It reached the Lake of Tanganyika in 1840, Uganda in 1852, Rwanda, Burundi and Eastern Congo by 1861. Islam reached the west coast of this region with trade with West Africa at the turn of twentieth century.

Thus, the Muslim population remained negligible during the first millennium and has likely reached 0.25 million or 2.3% in 1100 AD, then increasing to 0.59 million or 4.9% in 1200 AD, to 1.03 million or 7.6% in 1300 AD, to 1.54 million or 10.3% in 1400 AD, to 2.15 million or 13.1% in 1500 AD, to 3.19 million or 16.7% in 1600 AD, to 5.7 million or 25.5% in 1700 AD, to 6.3 million or 25.0% in 1800 AD, to 8.7 million or 28.0% in 1900, to 70 million or 29.4% in 2000, to 123 million or 29.5% in 2020, and is projected to reach 0.25 billion or 30% by 2050 and 0.46 billion or 31% by 2100.

A plot of centennial estimates of the Muslim population and its percentage with respect to the total population in this region from 600 to 2100 is provided in Fig. 3.4a. A zoom in of this plot, providing a plot of decennial estimates of the Muslim population and its percentage with respect to the total population in this region from 1900 to 2100 is provided in Fig. 3.4b. This shows that the Muslim population in this region was increasing slowly until 1940 and is increasing substantially afterwards towards the end of this century with an increasing rate. The percentage of Muslims on the other hand, was decreasing sharply from 31% in 1870, to almost 22% in 1940, but bounced back the following decade to 27%, and will exceed 31% by the end of this century.

The corresponding individual data for each country in this region is discussed below. In Section 3.4.13, the total population in each country in this region and the corresponding percentage and number of Muslims is presented centennially in Table 3.4a from 600 to 2100 and decennially in Tables 3.4b and 3.4c from 1790 to 2100.

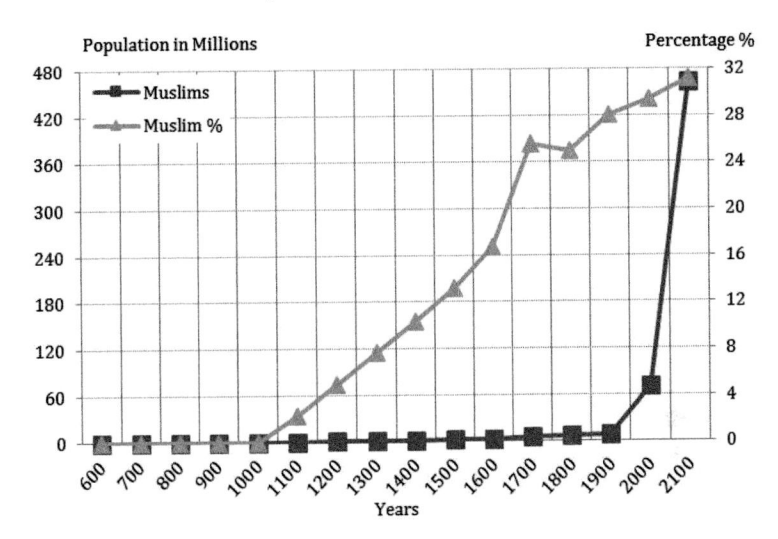

Figure 3.4a Plot of centennial estimates of the Muslim population and its percentage of the total population in Central Africa: 600–2100 AD (1–1500 H).

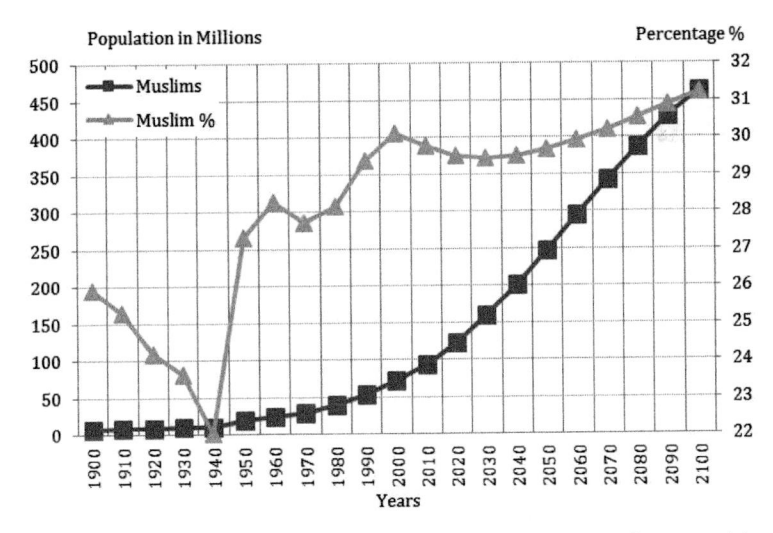

Figure 3.4b Plot of decennial estimates of the Muslim population and its percentage of the total population in Central Africa: 1900–2100 AD (1320–1520 H).

3.4.1 Burundi

The Republic of Burundi has an area of 27,830 sq. km and its map is presented in Fig. 3.4.1. It was occupied by the Germans in 1899, but lost it to the Belgians in 1916, and became part of Ruanda-Urundi; the predecessor of Rwanda and Burundi. They both gained independence from Belgium in 1962 as separate countries. Islam entered here during the nineteenth century through commerce. By 1861 these lands were under the Muslim Sultanate of Zanzibar.

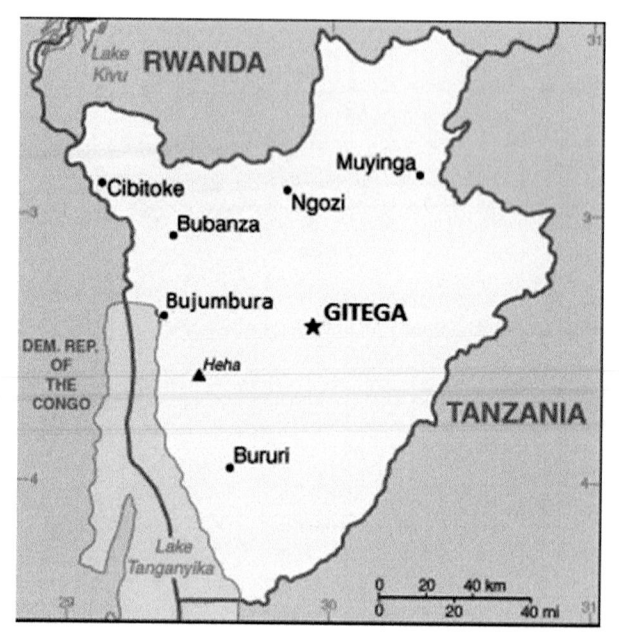

Figure 3.4.1 Map of the Republic of Burundi.

As shown in Table 3.4.1, the Muslim population was estimated at 8,220 or 0.5% in 1934, and 26,000 or 0.8% in 1965 (survey), then according to census data, it increased to 84,000 or 1.9% in 1990 and 201,000 or 2.7% in 2008. Thus, assuming that the percentage of Muslims will continue to increase by a quarter of a percentage point per decade, the Muslim population is expected to increase to one million or 3.8% by 2050 and 2.5 million or 5% by 2100.

Table 3.4.1 Evolution of the Muslim population in Burundi

Year	Population	Muslims	%	Source
1934	1,681,583	8,220	0.49	[ROD]es
1965	3,200,000	25,600	0.80	[BI65]s
1990	4,305,199	83,528	1.94	[UN04]c
2008	7,562,476	200,509	2.65	[UN]c
2050	25,325,000	950,000	3.75	es
2100	50,904,000	2,545,000	5.00	es

3.4.2 Cameroon

The Republic of Cameroon has an area of 475,440 sq. km and its map is presented in Fig. 3.4.2a. The former French Cameroon and part of British Cameroon merged in 1961 to form the present country. It was occupied by Germany in 1884, and then partitioned between France (most of current Cameroon) and Great Britain (west side) in 1919. Islam entered the northern parts in 1085 with the start of the Sayfawa Muslim Dynasty of the Kanem-Bornu Empire which was founded by Hummay bnu Abdul Jalil who ruled it until 1097. This Dynasty lasted until 1389 and was based in Kanem in west Chad, and controlled Lake Chad.

The Muslim population was estimated at half million or 23.8% in 1921. According to census data, the Muslim population increased from 0.34 million or 23.5% in 1953 to 3.7 million or 20.9% in 2005. A choropleth map illustrating the presence of Muslims per region is presented in Fig. 3.4.2b.

Thus, assuming that the percentage of Muslims remains constant at 20.9%, the Muslim population is expected to exceed ten million by 2050 and nineteen million by 2100. A summary of the data is provided in Table 3.4.2.

Table 3.4.2 Evolution of the Muslim population in Cameroon

Year	Population	Muslims	%	Source
1921	2,100,000	500,000	23.81	[MAS23]es
1953	1,439,870	338,820	23.53	[UN56]c
2005	17,463,836	3,651,688	20.91	[IPUMS]c
2050	50,573,000	10,575,000	20.91	es
2100	90,225,000	18,866,000	20.91	es

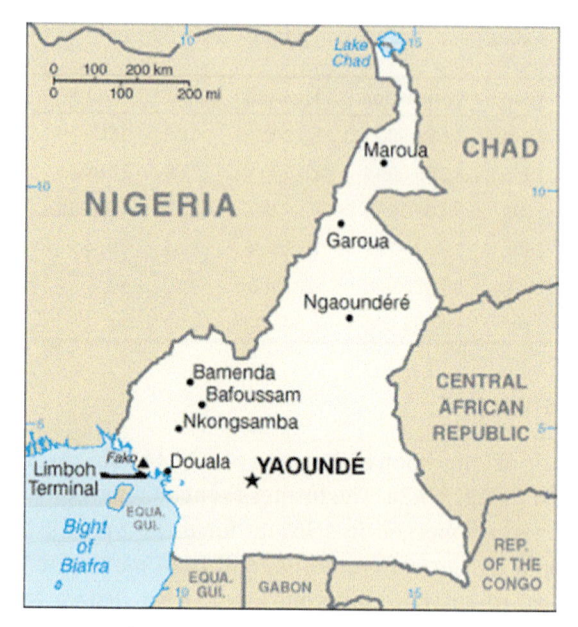

Figure 3.4.2a Map of the Republic of Cameroon.

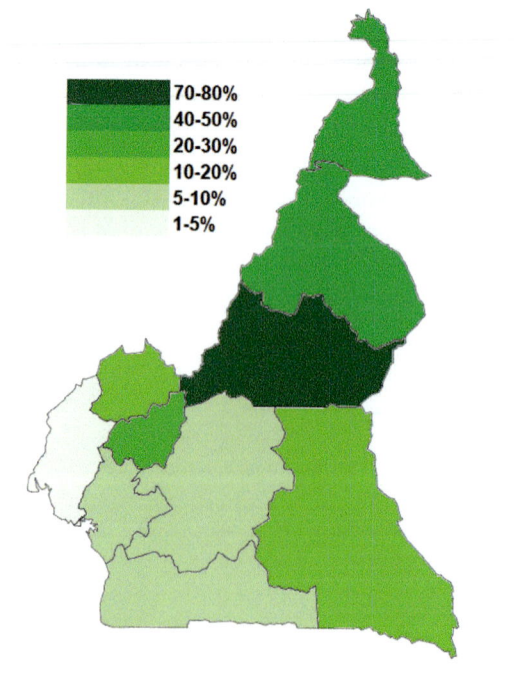

Figure 3.4.2b A choropleth map of Cameroon illustrating the presence of Muslims per region.

3.4.3 Central African Republic

It has an area of 622,984 sq. km and its map is presented in Fig. 3.4.3a. It was occupied by France in 1889 and upon independence from France in 1960 it changed its name from Ubangi-Shari. Islam entered here through trade in early nineteenth century. The number of Muslims increased from 16,000 or 2.5% in 1960 (survey) 318,000 or 10.3% in 2003 (census), most of whom live in the north of the country. A choropleth map illustrating the presence of Muslims per prefecture is presented in Fig. 3.4.3b.

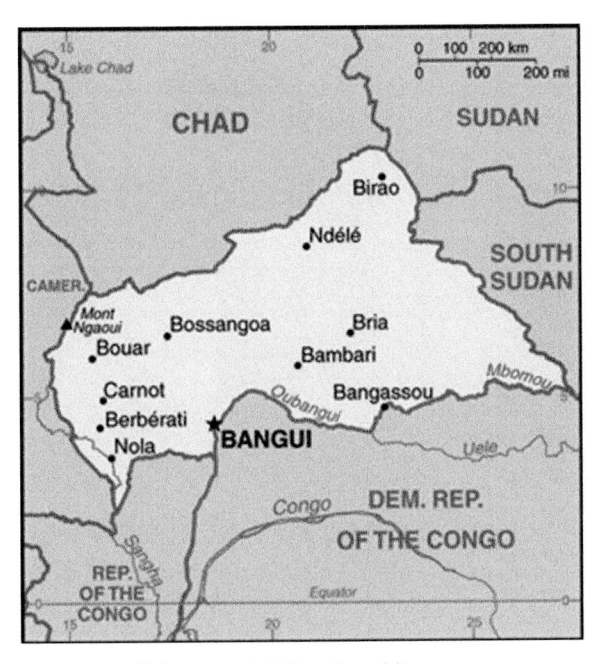

Figure 3.4.3a Map of the Central African Republic.

Thus, assuming that the percentage of Muslims will continue to increase by one percentage point per decade, the Muslim population is expected to reach 1.3 million or 15% by 2050 and 2.3 million or 20% by 2100. A summary of the data is provided in Table 3.4.3.

Table 3.4.3 Evolution of the Muslim population in the Central African Republic

Year	Population	Muslims	%	Source
1921	606,644	25,000	4.12	[MAS23]es
1949	1,000,000	30,000	3.00	[MAS55]es
1960	622,690	15,710	2.52	[CF60]s
2003	3,076,282	317,621	10.32	[CF03]c
2050	8,401,000	1,260,000	15.00	es
2100	11,631,000	2,326,000	20.00	es

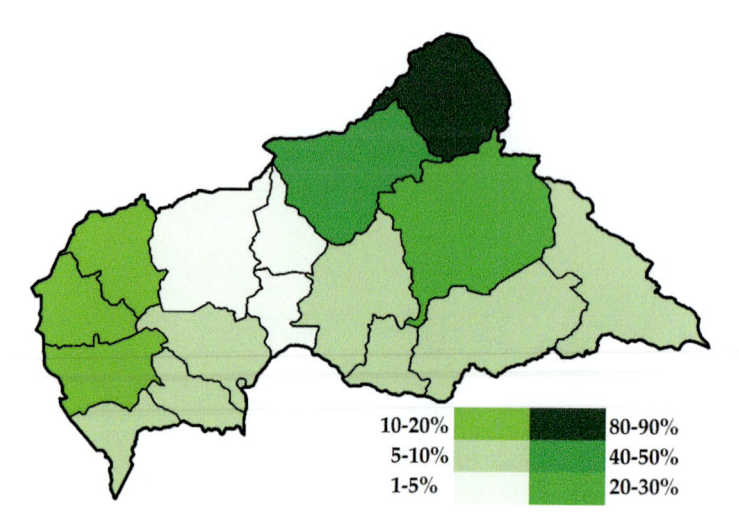

Figure 3.4.3b A choropleth map of Central African Republic illustrating the presence of Muslims per perfecture.

3.4.4 Congo–Brazzaville

The Republic of the Congo with its capital Brazzaville has a total area of 342,000 sq. km and its map is presented in Fig. 3.4.4. It was occupied by France in 1886 and gained independence from France in 1960. Islam entered here in mid nineteenth century with Muslim traders coming from North Africa and early twentieth century by immigrants from West Africa. The Muslim population was estimated at 2,000 or 0.3% in 1922. The first census recorded 4,343 Muslims or

Figure 3.4.4 Map of the Republic of the Congo.

Table 3.4.4 Evolution of the Muslim population in Congo–Brazzaville

Year	Population	Muslims	%	Source
1850	215,000	0	0.00	[MAS23]es
1922	581,143	2,000	0.34	[MAS23]es
1961	729,447	4,343	0.60	[CG61,CG61b]s
1974	1,319,790	40,000	3.03	[SYB80]es
1984	1,909,248	37,000	1.94	[KET86]es
2007	3,697,490	59,871	1.62	[CG07]c
2012	4,286,000	102,500	2.39	[DHS]s
2050	10,702,000	482,000	4.50	es
2100	20,962,000	1,467,000	7.00	es

0.6% of the population. This excluded the population of Point-Noire which was censused in 1958 with total population of 54,573 but did not include information on religious affiliation. The next census that collected such information was conducted in 2007 and recorded 60,000 or 1.6% of the total population was Muslim. According to DHS survey data, the percentage of Muslims was 1.61% in 2005, 1.25% in 2009 and 2.39% in 2012. Thus, assuming that the percentage of Muslims will continue to increase by half of a percentage point per decade, the Muslim population is expected to reach half a million or 4.5% by 2050 and 1.5 million or 7% by 2100. A summary of the data is provided in Table 3.4.4.

3.4.5 Congo–Kinshasa

The Democratic Republic of the Congo with its capital Kinshasa has a total area of 2,344,858 sq. km and its map is presented in Fig. 3.4.5. It was occupied by the Belgians in 1885 and named the Congo Free State, then it was made a Belgian colony in 1908 when it was renamed the Belgian Congo and it gained its independence from Belgium in 1960. Its name was Zaire between 1965 and 1997.

Islam entered here during the nineteenth century through commerce. By 1861 the eastern part as far as the city of Kindu on the Congo River was under the Muslim Sultanate of Zanzibar. Muslim population was estimated to be 1,000 or 0.01% in 1920. Based on census data, the Muslim population increased to 77,600 or 0.6% in 1957 and 415,000 or 1.4% in 1984. According to DHS, the percentage of Muslims was 1.49% in 2007 and 1.33% in 2014. Thus, assuming that the percentage of Muslims remains fixed at 1.5%, the Muslim population is expected to reach three million by 2050 and six million by 2100. A summary of the data is provided in Table 3.4.5.

Table 3.4.5 Evolution of the Muslim population in Congo–Kinshasa

Year	Population	Muslims	%	Source
1920	7,000,000	1,000	0.01	[MAS23]es
1957	12,183,661	77,600	0.64	[ROD]c
1984	29,671,407	415,000	1.40	[ROD]c
2007	59,835,000	890,000	1.49	[DHS]s
2014	74,877,000	999,000	1.33	[DHS]s
2050	194,489,000	2,898,000	1.49	es
2100	362,031,000	5,394,000	1.49	es

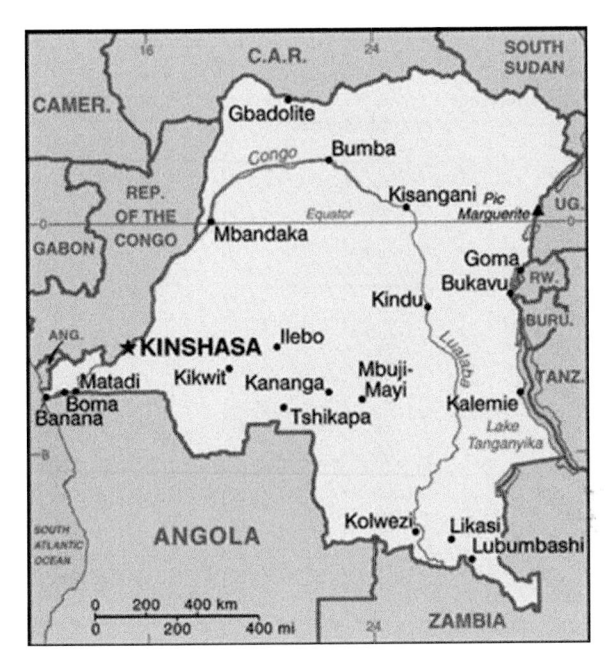

Figure 3.4.5 Map of the Democratic Republic of the Congo.

3.4.6 Equatorial Guinea

The Republic of Equatorial Guinea has an area of 28,051 sq. km, including five inhabited islands off its coast: Bioko (2,017 sq. km) where the capital Malabo is located, Annobón (18 sq. km), Corisco (14 sq. km), Great Elobey (2.3 sq. km) and Small Elobey (0.2 sq. km). A map of this country is presented in Fig. 3.4.6. It was occupied by Spain in 1844 and gained its independence in 1968. Islam entered here at the turn of the twentieth century and the number of Muslims was estimated at 1,000 or 0.3% in 1985. However, by 2011, DHS puts the percentage of Muslims at 3.69%. Many of the Muslims are coming from nearby Cameroon and Nigeria. Thus, assuming that the percentage of Muslims will increase by half of a percentage point per decade, the Muslim population is expected to reach 0.2 million by 2050 and 0.4 million by 2100. A summary of the data is provided in Table 3.4.6.

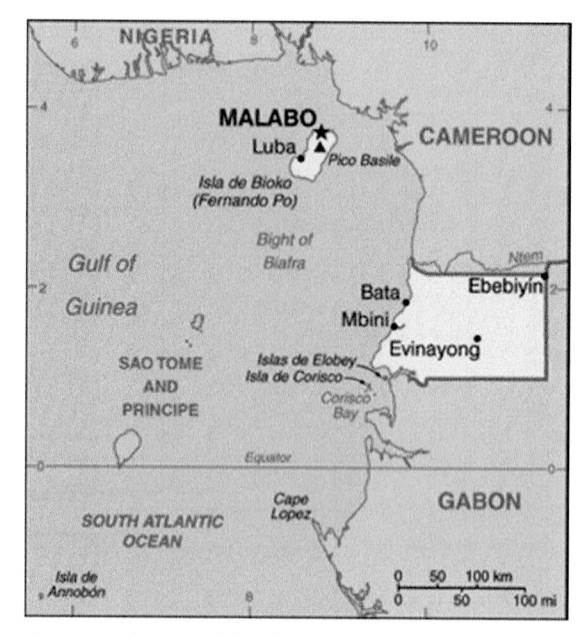

Figure 3.4.6 Map of the Republic of Equatorial Guinea.

Table 3.4.6 Evolution of the Muslim population in Equatorial Guinea

Year	Population	Muslims	%	Source
1985	312,772	1,000	0.32	[KET86]es
2011	751,000	27,700	3.69	[DHS]s
2050	2,821,000	155,000	5.50	es
2100	4,511,000	361,000	8.00	es

3.4.7 Gabon

The Gabonese Republic has an area of 267,667 sq. km and its map is presented in Fig. 3.4.7. It was occupied by France in 1885 and gained its independence in 1960. In 1967, Albert-Bernard Bongo became a president. He embraced Islam in 1973 and changed his name to Omar Bongo. He remained president of Gabon until his death in 2009, when his son Ali Bongo became the next president.

Islam entered here at the turn of the twentieth century with traders from West Africa. By 1922, the Muslim population consisted of about fifty individuals living in the Capital Libreville, or 0.01% of the total population. Estimates of the Muslim population increased to 2,000 or 0.2% in 1970, to 67,000 or 10.0% in 1982, to 0.12 million or 12.0% in 1993. Then according to DHS, the Muslim population increased to 0.14 million or 9.4% in 2003, and 0.18 million or 10.9% in 2013. Most of Muslims in Gabon however are immigrants from West Africa. Thus, assuming that the percentage of Muslims will increase by one percentage point per decade, then the Muslim population is expected to reach half a million or 15% by 2050 and one million or 20% by 2100. A summary of the data is provided in Table 3.4.7.

Table 3.4.7 Evolution of the Muslim population in Gabon

Year	Population	Muslims	%	Source
1922	388,778	50	0.01	[MAS23]es
1970	950,007	2,000	0.21	[SYB80]es
1982	670,000	67,000	10.00	[KET86]es
1993	1,014,976	121,800	12.00	[DOS00]es
2003	1,318,000	124,300	9.43	[DHS]s
2013	1,650,000	180,000	10.91	[DHS]s
2050	3,809,000	571,000	15.00	es
2100	5,859,000	1,172,000	20.00	es

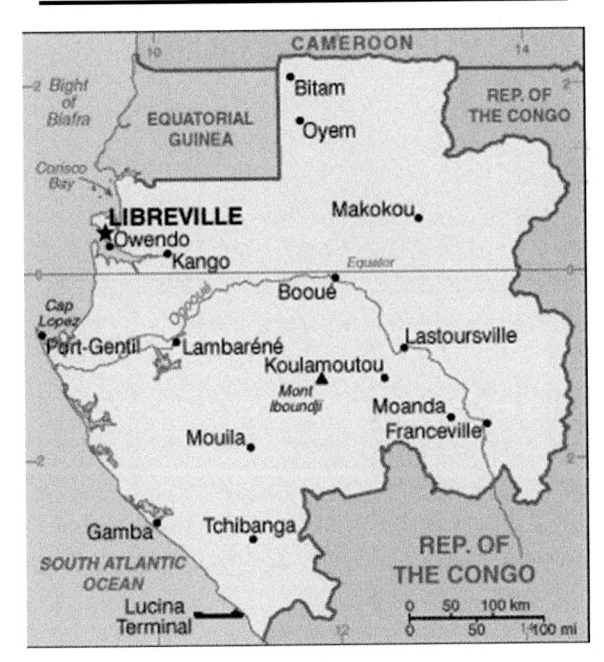

Figure 3.4.7 Map of the Gabonese Republic.

3.4.8 Nigeria

The Federal Republic of Nigeria has an area of 923,768 sq. km and is the most populous nation in Africa. Its map is presented in Fig. 3.4.8. It was occupied by the British in 1885 and gained its independence from the UK in 1960. Islam entered here from the northwest through the Za Dynasty of the Songhai Empire, whose ruler Kusoy Muslim Dam converted to Islam in 1010. Islam also entered from the northeast with the establishment of the Sayfawa Muslim Dynasty of the Kanem-Bornu Empire in 1085 by Hummay bnu Abdul Jalil.

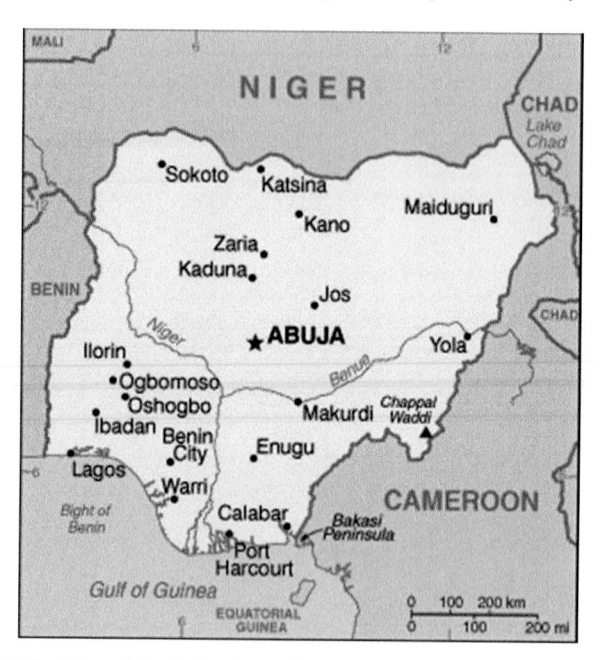

Figure 3.4.8 Map of the Federal Republic of Nigeria.

The majority of the northern half of Nigeria is Muslim. According to 1911 estimate, the Muslim population was eight million or 46.2% of the total population. Later census data indicate an increase to 27 million or 45.3% in 1953, and 26 million or 47.2% in 1963. According to DHS survey data, the percentage of Muslims was 47.53% in 1990, 44.42% in 1999, 50.43% in 2003, 44.73% in 2008 and 51.80% in 2013. Thus, assuming that the percentage of Muslims will continue to increase by a quarter of a percentage point per decade, the Muslim population is expected to reach 0.2 billion or 52.8% by 2050 and 0.4 billion or 54% by 2100. A summary of the data is provided in Table 3.4.8.

Table 3.4.8 Evolution of the Muslim population in Nigeria

Year	Population	Muslims	%	Source
1911	16,858,689	7,795,000	46.24	[MAS23]es
1953	59,431,466	26,912,046	45.28	[UN56]c
1963	55,670,055	26,276,496	47.20	[UN71]c
1990	95,617,000	45,447,000	47.53	[DHS]s
2003	132,581,000	66,860,000	50.43	[DHS]s
2013	172,817,000	89,520,000	51.80	[DHS]s
2050	401,315,000	211,694,000	52.75	es
2100	732,942,000	395,788,000	54.00	es

3.4.9 Rwanda

The Republic of Rwanda has an area of 26,338 sq. km and its map is presented in Fig. 3.4.9. It was occupied by the Germans in 1900, but lost it to the Belgians in 1916, and became part of Ruanda-Urundi; the predecessor of Rwanda and Burundi. They both gained independence from Belgium in 1962 as separate countries.

Islam entered here during the nineteenth century through commerce. By 1861 these lands were under the Muslim Sultanate of Zanzibar. According to census data, the Muslim population decreased from 302,000 or 8.5% in 1970, to 40,000 or 0.8% in 1978 due to turmoil between Tutsi and Hutu ethnicities between 1972 and 1973, which led to a military coup in mid-1973. The Muslim population then increased to 87,000 or 1.2% in 1991, to 145,000 or 1.8% in 2002 and 211,000 or 2.0% in 2012. Thus, assuming that the percentage of Muslims will continue to increase by a fifth of a percentage point per decade, the Muslim population is expected to reach 0.6 million or 2.8% by 2050 and 1.3 million or 3.8% by 2100. A summary of the data is provided in Table 3.4.9.

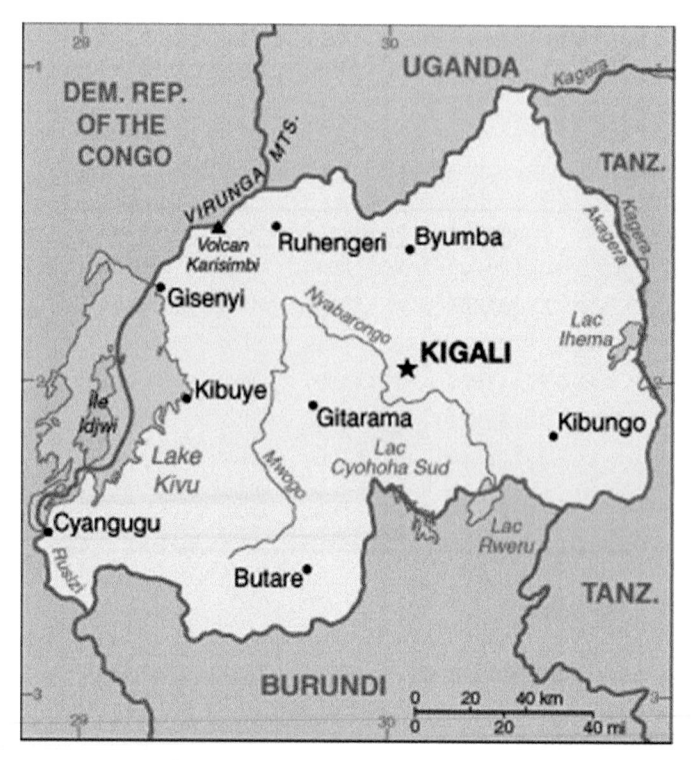

Figure 3.4.9 Map of the Republic of Rwanda.

Table 3.4.9 Evolution of the Muslim population in Rwanda

Year	Population	Muslims	%	Source
1970	3,572,550	302,300	8.46	[UN79]c
1978	4,730,979	39,676	0.84	[UN83]c
1991	7,157,040	86,550	1.21	[ROD]c
2002	7,925,140	144,968	1.83	[RW02]c
2012	10,375,603	211,011	2.03	[RW12]c
2050	23,048,000	645,000	2.80	es
2100	33,413,000	1,270,000	3.80	es

3.4.10 São Tomé and Príncipe

The Democratic Republic of São Tomé and Príncipe was uninhabited when it was discovered by the Portuguese in 1470. Consequently, the Portuguese occupied São Tomé in 1493 and Príncipe in 1500. The Islands gained their independence from Portugal in 1975. The country has an area of 1,001 sq. km, consisting of the two main islands of São Tomé (859 sq. km) with over 95% of the population and where the capital São Tomé is located and Príncipe (142 sq. km), and six much smaller islands. A map of these islands is presented in Fig. 3.4.10.

Islam entered here after independence with emigrants from nearby Cameroon and Nigeria. Hence, based on census data, the Muslim population increased from none in 1970 to 51 or 0.04% in 1991. According to a 2009 DHS, the numbers increased to 64 or 0.04%. Thus, assuming that the percentage of Muslims will remain fixed, the Muslim population is expected to remain under 300 for the rest of this century. A summary of the data is provided in Table 3.4.10.

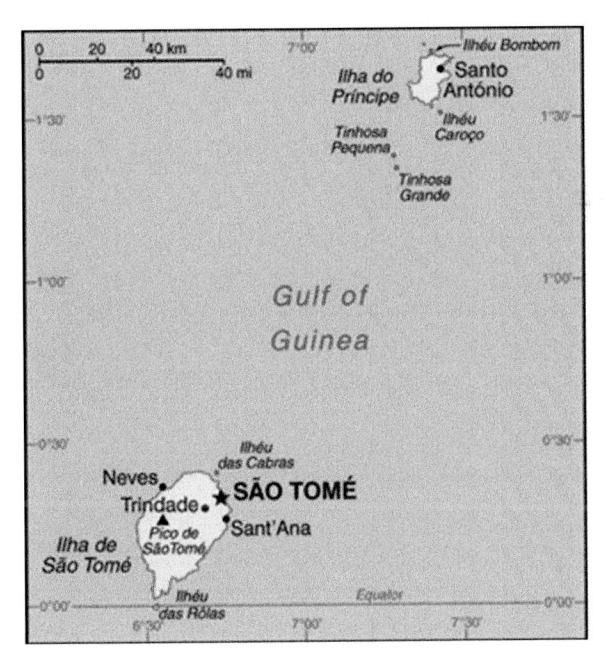

Figure 3.4.10 Map of the Democratic Republic of São Tomé and Príncipe.

Table 3.4.10 Evolution of the Muslim population in São Tomé and Príncipe

Year	Population	Muslims	%	Source
1891	49,000	0	0.00	[JAN]es
1970	73,800	0	0.00	[ST70]c
1991	115,787	51	0.04	[UN04]c
2009	167,000	60	0.04	[DHS]s
2050	394,000	200	0.04	es
2100	708,000	300	0.04	es

3.4.11 South Sudan

The Republic of South Sudan has an area of 619,745 sq. km and its map is presented in Fig. 3.4.11. It was occupied by the British in 1898 as Anglo-Egyptian Sudan and gained its independence from the UK in 1956, then from Sudan in 2011. Islam entered here in 1820 with the Egyptian occupation.

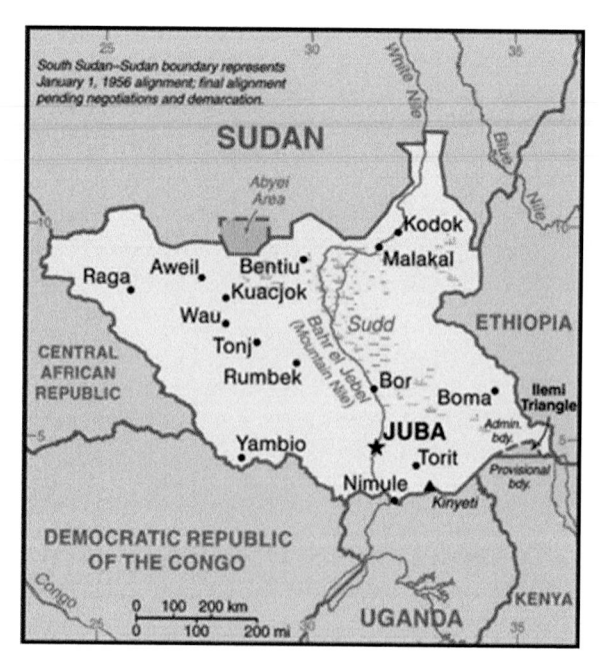

Figure 3.4.11 Map of the Republic of South Sudan.

According to a 1923 estimate, the total population was 1.5 million, none of which was Muslim. The 1956 census indicated 18% of the population Muslim, which was the only census ion record to inquire on religion in South Sudan. In 2012 it was estimated that one-tenth of the population is Muslim. Thus, assuming that the percentage of Muslims remains constant, the Muslim population is expected to increase to two million by 2050 and three million by 2100. The data is summarized in Table 3.4.11.

Table 3.4.11 Evolution of the Muslim population in South Sudan

Year	Population	Muslims	%	Source
1923	1,500,000	0	0.00	[MAS23]es
1956	2,782,000	500,760	18.00	[SS56]c
2008	8,260,000	826,000	10.00	[SS8]es
2050	19,963,000	1,996,000	10.00	es
2100	31,738,000	3,174,000	10.00	es

3.4.12 Uganda

The Republic of Uganda has an area of 241,038 sq. km and its map is presented in Fig. 3.4.12a. It was occupied by the British in 1888 and gained its independence from the UK in 1962. Islam entered here with Muslim traders coming from Zanzibar in 1852. Eventually, King Muteesa I embraced Islam in 1880 and made it the state religion. He was the ruler of the Kingdom of Buganda from 1856 to 1884.

Estimates of the Muslim population increased from none in 1850, to 50,000 in 1911, to 98,000 or 3.4% in 1921. According to census data, the Muslim population increased from 0.12 million or 3.5% in 1931, to 0.4 million or 5.5% in 1959, to 1.8 million or 10.6% in 1991, to 3.0 million or 12.1% in 2002, to 4.7 million or 13.7% in 2014. A choropleth map illustrating the presence of Muslims per administrative region is presented in Fig. 3.4.12b.

Thus, assuming that the percentage of Muslims will increase by one percentage point per decade, the Muslim population is expected to reach fifteen million or 17% by 2050 and thirty million or 22% by 2100. A summary of the data is provided in Table 3.4.12.

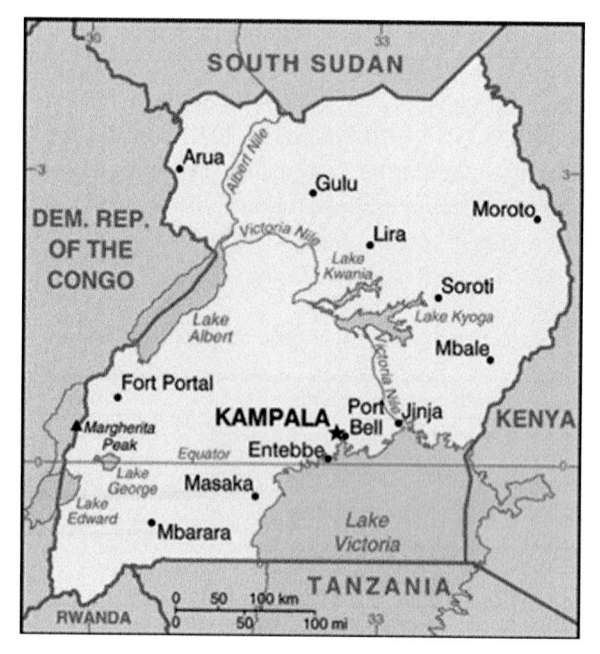

Figure 3.4.12a Map of the Republic of Uganda.

Table 3.4.12 Evolution of the Muslim population in Uganda

Year	Population	Muslims	%	Source
1850	2,750,000	0	0.00	[MAS55]es
1911	2,500,000	50,000	2.00	[ROD]es
1921	2,848,735	98,000	3.44	[ROD]es
1931	3,525,014	122,025	3.46	[ROD]c
1959	6,537,058	356,236	5.45	[UN63]c
1991	16,671,705	1,758,101	10.55	[UN04]c
2002	24,433,132	2,956,121	12.10	[UN]c
2014	34,124,155	4,663,204	13.67	[UG]c
2050	89,447,000	15,206,000	17.00	es
2100	136,785,000	30,093,000	22.00	es

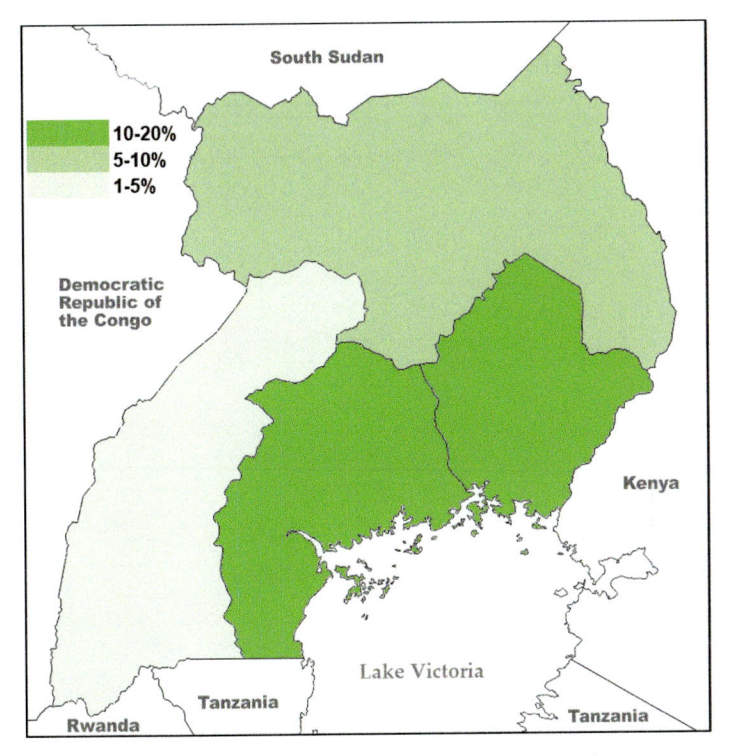

Figure 3.4.12b A choropleth map of Uganda illustrating the presence of Muslims per administrative region.

3.4.13 Regional Summary and Conclusion

Islam entered Central Africa in the eleventh century and spread through Sufi orders. Islam spread slowly and over a quarter of the population of this region is now Muslim. The Muslim population is expected to continue to increase in number and percentage for the next three centuries. The following tables present centennial data from 600 AD to 2100 AD (or approximately 1 H to 1500 H) in Table 3.4a and decennial data from 1790 AD to 2100 AD (or 1210 H to 1520 H) in Tables 3.4b and 3.4c for current countries in Central Africa. The data includes total population in thousands (P), the percentage of which is Muslim (M%), the corresponding Muslim population in thousands (M), and the annual population growth rate (APGR, or G%) of the total population in this region.

Table 3.4a Centennial estimates of the Muslim population (×1000) in Central Africa: 600–2100 AD (1–1500 H)

		600	700	800	900	1000	1100	1200	1300
Burundi	P	210	230	250	270	293	350	410	470
	M%	—	—	—	—	—	—	—	—
	M	—	—	—	—	—	—	—	—
Cameroon	P	910	980	1,070	1,160	1,250	1,500	1,750	2,000
	M%	—	—	—	—	—	1.00	5.00	10.00
	M	—	—	—	—	—	15	88	200
CAR	P	120	135	150	165	180	220	260	300
	M%	—	—	—	—	—	—	—	—
	M	—	—	—	—	—	—	—	—
Congo B.	P	36	42	48	54	70	85	100	115
	M%	—	—	—	—	—	—	—	—
	M	—	—	—	—	—	—	—	—
Congo K.	P	1,100	1,200	1,300	1,400	1,500	1,800	2,100	2,400
	M%	—	—	—	—	—	—	—	—
	M	—	—	—	—	—	—	—	—
Eq.Guinea	P	22	24	26	28	30	36	42	48
	M%	—	—	—	—	—	—	—	—
	M	—	—	—	—	—	—	—	—
Gabon	P	36	42	48	54	70	85	100	115
	M%	—	—	—	—	—	—	—	—
	M	—	—	—	—	—	—	—	—
Nigeria	P	3,400	3,600	3,800	4,000	4,200	4,600	5,000	5,500
	M%	—	—	—	—	—	5.00	10.00	15.00
	M	—	—	—	—	—	230	500	825
Rwanda	P	96	110	124	138	152	175	200	235
	M%	—	—	—	—	—	—	—	—
	M	—	—	—	—	—	—	—	—
STP	P	—	—	—	—	—	—	—	—
	M%	—	—	—	—	—	—	—	—
	M	—	—	—	—	—	—	—	—
S. Sudan	P	780	810	840	870	900	960	1,020	1,080
	M%	—	—	—	—	—	—	—	—
	M	—	—	—	—	—	—	—	—
Uganda	P	450	525	600	675	750	900	1,050	1,200
	M%	—	—	—	—	—	—	—	—
	M	—	—	—	—	—	—	—	—
Total	P	7,160	7,698	8,256	8,814	9,395	10,711	12,032	13,463
	M%	—	—	—	—	—	2.29	4.88	7.61
	M	—	—	—	—	—	245	588	1,025
	G%		0.072	0.070	0.065	0.064	0.131	0.116	0.112

Table 3.4a (Continued)

		1400	1500	1600	1700	1800	1900	2000	2100
Burundi	P	530	587	685	782	980	1,493	6,379	50,904
	M%	—	—	—	—	—	0.20	2.30	5.00
	M	—	—	—	—	—	3	147	2,545
Cameroon	P	2,250	2,500	2,660	2,800	3,160	3,290	15,514	90,225
	M%	15.00	20.00	25.00	25.00	23.81	23.81	20.91	20.91
	M	338	500	665	700	752	783	3,244	18,866
CAR	P	340	360	370	400	530	595	3,640	11,631
	M%	—	—	0.01	0.01	2.52	2.52	10.32	20.00
	M	—	—	0	0	13	15	376	2,326
Congo B.	P	130	140	150	160	190	240	3,127	20,962
	M%	—	—	—	—	—	0.34	1.62	7.00
	M	—	—	—	—	—	1	51	1,467
Congo K.	P	2,700	3,000	3,200	3,400	3,900	4,103	47,106	362,031
	M%	—	—	0.01	0.01	—	0.01	1.40	1.49
	M	—	—	0	0	—	0	659	5,394
Eq.Guinea	P	54	60	68	75	85	137	606	4,511
	M%	—	—	0.01	0.01	—	0.32	2.00	8.00
	M	—	—	0	0	—	0	12	361
Gabon	P	130	140	150	160	190	240	1,228	5,859
	M%	—	—	—	—	—	—	9.44	20.00
	M	—	—	—	—	—	—	116	1,172
Nigeria	P	6,000	6,600	8,400	10,800	12,050	15,589	122,284	732,942
	M%	20.00	25.00	30.00	35.00	46.24	46.24	50.43	54.00
	M	1,200	1,650	2,520	4,994	5,572	7,862	61,668	395,788
Rwanda	P	270	305	356	410	730	996	7,934	33,413
	M%	—	—	0.01	0.01	—	4.00	1.83	3.80
	M	—	—	0	0	—	40	145	1,270
STP	P	—	0	2	5	11	42	142	708
	M%	—	—	—	—	—	—	0.04	0.04
	M	—	—	—	—	—	—	0	0
S. Sudan	P	1,140	1,200	1,260	1,320	1,680	2,261	6,199	31,738
	M%	—	—	—	—	—	0.01	10.00	10.00
	M	—	—	—	—	—	0	620	3,174
Uganda	P	1,350	1,500	1,750	2,000	1,900	2,200	23,650	136,785
	M%	—	—	—	—	—	1.00	12.10	22.00
	M	—	—	—	—	—	22	2,862	30,093
Total	P	14,894	16,392	19,051	22,312	25,406	31,186	237,810	1,481,709
	M%	10.32	13.12	16.72	25.52	24.95	27.98	29.39	31.21
	M	1,538	2,150	3,185	5,694	6,338	8,727	69,899	462,457
	G%	0.101	0.096	0.150	0.158	0.130	0.205	2.032	1.829

Table 3.4b Decennial estimates of the Muslim population (×1000) in Central Africa: 1790–1940 AD (1210–1360 H)

		1790	1800	1810	1820	1830	1840	1850	1860
Burundi	P	960	980	1,000	1,016	1,080	1,140	1,200	1,260
	M%	—	—	—	—	—	—	0.10	0.10
	M	—	—	—	—	—	—	1	1
Cameroon	P	3,140	3,160	3,180	3,200	3,220	3,230	3,240	3,250
	M%	23.81	23.81	23.81	23.81	23.81	23.81	23.81	23.81
	M	748	752	757	762	767	769	771	774
CAR	P	520	530	540	550	560	565	570	575
	M%	2.52	2.52	2.52	2.52	2.52	2.52	2.52	2.52
	M	13	13	14	14	14	14	14	14
Congo B.	P	185	190	195	200	205	210	215	220
	M%	—	—	—	—	—	—	—	0.34
	M	—	—	—	—	—	—	—	1
Congo K.	P	3,850	3,900	3,950	4,000	4,020	4,040	4,050	4,060
	M%	—	—	—	—	—	—	0.01	0.01
	M	—	—	—	—	—	—	0	0
Eq.Guinea	P	80	85	90	95	100	105	110	115
	M%	—	—	—	—	—	—	—	—
	M	—	—	—	—	—	—	—	—
Gabon	P	185	190	195	200	205	210	215	220
	M%	—	—	—	—	—	—	—	—
	M	—	—	—	—	—	—	—	—
Nigeria	P	12,000	12,050	12,100	12,150	12,200	12,300	12,400	12,500
	M%	46.24	46.24	46.24	46.24	46.24	46.24	46.24	46.24
	M	5,549	5,572	5,595	5,618	5,641	5,688	5,734	5,780
Rwanda	P	720	730	750	760	770	780	800	810
	M%	—	—	—	—	—	—	0.01	0.10
	M	—	—	—	—	—	—	0	1
STP	P	10	11	12	14	15	16	13	15
	M%	—	—	—	—	—	—	—	—
	M	—	—	—	—	—	—	—	—
S. Sudan	P	1,660	1,680	1,700	1,719	1,780	1,840	1,900	1,960
	M%	—	—	—	—	0.01	0.01	0.01	0.01
	M	—	—	—	—	0	0	0	0
Uganda	P	1,880	1,900	1,920	1,940	1,960	1,980	1,960	1,980
	M%	—	—	—	—	—	—	—	0.10
	M	—	—	—	—	—	—	—	2
Total	P	25,190	25,406	25,632	25,844	26,115	26,416	26,673	26,965
	M%	25.05	24.95	24.84	24.74	24.59	24.50	24.45	24.38
	M	6,310	6,338	6,366	6,394	6,422	6,471	6,521	6,574
	G%		0.085	0.089	0.082	0.104	0.115	0.097	0.109

Table 3.4b (*Continued*)

		1870	1880	1890	1900	1910	1920	1930	1940
Burundi	P	1,326	1,379	1,435	1,493	1,650	1,822	2,013	2,223
	M%	0.10	0.10	0.10	0.20	0.30	0.40	0.49	0.60
	M	1	1	1	3	5	7	10	13
Cameroon	P	3,260	3,270	3,280	3,290	3,300	3,313	3,660	4,043
	M%	23.81	23.81	23.81	23.81	23.81	23.81	23.81	23.53
	M	776	779	781	783	786	789	871	951
CAR	P	580	585	590	595	600	607	933	1,206
	M%	2.52	2.52	2.52	2.52	2.52	2.52	2.52	2.52
	M	15	15	15	15	15	15	24	30
Congo B.	P	225	230	235	240	306	390	498	634
	M%	0.34	0.34	0.34	0.34	0.34	0.34	0.34	0.34
	M	1	1	1	1	1	1	2	2
Congo K.	P	4,070	4,080	4,090	4,103	5,101	6,341	7,884	10,356
	M%	0.01	0.01	0.01	0.01	0.01	0.01	0.10	0.20
	M	0	0	0	0	1	1	8	21
Eq.Guinea	P	120	125	132	137	178	168	185	205
	M%	—	—	—	0.32	0.32	0.32	0.32	0.32
	M	—	—	—	0	1	1	1	1
Gabon	P	225	230	235	240	259	389	387	389
	M%	—	—	—	—	—	—	0.01	0.01
	M	—	—	—	—	—	—	0	0
Nigeria	P	12,689	13,590	14,555	15,589	16,696	17,881	19,928	19,941
	M%	46.24	46.24	46.24	46.24	46.24	46.24	45.28	45.28
	M	5,868	6,284	6,730	7,208	7,720	8,097	9,023	9,029
Rwanda	P	820	840	856	996	1,158	1,347	1,567	1,823
	M%	1.00	2.00	3.00	4.00	5.00	6.00	7.00	8.00
	M	8	17	26	40	58	81	110	146
STP	P	22	26	49	42	66	59	55	61
	M%	—	—	—	—	—	—	—	—
	M	—	—	—	—	—	—	—	—
S. Sudan	P	2,040	2,111	2,185	2,261	2,364	2,421	2,497	2,576
	M%	0.01	0.01	0.01	0.01	0.01	0.01	1.00	5.00
	M	0	0	0	0	0	0	25	129
Uganda	P	2,000	2,050	2,100	2,200	2,466	2,855	3,542	4,200
	M%	0.20	0.40	0.50	1.00	2.00	3.44	3.46	4.00
	M	4	8	11	22	49	98	123	168
Total	P	27,377	28,517	29,742	31,186	34,145	37,593	43,150	47,656
	M%	24.38	24.92	25.44	25.89	25.29	24.18	23.63	22.01
	M	6,673	7,105	7,565	8,073	8,636	9,090	10,196	10,491
	G%	0.152	0.408	0.421	0.474	0.906	0.962	1.379	0.993

Table 3.4c Decennial estimates of the Muslim population (×1000) in Central Africa: 1950–2100 AD (1370–1520 H)

		1950	1960	1970	1980	1990	2000	2010	2020
Burundi	P	2,309	2,798	3,479	4,157	5,439	6,379	8,676	11,891
	M%	0.70	0.80	0.90	1.00	1.94	2.30	2.65	3.00
	M	16	22	31	42	106	147	230	357
Cameroon	P	4,307	5,177	6,520	8,621	11,780	15,514	20,341	26,546
	M%	23.53	23.53	22.00	20.91	20.91	20.91	20.91	20.91
	M	1,013	1,218	1,434	1,803	2,463	3,244	4,253	5,551
CAR	P	1,327	1,502	1,811	2,199	2,807	3,640	4,387	4,830
	M%	2.52	2.52	4.00	6.00	8.00	10.32	11.00	12.00
	M	33	38	72	132	225	376	483	580
Congo B.	P	827	1,018	1,327	1,778	2,357	3,127	4,274	5,518
	M%	0.60	0.60	3.03	1.94	1.62	1.62	2.39	3.00
	M	5	6	40	34	38	51	102	166
Congo K.	P	12,184	15,248	20,011	26,359	34,612	47,106	64,564	89,561
	M%	0.40	0.64	1.00	1.40	1.40	1.40	1.49	1.49
	M	49	98	200	369	485	659	962	1,334
Eq.Guinea	P	226	255	304	250	419	606	944	1,403
	M%	0.32	0.32	0.32	0.32	0.32	2.00	3.69	4.00
	M	1	1	1	1	1	12	35	56
Gabon	P	473	501	589	726	949	1,228	1,624	2,226
	M%	0.21	0.21	0.21	10.00	12.00	9.44	10.91	12.00
	M	1	1	1	73	114	116	177	267
Nigeria	P	37,860	45,138	55,982	73,424	95,212	122,284	158,503	206,140
	M%	45.28	47.20	47.20	47.20	47.53	50.43	51.80	52.00
	M	17,143	21,305	26,424	34,898	48,016	63,343	82,105	107,193
Rwanda	P	2,186	2,936	3,757	5,153	7,289	7,934	10,039	12,952
	M%	8.46	8.46	8.46	0.84	1.21	1.83	2.03	2.20
	M	185	248	318	43	88	145	204	285
STP	P	60	64	75	96	119	142	180	219
	M%	—	—	—	0.01	0.04	0.04	0.04	0.04
	M	—	—	—	0	0	0	0	0
S. Sudan	P	2,482	2,843	3,494	4,503	5,493	6,199	9,508	11,194
	M%	10.00	18.00	16.00	14.00	12.00	10.00	10.00	10.00
	M	248	512	559	630	659	620	951	1,119
Uganda	P	5,158	6,767	9,406	12,442	17,354	23,650	32,428	45,741
	M%	5.00	5.45	5.45	10.55	10.55	12.10	13.67	14.00
	M	258	369	513	1,313	1,831	2,862	4,433	6,404
Total	P	69,399	84,247	106,755	139,709	183,831	237,810	315,468	418,220
	M%	27.31	28.27	27.72	28.16	29.39	30.10	29.78	29.48
	M	18,952	23,818	29,594	39,338	54,025	71,574	93,934	123,311
	G%	3.759	1.939	2.368	2.690	2.745	2.575	2.826	2.819

Table 3.4c (Continued)

		2030	2040	2050	2060	2070	2080	2090	2100
Burundi	P	15,773	20,253	25,325	30,701	36,107	41,427	46,451	50,904
	M%	3.25	3.50	3.75	4.00	4.25	4.50	4.75	5.00
	M	513	709	950	1,228	1,535	1,864	2,206	2,545
Cameroon	P	33,766	41,873	50,573	59,427	68,127	76,377	83,821	90,225
	M%	20.91	20.91	20.91	20.91	20.91	20.91	20.91	20.91
	M	7,060	8,756	10,575	12,426	14,245	15,970	17,527	18,866
CAR	P	5,942	7,205	8,401	9,496	10,422	11,086	11,474	11,631
	M%	13.00	14.00	15.00	16.00	17.00	18.00	19.00	20.00
	M	772	1,009	1,260	1,519	1,772	1,995	2,180	2,326
Congo B.	P	7,020	8,777	10,702	12,737	14,867	17,017	19,070	20,962
	M%	3.50	4.00	4.50	5.00	5.50	6.00	6.50	7.00
	M	246	351	482	637	818	1,021	1,240	1,467
Congo K.	P	120,047	155,725	194,489	234,146	272,901	308,523	338,554	362,031
	M%	1.49	1.49	1.49	1.49	1.49	1.49	1.49	1.49
	M	1,789	2,320	2,898	3,489	4,066	4,597	5,044	5,394
Eq.Guinea	P	1,874	2,346	2,821	3,277	3,683	4,022	4,297	4,511
	M%	4.50	5.00	5.50	6.00	6.50	7.00	7.50	8.00
	M	84	117	155	197	239	282	322	361
Gabon	P	2,744	3,265	3,809	4,323	4,780	5,204	5,577	5,859
	M%	13.00	14.00	15.00	16.00	17.00	18.00	19.00	20.00
	M	357	457	571	692	813	937	1,060	1,172
Nigeria	P	262,977	329,067	401,315	476,130	550,375	620,457	681,978	732,942
	M%	52.25	52.50	52.75	53.00	53.25	53.50	53.75	54.00
	M	137,406	172,760	211,694	252,349	293,075	331,944	366,563	395,788
Rwanda	P	16,234	19,634	23,048	26,192	28,851	30,959	32,496	33,413
	M%	2.40	2.60	2.80	3.00	3.20	3.40	3.60	3.80
	M	390	510	645	786	923	1,053	1,170	1,270
STP	P	268	328	394	462	531	597	657	708
	M%	0.04	0.04	0.04	0.04	0.04	0.04	0.04	0.04
	M	0	0	0	0	0	0	0	0
S. Sudan	P	13,839	16,856	19,963	23,012	25,850	28,312	30,285	31,738
	M%	10.00	10.00	10.00	10.00	10.00	10.00	10.00	10.00
	M	1,384	1,686	1,996	2,301	2,585	2,831	3,029	3,174
Uganda	P	59,438	74,455	89,447	103,520	115,800	125,645	132,617	136,785
	M%	15.00	16.00	17.00	18.00	19.00	20.00	21.00	22.00
	M	8,916	11,913	15,206	18,634	22,002	25,129	27,850	30,093
Total	P	539,923	679,785	830,286	983,422	1,132,293	1,269,627	1,387,277	1,481,709
	M%	29.43	29.51	29.68	29.92	30.21	30.53	30.87	31.21
	M	158,916	200,588	246,432	294,257	342,073	387,624	428,191	462,457
	G%	2.554	2.303	2.000	1.693	1.410	1.145	0.886	0.659

3.5 Muslims in Southern Africa

This region consists of eleven countries and territories: Angola, Botswana, Eswatini (Swaziland), Lesotho, Malawi, Mozambique, Namibia, Saint Helena, South Africa, Zambia, and Zimbabwe. Islam entered this region from the northeast with the Kilwa Sultanate in 1140. It entered from the south after 1652 with the occupation of the Dutch of the South Africa, who expelled many Muslim from Java to here. Islam spread further in the east and south in the second half of the nineteenth century by Zanzibar Sultanate and with the arrival of the British, who brought settlers from India to help their exploitation of these lands. Islam entered the west of this region until the twentieth century with Muslim immigrants from West Africa.

Thus, the Muslim population remained negligible during the first millennium and has likely reached 3,000 or 0.1% in 1200 AD, then increasing to 6,000 or 0.2% in 1300 AD, to 10,000 or 0.2% in 1400 AD, to 15,000 or 0.3% in 1500 AD, to 22,000 or 0.4% in 1600 AD, to 33,000 or 0.6% in 1700 AD, to 49,000 or 0.7% in 1800 AD, to 0.14 million or 0.8% in 1900, to 5.4 million or 4.6% in 2000, to 10 million or 5.3% in 2020, and is projected to reach 21 million or 6% by 2050 and 42 million or 7% by 2100.

A plot of centennial estimates of the Muslim population and its percentage with respect to the total population in this region from 600 to 2100 is provided in Fig. 3.5a. A zoom in of this plot, providing a plot of decennial estimates of the Muslim population and its percentage with respect to the total population in this region from 1900 to 2100 is provided in Fig. 3.5b. This shows that the Muslim population in this region was increasing minimally until 1940, and then picked up a speed reaching the first million the following decade and is expected to continue increasing this century at a rate of two percentage points per decade.

The corresponding individual data for each country in this region is discussed below. In Section 3.5.12, the total population in each country in this region and the corresponding percentage and number of Muslims is presented centennially in Table 3.5a from 600 to 2100 and decennially in Tables 3.5b and 3.5c from 1790 to 2100.

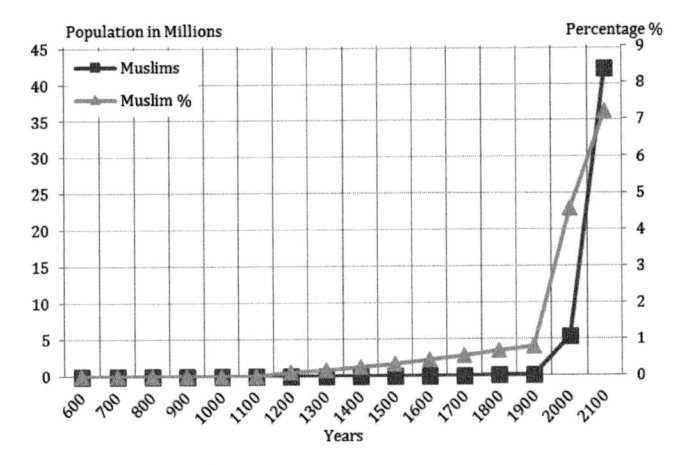

Figure 3.5a Plot of centennial estimates of the Muslim population and its percentage of the total population in Southern Africa: 600–2100 AD (1–1500 H).

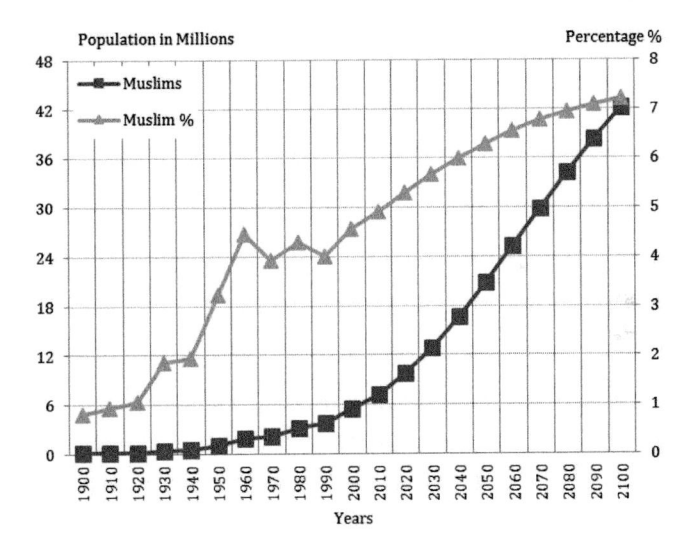

Figure 3.5b Plot of decennial estimates of the Muslim population and its percentage of the total population in Southern Africa: 1900–2100 AD (1320–1520 H).

3.5.1 Angola

The Republic of Angola has an area of 1,246,700 sq. km and its map is presented in Fig. 3.5.1. It was occupied by the Portuguese in 1483 and gained its independence from Portugal in 1975. Islam entered here towards the middle of the twentieth century with West African migrants. The Muslim population remained small throughout the twentieth century but increased dramatically at the turn of the twenty-first century. According to census data, the Muslim population increased from eleven in 1950 to 100,000 or 0.4% in 2014. Thus, assuming that the percentage of the Muslim population will increase by a tenth of a percentage point per decade, then the Muslim population is expected to reach 0.6 million or 0.8% by 2050 and 2.4 million or 1.3% by 2100. A summary of the data is provided in Table 3.5.1.

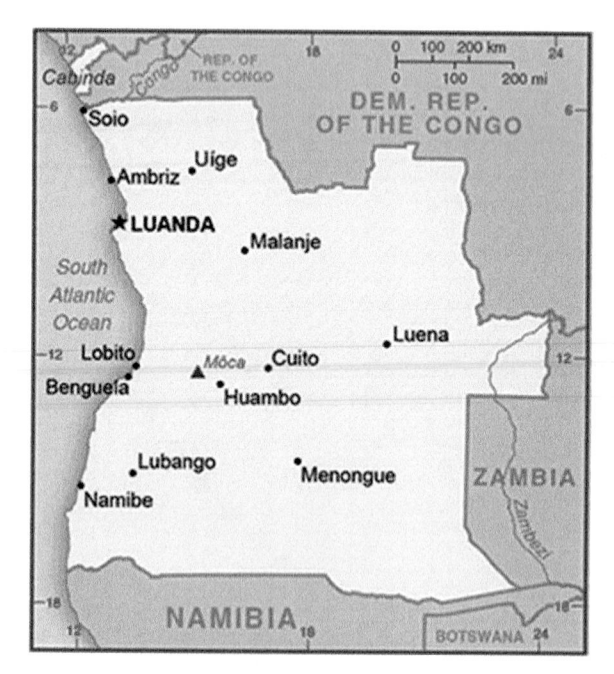

Figure 3.5.1 Map of the Republic of Angola.

Table 3.5.1 Evolution of the Muslim population in Angola

Year	Population	Muslims	%	Source
1950	4,145,266	11	0.00	[UN56]c
1982	8,808,000	8,000	0.09	[KET86]es
2014	25,789,024	100,000	0.39	[AO]c
2050	77,420,000	619,000	0.80	es
2100	188,283,000	2,448,000	1.30	es

3.5.2 Botswana

The Republic of Botswana has an area of 581,730 sq. km and its map is presented in Fig. 3.5.2. It was occupied by the British in 1885 and gained its independence from the UK in 1966 when it changed its name from Bechuanaland. Islam arrived here in 1890 with Indian traders and workers that were brought by the British. Based on census data, the Muslim population increased from 44 or 0.04% in 1911, to 98 or 0.03% in 1946, to 384 or 0.07% in 1964, to 3,628 or 0.3% in 1991, to 5,036 or 0.4% in 2001, to 10,148 or 0.7% in 2011. The last two censuses included only population twelve-years and older in the religion category. Thus, assuming that the percentage of Muslims will increase by a quarter of a percentage point per decade, the Muslim population is expected to reach 60,000 or 1.8% by 2050 and 0.13 million or 3.0% by 2100. A summary of the data is provided in Table 3.5.2.

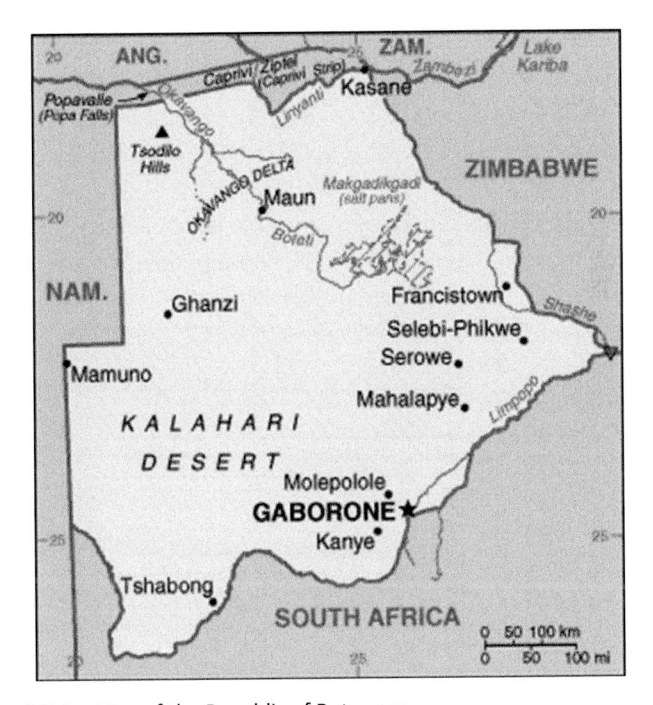

Figure 3.5.2 Map of the Republic of Botswana.

Table 3.5.2 Evolution of the Muslim population in Botswana

Year	Population	Muslims	%	Source
1885	110,000	0	0.00	[MAS23]es
1911	124,350	44	0.04	[ROD]c
1946	296,274	98	0.03	[UN56]c
1964	549,510	384	0.07	[ROD]c
1991	1,326,796	3,628	0.27	[ROD]c
2001	1,184,726	5,036	0.43	[BW01]c
2011	1,372,675	10,148	0.74	[BW11]c
2050	3,510,000	61,000	1.75	es
2100	4,166,000	125,000	3.00	es

3.5.3 Eswatini (Swaziland)

The Kingdom of Eswatini has an area of 17,364 sq. km and its map is presented in Fig. 3.5.3. It was occupied by the British in 1902 and gained its independence from the UK in 1968 and changed its name from Swaziland in 2018. Islam entered here in 1963 with the first Malawian workers in asbestos mines. The Muslim population reached 700 or 0.1% in 1986, and according to ABS survey, the percentage of Muslims increased to 0.17% in 2015. Thus, assuming that the percentage of Muslims will remain constant, the Muslim population is expected to remain less than 4,000 throughout this century. A summary of the data is provided in Table 3.5.3.

Table 3.5.3 Evolution of the Muslim population in Eswatini

Year	Population	Muslims	%	Source
1890	80,000	0	0.00	[JAN]es
1956	237,021	0	0.00	[KET86]es
1986	681,059	700	0.10	[KET86]es
2015	1,287,000	2,200	0.17	[ABS]s
2050	1,704,000	3,000	0.17	es
2100	2,145,000	4,000	0.17	es

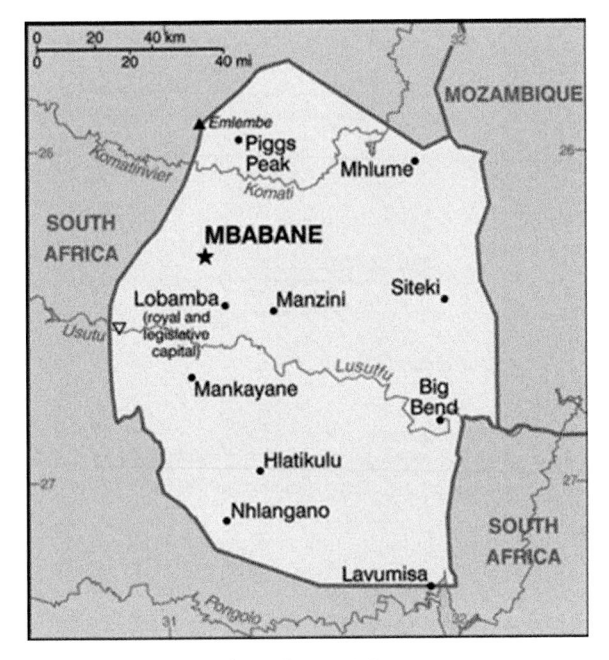

Figure 3.5.3 Map of the Kingdom of Eswatini.

3.5.4 Lesotho

The Kingdom of Lesotho is an enclave in the Republic of South Africa and has an area of 30,355 sq. km and its map is presented in Fig. 3.5.4. It was occupied by the British in 1870 and gained its independence from the UK in 1966, when it changed its name from Basutoland. Islam arrived here in towards the end of the nineteenth century with Indian traders and workers that were brought by the British. The first mosque was erected during the first decade of the twentieth century.

Based on census data, the Muslim population increased from none in 1891, to 119 or 0.03% in 1911, to 220 or 0.04% in 1946, to 331 or 0.05% in 1956. According to ABS survey data, the percentage of Muslims was 0.25% in 2003, 0.08% in 2006, 0.17% in 2012 and 0.08% in 2015. Thus, assuming that the percentage of Muslims will increase by 0.02 of a percentage point per decade, the Muslim population is expected to reach 4,000 or 0.2% by 2050 and 7,000 or 0.3% by 2100. A summary of the data is provided in Table 3.5.4.

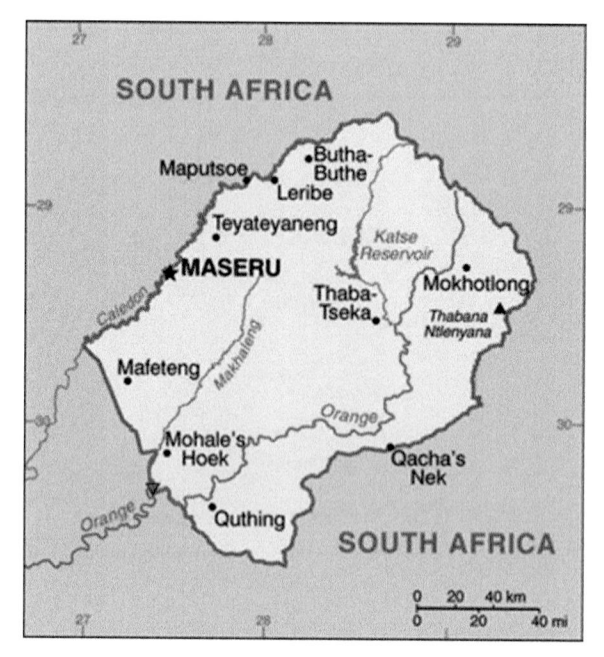

Figure 3.5.4 Map of the Kingdom of Lesotho.

Table 3.5.4 Evolution of the Muslim population in Lesotho

Year	Population	Muslims	%	Source
1891	218,902	0	0.00	[JAN]c
1911	404,507	119	0.03	[GB11]c
1946	563,854	220	0.04	[UN56]c
1956	641,674	331	0.05	[UN56]c
2006	1,940,000	1,500	0.08	[ABS]s
2016	2,007,201	1,700	0.08	[ABS]s
2050	2,665,000	4,000	0.16	es
2100	2,695,000	7,000	0.26	es

3.5.5 Malawi

The Republic of Malawi has an area of 118,484 sq. km and its map is presented in Fig. 3.5.5a. It was occupied by the British in 1891 and gained its independence from the UK in 1964. It used to be called British Central Africa from 1891 to 1907, then Nyasaland until 1963, and then Malawi afterwards. Islam arrived here in 1870 through the Sultanate of Zanzibar and by 1911 the size of the Muslim population was estimated at 50,000 or 5.2% of the total population. According to census data, the Muslim population increased from 0.13 million or 8.0% in 1931 to 1.3 million or 12.8% in 1998, to 1.7 million or 13.0% in 2008 and 2.4 million or 13.8% in 2018. A choropleth map illustrating the presence of Muslims per district is presented in Fig. 3.5.5b.

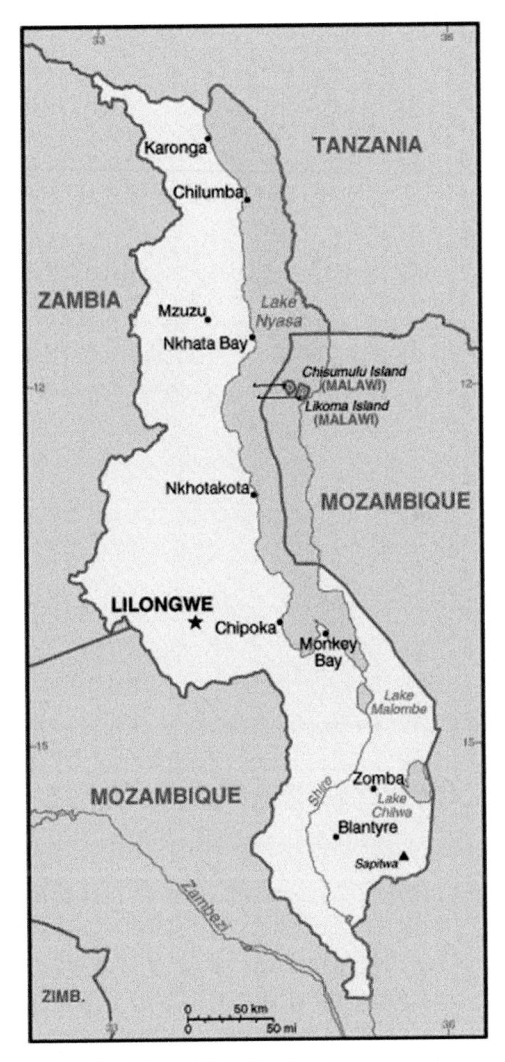

Figure 3.5.5a Map of the Republic of Malawi.

Thus, assuming that the percentage of Muslims will increase by 0.4 of a percentage point per decade, the Muslim population is expected to near six million or 15% by 2050 and exceed eleven million or 17% by 2100. A summary of the data is provided in Table 3.5.5.

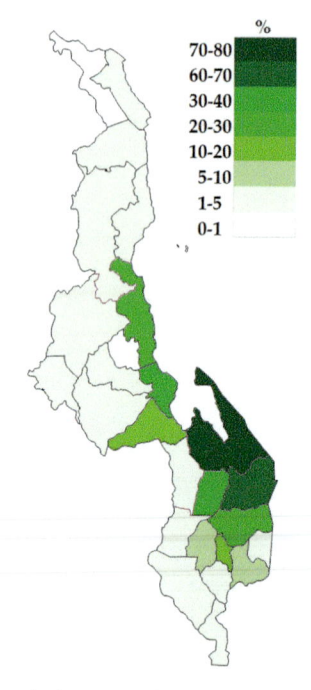

Figure 3.5.5b A choropleth map of Malawi illustrating the presence of Muslims per district.

Table 3.5.5 Evolution of the Muslim population in Malawi

Year	Population	Muslims	%	Source
1870	550,000	0	0.00	[MAS23]es
1911	970,430	50,150	5.17	[MWH]es
1931	1,599,888	128,000	8.00	[ROD]c
1998	9,933,963	1,272,429	12.81	[MW]c
2008	13,029,498	1,690,087	12.97	[MW]c
2018	17,563,749	2,426,754	13.82	[MW18]c
2050	38,143,000	5,721,000	15.00	es
2100	66,559,000	11,315,000	17.00	es

3.5.6 Mozambique

The Republic of Mozambique has an area of 799,380 sq. km and its map is presented in Fig. 3.5.6a. It was occupied by the Portuguese in 1498 and gained its independence from Portugal in 1975. Islam entered here in 1140, when the ruler of Kilwa Sultanate Dawud II ben Suleiman (ruled 1131 to 1170) conquered the costal line, including the cities of Quilimane and Sofala, and deeper inland including the city of Manica, on the border with Zimbabwe. It remained under Muslim control until the arrival of the Portuguese.

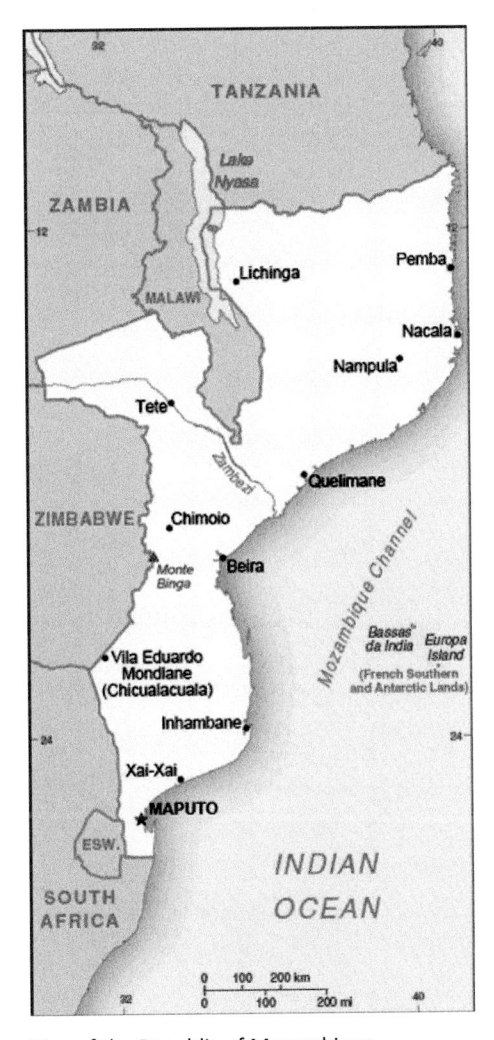

Figure 3.5.6a Map of the Republic of Mozambique.

Estimates of the Muslim population increased from 60,000 or 1.9% in 1919, to 0.25 million or 5.0% in 1940. However, based on census data, the Muslim population increased from 0.61 million or 10.7% in 1950, to 1.18 million or 17.8% in 1960, but decreased to 1.11 million or 13.6% in 1970, then continued to increase to 2.70 million or 18.1% in 1997, to 3.63 million or 18.0% in 2007. A choropleth map illustrating the presence of Muslims per district is presented in Fig. 3.5.6b.

Thus, assuming that the percentage of Muslims will increase by a quarter of a percentage point per decade, the Muslim population is expected to reach 12.4 million or 19% by 2050 and 25 million or 20.3% by 2100. A summary of the data is provided in Table 3.5.6.

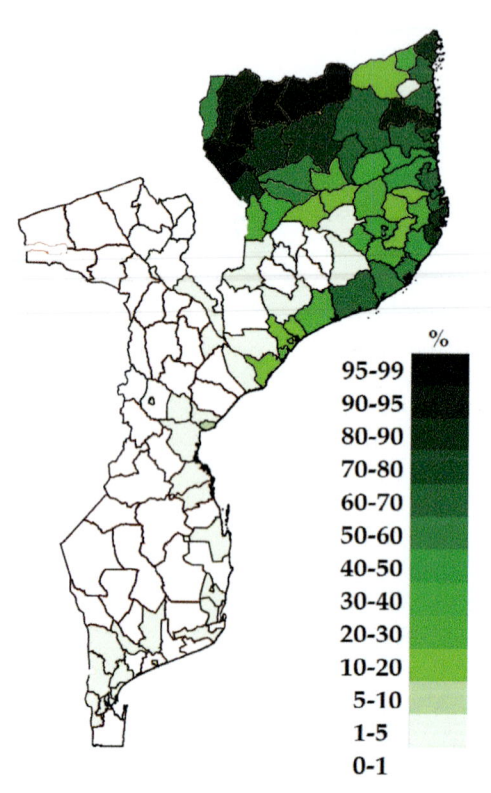

Figure 3.5.6b A choropleth map of Mozambique illustrating the presence of Muslims per district.

Table 3.5.6 Evolution of the Muslim population in Mozambique

Year	Population	Muslims	%	Source
1919	3,120,000	60,000	1.92	[MAS23]es
1940	5,085,630	252,000	4.96	[MAS55]es
1950	5,738,911	612,355	10.67	[UN56]c
1960	6,603,653	1,175,321	17.80	[ROD]c
1970	8,168,933	1,107,113	13.55	[ROD]c
1997	14,899,639	2,696,757	18.10	[UN]c
2007	20,121,129	3,628,913	18.04	[UN]c
2050	65,313,000	12,409,000	19.00	es
2100	123,647,000	25,039,000	20.25	es

3.5.7 Namibia

The Republic of Namibia has an area of 824,292 sq. km and its map is presented in Fig. 3.5.7. It was occupied by the Germans in 1884 as German South West Africa but taken from them by South Africa in 1915. Namibia gained its independence from South Africa in 1990, when it changed its name from South West Africa. Islam entered here in 1970s and gained attention when a prominent politician named Jacobs Salman Dhamir embraced Islam in 1980. The first mosque was erected in 1986 in Katutura. According to ABS survey data, the percentage of Muslims was 0.25% in 2003, 0.08% in 2006, 2012 and 2015. Thus, assuming that the percentage of Muslims will continue to increase by 0.02 of a percentage point per decade, the Muslim population is expected to reach 6,000 or 0.2% by 2050 and 14,000 or 0.3% by 2100. A summary of the data is provided in Table 3.5.7.

Table 3.5.7 Evolution of the Muslim population in Namibia

Year	Population	Muslims	%	Source
1970	761,010	0	0.00	es
2006	2,054,000	1,600	0.08	[ABS]s
2015	2,459,000	2,000	0.08	[ABS]s
2050	3,981,000	6,000	0.16	es
2100	5,374,000	14,000	0.26	es

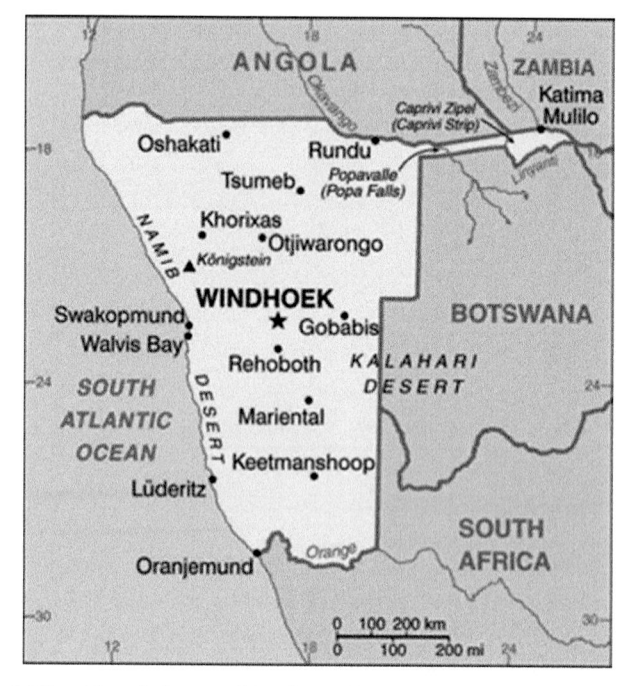

Figure 3.5.7 Map of the Republic of Namibia.

3.5.8 Saint Helena

Saint Helena, Ascension, and Tristan da Cunha is an Overseas Territory of the United Kingdom, which conquered it in 1657. The Territory has an area of 308 sq. km comprising three main islands spread over 2,500 km from Mid to South Atlantic Ocean. The largest of these islands is Saint Helena with area 122 sq. km and 80% of the population and where the capital Jamestown is located. It was captured by the British in 1657. About 1,300 km to its northwest is Ascension with 88 sq. km in total area, a seventh of the population, and which was captured by the British in 1815. About 2,500 km south of St. Helena is the Island of Tristan da Cunha with area 98 sq. km, which was captured by the British in 1816. A map of these islands is presented in Fig. 3.5.8.

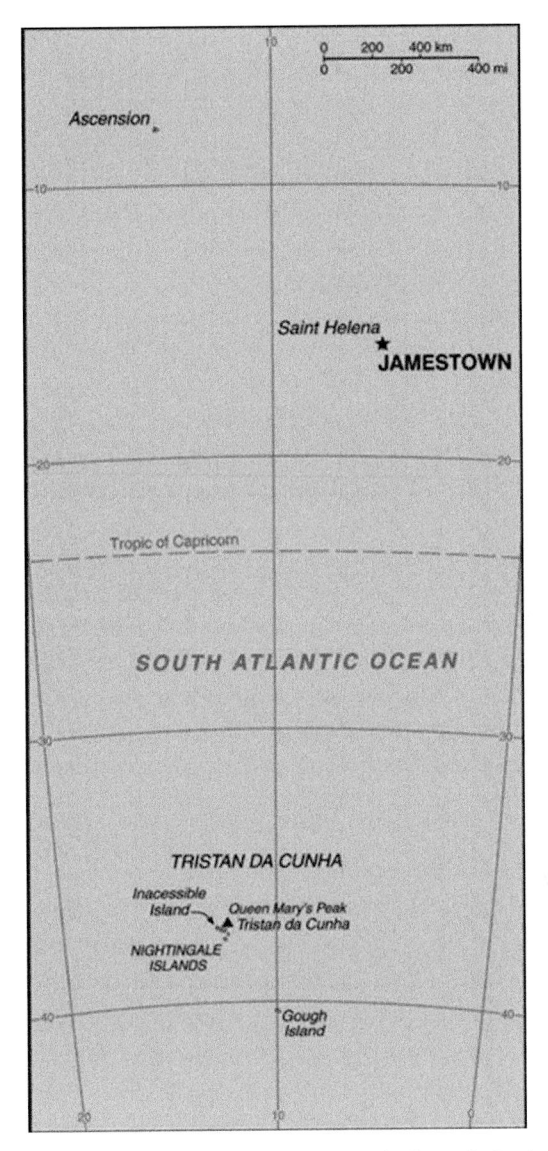

Figure 3.5.8 Map of Saint Helena, Ascension, and Tristan da Cunha.

The Islands were uninhabited when they were discovered by the Portuguese navigator João da Nova; the first was discovered in May 3rd, 1502, which coincided with the Catholic feast-day for the finding of the True Cross by Saint Helena in Jerusalem. The island of Ascension was discovered by the Portuguese navigator and naval general Afonso de Albuquerque on Ascension Day (May 21st) in 1503, from which fact it derives its name. Afonso was accompanied by a Portuguese nobleman named Fernão Lopes, who will convert to Islam in Goa, India, and return to St. Helena to live permanently. The island of Tristan da Cunha was discovered in 1506 by the Portuguese explorer Tristão da Cunha who named it after himself. However, rough seas prevented his landing.

For several centuries, the British have used the island of St. Helena as a place of exile, most notably for Napoleon Bonaparte's exile and death (1815–1821), the exile of Dinzizulu Kacetshwayo (1890–1897) from Zululand, and five and half thousand Boer prisoners of war from South Africa (1900–1902). In fact, the first permanent inhabitant of St. Helena was an exiled Muslim. His name is Fernão Lopes, a Portuguese nobleman who accompanied the Portuguese naval general Afonso de Albuquerque on his first voyage in 1503 to Goa on the west coast of India. Lopes converted to Islam and sided together with his troops in 1512 with Rasul Khan in a rebellion against Portuguese rule of Goa. Rasul Khan was a general that the Sultan of Bijapur, Ismael Adil Shan, sent with an army to reconquer Goa from the Portuguese. When Lopes was captured by the Portuguese, he was tortured, disfigured, and exiled to St. Helena where he lived in almost complete solitude from 1516 until his death in 1545.

Muslims in British colonies also took their share of exile and imprisonment in St. Helena. For example, the 1921 census recorded 32 Muslims or 0.9% of the total population, who were also listed as "African Prisoners". These Muslims consisted of Sheikh Sayyid Khalid bin Barghash Al-Busaid and his party who were exiled from Zanzibar to St Helena from June 1917 to April 1921. He ruled Zanzibar from August 25th to August 27th, 1896 seizing power after the sudden death of his cousin Hamad bin Thuwaini, who many suspect he was poisoned by Sheikh Khalid. Britain refused to recognize his claim to the throne, preferring as Sultan Hamud bin Muhammed who was more favorable to British interests. In accordance with a treaty signed in 1886 a condition for accession to the sultanate was that the candidate obtain the permission of the British consul and Sheikh Khalid had not fulfilled this requirement, so the British sent an ultimatum to him demanding that he surrender. He did not, barricading himself inside his heavily-fortified palace, which the British decided to take by force. Sheikh Khalid managed to evade the British forces and was smuggled out of the country to German East Africa where he lived as a Sultan for twenty years. The British continued to pursue him and on 27th February 1917, Sheikh Khalid was arrested in the Rufiji delta, 250 miles from Dar es Salaam. Four months later, on 22nd June he was escorted with his entourage to exile in St Helena.

On arrival Sheikh Khalid and his followers, seventeen of them, plus three political exiles from Kenya, were kept in military custody in the Jamestown Barracks. There is no information available on the prisoners; all newspapers and other records relating to Sheikh Khalid were censored during that period. It is known that they did not mix much with the local population.

The weather conditions and the lack of Muslims on the island did not suit Sheikh Khalid and his party. He requested to be moved to his relatives in Oman or to his property in Dar es Salaam, but this request was refused. However, in January 1921 it was decided to send Sheikh Khalid and his entourage to the Seychelles, where there were already held in exile political prisoners from the Gold Coast, Uganda, Nyasaland, and Somaliland. Sheikh Khalid and his entourage left St Helena at the end of April 1921 after four years on the island. He died on the 15th of March 1927 in Mombasa age 53.

The final exile by the British was that of three Bahrainis (1957–1961). The three, Abdali al Alaiwat, Abdulrahman al Bakir and Abdulaziz al Shamlan, had been prominent members of the National Union Committee in Bahrain and had been tried by the ruler of Bahrain for offences against the state and sentenced to fourteen years imprisonment. The ruler of Bahrain asked Britain for assistance in removing them to a British Territory and it was decided that they should be sent to St Helena. They were sent to England in 1961 after their four-year exile in St. Helena.

As shown in Table 3.5.8, with the exception of the 1921 census, no other official census recorded any Muslims in these islands. Thus, the situation is not expected to change during this century.

Table 3.5.8 Evolution of the Muslim population in Saint Helena

Year	Population	Muslims	%	Source
1516	1	1	100	[SH16]c
1545	1	1	100	[SH16]c
1871	6,241	0	0.00	[GB71]c
1911	3,697	0	0.00	[GB11]c
1921	3,747	32	0.85	[SH21]c
1931	3,995	0	0.00	[SH31]c
1946	4,748	0	0.00	[SH46]c
1956	4,642	0	0.00	[SH56]c
1966	4,649	0	0.00	[SH66]c
1976	5,147	0	0.00	[SH76]c
1987	6,222	0	0.00	[SH87]c
1998	6,190	0	0.00	[SH98]c
2008	3,935	0	0.00	[SH08]c
2016	3,640	0	0.00	[SH16]c
2050	5,800	0	0.00	es
2100	4,500	0	0.00	es

3.5.9 South Africa

The Republic of South Africa has an area of 1,219,090 sq. km and its map is presented in Fig. 3.5.9a. The Dutch arrived here in 1652, followed by the British in 1806. The Union of South Africa was formed from former British colonies: Cape Hope, Natal, Orange Free State, and Transvaal. It gained independence from the UK in 1931 and changed to republic in 1961. The first Muslims were brought here in 1658 by the Dutch soon after their arrival and consisted slaves and political prisoners mainly from current Indonesia. Then the British brought more Muslims from India starting the eighteenth century, consisting of indentured workers and businessmen with their families.

The census of 1891 found 15,000 Muslims living in the Cape Colony or 0.7% of the population of the Cape and Natal British Colonies, which was half the land and two-thirds of the population of current South Africa. Based on census data, the Muslim population increased from 15,000 or 0.7% in the colonies 33,000 or 0.6% in 1904, to 46,000 or 0.8% in 1911, to 50,000 or 0.7% in 1921, to 79,000 or 0.8% in 1936, to 0.11 million or 1.0% in 1946, to 0.15 million or 1.2% in 1951, to 0.19 million or 1.2% in 1960, to 0.27 million or 1.3% in 1970, to 0.35 million or 1.7% in 1980, but decreased to 0.34 million or 1.1% in 1991. It continued to increase since then to 0.55 million or 1.5% in 1996, to 0.65 million or 1.5% in 2001, to 0.89 million or 1.6% in 2016.

Figure 3.5.9a Map of the Republic of South Africa.

The 2016 data showed that over 88% of the Muslim population lives in three provinces: Western Cape (39%), Gauteng (28%) and KwaZulu-Natal (21%). The percentage of Muslims in these provinces is 5.6%, 1.9% and 1.7%, respectively. The main towns in these provinces are Cape Town, Pretoria/Johannesburg and Durban, respectively. In 1996, over 90% of Muslims lived in these three provinces, and the fraction of Muslims and makeup of the province was 48% and 6.7% for Western Cape, 19% and 1.5% for Gauteng and 23% and 1.5% for KwaZulu-Natal. A choropleth map illustrating the presence of Muslims per province is presented in Fig. 3.5.9b.

Thus, assuming that the percentage of Muslims will continue to increase by a tenth of a percentage point per decade, the Muslim population is expected to increase to 1.5 million or 2.0% by 2050 and two million or 2.5% by 2100. A summary of the data is provided in Table 3.5.9.

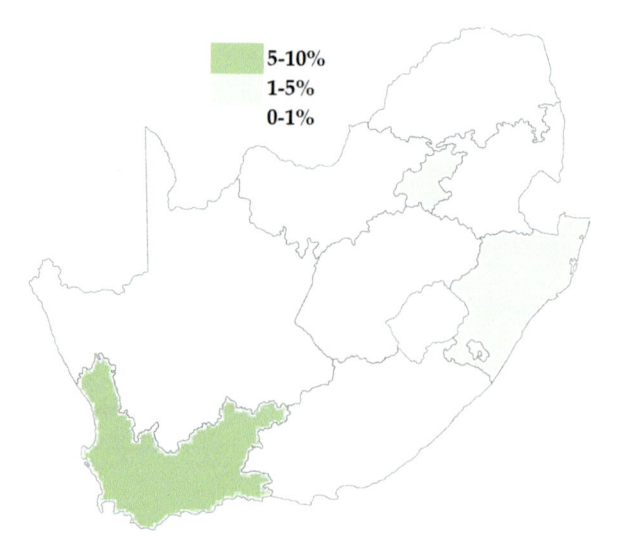

Figure 3.5.9b A choropleth map of South Africa illustrating the presence of Muslims per province.

Table 3.5.9 Evolution of the Muslim population in South Africa

Year	Population	Muslims	%	Source
1891	2,071,137	15,099	0.73	[SYB00]c
1904	5,175,824	32,773	0.63	[SYB10]c
1911	5,973,394	45,904	0.77	[GB11]c
1921	6,928,580	49,936	0.72	[SYB31]c
1936	9,589,898	79,088	0.82	[SYB50]c
1946	11,245,775	110,392	0.98	[UN56]c
1951	12,671,452	146,829	1.16	[ROD]c
1960	16,002,797	192,007	1.20	[UN63]c
1970	21,794,328	269,915	1.26	[UN79]c
1980	20,611,214	352,993	1.71	[UN88]c
1991	30,986,920	338,142	1.09	[ROD]c
1996	36,836,869	553,585	1.50	[UN]c
2001	44,208,807	654,064	1.48	[UN]c
2016	54,946,358	892,685	1.62	[ZA]s
2050	75,518,000	1,510,000	2.00	es
2100	79,191,000	1,980,000	2.50	es

3.5.10 Zambia

The Republic of Zambia has an area of 752,618 sq. km and its map is presented in Fig. 3.5.10a. It was occupied by the British in 1891 and gained its independence from the UK in 1964 when it changed its name from Northern Rhodesia. Islam entered the eastern parts in the twelfth century with the influence of the Kilwa Sultanate and trade. The Muslim population estimate increased from 3,000 or 0.4% in 1911, to 4,900 or 0.3% in 1948. Based on census data, the Muslim population aged five-years and older increased from 15,000 or 0.3% in 1980, to 42,000 or 0.5% in 2000, and 61,000 or 0.5% in 2010. A choropleth map illustrating the presence of Muslims per province is presented in Fig. 3.5.10b.

Thus, assuming that the percentage of Muslims will continue to increase by 0.05 of a percentage point per decade, then the Muslim population is expected to reach 0.3 million or 0.7% by 2050 and 0.8 million or 1.0% by 2100. A summary of the data is provided in Table 3.5.10.

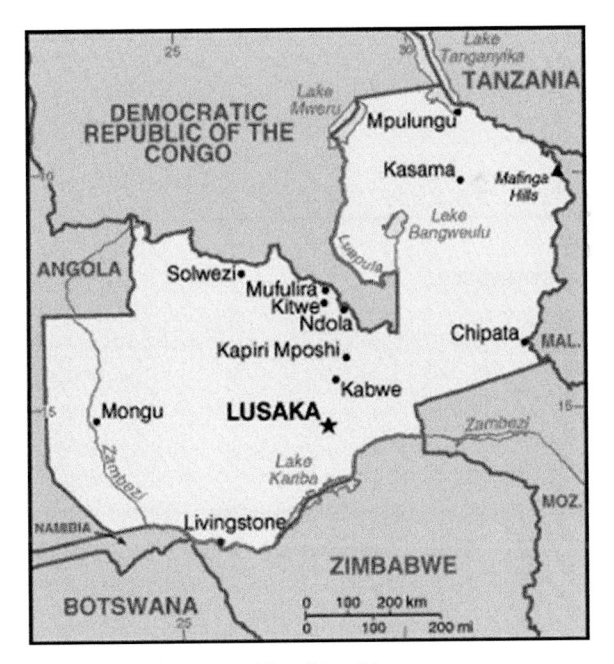

Figure 3.5.10a Map of the Republic of Zambia.

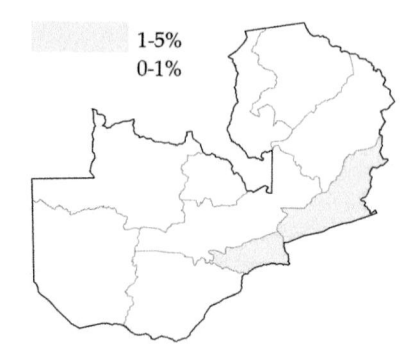

Figure 3.5.10b A choropleth map of Zambia illustrating the presence of Muslims per province.

Table 3.5.10 Evolution of the Muslim population in Zambia

Year	Population	Muslims	%	Source
1911	801,400	3,000	0.37	[ZMH]es
1948	1,721,000	4,900	0.29	[MAS55]es
1980	5,679,808	15,000	0.26	[ZM80]c
2000	9,337,425	41,932	0.45	[ZM00]c
2010	12,526,314	61,412	0.49	[ZM10]c
2050	39,121,000	274,000	0.70	es
2100	81,546,000	775,000	0.95	es

3.5.11 Zimbabwe

The Republic of Zimbabwe has an area of 390,757 sq. km and its map is presented in Fig. 3.5.11. It was occupied by the British in 1888 and gained its independence from the UK in 1980 when it changed its name from Southern Rhodesia. Islam entered the eastern parts in the twelfth century with the influence of the Kilwa Sultanate and trade. In 1890, Muslim from Malawi settled here to work on farms and mines for the British. The Muslim population was estimated in 1982 at 62,000 or 0.8% of the total population. According to DHS survey data, the percentage of Muslims was 0.86% in 1999, 0.88% in 2006, 0.53% in 2011 and 0.56% in 2015.

Thus, assuming that the percentage of Muslims will increase by a tenth of a percentage point per decade, the Muslim population is expected to increase to 0.2 million or 0.9% by 2050 and 0.4 million or 1.4% by 2100. A summary of the data is provided in Table 3.5.11.

Figure 3.5.11 Map of the Republic of Zimbabwe.

Table 3.5.11 Evolution of the Muslim population in Zimbabwe

Year	Population	Muslims	%	Source
1890	1,237,000	0	0.00	[KET86]es
1946	1,764,000	500	0.03	[MAS55]es
1982	7,546,071	62,000	0.82	[KET86]es
1999	12,374,000	106,000	0.86	[DHS]s
2006	13,128,000	116,000	0.88	[DHS]s
2011	14,256,000	76,000	0.53	[DHS]s
2015	15,603,000	87,000	0.56	[DHS]s
2050	23,948,000	216,000	0.90	es
2100	30,965,000	434,000	1.40	es

Table 3.5a Centennial estimates of the Muslim population (×1000) in Southern Africa: 600–2100 AD (1–1500 H)

		600	700	800	900	1000	1100	1200	1300
Angola	P	660	720	780	840	900	1,100	1,300	1,500
	M%	—	—	—	—	—	—	—	—
	M	—	—	—	—	—	—	—	—
Botswana	P	42	43	44	45	46	55	65	75
	M%	—	—	—	—	—	—	—	—
	M	—	—	—	—	—	—	—	—
Eswatini	P	4	4	5	5	6	7	8	9
	M%	—	—	—	—	—	—	—	—
	M	—	—	—	—	—	—	—	—
Lesotho	P	11	12	13	14	15	18	21	24
	M%	—	—	—	—	—	—	—	—
	M	—	—	—	—	—	—	—	—
Malawi	P	115	130	145	160	171	200	230	260
	M%	—	—	—	—	—	—	—	—
	M	—	—	—	—	—	—	—	—
Mozambique	P	200	225	250	275	300	440	580	720
	M%	—	—	—	—	—	—	0.50	0.75
	M	—	—	—	—	—	—	3	5
Namibia	P	50	51	52	53	54	70	85	90
	M%	—	—	—	—	—	—	—	—
	M	—	—	—	—	—	—	—	—
St.Helena	P	—	—	—	—	—	—	—	—
	M%	—	—	—	—	—	—	—	—
	M	—	—	—	—	—	—	—	—
S.Africa	P	220	240	260	280	300	350	400	450
	M%	—	—	—	—	—	—	—	—
	M	—	—	—	—	—	—	—	—
Zambia	P	95	110	125	140	155	170	205	240
	M%	—	—	—	—	—	—	0.10	0.10
	M	—	—	—	—	—	—	0	0
Zimbabwe	P	115	130	145	160	173	200	230	270
	M%	—	—	—	—	—	—	0.10	0.20
	M	—	—	—	—	—	—	0	1
Total	P	1,512	1,665	1,819	1,972	2,120	2,610	3,124	3,638
	M%	—	—	—	—	—	—	0.11	0.17
	M	—	—	—	—	—	—	3	6
	G%		0.096	0.088	0.081	0.072	0.208	0.180	0.152

Table 3.5a (*Continued*)

		1400	1500	1600	1700	1800	1900	2000	2100
Angola	P	1,700	1,800	1,900	2,000	2,400	4,790	16,395	188,283
	M%	—	—	—	—	—	—	0.30	1.30
	M	—	—	—	—	—	—	49	2,448
Botswana	P	85	92	92	92	74	120	1,643	4,166
	M%	—	—	—	—	—	0.02	0.43	3.00
	M	—	—	—	—	—	0	7	125
Eswatini	P	10	11	13	19	27	115	2,033	2,695
	M%	—	—	—	—	—	—	0.08	0.26
	M	—	—	—	—	—	—	2	7
Lesotho	P	27	30	35	50	73	349	11,149	66,559
	M%	—	—	—	—	—	0.03	12.81	17.00
	M	—	—	—	—	—	0	1,428	11,315
Malawi	P	300	343	377	411	340	737	17,712	123,647
	M%	—	—	—	—	—	4.00	17.65	20.25
	M	—	—	—	—	—	29	3,126	25,039
Mozambique	P	860	1,000	1,250	1,500	2,000	4,106	1,795	5,374
	M%	1.00	1.25	1.50	1.75	1.92	1.92	0.08	0.26
	M	9	13	19	26	38	79	1	14
Namibia	P	100	108	108	108	125	180	6	4
	M%	—	—	—	—	—	—	—	—
	M	—	—	—	—	—	—	—	—
St.Helena	P	—	—	0	1	3	10	44,968	79,191
	M%	—	—	—	—	—	—	1.48	2.50
	M	—	—	—	—	—	—	666	1,980
S.Africa	P	500	600	700	1,000	1,300	4,713	1,005	2,145
	M%	—	—	—	0.20	0.73	0.63	0.17	0.17
	M	—	—	—	2	9	30	2	4
Zambia	P	275	310	342	373	410	1,272	10,416	81,546
	M%	0.20	0.30	0.40	0.50	0.37	0.37	0.45	0.95
	M	1	1	1	2	2	5	47	775
Zimbabwe	P	310	347	382	416	459	1,422	11,881	30,965
	M%	0.30	0.50	0.60	0.70	0.01	0.03	0.86	1.40
	M	1	2	2	3	0	0	102	434
Total	P	4,167	4,641	5,199	5,970	7,211	17,814	119,003	584,576
	M%	0.24	0.33	0.43	0.55	0.69	0.80	4.56	7.21
	M	10	15	22	33	49	143	5,430	42,139
	G%	0.136	0.108	0.114	0.138	0.189	0.904	1.899	1.592

Table 3.5b Decennial estimates of the Muslim population (×1000) in Southern Africa: 1790–1940 AD (1210–1360 H)

		1790	1800	1810	1820	1830	1840	1850	1860
Angola	P	2,375	2,400	2,425	2,450	2,460	2,470	2,480	2,490
	M%	—	—	—	—	—	—	—	—
	M	—	—	—	—	—	—	—	—
Botswana	P	70	74	78	82	86	90	95	100
	M%	—	—	—	—	—	—	—	—
	M	—	—	—	—	—	—	—	—
Eswatini	P	26	27	28	29	35	41	47	53
	M%	—	—	—	—	—	—	—	—
	M	—	—	—	—	—	—	—	—
Lesotho	P	71	73	75	77	90	110	130	150
	M%	—	—	—	—	—	—	—	—
	M	—	—	—	—	—	—	—	—
Malawi	P	310	340	370	400	425	450	475	500
	M%	—	—	—	—	—	—	—	—
	M	—	—	—	—	—	—	—	—
Mozambique	P	1,950	2,000	2,050	2,096	2,300	2,500	2,700	2,900
	M%	1.92	1.92	1.92	1.92	1.92	1.92	1.92	1.92
	M	37	38	39	40	44	48	52	56
Namibia	P	120	125	130	135	140	145	150	155
	M%	—	—	—	—	—	—	—	—
	M	—	—	—	—	—	—	—	—
St.Helena	P	2	3	4	6	5	5	7	7
	M%	—	—	—	—	—	—	—	—
	M	—	—	—	—	—	—	—	—
S.Africa	P	1,200	1,300	1,400	1,550	1,800	2,000	2,200	2,400
	M%	0.73	0.73	0.73	0.73	0.73	0.73	0.73	0.73
	M	9	9	10	11	13	15	16	18
Zambia	P	406	410	414	418	500	580	660	740
	M%	0.37	0.37	0.37	0.37	0.37	0.37	0.37	0.37
	M	2	2	2	2	2	2	2	3
Zimbabwe	P	455	459	463	467	560	650	740	830
	M%	0.01	0.01	0.01	0.01	0.01	0.01	0.01	0.01
	M	0	0	0	0	0	0	0	0
Total	P	6,985	7,211	7,437	7,710	8,401	9,041	9,684	10,325
	M%	0.68	0.69	0.69	0.69	0.70	0.72	0.73	0.74
	M	48	49	51	53	59	65	70	76
	G%		0.318	0.308	0.359	0.860	0.735	0.687	0.641

Table 3.5b (*Continued*)

		1870	1880	1890	1900	1910	1920	1930	1940
Angola	P	2,497	3,510	4,934	4,790	4,650	4,515	4,383	3,738
	M%	—	—	—	—	—	—	—	—
	M	—	—	—	—	—	—	—	—
Botswana	P	105	110	115	120	124	153	234	272
	M%	—	—	0.00	0.02	0.04	0.04	0.03	0.03
	M	—	—	0	0	0	0	0	0
Eswatini	P	68	81	97	115	144	162	190	222
	M%	—	—	—	—	—	—	—	—
	M	—	—	—	—	—	—	—	—
Lesotho	P	184	254	219	349	401	466	543	631
	M%	—	—	—	0.03	0.03	0.03	0.03	0.03
	M	—	—	—	0	0	0	0	0
Malawi	P	550	600	650	737	970	1,202	1,603	1,865
	M%	—	1.00	2.00	4.00	5.17	7.00	8.00	9.00
	M	—	6	13	29	50	84	128	168
Mozambique	P	3,191	3,470	3,775	4,106	4,580	4,857	5,239	5,650
	M%	1.92	1.92	1.92	1.92	1.92	1.92	4.96	4.96
	M	61	67	72	79	88	93	260	280
Namibia	P	160	165	170	180	200	229	286	336
	M%	—	—	—	—	—	—	—	—
	M	—	—	—	—	—	—	—	—
St.Helena	P	6	5	4	10	4	4	4	5
	M%	—	—	—	—	—	0.85	—	—
	M	—	—	—	—	—	0	—	—
S.Africa	P	2,547	3,127	3,839	4,713	5,973	7,157	8,693	10,559
	M%	0.73	0.73	0.73	0.63	0.77	0.72	0.82	0.82
	M	19	23	28	30	46	52	71	87
Zambia	P	838	963	1,107	1,272	1,525	1,681	1,906	2,160
	M%	0.37	0.37	0.37	0.37	0.37	0.37	0.37	0.29
	M	3	4	4	5	6	6	7	6
Zimbabwe	P	936	1,076	1,237	1,422	1,704	1,879	2,130	2,414
	M%	0.01	0.01	0.01	0.03	0.03	0.03	0.03	0.03
	M	0	0	0	0	1	1	1	1
Total	P	11,082	13,361	16,147	17,814	20,275	22,305	25,212	27,852
	M%	0.75	0.74	0.73	0.80	0.94	1.06	1.85	1.95
	M	83	99	118	143	190	236	467	542
	G%	0.708	1.870	1.894	0.982	1.294	0.954	1.225	0.996

Table 3.5c Decennial estimates of the Muslim population (×1000) in Southern Africa: 1950–2100 AD (1370–1520 H)

		1950	1960	1970	1980	1990	2000	2010	2020
Angola	P	4,548	5,455	5,890	8,341	11,848	16,395	23,356	32,866
	M%	0.00	0.01	0.05	0.09	0.20	0.30	0.39	0.50
	M	0	1	3	8	24	49	91	164
Botswana	P	413	503	628	898	1,287	1,643	1,987	2,352
	M%	0.03	0.07	0.10	0.20	0.27	0.43	0.74	1.00
	M	0	0	1	2	3	7	15	24
Eswatini	P	273	337	431	588	822	1,005	1,065	1,160
	M%	—	—	0.10	0.10	0.10	0.17	0.17	0.17
	M	—	—	0	1	1	2	2	2
Lesotho	P	705	837	1,029	1,340	1,704	2,033	1,996	2,142
	M%	0.04	0.05	0.05	0.04	0.04	0.08	0.08	0.10
	M	0	0	1	1	1	2	2	2
Malawi	P	2,954	3,660	4,704	6,250	9,405	11,149	14,540	19,130
	M%	10.00	11.00	12.00	12.30	12.60	12.81	12.97	13.82
	M	295	403	564	769	1,185	1,428	1,886	2,644
Mozambique	P	5,959	7,185	9,023	11,630	12,987	17,712	23,532	31,255
	M%	10.67	17.80	13.55	15.00	16.00	17.65	17.92	18.25
	M	636	1,279	1,223	1,745	2,078	3,126	4,217	5,704
Namibia	P	514	634	817	1,058	1,433	1,795	2,119	2,541
	M%	—	—	—	0.01	0.04	0.08	0.08	0.10
	M	—	—	—	0	1	1	2	3
St.Helena	P	5	5	6	7	7	6	5	6
	M%	—	—	—	—	—	—	—	—
	M	—	—	—	—	—	—	—	—
S.Africa	P	13,628	17,100	22,070	28,557	36,801	44,968	51,217	59,309
	M%	1.16	1.20	1.26	1.71	1.09	1.48	1.62	1.70
	M	158	205	278	488	401	666	830	1,008
Zambia	P	2,310	3,071	4,179	5,852	8,037	10,416	13,606	18,384
	M%	0.29	0.29	0.26	0.26	0.26	0.45	0.49	0.55
	M	7	9	11	15	21	47	67	101
Zimbabwe	P	2,747	3,777	5,289	7,409	10,432	11,881	12,698	14,863
	M%	0.03	0.03	0.82	0.82	0.86	0.86	0.53	0.56
	M	1	1	43	61	90	102	67	83
Total	P	34,056	42,563	54,066	71,929	94,762	119,003	146,120	184,008
	M%	3.22	4.46	3.93	4.29	4.01	4.56	4.91	5.29
	M	1,097	1,898	2,124	3,088	3,804	5,430	7,177	9,735
	G%	2.011	2.230	2.392	2.855	2.757	2.278	2.053	2.306

Table 3.5c (Continued)

		2030	2040	2050	2060	2070	2080	2090	2100
Angola	P	44,835	59,782	77,420	97,343	119,149	142,116	165,392	188,283
	M%	0.60	0.70	0.80	0.90	1.00	1.10	1.20	1.30
	M	269	418	619	876	1,191	1,563	1,985	2,448
Botswana	P	2,774	3,168	3,510	3,773	3,958	4,081	4,148	4,166
	M%	1.25	1.50	1.75	2.00	2.25	2.50	2.75	3.00
	M	35	48	61	75	89	102	114	125
Eswatini	P	1,298	1,507	1,704	1,865	1,987	2,073	2,124	2,145
	M%	0.17	0.17	0.17	0.17	0.17	0.17	0.17	0.17
	M	2	3	3	3	3	4	4	4
Lesotho	P	2,325	2,507	2,665	2,775	2,829	2,831	2,785	2,695
	M%	0.12	0.14	0.16	0.18	0.20	0.22	0.24	0.26
	M	3	4	4	5	6	6	7	7
Malawi	P	24,849	31,317	38,143	45,060	51,684	57,612	62,614	66,559
	M%	14.20	14.60	15.00	15.40	15.80	16.20	16.60	17.00
	M	3,529	4,572	5,721	6,939	8,166	9,333	10,394	11,315
Mozambique	P	41,185	52,729	65,313	78,358	91,305	103,508	114,390	123,647
	M%	18.50	18.75	19.00	19.25	19.50	19.75	20.00	20.25
	M	7,619	9,887	12,409	15,084	17,804	20,443	22,878	25,039
Namibia	P	3,011	3,497	3,981	4,410	4,765	5,047	5,250	5,374
	M%	0.12	0.14	0.16	0.18	0.20	0.22	0.24	0.26
	M	4	5	6	8	10	11	13	14
St.Helena	P	6	6	6	5	5	5	5	4
	M%	—	—	—	—	—	—	—	—
	M	—	—	—	—	—	—	—	—
S.Africa	P	65,956	71,375	75,518	78,172	79,499	79,986	79,888	79,191
	M%	1.80	1.90	2.00	2.10	2.20	2.30	2.40	2.50
	M	1,187	1,356	1,510	1,642	1,749	1,840	1,917	1,980
Zambia	P	24,326	31,338	39,121	47,486	56,230	65,015	73,554	81,546
	M%	0.60	0.65	0.70	0.75	0.80	0.85	0.90	0.95
	M	146	204	274	356	450	553	662	775
Zimbabwe	P	17,596	20,864	23,948	26,477	28,533	30,014	30,774	30,965
	M%	0.70	0.80	0.90	1.00	1.10	1.20	1.30	1.40
	M	123	167	216	265	314	360	400	434
Total	P	228,161	278,092	331,329	385,725	439,945	492,286	540,924	584,576
	M%	5.66	5.99	6.29	6.55	6.77	6.95	7.09	7.21
	M	12,916	16,663	20,825	25,253	29,782	34,215	38,373	42,139
	G%	2.151	1.979	1.752	1.520	1.315	1.124	0.942	0.776

Table 3.6a Centennial estimates of the Muslim population (×1000) in Africa: 600–2100 AD (1–1500 H)

		600	700	800	900	1000	1100	1200	1300
North Africa	P	13,200	13,550	13,900	14,250	14,600	14,490	14,380	14,270
	M%	—	8.80	35.07	61.53	78.49	82.82	83.12	83.57
	M	—	1,193	4,874	8,769	11,460	12,000	11,952	11,926
West Africa	P	2,201	2,349	2,497	2,645	2,791	3,017	3,355	3,693
	M%	—	—	—	—	—	1.06	2.03	2.93
	M	—	—	—	—	—	32	68	108
East Africa	P	2,391	2,645	2,898	3,151	3,405	4,061	4,752	5,444
	M%	—	1.96	6.35	11.84	19.04	20.70	22.80	23.98
	M	—	52	184	373	648	841	1,084	1,306
Central Africa	P	7,160	7,698	8,256	8,814	9,395	10,711	12,032	13,463
	M%	—	—	—	—	—	2.29	4.88	7.61
	M	—	—	—	—	—	245	588	1,025
Southern Africa	P	1,512	1,665	1,819	1,972	2,120	2,610	3,124	3,638
	M%	—	—	—	—	—	—	0.11	0.17
	M	—	—	—	—	—	—	3	6
Total	P	26,465	27,908	29,371	30,833	32,312	34,889	37,643	40,508
	M%	—	4.46	17.22	29.65	37.47	37.60	36.38	35.48
	M	—	1,244	5,058	9,142	12,108	13,118	13,695	14,371
	G%		0.053	0.051	0.049	0.047	0.077	0.076	0.073

Table 3.6a (Continued)

		1400	1500	1600	1700	1800	1900	2000	2100
North Africa	P	14,160	14,100	17,290	16,080	17,487	31,614	204,587	829,102
	M%	84.19	83.79	85.44	85.15	86.34	88.28	95.26	96.80
	M	11,922	11,815	14,773	13,693	15,098	27,909	194,880	802,546
West Africa	P	4,046	4,389	5,594	7,200	8,108	11,594	87,551	488,455
	M%	5.08	7.23	9.42	14.69	20.12	26.57	46.34	59.88
	M	206	318	527	1,058	1,631	3,080	40,569	292,500
East Africa	P	6,231	7,115	8,067	9,270	11,348	21,404	162,033	896,285
	M%	25.61	27.36	28.27	30.37	23.98	28.27	28.77	36.34
	M	1,596	1,947	2,281	2,816	2,721	6,051	46,623	325,691
Central Africa	P	14,894	16,392	19,051	22,312	25,406	31,186	237,810	1,481,709
	M%	10.32	13.12	16.72	25.52	24.95	27.98	29.39	31.21
	M	1,538	2,150	3,185	5,694	6,338	8,727	69,899	462,457
Southern Africa	P	4,167	4,641	5,199	5,970	7,211	17,814	119,003	584,576
	M%	0.24	0.33	0.43	0.55	0.69	0.80	4.56	7.21
	M	10	15	22	33	49	143	5,430	42,139
Total	P	43,498	46,637	55,201	60,832	69,560	113,612	810,984	4,280,127
	M%	35.11	34.83	37.66	38.29	37.14	40.41	44.07	44.98
	M	15,271	16,244	20,788	23,293	25,838	45,909	357,401	1,925,333
	G%	0.071	0.070	0.169	0.097	0.134	0.491	1.965	1.663

Table 3.6b Decennial estimates of the Muslim population (×1000) in Africa: 1790–1940 AD (1210–1360 H)

		1790	1800	1810	1820	1830	1840	1850	1860
North Africa	P	16,710	17,487	18,246	18,797	19,827	20,074	21,177	22,130
	M%	86.07	86.34	86.59	86.74	87.05	87.13	87.44	87.67
	M	14,382	15,098	15,798	16,305	17,259	17,490	18,518	19,401
West Africa	P	7,931	8,108	8,286	8,461	8,724	8,995	9,287	9,591
	M%	20.04	20.12	20.19	20.27	20.75	21.13	21.63	22.24
	M	1,589	1,631	1,673	1,715	1,810	1,901	2,009	2,133
East Africa	P	11,183	11,348	11,514	11,712	12,509	13,395	14,371	15,461
	M%	24.02	23.98	23.93	23.84	24.54	25.14	25.58	25.88
	M	2,687	2,721	2,755	2,792	3,070	3,367	3,675	4,001
Central Africa	P	25,190	25,406	25,632	25,844	26,115	26,416	26,673	26,965
	M%	25.05	24.95	24.84	24.74	24.59	24.50	24.45	24.38
	M	6,310	6,338	6,366	6,394	6,422	6,471	6,521	6,574
Southern Africa	P	6,985	7,211	7,437	7,710	8,401	9,041	9,684	10,325
	M%	0.68	0.69	0.69	0.69	0.70	0.72	0.73	0.74
	M	48	49	51	53	59	65	70	76
Total	P	67,998	69,560	71,114	72,524	75,575	77,921	81,192	84,472
	M%	36.79	37.14	37.47	37.59	37.87	37.59	37.93	38.10
	M	25,015	25,838	26,644	27,259	28,620	29,294	30,794	32,184
	G%		0.227	0.221	0.196	0.412	0.306	0.411	0.396

Table 3.6b (*Continued*)

		1870	1880	1890	1900	1910	1920	1930	1940
North Africa	P	23,245	25,846	29,012	31,614	34,969	37,487	42,659	48,907
	M%	87.94	88.26	88.62	88.28	88.14	87.17	87.65	87.17
	M	20,442	22,812	25,712	27,909	30,821	32,677	37,389	42,632
West Africa	P	9,992	10,362	10,790	11,594	12,984	15,382	17,352	19,688
	M%	23.14	23.83	24.56	26.57	26.47	29.05	28.12	27.45
	M	2,312	2,469	2,650	3,080	3,437	4,469	4,879	5,404
East Africa	P	16,714	18,099	19,640	21,404	23,474	26,573	31,064	35,111
	M%	26.64	27.15	27.69	28.27	28.83	28.64	30.05	29.70
	M	4,453	4,914	5,439	6,051	6,769	7,610	9,334	10,429
Central Africa	P	27,377	28,517	29,742	31,186	34,145	37,593	43,150	47,656
	M%	24.38	24.92	25.44	25.89	25.29	24.18	23.63	22.01
	M	6,673	7,105	7,565	8,073	8,636	9,090	10,196	10,491
Southern Africa	P	11,082	13,361	16,147	17,814	20,275	22,305	25,211	27,852
	M%	0.75	0.74	0.73	0.80	0.94	1.06	1.85	1.95
	M	83	99	118	143	190	236	467	542
Total	P	88,410	96,185	105,332	113,612	125,847	139,339	159,436	179,214
	M%	38.42	38.88	39.38	39.83	39.61	38.81	39.05	38.78
	M	33,964	37,400	41,484	45,256	49,853	54,081	62,265	69,498
	G%	0.456	0.843	0.908	0.757	1.023	1.018	1.347	1.169

Table 3.6c Decennial estimates of the Muslim population (×1000) in Africa: 1950–2100 AD (1370–1520 H)

		1950	1960	1970	1980	1990	2000	2010	2020
North Africa	P	59,209	75,728	97,790	125,907	164,219	204,587	249,900	311,766
	M%	88.16	90.11	93.66	94.72	95.31	95.26	95.62	95.62
	M	52,199	68,240	91,588	119,253	156,514	194,880	238,955	298,114
West Africa	P	25,091	30,300	38,486	49,749	66,805	87,551	113,524	146,608
	M%	29.75	35.61	37.40	42.45	46.31	46.34	48.79	50.26
	M	7,463	10,789	14,395	21,120	30,940	40,569	55,384	73,685
East Africa	P	40,039	50,523	66,351	89,092	120,732	162,033	214,291	279,996
	M%	28.78	29.42	28.33	29.58	28.58	28.77	29.83	30.40
	M	11,525	14,865	18,797	26,356	34,508	46,623	63,927	85,109
Central Africa	P	69,399	84,247	106,755	139,709	183,831	237,810	315,468	418,220
	M%	27.31	28.27	27.72	28.16	29.39	30.10	29.78	29.48
	M	18,952	23,818	29,594	39,338	54,025	71,574	93,934	123,311
Southern Africa	P	34,056	42,563	54,066	71,929	94,762	119,003	146,120	184,008
	M%	3.22	4.46	3.93	4.29	4.01	4.56	4.91	5.29
	M	1,097	1,898	2,124	3,088	3,804	5,430	7,177	9,735
Total	P	227,794	283,361	363,448	476,386	630,350	810,984	1,039,304	1,340,598
	M%	40.05	42.21	43.06	43.90	44.39	44.28	44.20	44.01
	M	91,236	119,610	156,497	209,155	279,791	359,076	459,378	589,954
	G%	2.399	2.183	2.489	2.706	2.800	2.520	2.481	2.546

Table 3.6c (*Continued*)

		2030	2040	2050	2060	2070	2080	2090	2100
North Africa	P	377,894	449,013	523,779	596,797	665,793	728,894	784,174	829,102
	M%	95.66	95.73	95.84	95.98	96.15	96.35	96.57	96.80
	M	361,486	429,824	501,991	572,833	640,182	702,274	757,238	802,546
West Africa	P	185,879	229,914	276,970	324,951	371,962	416,037	455,297	488,455
	M%	51.63	52.96	54.24	55.45	56.61	57.73	58.82	59.88
	M	95,970	121,765	150,219	180,182	210,563	240,161	267,786	292,500
East Africa	P	356,464	439,946	526,911	614,083	697,535	773,727	840,396	896,285
	M%	31.14	31.91	32.69	33.47	34.23	34.96	35.67	36.34
	M	111,006	140,368	172,229	205,546	238,784	270,526	299,768	325,691
Central Africa	P	539,923	679,785	830,286	983,422	1,132,293	1,269,627	1,387,277	1,481,709
	M%	29.43	29.51	29.68	29.92	30.21	30.53	30.87	31.21
	M	158,916	200,588	246,432	294,257	342,073	387,624	428,191	462,457
Southern Africa	P	228,161	278,092	331,329	385,725	439,945	492,286	540,924	584,576
	M%	5.66	5.99	6.29	6.55	6.77	6.95	7.09	7.21
	M	12,916	16,663	20,825	25,253	29,782	34,215	38,373	42,139
Total	P	1,688,321	2,076,750	2,489,275	2,904,977	3,307,528	3,680,571	4,008,067	4,280,127
	M%	43.85	43.78	43.86	44.00	44.18	44.42	44.69	44.98
	M	740,295	909,208	1,091,695	1,278,072	1,461,385	1,634,799	1,791,356	1,925,333
	G%	2.306	2.071	1.812	1.544	1.298	1.069	0.852	0.657

3.5.12 Regional Summary and Conclusion

Southern Africa has the least concentration of Muslims among the five regions spanning Africa. Although Islam entered this region in the twelfth century, most of this region was never under Muslim control. The Muslim population has been increasing slowly. It is currently over 5% of the total population but is expected to remain less than 9% for the next three centuries. The following tables present centennial data from 600 AD to 2100 AD (or approximately 1 H to 1500 H) in Table 3.5a and decennial data from 1790 AD to 2100 AD (or 1210 H to 1520 H) in Tables 3.5b and 3.5c for current countries in Southern Africa. The data includes total population in thousands (P), the percentage of which is Muslim (M%), the corresponding Muslim population in thousands (M), and the annual population growth rate (APGR, or G%) of the total population in this region. The total population estimate in each country since 1950 is based on the United Nations' World Population Prospects [UNP] while pre-1950 data is based on [PSH, MAD, AVA]. Other estimates and census data is used to fill in missing data from the aforementioned sources.

3.6 Africa's Summary and Conclusion

Islam entered Africa after it started in Asia and by now over a quarter of the world Muslim population lives in Africa. This ratio is increasing and is expected to approach a half by the end of this century, largely due to rapid natural growth. The percentage of Muslims in Africa is currently 42% and is expected to remain around it for the next three centuries. The following tables present centennial data from 600 AD to 2100 AD (or approximately 1 H to 1500 H) in Table 3.6a and decennial data from 1790 AD to 2100 AD (or 1210 H to 1520 H) in Tables 3.6b and 3.6c for the five regions of Africa. The data includes total population in thousands (P), the percentage of which is Muslim (M%), the corresponding Muslim population in thousands (M), and the annual population growth rate (APGR, or G%) of the total population in Africa and each of its five regions.

References

[ABS] Afrobarometer (2015). *Afrobarometer Online Data Analysis.* Retrieved from http://www.afrobarometer-online-analysis.com/

[AO] Instituto Nacional de Estatística (INE) (2016). *Resultados Defenitivos do Recenseamento Geral da População e da Habitação de Angola 2014.* Luanda: INE.

[BF06] Institut National de la Statistique et de la Démographie (2006). Tableau 2.2: Répartition de la Population Résidente par Religion selon le Milieu de Résidence et le Sexe. *Principaux Tableaux du Recensement General de la Population et de l'Habitation 2006* [Online].

[BF61] Service de la Statistique et de la Mécanographie [Haute-Volta], Institut National de la Statistique et des Études Économiques [France] & Département de la Coopération [France] (1970). *Enquête démographique par sondage en République de Haute-Volta 1960–1961* (Tome 1). Paris: INSEE.

[BI65] Bihango, B. (1984). Enquêtes démographiques effectuées au Burundi Résumés méthodologiques et principaux résultats (pp. 11–14), *Séminaire sur l'utilisation des données du Recensement Général de la Population.* Bujumbura: Centre d'Etudes Démographiques pour le Developpement.

[BJ13] Institut National de la Statistique et de l'Analyse Economique (INSAE) (2016). *Principaux Indicateurs Socio Demographiques ey Economiques (RGPH-4, 2013).* Cotonou: INSAE.

[BJ02] Institut National de la Statistique et de l'Analyse Economique (INSAE) (2003). *Troisieme Recensement General de la Population et de l'Habitation - Fevrier 2002: Principaux Indicateure Socio-Demographiques.* Cotonou: Direction des Etudes Demographiques.

[BJ61] Némo, J., Cantrelle Pierre (collab.), Roussel, L. (collab.) (1964). Enquête démographique au Dahomey - 1961: résultats définitifs. Paris: Ministère de la Coopération; INSEE.

[BW11] Statistics Botswana (2014). *Population and Housing Census 2011 Analytical Report.* Gaborone: Statistics Botswana.

[BW01] Ntloedibe-Kuswani, G. S. (2003, September). African Religions and 2001 Population and Housing Census in Botswana. *Proceedings of the 2001 Population and Housing Census Dissemination Seminar,* Gaborone, Botswana. Gaborone: Central Statistics Office [Botswana].

[CF03] Institut Centrafricain des Statistiques et des Etudes Economiques et Sociales (ICASEES) (2008). Base de données du Recensement Général de la Population et de l'Habitatation (RGPH), 2003. Bangui: ICASEES. Retrieved from online database at http://celade.cepal.org/cgibin/RpWebEngine.exe/PortalAction?

[CF60] Service de la Statistique Générale [République Centrafricaine], Ministère de la Coopération [France] & Institut National de la Statistique et des Études Économiques [France] (1964). *Enquête Démographique en République Centrafricaine, 1959–1960: Résultats Définitifs*. Paris: Ministère de la Coopération & INSEE.

[CG07] Centre National de la Statistique et des Etudes Economiques (CNSEE) [Congo] (2010). *Le RGPH-2007 en quelques chiffres*. Brazzaville: CNSEE.

[CG61] Service de Statistique [Congo], Ministère de la Coopération [France] & Institut National de la Statistique et des Études Économiques [France] (1965). *Enquête Démographique, 1960–1961: Résultats Définitifs*. Paris: Ministère de la Coopération & INSEE.

[CG61b] Service de Statistique [Congo], Ministère de la Coopération [France] & Institut National de la Statistique et des Études Économiques [France] (1965). *Recensement de Brazzaville 1961: Résultats Définitifs*. Paris: Ministère de la Coopération & INSEE.

[CI98] Essoh, B. (2001). *Recensement général de la population et de l'habitat 1998*. Volume IV: Analyse des Résultats Définitifs. Tome 1: Etat et Structure de la Population. Abidjan: Institut National de la Statistique [Côte d'Ivoire].

[CIH] Miran, M. (2006). The Political Economy of Civil Islam in Côte d'Ivoire. In M. Bröning & H. Weiss (eds.), *Politischer Islam in Westafrika. Eine Bestandsaufnahme* (pp. 82–113). Berlin: Friedrich Ebert Stiftung et Lit Verlag.

[CV90] Inforpress (2009, December 23). Comunidade Islâmica na Praia Assinala Natal com Conferência Sobre Profetas [Islamic Community in Praia Notes About Christmas with the Prophets Conference]. *Inforpress*.

[CV10] Portuguese News Network (PNN) (2011, March 31). Cabo Verde: Últimos censos revelam população muito jovem [CaboVerde: Latest Census Shows Very Young Population]. *Journal Digital*.

[DHS] ICF International (1984–2018). *The Demographic and Health Surveys (DHS) Program*. Rockville, MD: ICF International. Retrieved from http://dhsprogram.com/

[DOS00] United States Department of State (DOS), Bureau of Democracy, Human Rights, and Labor (2000). *2000 Annual Report on International Religious Freedom*. Washington, DC: DOS. Retrieved from http://www.state.gov/j/drl/rls/irf/

[DOS05] United States Department of State (DOS), Bureau of Democracy, Human Rights, and Labor (2005). *2005 Annual Report on International Religious Freedom.* Washington, DC: DOS. Retrieved from http://www.state.gov/j/drl/rls/irf/

[DOS10] United States Department of State (DOS), Bureau of Democracy, Human Rights, and Labor (2010). *2010 Annual Report on International Religious Freedom.* Washington, DC: DOS. Retrieved from http://www.state.gov/j/drl/rls/irf/

[DOS15] United States Department of State (DOS), Bureau of Democracy, Human Rights, and Labor (2015). *2015 Annual Report on International Religious Freedom.* Washington, DC: DOS. Retrieved from http://www.state.gov/j/drl/rls/irf/

[DZH54] Good, D. (1961). Notes on the Demography of Algeria. *Population Index*, 27(1), 3–32.

[DZH66] Bahri, A. M., Negadi, G., Tabutin, D., Vallin, J., Bardinet, C., Dellouci, B., Mehani, M., Grangaud, M. F., Ouibrahim, L., Ribier, R., Von Allmen, F., & Von Allmen, M. (1974). *La Population de l'Algerie.* 1974 Country Monographs, CICRED (Comite International de Coordination de Recherches Nationales de Demographie) Series. Paris: CICRED. Retrieved from http://www.cicred.org/Eng/Publications/content/3MonographNational/Index.html

[EG60] Fargues, P. (1998). The Arab Christians of the Middle East: A Demographic Perspective. In A. Pacini (ed.), *Christian Communities in the Arab Middle East, the Challenge of the Future.* Oxford: Clarendon Press.

[EG868] Saleh, M. (2013). *On the Road to Heaven: Taxation, Conversions, and the Coptic-Muslim Socioeconomic Gap in Medieval Egypt.* Toulouse School of Economics Working Paper, No. 13-428.

[EJB] Houtsma, M. T. (1987). *E. J. Brill's First Encyclopedia of Islam 1913–1936.* Leiden: E. J. Brill.

[ER31] Instituto Centrale di Statistica del Regno d'Italia (1934). *Le Popolazioni delle Colonie e dei Possedimenti Italiani, Secondo il Censimento del 1931-IX.* Roma: Tipografia i. Failli

[ET07] Central Statistical Agency (CSA) (2008). *Summary and Statistical Report of the 2007 Population and Housing Census.* Addis Ababa: CSA.

[ET84] Central Statistical Authority (CSA) (1991). *The 1984 population and housing census of Ethiopia: analytical report at national level.* Addis Ababa: CSA.

[GB11] His Majesty's Stationary Office (HMSO) (1917). *Census of England and Wales. 1911.* (General Report with Appendices). London: Darling & Son.

[GB01] General Register Office (GRO) (1906) *Census of the British Empire. 1901. Report with Summary and Detailed Tables for the Several Colonies, &c., Area, Houses, and Population; also Population Classified by Ages, Condition as to Marriage, Occupation, Birthplaces, Religions, Degrees of Education, and Infirmities.* London: Darling & Son.

[GB871] Her Majesty's Stationary Office (HMSO) (1873). *Census of England and Wales for the Year 1871* (General Report. Vol. IV). London: George Edward Eyre and William Spottiswoode.

[GH10] Ghana Statistical Service (GSS) (2012). *2010 Population & Housing Census - Summary Report of Final Results.* Accra: GSS.

[GM13] Gambia Bureau of Statistics (GBOS) (2017). *2013 Population and Housing Census* (Volume 5: Spatial Distribution). Banjul: GBOS.

[GM03] Gambia Bureau of Statistics (GBOS) (2008). *Population and Housing Census of The Gambia 2003.* [Online Database]. Retrieved from http://www.gbos.gm/surveys/Census2003/survey0/dataSet/dataFiles/dataFile1/var53.html

[GM93] Gambia Bureau of Statistics (GBOS) (2008). *Population and Housing Census of The Gambia 1993.* [Online Database]. Retrieved from http://www.gbos.gm/surveys/Census1993/survey0/dataSet/dataFiles/dataFile0/var17.html

[GM83] Central Statistics Department (CSD) (1987). *Population and Housing Census 1983, General Report* (Volume I: Administrative abd Analytical Procedures). Banjul: CSD.

[GM63] Jallow, A. Y. (2010, April 19). Who is Kairaba? Harvard Graduate Jallow Critiques an African Statesman. *Senegambia News.*

[GN14] Bah, M. C., & Bangoura, M. A. (2017). *Le Troisième Recensement Général de la Population et de l'Habitation (RGPH-3), Rapport d'Analyse des Donnees du RGPH3* (Vol. 1: Etat et Structure de la Population). Conakry: Institut National de la Statistique [Guinea].

[GN96] Bah, M. C., & Bangoura, M. A. (2000). *Recensement Général de la Population et de l'Habitation de 1996, Rapport d'Analyse des Donnees du RGPH3* (Vol. 2: Etat de la Population). Conakry: Institut National de la Statistique [Guinea].

[GN83] Bureau National du Recensement [Guinea] (1989). *Recensement Général de la Population et de l'Habitation, Février 1983, Rapport* d'Analyse des Donnees du RGPH3 (Vol. 1: Résultats Définitifs). Conakry: Bureau National du Recensement [Guinea].

[GNH] Delafosse, M. (1922). L'Animisme Négre et sa Résistance a l'Islamisation en Afrique Occidentale. *Revue du Monde Musulman*, 49, pp. 121–163.

[GP91] Instituto Nacional de Estatistica e Censos (INEC) (1996). *Recenseamento Geral da Populacao d Habitacao 1991* (Volume I. Resultados Definitivos). Bissau: INEC.

[GP50] *Censo da Populacao de 1950* (Volume II. Populacao nao Civilizada). Lisbao: Tipografia Portuguesa, LDA.

[IML] Lapidus, I. M. (2002). *A History of Islamic Societies* (2nd ed.). Cambridge: Cambridge University Press.

[IPUMS] Minnesota Population Center (2013). *Integrated Public Use Microdata Series, International: Version 6.2* [Machine-readable database]. Minneapolis: University of Minnesota. Retrieved from https://international.ipums.org/

[JAN] Jansen, H. (1897). *Verbreitung des Islâms: mit angabe der verschiedenen riten, sekten und religiösen bruderschaften in den verschiedenen ländern der erde 1890 bis 1897. Mit benutzung der neuesten angaben (zählungen, berechnungen, schätzungen und vermutungen). [German for Dissemination of Islam: with number of various rites, sects and religious brotherhoods in the various countries of the earth 1890 to 1897. With use of information on the latest disclosures (censuses, calculations, estimates and guesses)]*. Berlin: Selbstverlag des Verfassers.

[KE09] Kenya National Bureau of Statistics (KNBS) (2010). *2009 Population and Housing Census Results*. Nairobi: KNBS.

[KE48] Watt, W. M. (1966). The Political Relevance of Islam in East Africa. *International Affairs*, 42(1), 35–44.

[KET86] Kettani, A. (1986). *Muslim Minorities in the World Today*. London: Mansell Publishing.

[LR08] Liberia Institute of Statistics and Geo-Information Services (LISGIS) (2009). *2008 Population and Housing Census Final Results*. Monrovia. LISGIS.

[LR84] Liberia Institute of Statistics and Geo-Information Services (LISGIS) (2009). 1984 Population and Housing Census (Summary Results in Graphics and Descriptive Analysis of som salient Demographic Characteristics. Monrovia. LISGIS.

[LY36] Pan, C. (1949). The Population of Libya. *Population Studies*, 3(1), 100–125.

[MAH] Institut National de Statistique et d'Economie Appliquée (INSEA) [Morocco] (1974). *La Population du Maroc*. 1974 Country Monographs, CICRED (Comite International de Coordination de Recherches Nationales de Demographie) Series. Rabat: INSEA. Retrieved http://www.cicred.org/Eng/Publications/content/3MonographNational/Index.html

[MAS23] Massignon, L. (1923). *Annuaire du Monde Musulman, Statistique, Historique, Social et Économique*. Paris: Ernest Leroux.

[MAS55] Massignon, L. (1955). *Annuaire du Monde Musulman, Statistique, Historique, Social et Économique* (4ème éd.). Paris: Presses Universitaires de France.

[ML09] Traore, S. M., Doumbia, A. G., Traore, V. & Tolno, D. F. (2011). *4ème Recensement General de la Population et de l'Habitat du Mali (RGPH-2009) - Analyse des Resultats Definitifs* (Theme 2: Etat et Structure de la Population). Bamako: Institut National de la Statistique (INSTAT).

[ML61] Service de Statistique [Mali], Secretariat d'Etat aux Affaires Etrangères [France], Institut Nationale des Statistiques et des Etudes Economiques [France], & Service de la Coopération [France] (1961). *Enquête démographique au Mali 1960–1961*. Bamako: Service de la Statistique, SEAE, INSEE, Service de la Coopération

[ML936] Imperato, P. J. & Imperato, E. M. (1982). *Mali: A Handbook of Historical Statistics*. Boston, MA: G.K. Hall.

[MU11] Statistics Mauritius (2012). *2011 Housing and Population Census* (Vol. II: Demographic and Fertility Characteristics). Port Louis: Statistics Mauritius.

[MU90] Central Statistical Office [Mauritius] (1991). *1990 Housing and Population Census of Mauritius* (Vol. II: Demographic and fertility Characteristics). Port Louis: CSO.

[MU83b] Central Statistical Office [Mauritius] (1985). *1983 Housing and Population Census of Mauritius* (Vol. V: Housing and Population Census Results (Island of Rodrigues)). Port Louis: CSO.

[MU83] Central Statistical Office [Mauritius] (1984). *1983 Housing and Population Census of Mauritius* (Vol. II: Demographic Characteristics (Island of Mauritius)). Port Louis: CSO.

[MU944] Koenig, M. (1945). *Final Report on the Census Enumeration Made in the Colony of Mauritius and Its Dependencies on 11th June, 1944*. Port Louis: Government Printer.

[MW18] National Statistical Office (NSO) (2019). *2018 Malawi Population and Housing Census - Main Report*. Zomba: NSO.

[MW08] National Statistical Office (NSO) (2009). *2008 Population and Housing Census - Final Results.* Zomba: NSO.

[MWH] Djinguiz, M. (1907). L'Islam dans l'Afrique Centrale Britannique. *Revue du Monde Musulman*, 3(11-12), 515-518.

[NE] Institut National de la Statistique du Niger (INS) (2012). *Données du Recensement Général de la Population et de l'Habitat 2001.* Niamey: INS. Retrieved from online database at http://celade.cepal.org/cgibin/RpWebEngine.exe/PortalAction?

[PEW] Pew Research Center (PRC) (2009). *Mapping the Global Muslim Population, A Report on the Size and Distribution of the World's Muslim Population.* Washington, DC: PRC.

[REH] Mourrégot, M. F. (2010). *L'Islam à l'Ile de la Réunion.* Paris: L'Harmattan.

[RE] A. Timol, A. (2005, February 14). La République et le Culte Musulman: l'Exemple de la Réunion. *Interventions.*

[ROD] Radončić, E. (2013). *World Almanac of the demographic history of Muslims: Rich cartographic representation of the geographical distribution of Muslim populations* (M. Durmić, Trans.). Sarajevo: Edin Radončić.

[RW12] National Institute of Statistics of Rwanda (NISR) (2014). *Fourth Population and Housing Census, Rwanda, 2012* (Final Results: Main Indicators Report). Kigali: NISR.

[RW02] National Census Service (NCS) (2003). *Final Results: Statistical Tables.* 3rd General Census of Population and Housing of Rwanda-August 2002. Kigali: NCS.

[SC] National Bureau of Statistics (NBS) (2011). *Population and Housing Census 2010–Report.* Victoria: NBS.

[SD93] Central Bureau of Statistics (CBS) (2008). *Statistical Year Book for the Year 2007.* Khartoum: CBS.

[SH16] St Helena Statistics Office (SHSO) (2016). *St Helena 2016 Population & Housing Census–Summary Report.* Jamestown: SHSO. And e-mail communication with Mr. B. Wahler from SHSO in October 2016.

[SH08] St Helena Statistics Office (SHSO) (2009). *The 2008 Population Census of St Helena.* Jamestown: SHSO. And e-mail communication with Mrs. D.M. Knipe from SHSO in October 2009.

[SH98] St Helena Statistics Office (SHSO) (1998). *The 1998 Population Census of St Helena.* Jamestown: SHSO.

[SH87] Overseas Development Administration (ODA) (1988). *St. Helena and Ascension Island Population Census, 1987.* Jamestown: ODA.

[SH76] St Helena Information Office (1977). *Census of the Population of St. Helena & Ascension Islands, 1976.* Jamestown: Government Printer.

[SH66] St Helena Education Office (1966). *Census of the Population of St. Helena Island and Ascension Island, 1966.* Jamestown: Government Printing Office.

[SH56] St Helena Education Office (1957). *Census of the Population of St. Helena Island and Ascension Island, 1956.* Jamestown: Government Printing Office.

[SH46] St Helena Statistics Office (SHSO) (1946). *Census of the Population of St. Helena Island and Ascension Island, 1946.* Jamestown: SHSO.

[SH31] Colonial Secretary's Office (1931). *Census of the Island of St. Helena in 1931.* Jamestown: The Government Printing Office by B. E. Grant.

[SH21] Colonial Secretary's Office (1921). *Census of the Island of St. Helena in 1921.* Jamestown: The Government Printing Office by B. E. Grant.

[SL15] Statistics Sierra Leone (SSL) (2017). *Sierra Leone 2015 Population and Housing Census 2004 With an Agricultural Module).* Freetown: SSL.

[SL04] Statistics Sierra Leone (SSL) (2006). *Sierra Leone Population and Housing Census: Thematic Report on Population Structure and Population Distribution.* Freetown: SSL.

[SN13] Agence Nationale de la Statistique et de la Démographie (ANSD) (2014). *Recensement Général de la Population et de l'Habitat, de l'Agriculture et de l'Elevage (RGPHAE 2013)* (Rapport Definitif). Dakar: ANSD.

[SN02] Agence Nationale de la Statistique et de la Démographie (ANSD) (2005). *Recensement Général de la Population et de l'Habitat 2002.* Dakar: ANSD.

[SN904] Becker, C., Martin, V., Schmitz, J. & Chastanet, M. (1983). *Les premiers recensements au Sénégal et l'évolution démographique.* Dakar: Office de la Recherche Scientifique et Technique Outre-Mer (ORSTOM).

[SNH] Delafosse, A. (1910). L'État Actuel de l'Islam dans l'Afrique Occedentale Française. *Revue du Monde Musulman,* 11(5), 32–53.

[SS8] Sudan Tribune (2012, August 17). S. Sudan's Kiir urges Muslims not to politicize Islam. *Sudan Tribune.*

[SS56] Agence France-Presse (2011, January 9). South Sudan's Muslims welcome secession. *The Daily Star.*

[ST70] Instituto Nacional de Estatistica (INE) (1971). Censo da Populacao 1970. Lisbao: INE

[SYB85] Paxton, J. (ed.) (1985). *The Statesman's Year-Book.* London: MacMillan.

[SYB80] Paxton, J. (ed.) (1980). *The Statesman's Year-Book.* London: MacMillan.

[SYB70] Paxton, J. (ed.) (1970). *The Statesman's Year-Book.* London: MacMillan.

[SYB50] Steinberg, S.H. (ed.) (1950). *The Statesman's Year-Book.* London: MacMillan.

[SYB31] Epstein, M. (ed.) (1931). *The Statesman's Year-Book.* London: MacMillan.

[SYB20] Scott-Keltie, J. & Epstein M. (eds.) (1920). *The Statesman's Year-Book.* London: MacMillan.

[SYB10] Scott-Keltie, J. (ed.) (1910). *The Statesman's Year-Book.* London: MacMillan.

[SYB00] Scott-Keltie, J. (ed.) (1900). *The Statesman's Year-Book.* London: MacMillan.

[SYB880] Martin, F. (ed.) (1880). *The Statesman's Year-Book.* London: MacMillan.

[TD] Institut National de la Statistique, des Études Économiques et Démographiques (INSEED) (2012). *Deuxieme Recensement General de la Population et de l'Habitat (RGPH2, 2009)* (Principaux Indicateurs Globaux, Issus de l'Analyse Thematique). N'Djamena: INSEED-TCHAD.

[TG10] Lamboni, M. (2016). *Analyses des Donn*ées RGPH4-Novembre 2010 (État Matrimonial et Nuptialité). Lomé: Institut National de la Statistique et des Études Économiques et Démographiques (INSEED-TOGO).

[TG81] Marguerat, Y. (1985). *La Population du Togo, Quatre Etudes sur le Recensement de 1981* (La Repartition Spatiale des Religions). Lomé: l'Office de la Recherche Scientifique et Technique Outre-Mer (ORSTOM).

[TN911] Direction Générale de l'Agriculture, du Commerce et de la Colonisation (1913). *Statistique Générale de la Tunisie, Année 1914.* Tunis: Société Anonyme de l'Imprimerie Rapide.

[TNH] Seklani, M. (1974). *La Population de la Tunisie.* 1974 Country Monographs, CICRED (Comite International de Coordination de Recherches Nationales de Demographie) Series. Paris: CICRED. Retrieved from http://www.cicred.org/Eng/Publications/content/3MonographNational/Index.html

[TZ67] Rubanza, Y. I. (2002). Religious Intolerance: The Tanzania Experience. *Proceedings of the Racism and Xenophobia Conference,* (pp. 1–23). Zanzibar, Tanzania.

[TZH] Chande, A. N. (1991). *Islam, Islamic Leadership and Community Development in Tanga, Tanzania.* (Unpublished doctoral dissertation). McGill University, Monteal, Canada.

[UG] Uganda Bureau of Statistics (UBS) (2016). *The National Population and Housing Census 2014–Main Report.* Kampala: UBS.

[UN] United Nations Statistics Division, UNSD Demographic Statistics (2018). Population by Religion, Sex and Urban/Rural Residence. *UNdata* [Online Database]. Retrieved from http://data.un.org

[UN04] United Nations, Department of International Economic and Social Affairs, Statistical Office (2006). Table 6 - Population by Religion, Sex and Urban/Rural Residence: Each Census, 1985–2004. *Demographic Yearbook Special Census Topics* (Vol. 2b: Ethnocultural Characteristics). New York, NY: United Nations.

[UN88] United Nations, Department of International Economic and Social Affairs, Statistical Office (1990). Table 29 - Population by Religion and Sex: Each Census, 1979–1988. *Demographic Yearbook 1988.* New York, NY: United Nations.

[UN83] United Nations, Department of International Economic and Social Affairs, Statistical Office (1985). Table 29 - Population by Religion and Sex: Each Census, 1974–1983. *Demographic Yearbook 1983.* New York, NY: United Nations.

[UN79] United Nations, Department of International Economic and Social Affairs, Statistical Office (1980). Table 29 - Population by Religion, Sex and Urban/Rural Residence: Each Census, 1970–1979. *Demographic Yearbook 1979.* New York, NY: United Nations.

[UN73] United Nations, Department of International Economic and Social Affairs, Statistical Office (1974). Table 31 - Population by Religion, Sex and Urban/Rural Residence: Each Census, 1965–1973. *Demographic Yearbook 1973.* New York, NY: United Nations.

[UN71] United Nations, Department of International Economic and Social Affairs, Statistical Office (1972). Table 17 - Population by Religion, Sex and Urban/Rural Residence: Each Census, 1962–1971. *Demographic Yearbook 1971.* New York, NY: United Nations.

[UN64] United Nations, Department of International Economic and Social Affairs, Statistical Office (1965). Table 32 - Population by Religion and Sex: Each Census, 1955–1964. *Demographic Yearbook 1964.* New York, NY: United Nations.

[UN63] United Nations, Department of International Economic and Social Affairs, Statistical Office (1964). Table 11 - Population by Religion and

Sex: Each Census, 1955–1963. *Demographic Yearbook 1963*. New York, NY: United Nations.

[UN56] United Nations, Department of International Economic and Social Affairs, Statistical Office (1957). Table 8 - Population by Religion and Sex: Each Census, 1945–1955. *Demographic Yearbook 1956*. New York, NY: United Nations.

[WVS] World Values Survey Association at the University of Aberdeen [The United Kingdom] (2015). *World Values Survey Online Data Analysis*. Retrieved from http://www.worldvaluessurvey.org/WVSOnline.jsp

[ZA] Statistics South Africa (STATS SA) (2016). *Community Survey 2016 in Brief*. Report 03-01-06. Pretoria: STATS SA.

[ZM10] Central Statistical Office (CSO) (2012). *Zambia 2010 Census of Population and Housing* (Vol. 11: National Descriptive Tables). Lusaka: CSO.

[ZM00] Jomo, F. (2006, July 19). Islam Making In-Roads in Zambia. *News from Africa*.

[ZM80] Phiri, F. J. (2008). Islam in Post-Colonial Zambia. In J.-B. Gewald, M. Hinfelaar, G. Macola (eds.). *One Zambia, Many Histories: Towards a History of Post-Colonial Zambia*, (pp. 164–184). Leiden: Brill.

[ZMH] Djinguiz, M. (1907). Les Musulmans de la Rhodesia Nord-Est. *Revue du Monde Musulman*, 2(5), 50–51.

Chapter 4

Islam in Europe

Europe was the third continent; after Asia and Africa, to which Islam has entered. Muslims crossed the Gibraltar strait in 711 AD, conquering all the Iberian Peninsula by 715 AD. They kept going north and conquered half of current France, reaching 100 Km southeast of Paris in 725 AD, until they were defeated in the Battle of Tours (Balat Ashuhada) in 732 AD. They were driven out of France by 759 AD but returned and conquered the Mediterranean coast of France from 891 AD to 973 AD. In the ninth century, Muslims controlled south of the Italian Peninsula for forty years and briefly controlled western coast of the Italian Peninsula. Muslims remained in control of southern Spain until the fall of Grenada in 1492 AD. Muslims also controlled East Europe under the Golden Horde Empire in 1313 AD. They remained in Crimea, in southern Ukraine, until 1796 AD, when the Muslim Crimean Khanate was captured by the Russian Empire.

Muslims controlled the Balkan Peninsula for several centuries, starting with the conquest of Istanbul in 1453, under the Ottoman Empire. They kept going north until they besieged Vienna in 1528 and 1683. Muslims were defeated in the second attempt and kept retreating south since then. However, there are several Muslim majority countries that remain in the Balkans today: Albania, Bosnia, East Thrace (Turkey) and Kosovo. All Mediterranean Islands were under Muslim control at some point: The Balearics (903–1232), Crete (827–961, 1645–1897), Corsica (806–930), Rhodes (653–658, 717–718, 1522–1912), Sardinia (809–1015), Sicily (831–1091),

---

The World Muslim Population: Spatial and Temporal Analyses
Houssain Kettani
Copyright © 2020 Jenny Stanford Publishing Pte. Ltd.
ISBN 978-981-4800-31-0 (Hardcover), 978-0-429-42253-9 (eBook)
www.jennystanford.com

and Malta (870–1091). As for Cyprus, it was listed under Asia, and controlled by Muslims 647–965, 1426–1460, 1518–1914, and the northern third of the island is under Muslim control since 1974.

Thus, the Muslim population has likely changed from 24,000 or 0.1% of the total European population in 700 AD, to 0.90 million or 2.5% in 800 AD, to 1.86 million or 5.2% in 900 AD, to 2.4 million or 6.6% in 1000 AD, to 2.7 million or 6.0% in 1100 AD, to 2.9 million or 5.3% in 1200 AD, to 3.0 million or 4.8% in 1300 AD, to 2.7 million or 3.7% in 1400 AD, to 2.7 million or 3.2% in 1500 AD, to 3.1 million or 2.9% in 1600 AD, to 3.6 million or 2.9% in 1700 AD, to 4.1 million or 2.2% in 1800 AD, to 8.7 million or 2.1% in 1900, to 33 million or 4.6% in 2000, to 47 million or 6.3% in 2020, and is projected to reach 58 million or 8% by 2050 and 74 million or 12% by 2100.

A plot of centennial estimates of the Muslim population and its percentage with respect to the total population in Europe from 600 to 2100 is provided in Fig. 4.0a. A zoom in of this plot, providing a plot of decennial estimates of the Muslim population and its percentage with respect to the total population in this continent from 1900 to 2100 is provided in Fig. 4.0b. This shows that the Muslim population in Europe was increasing slowly until 1950 but has increased substantially since then and towards the end of this century. The percentage of Muslims on the other hand was around 2% until 1960, and then started increasing by about one percentage point each decade as is expected to continue doing so towards the end of this century, reaching 12% by 2100.

Islam in Europe is a matter of diversity and necessity. Western Europe started having migrant workers from Muslim countries after WWII to stimulate its economy and cover for shortage of manpower. Thus, Muslims were instrumental in the economic prosperity of Western Europe and Europe at large. In addition, as more Europeans leave Christianity, Muslims present Europeans with another spiritual life, in which many Europeans find satisfaction and answers to their long puzzling questions. In this context, Islam may be a cause of stability and prosperity to Europe. Understanding the history and appreciating the numbers of Muslims in Western Europe may help in developing better relation between Christians and Muslims in this region and worldwide, away from political slogans, ethnic slurs or misuse of religion for worldly gains.

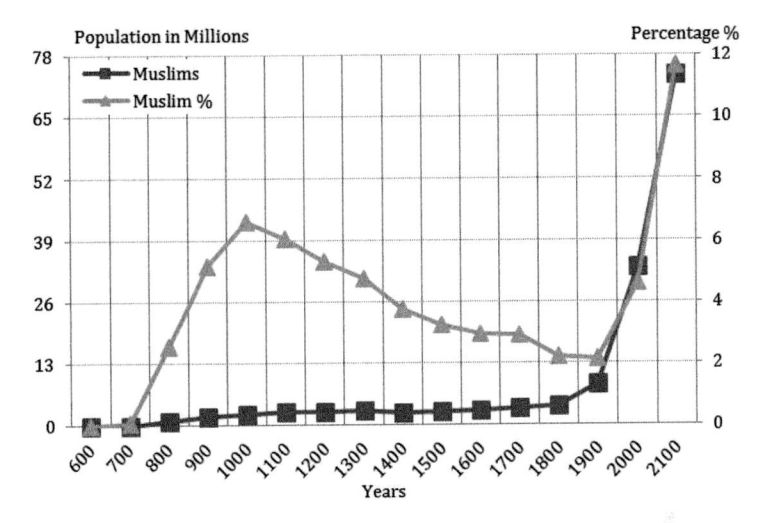

Figure 4.0a Plot of centennial estimates of the Muslim population and its percentage of the total population in Europe: 600–2100 AD (1–1500 H).

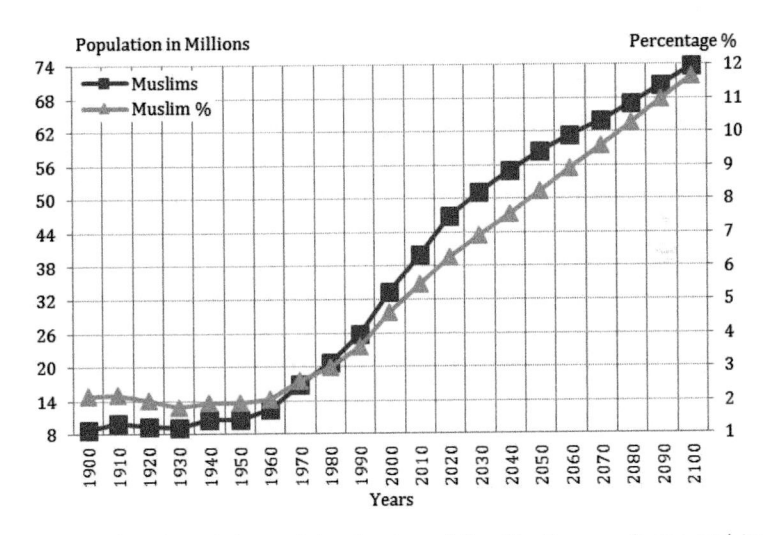

Figure 4.0b Plot of decennial estimates of the Muslim population and its percentage of the total population in Europe: 1900–2100 AD (1320–1520 H).

We divided Europe into five regions; the data for each is included in a separate section and are sorted in terms of the percentage of Muslims in descending order. These regions are the Balkan Peninsula

(Section 4.1), Eastern Europe (Section 4.2), Western Europe (Section 4.3), Northern Europe (Section 4.4), and Central Europe (Section 4.5). The country of Russia was included in Europe as most of its population lives in the European side of the country, although most of its territory is in Asia. In addition, East Thrace, which is the European side of Turkey was included in Asia and therefore excluded from Europe. In Section 4.6, the total population in each of the five European regions and the corresponding percentage and number of Muslims is presented centennially in Table 4.6a from 600 to 2100 and decennially in Tables 4.6b and 4.6c from 1790 to 2100.

4.1 Muslims in the Balkan Peninsula

This region consists of eleven countries: Albania, Bosnia and Herzegovina, Bulgaria, Croatia, Greece, Kosovo, Macedonia, Montenegro, Romania, Serbia, and Slovenia. Three of these countries (Albania, Kosovo, and Bosnia) are the only Muslim majority countries in Europe. Three other countries (Macedonia, Montenegro, and Bulgaria) have the largest Muslim minority in Europe in terms of percentage. This region has the largest concentration of Muslims in Europe. Islam is deeply rooted here, unlike the rest of Europe, where current Muslim population can only trace itself few decades ago at most to Muslim migrants that off-sat labor shortage in Europe after WWII. Hence, the case of Muslims in the Balkans provides Europe (a traditionally Christian majority region with various Christian denominations) with an interface and gate to understanding and collaborating with the Muslim world (a loose term referring to almost a fourth of the world population, spread worldwide with population majority in almost half of the world's land). Muslims in the Balkans were a significant minority for over half a millennium and are expected to remain so for at least the next several centuries. Muslims' relationship with Christians in this region does impact Muslim-Christian relationship worldwide to the better or to the worst. So far, the relationship has been tense, with several massacres against Muslims throughout the centuries. Understanding the history and appreciating the numbers of Muslims in the Balkans may help in developing better relation between Christians and Muslims in this region and worldwide, away from political slogans, ethnic slurs or misuse of religion for worldly gains.

The first Muslims settled here in the second half of the thirteenth century. They first consisted of Anatolian Turkoman settlers in Dobrudja, Romania. Then Tatars when the Golden Horde Empire, which controlled Bulgaria, Romania, and vast areas of Eastern Europe and Central Asia converted to Islam in 1313 under Sultan Mohammed Öz-Beg who reigned from 1313 to 1341. Islam expanded further with the advances of the Ottoman Empire. This started during the reign of Sultan Orhan I ben Osman I (ruled from 1324 to 1361), when Muslims conquered Gallipoli Peninsula in 1347, currently in East Thrace; the European side of Turkey. This part is not included here as it is treated under Asia. Muslims continued their advance conquering northern third of Greece and Macedonia in 1371, Albania in 1385, Kosovo in 1389, most of Bulgaria by 1393, Romania in 1411 to 1419, Istanbul in 1453, Serbia by 1459, rest of Greece by 1460, Bosnia and Herzegovina by 1482 and Croatia in 1493. Most of the Peninsula remained under Ottoman (Muslim) control for several centuries.

Thus, the Muslim population in this region remained non-existent until the thirteenth century. It then has likely increased from 16,000 or 0.3% in 1300 AD, to 0.10 million or 1.8% in 1400 AD, to 0.36 million or 5.8% in 1500 AD, to 0.91 million or 11.9% in 1600 AD, to 1.44 million or 17.3% in 1700 AD, then changed to 2.0 million or 13.6% in 1800 AD, to 3.7 million or 11.4% in 1900, to 8.2 million or 12.7% in 2000, to 7.7 million or 13.0% in 2020, and is projected to reach seven million or 14% by 2050 but decrease to below five million or 12% by 2100.

A plot of centennial estimates of the Muslim population and its percentage with respect to the total population in this region from 600 to 2100 is provided in Fig. 4.1a. A zoom in of this plot, providing a plot of decennial estimates of the Muslim population and its percentage with respect to the total population in this region from 1900 to 2100 is provided in Fig. 4.1b. The corresponding individual data for each country in this region is discussed below. In Section 4.1.12, the total population in each country in this region and the corresponding percentage and number of Muslims is presented centennially in Table 4.1a from 600 to 2100 and decennially in Tables 4.1b and 4.1c from 1790 to 2100.

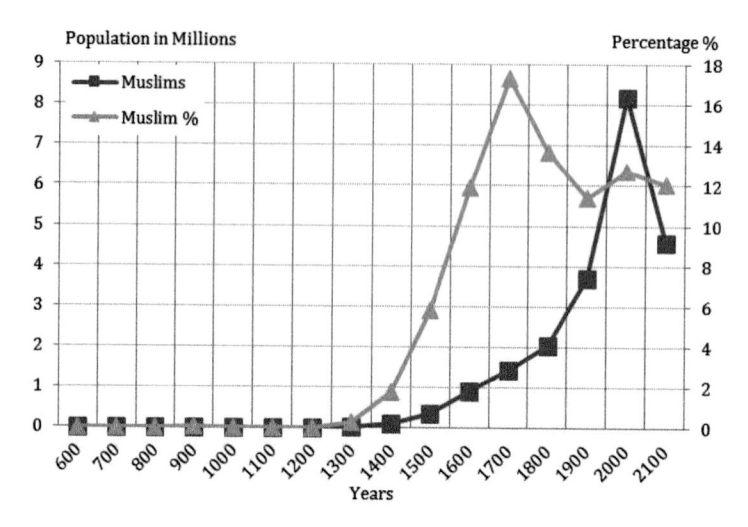

Figure 4.1a Plot of centennial estimates of the Muslim population and its percentage of the total population in the Balkan Peninsula: 600–2100 AD (1–1500 H).

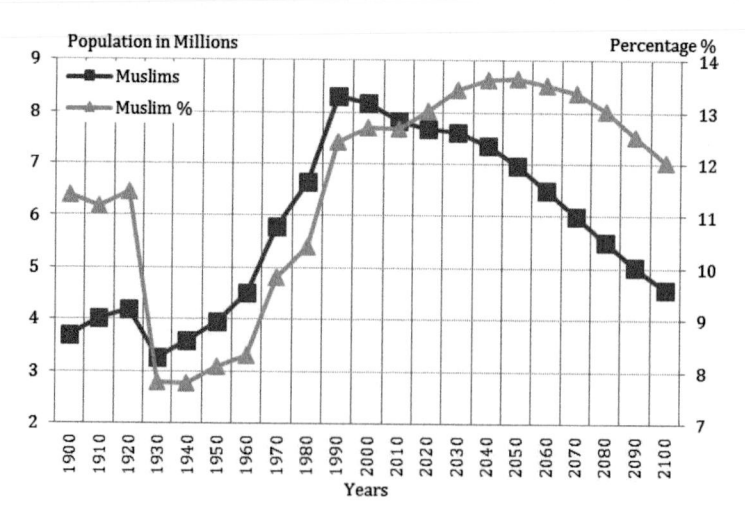

Figure 4.1b Plot of decennial estimates of the Muslim population and its percentage of the total population in the Balkan Peninsula: 1900–2100 AD (1320–1520 H).

4.1.1 Albania

Islam entered Albania when the Ottoman Empire conquered the lands in 1478 during the reign of Sultan Mohammed II el Fatih ben Murad II. Eventually the majority of Albanians chose Islam as their religion. Albania then declared its independence from the weak Ottoman Empire in 1912. It was then attacked by neighboring countries, which swallowed more than 60% of the new country. Today, historical Albania is contained in Kosovo, and parts of Greece, Serbia, Macedonia, and Montenegro. Currently, the Republic of Albania has an area of 28,748 sq. km and its map is presented in Fig. 4.1.1.

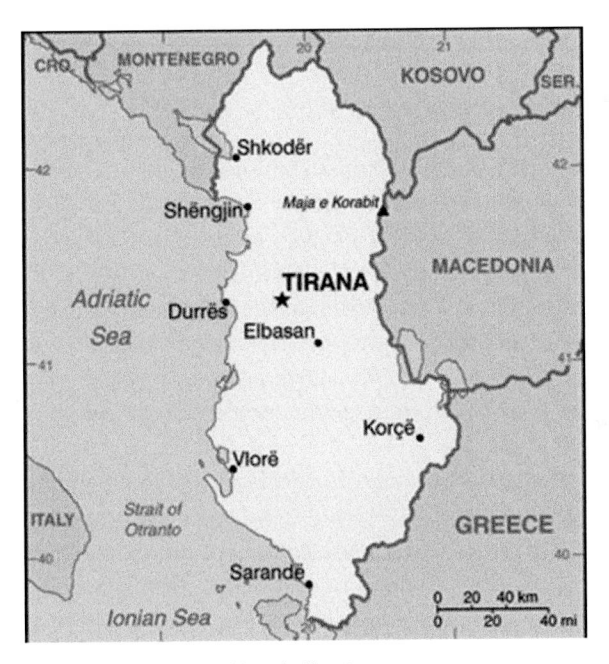

Figure 4.1.1   Map of the Republic of Albania.

According to census data, the Muslim population changed from 0.58 million or 70.3% in 1921, to 0.56 million or 68.6% in 1923, to 0.69 million or 68.6% in 1930, to 0.78 million or 69.1% in 1942, to 1.65 million or 70.2% in 2011. Thus, assuming that the percentage of Muslims will increase by one percentage point per decade, the Muslim population is expected to decrease to 1.8 million or 74%

by 2050 and 0.9 million or 79% by 2100. A summary of the data is provided in Table 4.1.1.

Table 4.1.1 Evolution of the Muslim population in Albania

Year	Population	Muslims	%	Source
1921	831,877	584,675	70.28	[MAS23]c
1923	817,378	560,348	68.55	[ROD]c
1930	1,003,068	688,280	68.62	[SYB31]c
1942	1,128,143	779,417	69.09	[ROD]c
2011	2,346,092	1,646,236	70.17	[AL11]c
2050	2,424,000	1,794,000	74.00	es
2100	1,088,000	860,000	79.00	es

4.1.2 Bosnia and Herzegovina

Islam entered it when the Ottoman Empire conquered Bosnia in 1463 during the reign of Sultan Mohammed el Fatih, and Herzegovina (southern part) in 1482. This was in response to an appeal by Bogomil residents to avoid religious prosecution by Catholic and Orthodox Churches. When Bosnians were introduced to Islam, they found many similarities in it with their belief, thus they embraced Islam. As the Ottoman Empire started getting weaker, it was forced to abandon Bosnia and Herzegovina to the Austria-Hungary Empire in 1878.

This country was able to declare its independence from crumbling Yugoslavia in 1992. But this decision was met by a bloody war, in which Serbia and Croatia wanted to divide the new country between them. This resulted in systematic ethnic cleansing of the Muslim majority, leading to millions of deaths and refugees. The massacres stopped by 1995 following the Dayton Accord, and the country is now independent with Muslims constituting the largest religious minority. Currently, Bosnia and Herzegovina have an area of 51,209 sq. km. It consists of two governing entities: Federation of Bosnia and Herzegovina (26,111 sq. km or 51% of the total area), and Republika Srpska (24,606 sq. km or 48%), with a third region, the Brčko District (493 sq. km or 1%) being administered by both. A map of the country is presented in Fig. 4.1.2.

Figure 4.1.2 Map of Bosnia and Herzegovina.

Based on census data as shown in Table 4.1.2a, the Muslim population changed from 0.33 million or 35.8% in 1851, to 0.42 million or 32.8% in 1865, to 0.54 million or 51.9% in 1871, to 0.48 million or 45.6% in 1876, to 0.45 million or 38.7% in 1879, to 0.49 million or 36.9% in 1885, to 0.55 million or 35.0% in 1895, to 0.61 million or 32.3% in 1910, to 0.59 million or 31.1% in 1921, to 0.72 million or 30.9% in 1931. This is due to massacres of some Muslims, and migration of others to the Ottoman Empire and Turkey. The percentage then increased to 0.89 million or 33.4% in 1948, decreased to 0.89 million or 32.2% in 1953, then increased constantly since then to 1.12 million or 34.1% in 1961, to 1.48 million or 39.6% in 1971, to 1.63 million or 39.5% in 1981, reaching 1.91 million or 43.8% in the last census of 1991 and 1.79 million or 51.3% in 2013. In 1961, the number of Muslims was taken as those declaring themselves Muslim by nationality (842,248) and those declaring themselves as Yugoslavs (275,883). In 1971 to 1991, only those declaring themselves Muslim by nationality were considered Muslim.

Table 4.1.2a Evolution of the Muslim population in Bosnia and Herzegovina

Year	Population	Muslims	%	Source
1851	916,000	328,000	35.81	[BA876]c
1865	1,278,850	419,628	32.81	[BA876]c
1871	1,042,000	541,000	51.92	[BA876]c
1876	1,053,700	480,000	45.55	[BA876]c
1879	1,158,164	448,613	38.73	[BA91]c
1885	1,336,091	492,710	36.88	[BA91]c
1895	1,568,092	548,632	34.99	[BA91]c
1910	1,898,044	612,137	32.25	[BA91]c
1921	1,889,929	588,247	31.13	[BA91]c
1931	2,323,555	718,079	30.90	[YU31]c
1948	2,666,968	890,094	33.37	[BA91]c
1953	2,847,790	917,720	32.23	[SI53]c
1961	3,277,948	1,118,131	34.11	[BA91]e
1971	3,746,111	1,482,430	39.57	[BA91]c
1981	4,124,256	1,630,033	39.52	[BA91]c
1991	4,354,911	1,905,829	43.76	[BA91]c
2013	3,491,871	1,790,454	51.27	[BA13]c
2050	2,685,000	1,584,000	59.00	es
2100	1,641,000	1,132,000	69.00	es

The independence war in 1992–1995 decreased the total population, but the 2013 census showed that Serbs and Croats were decreased more than Bosnians. The percentage of decrease was 27.3% for Croats, 9.3% for Serbs and 3.5% for Bosnians. Table 4.1.2b shows the distribution of Muslims in the three autonomous regions of Bosnia since 1961. The 2013 census showed that 88% of Muslims live in Federation of Bosnia and Herzegovina (FBiH) which is an increase from 52% in 1991 and 43% in 1961. The ratio of Muslims living in FBiH of total Muslims in BiH increased from 69% in 1961, to 75% in 1991, to 88% in 2013. On the other hand, in 2013 Muslims made up only 14% of the population in Republika Srpska (RS), which is a decrease from 28% in 1991. The ratio of Muslims living in RS decreased from 29% in 1961, to 24% in 1981, to 23% in 1991, to 10% in 2013.

Table 4.1.2b Evolution of the Muslim population since 1961 in the three autonomous regions comprising Bosnia and Herzegovina (BiH): Federation of Bosnia and Herzegovina (FBiH), Republika Srpska (RS) and Brčko District (BD). *P*: Total population, *M*: Muslim population, *M%*: Percentage of Muslim population, *MR%*: Muslim Ratio in percentage

		1961	1971	1981	1991	2013
FBiH	P	1,795,950	2,191,906	2,370,977	2,720,074	2,191,256
	M	774,318	1,081,214	1,212,555	1,423,593	1,581,868
	M%	43.11	49.33	51.14	52.34	72.19
	MR%	69.25	72.94	74.39	74.81	88.35
RS	P	1,419,046	1,479,434	1,670,511	1,569,332	1,217,832
	M	327,329	371,035	385,044	440,746	172,742
	M%	23.07	25.08	23.05	28.08	14.18
	MR%	29.27	25.03	23.62	23.16	9.65
BD	P	62,952	74,771	82,768	87,627	82,783
	M	16,484	30,181	32,434	38,617	35,844
	M%	26.19	40.36	39.19	44.07	43.30
	MR%	1.47	2.04	1.99	2.03	2.00
BiH	P	3,277,948	3,746,111	4,124,256	4,377,033	3,491,871
	M	1,118,131	1,482,430	1,630,033	1,902,956	1,790,454
	M%	34.11	39.57	39.52	43.48	51.27

Thus, assuming that the percentage of Muslims in Bosnia and Herzegovina will increase by two percentage points per decade, the Muslim population is expected to reach 1.6 million or 59% by 2050, and 1.1 million or 69% by 2100.

4.1.3 Bulgaria

Islam entered Bulgaria when the capital Sofia was conquered by the Ottoman Empire in 1385, then all Bulgaria was under the Ottomans by 1393. The Muslim rule of Bulgaria remained until it declared its independence from the weakened Ottoman Empire in 1908. Currently, the Republic of Bulgaria has an area of 110,879 sq. km and its map is presented in Fig. 4.1.3.

Figure 4.1.3 Map of the Republic of Bulgaria.

As shown in Table 4.1.3, the percentage of the Muslim population decreased by more than a half from 29% in 1881, to 14% in 1910. Muslims continued to decrease due to persecution and migration to Turkey, reaching 11% in 1956. This percentage bounced to 13% in 1992 but decreased to 12% in 2001. The data for 1956–1975 is based on the assumption that all and only Turks and Roma (Gypsy) are Muslim. The number of Turks increased from 656,025 in 1956, to 780,928 in 1965, and then decreased to 730,728. The number of Roma decreased from 197,865 in 1956, to 148,874 in 1965, to 18,323 in 1975.

Indeed, the Muslim population changed from 0.58 million or 28.8% in 1881, to 0.68 million or 21.4% in 1888, to 0.64 million or 19.4% in 1893, to 0.64 million or 17.2% in 1900, to 0.60 million or 15.0% in 1905, to 0.60 million or 13.9% in 1910, to 0.69 million or 14.3% in 1920, to 0.79 million or 14.4% in 1926, to 0.82 million or 13.5% in 1934, to 0.93 million or 13.3% in 1946, to 0.85 million or 11.2% in 1956, to 0.93 million or 11.3% in 1965, to 0.75 million or 8.6% in 1975, to 1.11 million or 13.1% in 1992, to 0.97 million or 12.6% in 2001, to 0.58 million or 10.8% in 2011.

In the 2011 census, more than a fifth of the population did not respond to the religion question. This explains the decrease in total population in Table 4.1.3, which is only the number of those responding to the religion question. The distribution of Muslims among ethnic groups of those who responded to the religion question in the 2011 census was as follows: 444,434 or 88% of Turks, 42,201 or 18% of Roma and 67,350 or 2% of Bulgars. In other words, 77% of Muslims were Turks, 12% were Bulgar, and 7% were Roma. Thus, assuming that the percentage of Muslims will remain fixed at 10.8%, the Muslim population is expected to continue to decrease to 0.6 million by 2050 and 0.4 million by 2100.

Table 4.1.3 Evolution of the Muslim population in Bulgaria

Year	Population	Muslims	%	Source
1881	2,007,919	578,060	28.79	[BG]c
1888	3,154,375	676,215	21.44	[BG]c
1893	3,310,713	643,258	19.43	[ROD]c
1900	3,744,283	643,300	17.18	[BG]c
1905	4,035,575	603,867	14.96	[BG]c
1910	4,337,513	602,078	13.88	[BG]c
1920	4,846,954	690,734	14.25	[BG]c
1926	5,478,740	789,296	14.41	[BG]c
1934	6,077,939	821,298	13.51	[BG]c
1946	7,029,349	934,418	13.29	[BG]c
1956	7,613,709	853,890	11.22	[BG11]e
1965	8,227,046	929,802	11.30	[BG11]e
1975	8,727,771	749,051	8.58	[BG11]e
1992	8,487,317	1,112,331	13.11	[BG]c
2001	7,645,592	966,978	12.65	[UN]c
2011	5,348,403	577,139	10.79	[BG11]c
2050	5,385,000	581,000	10.79	es
2100	3,588,000	387,000	10.79	es

4.1.4 Croatia

Islam entered Croatia when it was conquered by the Ottoman Empire in 1526 after the Battle of Mohács. But Muslims lost control of it to the Austrian Empire in 1593 after the Battle of Sisak. Between 1868 and 1918 it formed a nominally autonomous kingdom of Croatia-Slovenia that included current Croatia without the Dalmatia region (coastal area) which was mostly controlled by Austria, and the eastern part of bordering Serbia and half of the border with Hungary, which was controlled by Hungary until 1920. It then joined the Kingdom of Serbs, Croats and Slovenes, which became Kingdom of Yugoslavia in 1929. The Republic of Croatia gained its independence from Yugoslavia in 1991 and currently has an area of 56,594 sq. km and its map is presented in Fig. 4.1.4.

Figure 4.1.4 Map of the Republic of Croatia.

As shown in Table 4.1.3 and based on census data, the Muslim population in the Kingdom of Croatia-Slovenia increased from none in 1857, to 19 in 1869 to 73 in 1880 but decreased to 39 in 1890. In addition, Austrian census recorded several Muslims living in areas currently in Croatia but were outside Croatia-Slovenia Kingdom. These are four in the region of Istria in 1880, but none in the other censuses, while Muslim population in Dubrovnik was none in 1869, four in 1880, two in 1890 and nine in 1900. In the Croatian part that was under Hungary, twelve Muslims were recorded in 1869 (one in Vinkovci, five in Našice and six in Osijek), then none in 1880.

In 1910, the Muslim population increased to twenty in Croatian Silesia, seventy in Croatian Dalmatia and fourteen in Fiume district (Rijeka) which was under Hungary. This makes the total of Muslims in current lands occupied by Croatia 308 or 0.01% in 1910.

Table 4.1.4 Evolution of the Muslim population in Croatia

Year	Population	Muslims	%	Source
1857	1,628,890	—	0.00	[EUH]c
1869	1,838,198	31	0.00	[EUH,AT890,HU870]c
1880	2,506,228	81	0.00	[EUH, AT880]c
1890	2,854,558	41	0.00	[EUH. AT890]c
1910	3,460,584	308	0.01	[HRH,AT910,HU910]c
1921	3,360,320	3,015	0.09	[YU21]c
1931	3,430,270	4,750	0.14	[HRH]c
1948	3,779,858	1,712	0.05	[HR81]e
1953	3,918,800	16,185	0.41	[HRH]c
1961	4,159,696	20,799	0.50	[HR81]e
1971	4,426,221	22,632	0.51	[HR81]e
1981	4,601,469	29,746	0.65	[HR81]e
1991	4,699,097	54,814	1.17	[HR01]c
2001	4,411,586	56,777	1.29	[HR01]c
2011	4,272,429	62,977	1.47	[HR11]c
2050	3,365,000	77,000	2.30	es
2100	2,183,000	72,000	3.30	es

Accordingly, the Muslim population in current Croatia increased from 294 or 0.01% in 1910, 3,000 or 0.1% in 1921, to 4,750 or 0.1% in 1931, then decreased to 0.05% in 1948. Indeed, the Muslim population increased from 200 or 0.01% in 1910, to 3,000 or 0.1% in 1921, to 5,000 or 0.1% in 1931, to 2,000 or 0.1% in 1948. It continued its steady increase since then to 16,000 or 0.4% in 1953, to 21,000 or 0.5% in 1961, to 23,000 or 0.5% in 1971, to 30,000 or 0.7% in 1981, to 55,000 or 1.2% in 1991, to 57,000 or 1.3% in 2001, to 63,000 or 1.5% in 2011. Thus, assuming that the percentage of Muslims will continue to increase by a fifth of a percentage point per decade, the Muslim population is expected to remain less than 0.1 million throughout this century, comprising 2.3% of the total population by 2050 and 3.3% by 2100.

When inferring religious adherence from ethnicity in 1948, 1961, 1971 and 1981, the Muslims by nationality, Albanians and Yugoslav (1961 only) ethnicities were assumed Muslim. The distributions are as follows:

- 1948: 1,077 Muslims and 633 Albanians;
- 1961: 3,113 Muslims, 2,126 Albanians, and 16,964 Yugoslav;
- 1971: 18,457 Muslims and 4,175 Albanians;
- 1981: 23,740 Muslims and 6,006 Albanians.

The 2011 census included for the first time a mapping between ethnicity and religion. It indicated that Muslims comprised of 27,959 or 88.82% of ethnic Bosniacs, 9,647 or 0.25% of ethnic Croats, 9,647 or 54.78% of ethnic Albanians, 5,039 or 29.68% of ethnic Roma and 343 or 93.46% of ethnic Turks.

4.1.5 Greece

Muslims controlled Rhodes Island between 653 and 658 during the reign of Caliph Muawiya bnu Abi Sufyan, in their failed attempt to conquer Constantinople. It was also captured from 717 to 718. Muslims conquered Crete from the Byzantine Empire from 827 to 961 by expelled Muslims from Cordoba. They were expelled by an Umayyad Caliph in al-Andalus, and they were led by Abu Hafs Omar el Ballooti. During this time, the majority of the Island was Muslim. However, after the Muslims' defeat, the Muslim population was forced to migrate out of the Island or become Christian, and therefore no Muslims left in the Island.

Current Greece fell under the Ottoman Empire between 1380 and 1718. The Macedonian (Northern) part was conquered in 1380, Thessaly (Central part) in 1393, and the rest of the mainland in 1500, then the islands of Rhodes in 1522, then Crete between 1645 and 1718. Greece then gained its independence in 1828, which included the middle part, Peloponnese (Morea) peninsula, and some islands, with a total area of 47,516 sq. km. At this time Muslims constituted 30% of the total population. This percentage reduced to 1% by 1889 due to migration of Muslims to the Ottoman Empire fearing harassments and discrimination by Greek Orthodox. The new country expanded its area to 50,211 sq. km by annexing the Ionian Islands including Corfu in 1864 (ceded by the UK), to 63,606 sq. km by annexing Thessaly in 1881, to 71,942 sq. km by annexing Crete in 1898, to 129,281 sq. km by annexing parts of Macedonia, Epirus, Thrace (west) and Aegean islands in 1913, to 131,944 by annexing the Dodecanese Islands including Rhodes in 1947 (ceded by Italy). This doubled the size and population of Greece and increased the Muslims to 25% by 1920 census.

Following the treaty of Lausanne, a massive population exchange took place between Greece and Turkey based on their religious belief and regardless of their ethnicity and mother tongue. This resulted in over a million Muslims moving from Greece to Turkey, and similar number of Greek Orthodox moved from Turkey to Greece. This reduced the percentage of Muslims in Greece to 2% by 1928 census. The treaty excluded Istanbul and West Thrace from the exchange, and that is why the Muslim presence in Greece was not reduced to zero. In 1900, the island of Crete had a total population of 303,543, out of which 33,496 or 11% was Muslim, but they were expelled to Turkey, reducing its Muslim population to zero. Currently, the Hellenic Republic has an area of 131,957 sq. km and its map is presented in Fig. 4.1.5.

Thus, the Muslim population was reduced from 0.23 million or 30% in 1828, to 1,000 or 0.1% in 1870, increased to 25,000 or 1.3% in 1882 after the annexation of Thessaly, then decreased to 24,000 or 1.1% in 1889, then increased to 1.4 million or 25.3% in 1920 with the annexation of new lands, then reduced to 0.13 million or 2.0% in 1928 as a result of the deportation of Muslims to Turkey, and 0.14 million or 1.9% in 1940. The Muslim population then remained at 0.11 million, reducing in percentage with respect to the total population to 1.5% in 1951 and 1.4% in 1961, which was the last census to inquire on the religious affiliation of the population. According to the European Values Survey (EVS), the percentage of Muslims was 1.78% in 2008. Thus, assuming that the percentage of Muslims will continue to increase by a tenth of a percentage point per decade, the Muslim population is expected to remain less than 0.2 million throughout this century. A summary of the data is provided in Table 4.1.5.

Figure 4.1.5 Map of the Hellenic Republic.

Table 4.1.5 Evolution of the Muslim population in Greece

Year	Population	Muslims	%	Source
1828	753,400	226,000	30.00	[KET86]c
1870	1,457,026	917	0.06	[EUH]c
1879	1,679,775	740	0.04	[EUH]c
1882	1,979,305	25,243	1.28	[SYB85]c
1889	2,188,008	24,165	1.10	[SYB10]c
1920	5,541,474	1,400,000	25.26	[KET76]c
1928	6,204,684	126,017	2.03	[GR51]c
1940	7,460,203	141,090	1.89	[GR51]c
1951	7,632,801	112,665	1.48	[GR51]c
1961	8,388,553	114,955	1.37	[KET76]c
2008	11,162,000	199,000	1.78	[EVS]s
2050	9,029,000	199,000	2.20	es
2100	6,583,000	178,000	2.70	es

4.1.6 Kosovo

Islam entered Kosovo when it was conquered by the Ottoman Empire in 1389 under the reign of Sultan Murad ben Orhan (ruled from 1363 to 1389 and was the third king of the empire). It was then conquered by Serbs in 1912, following the declaration of independence of Albania from the Ottomans. As the Yugoslav Federation started disintegrating, Kosovars yearned for independence. Following ethnic cleansings by Orthodox Serbs, and international intervention, Kosovo was able to obtain its independence in 2008, and is now a majority Muslim country. Currently, the Republic of Kosovo has an area of 10,887 sq. km and its map is presented in Fig. 4.1.6.

Figure 4.1.6 Map of the Republic of Kosovo.

As shown in Table 4.1.6a and based on ethnic census data, the Muslim population increased in number, but its percentage fluctuated due to migration of Orthodox Serbs to Kosovo. Accordingly, Muslims increased 0.33 million or 75.1% in 1921, to 0.38 million or 68.8% in 1931, to 0.51 million or 70.0% in 1948, to 0.54 million or 67.3% in 1953, to 0.68 million or 70.6% in 1961. However, the

Muslims population continued to increase constantly since then in both number and percentage to 0.95 million or 76.8% in 1971, to 1.30 million or 81.9% in 1981, to 1.67 million or 85.5% in 1991, just before the breakup of Yugoslavia. By 2011, Muslims reached 1.66 million or 96.1% of the total population due to the migration of Serbs back to Serbia and higher birth rate by Muslims.

When inferring religious adherence from ethnicity in 1948, 1961 and onward, the following were assumed Muslims: Albanians, Turks, Gypsy (Roma) and Muslims by nationality or Bosniac. The numbers of these ethnicities recorded in each census is provided in Table 4.1.6b. Thus, assuming that the percentage of Muslims will continue to increase by a tenth of a percentage point per decade, the Muslim population is expected to reach 1.6 million or 96.5% by 2050 and 1.0 million or 97.0% by 2100.

Table 4.1.6a Evolution of the Muslim population in Kosovo

Year	Population	Muslims	%	Source
1921	439,010	329,502	75.06	[YU]c
1931	552,064	379,981	68.83	[YU31]c
1948	727,820	509,236	69.97	[YU]e
1953	808,141	544,182	67.34	[YU]c
1961	963,988	680,595	70.60	[YU]e
1971	1,243,693	954,769	76.77	[YU]e
1981	1,584,440	1,297,811	81.91	[YU]e
1991	1,956,196	1,672,707	85.51	[YU]e
2011	1,730,117	1,663,412	96.14	[KO]c
2050	1,666,000	1,607,000	96.50	es
2100	1,016,000	985,000	97.00	es

Table 4.1.6b Evolution of the largest Muslim ethnicities in Kosovo since 1948

	1948	1953	1961	1971	1981	1991	2011
Albanian	498,242	524,559	646,605	916,168	1,226,736	1,596,072	1,616,869
Muslim/Bosniac	9,679	6,241	8,226	26,357	58,562	66,189	27,533
Roma	11,230	11,904	3,202	14,593	34,126	45,745	8,824
Turk	1,315	34,583	25,764	12,244	12,513	10,446	18,738
Total	520,466	577,287	683,797	969,362	1,331,937	1,718,452	1,671,964

4.1.7 Montenegro

Islam entered Montenegro when the Ottoman Empire conquered the lands in 1452. Muslims lost control of Montenegro when its residents rebelled and declared independence in 1697. Currently, Montenegro has an area of 13,812 sq. km and its map is presented in Fig. 4.1.7.

Figure 4.1.7 Map of Montenegro.

As shown in Table 4.1.7a, and based on census data, the Muslim population increased from 14,000 or 6.1% in 1895, to 56,000 or 18.0% in 1921, to 61,000 or 17.0% in 1931, but decreased to 20,000 or 5.3% in 1948, but bounced back to 74,000 or 17.7% in 1953, then decreased to 56,000 or 12.0% in 1961, then peaked in percentage to 106,000 or 20.0% in 1971 then continued to decrease in percentage to 116,000 or 19.8% in 1981, to 118,000 or 20.1% in 1991, to 110,000 or 17.9% in 2003, and then increased to 118,000 or 20.2% in 2011.

The censuses of 1895, 1921, 1931, 1953, 1991, and on did inquire on the religion. The decrease of the percentage of Muslims in 1948 and 1961 ethnic censuses is due the fact that many Muslims declared themselves as Montenegrins. Muslim ethnicity was introduced in 1961 census and onward, which makes deduction of religious affiliation from ethnic affiliation more accurate.

Table 4.1.7a Evolution of the Muslim population in Montenegro

Year	Population	Muslims	%	Source
1895	227,831	13,840	6.07	[SYB00]c
1921	311,341	55,978	17.98	[VV]c
1931	360,044	61,038	16.95	[YU31]c
1948	377,189	19,812	5.25	[YU]e
1953	419,873	74,140	17.66	[YU]c
1961	471,894	56,468	11.97	[YU]e
1971	529,604	105,907	20.00	[YU]e
1981	584,310	115,815	19.82	[YU]e
1991	585,775	118,016	20.15	[YU]c
2003	615,136	110,034	17.89	[ME03]c
2011	587,442	118,477	20.17	[ME11]c
2050	589,000	124,000	21.00	es
2100	454,000	100,000	22.00	es

When inferring religious adherence from ethnicity in 1948, 1961, 1971, and 1981, Albanians and Muslims by nationality were assumed Muslim. The numbers of these ethnicities recorded in each census is provided in Table 4.1.7b. In Yugoslavia times, Bosnians used to declare themselves as Muslim by nationality, and Bosnian ethnicity was first collected in the 2003 census. So as the number of declared Bosnians increased, the number of Muslim nationals decreased. Table 4.1.7b also shows that the difference between inferring religious adherence based on these ethnic affiliations and the result of the religion question in post-independence censuses. But it seems from 1953 and 1991 censuses which collected religious information; that Muslims made up between 18% and 19% of the total population throughout the second half of the twentieth century. Thus, assuming that the percentage of Muslims will increase by a fifth of a percentage point per decade, the Muslim population is expected to remain around 0.1 million throughout this century, comprising 21% of the total population by 2050 and 22% by 2100.

Table 4.1.7b Evolution of the largest Muslim ethnicities in Montenegro since 1948

	1948	1953	1961	1971	1981	1991	2003	2011
Bosniac	NA	NA	NA	NA	NA	NA	48,184	54,396
Albanian	19,425	23,460	25,803	35,671	37,735	40,415	31,163	30,439
Muslims	387	6,424	30,665	70,236	78,080	89,614	24,625	20,871
Total		19,812	29,884	56,468	105,907	115,815	130,029	103,972	105,706

4.1.8 North Macedonia

Islam entered Macedonia when it was conquered by the Ottoman Empire between 1371 and 1395 mostly under the reign of Sultan Murad ben Orhan (ruled from 1363 to 1389 and was the third king of the empire). The Muslim rule ended in 1913 after the Balkan Wars. However, it was then divided among Bulgaria, Greece and Serbia. Eventually parts of Macedonia became one of the six socialist republics forming the Socialist Federal Republic of Yugoslavia in 1963. It then gained its independence in 1992 forming the Republic of Macedonia with an area of 25,713 sq. km; which is less than half of historical Macedonia. The country changed its name in 2019 to the Republic of North Macedonia. Its map is presented in Fig. 4.1.8.

As shown in Table 4.1.8a and based on census data, the Muslim population decreased from 32.4% in 1921, to 24.1% in 1961, due to wars and hardships targeting the Muslim population. The situation improved after the formation of Yugoslavia, prompting a constant increase to 25.1% in 1971, to 28.6% in 1981, to 30.9% in 1991 just before the breakup of Yugoslavia. The percentage of Muslims continued to increase after independence to 30.4% in 1994, and reaching 33.3% in 2002, due to the settlement of Muslim refugees from Bosnia and Kosovo, who fled the ethnic cleansing by Orthodox Serbs. This was the percentage of Muslims ninety years earlier, just before the Ottoman Empire lost control over Macedonia.

The data for 1921, was calculated as follows: total population from [MK], Muslim population from data in [YU21] for Serbia and Montenegro (which included current Kosovo and Macedonia), minus data in [VV] for current Serbia, Kosovo and Montenegro. When inferring religious adherence from ethnicity in 1948 to 1991, the following were assumed Muslims: Albanians, Turks,

Gypsy (Roma) and Muslims by nationality. The numbers of these ethnicities recorded in each census is provided in Table 4.1.8b. In Yugoslavia times, Bosnians used to declare themselves as Muslim by nationality, and Bosnian ethnicity was first collected in the 1994 census. So as the number of declared Bosnians increased, the number of Muslim nationals increased. Table 4.1.8b also shows that the difference is minimal between inferring religious adherence based on these ethnic affiliations and the result of the religion question in post-independence censuses. Thus, assuming that the percentage of Muslims will continue to increase by half of a percentage point per decade, the Muslim population is expected to reach 0.7 million or 35.5% by 2050 and 0.5 million or 38.0% by 2100.

Table 4.1.8a Evolution of the Muslim population in North Macedonia

Year	Population	Muslims	%	Source
1921	808,724	261,804	32.37	[VV,YU21]c
1931	949,958	282,820	29.77	[YU31]c
1948	1,152,986	314,603	27.29	[MK]e
1953	1,304,514	390,949	29.97	[ROD]c
1961	1,406,003	338,200	24.05	[MK]e
1971	1,647,308	414,176	25.14	[MK]e
1981	1,909,136	546,437	28.62	[MK]e
1991	1.997,209	611,326	30.92	[UN04]c
1994	1,910,339	581,203	30.42	[UN04]c
2002	2,022,547	674,015	33.33	[UN]c
2050	1,857,000	659,000	35.50	es
2100	1,249,000	475,000	38.00	es

Table 4.1.8b Evolution of the largest Muslim ethnicities in North Macedonia since 1948

	1948	1953	1961	1971	1981	1991	1994	2002
Albanian	197,389	162,524	183,108	279,871	377,208	441,987	441,104	509,083
Turkish	95,940	203,938	131,484	108,552	86,591	77,080	78,019	77,959
Roma	19,500	20,462	20,606	24,505	43,125	52,103	43,707	53,879
Bosniac	NA	NA	NA	NA	NA	NA	6,829	17,018
Muslims	1,560	1,591	3,002	1,248	39,513	31,356	15,418	2,553
Total	314,389	388,515	338,200	414,176	546,437	602,526	585,077	660,492

Figure 4.1.8 Map of the Republic of North Macedonia.

4.1.9 Romania

The first Muslims settled down in Dobruja in 1263, which is an area between the Black Sea and the Danube River shared between Romania and Bulgaria. The group consisted of over 10,000 Anatolian Turkomans led by Sari Saltik [ROH]. The first group of Tatars settled here as part of the Golden Horde Empire during the time of its leader Noghai (1280–1313). The whole empire came under Muslim control when Sultan Mohammed Öz-Beg accepted Islam before taking the throne of the Golden Horde and ruled it from 1313 to 1341. He adopted Islam as the state's religion and continued the spread Islam among the Turkic people. The leaders of the Golden Horde remained Muslim afterwards. The Ottoman Empire conquered Dobruja in 1411, then Wallachia (which includes the current Capital Bucharest) in 1416, then Transylvania (Northwest of current Romania) in 1419. Romania remained under Muslim control until it declared its independence from the ailing Ottoman Empire in 1878. Currently,

Romania has an area of 238,391 sq. km and its map is presented in Fig. 4.1.9.

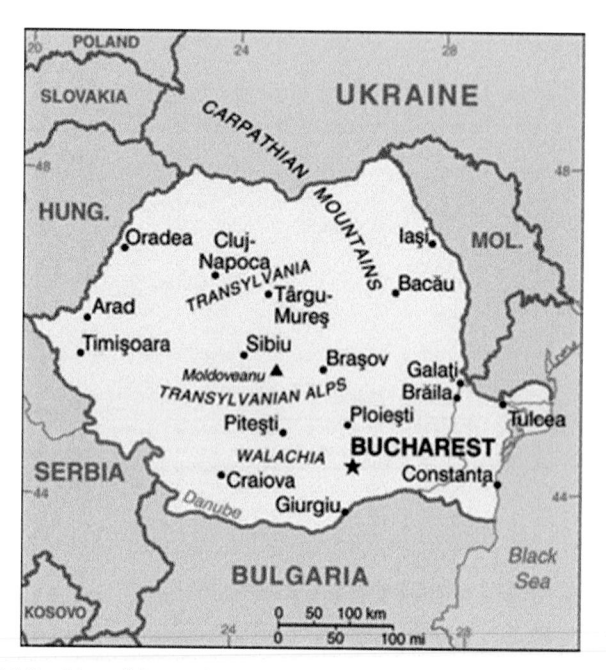

Figure 4.1.9 Map of Romania.

The birth of current Romania started with the union of Wallachia (71,225 sq. km) and Moldavia (46,988 sq. km) in 1859, covering less than half of current Romania, and had 1,323 Muslims or 0.03% of its population in 1859. The Muslim population increased to 45,000 or 0.3% in 1899 after the annexation of Northern Dobruja (15,500 sq. km) in 1878; the Romanian lands on the Black Sea and referred to in Romania as Dobrogea. The Muslim population then increased to 46,000 or 0.6% in 1912. Romania occupied Southern Dobruja (7,565 sq. km) from Bulgaria between 1913 and 1940. By the end of WWI in 1918, Romania almost doubled its size by acquiring Southern Bukovina (10,438 sq. km) from Austria (northern part is with Ukraine), Banat (28,513 sq. km), Transylvania (57,788 sq. km), eastern part of Crișana (20,818 sq. km) and southern part of Maramureș (16,208 sq. km) from Hungary and Bessarabia (44,408 sq. km) from Russia. The census of 1930 counted 185,486 Muslims or 1.0% out of the total population of 18,057,028. Most of these or 129,025 were living in Southern Dobruja which is part of Bulgaria since 1940 and is referred to in Bulgaria as Dobrudzha. Another 148 was in Basarabia, which is now Moldova and parts of Ukraine. Northern Dobruja had 38,132 Muslims. Thus, in current Romania, there were 43,000 or 0.3% in 1930.

Table 4.1.9a shows the census data on the number of Muslims in the historical regions of Romania. These historical regions that were ruled by different countries are:

- Dobrogea or Southern Dobruja is now organized as the counties of Constanța and Tulcea, with a total area of 15,570 sq. km. The vast majority of Muslims live in the south of this region or Constanța County. The Muslim population here changed from 48,000 or 34% in 1880, to 42,000 or 22% in 1890, to 40,000 or 16% in 1900, to 41,000 or 11% in 1912, to 38,000 or 9% in 1930, to 52,000 or 5% in 1992 and 2002, to 47,000 or 6% in 2011. Thus, the Muslim population remained similar in number but decreased significantly in percentage with respect to the rest of the population in Dobrogea.

- Wallachia is now organized as Bucharest Municipality and the counties of Argeș, Brăila, Buzău, Călărași, Dâmbovița, Dolj, Giurgiu, Gorj, Ialomița, Ilfov, Mehedinți, Olt, Prahova, Teleorman and Vâlcea, with a total area of 76,381 sq. km. The number of Muslims here changed from 4,000 in 1930, to 2,800 in 1992, 11,500 in 2002, to 13,000 in 2011. Most are living in the capital city of Bucharest.

- Moldavia is now organized as the counties Bacău, Botoșani, Galați, Iași, Neamț, Vaslui, and Vrancea, with a total area of 37,620 sq. km. The number of Muslims in Wallachia and Moldavia remained negligible in the nineteenth century, increasing from 1,300 in 1860, to 4,100 in 1890, to 4,300 in 1900, to 5,000 in 1912. The number of Muslim changed from 500 in 1930, to 300 in 1992, to 1,100 in 2002, to 1,200 in 2011.

- Transylvania (including Banat, Eastern Crișana and Southern Maramureș) is now organized as the counties of Alba, Arad,

Bihor, Bistriţa-Năsăud, Braşov, Caraş-Severin, Cluj, Covasna, Harghita, Hunedoara, Maramureş, Mureş, Satu Mare, Sălaj, Sibiu, and Timiş with a total area of 100,287 sq. km. The Hungarian census showed that the Muslim population here increased from none in 1850, to 22 in 1870, to four in 1880, to 110 in 1910. The presence of Muslims remains negligible in this region, however. The Romanian census showed an increase to 419 in 1930, then 534 in 1992, then 2,425 in 2002 and 3,416 in 2011.

- Southern Bukovina is mostly under Soceava County with a total area of 8,553 sq. km. The Austrian census showed that the Muslim population here increased from none in 1850, to twelve in 1880, to three in 1890 and 1900, to six in 1910. The presence of Muslims remains negligible in this region, however. The Romanian census showed a slight increase to 49 in 1930, thirteen in 1992, eighty in 2002 and 81 in 2011.

The combined data for Romania in its current borders is summarized in Table 4.1.9b. Accordingly, the Muslim population in Romania changed from 49,000 or 0.6% in 1880, to 46,000 or 0.5% in 1890, to 45,000 in 1900, to 47,000 or 0.4% in 1912, to 43,000 or 0.3% in 1930. Due to communism rule, religious affiliation was not inquired in censuses between 1948 and 1977, but ethnic affiliation was collected. Thus, we infer religious adherence from ethnicity in 1948 to 1977, by assuming all and only Turks and Tatar were Muslim. The numbers of both ethnicities per census year are summarized in Table 4.1.9c. Accordingly, Muslims decreased further to 0.2% from 1948 to 1977, numbering between 29,000 and 47,000. They remained at 0.3% since 1992, numbering 56,000 in 1992, then 67,000 in 2002, and 64,000 in 2011.

The 2002 and 2011 Censuses showed that 97% of Turks and 99% or of Tatar are Muslim. In addition, over two-thirds of all Muslims live in Constanta County on the Black Sea, 72% in 2002, and 67% in 2011. Thus, assuming that the percentage of Muslims will continue to increase by 0.01 of a percentage point per decade, the Muslim population is expected to remain just over 50,000 or around 0.4% of the total population throughout the second half of this century.

Muslims in the Balkan Peninsula | 435

Table 4.1.9a Evolution of the Muslim population in historical regions of Romania

Region		1850	1860	1870	1880	1890	1900	1910	1930	1992	2002	2011
Northern Dobruja (Dobrogea)	P				139,671	192,000	258,242	380,430	447,810	1,019,766	971,392	824,360
	M				48,100	42,000	40,383	41,442	38,132	52,303	52,158	46,569
	M%				34.44	21.88	15.64	10.89	8.52	5.13	5.37	5.65
Wallachia & Moldavia	P		4,424,961			4,846,342	5,654,278	6,854,890	5,473,307	9,280,754	8,802,513	7,701,358
	M		1,323			4,082	4,349	4,964	3,905	2,753	11,520	13,037
	M%		0.03			0.08	0.08	0.07	0.07	0.03	0.13	0.17
Moldavia	P								2,367,415	4,084,372	3,992,409	3,330,738
	M								477	325	1,074	1,234
	M%								0.02	0.01	0.03	0.04
Transylvania	P	3,646,424		4,224,436	4,032,851	4,429,564	4,840,722	5,262,495	5,549,806	7,723,313	7,214,749	6,394,745
	M	—		22	4			110	419	534	2,425	3,416
	M%	0.00		0.00	0.00			0.00	0.01	0.01	0.03	0.05
Southern Bukovina (Suceava)	P	210,281			259,384	291,215	330,708	381,541	472,111	701,830	688,177	610,701
	M	—			12	3	3	6	49	13	80	81
	M%	0.00			0.00	0.00	0.00	0.00	0.01	0.00	0.01	0.01

Table 4.1.9b Evolution of the Muslim population in Romania

Year	Population	Muslims	%	Source
1880	8,856,867	49,439	0.56	[EUH,HU880,AT880]c
1890	9,759,121	46,085	0.47	[EUH,HU890,AT890]c
1899	11,083,950	44,735	0.40	[EUH,HU900,AT900]c
1912	12,879,356	46,522	0.36	[RO11,HU910,AT910]c
1930	14,280,729	42,982	0.30	[RO30]c
1948	15,872,624	28,782	0.18	[RO77]e
1956	17,489,450	34,798	0.20	[RO77]e
1966	19,103,163	40,191	0.21	[RO77]e
1977	21,559,910	46,791	0.22	[RO77]e
1992	22,791,565	55,928	0.25	[RO11]c
2002	21,669,240	67,257	0.31	[RO11]c
2011	18,861,902	64,337	0.34	[RO11]c
2050	16,260,000	59,000	0.36	es
2100	11,878,000	49,000	0.41	es

Table 4.1.9c Evolution of the largest Muslim ethnicities in Romania since 1930 (current borders)

	1930	1956	1966	1977	1992	2002	2011
Turk	26,080	14,329	18,040	23,422	29,832	32,596	27,698
Tatar	15,580	20,469	22,151	23,369	24,596	24,137	20,282
Total	41,660	34,798	40,191	46,791	54,428	56,733	47,980

4.1.10 Serbia

Islam entered Serbia when the Ottoman Empire conquered Belgrade in 1452. As the Ottoman Empire started getting weaker, Muslims lost control of Serbia to the Austrian Empire in 1718 but was regained in 1738. Muslims then lost control for good when Serbs rebelled and declared their independence in 1830. The Vojvodina region was under Hungary between 1848 and 1918, after which it was annexed by Serbia.

Currently, the Republic of Serbia has an area of 77,474 sq. km and its map is presented in Fig. 4.1.10. It consists of Proper Serbia (55,968 sq. km) and the Autonomous Province of Vojvodina (21,506 sq. km) in the north. Censuses in both territories were conducted regularly since mid-eighteenth century. For Vojvodina (Table 4.1.10a), Muslim adherence was collected by the Hungarians until the 1910 census, and later collected by Yugoslavia in the censuses of 1921, 1931, 1953, 1991 and onward by the Serbs. For Central Serbia or Serbia Proper (Table 4.1.10c), Islam was asked on the censuses between 1884 and 1921, and in 1953, 1991, and 2002. When inferring religious adherence from ethnicity in 1948, 1961, 1971, and 1981, the Albanians and Muslims by nationality were assumed Muslim. The recorded numbers of both ethnicities in censuses since 1948 are summarized in Table 4.1.10b for Vojvodina and Table 4.1.10d for Serbia Proper.

Figure 4.1.10 Map of Republic of Serbia.

The presence of Muslims in Vojvodina was negligible until the end of WWI and its subsequent succession from Hungary and joining Yugoslavia. Accordingly, the number of Muslims increased from none in 1870, to one in 1880, to thirteen in 1910. The Muslim percentage of the population remained low though, increasing from around 0.1% in 1921–1948, to 0.2% in 1953, to 0.3% in 1961, to 0.4% in 1971 and 1981, to 0.5% in 1991, and then decreased to 0.4% in 2002, but increased again to 0.8% in 2011.

Table 4.1.10a Evolution of the Muslim population in Vojvodina

Year	Population	Muslims	%	Source
1870	1,100,000	—	0.00	[HU870]c
1880	1,172,729	1	0.00	[HU880]c
1890	1,331,143	1	0.00	es
1900	1,432,748	1	0.00	es
1910	1,515,304	13	0.00	[HU910]c
1921	1,535,794	1,870	0.12	[YU]c
1931	1,624,158	1,654	0.10	[YU]c
1948	1,663,212	1,530	0.09	[VV]e
1953	1,712,619	3,254	0.19	[YU]c
1961	1,854,956	3,624	0.20	[VV]e
1971	1,952,533	6,577	0.34	[VV]e
1981	2,034,772	8,742	0.43	[VV]e
1991	1,802,544	9,775	0.54	[YU]c
2002	1,887,972	8,073	0.43	[YU]c
2011	1,806,864	14,206	0.79	[RS]c

As for Serbia Proper, The Muslim population increased from 0.8% to 0.5% between 1884 and 1910, to 3% in 1921, decreased again to 1% in 1948, then remained between 3% and 4% from 1953 to 1991, and reached 4.4% in 2002, but decreased to 4.1% in 2011.

Thus, combining the data for Vojvodina and Central Serbia, we get the data in Table 4.1.10e for the Republic of Serbia in its current borders. Accordingly, the Muslim population increased from around 6,000 or 0.3% in 1874, to 15,000 or 0.5% in 1884, to 17,000 or 0.5% in 1891, but decreased to 15,000 or 0.4% in 1900 and 14,000 or 0.3% in 1910. The Muslim population then increased to 98,000 or

2.2% in 1921, to 114,000 or 2.2% in 1931, then decreased to 41,000 or 0.7% in 1948, then remained between 2.4% and 3.0% from 1953 to 1991, numbering between 0.12 million and 0.23 million, and reached 0.24 million or 3.3% in 2002, but decreased to 0.22 million or 3.2% in 2011. Thus, assuming that the percentage of Muslims will increase by a tenth of a percentage point per decade, the Muslim population is expected to increase from 0.2 million or 3.5% in 2050 to 0.3 million or 4% by 2100.

Table 4.1.10b Evolution of the largest Muslim ethnicities in Vojvodina since 1948

	1948	1953	1961	1971	1981	1991	2002	2011
Bosniac							417	780
Albanian	480	965	1,994	3,086	3,812	2,556	1,695	2,251
Muslims	1,050	10,537	1,630	3,491	4,930	5,851	3,634	3,360
Total	1,530	11,502	3,624	6,577	8,742	8,407	5,746	6,391

Table 4.1.10c Evolution of the Muslim population in Serbia Proper

Year	Population	Muslims	%	Source
1874	1,353,898	6,176	0.46	[EUH]c
1884	1,937,172	14,569	0.75	[SYB90]c
1891	2,161,905	16,764	0.78	[SYB00]c
1900	2,492,882	14,745	0.59	[SYB10]c
1910	2,911,701	14,435	0.50	[SYB20]c
1921	2,856,897	95,819	3.35	[YU]c
1931	3,503,925	111,967	3.20	[YU31]c
1948	4,136,934	39,875	0.96	[YU]e
1953	4,458,394	152,403	3.42	[YU]c
1961	4,823,208	134,984	2.80	[YU]e
1971	5,250,170	189,989	3.62	[YU]e
1981	5,694,464	224,158	3.94	[YU]e
1991	5,527,467	214,345	3.88	[YU]c
2002	5,275,707	231,585	4.39	[YU]c
2011	5,059,549	208,622	4.12	[RS]c

Table 4.1.10d Evolution of the largest Muslim ethnicities in Serbia Proper since 1948

	1948	1953	1961	1971	1981	1991	2002	2011
Bosniac							135,670	144,498
Albanian	33,289	39,989	51,173	65,507	72,484	75,725	59,952	3,558
Muslims	6,586	64,303	83,811	124,482	151,674	174,371	15,869	18,941
Total	39,875	104,292	134,984	189,989	224,158	250,096	211,491	166,997

Table 4.1.10e Evolution of the Muslim population in Serbia

Year	Population	Muslims	%	Source
1874	2,453,898	6,176	0.25	[EUH]c
1884	3,109,901	14,570	0.47	[SYB90]c
1891	3,493,048	16,765	0.48	[SYB00]c
1900	3,925,630	14,746	0.38	[SYB10]c
1910	4,427,005	14,448	0.33	[SYB20]c
1921	4,392,691	97,689	2.22	[YU]c
1931	5,128,083	113,621	2.22	[YU31]c
1948	5,800,146	41,405	0.71	[YU]e
1953	6,171,013	155,657	2.52	[YU]c
1961	6,678,164	138,608	2.08	[YU]e
1971	7,202,703	196,566	2.73	[YU]e
1981	7,729,236	232,900	3.01	[YU]e
1991	7,330,011	224,120	3.06	[RS]c
2002	7,163,679	239,658	3.35	[RS]c
2011	6,866,413	222,828	3.25	[RS]c
2050	5,891,000	206,000	3.50	es
2100	6,819,000	273,000	4.00	es

4.1.11 Slovenia

The Ottoman Empire reached Slovenia in 1566 but did not conquer it, except for the northeastern most Prekmurje region, which was under Ottoman control from 1566 to 1688. It was under Austrian rule from 1797 until 1918 when it joined the Kingdom of Serbs, Croats and Slovenes, which became Kingdom of Yugoslavia in 1929. The Republic of Slovenia gained its independence from Yugoslavia in 1991 and currently has an area of 20,273 sq. km and its map is presented in Fig. 4.1.11.

Figure 4.1.11 Map of Republic of Slovenia.

The 1900 Austrian census recorded one Muslim in the lands currently occupied by Slovenia, which was mostly the region of Carniola (Krain) and the southern part of Styria (Steiermark) region, while the previous censuses showed no Muslims in these lands. The 1910 census recorded thirteen Muslims in the city of Koper and two Muslims in the city of Rudolfswert.

The Muslim population in current Slovenia remained at 0.1% or less until 1971, when it increased to 5,000 or 0.3% due to migration from other Yugoslav states for economic reasons. It continued to increase since then to 15,000 or 0.8% in 1981, to 29,000 or 1.9% in 1991, and 47,000 or 3.1% in 2002, doubling in size each decade. In 2011, Slovenia switched to register-based census, which does not collect religious information. The census data is summarized in Table 4.1.11a.

Table 4.1.11a Evolution of the Muslim population in Slovenia

Year	Population	Muslims	%	Source
1869	1,128,768	—	0.00	[AT890]c
1880	1,182,223	—	0.00	[AT880]c
1890	1,234,056	—	0.00	[AT890]c
1900	1,268,055	1	0.00	[AT900]c
1910	1,321,098	15	0.00	[AT910]c
1921	1,054,919	649	0.06	[YU21]c
1931	1,144,298	927	0.08	[YU31]c
1948	1,391,873	395	0.03	[SI91]e
1953	1,466,425	668	0.05	[SI53]c
1961	1,591,523	747	0.05	[SI91]e
1971	1,679,051	4,512	0.27	[SI91]e
1981	1,838,381	15,410	0.84	[SI91]e
1991	1,553,486	29,361	1.89	[SI02]c
2002	1,516,966	47,488	3.13	[SI02]c
2050	1,940,000	81,000	4.20	es
2100	1,676,000	87,000	5.20	es

When inferring religious adherence from ethnicity in 1948, 1961, 1971, and 1981, the Albanians and Muslims by nationality were assumed Muslim. The recorded numbers of both ethnicities in censuses since 1948 are summarized in Table 4.1.11b. Thus, assuming that the percentage of Muslims will continue to increase by a fifth of a percentage point per decade, the Muslim population is expected to remain below 0.1 million throughout this century, comprising 4.2% of the total population by 2050 and 5.2% by 2100.

Table 4.1.11b Evolution of the largest Muslim ethnicities in Slovenia since 1948

	1948	1953	1961	1971	1981	1991	2002
Bosniac							21,542
Albanian	216	169	282	1,266	1,933	3,534	6,186
Muslims	179	1,617	465	3,197	13,339	26,577	10,467
Total	395	1,786	747	4,463	15,272	30,111	38,195

4.1.12 Regional Summary and Conclusion

The Balkan Peninsula was the second European region to come under Muslim control, as early as the fourteenth century, and most of it remained under Muslim control for over five centuries. The Balkans has the largest concentration of Muslim in Europe, with the only three majority Muslim countries in Europe and a fourth of which of third is Muslim. However, the ratio of Muslims is expected to remain around one eighth of the total population throughout this century. The following tables present centennial data from 600 AD to 2100 AD (or approximately 1 H to 1500 H) in Table 4.1a and decennial data from 1790 AD to 2100 AD (or 1210 H to 1520 H) in Tables 4.1b and 4.1c for current countries in the Balkan Peninsula. The data includes total population in thousands (P), the percentage of which is Muslim (M%), the corresponding Muslim population in thousands (M), and the annual population growth rate (APGR, or G%) of the total population in this region.

Table 4.1a Centennial estimates of the Muslim population (×1000) in the Balkan Peninsula: 600–2100 AD (1–1500 H)

		600	700	800	900	1000	1100	1200	1300
Albania	P	200	200	200	200	200	200	200	200
	M%	—	—	—	—	—	—	—	—
	M	—	—	—	—	—	—	—	—
Bosnia	P	270	274	278	282	286	300	315	330
	M%	—	—	—	—	—	—	—	—
	M	—	—	—	—	—	—	—	—
Bulgaria	P	680	710	740	770	800	800	800	800
	M%	—	—	—	—	—	—	—	—
	M	—	—	—	—	—	—	—	—
Croatia	P	388	394	400	406	412	430	450	470
	M%	—	—	—	—	—	—	—	—
	M	—	—	—	—	—	—	—	—
Greece	P	1,400	1,300	1,200	1,100	1,000	1,000	1,000	1,000
	M%	—	—	—	0.10	—	—	—	—
	M	—	—	—	1	—	—	—	—
Kosovo	P	80	81	81	82	82	85	90	95
	M%	—	—	—	—	—	—	—	—
	M	—	—	—	—	—	—	—	—
Montenegro	P	41	42	42	43	43	46	49	52
	M%	—	—	—	—	—	—	—	—
	M	—	—	—	—	—	—	—	—
N. Macedonia	P	130	131	131	132	132	140	148	156
	M%	—	—	—	—	—	—	—	—
	M	—	—	—	—	—	—	—	—
Romania	P	800	800	800	800	800	1,050	1,300	1,550
	M%	—	—	—	—	—	—	—	1.00
	M	—	—	—	—	—	—	—	16
Serbia	P	594	603	612	621	630	660	690	720
	M%	—	—	—	—	—	—	—	—
	M	—	—	—	—	—	—	—	—
Slovenia	P	150	152	154	156	158	167	176	185
	M%	—	—	—	—	—	—	—	—
	M	—	—	—	—	—	—	—	—
Total	P	4,733	4,687	4,638	4,592	4,543	4,878	5,218	5,558
	M%	0.00	0.00	0.00	0.02	0.00	0.00	0.00	0.28
	M	—	—	—	1	—	—	—	16
	G%		−0.010	−0.011	−0.010	−0.011	0.071	0.067	0.063

Table 4.1a (*Continued*)

		1400	1500	1600	1700	1800	1900	2000	2100
Albania	P	200	200	200	300	410	800	3,129	1,088
	M%	—	20.00	40.00	60.00	70.28	70.28	70.10	79.00
	M	—	40	80	180	288	562	2,194	860
Bosnia	P	345	367	449	449	700	1,568	3,751	1,641
	M%	—	10.00	20.00	30.00	38.73	34.99	48.00	69.00
	M	—	37	90	135	271	549	1,801	1,132
Bulgaria	P	800	800	1,250	1,250	2,050	4,315	7,998	3,588
	M%	—	10.00	20.00	30.00	28.79	17.18	12.65	10.79
	M	—	80	250	375	590	741	1,012	387
Croatia	P	490	530	647	647	1,450	3,162	4,428	2,183
	M%	—	—	0.01	0.01	—	0.00	1.29	3.30
	M	—	—	—	—	—	—	57	72
Greece	P	1,000	1,000	1,500	1,500	2,175	4,962	11,082	6,583
	M%	5.00	10.00	20.00	30.00	30.00	25.26	1.70	2.70
	M	50	100	300	450	652	1,253	188	178
Kosovo	P	100	105	129	129	100	349	1,900	1,016
	M%	10.00	20.00	40.00	60.00	75.06	75.06	96.14	97.00
	M	10	21	52	77	75	262	1,827	985
Montenegro	P	55	55	67	67	120	300	614	454
	M%	—	1.00	2.00	3.00	6.07	6.07	17.89	22.00
	M	—	1	1	2	7	18	110	100
N. Macedonia	P	164	170	208	208	390	787	2,035	1,249
	M%	5.00	10.00	20.00	30.00	32.37	32.37	33.33	38.00
	M	8	17	42	62	126	255	678	475
Romania	P	1,800	2,000	2,000	2,500	5,500	11,000	22,137	11,878
	M%	2.00	3.00	4.00	5.00	0.56	0.40	0.31	0.41
	M	36	60	80	125	31	44	69	49
Serbia	P	750	810	990	990	1,450	4,000	5,335	6,819
	M%	—	1.00	2.00	3.00	0.25	0.38	3.35	4.00
	M	—	8	20	30	4	15	179	273
Slovenia	P	194	203	248	248	600	1,268	1,988	1,676
	M%	—	—	0.01	0.01	—	0.00	3.13	5.20
	M	—	—	—	—	—	—	62	87
Total	P	5,898	6,240	7,688	8,288	14,945	32,511	64,397	38,177
	M%	1.77	5.82	11.89	17.33	13.68	11.38	12.70	12.04
	M	104	363	914	1,436	2,045	3,700	8,176	4,598
	G%	0.059	0.056	0.209	0.075	0.590	0.777	0.711	−0.646

Table 4.1b Decennial estimates of the Muslim population (×1000) in the Balkan Peninsula: 1790–1940 AD (1210–1360 H)

		1790	1800	1810	1820	1830	1840	1850	1860
Albania	P	400	410	420	437	450	470	500	550
	M%	70.28	70.28	70.28	70.28	70.28	70.28	70.28	70.28
	M	281	288	295	307	316	330	351	387
Bosnia	P	690	700	710	720	800	900	1,100	1,220
	M%	38.73	38.73	38.73	38.73	38.73	38.73	38.73	38.73
	M	267	271	275	279	310	349	426	473
Bulgaria	P	2,000	2,050	2,100	2,187	2,200	2,300	2,500	2,550
	M%	28.79	28.79	28.79	28.79	28.79	28.79	28.79	28.79
	M	576	590	605	630	633	662	720	734
Croatia	P	1,400	1,450	1,500	1,550	1,600	1,650	1,700	2,182
	M%	—	—	—	—	—	—	—	—
	M	—	—	—	—	—	—	—	—
Greece	P	2,011	2,175	2,207	2,312	2,534	2,777	3,044	3,336
	M%	30.00	30.00	30.00	30.00	30.00	30.00	30.00	30.00
	M	603	652	662	694	760	833	913	1,001
Kosovo	P	90	100	110	120	130	140	160	180
	M%	75.06	75.06	75.06	75.06	75.06	75.06	75.06	75.06
	M	68	75	83	90	98	105	120	135
Montenegro	P	110	120	130	140	150	160	170	180
	M%	6.07	6.07	6.07	6.07	6.07	6.07	6.07	6.07
	M	7	7	8	8	9	10	10	11
N. Macedonia	P	380	390	400	410	420	440	400	450
	M%	32.37	32.37	32.37	32.37	32.37	32.37	32.37	32.37
	M	123	126	129	133	136	142	129	146
Romania	P	5,000	5,500	6,000	6,389	7,000	7,500	8,000	8,500
	M%	0.56	0.56	0.56	0.56	0.56	0.56	0.56	0.56
	M	28	31	34	36	39	42	45	48
Serbia	P	1,400	1,450	1,500	1,550	1,600	1,650	1,700	2,278
	M%	0.25	0.25	0.25	0.25	0.25	0.25	0.25	0.25
	M	4	4	4	4	4	4	4	6
Slovenia	P	550	600	650	700	750	800	900	1,102
	M%	—	—	—	—	—	—	—	—
	M	—	—	—	—	—	—	—	—
Total	P	14,031	14,945	15,727	16,515	17,634	18,787	20,174	22,528
	M%	13.94	13.68	13.32	13.20	13.07	13.19	13.48	13.05
	M	1,956	2,045	2,094	2,180	2,306	2,478	2,719	2,939
	G%		0.631	0.510	0.489	0.656	0.633	0.712	1.103

Table 4.1b (Continued)

		1870	1880	1890	1900	1910	1920	1930	1940
Albania	P	603	640	726	800	874	932	982	1,088
	M%	70.28	70.28	70.28	70.28	70.28	70.28	68.62	69.09
	M	424	450	510	562	614	655	674	752
Bosnia	P	1,260	1,158	1,336	1,568	1,898	1,890	2,324	2,450
	M%	38.73	38.73	36.88	34.99	32.25	31.13	30.90	32.23
	M	488	449	493	549	612	588	718	790
Bulgaria	P	2,586	3,155	3,762	4,315	4,980	5,038	5,998	6,624
	M%	28.79	28.79	21.44	17.18	13.88	14.25	14.41	13.51
	M	745	908	807	741	691	718	864	895
Croatia	P	2,398	2,479	2,855	3,162	3,375	3,427	3,789	3,780
	M%	0.00	0.00	0.00	0.00	0.01	0.09	0.14	0.14
	M	—	—	—	—	—	3	5	5
Greece	P	3,657	4,059	4,482	4,962	5,320	5,700	6,351	7,280
	M%	30.00	30.00	30.00	25.26	25.26	25.26	2.03	1.89
	M	1,097	1,218	1,345	1,253	1,344	1,440	129	138
Kosovo	P	200	240	300	349	475	439	552	660
	M%	75.06	75.06	75.06	75.06	75.06	75.06	68.83	67.34
	M	150	180	225	262	357	330	380	444
Montenegro	P	190	207	250	300	344	311	360	370
	M%	6.07	6.07	6.07	6.07	17.98	17.98	16.95	17.66
	M	12	13	15	18	62	56	61	65
N. Macedonia	P	500	528	776	787	876	809	950	1,050
	M%	32.37	32.37	32.37	32.37	32.37	32.37	29.77	29.78
	M	162	171	251	255	284	262	283	313
Romania	P	9,179	9,758	10,373	11,000	11,866	12,340	14,141	15,920
	M%	0.56	0.56	0.47	0.40	0.36	0.30	0.30	0.30
	M	51	55	49	44	43	37	42	48
Serbia	P	2,696	3,083	3,500	4,000	4,673	4,380	5,174	5,500
	M%	0.25	0.47	0.48	0.38	0.33	2.22	2.22	2.52
	M	7	14	17	15	15	97	115	139
Slovenia	P	1,129	1,182	1,234	1,268	1,321	1,305	1,398	1,450
	M%	0.00	0.00	0.00	0.00	0.01	0.06	0.08	0.05
	M	—	—	—	—	—	1	1	1
Total	P	24,398	26,489	29,594	32,511	36,002	36,572	42,018	46,172
	M%	12.85	13.05	12.54	11.38	11.17	11.45	7.79	7.77
	M	3,135	3,457	3,711	3,700	4,022	4,187	3,273	3,589
	G%	0.798	0.822	1.108	0.940	1.020	0.157	1.388	0.943

Table 4.1c Decennial estimates of the Muslim population (×1000) in the Balkan Peninsula: 1950–2100 AD (1370–1520 H)

		1950	1960	1970	1980	1990	2000	2010	2020
Albania	P	1,263	1,636	2,151	2,683	3,286	3,129	2,948	2,878
	M%	69.20	69.40	69.60	69.80	70.00	70.10	70.17	71.00
	M	874	1,135	1,497	1,873	2,300	2,194	2,069	2,043
Bosnia	P	2,661	3,226	3,761	4,180	4,463	3,751	3,705	3,281
	M%	32.23	34.11	39.57	39.52	43.76	48.00	51.27	53.00
	M	858	1,100	1,488	1,652	1,953	1,801	1,900	1,739
Bulgaria	P	7,251	7,886	8,508	8,879	8,841	7,998	7,425	6,948
	M%	13.29	11.22	11.30	8.58	13.11	12.65	10.79	10.79
	M	964	885	961	762	1,159	1,012	801	750
Croatia	P	3,850	4,193	4,423	4,598	4,776	4,428	4,328	4,105
	M%	0.41	0.50	0.51	0.65	1.17	1.29	1.47	1.70
	M	16	21	23	30	56	57	64	70
Greece	P	7,669	8,274	8,664	9,627	10,226	11,082	10,888	10,423
	M%	1.48	1.37	1.40	1.50	1.60	1.70	1.78	1.90
	M	113	113	121	144	164	188	194	198
Kosovo	P	733	964	1,244	1,584	1,956	1,900	1,730	1,804
	M%	67.34	70.60	76.77	81.91	85.51	96.14	96.14	96.20
	M	494	681	955	1,298	1,673	1,827	1,663	1,735
Montenegro	P	395	487	520	581	615	614	624	628
	M%	17.66	11.97	20.00	19.82	20.15	17.89	20.17	20.40
	M	70	58	104	115	124	110	126	128
N. Macedonia	P	1,254	1,489	1,721	1,924	1,996	2,035	2,071	2,083
	M%	30.00	24.05	25.14	28.62	30.92	33.33	33.50	34.00
	M	376	358	433	551	617	678	694	708
Romania	P	16,236	18,614	20,549	22,616	23,489	22,137	20,472	19,238
	M%	0.18	0.20	0.21	0.22	0.25	0.31	0.34	0.33
	M	29	37	43	50	59	69	70	63
Serbia	P	5,999	5,889	5,720	5,479	5,197	5,335	5,579	5,573
	M%	2.52	2.08	2.73	3.01	3.06	3.35	3.25	3.20
	M	151	122	156	165	159	179	181	178
Slovenia	P	1,473	1,587	1,670	1,836	2,006	1,988	2,043	2,079
	M%	0.05	0.05	0.27	0.84	1.89	3.13	3.40	3.60
	M	1	1	5	15	38	62	69	75
Total	P	48,785	54,244	58,928	63,986	66,853	64,397	61,813	59,040
	M%	8.09	8.32	9.82	10.40	12.42	12.70	12.67	13.02
	M	3,946	4,512	5,785	6,654	8,302	8,176	7,830	7,688
	G%	0.551	1.061	0.828	0.824	0.438	−0.374	−0.409	−0.459

Table 4.1c (Continued)

		2030	2040	2050	2060	2070	2080	2090	2100
Albania	P	2,787	2,634	2,424	2,191	1,942	1,660	1,359	1,088
	M%	72.00	73.00	74.00	75.00	76.00	77.00	78.00	79.00
	M	2,007	1,923	1,794	1,643	1,476	1,278	1,060	860
Bosnia	P	3,127	2,923	2,685	2,439	2,213	2,002	1,808	1,641
	M%	55.00	57.00	59.00	61.00	63.00	65.00	67.00	69.00
	M	1,720	1,666	1,584	1,488	1,394	1,302	1,212	1,132
Bulgaria	P	6,417	5,873	5,385	4,923	4,484	4,122	3,847	3,588
	M%	10.79	10.79	10.79	10.79	10.79	10.79	10.79	10.79
	M	692	634	581	531	484	445	415	387
Croatia	P	3,877	3,629	3,365	3,095	2,830	2,584	2,371	2,183
	M%	1.90	2.10	2.30	2.50	2.70	2.90	3.10	3.30
	M	74	76	77	77	76	75	74	72
Greece	P	9,917	9,509	9,029	8,400	7,751	7,249	6,895	6,583
	M%	2.00	2.10	2.20	2.30	2.40	2.50	2.60	2.70
	M	198	200	199	193	186	181	179	178
Kosovo	P	1,821	1,768	1,666	1,510	1,370	1,240	1,119	1,016
	M%	96.30	96.40	96.50	96.60	96.70	96.80	96.90	97.00
	M	1,754	1,704	1,607	1,458	1,325	1,200	1,085	985
Montenegro	P	624	610	589	566	540	510	481	454
	M%	20.60	20.80	21.00	21.20	21.40	21.60	21.80	22.00
	M	129	127	124	120	115	110	105	100
N. Macedonia	P	2,051	1,967	1,857	1,734	1,597	1,461	1,347	1,249
	M%	34.50	35.00	35.50	36.00	36.50	37.00	37.50	38.00
	M	708	688	659	624	583	541	505	475
Romania	P	18,306	17,307	16,260	15,175	14,126	13,256	12,535	11,878
	M%	0.34	0.35	0.36	0.37	0.38	0.39	0.40	0.41
	M	62	61	59	56	54	52	50	49
Serbia	P	5,618	5,731	5,891	6,104	6,300	6,486	6,661	6,819
	M%	3.30	3.40	3.50	3.60	3.70	3.80	3.90	4.00
	M	185	195	206	220	233	246	260	273
Slovenia	P	2,056	2,006	1,940	1,859	1,779	1,720	1,692	1,676
	M%	3.80	4.00	4.20	4.40	4.60	4.80	5.00	5.20
	M	78	80	81	82	82	83	85	87
Total	P	56,602	53,958	51,091	47,996	44,931	42,291	40,116	38,177
	M%	13.44	13.63	13.65	13.53	13.37	13.03	12.53	12.04
	M	7,607	7,354	6,972	6,493	6,008	5,512	5,028	4,598
	G%	−0.422	−0.478	−0.546	−0.625	−0.660	−0.606	−0.528	−0.495

4.2 Muslims in Eastern Europe

This region consists of seven countries: Belarus, Estonia, Latvia, Lithuania, Moldova, Russia, and Ukraine. Islam entered the Caucasus in 22 H or 643 AD during the time of Caliph Omar. Islam then continued spreading among the Turkic peoples, including the conversion of the king of Volga Bulgaria or Khanate of Volga in 310 H or 922 AD, and King of the Golden Horde Empire in 1313. By then, significant parts of current Russia (the region west of a line extending from the short Russian-Chinese border between Kazakhstan and Mongolia to the Arctic Ocean), Moldova, and Ukraine, were under Muslim control. This lasted several centuries until Moscow got stronger and started conquering a Muslim Khanate after another. It annexed most of current Russia by mid to end of the sixteenth century. The last territory held by Muslims was the Crimean Khanate (Southern Ukraine), which fell to Russian control in 1783. Between the sixteenth and mid twentieth centuries, Muslims in this region faced vicious forms of ethnic and religious cleansing by the Russians. This included mass deportations, starvation, and Christianization. Their numbers decreased significantly, but surprisingly they were not annihilated. Islam in this region gained a momentum after the fall of the Soviet Union in 1991.

Thus, the Muslim population has likely increased from 24,000 or 0.7% of the total population of this region in 700 AD, to 50,000 or 1.3% in 800 AD, to 79,000 or 2.0% in 900 AD, to 0.11 million or 2.6% in 1000 AD, to 0.18 million or 3.1% in 1100 AD, to 0.29 million or 3.8% in 1200 AD, to 0.41 million or 4.7% in 1300 AD, to 0.56 million or 5.6% in 1400 AD, to 0.88 million or 7.0% in 1500 AD, to 1.29 million or 8.2% in 1600 AD, to 1.98 million or 9.4% in 1700 AD, to 2.1 million or 5.0% in 1800 AD, to 5.0 million or 4.6% in 1900, to 15 million or 6.9% in 2000, to 17 million or 7.9% in 2020, and is projected to reach 18 million or 9% by 2050 and 19 million or 12% by 2100.

A plot of centennial estimates of the Muslim population and its percentage with respect to the total population in this region from 600 to 2100 is provided in Fig. 4.2a. A zoom in of this plot, providing a plot of decennial estimates of the Muslim population and its percentage with respect to the total population in this region from 1900 to 2100 is provided in Fig. 4.2b. The corresponding individual data for each country in this region is discussed below. In Section 4.2.8, the total population in each country in this region and the corresponding percentage and number of Muslims is presented centennially in Table 4.2a from 600 to 2100 and decennially in Tables 4.2b and 4.2c from 1790 to 2100.

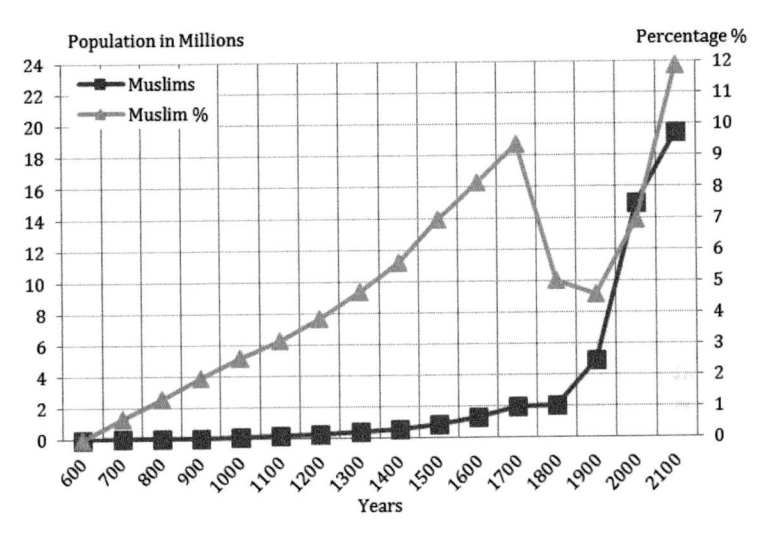

Figure 4.2a Plot of centennial estimates of the Muslim population and its percentage of the total population in Eastern Europe: 600–2100 AD (1–1500 H).

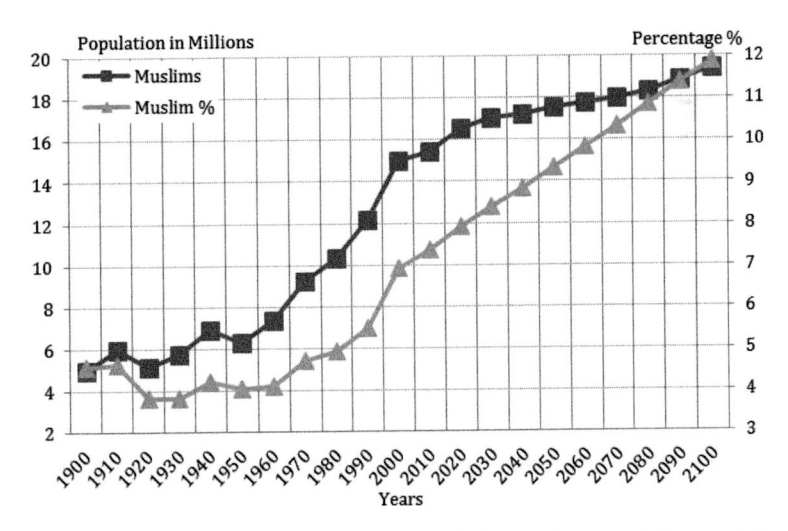

Figure 4.2b Plot of decennial estimates of the Muslim population and its percentage of the total population in Eastern Europe: 1900–2100 AD (1320–1520 H).

4.2.1 Belarus

The Republic of Belarus gained its independence from the Soviet Union in 1991. It has an area of 207,600 sq. km and its map is presented in Fig. 4.2.1. Islam entered Belarus through Muslim Tatars from the Golden Horde Empire in the thirteenth century but remained a small minority of tens of thousands by the end of the sixteenth century.

Figure 4.2.1 Map of the Republic of Belarus.

According to the 1897 Russian Empire census, the Muslim population in the Governorates of Grodno, Minsk, Mogilev, Vilna (Wilensky) and Vitebsk was 13,570 or 0.16% of the total population. About 5% of Minsk and Mogilev governorates lie in Ukraine and Russia, respectively, while the rest lies in Belarus. About a quarter of Grodno Governorate lies in Poland, while the rest lies in Belarus. Vilna is split between Belarus and Lithuania, while Vitebsk is divided equally between Belarus, Latvia, and Russia. Moreover, Belarus is contained in these five governorates. Data of the 1897 census for these governorates is summarized in Table 4.2.1a. For comparison, the table also contains an estimate based on the 1897 of the total population living in the current border of Belarus and the current area of the country.

Censuses since 1926 collected ethnic affiliation demography of the population. The top twenty Muslim ethnicities in Belarus from largest in number are Tatar, Azerbaijani, Uzbek, Turkmen, Kazakh, Arab, Tajik, Bashkir, Lezgin, Persian, Chechen, Avar, Kyrgyz, Darghin, Ingush, Kabardian, Lak, Tabasaran, Adyghei and Karakalpak. The number of members of each of these ethnicities since 1926 is provided in Table 4.2.1b. Based on ethnic censuses, the Muslim population decreased to 3,800 or 0.1% in 1926, then bounced back to 12,300 or 0.2% in 1939, and continued its slow increase to 12,900 or 0.2% in 1959, to 15,900 or 0.2% in 1970, to 20,100 or 0.2% in 1979, to 29,300 or 0.3% in 1989. The Muslim population decreased since then after the breakup of the Soviet Union in 1991 as various ethnicities started returning to their country of origin. Thus, the Muslim population changed to 21,800 or 0.2% in 1999, to 23,000 or 0.2% in 2009. Thus, assuming that the percentage of Muslims will continue to increase by 0.02 of a percentage point, the Muslim population is expected to remain around 30,000 throughout this century, comprising 0.3% of the total population by 2050 and 0.4% by 2100. The data is summarized in Table 4.2.1c.

Table 4.2.1a 1897 Census data for territory covering current Belarus

Governorate	Population	Muslims	% Muslims	Area (km^2)
Grodno	1,603,409	3,731	0.23	38,671
Minsk	2,147,621	4,619	0.22	91,408
Mogilev	1,686,764	184	0.01	48,047
Vilna	1,591,207	4,375	0.27	42,530
Vitebsk	1,489,246	661	0.04	45,169
Total	8,518,247	13,570	0.16	265,826
Belarus	5,803,000			207,600

Table 4.2.1b Evolution of the largest Muslim ethnic populations in Belarus since 1926

	1926	1939	1959	1970	1979	1989	1999	2009
Tatar	3,777	7,664	8,654	10,031	10,911	12,552	10,146	7,316
Azeri	NA	633	1,402	1,335	2,654	5,009	6,362	5,567
Uzbek	NA	885	886	1,606	2,333	3,537	1,571	1,593
Turkmen	1	215	219	197	170	777	921	2,685
Kazakh	NA	1,423	633	1,062	1,355	2,266	1,239	1,355
Arab	—	—	—	24	35	101	490	1,330
Tajik	NA	148	246	256	383	920	NA	871
Bashkir	8	764	346	673	772	1,252	1,091	607
Lezgin	5	83	110	165	244	652	NA	404
Persian	2	2	4	10	12	127	NA	323
Chechen	7	71	53	78	140	298	NA	265
Avar	NA	27	63	76	163	293	NA	149
Kyrgyz	1	142	113	93	458	564	NA	95
Darghin	NA	23	NA	35	60	162	NA	88
Ingush	1	29	19	53	57	116	NA	88
Kabardian	3	106	70	87	118	178	NA	69
Lak	NA	NA	15	47	49	120	NA	64
Tabasaran	NA	1	13	20	36	102	NA	57
Adyghei	NA	113	45	57	57	139	NA	34
Karakalpak	NA	6	43	35	48	152	NA	8
Total	3,805	12,335	12,934	15,940	20,055	29,317	21,820	22,968

Table 4.2.1c Evolution of the Muslim population in Belarus

Year	Population	Muslims	%	Source
1897	8,518,247	13,570	0.16	[SU]c
1926	4,983,240	3,805	0.08	[SU]e
1939	5,568,994	12,335	0.22	[SU]e
1959	8,055,714	12,934	0.16	[BY]e
1970	9,002,338	15,940	0.18	[BY]e
1979	9,532,516	20,055	0.21	[BY]e
1989	10,151,806	29,317	0.29	[BY]e
1999	10,045,237	21,820	0.22	[BY]e
2009	9,503,807	20,453	0.24	[BY]e
2050	8,634,000	28,000	0.32	es
2100	7,430,000	31,000	0.42	es

4.2.2 Estonia

The Republic of Estonia gained its independence from the Soviet Union in 1991 and has an area of 45,225 sq. km and its map is presented in Fig. 4.2.2. Because of the Russian occupation, some Tatar Muslims moved in to Estonia. According to the 1881 census, there were eleven Tatars in the country.

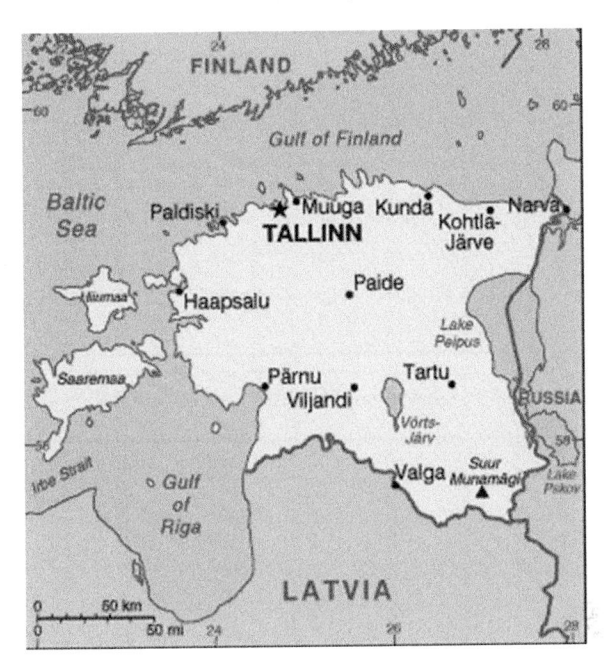

Figure 4.2.2 Map of the Republic of Estonia.

According to the 1897 Russian Empire census, the Muslim population in the Governorates of Estland and Lifland was 611 or 0.04% of the total population. The first governorate lies in current Estonia, while the second is split between Estonia and Latvia. Data of the 1897 census for the two governorates is summarized in Table 4.2.2a. For comparison, the table also contains an estimate based on the 1897 of the total population living in the current border of Estonia and the current area of the country.

Based on ethnic data, the number of Muslims dropped to zero in 1922 as a result of Estonia's independence from the Russian Empire in 1918. The number of Muslims increased later to 182 or 0.02%

in 1934, of which 166 are Tatar and 16 are Turks. Censuses since 1959 collected ethnic affiliation demography of the population. The top twelve Muslim ethnicities in Estonia from largest in number are Tatar, Azerbaijani, Bashkir, Uzbek, Kazakh, Lezghin, Chechen, Turkmen, Tajik, Avar, Kyrgyz and Turk. The number of members of each of these ethnicities since 1959 is provided in Table 4.2.2b. Based on ethnic censuses, the Muslim population increased to 2,972 or 0.25% in 1959, to 3,359 or 0.25% in 1970, to 5,016 or 0.34% in 1979, and peaked at 7,313 or 0.47% in 1989.

Table 4.2.2a   1897 Census data for territory covering current Estonia

Governorate	Population	Muslims	% Muslims	Area (km^2)
Estland	412,716	75	0.02	20,249
Lifland	1,299,365	536	0.04	47,029
Total	1,712,081	611	0.04	67,278
Estonia	676,000			45,225

Table 4.2.2b   Evolution of the largest Muslim ethnic populations in Estonia

	1959	1970	1979	1989	2000	2011
Tatar	1,535	2,205	3,199	4,070	2,582	1,945
Azeri	422	264	543	1,238	880	923
Bashkir	85	126	332	371	152	112
Uzbek	465	224	397	595	132	116
Kazakh	141	195	226	424	127	103
Lezghin	30	52	67	178	121	102
Chechen	34	11	38	45	48	51
Turkmen	7	149	66	106	36	29
Tajik	207	45	52	113	35	33
Avar	19	16	28	69	30	28
Kyrgyz	21	49	46	81	26	20
Turk	6	23	22	23	24	82
Total	2,972	3,359	5,016	7,313	4,193	3,544

According to the 2000 census, which inquired on religious adherence of population aged 15 and older after the independence of Estonia from the Soviet Union in 1991, the Muslim population dropped to 1,387 or 0.2% in 2000 and then increased to 1,508 or 0.2% in 2011. These censuses also showed that only 754 or 29% in 2000 and 604 or 31% in 2011 of Tatars were Muslim. The next largest ethnic affiliation of Muslims was Estonian 83 in 2000 and 148 in 2011, then Russian: 79 in 2000 and 107 in 2011. Thus, assuming that the percentage of Muslims will continue to increase by 0.01 of a percentage point per decades, the Muslim population is expected to remain less than 3,000 or 0.3% throughout this century. A summary of the data is provided in Table 4.2.2c.

Table 4.2.2c Evolution of the Muslim population in Estonia

Year	Population	Muslims	%	Source
1881	880,317	11	0.00	[EEH]e
1897	1,712,081	611	0.04	[SU]c
1922	1,094,859	—	0.00	[EEH]e
1934	1,128,658	182	0.02	[EEH]e
1959	1,196,791	2,972	0.25	[SU]e
1970	1,356,079	3,359	0.25	[SU]e
1979	1,464,476	5,016	0.34	[SU]e
1989	1,565,662	7,313	0.47	[SU]e
2000	778,290	1,387	0.18	[UN]c
2011	911,656	1,508	0.17	[EE]c
2050	1,158,000	2,400	0.22	es
2100	838,000	2,200	0.26	es

4.2.3 Latvia

The Republic of Latvia gained its independence from the Soviet Union in 1991 and has an area of 64,589 sq. km and its map is presented in Fig. 4.2.3. There were only two Muslims in 1863 in the Kurland Governorate. The number of Muslims increased later due to Turkish prisoners of war brought by the Russian Empire from the Russian Turkish war of 1877–1878.

According to the 1897 Russian Empire census, the Muslim population in the Governorates of Kurland, Lifland and Vitebsk was 1,793 or 0.05% of the total population. The first governorate lies in current Latvia; the second is split between Estonia and Latvia, while the third is divided equally between Latvia, Belarus, and Russia. Data of the 1897 census for the three governorates is summarized in Table 4.2.3a. For comparison, the table also contains an estimate based on the 1897 of the total population living in the current border of Latvia and the current area of the country.

Figure 4.2.3 Map of the Republic of Latvia.

Table 4.2.3a 1897 Census data for territory covering current Latvia

Governorate	Population	Muslims	% Muslims	Area (km^2)
Kurland	674,034	596	0.09	27,286
Lifland	1,299,365	536	0.04	47,029
Vitebsk	1,489,246	661	0.04	45,169
Total	3,462,645	1,793	0.05	119,484
Latvia	1,075,000			64,589

The number of Muslims dropped as a result of Latvian independence from the Russian Empire in 1918 and remained less than 0.01% throughout its independence. Accordingly, the number of

Muslims decreased from 115 in 1920, to 103 in 1925, to 81 in 1930, and to 66 in 1935. The numbers for 1920 and 1925 are based on ethnic data while data for 1930 and 1935 are the number of Muslims that was recorded in the census. The ethnic data is as follows: The number of ethnic Tatars was 115 in 1920; 72 in 1925; 43 in 1930; and 39 in 1935. The number of ethnic Turks was 24 in 1925; 28 in 1930; and 38 in 1935. The number of Persians was seven in 1925; eleven in 1930; and seventeen in 1935.

Table 4.2.3b Evolution of the largest Muslim ethnic populations in Latvia

	1959	1970	1979	1989	2000	2011
Tatar	1,836	2,688	3,772	4,888	3,168	2,223
Azeri	324	558	954	2,765	1,700	1,657
Uzbek	776	402	482	925	306	339
Kazakh	143	523	447	1,044	258	241
Bashkir	115	270	476	629	304	215
Lezghin	36	55	113	348	266	193
Chechen	53	56	39	158	135	139
Tajik	88	83	116	343	93	119
Turk	18	12	3	9	NA	81
Kyrgyz	44	31	109	189	NA	52
Avar	20	61	52	121	73	51
Turkmen	67	126	90	228	45	49
Lak	14	11	25	45	42	37
Afghan	NA	1	2	9	39	26
Abkhaz	11	46	26	90	39	25
Total	3,545	4,923	6,706	11,791	6,468	5,447

The Muslim population increased as a result of the Soviet occupation from 1944 to 1991. The Soviets carried out censuses since 1959 that collected ethnic affiliation demography of the population. The top fifteen Muslim ethnicities in Latvia from largest in number are Tatar, Azerbaijani, Uzbek, Kazakh, Bashkir, Lezghin, Chechen, Tajik, Turk, Kyrgyz, Avar, Turkmen, Lak, Afghan and Abkhaz. The number of members of each of these ethnicities since 1959 is provided in Table 4.2.3b. Based on ethnic censuses, the Muslim population increased to 3,545 or 0.2% in 1959, to 4,923 or 0.2% in 1970, to 6,706 or 0.3% in 1979, to 11,791 or 0.4% in 1989. The Muslim population continued to decrease since then after Latvia's independence from the Soviet Union in 1991. Accordingly, the Muslim population decreased to 6,468 or 0.3% in 2000, to 5,447 or 0.26% in 2011. Thus, assuming that the percentage of Muslims will increase by 0.01 of a percentage point per decade, the Muslim population is expected to remain around 4,000 or less than 0.4% of the total population throughout this century. A summary of the data is provided in Table 4.2.3c.

Table 4.2.3c Evolution of the Muslim population in Latvia

Year	Population	Muslims	%	Source
1863	1,499,200	2	0.00	[LVH]c
1897	3,462,645	1,793	0.05	[SU]c
1920	1,596,131	115	0.01	[LVH]e
1925	1,844,805	103	0.01	[LVH]e
1930	1,900,045	81	0.00	[LVH]c
1935	1,950,502	66	0.00	[LVH]c
1959	2,093,458	3,545	0.17	[SU]e
1970	2,364,127	4,923	0.21	[SU]e
1979	2,502,816	6,706	0.27	[SU]e
1989	2,666,567	11,791	0.44	[SU]e
2000	2,377,383	6,468	0.27	[UNE]e
2011	2,070,371	5,447	0.26	[LV]e
2050	1,479,000	4,400	0.30	es
2100	1,114,000	3,900	0.35	es

4.2.4   Lithuania

The Republic of Lithuania gained its independence from the Soviet Union in 1991 and has an area of 65,300 sq. km and its map is presented in Fig. 4.2.4. As the medieval Grand Duchy of Lithuania expanded south in Europe from the thirteenth to the fifteenth centuries, it occupied lands where some Muslim Tatar lived, who came with the Golden Horde Empire. Some Tatar Muslims moved north to the lands of current Lithuania. Under the rule of Grand

Duke Vytautas (reigned 1392 to 1430), more Muslim Tatars were welcomed in the north. They remained welcomed there until the Russian occupation towards the end of the eighteenth century.

Figure 4.2.4 Map of the Republic of Lithuania.

According to the 1897 Russian Empire census, the Muslim population in the Governorates of Kovno, Suvalki and Vilna was 7,081 or 0.19% of the total population. Kovno governorate lies in current Lithuania, with 5% in Belarus and 1% in Latvia. About two-thirds of Suvalki governorate lays in Lithuania, while the rest of it lays in Poland. Vilna Governorate is split between Lithuania and Belarus. The rest of Lithuania is Klaipėdia Region, which was annexed from Prussia in 1923 and therefore the 1897 census was not carried there. It had an area of 2,848 sq. km, and population around 100,000. Data of the 1897 census for the three governorates is summarized in Table 4.2.4a. For comparison, the table also contains an estimate based on the 1897 of the total population living in the current border of Latvia and the current area of the country.

The 1923 Census recorded 1,107 Muslims or 0.05% of the total population. There were 973 Tatars of whom 961 were Muslim, and three Turks, all Christian. Other Muslims were 117 ethnic Lithuanian, twelve ethnic Polish and two ethnic Russian.

Table 4.2.4a 1897 Census data for territory covering current Lithuania

Governorate	Population	Muslims	% Muslims	Area (km²)
Kovno	1,544,564	1,920	0.12	40,642
Suvalki	582,913	786	0.13	12,551
Vilna	1,591,207	4,375	0.27	42,530
Total	3,718,684	7,081	0.19	95,723
Lithuania	1,876,000			65,300

The Muslim population increased because of the Soviet occupation from 1944 to 1991. The Soviets carried out censuses since 1959 that collected ethnic affiliation demography of the population. The top ten Muslim ethnicities in Lithuania from largest in number are Tatar, Azerbaijani, Uzbek, Kazakh, Bashkir, Lezghin, Chechen, Tajik, Turkmen and Kyrgyz. The number of members of each of these ethnicities since 1959 is provided in Table 4.2.4b. Based on ethnic censuses, the Muslim population increased to 4,470 or 0.16% in 1959, to 4,911 or 0.16% in 1970, to 8,651 or 0.26% in 1979 and peaked at 10,055 or 0.27% in 1989, just before the breakup of the Soviet Union. According to the 2001 census which inquired on religious adherence, after the independence of Lithuania from the Soviet Union in 1991, the Muslim population dropped to 2,860 or 0.09% in 2001. This census also showed that only 1,679 or 51.9% of Tatars, and 362 or 45.9% of Azeris are Muslim. The ethnic affiliation of the rest of Muslims was 185 Lithuanian, 74 Russians, and the rest are from other ethnicities. The number of Muslims decreased to 2,700 or 0.10% in 2011. Thus, assuming that the percentage of Muslims will increase by 0.01 of a percentage point per decade, the Muslim population is expected to remain around 3,000 or less than 0.2% of the total population throughout this century. A summary of the data is provided in Table 4.2.4c.

Table 4.2.4b Evolution of the largest Muslim ethnic populations in Lithuania

	1959	1970	1979	1989	2001	2011
Tatar	3,023	3,460	4,006	5,188	3,235	2,793
Azeri	500	711	1,078	1,314	788	648
Uzbek	548	252	2,011	1,453	159	
Kazakh	112	200	567	663	145	
Bashkir	101	171	293	420	136	
Lezghin	35	32	49	112	82	
Chechen	6	11	74	72	54	
Tajik	75	31	207	522	65	
Turkmen	52	28	143	193	NA	
Kyrgyz	18	15	223	118	NA	
Total	4,470	4,911	8,651	10,055	4,664	3,441

Table 4.2.4c Evolution of the Muslim population in Lithuania

Year	Population	Muslims	%	Source
1897	3,718,684	7,081	0.19	[SU]c
1923	2,028,971	1,107	0.05	[LT23]c
1959	2,711,445	4,470	0.16	[SU]e
1970	3,128,236	4,911	0.16	[SU]e
1979	3,391,490	8,651	0.26	[SU]e
1989	3,674,802	10,055	0.27	[SU]e
2001	3,297,525	2,860	0.09	[UN]c
2011	2,735,672	2,727	0.10	[LT]c
2050	2,121,000	3,000	0.14	es
2100	1,524,000	2,900	0.19	es

4.2.5 Moldova

The Republic of Moldova gained its independence from the Soviet Union in 1991 and has an area of 33,851 sq. km, including the Autonomous Territorial Unit of Gagauzia (1,832 sq. km) to the south and the de facto independent territory of Transnistria (4,163 sq. km) to the east. The map of Moldova is presented in Fig. 4.2.5. The lands of current Moldova came under control of the Golden Horde Empire in the thirteenth and fourteenth centuries, then Ottoman Empire from 1538 to 1812, after which it was ceded to the Russian Empire in the Treaty of Bucharest. However, Islam did not spread in these lands. In 1991 Moldova gained its independence from the Soviet Union but the Moldovan land east of the Dniester River, known as Transnistria, declared its independence as the Pridnestrovian Moldavian Republic. PMR however is internationally unrecognized and is still claimed, but uncontrolled by Moldova.

Figure 4.2.5 Map of the Republic of Moldova.

According to the 1897 Russian Empire census, the Muslim population in the Governorates of Bessarabia was 617 or 0.03% of the total population. The governorate had an area of 45,633 sq. km and its northern two thirds, which included more than 80% of the population, constituted current Moldova and Transnistria, while the southern third belongs to Ukraine. Bessarabia was under Romania from 1918 to 1940, which held a census in 1930 that inquired on religious adherence. Bessarabia then included nine counties; all are within current Moldova except Ismail, Cetatea Albă and the northern half of Hotin. The result of census is 119 Muslims or 0.01% of the total population of current Moldova. Another 29 Muslims lived in lands currently under Ukraine.

The Soviets carried out censuses since 1926 that collected ethnic affiliation demography of the population. The top fifteen Muslim ethnicities in Moldova from largest in number are Tatar, Azerbaijani, Uzbek, Kazakh, Turk, Arab, Kazakh, Turkmen, Tajik, Bashkir, Chechen, Albanian, Lezghin, Kyrgyz, Avar and Karachay. The number of members of each of these ethnicities since 1926 is provided in Table 4.2.5a. However, in 1926 the Moldovan Autonomous Soviet Socialist Republic was mostly in current Ukraine and included Transnistria.

Based on ethnic censuses, the Muslim population increased from 2,350 or 0.08% in 1959, to 3,484 or 0.10% in 1970, to 6,135 or 0.16% in 1979, to 11,196 or 0.26% in 1989. Census data after the independence started collecting information on religious adherence. Accordingly, the Muslim population increased from 1,667 or 0.05% in 2004 to 2,009 or 0.08% in 2014. The decrease from 1989 to 2004 is due to some leaving the country after its independence in 1991, but also due to the discrepancy when trying to infer religious data from ethnic data. For example, the ethnic data of the 2004 census shows 3,896 people from traditionally Muslim ethnicities, which is more than double of the number of those declaring themselves as Muslims in the same census. Thus, assuming that the percentage of Muslims will increase by 0.05 of a percentage point per decade, the Muslim population is expected to remain less than 10,000 or 0.5% of the total population throughout this century. A summary of the data is provided in Table 4.2.5b.

Table 4.2.5a Evolution of the largest Muslim ethnic populations in Moldova (excludes Transnistria in 2004)

	1926	1930	1959	1970	1979	1989	2004
Tatar	104	12	1,047	1,859	2,637	3,477	974
Azeri	—		373	425	1,062	2,642	891
Uzbek	—		405	365	605	1,391	416
Turk	1	147	10	26	20	14	269
Arab	1		8	4	26	26	259
Kazakh	—		154	234	533	1,108	256
Turkmen	—		25	51	165	337	220
Tajik	—		44	91	194	592	211
Bashkir	1		131	164	342	610	112
Chechen	2		16	35	33	150	108
Albanian	—	31	52	95	157	204	106
Lezghin	—		31	73	170	218	74
Kyrgyz	—		39	20	97	221	NA
Avar	—		14	25	78	120	NA
Karachay	—		1	17	16	86	NA
Total	109	190	2,350	3,484	6,135	11,196	3,896

Table 4.2.5b Evolution of the Muslim population in Moldova (excludes Transnistria in 2004 and 2014)

Year	Population	Muslims	%	Source
1897	1,935,412	617	0.03	[SU]c
1926	572,339	109	0.02	[SU]e
1930	2,240,300	119	0.01	[RO30]c
1959	2,884,477	2,350	0.08	[SU]e
1970	3,568,873	3,484	0.10	[SU]e
1979	3,949,756	6,135	0.16	[SU]e
1989	4,335,360	11,196	0.26	[SU]e
2004	3,307,605	1,667	0.05	[MD04]c
2014	2,611,759	2,009	0.08	[MD14]c
2050	3,360,000	8,000	0.25	es
2100	2,012,000	10,000	0.50	es

As for Transnistria, recent censuses indicate that there are no Muslims in the country as shown in Table 4.2.5c. This is expected to remain the case throughout this century.

Table 4.2.5c Evolution of the Muslim population in Transnistria

Year	Population	Muslims	%	Source
1926	335,000	—	0.00	[SU]e
1979	578,000	—	0.00	[PV]c
1989	679,000	—	0.00	[PV]c
2004	555,347	—	0.00	[PV]c
2015	475,665	—	0.00	es

4.2.6   Russia

Islam entered the Northern Caucasus in 22 H or 643 AD during the time of Caliph Omar. The Muslim troops led by the Prophet's companion Bakeer bnu Abdellah al-Ashaj, followed by troops from the western side led by the Prophet companion Utba bnu Farqad Assulami, until they conquered Derbent city (Babel-Abwab) in Southern Dagestan on the west coast of the Caspian Sea. Islam then continued spreading among the Turkic peoples, including the conversion of the king of Volga Bulgaria or Khanate of Volga in 310 H or 922 AD. His name was Almış, which he replaced by adopting the Arabic name Jaafar bnu Abdillah, after the Abbasid King Jaafar al-Muqtader, who encouraged him to Islam.

This Muslim Khanate was taken by the Pagan Tatar Golden Horde Empire in 1236 AD, led by Batu Khan who reigned from 1242 to 1255, and was a grandson of Genghis Khan. Batu's grandson, Sultan Mohammed Öz-Beg accepted Islam before taking the throne and reigned from 1313 to 1341. He adopted Islam as the state's religion and continued the spread of Islam among Tatars. By then the Empire controlled significant parts of current Russia (the region west of a line extending from the short Russian-Chinese border between Kazakhstan and Mongolia to the Arctic Ocean), Kazakhstan, Ukraine, Moldova, eastern Poland, Romania and Bulgaria.

The Golden Horde collapsed few years after it was devastated by Timur the Lame in 1395, who was the ruler of the Timurid Empire, another Tatar Muslim Empire in Central Asia. The Golden Horde disintegrated into several Muslim Khanates:

- Qasim, with its capital Kasimov, 200 Km southeast of Moscow, established in 1452 and was a vassal state to Christian Orthodox Russia, until it was abolished and annexed by Russia in 1681.
- Kazan, with its same name capital 700 Km east of Moscow, established in 1438 and was bitterly destroyed by and annexed by Russia in 1552, under the reign of the Russian Tsar, Ivan IV the Terrible who ruled Russia from 1533 to 1584.
- Astrakhan, north of the Caspian Sea with its same name capital. The Khanate was established in 1466 and was bitterly destroyed by and annexed by Russia in 1556 under the rule of Ivan IV the Terrible.
- Sibir, with its same name ruined capital 2000 km east of Moscow, and 1000 km north of Astana, the capital of current Kazakhstan. The Khanate was established in 1490 and was bitterly destroyed by and annexed by Russia in 1598, under the reign of the Russian Tsar son of Ivan IV, Feodor I who ruled Russia from 1584 to 1598.
- Crimea, with its capital Bahçeseray, south of the Crimean Peninsula. The Khanate was established in 1441, it controlled the peninsula, south Ukraine Proper and northwest Caucasus, by which having full control over the Azov Sea. Eventually, the Khanate was conquered by the Russian Empire in 1783 under the reign of Tsar Catherine II, who ruled Russia from 1762 to 1796.
- Kazakh, with its capital Turkistan located south of current Kazakhstan. The Khanate was established in 1456 and was conquered by the Russian Empire in 1847 under the reign of Tsar Nicholas I, who ruled Russia from 1796 to 1855.

Currently, the Russian Federation is the largest country in the world with total area of 17,098,242 sq. km and its map is presented in Fig. 4.2.6a. It spans two continents with four million sq. km in Europe and thirteen million sq. km in Asia. However, less than quarter of the population lives in the Asian part. In other words, while 23% of Russia is in Europe, only 23% of the Russian population lives in the Asian part. The Muslim population is spread all over the country but is concentrated in the Caucasus and central Asian part. The Russian territory is organized in eight federal districts, three of which lay in the Asian continent (Ural, Siberian and Far Eastern), while the rest are in the European continent. A ninth federal district was established in 2014 that consists of the Crimean Peninsula that was annexed from Ukraine and remains internationally recognized as part of the later. Thus, almost a quarter of the Russian land and almost three-quarter of its population are in the European side. The eight federal districts are:

- Central: Has an area of 652,800 sq. km (4%) and over a quarter of the Russian population, It includes the capital and Russia's largest city of Moscow and seventeen oblasts: Belgorod, Bryansk, Vladimir, Voronezh, Ivanovo, Kaluga, Kostroma, Kursk, Lipetsk, Moscow, Oryol, Ryazan, Smolensk, Tambov, Tver, Tula and Yaroslavl. In 1897, this area was covered by the governorates of Chernihiv (Chernigov), Kaluga, Kostroma, Kursk, Tver, Moscow, Orel (Oryol), Ryazan, Smolensk, Tambov, Tula, Voronezh, Vladimir and Yaroslavl.
- Northwestern: Has an area of 1,677,900 sq. km (10%) and a tenth of the population. It includes Russia's second largest city of Saint Petersburg, the republics of Karelia and Komi, and seven oblasts: Arkhangelsk, Vologda, Kaliningrad, Leningrad, Murmansk, Novgorod and Pskov. In 1897, this area was covered by the governorates of Arkhangelsk, Novgorod, Olonetsk (Olonets), Pskov, St. Petersburg, Vitebsk and Vologda.
- Southern: Has an area of 418,500 sq. km (2%) and a tenth of the population. It includes Krasnodar Krai, the republics of Adygea and Kalmykia, and three oblasts: Astrakhan, Volgograd and Rostov. In 1897, this area was covered by the governorates of Astrakhan, Black Sea, Don and Kuban.

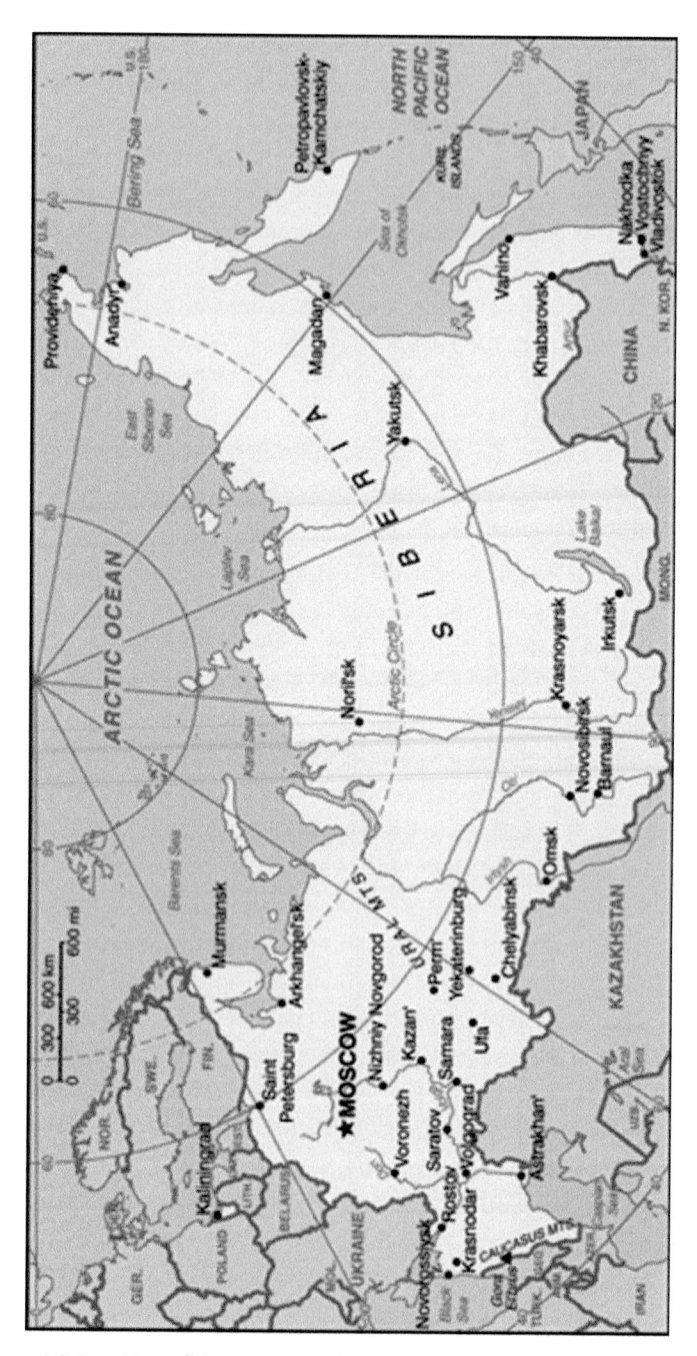

Figure 4.2.6a Map of the Russian Federation.

- North Caucasian: Has an area of 170,700 sq. km (1%) and 7% of the population. It includes Stavropol Krai and six republics: Dagestan, Ingushetia, Kabardino-Balkar, Karachay-Cherkess, North Ossetia-Alania and Chechenia. In 1897, this area was covered by the governorates of Dagestan, Stavropol and Terek.
- Volga: Has an area of 1,038,000 sq. km (6%) and a fifth of the population. It includes Perm Krai, six republics: Bashkortostan, Mari El, Mordovia, Tatarstan, Udmurt and Chuvash, and seven oblasts: Kirov, Nizhny Novgorod, Orenburg, Penza, Samara, Saratov and Ulyanovsk. In 1897, this area was covered by the governorates of Kazan, Penza, Perm, Nizhny Novgorod, Orenburg, Samara, Saratov, Simbirskaya (Simbirsk), Vyatka and Ufa.
- Ural: Has an area of 1,788,900 sq. km (10%) and 8% of the population. It includes four oblasts: Kurgan, Sverdlovsk, Tyumen and Chelyabinsk. In 1897, this area was covered by Tobolsk governorate.
- Siberian: Has an area of 5,114,800 sq. km (30%) and 13% of the population. It includes four republics: Altai, Buryatia, Tyva and Khakassia, and three krais: Altai, Transbaikal (Zabaykalsky) and Krasnoyarsk, and five oblasts: Irkutsk, Kemerovo, Novosibirsk, Omsk and Tomsk. In 1897, this area was covered by the governorates of Irkutsk, Tomsk, Transbaikal and Yenisei.
- Far Eastern: Has an area of 6,215,900 sq. km (36%) and 4% of the population. It includes Sakha (Yakutia) Republic, three krais: Kamchatka, Primorsky and Khabarovsk, and three oblasts: Amur, Magadan and Sakhalin, and Jewish Autonomous Oblast and Chukotka Autonomous Okrug. In 1897, this area was covered by the governorates of Amur, Maritime (Primorskaya), Sakhalin Island and Yakut.

According to an 1867 estimate, the Muslim population was 5.66 million or 7.4% of the total population. The 1897 census of the Russian Empire found 13,906,972 Muslims or 11.07% of the total population of 125,640,021. From this, in current Russia lived 4.6 million or 6.7% of the total population in the lands currently controlled by the Russian Federation.

Table 4.2.6a Evolution of the largest Muslim ethnic populations in Russia

	1926	1939	1959	1970	1979	1989	2002	2010
Tatar	2,926,053	3,682,956	4,074,669	4,757,913	5,010,922	5,543,371	5,558,732	5,310,649
Bashkir	738,861	824,537	953,801	1,180,913	1,290,994	1,345,273	1,673,389	1,584,554
Chechen	318,361	400,325	261,311	572,220	712,161	898,999	1,360,253	1,431,360
Avar	178,263	235,715	249,529	361,613	438,306	544,016	814,473	912,090
Kazakh	136,501	356,646	382,431	477,820	518,060	635,865	653,962	647,732
Azeri	24,335	43,014	70,947	95,689	152,421	335,889	621,840	603,070
Dargin	125,759	152,007	152,563	224,172	280,444	353,348	510,156	589,386
Kabardian	139,864	161,216	200,634	277,435	318,822	386,055	519,958	516,826
Kumyk	94,509	110,299	132,896	186,690	225,800	277,163	422,409	503,060
Lezgin	92,937	100,328	114,210	170,494	202,854	257,270	411,535	473,722
Ingush	72,137	90,980	55,799	137,380	165,997	215,068	413,016	444,833
Uzbek	942	16,166	29,512	61,588	72,385	126,899	122,916	289,862
Karachai	55,116	74,488	70,537	106,831	125,792	150,332	192,182	218,403
Tajik	52	3,315	7,027	14,108	17,863	38,208	120,136	200,303
Lak	40,243	54,348	58,397	78,625	91,412	106,245	156,545	178,630

Tabasaran	31,983	33,471	34,228	54,047	73,433	93,587	131,785	146,360
Adyghe	64,959	85,588	78,561	98,461	107,239	122,908	131,759	128,717
Balkar	33,298	41,949	35,249	52,969	61,828	78,341	108,426	112,924
Turk	1,846	2,668	1,377	1,568	3,561	9,890	95,672	109,883
Nogai	36,089	36,088	37,656	51,159	58,639	73,703	90,666	103,660
Kyrgyz	285	6,328	4,701	9,107	15,011	41,734	31,808	103,422
Cherkes	NA	NA	28,986	38,356	44,572	50,764	60,517	73,184
Abaza	13,825	14,739	19,059	24,892	28,800	32,983	37,942	43,341
Turkmen	7,849	12,927	11,631	20,040	22,979	39,739	33,053	36,885
Rutul	10,333	10,000	6,703	11,904	14,835	19,503	29,929	35,240
Agul	7,653	7,000	6,460	8,751	11,752	17,728	28,297	34,160
Kurd	164	387	855	1,015	1,631	4,724	19,607	23,232
Tasakhur	3,533	3,962	4,437	4,730	4,774	6,492	10,366	12,769
Arab	980	95	649	2,555	2,339	2,704	10,811	9,583
Afghan	107	190	175	561	184	858	9,800	5,350
Total	5,156,837	6,561,732	7,084,990	9,083,606	10,075,810	11,809,659	14,381,940	14,883,190

Table 4.2.6b Evolution of the Muslim population in Russia per federal district: 1897–2010

		1897	1926	1939	1959	1970
Central	P	25,971,498	31,695,660	36,171,775	33,487,258	35,649,787
	M	32,038	52,196	208,026	218,515	298,987
	M%	0.12	0.16	0.58	0.65	0.84
	MR%	0.69	1.02	3.17	3.10	3.32
North Western	P	8,143,095	9,028,151	14,439,399	11,474,054	12,888,896
	M	7,516	16,151	154,085	68,412	99,805
	M%	0.09	0.18	1.07	0.60	0.77
	MR%	0.16	0.32	2.35	0.97	1.11
Southern	P	5,544,139	7,242,012	8,574,077	9,815,005	11,799,455
	M	416,739	186,771	234,815	290,522	405,001
	M%	7.52	2.58	2.74	2.96	3.43
	MR%	8.96	3.66	3.57	4.12	4.50
North Caucasian	P	2,378,391	3,969,976	4,266,736	4,526,503	5,939,575
	M	1,068,537	1,254,253	1,597,780	1,511,535	2,301,641
	M%	44.93	31.59	37.45	33.39	38.75
	MR%	22.99	24.57	24.32	21.46	25.60
Volga	P	21,732,846	22,609,662	26,880,285	27,652,383	29,753,951
	M	2,997,167	3,156,331	3,738,053	3,923,488	4,648,754
	M%	13.79	13.96	13.91	14.19	15.62
	MR%	64.48	61.83	56.91	55.69	51.70
Ural	P	1,433,043	6,786,339	5,452,276	9,112,337	10,100,203
	M	64,880	262,150	317,016	606,347	743,174
	M%	4.53	3.86	5.81	6.65	7.36
	M%	1.40	5.13	4.83	8.61	8.26
Siberian	P	3,684,144	9,179,175	9,706,754	16,632,629	18,166,834
	M	56,009	169,244	276,068	357,584	405,318
	M%	1.52	1.84	2.84	2.15	2.23
	MR%	1.20	3.32	4.20	5.08	4.51
Far Eastern	P	641,635	2,170,436	2,779,732	4,834,146	5,780,509
	M	5,685	8,074	42,860	68,581	89,428
	M%	0.89	0.37	1.54	1.42	1.55
	MR%	0.12	0.16	0.65	0.97	0.99

Table 4.2.6b (Continued)

		1979	1989	2002	2010
Central	P	36,677,954	37,939,772	37,264,631	36,483,008
	M	334,000	478,876	706,159	761,594
	M%	0.91	1.26	1.89	2.09
	MR%	3.31	4.04	4.90	5.11
North Western	P	14,058,201	15,236,761	13,521,273	12,500,673
	M	137,517	215,751	195,299	227,019
	M%	0.98	1.42	1.44	1.82
	MR%	1.36	1.82	1.36	1.52
Southern	P	12,510,593	13,252,223	13,939,028	13,515,784
	M	483,974	642,158	749,266	758,985
	M%	3.87	4.85	5.38	5.62
	MR%	4.79	5.42	5.20	5.09
North Caucasian	P	6,539,174	7,283,925	8,929,300	9,365,804
	M	2,873,885	3,418,954	5,110,448	5,641,508
	M%	43.95	46.94	57.23	60.24
	MR%	28.47	28.87	35.48	37.85
Volga	P	30,684,010	31,765,311	31,076,209	29,128,264
	M	4,879,056	5,298,018	5,840,345	5,734,793
	M%	15.90	16.68	18.79	19.69
	MR%	48.33	44.74	40.55	38.48
Ural	P	10,851,080	12,525,837	12,304,762	11,540,584
	M	834,358	1,078,322	1,148,190	1,137,258
	M%	7.69	8.61	9.33	9.85
	MR%	8.27	9.11	7.97	7.63
Siberian	P	19,243,841	21,068,035	20,021,659	18,695,220
	M	437,530	539,242	536,359	526,101
	M%	2.27	2.56	2.68	2.81
	MR%	4.33	4.55	3.72	3.53
Far Eastern	P	6,879,919	7,950,005	6,649,118	5,997,770
	M	114,248	169,859	116,233	117,694
	M%	1.66	2.14	1.75	1.96
	MR%	1.13	1.43	0.81	0.79

Table 4.2.6c Evolution of the Muslim population in Russia

Year	Population	Muslims	%	Source
1867	77,049,000	5,660,000	7.35	[SYB70]es
1897	69,528,791	4,648,571	6.69	[SU]c
1926	92,395,448	5,156,837	5.58	[SU]e
1939	108,271,034	6,561,732	6.06	[SU]e
1959	117,534,315	7,084,990	6.03	[SU]e
1970	130,079,210	9,083,606	6.98	[SU]e
1979	137,409,921	10,075,810	7.33	[SU]e
1989	147,021,869	11,809,659	8.03	[SU]e
2002	145,166,731	14,381,940	9.91	[RU]e
2010	142,905,200	14,883,190	10.41	[RU10]e
2050	135,824,000	16,978,000	12.50	es
2100	126,143,000	18,921,000	15.00	es

Censuses in Russia since 1926 include ethnic demography, from which we can deduce religion. The top thirty Muslim ethnicities in Russia from largest in number are Tatar, Bashkir, Chechen, Avar, Kazakh, Azeri, Dargin, Kumyk, Lezgin, Ingush, Uzbek, Karachai, Tajik, Lak, Tabasaran, Adyghe, Balkar, Turk, Nogai, Kyrgyz, Cherkes, Abaza, Turkmen, Rutul, Agul, Kurd, Tasakhur, Arab and Afghan. The number of members of each of these ethnicities since 1926 is provided in Table 4.2.6a. Based on ethnic censuses, the Muslim population increased from 4.65 million or 6.7% in 1897, to 5.16 million or 5.6% in 1926, to 6.56 million or 6.1% in 1939, to 7.08 million or 6.0% in 1959, to 9.08 million or 7.0% in 1970, to 10.08 million or 7.3% in 1979, to 11.81 million or 8.0% in 1989, to 14.38 million or 9.9% in 2002, to 14.88 million or 10.4% in 2010. Thus, assuming that the percentage of Muslims will continue to increase by half of a percentage point per decade, the Muslim population is expected to increase to seventeen million or 12.5% by 2050 and nineteen million or 15% by 2100. A summary of the data is provided in Table 4.2.6c.

Table 4.2.6b present the distribution of Muslims and the total population in the federal districts. About a third of the Muslims population lives North Caucasian Federal District, where their share of the population increased to 60%. Another third lives in Volga FD, where they constitute a fifth of the population. Almost a tenth of the Muslim population lives in Ural FD, where they make up a tenth of the population. The share of Muslims has been increasing in every federal district since 1950s, however 88% of Russian Muslim population lives in the European part of Russia, which has increased from 30% in 1959.

A choropleth map of Russia illustrating the presence of Muslims per federal district is presented in Fig. 4.2.6b, where white is less than 1% of the district's population and the darker the region, the more is the percentage of Muslims.

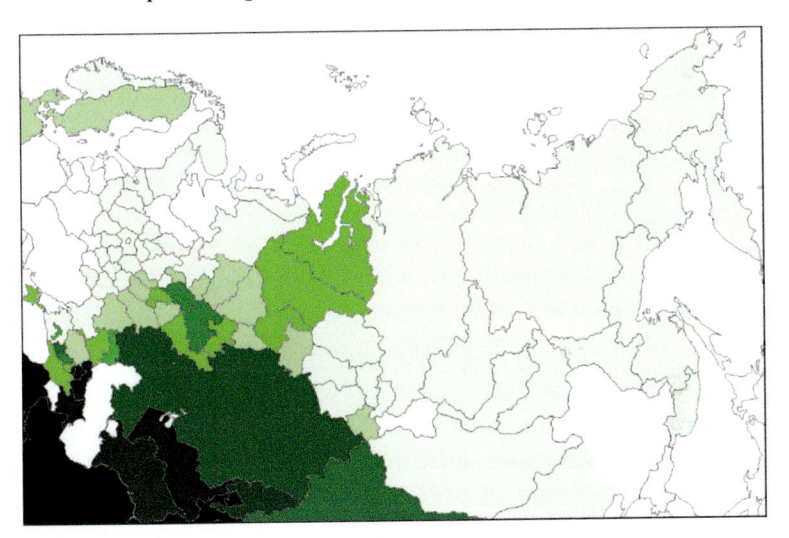

Figure 4.2.6b A choropleth map of Russia illustrating the presence of Muslims per federal district.

4.2.7 Ukraine

Islam entered Ukraine in the thirteenth century as Tatars of the Golden Horde Empire started embracing Islam. The southern parts of Ukraine remained under Muslim rule through the Crimean Khanate, a successor of the Golden Horde in this region, until it was annexed by the Russian Empire in 1796. Thus, significant parts of

Ukraine, especially the southern part, were under Muslim control for 483 years. For the next century and a half following the fall of the Crimean Khanate, the Muslims were subjugated to various forms of persecution and discrimination by the Russians. Before its fall, the Muslim Crimean Tatars constituted the vast majority (over 95%) of the population. However, due to Russian wars, many fled to the Ottoman Empire, or deported to Central Asia and Siberia.

In 1802, part of the Crimean Khanate became known as Turida (Tavrisheskaya) governorate, which included the Crimean Peninsula and about the same size into Ukraine Proper. By 1897, Muslims numbered 190,800 or 13% of the population of the Governorate. In 1921, Crimean Autonomous Soviet Socialist Republic was created by the Soviet Union within the Crimean Peninsula. Its Muslim population, almost all Crimean Tatars, increased in number but decreased in percentage from 184,355 or 26% in 1926, to 222,359 or 20% in 1939. The decrease is due to the Russian policy of settling many Russians and other non-Muslim ethnic groups in the peninsula, to change its demography.

In 1944, Joseph Stalin, who ruled the Soviet Union from 1922 to 1953, deported all Crimean Tatars from Crimea to Siberia and Uzbekistan, about half of whom perished in this process. Stalin falsely accused them all of collaborating with Nazi Germany. They were vindicated by the Russian state in 1956 but were not compensated or allowed to return to their homeland, safe few thousands by the end of 1960s. Crimea was abolished in 1945 and given to Ukraine in 1954. Thus, by 1959, the Muslim population was reduced to zero, and then increased to 6,479 or 0.4% in 1979, to 15,100 or 2% in 1989 as the Soviet Union neared its dissolution.

The Crimean Autonomous Republic within the Ukraine was re-established in the Crimean Peninsula in 1991, excluding the city of Sevastopol due to its strategic importance as a naval base. Crimean Tatars continued returning to their homeland, increasing the number of Muslims in the peninsula to 2.66 million or 11.1% in 2001 and 2.87 million or 12.5% in 2014. A summary of the change of the Muslim population in the Crimean Peninsula including Sevastopol is provided in Table 4.2.7a.

Table 4.2.7a Evolution of the Muslim population in the Crimean Peninsula (ARC & Sevastopol)

Year	Population	Muslims	%	Source
1750	467,000	444,000	95.07	[CPH]e
1770	454,700	421,000	92.59	[CPH]e
1793	127,800	112,200	87.79	[CPH]e
1795	156,400	137,000	87.60	[CPH]e
1816	212,600	182,600	85.89	[CPH]e
1835	279,400	233,300	83.50	[CPH]e
1850	343,500	267,200	77.79	[CPH]e
1858	331,300	242,000	73.05	[CPH]e
1864	198,700	100,000	50.33	[CPH]e
1897	546,592	190,800	34.88	[SU]c
1917	749,800	220,500	29.41	[CPH]e
1920	718,900	187,000	26.01	[CPH]e
1926	713,823	184,355	25.83	[SU]e
1934	832,000	198,000	23.80	[CPH]e
1937	996,800	206,500	20.72	[CPH]e
1939	1,126,429	222,359	19.74	[SU]e
1944	379,000	—	0.00	[CPH]e
1959	1,201,517	—	0.00	[CPH]e
1970	1,813,502	6,479	0.36	[CPH]e
1979	2,135,900	15,100	0.71	[CP]e
1989	2,430,495	52,172	2.15	[CP]e
2001	2,401,209	265,900	11.07	[CP]e
2014	2,284,769	285,590	12.50	[CP14]e

The Ukraine gained its independence from the Soviet Union in 1991 and has an area of 603,550 sq. km, including the Crimean Peninsula, which consists of the Autonomous Republic of Crimea (26,100 sq. km), and the Sevastopol City Council (1,079 sq. km). In 2014, the latter two declared their independence and joined

Russia as its ninth federal district. The move was not internationally recognized, and the Crimean Peninsula is still claimed but uncontrolled by Ukraine. A map of the Ukraine is presented in Fig. 4.2.7.

Figure 4.2.7 Map of the Ukraine.

According to the 1897 Russian Empire census, the Muslim population in ten governorates covering current Ukraine and Crimea was 209,672 or 0.83% of the total population. These governorates were Bessarabia (southern one third laid in Ukraine while the rest made up Moldova), Cherigov (northern one third in Ukraine, while the rest in Russia), Katerynoslav (Yekaterinoslav), Kharkiv, Kherson, Kiev, Podolia (Podolsk or Podolsky), Taurida (Tavrida or Taurian) and Volynia (Volyn). Data of the 1897 census for the three governorates is summarized in Table 4.2.8b. For comparison, the table also contains an estimate based on the 1897 of the total population living in the current border of Ukraine and the current area of the country.

Censuses in Ukraine since 1926 include ethnic demography, from which we can deduce religion. The largest thirty Muslim ethnicities in Ukraine starting with the largest in number are Tatar, Azeri, Uzbek, Turk, Arab, Kazakh, Lezghin, Tajik, Bashkir, Turkmen, Albanian, Chechen, Kurd, Darghin, Avar, Abkhaz, Kyrgyz, Lak, Afghan, Tabasaran, Kumyk, Kabardian, Ingush, Persian, Nogai, Adyghei, Balkar, Cherkes, Uighur and Karachay. The number of members of each of these ethnicities since 1926 is provided in Table 4.2.8c. Based on ethnic censuses, the Muslim population increased from 0.21 million or 0.7% in 1926, to 0.32 million or 1.0% in 1939, but decreased to 0.10 million or 0.2% in 1959 because of Stalin's deportation of Crimean Tatars to Uzbekistan and Siberia after WWII. The Muslim population continued growing since then, increasing to 0.13 million or 0.3% in 1970, to 0.15 million or 0.3% in 1979, to 0.24 million or 0.5% in 1989, to 0.44 million or 0.9% in 2001. According to a 2007 DHS, this percentage increased to 1.0%. Thus, assuming that the percentage of Muslims will continue to increase by a tenth of a percentage point per decade, the Muslim population is expected to increase to remain close to half a million for the rest of this century, increasing in percentage from 1% of the total population to 2% by 2100. A summary of the data is provided in Table 4.2.7d.

Table 4.2.7b 1897 Census data for territory covering current Ukraine (including Crimea)

Governorate	Population	Muslims	% Muslims	Area (km^2)
Bessarabia	1,935,412	617	0.03	45,633
Chernigov	2,297,854	528	0.02	52,403
Katerynoslav	2,113,674	2,090	0.10	63,398
Kharkiv	2,492,316	1,362	0.05	54,496
Kherson	2,733,612	2,367	0.09	71,284
Kiev	3,559,229	2,931	0.08	50,999
Podolia	3,018,299	3,460	0.11	42,020
Poltava	2,778,151	640	0.02	49,896
Taurida	1,447,790	190,800	13.18	63,447
Volynia	2,989,482	4,877	0.16	71,854
Total	25,365,819	209,672	0.83	565,431
Ukraine	26,520,000			603,550

Table 4.2.7c Evolution of the largest Muslim ethnic populations in Ukraine (including Crimea)

	1926	1939	1959	1970	1979	1989	2001
Tatar	201,375	274,335	61,527	76,212	90,542	133,682	321,497
Azeri	NA	4,735	6,680	10,769	17,235	36,961	45,176
Uzbek	26	13,063	8,472	10,563	9,862	20,333	12,353
Turk	596	1,121	284	226	257	262	9,180
Arab	4	27	30	796	1,352	1,240	6,575
Kazakh	NA	11,415	4,694	7,555	7,171	10,505	5,526
Lezghin	268	1,010	1,484	1,708	2,354	4,810	4,349
Tajik	NA	1,115	2,201	2,473	2,415	4,447	4,255
Bashkir	121	3,823	3,345	3,672	5,367	7,402	4,253
Turkmen	25	3,770	1,745	1,045	1,696	3,399	3,709
Albanian	3,028	1,661	3,809	3,972	3,874	3,343	3,308
Chechen	59	2,045	424	939	1,046	1,844	2,877
Kurd	15	90	65	117	122	238	2,088
Darghin	NA	140	NA	634	595	1,550	1,610
Avar	3	426	714	893	1,211	2,677	1,496
Abkhaz	9	265	633	476	941	990	1,458
Kyrgyz	49	1,693	1,301	1,576	2,370	2,297	1,128
Lak	NA	NA	623	574	662	1,035	1,019
Afghan	NA	70	3	295	148	360	1,008
Tabasaran	NA	7	120	118	300	932	977
Kumyk	NA	267	295	324	313	868	718
Kabardian	19	867	680	554	673	959	473
Ingush	10	477	148	285	306	466	455
Persian	328	831	250	301	171	228	419
Nogai	3	132	87	84	248	331	385
Adyghei	NA	679	381	426	458	688	338
Balkar	NA	149	310	104	152	244	206
Cherkes	101		289	292	463	447	199
Uigur	NA	185	173	371	135	194	197
Karachay	NA	293	105	269	203	342	190
Total	206,039	324,691	100,872	127,623	152,642	243,074	437,422

Table 4.2.7d Evolution of the Muslim population in Ukraine (including Crimea)

Year	Population	Muslims	%	Source
1897	25,365,819	209,672	0.83	[SU]c
1926	29,159,671	206,039	0.71	[SU]e
1939	32,072,647	324,691	1.01	[SU]e
1959	41,869,046	100,872	0.24	[SU]e
1970	47,126,517	127,623	0.27	[SU]e
1979	49,609,333	152,642	0.31	[SU]e
1989	51,452,034	243,074	0.47	[SU]e
2001	48,240,902	437,422	0.91	[UNE]e
2007	46,249,000	455,000	0.98	[DHS]s
2050	35,219,000	493,000	1.40	es
2100	24,413,000	464,000	1.90	es

4.2.8   Regional Summary and Conclusion

Islam entered the Caucasus two decades after the death of Prophet Mohammed peace and blessings upon him. Islam spread to Eastern Europe in the fourteenth century with the Golden Horde Empire. Significant parts remained under Muslim control until the end of the eighteenth century with the fall of the Crimean Khanate. Muslim now constitute 7% of the total population and are increasing to 11% by the end of this century. The following tables present centennial data from 600 AD to 2100 AD (or approximately 1 H to 1500 H) in Table 4.2a and decennial data from 1790 AD to 2100 AD (or 1210 H to 1520 H) in Tables 4.2b and 4.2c for current countries in Eastern Europe. The data includes total population in thousands (P), the percentage of which is Muslim (M%), the corresponding Muslim population in thousands (M), and the annual population growth rate (APGR, or G%) of the total population in this region.

Table 4.2a Centennial estimates of the Muslim population (×1000) in Eastern Europe: 600–2100 AD (1–1500 H)

		600	700	800	900	1000	1100	1200	1300
Belarus	P	190	200	210	220	230	320	410	500
	M%	—	—	—	—	—	—	—	—
	M	—	—	—	—	—	—	—	—
Estonia	P	23	24	25	26	27	38	49	60
	M%	—	—	—	—	—	—	—	—
	M	—	—	—	—	—	—	—	—
Latvia	P	40	42	44	46	48	60	80	100
	M%	—	—	—	—	—	—	—	—
	M	—	—	—	—	—	—	—	—
Lithuania	P	59	63	67	71	75	100	125	150
	M%	—	—	—	—	—	—	—	—
	M	—	—	—	—	—	—	—	—
Moldova	P	51	54	57	60	63	95	120	145
	M%	—	—	—	—	—	—	—	0.50
	M	—	—	—	—	—	—	—	1
Russia	P	2,250	2,375	2,500	2,625	2,750	3,500	4,750	5,500
	M%	—	1.00	2.00	3.00	4.00	5.00	6.00	7.00
	M	—	24	50	79	110	175	285	385
Ukraine	P	850	900	950	1,000	1,053	1,500	1,900	2,300
	M%	—	—	—	—	—	—	—	1.00
	M	—	—	—	—	—	—	—	23
Total	P	3,463	3,658	3,853	4,048	4,246	5,613	7,434	8,755
	M%	0.00	0.65	1.30	1.95	2.59	3.12	3.83	4.67
	M	—	24	50	79	110	175	285	409
	G%		0.055	0.052	0.049	0.048	0.279	0.281	0.164

Table 4.2a (Continued)

		1400	1500	1600	1700	1800	1900	2000	2100
Belarus	P	590	691	864	1,152	2,300	5,803	9,872	7,430
	M%	1.00	2.00	3.00	4.00	0.16	0.16	0.22	0.42
	M	6	14	26	46	4	9	22	31
Estonia	P	71	81	101	134	250	676	1,399	838
	M%	—	—	—	—	—	0.04	0.18	0.26
	M	—	—	—	—	—	—	3	2
Latvia	P	120	144	180	240	490	1,210	2,384	1,114
	M%	—	—	—	—	—	0.05	0.27	0.35
	M	—	—	—	—	—	1	6	4
Lithuania	P	180	224	279	373	770	1,876	3,502	1,524
	M%	0.50	1.00	1.00	1.00	0.19	0.19	0.09	0.19
	M	1	2	3	4	1	4	3	3
Moldova	P	160	189	236	314	630	1,583	4,203	2,012
	M%	1.00	1.50	2.00	2.50	0.03	0.03	0.05	0.50
	M	2	3	5	8	—	—	2	10
Russia	P	6,250	8,183	10,195	13,616	26,000	70,628	146,405	126,143
	M%	8.00	9.00	10.00	11.00	7.35	6.69	9.91	15.00
	M	500	736	1,020	1,498	1,911	4,725	14,509	18,921
Ukraine	P	2,700	3,160	3,950	5,267	10,500	26,520	48,838	24,413
	M%	2.00	4.00	6.00	8.00	1.30	0.83	0.91	1.90
	M	54	126	237	421	137	220	444	464
Total	P	10,071	12,672	15,805	21,096	40,940	108,296	216,602	163,474
	M%	5.58	6.96	8.16	9.37	5.01	4.58	6.92	11.89
	M	562	882	1,290	1,977	2,053	4,959	14,989	19,435
	G%	0.140	0.230	0.221	0.289	0.663	0.973	0.693	−0.281

Table 4.2b Decennial estimates of the Muslim population (×1000) in Eastern Europe: 1790–1940 AD (1210–1360 H)

		1790	1800	1810	1820	1830	1840	1850	1860
Belarus	P	2,200	2,300	2,400	2,545	2,900	3,200	3,311	3,535
	M%	0.16	0.16	0.16	0.16	0.16	0.16	0.16	0.16
	M	4	4	4	4	5	5	5	6
Estonia	P	230	250	263	296	330	360	400	440
	M%	—	—	—	—	—	—	—	—
	M	—	—	—	—	—	—	—	—
Latvia	P	470	490	510	531	590	650	710	770
	M%	—	—	—	—	—	—	—	0.00
	M	—	—	—	—	—	—	—	—
Lithuania	P	740	770	800	823	900	1,000	1,100	1,200
	M%	0.19	0.19	0.19	0.19	0.19	0.19	0.19	0.19
	M	1	1	2	2	2	2	2	2
Moldova	P	600	630	660	694	780	860	940	1,026
	M%	0.03	0.03	0.03	0.03	0.03	0.03	0.03	0.03
	M	—	—	—	—	—	—	—	—
Russia	P	24,000	26,000	28,000	30,588	34,000	37,000	40,000	45,000
	M%	7.35	7.35	7.35	7.35	7.35	7.35	7.35	7.35
	M	1,764	1,911	2,058	2,248	2,499	2,720	2,940	3,308
Ukraine	P	10,000	10,500	11,000	11,629	13,000	14,400	15,800	17,200
	M%	1.20	1.30	1.50	1.60	1.70	1.70	1.70	1.40
	M	120	137	165	186	221	245	269	241
Total	P	38,240	40,940	43,633	47,106	52,500	57,470	62,261	69,171
	M%	4.94	5.01	5.11	51.8	5.19	5.17	5.17	5.14
	M	1,889	2,053	2,229	2,440	2,727	2,972	3,216	3,557
	G%		0.682	0.637	0.766	1.084	0.904	0.801	1.052

Table 4.2b (Continued)

		1870	1880	1890	1900	1910	1920	1930	1940
Belarus	P	4,133	4,617	5,158	5,803	7,100	7,206	8,018	8,786
	M%	0.16	0.16	0.16	0.16	0.16	0.08	0.08	0.22
	M	7	7	8	9	11	6	6	19
Estonia	P	482	538	601	676	760	840	934	1,024
	M%	—	0.00	0.04	0.04	0.01	0.01	0.02	0.02
	M	—	—	—	—	—	—	—	—
Latvia	P	862	963	1,075	1,210	1,350	1,502	1,672	1,832
	M%	0.00	0.00	0.05	0.05	0.72	0.01	0.01	0.01
	M	—	—	1	1	10	—	—	—
Lithuania	P	1,336	1,493	1,668	1,876	2,100	2,330	2,593	2,841
	M%	0.19	0.19	0.19	0.19	0.19	0.05	0.05	0.05
	M	3	3	3	4	4	1	1	1
Moldova	P	1,128	1,493	1,407	1,583	1,800	1,966	2,187	2,397
	M%	0.03	0.03	0.03	0.03	0.01	0.01	0.01	0.01
	M	—	—	—	—	—	—	—	—
Russia	P	50,303	56,196	62,779	70,628	85,432	87,707	98,065	106,933
	M%	7.35	7.35	6.69	6.69	6.69	5.61	5.61	6.10
	M	3,697	4,130	4,200	4,725	5,715	4,920	5,501	6,523
Ukraine	P	18,888	21,101	23,572	26,520	30,837	32,933	36,641	40,152
	M%	0.60	0.70	0.80	0.83	0.83	0.71	0.71	1.01
	M	113	148	189	220	256	234	260	406
Total	P	77,132	86,401	96,260	108,296	129,379	134,484	150,110	163,965
	M%	4.95	4.96	4.57	4.58	4.63	3.84	3.84	4.24
	M	3,820	4,289	4,401	4,959	5,997	5,162	5,770	6,950
	G%	1.089	1.135	1.081	1.178	1.779	0.387	1.099	0.883

Table 4.2c Decennial estimates of the Muslim population (×1000) in Eastern Europe: 1950–2100 AD (1370–1520 H)

		1950	1960	1970	1980	1990	2000	2010	2020
Belarus	P	7,745	8,125	8,914	9,570	10,151	9,872	9,421	9,449
	M%	0.16	0.16	0.18	0.21	0.29	0.22	0.24	0.26
	M	12	13	16	20	29	22	23	25
Estonia	P	1,101	1,217	1,360	1,474	1,565	1,399	1,332	1,327
	M%	0.26	0.25	0.25	0.34	0.47	0.18	0.17	0.18
	M	3	3	3	5	7	3	2	2
Latvia	P	1,927	2,130	2,378	2,521	2,664	2,384	2,119	1,886
	M%	0.17	0.17	0.21	0.27	0.44	0.27	0.26	0.27
	M	3	4	5	7	12	6	6	5
Lithuania	P	2,567	2,770	3,137	3,431	3,696	3,502	3,124	2,722
	M%	0.16	0.16	0.16	0.26	0.27	0.09	0.10	0.11
	M	4	4	5	9	10	3	3	3
Moldova	P	2,341	3,004	3,596	4,011	4,366	4,203	4,086	4,034
	M%	0.08	0.08	0.10	0.16	0.26	0.05	0.08	0.10
	M	2	2	4	6	11	2	3	4
Russia	P	102,799	119,872	130,149	138,053	147,532	146,405	143,479	145,934
	M%	6.03	6.03	6.98	7.33	8.03	9.91	10.41	11.00
	M	6,199	7,228	9,084	10,119	11,847	14,509	14,936	16,053
Ukraine	P	37,298	42,665	47,089	49,966	51,463	48,838	45,792	43,734
	M%	0.24	0.24	0.27	0.31	0.47	0.91	0.98	1.10
	M	90	102	127	155	242	444	449	481
Total	P	155,777	179,782	196,621	209,027	221,437	216,602	209,353	209,087
	M%	4.05	4.09	4.70	4.94	5.49	6.92	7.37	7.93
	M	6,313	7,357	9,245	10,321	12,159	14,989	15,422	16,573
	G%	−0.512	1.433	0.895	0.612	0.577	−0.221	−0.340	−0.013

Table 4.2c (*Continued*)

		2030	2040	2050	2060	2070	2080	2090	2100
Belarus	P	9,265	8,938	8,634	8,333	8,002	7,728	7,569	7,430
	M%	0.28	0.30	0.32	0.34	0.36	0.38	0.40	0.42
	M	26	27	28	28	29	29	30	31
Estonia	P	1,280	1,219	1,158	1,092	1,017	946	887	838
	M%	0.19	0.20	0.21	0.22	0.23	0.24	0.25	0.26
	M	2	2	2	2	2	2	2	2
Latvia	P	1,720	1,584	1,479	1,384	1,290	1,218	1,164	1,114
	M%	0.28	0.29	0.30	0.31	0.32	0.33	0.34	0.35
	M	5	5	4	4	4	4	4	4
Lithuania	P	2,485	2,284	2,121	1,969	1,820	1,703	1,612	1,524
	M%	0.12	0.13	0.14	0.15	0.16	0.17	0.18	0.19
	M	3	3	3	3	3	3	3	3
Moldova	P	3,886	3,641	3,360	3,059	2,732	2,426	2,195	2,012
	M%	0.15	0.20	0.25	0.30	0.35	0.40	0.45	0.50
	M	6	7	8	9	10	10	10	10
Russia	P	143,348	139,031	135,824	132,692	129,229	127,150	126,688	126,143
	M%	11.50	12.00	12.50	13.00	13.50	14.00	14.50	15.00
	M	16,485	16,684	16,978	17,250	17,446	17,801	18,370	18,921
Ukraine	P	40,882	38,002	35,219	32,423	29,728	27,503	25,851	24,413
	M%	1.20	1.30	1.40	1.50	1.60	1.70	1.80	1.90
	M	491	494	493	486	476	468	465	464
Total	P	202,865	194,699	187,797	180,952	173,818	168,673	165,966	163,474
	M%	8.39	8.85	9.33	9.83	10.34	10.86	11.38	11.89
	M	17,018	17,222	17,517	17,783	17,969	18,317	18,884	19,435
	G%	−0.302	−0.411	−0.361	−0.371	−0.402	−0.300	−0.162	−0.151

4.3 Muslims in Western Europe

Western Europe consists of nine countries and territories: Andorra, Belgium, France, Gibraltar, Luxembourg, Monaco, the Netherlands, Portugal and Spain. It was the first European region that came under Muslim control. Indeed, Islam entered this region in 711 and remained until 1492. For a period of time, Muslims controlled most of the region: all of the Iberian Peninsula, West Mediterranean Islands, and half of France. Muslims later were driven out from the whole region and their number was reduced to zero. However, the current Muslim population in this region is not a continuum of these previous presences. Muslims started trickling back by the end of the nineteenth century from colonies of this region, many of which had majority Muslim population. Nevertheless, the Muslim population remained small and only picked up momentum after WWII, when the region needed more manpower for its economic prosperity.

Thus, the Muslim population has likely increased from none in 700 AD, to 0.85 million or 7.5% of the total population of this region in 800 AD, to 1.72 million or 14.8% in 900 AD, to 2.2 million or 18.5% in 1000 AD, to 2.5 million or 17.3% in 1100 AD, then decreased to 2.5 million or 14.8% in 1200 AD, to 2.4 million or 12.7% in 1300 AD, to 2.0 million or 8.9% in 1400 AD, to 1.41 million or 5.6% in 1500 AD, to 0.84 million or 2.7% in 1600 AD, to 90,000 or 0.2% in 1700, to almost none in 1800, then increased to 5,000 or 0.01% in 1900, to 4.3 million or 3.2% in 2000, to 9.5 million or 6.3% in 2020, and is projected to reach 14 million or 9% by 2050 and 18 million or 13.7% by 2100.

A plot of centennial estimates of the Muslim population and its percentage with respect to the total population in this region from 600 to 2100 is provided in Fig. 4.3a. A zoom in of this plot, providing a plot of decennial estimates of the Muslim population and its percentage with respect to the total population in this region from 1900 to 2100 is provided in Fig. 4.3b. The corresponding individual data for each country in this region is discussed below. In Section 4.3.10, the total population in each country in this region and the corresponding percentage and number of Muslims is presented centennially in Table 4.3a from 600 to 2100 and decennially in Tables 4.3b and 4.3c from 1790 to 2100.

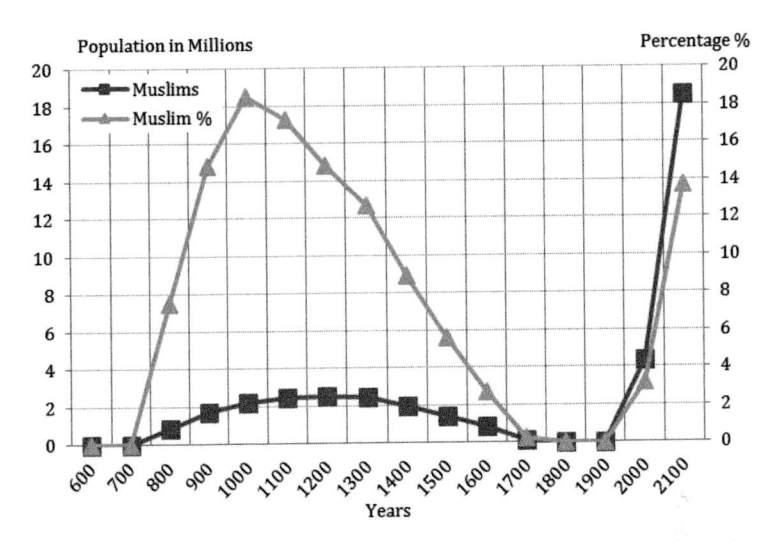

Figure 4.3a Plot of centennial estimates of the Muslim population and its percentage of the total population in the Western Europe: 600–2100 AD (1–1500 H).

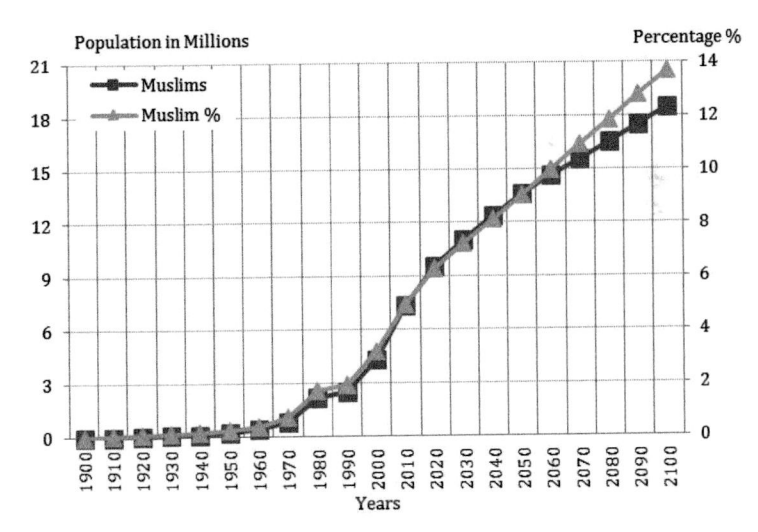

Figure 4.3b Plot of decennial estimates of the Muslim population and its percentage of the total population in the Western Europe: 1900–2100 AD (1320–1520 H).

4.3.1 Andorra

The Principality of Andorra is a small country with area 468 sq. km located in the eastern Pyrenees Mountains, between Spain and France. Its map is presented in Fig. 4.3.1. It was conquered by Muslims in 715 as they advanced in the Iberian Peninsula. Muslims lasted their grip on it in 800. Thus, it was under Muslim control for 85 years. There were no Muslims in Andorra in 1970. Later estimates are based on the number of Moroccan nationals living in the country as summarized in Table 4.3.1. Accordingly, the Muslim (Moroccan) population increased from 174 or 0.4% in 1982, to 211 or 0.5% in 1985, to 296 or 0.5% in 1990, to 406 or 0.6% in 1995, to 469 or 0.7% in 2000, to 504 or 0.6% in 2005, to 508 or 0.6% in 2010, but decreased to 449 or 0.6% in 2015 and 456 or 0.6% in 2018. According to WVS survey data, the percentage of Muslims was 1.10% in 2005. Thus, assuming that the percentage of Muslims will remain constant at 0.6%, the Muslim population is expected to remain less than 500 throughout this century.

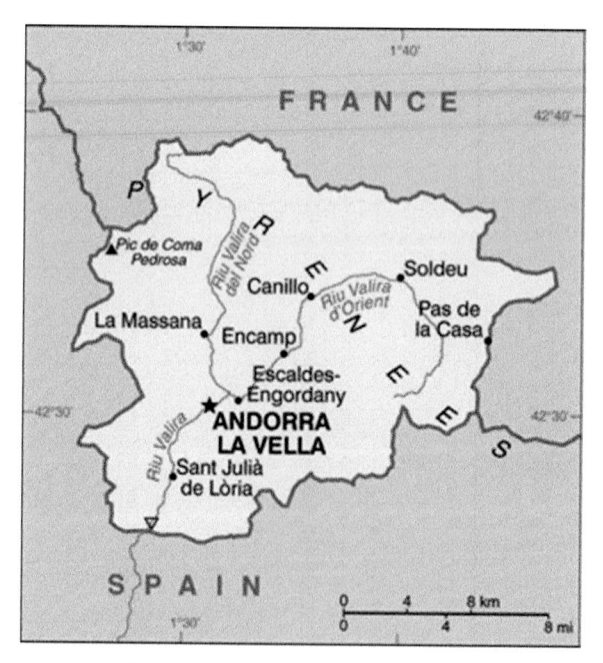

Figure 4.3.1 Map of the Principality of Andorra.

Table 4.3.1 Year-end evolution of the Muslim population in Andorra

Year	Population	Muslims	%	Source
1970	19,545	—	0.00	[KET76]es
1980	35,460	163	0.46	[AD]e
1985	44,596	174	0.39	[AD]e
1990	54,507	251	0.46	[AD]e
1995	63,859	317	0.50	[AD]e
2000	65,844	420	0.64	[AD]e
2005	77,081	469	0.61	[AD]e
2010	85,015	508	0.60	[AD]e
2015	78,014	449	0.58	[AD]e
2017	80,209	461	0.57	[AD]e
2050	76,000	450	0.60	es
2100	62,000	370	0.60	es

4.3.2 Belgium

The Kingdom of Belgium has an area of 30,528 sq. km and its map is presented in Fig. 4.3.2. According to census data, in 1890 there were no people from a Muslim majority country. The number of Turkish citizens between 1900 and 1947 summarized in Table 4.3.2a, which shows during that period the Muslim population constituted 0.01% of the total population. The largest population from countries with well over 90% Muslim majority (excluding Iran) between 1961 and 1991 is summarized in Table 4.3.2b and is used to estimate the number of Muslims during that period. Accordingly, a dramatic increase in the number of Muslims occurred in the 1960s when migration agreements were signed with Morocco and Turkey and then at the end of the 1960s with Algeria and Tunisia, to help with work force shortage in Belgium. This caused the Muslim population to increase from 1,200 or 0.01% in 1961, to 70,000 or 0.7% in 1970, then 0.19 million or 1.9% in 1981 and 0.25 million or 2.5% in 1991.

Towards the end of the second millennium more Muslims obtained Belgian citizenship, rendering estimates based on citizenship from Muslim-majority countries inaccurate. For example, in 1991 the number of up to third generation Moroccan descents was 153,288, while that of Turkish descent was 88,631. Thus by 1991, almost all Turkish descents and 93% of Moroccan descents retained their nationality and did not obtain Belgian citizenship. However, those who declared themselves Moroccans decrease to 140,303 in 1996 and 94,264 in 2001. Similarly, those declaring Turkish nationality decreased to 81,744 in 1996 and 47,725 in 2001. Thus, towards the end of the 1990s a substantial portion of these nationalities obtained Belgian citizenship.

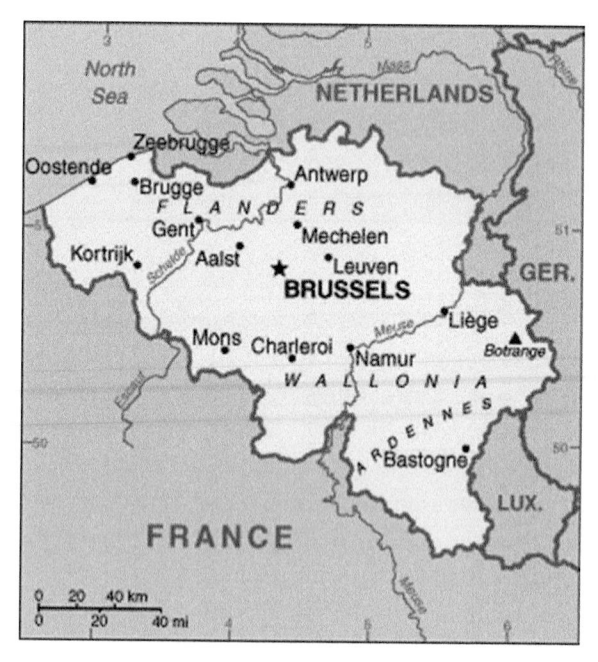

Figure 4.3.2 Map of the Kingdom of Belgium.

According to the European Values Survey (EVS), the percentage of Muslims was 0.64% in 1981, 0.32% in 1990, 2.81% in 1999 and 5.58% in 2009. Thus, assuming that the percentage of Muslims will continue to increase by one percentage point per decade, the Muslim population is expected to reach over 1.2 million or 10% by 2050 and 1.9 million or 15% by 2100. A summary of the data is provided in Table 4.3.2c.

Table 4.3.2a Evolution of the Turkish citizens in Belgium: 1890–1947

	1890	1900	1910	1920	1930	1947
Turkey	—	414	759	428	1,350	590

Table 4.3.2b Evolution of the largest Muslim nationalities in Belgium: 1961–1991

	1961	1970	1981	1991
Morocco	461	39,294	105,133	142,125
Turkey	320	20,312	63,587	88,688
Algeria	202	6,621	10,796	10,701
Tunisia	204	2,201	6,871	6,324
Pakistan	17	125	1,291	1,842
Syria	—	—	409	865
Egypt	—	339	314	530
Senegal	—	—	406	488
Iraq	—	482	159	234
Total	1,204	69,374	188,966	251,797

Table 4.3.2c Evolution of the Muslim population in Belgium

Year	Population	Muslims	%	Source
1890	6,069,321	—	0.00	[BEH]e
1900	6,693,548	414	0.01	[BEH]e
1910	7,423,784	759	0.01	[BEH]e
1920	7,465,782	428	0.01	[BEH]e
1930	8,411,234	1,350	0.02	[BEH]e
1947	8,879,814	590	0.01	[BEH]e
1961	9,189,741	1,204	0.01	[BEH]e
1970	9,650,944	69,374	0.72	[BEH]e
1981	9,963,374	188,966	1.90	[BEH]e
1991	9,978,681	251,797	2.52	[EU]e
1999	10,237,000	288,000	2.81	[EVS]s
2009	10,854,000	606,000	5.58	[EVS]s
2050	12,221,000	1,222,000	10.00	es
2100	12,493,000	1,874,000	15.00	es

4.3.3 France

After conquering the Iberian Peninsula in 715, the Muslim forces continued to conquer southern France. In 719 they conquered Carcassonne and Narbonne, and advanced until they besieged Toulouse. The incursions were led by the governor of al-Andalus, Assamh bnu Malik al-Kholani, who died at the Battle of Toulouse in 721. He was succeeded by Anbasa bnu Sahim al-Kalbi, who continued the conquest by capturing in 725, Nîmes, Lyon, Dijon, Autun, Langres, and Sens (100 km southeast of Paris). He was then succeeded by Abdul Rahman Bnu Abdillah al-Ghafiqi, who continued the conquest in 732 from the southwest region of France, by capturing Bayonne, Bordeaux, and Poitiers. In October 732, the Muslims were defeated just south of Châtellerault, about 20 km northeast of Poitiers, in the Battle of Tours/Poitiers, or the Battle of Balat Ashuhada (Arabic for Courts of the Martyrs), in which al-Ghafiqi lost his life. In 734, the Muslim governor of Narbonne, Yusuf ben Abdul Rahman, conquered Arles and Avignon, between Nîmes and Marseille.

The Muslims lacked demographic power to retain the new acquired vast land. Thus, in 736 the Christian forces then regained control over Bordeaux, Avignon in 737, Narbonne in 751, Nîmes in 754, and Carcassonne in 759. Thus, Muslims were able to conquer about two thirds of current France and were able to stay there up to forty years. There were other successful attempts by Muslim Andalusian seamen, who settled in western coast, between Nice and Toulon, starting with Saint Tropez in 891. Their state eventually stretched from Marseille to Genoa, Italy in 934, but it was toppled by the Christian forces by 973. As far as Corsica is concerned, it was conquered by the Aghlabid Muslim Dynasty in 806 after conquering Sicily and Sardinia. Muslims lost the island during the Fatimid Dynasty in 930. Thus, it was under Muslim control for 126 years.

Currently, the European region of the French Republic has an area of 551,500 sq. km, which includes the West Mediterranean Island of Corsica (8,680 sq. km). A map of France is presented in Fig. 4.3.3. The Muslim presence in France was not tolerated until France occupied Algeria in 1830. Information on the religious affiliation of the population was collected in the censuses of 1851 to 1872. The number of those following a religion other than Christianity and Judaism increased from 1,295 in 1861, to 1,400 in 1866, to 3,071 in 1872, remaining less than 0.01% of the total population.

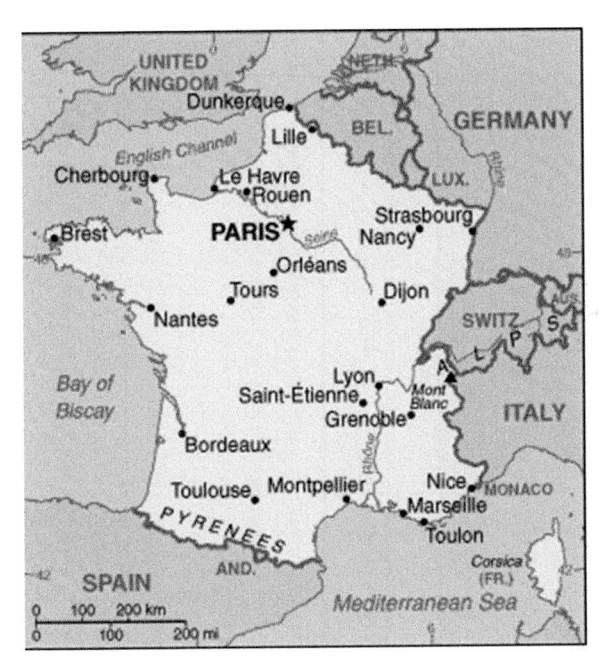

Figure 4.3.3 Map of the French Republic.

Table 4.3.3a presents the census data on the number of Turks and Africans recorded in the corresponding censuses. Estimating the number of Muslims based on these ethnicities indicates that the Muslim population remained less than 0.01% of the total population throughout the nineteenth century. It increased in number from 438 in 1861, to 565 in 1866, to 1,173 in 1872, to 1,174 in 1876, to 1,494 in 1881, to 1,612 in 1886, to 2,664 in 1891, to 3,054 or 0.01% in 1896, to 11,278 or 0.03% in 1911, to 42,706 or 0.1% in 1921. The African population then increased to 72,000 or 0.2% in 1926, to 105,000 or 0.3% in 1931, and then decreased to 87,000 or 0.2% in 1936.

Table 4.3.3a Evolution of the largest Muslim ethnicities in France from 1861 to 1921

	1861	1866	1872	1876	1881	1886	1891	1896	1911	1921
Turks	438	565	1,173	1,174	1,494	1,612	1,851	2,362	8,158	5,040
Africans							813	692	3,120	37,666
Total	438	565	1,173	1,174	1,494	1,612	2,664	3,054	11,278	42,706

Table 4.3.3b Estimate of the Muslim population in France based on the evolution of the largest Muslim nationalities from 1946 to 2011

	1946	1954	1962	1968	1975	1982	1990	1999	2006	2011
Algeria	22,114	211,675	350,484	473,812	710,690	805,116	614,207	475,216	481,211	465,849
Morocco	16,458	10,734	3,320	84,236	260,025	441,308	572,652	506,305	460,708	433,026
Tunisia	1,916	4,800	26,569	61,028	139,735	190,800	206,336	153,574	145,990	150,109
Other Africa	13,517	2,296	17,787	33,020	81,850	157,548	239,947	282,736	414,432	515,412
Turkey	7,770	5,273	6,000	7,628	50,860	122,260	197,712	205,589	223,629	219,534
Total	61,775	234,778	404,160	659,724	1,243,160	1,717,032	1,830,854	1,623,420	1,725,970	1,783,930

Table 4.3.3c Evolution of the Muslim population in France

Year	Population	Muslims	%	Source
1861	37,386,313	438	0.00	[FR21]e
1866	38,067,064	565	0.00	[FR21]e
1872	36,102,921	1,173	0.00	[FR21]e
1876	36,905,788	1,174	0.00	[FR21]e
1881	37,672,048	1,494	0.00	[FR21]e
1886	38,218,903	1,612	0.00	[FR21]e
1891	38,342,948	2,664	0.01	[FR21]e
1896	38,517,975	3,054	0.01	[FR21]e
1911	39,601,509	11,278	0.03	[FR21]e
1921	39,209,158	42,706	0.11	[FR21]e
1926	40,743,874	72,000	0.18	[FR99]e
1931	41,834,923	105,000	0.25	[FR99]e
1936	41,907,056	87,000	0.21	[FR99]e
1946	40,502,513	61,775	0.15	[FR99]e
1954	42,777,162	234,778	0.55	[FR99]e
1962	46,520,271	404,160	0.87	[FR99]e
1968	49,778,540	659,724	1.33	[FR99]e
1975	52,655,864	1,243,160	2.36	[FR99]e
1982	54,334,871	1,717,032	3.16	[FR99]e
1990	56,615,155	1,830,854	3.23	[FR99]e
2008	62,309,000	4,195,000	6.73	[EVS]s
2050	67,587,000	7,435,000	11.00	es
2100	65,498,000	10,480,000	16.00	es

After World War II, France needed more manpower, many of whom came from northwestern Africa, being a French colony. To estimate the number of Muslims post WWII, we use census data which provides citizenship data on foreign population residing in France. We assume that Algerians, Moroccans, Tunisians, Turks and Africans are Muslim (most of the French African colonies were Muslim majority). Table 4.3.3b presents the data between 1946 and 2011. Accordingly, the Muslim population increased from 0.06 million or 0.2% in 1946, to 0.23 million or 0.6% in 1954, to 0.40 million or 0.9% in 1962, to 0.66 million or 1.3% in 1968, to 1.24 million or 2.4% in 1975, to 1.71 million or 3.2% in 1982, to 1.83 million or 3.2% in 1990. Since 1980s however, more Muslims obtained French citizenship, rendering estimates based on citizenship from Muslim-majority countries inaccurate. According to the European values Survey (EVS), the percentage of Muslims was 6.73% in 2008. Thus, assuming that the percentage of Muslims will continue to increase by one percentage point per decade, the Muslim population is expected to increase to 7.4 million or 11% by 2050 and 10.5 million or 16% by 2100. The evolution of Muslims in France since 1861 to 2100 is summarized in Table 4.3.3c.

4.3.4 Gibraltar

This overseas territory of the United Kingdom has an area of 7 sq. km and its map is presented in Fig. 4.3.4. It is named after the Muslim General Tariq bnu Zeyad, who crossed from Tangier in north Morocco, and started its campaign of conquering the Iberian Peninsula, by capturing this Mountain in 711. It was named Jabal Tariq, or Traiq's mountain, while the opposite mountain in Morocco across the Strait was named Jabal Musa, after his superior Musa bnu Nusayr, who later joined him in their successful conquest of the Iberian Peninsula. The Mountain fell to the Christians between 1309 and 1331, but was recaptured by Muslims, until they lost it for good to the Christian Castilians in 1462 as the Nasrid Dynasty was losing territory. The Castilians then expelled all Muslims from this area. Thus, the Mountain was under Muslim rule for 729 years. It was seized by the British in 1702 and is under their control since then.

As summarized in Table 4.3.4, British censuses in the eighteenth century show the presence of no Muslims in the Peninsula. In fact, until the 1960s, the presence of Muslims was negligible, peaking at fifty or 0.3% in 1911. Post WWII censuses show the increase of Muslims from none in 1951, to six in 1961, to 1,989 or 8.1% in 1970. The significant increase is due to Spain's closure of the border, which caused shortage in labor force. Thus, Spanish workforce was replaced by Moroccan ones, consequently the vast majority of Muslims in Gibraltar are of Moroccan descent. The Muslim population then increased to 2,124 or 8.0% in 1981, but decreased to 1,850 or 6.9%

in 1991, to 1,102 or 4.0% in 2001 and 1,066 or 3.6% in 2012. Thus, assuming that the percentage of Muslims will remain constant, the Muslim population is expected to continue to decrease in number and remain around one thousand throughout this century.

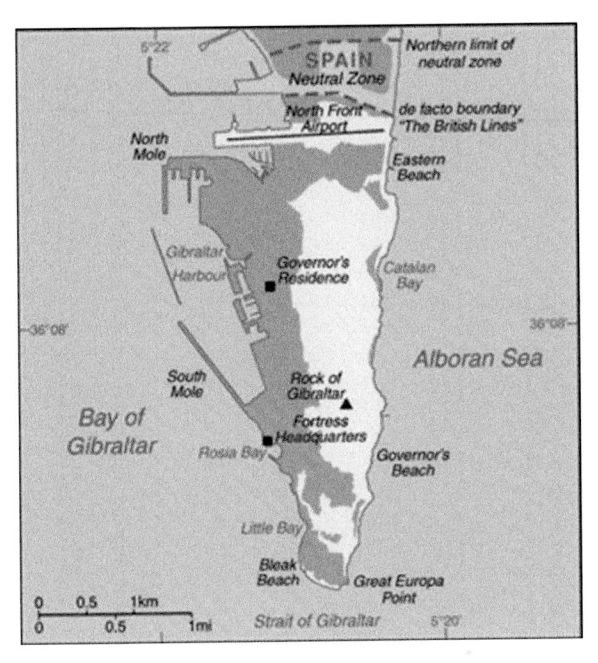

Figure 4.3.4 Map of Gibraltar.

4.3.5   Luxembourg

The Grand Duchy of Luxembourg has an area of 2,586 sq. km and its map is presented in Fig. 4.3.5. Census data show that the number of adherents to religions; other than Christianity and Judaism, were eight or 0.00% in 1871 and 1905.

Censuses in Luxembourg gathered information on the nationality of foreign citizens living in the country. Based on these numbers the Muslim population increased from 54 (Turks) or 0.02% in 1970, to 291 (147 Turks, 74 Moroccans, 70 Tunisians) or 0.1% in 1981, to 513 (190 Turks, 120 Moroccans, 109 Tunisians, 94 Algerians) or 0.1% in 1991. A dramatic increase in the number of Muslims happened

Table 4.3.4 Evolution of the Muslim population in Gibraltar

Year	Population	Muslims	%	Source
1753	1,816	—	0.00	[GI12]c
1767	2,710	—	0.00	[GI12]c
1771	3,201	—	0.00	[GI791]c
1777	3,201	—	0.00	[GI12]c
1783	2,890	—	0.00	[GI12]c
1787	3,386	—	0.00	[GI12]c
1791	2,874	—	0.00	[GI792]c
1814	10,136	19	0.19	[GI814]c
1816	11,424	21	0.18	[GI816]c
1834	15,008	3	0.02	[GI834]c
1868	16,934	15	0.09	[GI868]c
1871	18,695	26	0.14	[GI871]c
1878	18,008	44	0.24	[GI901]c
1881	18,381	35	0.19	[GI881]c
1911	19,825	50	0.25	[GI911]c
1921	18,540	42	0.23	[GI921]c
1931	17,613	16	0.09	[GI931]c
1951	21,314	—	0.00	[GI951]c
1961	21,636	6	0.03	[GI970]c
1970	24,672	1,989	8.06	[GI12]c
1981	26,479	2,124	8.02	[GI12]c
1991	26,703	1,850	6.93	[GI12]c
2001	27,495	1,102	4.01	[GI12]c
2012	32,150	1,166	3.63	[GI12]c
2050	34,000	1,200	3.63	es
2100	27,000	1,000	3.63	es

afterwards, including an influx of refugees from Bosnia and Kosovo following the breakup of former Yugoslavia. By 2011, numbers of nationals from overwhelmingly Muslim majority countries exceeded 6,000, with over a third from Bosnia and a fourth from Kosovo. According to the European Values Survey (EVS), the percentage of Muslims was 1.00% in 1999 and 3.11% in 2008. Thus, assuming that the percentage of Muslims will continue to increase by half of a percentage point per decade, the Muslim population is expected to reach 40,000 or 5.5% by 2050 and 80,000 or 8% by 2100. The data is summarized in Table 4.3.5.

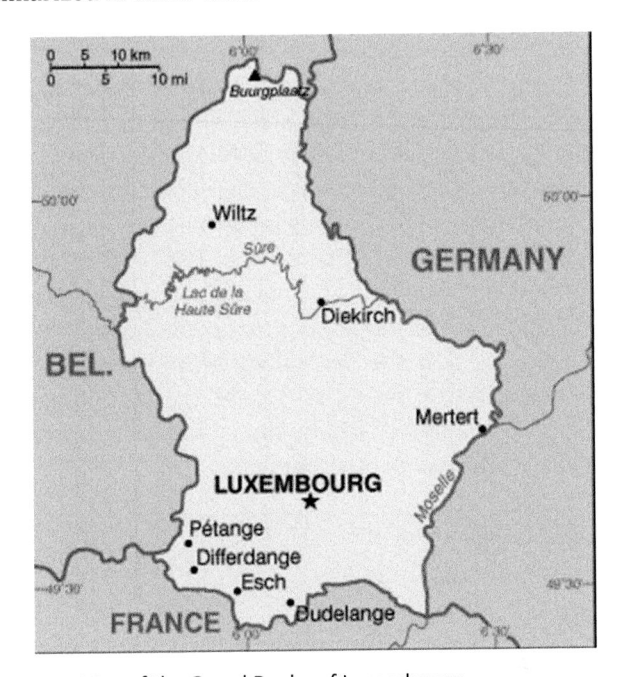

Figure 4.3.5 Map of the Grand Duchy of Luxembourg.

Table 4.3.5 Evolution of the Muslim population in Luxembourg

Year	Population	Muslims	%	Source
1871	204,028	<8	0.00	[LUH]c
1905	222,000	<8	0.00	[LUH]c
1970	339,848	54	0.02	[UNE]e
1981	364,602	291	0.08	[UNE]e
1991	384,634	513	0.13	[EU]e
1999	431,000	4,300	1.00	[EVS]s
2008	485,000	15,100	3.11	[EVS]s
2050	790,000	43,000	5.50	es
2100	985,000	79,000	8.00	es

4.3.6 Monaco

The Principality of Monaco has a total area of 2 sq. km and its map is presented in Fig. 4.3.6. It was under Muslim control from 891 to 973 by Muslim Andalusians who controlled southeast current France. Censuses here gather information on the nationality of foreign citizens living in the country. Estimate of Muslims from 1908 to 1951 is taken to be the number of Turks recorded in the corresponding census. The largest ten nationalities of Muslim majority countries since 1956 is provided in Table 4.3.6a. Based on these numbers and as shown in Table 4.3.6b, the Muslim population increased from double digits until 1961, to 567 or 2.1% in 1982 and peaked at 905 or 3.0% in 1990, but remained at triple digits, changing from 886 or 2.0% in 2000, to 706 or 2.0% in 2008, to 865 or 2.3% in 2016.

Thus, assuming that the percentage of Muslims will increase by half of a percentage point per decade, the Muslim population is expected to near 2,000 or 4% by the middle of this century and 4,000 or 6.5% by the end of this century.

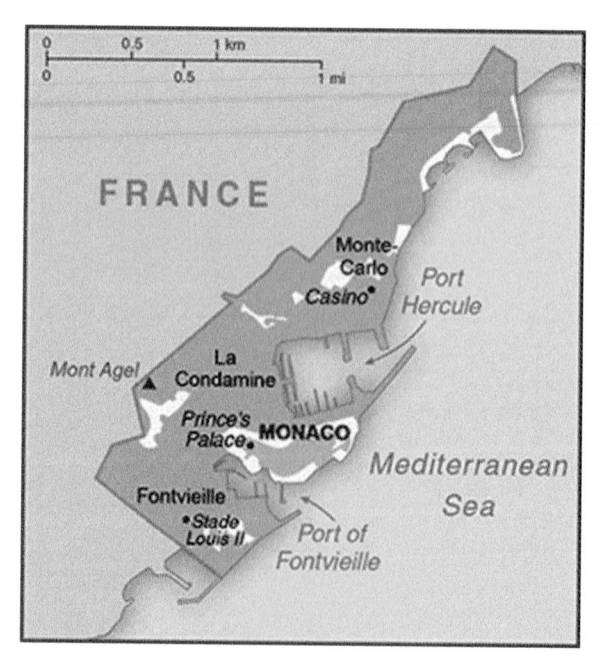

Figure 4.3.6 Map of the Principality of Monaco.

Table 4.3.6a Evolution of the largest Muslim nationality populations in Monaco since 1956

	1956	1961	1982	1990	2000	2008	2016
Morocco		6	151	252	307	242	239
Lebanon		10	129	262	186	145	195
Tunisia		11	60	103	103	100	98
Turkey	25	17	98	77	70	57	90
Algeria			22	29	28	27	72
Iran		17	41	68	70	46	60
Egypt		6	48	80	80	53	54
Senegal			13	15	24	16	33
Syria			5	19	18	20	17
Iraq	15						7
Total	40	67	567	905	886	706	865

Table 4.3.6b Evolution of the Muslim population in Monaco

Year	Population	Muslims	%	Source
1908	19,121	16	0.08	[MC61]e
1913	22,956	28	0.12	[MC61]e
1918	15,960	16	0.10	[MC61]e
1923	22,153	51	0.23	[MC61]e
1928	24,927	53	0.21	[MC61]e
1933	22,994	34	0.15	[MC61]e
1938	23,956	22	0.08	[MC61]e
1946	19,242	25	0.13	[MC61]e
1951	20,202	20	0.10	[MC61]e
1956	20,422	40	0.20	[UN63]e
1961	22,297	67	0.30	[MC61]e
1982	27,063	567	2.10	[MC08]e
1990	29,972	905	3.02	[MC08]e
2000	32,020	886	2.77	[MC08]e
2008	35,352	706	2.00	[MC08]e
2016	37,308	865	2.32	[MC16]e
2050	46,000	1,800	4.00	es
2100	64,000	4,200	6.50	es

4.3.7 The Netherlands

The Kingdom of the Netherlands has an area of 41,543 sq. km and its map is presented in Fig. 4.3.7. The first Muslims in this country came from Indonesia, which was a Dutch colony under the name Dutch East Indies from 1800 to 1949. Censuses inquiring on religious affiliation were held regularly from 1879 until 1970. According to theses censuses, the number of Muslims remained at 0.00 until the second half of the twentieth century, increasing from double digits between 1879 and 1909, to triple digits between 1920 and 1947, then 1,400 or 0.01% in 1960 and 54,000 or 0.4% in 1970.

Figure 4.3.7 Map of the Kingdom of the Netherlands.

Table 4.3.7a shows the evolution of first-generation population from Turkey and Morocco from 1899 to 1991, which we use to estimate the Muslim population in the Netherlands. Accordingly, the Muslim population increased 0.12 million or 0.9% in 1976, to 0.22 million or 1.6% in 1981, to 0.27 million or 1.9% in 1986, to 0.36 million or 2.4% in 1991.

Table 4.3.7a Estimate of Muslim population in the Netherlands based on foreign background: 1899–1991 [NLE]

	1899	1909	1920	1930	1947	1956	1961
Turkey	—	100	100	100	—	—	100
Morocco	—	—	—	—	—	—	200
Total	—	100	100	100	—	—	300

Table 4.3.7a (Continued)

	1966	1971	1976	1981	1986	1991
Turkey	8,700	31,800	76,500	138,500	156,400	203,500
Morocco	5,800	23,000	42,200	83,400	116,400	156,900
Total	14,500	54,800	118,700	221,900	272,800	360,400

Thus, the Muslim population remained negligible until the 1960s, when tens of thousands of Turkish and Moroccan 'guest workers' arrived in the Netherlands to work in the country's industrial sector. The number of Muslim increased further in the 1970s as more former Dutch colonies gained independence, including Indonesia and Suriname. The number continued to increase in the 1980s as the migrant workers were joined by their families. From the late 1980s, the number of Muslims also increased as a result of a growing influx of political refugees and asylum seekers from the Middle East and East Africa.

Since 1996, Statistics Netherlands provides data on the population with foreign background, which it defines as follows: "Characteristic showing with which country someone actually is closely related given their own country of birth and that of their parents. For someone with a first-generation foreign background the origin is indicated as the country of birth of that person. For someone with a second-generation foreign background the origin is indicated as the country of birth of the mother of that person. If the mother's country of birth is the Netherlands than the origin is indicated as the father's country of birth." Statistics Netherlands also held religious survey (Permanent Quality of Life Survey or POLS) in 2006 that found the percentage of Muslims in those of foreign background as almost 100% for Pakistanis, Somalis, around 90% of Turks, Moroccans, around 80% of Afghanis, 60% of Egyptians, Iraqis, Tunisians, 40% of Iranians, 10% of Surinamese and 2% of Indonesians.

Table 4.3.7b Beginning of year estimate of the Muslim population in the Netherlands based on country of origin since 1996 [NLO]

	1996			2001			2006			2011			2016		
	1st Gen	2nd Gen	Total	1st Gen	2nd Gen	Total	1st Gen	2nd Gen	Total	1st Gen	2nd Gen	Total	1st Gen	2nd Gen	Total
Turkey	167,248	104,266	271,514	181,595	138,005	319,600	195,711	168,622	364,333	197,042	191,925	388,967	190,621	206,850	397,471
Morocco	140,572	84,516	225,088	155,669	117,083	272,752	168,504	154,735	323,239	167,607	188,276	355,883	168,336	217,425	385,761
Syria	2,846	758	3,604	4,938	1,550	6,488	6,620	2,571	9,191	7,093	3,566	10,659	38,487	5,351	43,838
Iraq	10,148	1,130	11,278	33,685	4,506	38,191	35,246	8,511	43,757	40,938	11,920	52,858	40,831	15,438	56,269
Afghan.	4,537	379	4,916	24,254	2,140	26,394	31,987	5,259	37,246	31,823	8,241	40,064	33,030	11,309	44,339
Somalia	17,156	2,904	20,060	21,705	7,926	29,631	13,691	6,202	19,893	23,177	8,060	31,237	26,803	12,662	39,465
Egypt	7,726	3,872	11,598	9,359	5,806	15,165	11,131	7,864	18,995	11,919	9,154	21,073	13,066	10,132	23,198
Pakistan	9,628	4,499	14,127	10,649	6,138	16,787	10,828	7,356	18,184	11,131	8,277	19,408	12,093	9,354	21,447
Tunisia	3,345	2,453	5,798	3,754	3,154	6,908	4,126	3,905	8,031	4,198	4,492	8,690	4,18	4,997	9,415
Algeria	2,861	2,018	4,879	3,914	2,754	6,668	3,888	3,487	7,375	3,881	3,954	7,835	3,899	4,274	8,173
Sudan	806	137	943	4,818	679	5,497	5,290	1,623	6,913	4,324	1,884	6,208	4,664	2,047	6,711
Total	366,873	206,932	573,805	454,340	289,741	744,081	487,022	370,135	857,157	503,133	439,749	942,882	536,248	499,839	1,036,087

We assume that the largest ten populations of origin from a 95% or more Muslim-majority country are Muslim (except Iran), and use their numbers to estimate the number of Muslims in the Netherlands. So, we exclude Indonesia, Iran, Malaysia, Sierra Leone, Lebanon, Guinea, etc. The data and aggregation steps are shown in Table 4.3.7b, which also shows that descents of Turkey and Morocco each constitute more than a third of the Muslim population in the Netherlands. Furthermore, about half of the Muslim population is second generation, versus a third before 1996. Accordingly, the Muslim population continued to increase to 0.57 million or 3.7% in 1996, to 0.74 million or 4.7% in 2001, to 0.86 million or 5.3% in 2006, to 0.94 million or 5.7% in 2011, to 1.04 million or 6.1% in 2016 and 1.11 million or 6.5% in 2018. Thus, assuming that the percentage of Muslims will continue to increase by half of a percentage point per decade, the Muslim population is expected to reach 1.4 million or 8% by 2050 and 1.7 million or 10.5% by 2100. A summary of the data is provided in Table 4.3.7c.

Table 4.3.7c Evolution of the Muslim population in the Netherlands

Year	Population	Muslims	%	Source
1879	4,012,693	49	0.00	[ROD]c
1889	4,511,415	47	0.00	[ROD]c
1899	5,104,137	29	0.00	[ROD]c
1909	5,858,175	54	0.00	[ROD]c
1920	6,865,314	203	0.00	[ROD]c
1930	7,935,565	445	0.00	[ROD]c
1947	9,625,499	300	0.00	[NL971]c
1960	11,461,964	1,399	0.01	[NL971]c
1970	13,060,115	53,975	0.41	[NL971]c
1976	13,733,578	118,700	0.86	[NLE]e
1981	14,208,586	221,900	1.56	[NLE]e
1986	14,529,430	272,800	1.88	[NLE]e
1991	15,010,445	360,400	2.40	[NLE]e
1996	15,493,889	573,805	3.70	[NLO]e
2001	15,987,075	744,081	4.65	[NLO]e
2006	16,334,210	857,157	5.25	[NLO]e
2011	16,665,799	942,882	5.66	[NLO]e
2016	16,979,120	1,036,087	6.10	[NLO]e
2018	17,081,507	1,114,293	6.52	[NLO]e
2050	17,165,000	1,373,000	8.00	es
2100	15,760,000	1,655,000	10.50	es

4.3.8 Portugal

Islam first entered in 711 as Tariq bnu Ziyad troops conquered the Iberian Peninsula. It was part of the Umayyad Empire in al-Andalus. As Muslims lost their grip on the northwestern part of the Iberian Peninsula, the Christian forces continued their slow advance towards the south. They conquered Porto (north) in 868, then Viseu in 1058, then Lisbon (west) in 1147, Évora (middle) in 1166, and finally Faro (south) in 1249. Thus, Portugal was under Muslim control between 157 years in the north and 538 years in the south. The Christians enslaved the Muslim population and confiscated their lands and houses by giving it to Christian settlers from the north. The Muslim population was expelled from Portugal in 1540, most of who settled in north Morocco.

Currently, the Portuguese Republic has a total area of 92,225 sq. km, which includes two autonomous regions of the Azores (2,322 sq. km) and Madeira (801 sq. km). The mainland of Portugal is organized into five regions: Northern (21,286 sq. km), Central (28,199 sq. km), Lisbon (3,015 sq. km), Alentejo (31,605 sq. km) and Algarve Region (4,997 sq. km). A map of the mainland Portugal is presented in Fig. 4.3.8a.

The Azores archipelago is in the northern Atlantic Ocean, 1,370 km west of Portugal mainland. The islands were uninhabited when they were annexed by Portugal in 1427. There are nine major Azorean islands and an islet cluster, in three main groups. These are Flores (141 sq. km) and Corvo (17 sq. km), to the west; Graciosa (61 sq. km), Terceira (402 sq. km), São Jorge (244 sq. km), Pico (445 sq. km), and Faial (173 sq. km) in the center; and São Miguel (745 sq. km), Santa Maria (97 sq. km), and the Formigas Reef (9 sq. km) to the east. They extend for more than 600 km (370 mi) and lie in a northwest-southeast direction. A map of the Azores is presented in Fig. 4.3.8b.

The Madeira archipelago is also located in the northern Atlantic Ocean, 780 km southwest of Portugal mainland, 660 km west of Morocco and 410 km north of the Canary Islands. The islands were uninhabited when they were annexed by Portugal in 1419. The archipelago consists of the main island of Madeira (741 sq. km)

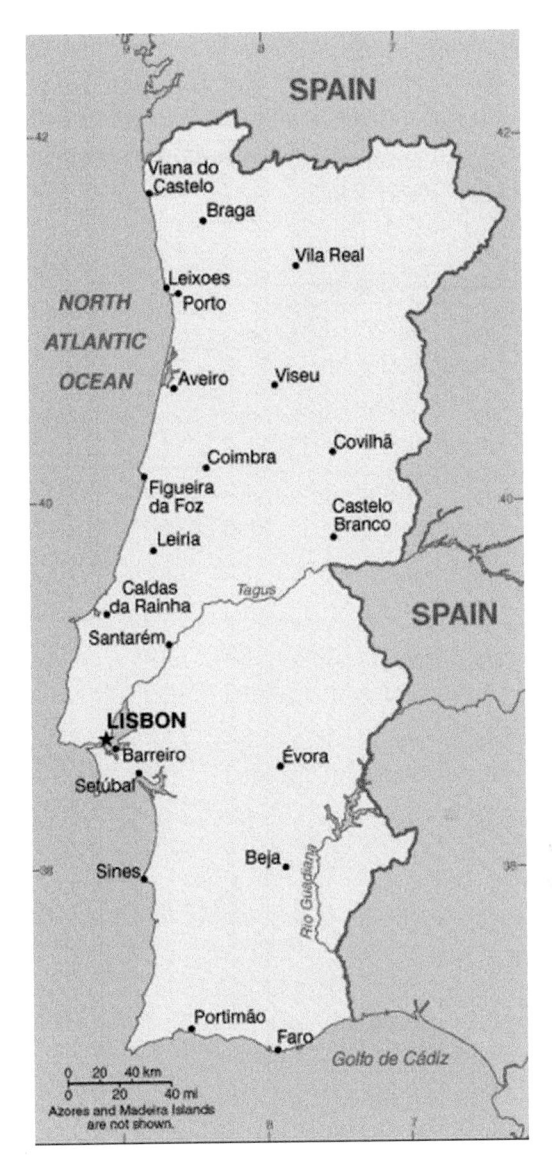

Figure 4.3.8a Map of the Portuguese Republic.

where almost all the population lives, Porto Santo (42 sq. km), and the Desertas. The latter is uninhabited nature reserve that consists of Deserta Grande (10 sq. km), Bugio (3 sq. km) and Chão Islet (1 sq. km). A map of Madeira is presented in Fig. 4.3.8c. Madeira is administered together with the separate archipelago of the Savage (Selvagens) Islands (3 sq. km), located 160 km north of the Canary Islands and 300 km south of Madeira Island, and is also uninhabited nature reserve. A map of the Savages is presented in Fig. 4.3.9b together with the Canary Islands.

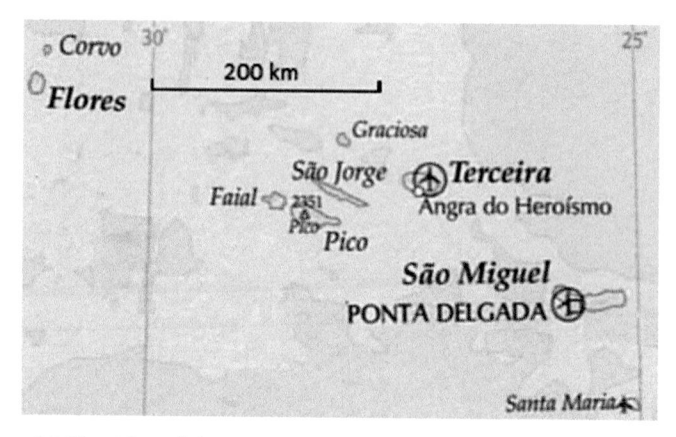

Figure 4.3.8b Map of the Azores.

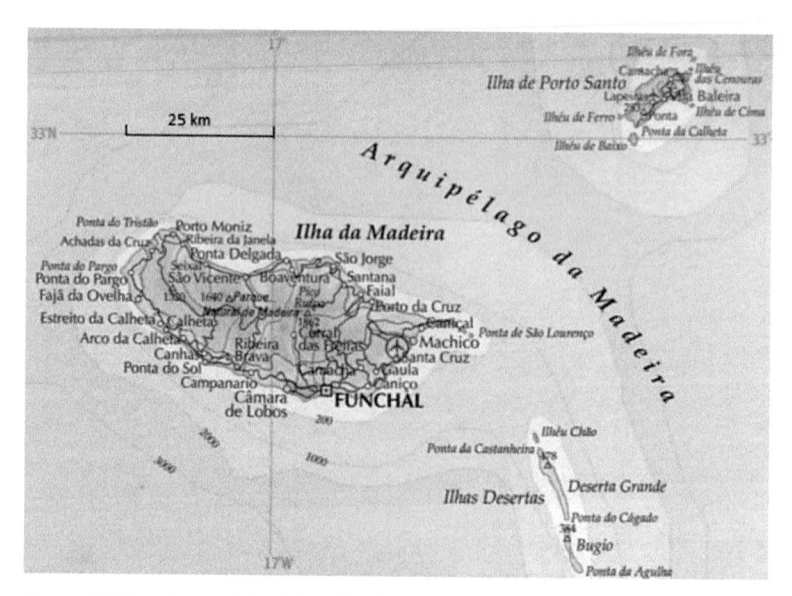

Figure 4.3.8c Map of Madeira Islands.

As shown in Table 4.3.8a and based on census data, the Muslim population in Portugal increased from none in 1890, to 365 or 0.0% in 1970, to 4,335 or 0.1% in 1981, to 9,134 or 0.1% in 1991, to 12,014 or 0.2% in 2001, to 20,640 or 0.3% in 2011. The data was collected from people fifteen years old and above in 2001 and later, and twelve years old in 1991 and before. The difference in 1991 between the two age groups is only 25 Muslims and 4,107 in total population. Thus, assuming that the percentage of Muslims will continue to increase by a tenth of a percentage point per decade, the Muslim population is expected to remain less than 0.1 million throughout this century, comprising 0.6% of the total population by 2050 and 1.1% by 2100.

Table 4.3.8a Evolution of the Muslim population in Portugal

Year	Population	Muslims	%	Source
1890	5,049,729	—	0.00	[EUH]c
1970	8,458,425	365	0.00	[UN79]c
1981	6,721,889	4,335	0.06	[UN88]c
1991	6,900,097	9,134	0.13	[UN04]c
2001	7,912,693	12,014	0.15	[UN]c
2011	8,244,975	20,640	0.25	[UN]c
2050	9,085,000	55,000	0.60	es
2100	6,985,000	77,000	1.10	es

Table 4.3.8b shows the evolution of Muslim population per region since 1981. Over two-thirds of the Muslim population lives in the capital Lisbon, where they remain less than 1% of the total population. The fraction of Muslims living in the capital is decreasing as more Muslims live in the other regions. On the islands, Muslims remain less than 0.1% of the total population. Distribution of Muslims per island is presented in Table 4.3.8c.

Table 4.3.8b Evolution of the Muslim population in Portugal per region since 1981. *P*: Total population, *M*: Muslim population, *M%*: Percentage of Muslim population, *MR%*: Muslim Ratio in percentage

Region		1981	1991	2001	2011
Northern	P	2,107,330	2,534,160	2,864,615	2,953,847
	M%	0.02	0.09	0.03	0.08
	M	405	2,257	835	2,263
	MR%	9.34	24.71	6.95	10.96
Central	P	1,689,268	1,251,955	1,858,740	1,863,346
	M%	0.01	0.05	0.03	0.09
	M	223	564	638	1,611
	MR%	5.14	6.17	5.31	7.81
Lisbon	P	1,384,917	2,204,272	1,952,112	2,094,657
	M%	0.20	0.26	0.49	0.68
	M	2,710	5,792	9,600	14,202
	MR%	62.51	63.41	79.91	68.81
Alentejo	P	980,777	334,955	580,452	585,170
	M%	0.10	0.07	0.04	0.09
	M	935	226	219	552
	MR%	21.57	2.47	1.82	2.67
Algarve	P	199,838	220,084	289,637	340,770
	M%	0.02	0.10	0.22	0.50
	M	33	219	630	1,709
	MR%	0.76	2.40	5.24	8.28
Azores	P	176,919	176,087	182,691	193,193
	M%	0.01	0.03	0.01	0.07
	M	9	47	19	136
	MR%	0.21	0.51	0.16	0.66
Madeira	P	182,840	178,572	184,446	213,992
	M%	0.01	0.02	0.04	0.08
	M	20	29	73	167
	MR%	0.46	0.32	0.61	0.81

Table 4.3.8c Evolution of the Muslim population in the Portuguese Islands since 1981. *P*: Total population, *M*: Muslim population, *M%*: Percentage of Muslim population

Island		1981	1991	2001	2011
Corvo	P	311	331	369	362
	M%	0.00	0.00	0.00	4.42
	M	—	—	—	16
Flores	P	3,461	3,391	3,167	3,058
	M%	0.00	0.00	0.00	0.23
	M	—	—	—	7
Faial	P	11,667	11,112	11,808	11,934
	M%	0.01	0.03	0.01	0.04
	M	1	3	1	5
Pico	P	12,326	11,589	11,908	11,383
	M%	0.01	0.01	0.02	0.05
	M	1	1	2	6
São Jorge	P	7,861	7,229	7,614	7,560
	M%	0.00	0.00	0.00	0.00
	M	—	—	—	—
Graciosa	P	4,236	4,080	3,871	3,623
	M%	0.00	0.00	0.00	0.03
	M	—	—	—	1
Terceira	P	40,382	42,196	43,455	45,058
	M%	0.00	0.02	0.01	0.07
	M	2	10	5	30
São Miguel	P	91,982	91,639	96,385	105,848
	M%	0.01	0.03	0.01	0.07
	M	5	31	10	69
Santa Maria	P	4,692	4,520	4,114	4,367
	M%	0.00	0.04	0.02	0.05
	M	—	2	1	2
Madeira	P	179,638	175,063	194,431	209,621
	M%	0.01	0.02	0.04	0.08
	M	19	28	73	165
Porto Santo	P	3,202	3,509	3,679	4,371
	M%	0.03	0.00	0.00	0.05
	M	1	1	—	2

4.3.9 Spain

Islam entered Spain when its residents requested the help of Muslims who just conquered North Africa. The residents of the Iberian Peninsula were under the wrath of the German Visigothic Kingdom. The Muslim conquest occurred during the Umayyad Empire, within the reign of al-Walid I bnu Abdel Malik, who reigned from 705 to 715, and was the sixth Umayyad King. The Muslim troops in northwest Africa were under the leadership of Musa bnu Nusayr, who sent troops under the leadership of Tariq bnu Ziyad to conquer the Peninsula. The mission was a great success and the whole peninsula fell to Muslim control between 711 and 718, after the decisive Battle of Guadalete. Accordingly, Muslims captured Toledo in 712, which was the capital of the Visigothic Kingdom, and by 715, Barcelona in the northeastern shore and the rest of the Peninsula was under Muslim control. The Balearic Islands were captured in 903.

While the Umayyad Empire fell down in the East in 750, and replaced by the Abbasid Empire, Abdul Rahamn I Addakhil (the Enterer) bnu Muawiya bnu Hisham, continued in 756 the Empire in the Iberian Peninsula, or al-Andalus, with Cordoba as its capital. The dynasty lasted until 1031, when the country fell into 23 small Muslim emirates, which were easy prey for Christian forces from the north.

Many Christians, who were not happy with the Muslim conquest, migrated north to treacherous mountains. Muslims did not hold the northwest part of the Peninsula for long. Pelayo ben Favila, a nobleman who escaped from Toledo as it was conquered by Muslims, triggered a revolt in the city of Gijon in 718, which started the Christian Kingdom of Asturias, with its capital in Cangas de Onis. It was then eventually disintegrated into three other Christian Kingdoms: Castile, Galicia, and Leon, all advancing and gaining territory from the Muslims in a span of the following seven centuries, together with other Christian Kingdoms coming from the Pyrenees region. After the defeat of the Umayyad Muslims in the Battle of Covadonga in 722, the Asturians conquered Galicia in 739 and León in 754. Muslims lost Barcelona in 801 and Salamanca (northeast of Portugal) in 939.

After the disintegration of the Umayyads in 1031, the Christian troops kept moving south, conquering one emirate after another, until they captured Toledo in 1089, which is located in the middle of the Peninsula, just south of Madrid, the current capital of Spain. The Andalusian Muslims then requested the help of Almoravid Muslim Dynasty with their Capital Marrakech in current Morocco. The ruler of the dynast was Yusuf bnu Tashfin, who ruled between 1061 and 1106. He stopped the advance of the Christians at the Battle of Zallaqa (Arabic for slippery) or the Battle of Sagrajas in 1086.

The Almoravids got weaker and then replaced by the Almohad Dynasty by 1147. By then Muslims lost Zaragoza in 1139, located in the northeast of the Peninsula, Lisbon in 1147, located in the southwest of the Peninsula and current capital of Portugal. The Almohads continued helping Muslims and fighting the Christians in the Peninsula, until the Almohads had a terrible defeat in the Battle of Oqab (crow in Arabic) or the Battle of Las Navas de Tolosa in 1212. This was followed by the fall of the Balearic Islands from 1228 to 1232, Badajoz in 1229, Cordoba in 1236, Valencia in 1238, Murcia in 1243, Cartagena in 1245, Seville in 1248, Huelva in 1250 Arcos, Medina-Sidona, Jerez, and Cadiz in 1262, and Tarifa in 1294.

The Nasrid Muslim Dynasty was established in 1232 with its capital Grenada in the southeast of the Peninsula. The latter continued to shrink with the fall of Baeza, Malaga, and Almeria in 1487, then Almuñécar and Salobreña in 1489. By 1491, the Dynasty consisted of only the city of Grenada which eventually was captured by the Christians in 1492 (897 H), which marks the end of Muslim rule in the Iberian Peninsula. Thus, Spain was under Muslim control for up to 781 years. The Canary Islands were captured by the Castilians between 1402 and 1405, then the two cities in north Morocco: Ceuta in 1415, Melilla in 1497.

Muslims continued to decrease in Iberia by fleeing or being forced to accept Christianity, including by burning them alive. In 1609–1614 (1018–1023 H), following a decree by King Felipe III (reined

1598–1621), a third of a million of Muslims were forcibly expelled mostly to North Africa, after all their possessions were confiscated and their children less than five years old were kidnapped for Christianization, and all mosques were converted to churches. Many Muslims remained in the Peninsula, keeping their religion secret and facing horrible death whenever discovered. From time to time, some managed to escape to Muslim lands.

Figure 4.3.9a Map of the Kingdom of Spain.

Currently, the Kingdom of Spain has an area of 505,370 sq. km, including the Balearic Islands (4,992 sq. km) west of the mainland, the Canary Islands (7,493 sq. km) west of Morocco and other smaller Islands off the northern coast of Morocco and exclaves in North Morocco with total area of 31 sq. km including the exclaves of Ceuta (18 sq. km) and Melilla (12 sq. km). A map of Spain is presented in Fig. 4.3.9a and that of the Canary Islands in Fig. 4.3.9b, while Spanish possessions in northern Morocco are illustrated in Fig. 4.3.9c.

The censuses of 1877 and 1897 showed non-Christian population of 83 and 510, respectively. In 1967, Spain allowed freedom of religions, allowing non-Christians to build their places of worship and confess their religion openly. Islam was officially recognized as a religion only in 1989. Until 1980s, the vast majority of Muslims was located in the Spanish enclaves of Ceuta and Mililla who were restricted from passing the strait to the Iberian Peninsula and only in 1986 were allowed to obtain Spanish citizenship when Spain joined the European Union. The vast majority of Muslim in Spain are Moroccans or Moroccan origin.

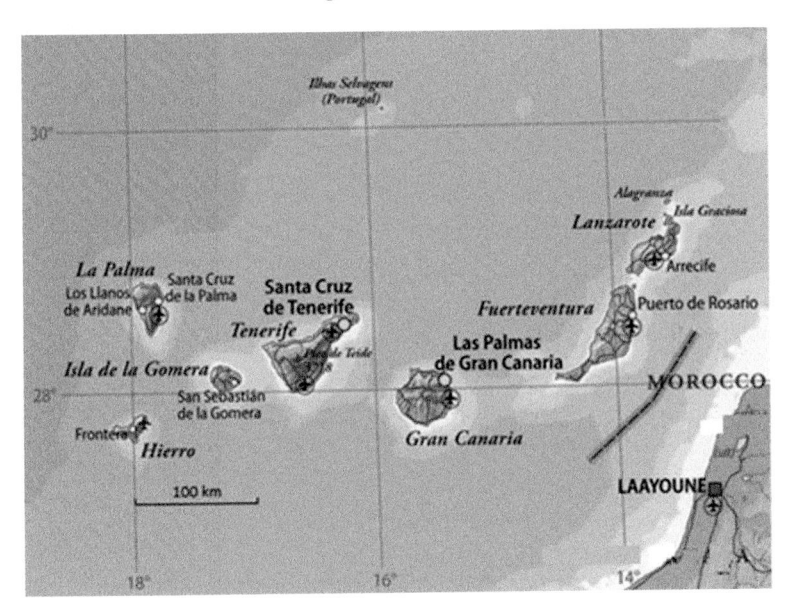

Figure 4.3.9b Map of the Canary Islands.

We estimate Muslim population based on total nationalities from Muslim majority countries. Thus, as shown in Table 4.3.9a, until 1991 Muslims constituted less than 0.2% of the total population.

However, Muslims increased dramatically from 0.07 million or 0.2% in 1991, to 0.17 million or 0.4% in 1996, to 0.34 million or 0.8% in 2001, to 1.1 million or 2.4% in 2006, to 1.5 million or 3.2% in 2011, to 1.9 million or 4.1% in 2016 and 2.0 million or 4.3% in 2019. Thus, assuming that the percentage of Muslims will continue to increase by one percentage point per decade, the Muslim population in Spain is expected to increase to 3.5 million or 8% by 2050 and 4.3 million or 13% by 2100.

Table 4.3.9a Evolution of the Muslim population in Spain

Year	Population	Muslims	%	Source
1877	18,341,281	<83	0.00	[EUH]c
1897	18,132,475	<510	0.00	[SYB10]c
1950	28,117,873	10,882	0.04	[ES1]e
1960	30,583,466	15,556	0.05	[ES1]e
1970	34,040,643	29,234	0.09	[ES1]e
1981	37,764,460	30,338	0.08	[ES1]e
1986	38,480,822	45,582	0.12	[ES1]e
1991	38,870,852	65,774	0.17	[ES1]e
1996	39,429,598	174,832	0.44	[ES1]e
1999	40,393,000	125,000	0.31	[EVS]s
2001	40,476,723	337,389	0.83	[ES1]e
2006	44,009,969	1,053,107	2.39	[ES2]e
2008	45,817,000	900,000	1.96	[EVS]s
2011	46,667,175	1,498,707	3.21	[ES2]e
2016	46,440,110	1,887,906	4.07	[ES2]e
2019	46,733,038	1,993,675	4.27	[ES2]e
2050	43,637,000	3,491,000	8.00	es
2100	33,210,000	4,317,000	13.00	es

Table 4.3.9b, presents the distribution of Muslims per autonomous region. More than fourth of Muslims live in Catalonia region, and almost a third is split between Madrid and Andalucia. The largest concentration of Muslims is in the Mililla and Ceuta where they constitute around half of the population.

Figure 4.3.9c Map of the Spanish Possessions in Northern Morocco.

4.3.10   Regional Summary and Conclusion

Western Europe was the first part of Europe to come under Muslim control at the start of the eighth century, and large part of it was under Muslim control for almost eight centuries. By the beginning of the seventeenth century, Islam was exterminated by brutal force under the name of Christianity and blessings of Church and the Pope. Islam returned after WWII as Europe needed more workforce and the Muslim population now consists mainly of immigrants from majority Muslim countries and their descendants. The number of Muslims is increasing rapidly, now at 5% of the total population, and is expected to exceed 11% by the end of this century. The following tables present centennial data from 600 AD to 2100 AD (or approximately 1 H to 1500 H) in Table 4.3a and decennial data from 1790 AD to 2100 AD (or 1210 H to 1520 H) in Tables 4.3b and 4.3c for current countries in Western Europe. The data includes total population in thousands (P), the percentage of which is Muslim (M%), the corresponding Muslim population in thousands (M), and the annual population growth rate (APGR, or G%) of the total population in this region.

Table 4.3.9b Beginning of year evolution of the Muslim population in Spain per autonomous region since 1950. *P*: Total population, *M*: Muslim population, *M%*: Percentage of Muslim population

Region		1950	1960	1970	1981	1986	1991	1996	2001	2006	2011	2016
Galicia	P	2,604,200	2,602,962	2,676,330	2,812,775	2,795,890	2,731,669	2,711,516	2,697,288	2,730,097	2,773,415	2,720,102
	M	—	8	89	138	138	395	927	2,592	5,784	12,233	17,166
	M%	—	0.00	0.00	0.00	0.00	0.01	0.03	0.10	0.21	0.44	0.63
Asturias	P	888,149	989,344	1,052,039	1,130,074	1,121,785	1,093,937	1,079,808	1,063,195	1,063,488	1,075,877	1,040,925
	M	—	—	11	106	101	185	305	738	2,149	5,628	7,764
	M%	—	—	0.00	0.01	0.01	0.02	0.03	0.07	0.20	0.52	0.75
Cantabria	P	404,921	432,132	469,079	514,403	525,170	527,326	527,558	532,958	561,042	590,874	582,504
	M	—	—	—	—	29	103	199	677	1,888	3,928	5,059
	M%	—	—	—	0.01	0.01	0.02	0.04	0.13	0.34	0.66	0.87
Basque	P	1,061,240	1,371,654	1,867,274	2,144,071	2,150,012	2,104,041	2,078,931	2,076,441	2,127,670	2,183,134	2,164,066
	M	—	1	18	211	344	864	1,772	3,328	12,983	30,319	51,654
	M%	—	0.00	0.00	0.01	0.02	0.04	0.09	0.16	0.61	1.39	2.39
Navarre	P	382,932	402,042	466,593	510,055	518,350	519,277	533,961	551,417	592,146	637,099	637,486
	M	—	—	89	26	58	128	598	3,575	10,256	17,473	23,601
	M%	—	—	0.02	0.01	0.01	0.02	0.11	0.65	1.73	2.74	3.70

Region													
La Rioja	P	229,791	229,852	234,620	254,915	261,373	263,434	263,331	270,991	302,697	320,850	312,810
	M	—	—	—	7	41	93	659	3,181	9,307	14,516	18,359
	M%	—	—	—	0.00	0.02	0.04	0.25	1.17	3.07	4.52	5.87
Castilla & León	P	2,864,378	2,848,352	2,668,279	2,585,184	2,597,404	2,545,926	2,506,764	2,458,661	2,494,676	2,545,286	2,454,454
	M	5	33	163	205	602	778	1,444	4,137	12,648	29,702	37,775
	M%	0.00	0.00	0.01	0.01	0.02	0.03	0.06	0.17	0.51	1.17	1.54
Aragón	P	1,094,002	1,105,498	1,153,053	1,198,189	1,201,098	1,188,817	1,196,298	1,198,318	1,273,568	1,344,396	1,318,571
	M	1	—	367	247	210	545	2,056	8,062	28,454	42,543	55,105
	M%	0.00	—	0.03	0.02	0.02	0.05	0.17	0.67	2.23	3.16	4.18
Cataluña	P	3,240,313	3,925,779	5,107,469	5,964,918	6,014,714	6,059,494	6,117,984	6,285,640	7,010,868	7,493,252	7,408,290
	M	103	100	727	1,577	2,675	9,808	41,709	104,497	245,222	387,582	510,481
	M%	0.00	0.00	0.01	0.03	0.04	0.16	0.68	1.66	3.50	5.17	6.89
Valencia	P	2,307,068	2,480,879	3,077,955	3,658,332	3,766,803	3,857,234	3,916,778	4,073,263	4,681,039	4,999,211	4,932,347
	M	10	11	793	812	1,387	2,949	10,079	26,367	113,595	157,992	200,572
	M%	0.00	0.00	0.03	0.02	0.04	0.08	0.26	0.65	2.43	3.16	4.07

Table 4.3.9b (Continued)

Region		1950	1960	1970	1981	1986	1991	1996	2001	2006	2011	2016
Castilla-La Mancha	P	2,030,598	1,975,539	1,732,601	1,650,564	1,669,415	1,657,032	1,701,751	1,740,501	1,917,447	2,099,057	2,048,900
	M	1	—	4	29	57	165	2,324	7,039	32,960	49,226	62,619
	M%	0.00	—	0.00	0.00	0.00	0.01	0.14	0.40	1.72	2.35	3.06
Madrid	P	1,926,311	2,606,254	3,761,320	4,702,568	4,835,649	4,947,555	5,055,266	5,274,180	5,953,604	6,394,239	6,424,275
	M	17	90	2,325	1,696	3,200	5,375	20,268	63,930	195,254	241,372	278,976
	M%	0.00	0.00	0.06	0.04	0.07	0.11	0.40	1.21	3.28	3.77	4.34
Extremadura	P	1,364,857	1,378,777	1,169,355	1,065,609	1,077,591	1,061,852	1,063,101	1,059,901	1,074,930	1,102,300	1,084,969
	M	—	—	2	21	92	251	2,206	6,306	15,536	15,426	19,064
	M%	—	—	0.00	0.00	0.01	0.02	0.21	0.59	1.45	1.40	1.76
Andalucía	P	5,605,857	5,893,396	5,990,874	6,462,979	6,726,619	6,940,522	7,125,599	7,303,603	7,865,775	8,332,087	8,403,774
	M	—	6	836	762	1,825	6,464	21,588	72,824	176,177	235,098	300,460
	M%	—	0.00	0.01	0.01	0.03	0.09	0.30	1.00	2.24	2.82	3.58
Murcia	P	756,721	800,463	832,005	959,392	1,007,214	1,045,601	1,098,386	1,169,047	1,351,109	1,459,076	1,466,474
	M	—	—	4	32	188	404	3,943	13,312	50,819	80,817	95,387
	M%	—	—	0.00	0.00	0.02	0.04	0.36	1.14	3.76	5.54	6.50

Baleares Islands	P	422,089	443,327	532,942	658,350	667,187	709,136	753,620	849,986	970,288	1,091,656	1,135,527
	M	—	18	54	151	562	1,144	2,338	5,284	22,585	40,109	51,785
	M%	—	0.00	0.01	0.02	0.08	0.16	0.31	0.62	2.33	3.67	4.56
Ceuta	P	59,936	73,182	62,607	65,340	65,151	67,615	69,959	71,336	72,806	81,627	84,663
	M	4,459	7,102	10,000	12,027	15,002	15,002	27,956	28,821	30,110	33,961	36,181
	M%	7.44	9.70	15.97	18.41	23.03	22.19	39.96	40.40	41.36	41.61	42.74
Melilla	P	81,182	79,586	60,815	53,628	52,388	56,600	61,540	65,838	67,110	78,863	84,764
	M	6,277	7,626	12,933	11,105	17,824	17,824	26,673	26,139	32,744	37,446	43,981
	M%	7.73	9.58	21.27	20.71	34.02	31.49	43.34	39.70	48.79	47.48	51.89
Canary Islands	P	793,328	944,448	1,125,433	1,373,114	1,427,009	1,493,784	1,567,447	1,734,159	1,899,609	2,064,876	2,135,209
	M	9	561	819	1,186	1,247	3,297	7,788	21,418	54,636	63,336	71,917
	M%	0.00	0.06	0.07	0.09	0.09	0.22	0.50	1.24	2.88	3.07	3.37

Table 4.3a Centennial estimates of the Muslim population (×1000) in Western Europe: 600–2100 AD (1–1500 H)

		600	700	800	900	1000	1100	1200	1300
Andorra	P	—	—	1	1	2	2	2	3
	M%	—	—	—	—	—	—	—	—
	M	—	—	—	—	—	—	—	—
Belgium	P	360	370	380	390	400	600	800	1,000
	M%	—	—	—	—	—	—	—	—
	M	—	—	—	—	—	—	—	—
France	P	5,900	6,050	6,200	6,350	6,500	8,000	9,500	11,000
	M%	—	—	0.10	0.20	—	—	—	—
	M	—	—	6	13	—	—	—	—
Gibraltar	P	—	—	1	1	2	2	3	3
	M%	—	—	50.00	60.00	70.00	80.00	80.00	80.00
	M	—	—	1	1	1	2	2	2
Luxembourg	P	—	—	—	—	1	5	10	20
	M%	—	—	—	—	—	—	—	—
	M	—	—	—	—	—	—	—	—
Monaco	P	—	—	—	—	—	—	—	1
	M%	—	—	—	1.00	—	—	—	—
	M	—	—	—	—	—	—	—	—
Netherlands	P	260	270	280	290	300	430	560	690
	M%	—	—	—	—	—	—	—	—
	M	—	—	—	—	—	—	—	—
Portugal	P	520	540	560	580	600	675	750	825
	M%	—	—	10.00	20.00	30.00	25.00	20.00	15.00
	M	—	—	56	116	180	169	150	124
Spain	P	3,900	3,925	3,950	3,975	4,000	4,600	5,200	5,800
	M%	—	—	20.00	40.00	50.00	50.00	45.00	40.00
	M	—	—	790	1,590	2,000	2,300	2,340	2,320
Total	P	10,942	11,157	11,373	11,588	11,805	14,314	16,825	19,342
	M%	0.00	0.00	7.50	14.84	18.48	17.26	14.81	12.65
	M	—	—	853	1,719	2,181	2,470	2,492	2,446
	G%		0.019	0.019	0.019	0.019	0.193	0.162	0.139

Table 4.3a (*Continued*)

		1400	1500	1600	1700	1800	1900	2000	2100
Andorra	P	3	3	4	4	5	5	65	62
	M%	—	—	—	—	—	—	0.71	0.60
	M	—	—	—	—	—	—	—	—
Belgium	P	1,200	1,400	1,600	2,000	3,100	6,719	10,282	12,493
	M%	—	—	—	—	—	0.01	2.81	15.00
	M	—	—	—	—	—	1	289	1,874
France	P	13,000	15,000	18,500	21,471	27,349	38,962	59,015	65,498
	M%	—	—	—	—	0.00	0.01	5.00	16.00
	M	—	—	—	—	—	4	2,951	10,480
Gibraltar	P	4	4	5	6	5	20	31	27
	M%	80.00	10.00	—	—	—	0.25	4.01	3.63
	M	3	—	—	—	—	—	1	1
Luxembourg	P	30	40	50	60	125	235	436	985
	M%	—	—	—	—	—	—	1.00	8.00
	M	—	—	—	—	—	—	4	79
Monaco	P	1	1	1	1	1	15	32	64
	M%	—	—	—	—	—	0.08	2.77	6.50
	M	—	—	—	—	—	—	1	4
Netherlands	P	820	950	1,500	1,900	1,982	5,104	15,926	15,760
	M%	—	—	—	—	—	0.00	4.65	10.50
	M	—	—	—	—	—	—	741	1,655
Portugal	P	900	1,000	1,100	2,000	3,345	5,423	10,297	6,985
	M%	10.00	5.00	1.00	—	—	—	0.15	1.10
	M	90	50	11	—	—	—	15	77
Spain	P	6,300	6,800	8,240	8,770	10,541	18,594	40,825	33,210
	M%	30.00	20.00	10.00	1.00	0.00	0.00	0.83	13.00
	M	1,890	1,360	824	88	—	—	339	4,317
Total	P	22,258	25,198	31,000	36,212	46,454	75,078	136,910	135,084
	M%	8.91	5.60	2.69	0.24	0.00	0.01	3.17	13.69
	M	1,983	1,410	835	88	—	5	4,342	18,487
	G%	0.140	0.124	0.207	0.155	0.249	0.480	0.601	−0.013

Table 4.3b Decennial estimates of the Muslim population (×1000) in Western Europe: 1790–1940 AD (1210–1360 H)

		1790	1800	1810	1820	1830	1840	1850	1860
Andorra	P	5	5	5	5	5	5	6	6
	M%	—	—	—	—	—	—	—	—
	M	—	—	—	—	—	—	—	—
Belgium	P	3,000	3,100	3,200	3,434	3,750	4,080	4,449	4,740
	M%	—	—	—	—	—	—	—	—
	M	—	—	—	—	—	—	—	—
France	P	28,100	27,349	29,280	30,250	32,370	34,080	35,630	36,510
	M%	0.00	0.00	0.00	0.00	0.00	0.00	0.00	0.00
	M	—	—	—	—	—	—	—	—
Gibraltar	P	3	5	11	11	15	16	16	15
	M%	—	—	0.19	0.18	0.11	0.11	0.09	0.09
	M	—	—	—	—	—	—	—	—
Luxembourg	P	120	125	130	134	150	175	190	198
	M%	—	—	—	—	—	—	—	—
	M	—	—	—	—	—	—	—	—
Monaco	P	1	1	1	1	1	1	1	1
	M%	—	—	—	—	—	—	—	—
	M	—	—	—	—	—	—	—	—
Netherlands	P	1,963	1,982	2,194	2,354	2,626	2,874	3,084	3,312
	M%	—	—	—	—	—	—	—	—
	M	—	—	—	—	—	—	—	—
Portugal	P	2,960	3,345	3,350	3,356	3,396	3,737	3,844	4,035
	M%	—	—	—	—	—	—	—	—
	M	—	—	—	—	—	—	—	—
Spain	P	10,062	10,541	11,500	12,500	13,698	12,387	14,216	15,656
	M%	0.00	0.00	0.00	0.00	0.00	0.00	0.00	0.00
	M	—	—	—	—	—	—	—	—
Total	P	46,213	46,454	49,671	52,046	56,011	57,356	61,435	64,473
	M%	0.00	0.00	0.00	0.00	0.00	0.00	0.00	0.00
	M	—	—	—	—	—	—	1	1
	G%		0.052	0.670	0.467	0.734	0.237	0.687	0.483

Table 4.3b (*Continued*)

		1870	1880	1890	1900	1910	1920	1930	1940
Andorra	P	6	6	6	5	5	4	6	6
	M%	—	—	—	—	—	—	—	—
	M	—	—	—	—	—	—	—	—
Belgium	P	5,096	5,541	6,069	6,719	7,498	7,552	8,076	8,346
	M%	—	—	—	0.01	0.01	0.01	0.02	0.01
	M	—	—	—	1	1	1	2	1
France	P	36,103	37,672	38,343	38,962	39,605	39,108	41,524	41,000
	M%	0.00	0.00	0.01	0.01	0.03	0.11	0.25	0.31
	M	—	2	4	4	12	43	104	127
Gibraltar	P	19	18	19	20	19	18	17	18
	M%	0.14	0.19	0.19	0.25	0.25	0.23	0.09	0.00
	M	—	—	—	—	—	—	—	—
Luxembourg	P	204	211	211	235	259	263	300	284
	M%	—	—	—	—	—	—	—	—
	M	—	—	—	—	—	—	—	—
Monaco	P	3	6	10	15	19	23	25	24
	M%	—	—	—	0.08	0.08	0.10	0.21	0.09
	M	—	—	—	—	—	—	—	—
Netherlands	P	3,580	4,016	4,511	5,104	5,858	6,865	7,936	8,834
	M%	—	0.00	0.00	0.00	0.00	0.00	0.00	0.00
	M	—	—	—	—	—	—	—	—
Portugal	P	4,188	4,551	5,050	5,423	5,960	6,033	6,826	7,722
	M%	—	—	—	—	—	—	—	—
	M	—	—	—	—	—	—	—	—
Spain	P	16,835	17,000	17,800	18,594	19,927	21,303	23,564	25,878
	M%	0.00	0.00	0.00	0.00	0.00	0.00	0.01	0.02
	M	—	—	—	—	—	—	2	5
Total	P	66,034	69,021	72,019	75,078	79,151	81,170	88,273	92,112
	M%	0.00	0.00	0.01	0.01	0.02	0.05	0.12	0.14
	M	1	2	4	5	13	44	108	133
	G%	0.239	0.442	0.425	0.416	0.528	0.251	0.840	0.426

Table 4.3c Decennial estimates of the Muslim population (×1000) in Western Europe: 1950–2100 AD (1370–1520 H)

		1950	1960	1970	1980	1990	2000	2010	2020
Andorra	P	6	13	24	36	55	65	84	77
	M%	—	—	—	0.44	0.54	0.71	0.60	0.57
	M	—	—	—	—	—	—	1	—
Belgium	P	8,638	9,167	9,632	9,869	10,007	10,282	10,939	11,590
	M%	0.01	0.01	0.72	1.90	2.52	2.81	5.58	7.00
	M	1	1	69	188	252	289	610	811
France	P	41,834	45,673	50,764	53,868	56,667	59,015	62,880	65,274
	M%	0.55	0.87	1.33	3.16	3.23	5.00	6.73	8.00
	M	230	397	675	1,702	1,830	2,951	4,232	5,222
Gibraltar	P	22	23	29	30	29	31	34	34
	M%	0.00	0.03	8.06	8.02	6.93	4.01	3.63	3.63
	M	—	—	2	2	2	1	1	1
Luxembourg	P	296	315	340	364	382	436	508	626
	M%	—	0.01	0.02	0.08	0.13	1.00	3.11	4.00
	M	—	—	—	—	—	4	16	25
Monaco	P	20	22	23	27	29	32	36	39
	M%	0.10	0.25	0.50	2.10	3.02	2.77	2.00	2.32
	M	—	—	—	1	1	1	1	1
Netherlands	P	10,042	11,449	13,002	14,148	14,965	15,926	16,683	17,135
	M%	0.00	0.01	0.41	1.56	2.40	4.65	5.66	6.50
	M	—	1	53	221	359	741	944	1,114
Portugal	P	8,417	8,845	8,651	9,751	9,895	10,297	10,596	10,197
	M%	0.00	0.00	0.00	0.06	0.13	0.15	0.25	0.30
	M	—	—	—	6	13	15	26	31
Spain	P	28,070	30,402	33,884	37,698	39,203	40,825	46,931	46,755
	M%	0.04	0.05	0.09	0.08	0.17	0.83	3.21	5.00
	M	11	15	30	30	67	339	1,506	2,338
Total	P	97,344	105,910	116,349	125,791	131,232	136,910	148,690	151,726
	M%	0.25	0.39	0.71	1.71	1.92	3.17	4.93	6.29
	M	243	415	831	2,150	2,525	4,342	7,338	9,543
	G%	0.552	0.843	0.940	0.780	0.423	0.424	0.825	0.202

Table 4.3c (Continued)

		2030	2040	2050	2060	2070	2080	2090	2100
Andorra	P	78	78	76	72	67	64	63	62
	M%	0.60	0.60	0.60	0.60	0.60	0.60	0.60	0.60
	M	—	—	—	—	—	—	—	—
Belgium	P	11,904	12,121	12,221	12,230	12,251	12,310	12,390	12,493
	M%	8.00	9.00	10.00	11.00	12.00	13.00	14.00	15.00
	M	952	1,091	1,222	1,345	1,470	1,600	1,735	1,874
France	P	66,696	67,571	67,587	67,083	66,596	66,295	65,938	65,498
	M%	9.00	10.00	11.00	12.00	13.00	14.00	15.00	16.00
	M	6,003	6,757	7,435	8,050	8,657	9,281	9,891	10,480
Gibraltar	P	34	34	34	33	32	30	29	27
	M%	3.63	3.63	3.63	3.63	3.63	3.63	3.63	3.63
	M	1	1	1	1	1	1	1	1
Luxembourg	P	690	744	790	828	864	901	940	985
	M%	4.50	5.00	5.50	6.00	6.50	7.00	7.50	8.00
	M	31	37	43	50	56	63	71	79
Monaco	P	42	44	46	49	52	56	60	64
	M%	3.00	3.50	4.00	4.50	5.00	5.50	6.00	6.50
	M	1	2	2	2	3	3	4	4
Netherlands	P	17,450	17,461	17,165	16,787	16,510	16,279	16,002	15,760
	M%	7.00	7.50	8.00	8.50	9.00	9.50	10.00	10.50
	M	1,222	1,310	1,373	1,427	1,486	1,546	1,600	1,655
Portugal	P	9,913	9,558	9,085	8,514	7,964	7,553	7,249	6,985
	M%	0.40	0.50	0.60	0.70	0.80	0.90	1.00	1.10
	M	40	48	55	60	64	68	72	77
Spain	P	46,230	45,225	43,637	41,046	38,027	35,720	34,389	33,210
	M%	6.00	7.00	8.00	9.00	10.00	11.00	12.00	13.00
	M	2,774	3,166	3,491	3,694	3,803	3,929	4,127	4,317
Total	P	153,037	152,836	150,641	146,642	142,363	139,208	137,061	135,084
	M%	7.20	8.12	9.04	9.98	10.92	11.85	12.77	13.69
	M	11,024	12,412	13,622	14,629	15,540	16,493	17,500	18,487
	G%	0.086	−0.013	−0.145	−0.269	−0.296	−0.224	−0.155	−0.145

4.4 Muslims in Northern Europe

This region consists of eleven countries and territories: Denmark, Faroe Islands, Finland, Guernsey, Iceland, Ireland, Isle of Man, Jersey, Norway, Sweden, and the United Kingdom. Islam entered this region as few Tatars moved to Finland after its conquest by the Russian Empire in early nineteenth century, and Muslims moved from former British colonies to the UK. However, the Muslim population remained small until after WWII, when more Muslims moved to this region for economic, education, and political reasons.

Thus, the Muslim population remained negligible until the twentieth century. It has likely increased from less than a thousand in 1900, to 2.2 million or 2.6% in 2000, to 5.5 million or 5.6% in 2020, and is projected to reach ten million or 8.8% by 2050 and 17 million or 14.4% by 2100.

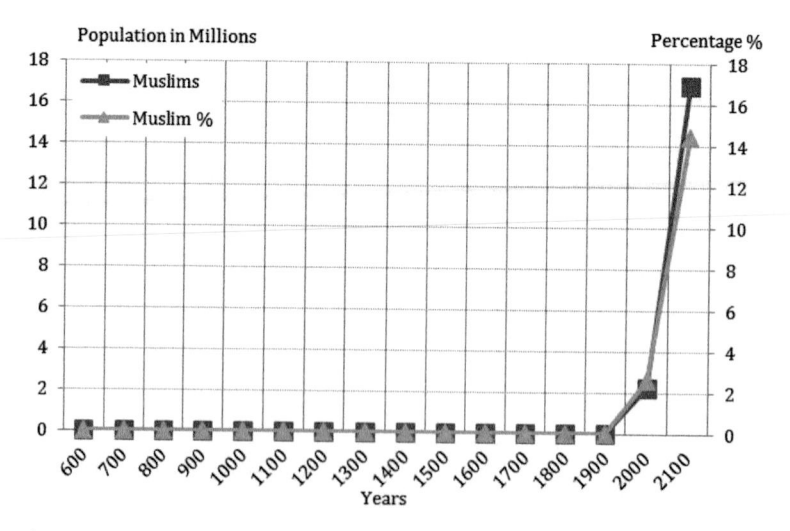

Figure 4.4a Plot of centennial estimates of the Muslim population and its percentage of the total population in Northern Europe: 600–2100 AD (1–1500 H).

A plot of centennial estimates of the Muslim population and its percentage with respect to the total population in this region from 600 to 2100 is provided in Fig. 4.4a. A zoom in of this plot, providing a plot of decennial estimates of the Muslim population and its percentage with respect to the total population in this region from

1900 to 2100 is provided in Fig. 4.4b. The corresponding individual data for each country in this region is discussed below. In Section 4.4.12, the total population in each country in this region and the corresponding percentage and number of Muslims is presented centennially in Table 4.4a from 600 to 2100 and decennially in Tables 4.4b and 4.4c from 1790 to 2100.

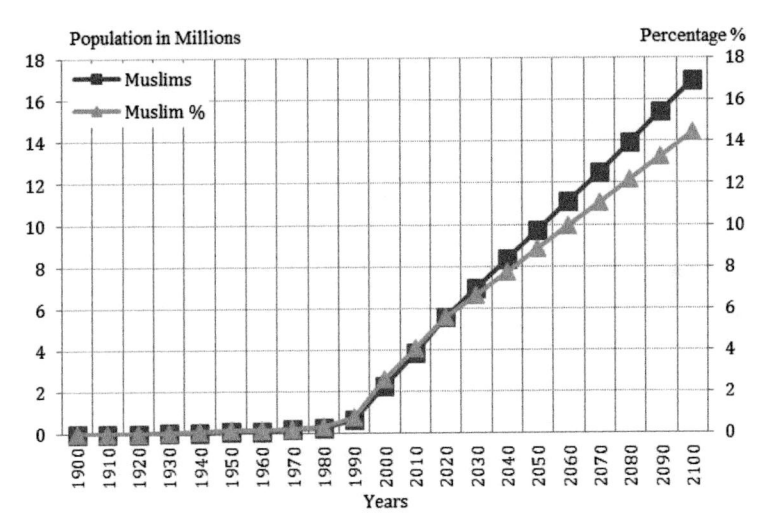

Figure 4.4b Plot of decennial estimates of the Muslim population and its percentage of the total population in Northern Europe: 1900–2100 AD (1320–1520 H).

4.4.1 Denmark

The Kingdom of Denmark (excluding the Faroe Islands and Greenland) has an area of 43,094 sq. km and its map is presented in Fig. 4.4.1. Islam is relatively new in this country, as Muslims started migrating after 1968 to supply the country's need of manpower. Censuses in Denmark do not collect information on religious affiliations of the population. Nevertheless, Statistics Denmark does collect information on the citizenship, country of birth and country of origin of the population in Denmark. From this data we can infer the number of Muslims in Denmark. The number of Turkish citizens who had residence or work permit increased from 41 in 1960, to 85 in 1965 to 1,852 in 1970. At the same period other Muslim nationalities were not present in Denmark. In 1975, there were 1,292 Moroccans, 4,982 Pakistanis, and 8,129 Turks.

Figure 4.4.1 Map of the Kingdom of Denmark.

Since 1980, Statistics Denmark started publishing on yearly basis the country of origin of its population, which it defined as follows:

- If none of the parents are known, the country of origin is defined from the information on the person in question. If the person is an immigrant, the country of origin is the same as the country of birth. If the person is a descendant, the country of origin is the same as the country of citizenship.
- If only one of the parents is known, the country of origin is defined from by person's country of birth.
- If both parents are known, the country of origin is defined by the country of birth or the country of citizenship of the mother.

Thus, with the assumption that those of Muslim-majority countries origin are Muslim, we construct Table 4.4.1a, summarizing the top countries of origin with Muslim majority sorted by the

Table 4.4.1a Beginning of year estimate of the Muslim population in Denmark based on country of origin

	1980	1985	1990	1995	2000	2005	2010	2015	2019
Turkey	14,086	18,989	29,431	39,222	48,773	54,859	59,216	61,634	63,819
Syria	213	252	625	1,493	2,284	3,046	3,707	14,093	42,467
Iraq	160	266	2,423	6,415	14,902	26,351	29,264	30,994	33,089
Lebanon	222	272	7,938	15,110	19,011	22,232	23,775	25,572	27,076
Pakistan	7,845	9,489	12,006	14,692	17,509	19,301	20,392	23,770	25,661
Bosnia	—	—	—	119	19,727	20,875	22,221	22,697	23,225
Somalia	133	168	531	5,280	14,856	16,952	16,831	19,707	21,432
Iran	241	984	8,591	11,157	12,980	14,289	15,209	18,572	21,227
Afghanistan	35	100	342	1,183	3,275	10,876	12,630	16,637	19,088
Morocco	2,104	2,968	4,267	5,955	7,813	8,974	9,831	10,699	11,527
Others	3,040	3,485	4,703	7,603	10,388	13,339	16,477	21,330	25,501
Total	28,079	36,973	70,857	108,229	171,518	211,094	229,553	265,705	314,112
Muslim %	0.55	0.72	1.38	2.08	3.22	3.90	4.15	4.69	5.41

number of descendants from such countries. Out of these numbers, the percentage of those born in a Muslim majority country decreased from 91% in 1980, to 77% in 1990, to 72% in 1995, to 71% in 2000, to 67% in 2005, to 64% in 2010, to 63% in 2015.

All in all, and as shown in Table 4.4.1b, the Muslim population increased from none before WWII, to 41 in 1960, to 85 in 1965, to 2,000 or 0.04% in 1970, to 14,000 or 0.3% in 1975, to 28,000 or 0.6% in 1980, to 37,000 or 0.7% in 1985, to 71,000 or 1.4% in 1990, to 0.11 million or 2.1% in 1995, to 0.17 million or 3.2% in 2000, to 0.21 million or 3.9% in 2005, to 0.23 million or 4.2% in 2010, to 0.27 million or 4.7% in 2015 and 0.31 million or 5.4% in 2019. Thus, assuming that the percentage of Muslims will continue to increase by two percentage points per decade, the Muslim population is expected to reach 0.7 million or 12% by 2050 and 1.5 million or 22% by 2100.

Table 4.4.1b Evolution of the Muslim population in Denmark

Year	Population	Muslims	%	Source
1850	1,497,747	—	0.00	[EUH]c
1860	1600,551	—	0.00	[EUH]c
1870	1,784,741	—	0.00	[EUH]c
1880	1,969,839	—	0.00	[EUH]c
1890	2,172,380	—	0.00	[EUH]c
1960	4,585,256	41	0.00	[DKH]e
1965	4,767,597	85	0.00	[DKH]e
1970	4,937,579	1,852	0.04	[DKH]e
1975	5,054,410	14,403	0.28	[DKH]e
1980	5,122,065	28,182	0.55	[DK]e
1985	5,111,108	37,020	0.72	[DK]e
1990	5,135,409	70,943	1.38	[DK]e
1995	5,215,718	108,310	2.08	[DK]e
2000	5,330,020	171,573	3.22	[DK]e
2005	5,411,405	211,158	3.90	[DK]e
2010	5,534,738	229,612	4.15	[DK]e
2015	5,659,715	265,714	4.69	[DK]e
2019	5,806,081	314,112	5.41	[DK]e
2050	6,245,000	749,000	12.00	es
2100	6,873,000	1,512,000	22.00	es

4.4.2 Faroe Islands

It is a self-governing overseas administrative division of the Kingdom of Denmark since 1948. The Islands belonged to Norway until 1380, when they became under Danish control. It has an area of 1,393 sq. km and its map is presented in Fig. 4.4.2. It consists mainly of eighteen islands: Streymoy (374 sq. km) where the capital is located and where almost half of the population lives, Eysturoy (286 sq. km) where almost a quarter of the population lives, Vágar (176 sq. km), Suðuroy (165 sq. km), Sandoy (111 sq. km), Borðoy (95 sq. km), Viðoy (41 sq. km), Kunoy (35 sq. km), Kalsoy (31 sq. km), Svínoy (27 sq. km), Fugloy (11 sq. km), Nólsoy (10 sq. km), Mykines (10 sq. km), Skúvoy (10 sq. km), Hestur (6 sq. km), Stóra Dímun (3 sq. km), Koltur (2 sq. km) and Lítla Dímun (1 sq. km), which is uninhabited.

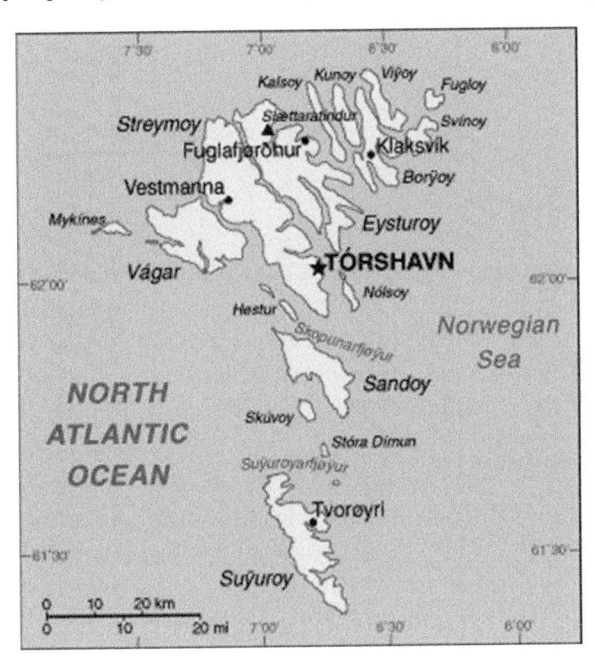

Figure 4.4.2 Map of the Faroe Islands.

Muslims moved here from Denmark for economic reasons towards the end of the twentieth century. The first census inquiring on religious adherence was held in 2011 and collected information from population aged fifteen years and older. It also provided information on when the person came to the islands and their place of birth. From such data we deduce older statistics. Accordingly, the Muslim population increased from one or 0.00% in 1979, to thirteen or 0.04% in 1999, to 23 or 0.07% in 2011. Thus, assuming that the percentage of Muslims will continue to increase by 0.05 of a percentage point per decade, the Muslim population is expected to reach 150 or 0.3% by 2050 and 300 or 0.5% by 2100. The data is summarized in Table 4.4.2.

Table 4.4.2 Evolution of the Muslim population in the Faroe Islands

Year	Population	Muslims	%	Source
1979	33,762	1	0.00	[FO]c
1999	34,538	13	0.04	[FO]c
2011	34,595	23	0.07	[FO]c
2020	49,000	50	0.10	es
2050	53,000	150	0.25	es
2100	55,000	300	0.50	es

4.4.3 Finland

Muslims started migrating to Finland after it was conquered by the Russian Empire in 1809. They were merchant Tatars from Kazan, the capital of the Republic of Tatarstan, Russia. Finland gained its independence in 1917, and recognized Islam as a religion in 1925. Currently, the Republic of Finland has an area of 338,145 sq. km and its map is presented in Fig. 4.4.3. The area includes the autonomous region of Åland Islands with area 13,517 sq. km.

Based on census data, the Muslim population increased from none in the nineteenth century to 262 or 0.01% in 1930. Subsequent censuses show that the Muslim population remained at 0.02% between 1934 and 2000. However, largely due to immigration, the Muslim population increased from 1,000 or 0.02% in 2000, to 4,000 or 0.1% in 2005, to 9,000 or 0.2% in 2010, to 13,000 or 0.2% in 2015, to 16,000 or 0.3% in 2018. Thus, assuming that the percentage of Muslims will increase by a tenth of a percentage point each decade, the Muslim population is expected to reach 30,000 or 0.6% by 2050 and 60,000 or 1.1% by 2100. The data is summarized in Table 4.4.3.

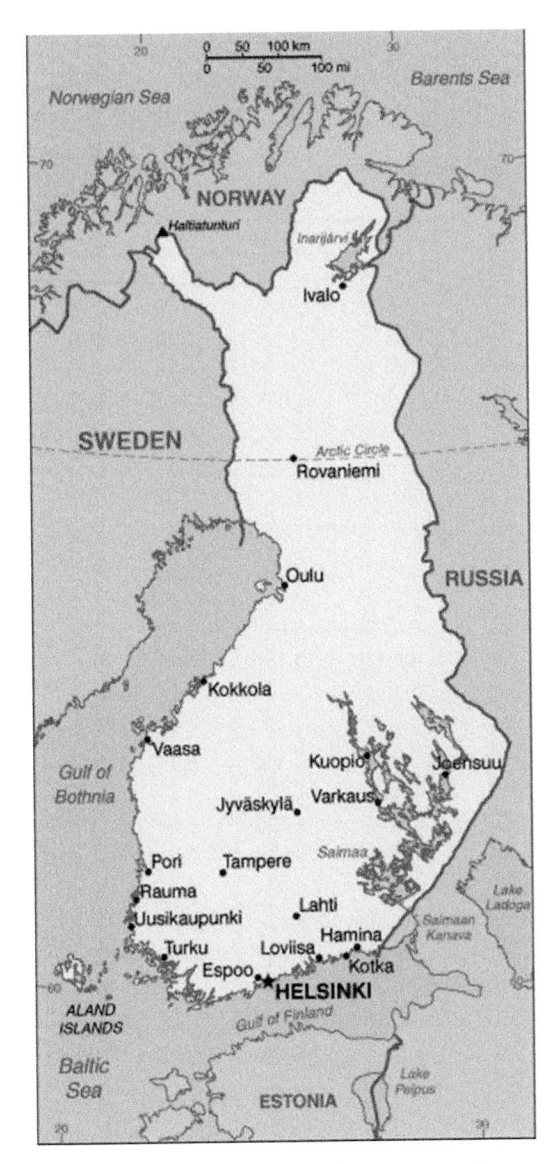

Figure 4.4.3 Map of the Republic of Finland.

These figures include only those persons, who are a member of Islamic congregations which are registered in Finland. Therefore, actual number might be bigger. The information is gathered by

Statistics Finland on religion from Population Register Centre and a person must be a member of a congregation registered in Finland in order to have religion included in these statistics.

Table 4.4.3   Year-end evolution of the Muslim population in Finland

Year	Population	Muslims	%	Source
1860	1,746,725	—	0.00	[EUH]c
1870	1,768,769	—	0.00	[EUH]c
1880	2,058,782	—	0.00	[EUH]c
1890	2,389,149	—	0.00	[EUH]c
1930	3,634,047	262	0.01	[SYB31]c
1934	3,561,600	648	0.02	[MAS55]c
1950	4,029,803	813	0.02	[UN56]c
1970	4,598,336	823	0.02	[UN73]c
1980	4,787,778	786	0.02	[FIH]c
1985	4,910,664	787	0.02	[UN04]c
1990	4,998,478	810	0.02	[UN04]c
1995	5,116,826	946	0.02	[FIH]c
2000	5,181,115	1,199	0.02	[FI]c
2001	5,194,901	2,104	0.04	[FI]c
2002	5,206,295	2,320	0.04	[FI]c
2003	5,219,732	2,748	0.05	[FI]c
2004	5,236,611	2,833	0.05	[FI]c
2005	5,255,580	4,239	0.08	[FI]c
2006	5,276,955	4,944	0.09	[FI]c
2007	5,300,484	5,689	0.11	[FI]c
2008	5,326,314	6,822	0.13	[FI]c
2009	5,351,427	8,230	0.15	[FI]c
2010	5,375,276	9,393	0.17	[FI]c
2011	5,401,267	10,088	0.19	[FI]c
2012	5,426,674	10,596	0.20	[FI]c
2013	5,451,270	11,125	0.20	[FI]c
2014	5,471,753	12,313	0.23	[FI]c
2015	5,487,308	13,289	0.24	[FI]c
2016	5,503,297	14,141	0.26	[FI]c
2017	5,513,130	15,359	0.28	[FI]c
2018	5,517,919	16,057	0.29	[FI]c
2050	5,486,000	33,000	0.60	es
2100	5,254,000	58,000	1.10	es

The information on religious community does not represent foreigners accurately. Not all their religious communities are included in the register of the National Board of Patents and Registration and not all those practicing a religion belong to parishes. At the end of 2013, 84% of foreign-language speakers did not belong to any registered religious community according to the Population Information System. Since most Muslims in Finland are expected to have a foreign background, this shows a major underestimate of the Muslim population. The recorded number of people born or one of their parents was born in over 90% Muslim-majority country (excluding Iran) increased from 2,000 or 0.04% in 1990, to 10,000 or 0.19% in 1995, to 17,000 or 0.33% in 2000, to 26,000 or 0.49% in 2005, to 42,000 or 0.78% in 2010, to 54,000 or 0.99% in 2013. These numbers are about four times the estimated number of Muslims given in Table 4.4.3.

4.4.4 Guernsey

The Bailiwick of Guernsey has an area of 78 sq. km and a British Crown Dependency in the English Channel, off the French coast of Normandy, a map of which is presented in Fig. 4.4.4. It is an archipelago consisting of the main island of Guernsey (65 sq. km) and much smaller islands. Together with the Bailiwick of Jersey, this archipelago is referred to as the Channel Islands.

Muslims started moving in here in the 1970s, and the Muslim population doubled from forty or 0.1% in 2006 to eighty or 0.1% in 2010. Thus, assuming that the percentage of Muslim will continue to increase by 0.05 of a percentage point per decade, the Muslim population is expected to remain less than 500 or 0.6% throughout this century. The data is summarized in Table 4.4.4.

Table 4.4.4 Evolution of the Muslim population in Guernsey

Year	Population	Muslims	%	Source
1911	41,858	—	0.00	[UKH2]es
2006	61,029	40	0.07	[GG]es
2010	62,431	80	0.13	[GG]es
2050	80,000	300	0.35	es
2100	89,000	500	0.60	es

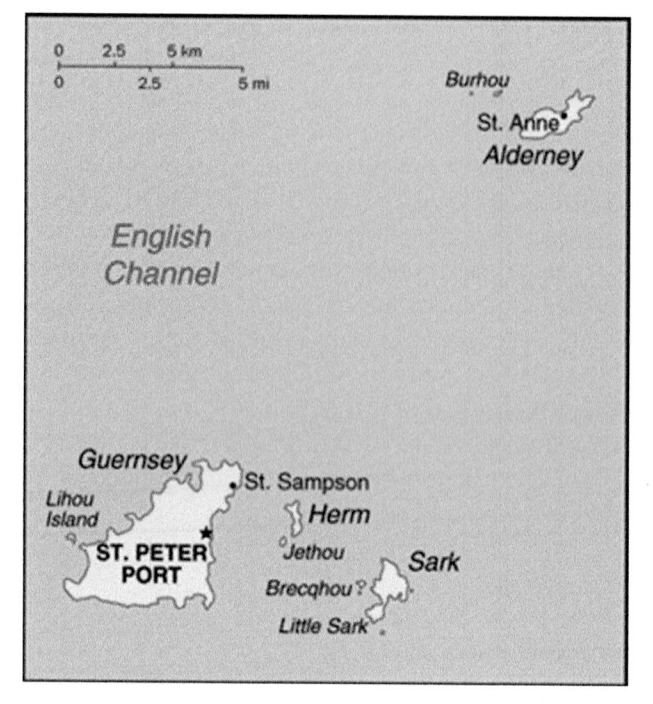

Figure 4.4.4 Map of the Bailiwick of Guernsey.

4.4.5 Iceland

The Republic of Iceland has an area of 103,000 sq. km and its map is presented in Fig. 4.4.5. Historically, the island belonged to Denmark, from which Iceland gained its independence in 1944. The 1973 Census recorded the presence of 14 Moroccans, which we use as an estimate of the Muslim population in Iceland in 1973, making 0.01% of the total population.

Statistics Iceland collects annual information on the number of religious affiliates in the country. The Muslim Association of Iceland was founded in 1997, while the Muslim Cultural Centre in Iceland was founded in 2009. Thus, the number of Muslims started being recorded in 1997. Most of the Muslim population resides in the capital Reykjavík.

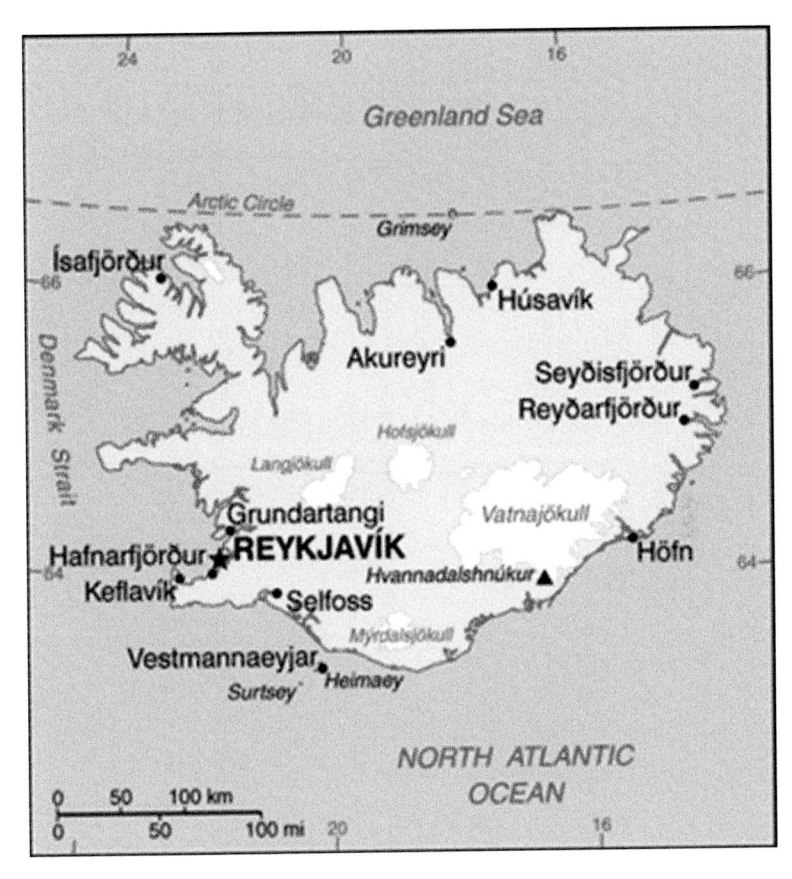

Figure 4.4.5 Map of the Republic of Iceland.

As shown in Table 4.4.5, the number of Muslims increased from 14 or 0.01% of the total population in 1973 (based on citizenship data), to 74 or 0.03% in 1997, to 134 or 0.05% in 2000, to 318 or 0.1% in 2005, to 591 or 0.2% in 2010, to 875 or 0.3% in 2015, then 1,132 or 0.3% in 2019. Thus, assuming that the percentage of Muslims will continue to increase by a fifth of a percentage point per decade, the Muslim population is expected to reach 4,000 or 1% by 2050 and 8,000 or 2% by 2100.

Table 4.4.5 Beginning of year evolution of the Muslim population in Iceland

Year	Population	Muslims	%	Source
1973	213,499	14	0.01	[UNE]e
1997	269,874	74	0.03	[IS]c
1998	272,381	78	0.03	[IS]c
1999	275,712	89	0.03	[IS]c
2000	279,049	134	0.05	[IS]c
2001	283,361	165	0.06	[IS]c
2002	286,575	179	0.06	[IS]c
2003	288,471	229	0.08	[IS]c
2004	290,570	292	0.10	[IS]c
2005	293,577	318	0.11	[IS]c
2006	299,891	340	0.11	[IS]c
2007	307,672	352	0.11	[IS]c
2008	315,459	373	0.12	[IS]c
2009	319,368	404	0.13	[IS]c
2010	317,630	591	0.19	[IS]c
2011	318,452	644	0.20	[IS]c
2012	319,575	694	0.22	[IS]c
2013	321,857	770	0.24	[IS]c
2014	325,671	841	0.26	[IS]c
2015	329,100	875	0.27	[IS]c
2016	332,529	865	0.26	[IS]c
2017	338,349	948	0.28	[IS]c
2018	348,450	1,051	0.30	[IS]c
2019	356,991	1,134	0.32	[IS]c
2050	390,000	4,000	1.00	es
2100	383,000	8,000	2.00	es

4.4.6 Ireland

The Republic of Ireland gained its independence from the United Kingdom in 1922 and has an area of 70,273 sq. km and its map is presented in Fig. 4.4.6. Historical census data show that the Muslim population increased from none in 1861 to two in 1871 to six in 1901 and seven in 1911. Thus, no noticeable presence of Muslims existed until the 1950s. Accordingly, as shown in Table 4.4.6, the estimate of the Muslim population increased to 2,000 or 0.1% in 1971, while EVS put it at 0.11% in 1981.

Figure 4.4.6 Map of the Republic of Ireland.

The Muslim population increased sharply in the 1990s and 2000s as a result of the Irish economic boom and asylum seekers from diverse Muslim countries. Consequently, later censuses showed that the Muslim population quintupled from 4,000 or 0.1% in 1991, to 19,000 or 0.5% in 2002, then doubled to 33,000 or 0.8% in 2006, then increased by half to 49,000 or 1.1% in 2011 and 63,000 or 1.4% in 2016. Thus, assuming that the percentage of Muslims will continue to increase by half of a percentage point per decade, the Muslim population is expected to reach 0.2 million or 3.0% by 2050 and 0.3 million or 5.5% by 2100.

Table 4.4.6 Evolution of the Muslim population in Ireland

Year	Population	Muslims	%	Source
1861	4,402,111	—	0.00	[IE861]c
1871	4,053,187	2	0.00	[IE871]c
1881	3,870,020	1	0.00	[IE881]c
1891	3,468,694	3	0.00	[IE891]c
1901	3,221,823	6	0.00	[GB901]c
1911	3,139,688	7	0.00	[GB911]c
1971	2,978,248	2,000	0.07	[KET76]es
1981	3,372,429	3,710	0.11	[EVS]s
1991	3,442,344	3,875	0.11	[UN04]c
2002	3,838,109	19,147	0.50	[IE06]c
2006	4,169,526	32,539	0.78	[IE06]c
2011	4,515,338	49,204	1.09	[IE16]c
2016	4,636,600	63,400	1.37	[IE16]c
2050	5,678,000	170,000	3.00	es
2100	5,685,000	313,000	5.50	es

4.4.7 Isle of Man

This is a British Crown Dependency, located in the Irish Sea between the islands of Great Britain and Ireland, with a total area of 572 sq. km. Its map is presented in Fig. 4.4.7. As shown in Table 4.4.6, estimates of the Muslim population increase from ten or 0.02% in 1901, to forty or 0.05% in 2001, to seventy or 0.08% in 2006. Thus, assuming that the percentage of Muslims increases by 0.05 of a percentage point per decade, the Muslim population is expected to remain less than 500 or 0.6% of the total population throughout this century.

Table 4.4.7 Evolution of the Muslim population in the Isle of Man

Year	Population	Muslims	%	Source
1911	52,016	10	0.02	[UKH2]es
2001	76,315	40	0.05	[IM]es
2011	84,497	70	0.08	[IM]es
2050	91,000	300	0.30	es
2100	96,000	500	0.55	es

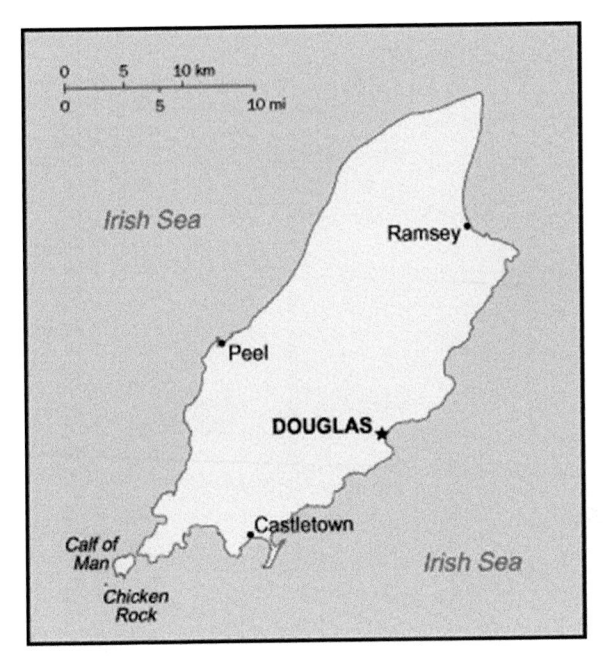

Figure 4.4.7 Map of the Isle of Man.

4.4.8 Jersey

The Bailiwick of Jersey has an area of 116 sq. km and a British Crown Dependency in the English Channel, off the French coast of Normandy, a map of which is presented in Fig. 4.4.8. It is an archipelago consisting of the main island of Jersey and little uninhabited nearby islands. Together with the Bailiwick of Guernsey, this archipelago is referred to as the Channel Islands.

Muslims started moving in here in the 1970s, and their population size increased to 52 or 0.1% in 2001 and to 128 or 0.1% in 2011. The data for the 2001 and 2011 was taken from the corresponding census with the assumption that the Muslim population consists of Pakistanis (21 in 2001 and 53 in 2011) and Bangladeshis (31 in 2001 and 74 in 2011). Thus, assuming that the percentage of Muslims will continue to increase by 0.05 of a percentage point per decade, the Muslim population is expected to remain less than 800 or 0.6% of the total population throughout this century. The data is summarized in Table 4.4.8.

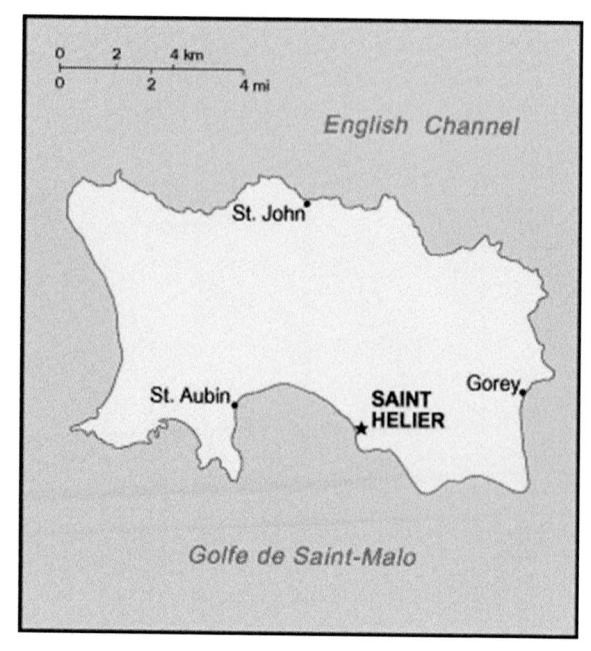

Figure 4.4.8 Map of the Bailiwick of Jersey.

Table 4.4.8 Evolution of the Muslim population in Jersey

Year	Population	Muslims	%	Source
1911	51,898	—	0.00	[UKH2]es
2001	87,186	52	0.06	[JE]e
2011	97,857	128	0.13	[JE]e
2050	120,000	400	0.35	es
2100	134,000	800	0.60	es

4.4.9 Norway

The Kingdom of Norway has an area of 385,186 sq. km including the archipelago of Svalbard and the island of Jan Mayen. Svalbard has an area of 61,022 sq. km including 36,500 sq. km of glaciers. Its population constantly decreased from its peak of 4,012 in 1982, to 3,017 in 1993, to 2,310 in 2018. No Muslims have lived so far in Svalbard. Jan Mayen has a total area of 377 sq. km including 125 sq.

km of glaciers and is almost uninhabited. Maps of the mainland of Norway, Svalbard and Jan Mayen are presented in Figs. 4.4.9a, 4.4.9b, and 4.4.9c, respectively.

Figure 4.4.9a Map of the Kingdom of Norway.

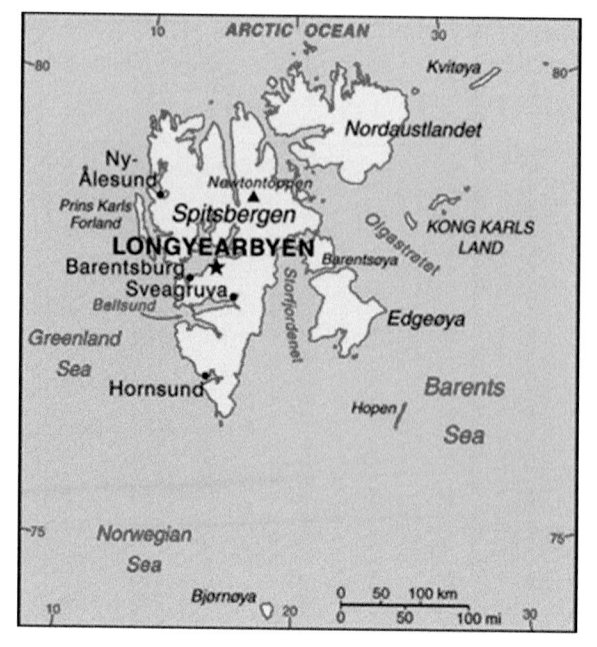

Figure 4.4.9b Map of Svalbard.

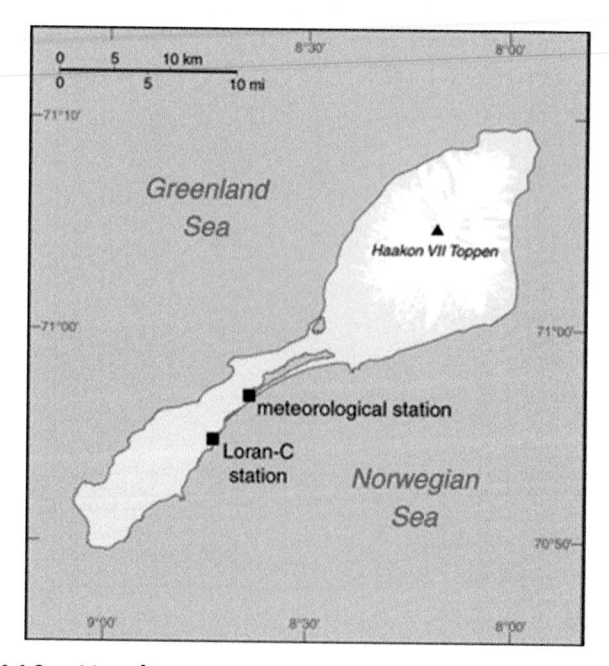

Figure 4.4.9c Map of Jan Mayen.

Muslims started migrating to Norway in the 1960s as the country needed more labor workforce. Based on place of birth, until 1970 the Muslim population remained at 0.00% of the total population and comprised of few Turks. Then in 1970, there were 244 born in Turkey, 407 born in Morocco and 170 born in Pakistan; bringing the total number of Muslims to 821 or 0.02% of the total population. Since 1969, Statistics Norway collects annual information on the number of religious affiliates in the country. This shows an increase of the Muslim population from 1,006 or 0.02% in 1980, to 8,214 or 0.2% in 1985, to 19,189 or 0.5% in 1990, to 40,550 or 0.9% in 1995, to 56,458 or 1.3% in 2000, to 76,621 or 1.7% in 2005, to 98,953 or 2.0% in 2010, to 141,027 or 2.7% in 2015 and 166,861 or 3.2% in 2018. A summary of the data is provided in Table 4.4.9a. Over 80% of the Muslim population in Norway used to live in the capital Oslo in 2000, but this percentage is reduced to 47% in 2006 and 37% in 2018. Per county distribution of the Muslim population in Norway since 2000 is presented in Table 4.4.9b.

The number of Muslims increased in the 1990s due to the incoming of refugees from Bosnia and Somalia, and in the 2000s due to refugees from Iraq. Thus, assuming that the percentage of Muslims will continue to increase by one percentage point per decade, the Muslim population is expected to increase to half a million or 6% by 2050 and one million or 11% by 2100.

Table 4.4.9a Beginning of year evolution of the Muslim population in Norway

Year	Population	Muslims	%	Source
1865	1,701,756	—	0.00	[EUH]c
1875	1,813,424	—	0.00	[EUH]c
1891	2,000,917	—	0.00	[EUH]c
1900	2,240,032	9	0.00	[NOC]e
1910	2,391,782	—	0.00	[NOC]e
1920	2,649,775	17	0.00	[NOC]e
1930	2,814,194	9	0.00	[NOC]e
1950	3,278,546	—	0.00	[NOC]e
1960	3,591,234	—	0.00	[NOC]e
1970	3,874,133	821	0.02	[NOC]e

Table 4.4.9a (Continued)

Year	Population	Muslims	%	Source
1980	4,078,900	1,006	0.02	[NO]c
1983	4,122,511	3,540	0.09	[NO]c
1984	4,134,353	4,432	0.11	[NO]c
1985	4,145,845	8,214	0.20	[NO]c
1986	4,159,187	10,520	0.25	[NO]c
1987	4,175,521	12,090	0.29	[NO]c
1988	4,198,289	14,727	0.35	[NO]c
1989	4,220,686	16,322	0.39	[NO]c
1990	4,233,116	19,189	0.45	[NO]c
1991	4,249,830	22,158	0.52	[NO]c
1992	4,273,634	28,906	0.68	[NO]c
1993	4,299,167	21,685	0.50	[NO]c
1994	4,324,815	32,811	0.76	[NO]c
1995	4,348,410	40,550	0.93	[NO]c
1996	4,369,959	42,134	0.96	[NO]c
1997	4,392,714	47,438	1.08	[NO]c
1998	4,417,599	46,634	1.06	[NO]c
1999	4,445,329	51,193	1.15	[NO]c
2000	4,478,497	56,458	1.26	[NO]c
2001	4,520,947	62,753	1.39	[NO]c
2002	4,524,066	70,487	1.56	[NO]c
2003	4,552,252	75,761	1.66	[NO]c
2004	4,577,457	80,838	1.77	[NO]c
2005	4,606,363	76,621	1.66	[NO]c
2006	4,640,219	72,023	1.55	[NO]c
2007	4,681,134	79,068	1.69	[NO]c
2008	4,737,171	83,684	1.77	[NO]c
2009	4,799,252	92,744	1.93	[NO]c
2010	4,858,199	98,953	2.04	[NO]c
2011	4,920,300	106,735	2.17	[NO]c
2012	4,985,870	112,236	2.25	[NO]c
2013	5,051,275	120,882	2.39	[NO]c
2014	5,109,056	132,135	2.59	[NO]c
2015	5,165,802	141,027	2.73	[NO]c
2016	5,213,985	148,189	2.84	[NO]c
2017	5,258,317	153,067	2.91	[NO]c
2018	5,295,619	166,861	3.15	[NO]c
2050	6,600,000	396,000	6.00	es
2100	7,953,000	875,000	11.00	es

Table 4.4.9b Beginning of year county distribution of the Muslim population in Norway since 2000. Oslo is included in Akershus in 2000–2004

	2000	2002	2004	2006	2008	2010	2012	2014	2016	2018
Østfold	1,846	2,138	1,954	5,019	5,742	7,460	8,280	9,850	10,840	12,895
Akershus	39,811	53,224	60,151	6,512	8,465	10,369	12,557	15,445	18,238	21,333
Oslo				33,895	39,263	44,583	49,656	55,284	60,123	62,462
Hedmark	—	579	112	1,029	1,130	1,438	1,730	2,258	2,533	3,003
Oppland	—	—	477	1,069	1,136	1,349	1,627	1,987	2,524	2,999
Buskerud	2,723	2,861	3,892	5,448	6,790	8,097	8,965	10,386	11,240	12,592
Vestfold	108	748	1,358	2,341	2,697	3,276	3,859	4,588	4,893	5,544
Telemark	770	667	1,143	1,997	2,194	2,524	2,909	3,334	3,751	4,429
Aust-Agder	324	315	383	632	585	821	970	1,027	1.300	1,856
Vest-Agder	927	1,338	1,479	1,638	1,813	2,212	2,420	3,140	3,598	4,267
Rogaland	2,181	3,855	4,006	4,973	5,661	6,565	7,444	8,396	9,518	11,127
Hordaland	943	1,841	2,079	2,549	2,716	3,721	3,511	5,019	6,135	7,460
Sogn & Fjordane	—	191	295	268	314	387	339	666	805	1,256
Møre & Romsdal	—	—	188	907	887	982	1,105	1,431	1,890	2,641
Sør-Trøndelag	—	—	1,827	2,088	2,385	2,679	3,157	3,941	4,376	6,818
Nord-Trøndelag	—	—	483	376	403	508	837	1,240	1,459	
Nordland	—	—	437	588	699	1,030	1,526	2,079	2,292	2,714
Troms	—	342	574	614	627	717	826	1,368	1,622	2,022
Finnmark	—	—	—	80	177	235	368	509	788	957
Svalbard	—	—	—	—	—	—	—	—	—	—

4.4.10 Sweden

Currently, the Kingdom of Sweden has an area of 450,295 sq. km and its map is presented in Fig. 4.4.10. Like Finland, the first Muslims to come to Sweden were Tatars, who migrated in the seventeenth century. According to census data, the number of Muslims increased

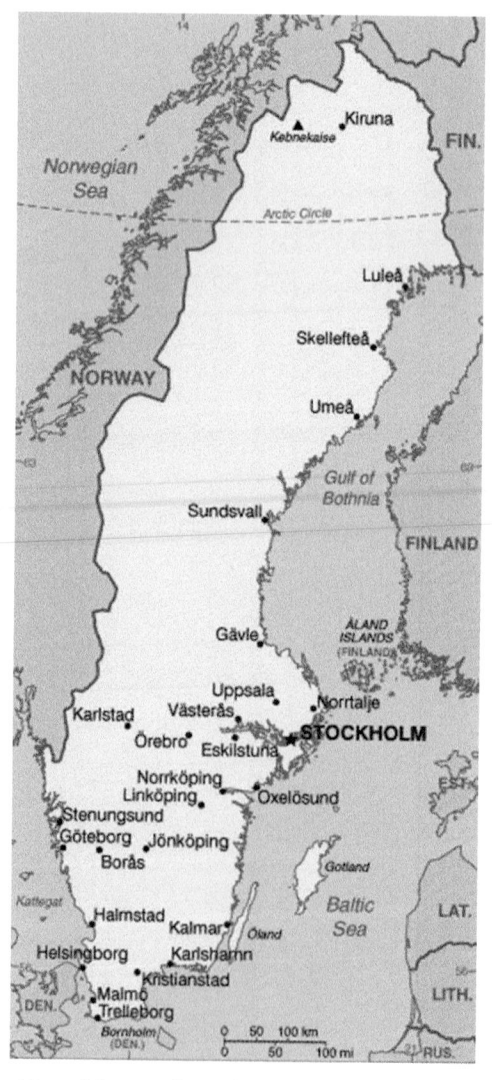

Figure 4.4.10 Map of the Kingdom of Sweden.

from one in 1910, to six in 1920 to fifteen in 1930. Later censuses in Sweden did not inquire on the number of Muslims in the country. Nevertheless, Statistics Sweden does collect information on the country of birth of the population in Sweden. From this data, we can infer the number of Muslims in Sweden with the assumption that those born in Muslim-majority countries are Muslim. The number of people born in such countries is summarized in Table 4.4.10a.

Table 4.4.10a Year-end estimate of the Muslim population in Sweden based on country of birth

	1900	1930	1950	1960	1970	1975	1980	1985
Syria	—	—	—	6	100	334	1,606	2,847
Iraq	—	—	5	16	108	161	631	4,075
Iran	2	8	110	115	411	998	3,348	8,900
Somalia	—	—	—	—	16	26	100	258
Afghanistan	—	—	58	17	28	27	31	130
Turkey	15	22	87	202	3,768	6,143	14,357	19,264
Lebanon	—	—	—	15	240	486	2,170	4,709
Pakistan	—	—	—	11	188	469	1,127	1,590
Morocco	—	—	—	22	775	999	1,584	1,952
Others	7	19	57	305	2,472	2,971	5,447	7,370
Total	24	49	317	709	8,106	12,614	30,401	51,095
Muslim %	0.00	0.00	0.00	0.01	0.10	0.15	0.37	0.61

Table 4.4.10a (Continued)

	1990	1995	2000	2005	2010	2015	2018
Syria	5,874	9,403	14,162	16,772	20,758	98,216	185,991
Iraq	9,818	26,361	49,372	72,553	121,761	131,888	144,035
Iran	40,084	49,040	51,101	54,470	62,120	69,067	77,386
Somalia	1,441	10,377	13,082	16,045	37,846	60,623	68,678
Bosnia	—	45,602	51,526	54,813	56,183	57,705	59,395
Afghanistan	534	1,544	4,287	8,345	14,420	31,267	51,979
Turkey	25,528	29,761	31,894	35,853	42,527	46,373	49,948
Lebanon	15,986	21,555	20,038	21,441	24,116	26,159	28,119
Pakistan	2,291	2,626	3,100	4,703	10,265	11,469	16,185
Morocco	2,720	3,818	4,492	5,547	7,391	9,480	11,025
Others	11,617	16,655	25,399	35,309	56,654	86,299	122,113
Total	115,893	216,742	268,453	325,851	454,041	628,546	814,854
Muslim %	1.35	2.45	3.02	3.60	4.82	6.38	7.97

Table 4.4.10b Year-end evolution of the Muslim population in Sweden

Year	Population	Muslims	%	Source
1850	3,482,541	—	0.00	[EUH]c
1860	3,859,728	—	0.00	[EUH]c
1870	4,169,525	—	0.00	[EUH]c
1880	4,555,668	—	0.00	[EUH]c
1890	4,784,981	—	0.00	[EUH]c
1900	5,136,441	—	0.00	[SE910]c
1910	5,169,678	1	0.00	[SE910]c
1920	5,509,497	6	0.00	[SE920]c
1930	6,142,191	15	0.00	[SE930]c
1950	7,041,829	317	0.00	[SE95]e
1960	7,041,829	709	0.01	[SE95]e
1970	8,081,142	8,106	0.10	[SE95]e
1975	8,208,442	12,614	0.15	[SE95]e
1980	8,317,937	30,401	0.37	[SE95]e
1985	8,358,139	51,095	0.61	[SE95]e
1990	8,590,630	115,893	1.35	[SE95]e
1995	8,837,496	216,742	2.45	[SE95]e
2000	8,882,792	268,453	3.02	[SE]e
2005	9,047,752	325,851	3.60	[SE]e
2010	9,415,570	454,041	4.82	[SE]e
2015	9,851,017	628,546	6.38	[SE]e
2018	10,230,185	814,854	7.97	[SE]e
2050	11,389,000	1,708,000	15.00	es
2100	13,023,000	3,256,000	25.00	es

Thus, the percentage of Muslims in Sweden was less than 0.01% until 1960; increasing in number from none in the nineteenth century, to one in 1910, to six in 1920, to fifteen in 1930, to 300 in 1950, to 700 or 0.01% in 1960. The increase of Muslims picked up momentum afterwards due to the country's need for immigrant workers as its economy improved. Accordingly, the Muslim population increased to 8,000 or 0.1% in 1970, to 13,000 or 0.2% in 1975, to 30,000 or 0.4% in 1980, to 51,000 or 0.6% in 1985. The number of Muslims continued to increase to 0.12 million or 1.4% in 1990, to 0.22 million or 2.5% in 1995, to 0.27 million or 3.0% in 2000, to 0.33 million or 3.6% in 2005, to 0.45 million or 4.8% in 2010, to 0.63 million or 6.4% in 2015 and 0.81 million or 8.0% in 2018. A significant portion of the increase is due to an influx of refugees from Muslim majority war torn countries such as Bosnia, Iraq, Somalia and Syria. Thus, assuming that the percentage of Muslims will continue to increase by two percentage points per decade, the Muslim population is expected to reach close to two million or 15% by 2050 and over three million or 25% by 2100. A summary of the data is provided in Table 4.4.10b.

4.4.11 The United Kingdom

The United Kingdom of Great Britain and Northern Ireland is an island nation with a total area of 243,610 sq. km, excluding fourteen overseas territories, and the three British Crown Dependencies: Isle of Man, Jersey and Guernsey. The UK, for short, consists of the following four countries: England (130,395 sq. km), Northern Ireland (13,843 sq. km), Scotland (78,772 sq. km), and Wales (20,779 sq. km). A map of the UK is presented in Fig. 4.4.11.

The first notable Muslim presence in the UK was in the second half of the nineteenth century by Yemeni migrants from the city port of Aden, which was a British colony. By the twentieth century, Muslims settled from India near London. The Muslim population pre-WWI remained at 0.00% of the total population. It increased from 543 in 1901, to 715 in 1911. In the latter, it was distributed as follows: 681 in England, 25 in Scotland, nine in Wales. In Northern Ireland, census data indicate that the number of Muslims increased from none in the nineteenth century to three in 1901, to one in 1911 to less than ten in 1926, to fifty in 1937.

During the first half of the twentieth century, migrants settled in the UK from former British colonies with substantial Muslim population, including Cyprus, Egypt, and Iraq. Thus, the Muslim population reached 50,000 or 0.1% in 1939, just before the start of WWII.

Figure 4.4.11 Map of the United Kingdom.

The Muslim population picked up momentum after WWII, as the country needed more manpower, which was fulfilled with migration from former colonies that were gaining independence. Consequently, the Muslim population doubled to 100,000 or 0.2% in 1951, and again to 0.2 million or 0.4% in 1981 and 0.4 or 0.8% million in 1991. The 2001 census asked for the first time about religious affiliation.

Thus, the number of Muslims reached 1.59 million or 2.9% in 2001 and 2.79 million or 4.8% in 2011. According to European Values Survey (EVS), the percentage of Muslims was 0.36% in 1981, 0.75% in 1990, 1.20% in 1999 and 4.43% in 2009. A summary of the data is provided in Table 4.4.11a. Thus, assuming that the percentage of Muslims will increase by one percentage point per decade, the Muslim population is expected to reach close to seven million or 9% by 2050 and eleven million or 14% by 2100. Table 4.4.11b presents the distribution of Muslims per country within the union. According to census results, the vast majority of Muslim population, over 95% is concentrated in England.

Table 4.4.11a Evolution of the Muslim population in the United Kingdom

Year	Population	Muslims	%	Source
1801	8,892,536	—	0.00	[GB851]c
1811	10,364,256	—	0.00	[GB851]c
1821	12,000,236	—	0.00	[GB851]c
1831	13,896,797	—	0.00	[GB851]c
1841	15,914,148	—	0.00	[GB851]c
1851	17,927,609	—	0.00	[GB851]c
1901	37,518,100	546	0.00	[UKH1]es
1911	41,126,000	716	0.00	[UKH2]es
1939	47,760,000	50,000	0.10	[KET96]es
1951	50,562,000	100,000	0.20	[KET96]es
1981	54,814,574	200,000	0.36	[EVS]s
1991	56,206,521	420,000	0.75	[EVS]s
2001	54,266,622	1,591,126	2.93	[EW,SQ,ND]c
2011	58,653,855	2,786,635	4.75	[EW2,SQ2,ND2]c
2050	74,082,000	6,667,000	9.00	es
2100	78,053,000	10,927,000	14.00	es

Table 4.4.11b Evolution of the Muslim population in the United Kingdom per country. P: Total population, M: Muslim population, M%: Percentage of Muslim population, MR%: Muslim Ratio

Country		1901	1911	2001	2011
England	P	30,072,180	33,561,235	45,362,316	49,208,352
	M%	0.00	0.00	3.36	5.41
	M	520	681	1,524,887	2,660,116
	MR%	95.24	95.11	95.84	95.46
Scotland	P	4,472,103	4,760,904	4,783,950	4,927,364
	M%	0.00	0.00	0.89	1.56
	M	17	25	42,557	76,737
	MR%	3.11	3.49	2.67	2.75
Wales	P	2,012,876	2,420,921	2,668,942	2,829,528
	M%	0.00	0.00	0.81	1.62
	M	6	9	21,739	45,950
	MR%			1.37	1.65
Northern Ireland	P	1,236,952	1,250,531	1,451,414	1,688,611
	M%	0.00	0.00	0.13	0.23
	M	3	1	1,943	3,832
	MR%	0.55	0.14	0.12	0.14

4.4.12 Regional Summary and Conclusion

Northern Europe has the second least concentration of Muslims among the five regions spanning the European Continent. The Muslim population in this region was almost nonexistent prior to the twentieth century and only picked up momentum in its second half. The increase was at first through migration as countries in this region needed to compensate for their labor shortage in post WWII era. The Muslim population continued to increase significantly towards the end of last century and is expected to continue this trend, reaching over one eighth of the total population by the end of this century. The main cause of increase is due to migration of Muslims, then their natural increase and finally conversion to Islam by natives. The following tables present centennial data from 600 AD to 2100 AD (or approximately 1 H to 1500 H) in Table 4.4a and decennial data from 1790 AD to 2100 AD (or 1210 H to 1520 H) in Tables 4.4b and 4.4c for current countries in Northern Europe. The data includes total population in thousands (P), the percentage of which is Muslim (M%), the corresponding Muslim population in thousands (M), and the annual population growth rate (APGR, or G%) of the total population in this region.

4.5 Muslims in Central Europe

This region consists of eleven countries: Austria, Czechia, Germany, Hungary, Italy, Liechtenstein, Malta, Poland, San Marino, Slovakia, and Switzerland. Islam entered this region from the south starting the eighth century with the capture of Mediterranean islands and parts of the Italian peninsula during the ninth century. Muslims lost control over their captured regions in the Italian Peninsula in the ninth century, and all Mediterranean islands in this region by the end of the eleventh century. Muslims entered Poland in the thirteenth century with the Golden Horde Empire and Muslim Tatars remain a small minority in Poland till this day. However, larger waves of Muslims came after WWII as the region needed more manpower.

Thus, the Muslim population has likely changed from a thousand in 700 AD and 800 AD, to 57,000 or 0.5% in 900 AD, to 111,000 or 0.9% in 1000 AD, to 68,000 or 0.4% in 1100 AD, to 100,000 or 0.5% in 1200 AD, to 110,000 or 0.5% in 1300 AD, to 35,000 or 0.1% in 1400 AD, to 60,000 or 0.2% in 1500 AD, to 100,000 or 0.2% in 1600 AD, to 60,000 or 0.1% in 1700 AD, to 4,000 or 0.01% in 1800 AD, to 14,000 or 0.01% in 1900, to 3.6 million or 1.6% in 2000, to 7.5 million or 3.3% in 2020, and is projected to near eleven million or 5% by 2050 and near fifteen million or 8% by 2100.

A plot of centennial estimates of the Muslim population and its percentage with respect to the total population in this region from 600 to 2100 is provided in Fig. 4.5a. A zoom in of this plot, providing a plot of decennial estimates of the Muslim population and its percentage with respect to the total population in this region from 1900 to 2100 is provided in Fig. 4.5b. The corresponding individual data for each country in this region is discussed below. In Section 4.5.12, the total population in each country in this region and the corresponding percentage and number of Muslims is presented centennially in Table 4.5a from 600 to 2100 and decennially in Tables 4.5b and 4.5c from 1790 to 2100.

Table 4.4a Centennial estimates of the Muslim population (×1000) in Northern Europe: 600–2100 AD (1–1500 H)

		600	700	800	900	1000	1100	1200	1300
Denmark	P	280	300	320	340	360	400	450	500
	M%	—	—	—	—	—	—	—	—
	M	—	—	—	—	—	—	—	—
Faroes	P	1	1	1	1	1	1	1	1
	M%	—	—	—	—	—	—	—	—
	M	—	—	—	—	—	—	—	—
Finland	P	32	34	36	38	40	100	150	200
	M%	—	—	—	—	—	—	—	—
	M	—	—	—	—	—	—	—	—
Guernsey	P	1	1	1	1	1	1	1	1
	M%	—	—	—	—	—	—	—	—
	M	—	—	—	—	—	—	—	—
Iceland	P	2	3	4	6	8	10	15	20
	M%	—	—	—	—	—	—	—	—
	M	—	—	—	—	—	—	—	—
Ireland	P	230	245	260	275	289	400	500	600
	M%	—	—	—	—	—	—	—	—
	M	—	—	—	—	—	—	—	—
Isle of Man	P	1	1	2	3	4	5	10	15
	M%	—	—	—	—	—	—	—	—
	M	—	—	—	—	—	—	—	—
Jersey	P	1	1	1	1	1	2	3	4
	M%	—	—	—	—	—	—	—	—
	M	—	—	—	—	—	—	—	—
Norway	P	160	170	180	190	200	220	240	260
	M%	—	—	—	—	—	—	—	—
	M	—	—	—	—	—	—	—	—
Sweden	P	320	340	360	380	400	425	450	475
	M%	—	—	—	—	—	—	—	—
	M	—	—	—	—	—	—	—	—
UK	P	1,520	1,640	1,760	1,880	2,000	2,400	2,800	3,200
	M%	—	—	—	—	—	—	—	—
	M	—	—	—	—	—	—	—	—
Total	P	2,548	2,736	2,925	3,115	3,304	3,964	4,620	5,276
	M%	0.00	0.00	0.00	0.00	0.00	0.00	0.00	0.00
	M	—	—	—	—	—	—	—	—
	G%		0.071	0.067	0.063	0.059	0.182	0.153	0.133

Table 4.4a (*Continued*)

		1400	1500	1600	1700	1800	1900	2000	2100
Denmark	P	550	600	650	700	929	2,418	5,341	6,873
	M%	—	—	—	—	—	—	3.22	22.00
	M	—	—	—	—	—	—	172	1,512
Faroes	P	1	2	3	4	5	15	47	55
	M%	—	—	—	—	—	—	0.04	0.50
	M	—	—	—	—	—	—	—	—
Finland	P	250	300	400	400	833	2,635	5,188	5,254
	M%	—	—	—	—	—	—	0.02	1.10
	M	—	—	—	—	—	—	1	58
Guernsey	P	1	2	5	10	17	40	59	89
	M%	—	—	—	—	—	—	0.07	0.60
	M	—	—	—	—	—	—	—	1
Iceland	P	25	30	35	40	46	78	280	353
	M%	—	—	—	—	—	—	0.05	2.00
	M	—	—	—	—	—	—	—	7
Ireland	P	700	800	1,000	1,925	5,220	3,222	3,783	5,685
	M%	—	—	—	—	—	0.00	0.50	5.50
	M	—	—	—	—	—	—	19	313
Isle of Man	P	20	25	30	35	38	55	77	96
	M%	—	—	—	—	—	0.01	0.05	0.55
	M	—	—	—	—	—	—	—	1
Jersey	P	5	10	15	20	23	53	89	134
	M%	—	—	—	—	—	—	0.06	0.60
	M	—	—	—	—	—	—	—	1
Norway	P	280	300	400	500	884	2,240	4,499	7,953
	M%	—	—	—	—	—	0.00	1.26	11.00
	M	—	—	—	—	—	—	57	875
Sweden	P	500	550	760	1,260	2,347	5,136	8,882	13,023
	M%	—	—	—	—	—	0.00	2.98	25.00
	M	—	—	—	—	—	—	265	3,256
UK	P	3,600	3,942	6,170	8,565	10,501	35,406	58,923	78,053
	M%	—	—	—	—	—	0.00	2.93	14.00
	M	—	—	—	—	—	—	1,726	10,927
Total	P	5,932	6,561	9,468	13,459	20,843	51,298	87,169	117,568
	M%	0.00	0.00	0.00	0.00	0.00	0.00	2.57	14.42
	M	—	—	—	—	—	—	2,240	16,950
	G%	0.117	0.101	0.367	0.352	0.437	0.901	0.530	0.299

Table 4.4b Decennial estimates of the Muslim population (×1000) in Northern Europe: 1790–1940 AD (1210–1360 H)

		1790	1800	1810	1820	1830	1840	1850	1860
Denmark	P	842	929	1,003	1,155	1,273	1,357	1,417	1,601
	M%	—	—	—	—	—	—	—	—
	M	—	—	—	—	—	—	—	—
Faroes	P	5	5	6	6	7	7	8	9
	M%	—	—	—	—	—	—	—	—
	M	—	—	—	—	—	—	—	—
Finland	P	706	833	863	1,178	1,372	1,441	1,621	1,726
	M%	—	—	—	—	—	—	—	—
	M	—	—	—	—	—	—	—	—
Guernsey	P	15	17	19	20	25	27	30	30
	M%	—	—	—	—	—	—	—	—
	M	—	—	—	—	—	—	—	—
Iceland	P	40	46	48	48	51	57	60	68
	M%	—	—	—	—	—	—	—	—
	M	—	—	—	—	—	—	—	—
Ireland	P	4,200	5,220	5,960	6,802	7,767	6,529	5,112	4,402
	M%	—	—	—	—	—	—	—	—
	M	—	—	—	—	—	—	—	—
Isle of Man	P	37	38	39	40	41	48	52	52
	M%	—	—	—	—	—	—	—	—
	M	—	—	—	—	—	—	—	—
Jersey	P	20	23	26	29	37	48	57	56
	M%	—	—	—	—	—	—	—	—
	M	—	—	—	—	—	—	—	—
Norway	P	800	884	885	885	1,051	1,195	1,328	1,490
	M%	—	—	—	—	—	—	—	—
	M	—	—	—	—	—	—	—	—
Sweden	P	2,188	2,347	2,396	2,585	2,888	3,139	3,483	3,860
	M%	—	—	—	—	—	—	—	—
	M	—	—	—	—	—	—	—	—
UK	P	9,500	10,501	11,970	14,092	16,261	18,534	20,817	23,129
	M%	—	—	—	—	—	—	—	—
	M	—	—	—	—	—	—	—	—
Total	P	18,352	20,843	23,216	26,840	30,773	32,381	33,984	36,422
	M%	0.00	0.00	0.00	0.00	0.00	0.00	0.00	0.00
	M	—	—	—	—	—	—	—	—
	G%		1.273	1.078	1.451	1.368	0.509	0.483	0.693

Table 4.4b (Continued)

		1870	1880	1890	1900	1910	1920	1930	1940
Denmark	P	1,785	1,968	2,170	2,418	2,722	3,061	3,531	3,826
	M%	—	—	—	—	—	—	—	—
	M	—	—	—	—	—	—	—	—
Faroes	P	10	11	13	15	18	21	24	27
	M%	—	—	—	—	—	—	—	—
	M	—	—	—	—	—	—	—	—
Finland	P	1,740	2,033	2,348	2,635	2,915	3,118	3,435	3,700
	M%	—	—	—	—	—	0.01	0.01	0.02
	M	—	—	—	—	—	—	—	1
Guernsey	P	31	33	35	40	42	38	41	44
	M%	—	—	—	—	—	—	—	—
	M	—	—	—	—	—	—	—	—
Iceland	P	69	72	71	78	85	93	106	120
	M%	—	—	—	—	—	—	—	—
	M	—	—	—	—	—	—	—	—
Ireland	P	4,035	3,870	3,469	3,222	3,140	3,000	2,970	2,960
	M%	—	—	—	0.00	0.00	0.00	0.00	0.00
	M	—	—	—	—	—	—	—	—
Isle of Man	P	54	54	56	55	52	60	49	52
	M%	—	—	—	0.01	0.02	0.02	0.02	0.02
	M	—	—	—	—	—	—	—	—
Jersey	P	57	52	55	53	52	50	50	50
	M%	—	—	—	—	—	—	—	—
	M	—	—	—	—	—	—	—	—
Norway	P	1,702	1,813	2,001	2,240	2,392	2,650	2,814	3,157
	M%	—	—	—	—	—	0.00	0.00	—
	M	—	—	—	—	—	—	—	—
Sweden	P	4,169	4,566	4,785	5,136	5,522	5,904	6,142	6,371
	M%	—	—	—	0.00	0.00	0.00	0.00	0.00
	M	—	—	—	—	—	—	—	—
UK	P	26,072	29,709	33,400	35,406	41,126	43,717	44,937	48,220
	M%	—	0.00	0.00	0.00	0.00	0.01	0.05	0.10
	M	—	—	—	—	—	4	22	48
Total	P	39,723	44,181	48,402	51,298	58,065	61,712	64,100	68,528
	M%	0.00	0.00	0.00	0.00	0.00	0.01	0.04	0.07
	M	—	1	1	1	1	5	23	49
	G%	0.868	1.064	0.913	0.581	1.239	0.609	0.380	0.668

Table 4.4c Decennial estimates of the Muslim population (×1000) in Northern Europe: 1950–2100 AD (1370–1520 H)

		1950	1960	1970	1980	1990	2000	2010	2020
Denmark	P	4,268	4,581	4,931	5,124	5,141	5,341	5,555	5,792
	M%	—	0.00	0.04	0.55	1.38	3.22	4.15	6.00
	M	—	—	2	28	71	172	231	348
Faroes	P	32	35	39	43	47	47	48	49
	M%	—	—	—	0.00	0.02	0.04	0.07	0.10
	M	—	—	—	—	—	—	—	—
Finland	P	4,008	4,435	4,612	4,788	4,996	5,188	5,366	5,541
	M%	0.02	0.02	0.02	0.02	0.02	0.02	0.17	0.30
	M	1	1	1	1	1	1	9	17
Guernsey	P	41	44	48	51	56	59	64	70
	M%	—	—	—	0.03	0.07	0.07	0.13	0.14
	M	—	—	—	—	—	—	—	—
Iceland	P	143	176	204	228	255	280	320	341
	M%	—	0.01	0.01	0.01	0.01	0.05	0.19	0.40
	M	—	—	—	—	—	—	1	1
Ireland	P	2,913	2,791	2,908	3,385	3,511	3,783	4,554	4,938
	M%	0.00	0.00	0.07	0.11	0.11	0.50	1.09	1.37
	M	—	—	2	4	4	19	50	68
Isle of Man	P	55	48	55	64	70	77	85	85
	M%	0.02	0.02	0.02	0.02	0.05	0.05	0.08	0.15
	M	—	—	—	—	—	—	—	—
Jersey	P	61	66	73	77	84	89	96	104
	M%	—	—	—	0.01	0.03	0.06	0.13	0.20
	M	—	—	—	—	—	—	—	—
Norway	P	3,265	3,583	3,876	4,086	4,247	4,499	4,886	5,421
	M%	—	—	0.02	0.02	0.45	1.26	2.04	3.00
	M	—	—	1	1	19	57	100	163
Sweden	P	7,010	7,483	8,055	8,316	8,567	8,882	9,390	10,099
	M%	0.00	0.01	0.10	0.37	1.35	2.98	4.76	9.00
	M	—	1	8	31	116	265	447	909
UK	P	50,616	52,371	55,573	56,209	57,134	58,923	63,460	67,886
	M%	0.20	0.20	0.30	0.36	0.75	2.93	4.75	6.00
	M	101	105	167	202	429	1,726	3,014	4,073
Total	P	72,412	75,611	80,377	82,372	84,111	87,169	93,824	100,326
	M%	0.14	0.14	0.22	0.32	0.76	2.57	4.10	5.56
	M	102	106	181	267	639	2,240	3,851	5,578
	G%	0.551	0.432	0.611	0.245	0.209	0.357	0.736	0.670

Table 4.4c (*Continued*)

		2030	2040	2050	2060	2070	2080	2090	2100
Denmark	P	6,002	6,147	6,245	6,355	6,490	6,630	6,747	6,873
	M%	8.00	10.00	12.00	14.00	16.00	18.00	20.00	22.00
	M	480	615	749	890	1,038	1,193	1,349	1,512
Faroes	P	51	53	53	54	55	56	56	55
	M%	0.15	0.20	0.25	0.30	0.35	0.40	0.45	0.50
	M	—	—	—	—	—	—	—	—
Finland	P	5,581	5,552	5,486	5,422	5,386	5,344	5,290	5,254
	M%	0.40	0.50	0.60	0.70	0.80	0.90	1.00	1.10
	M	22	28	33	38	43	48	53	58
Guernsey	P	74	77	80	81	83	85	87	89
	M%	0.25	0.30	0.35	0.40	0.45	0.50	0.55	0.60
	M	—	—	—	—	—	—	—	1
Iceland	P	360	372	377	376	374	369	361	353
	M%	0.60	0.80	1.00	1.20	1.40	1.60	1.80	2.00
	M	2	3	4	5	5	6	6	7
Ireland	P	5,248	5,489	5,678	5,739	5,710	5,698	5,717	5,685
	M%	2.00	2.50	3.00	3.50	4.00	4.50	5.00	5.50
	M	105	137	170	201	228	256	286	313
Isle of Man	P	88	90	91	91	92	94	95	96
	M%	0.20	0.25	0.30	0.35	0.40	0.45	0.50	0.55
	M	—	—	—	—	—	—	—	1
Jersey	P	111	116	120	122	124	127	130	134
	M%	0.25	0.30	0.35	0.40	0.45	0.50	0.55	0.60
	M	—	—	—	—	1	1	1	1
Norway	P	5,876	6,271	6,600	6,892	7,175	7,452	7,702	7,953
	M%	4.00	5.00	6.00	7.00	8.00	9.00	10.00	11.00
	M	235	314	396	482	574	671	770	875
Sweden	P	10,630	11,008	11,389	11,739	12,056	12,380	12,679	13,023
	M%	11.00	13.00	15.00	17.00	19.00	21.00	23.00	25.00
	M	1,169	1,431	1,708	1,996	2,291	2,600	2,916	3,256
UK	P	70,485	72,487	74,082	75,060	75,815	76,535	77,220	78,053
	M%	7.00	8.00	9.00	10.00	11.00	12.00	13.00	14.00
	M	4,934	5,799	6,667	7,506	8,340	9,184	10,039	10,927
Total	P	104,505	107,662	110,201	111,932	113,360	114,767	116,084	117,568
	M%	6.65	7.73	8.83	9.93	11.05	12.16	13.28	14.42
	M	6,949	8,327	9,729	11,118	12,521	13,960	15,422	16,950
	G%	0.408	0.298	0.233	0.156	0.127	0.123	0.114	0.127

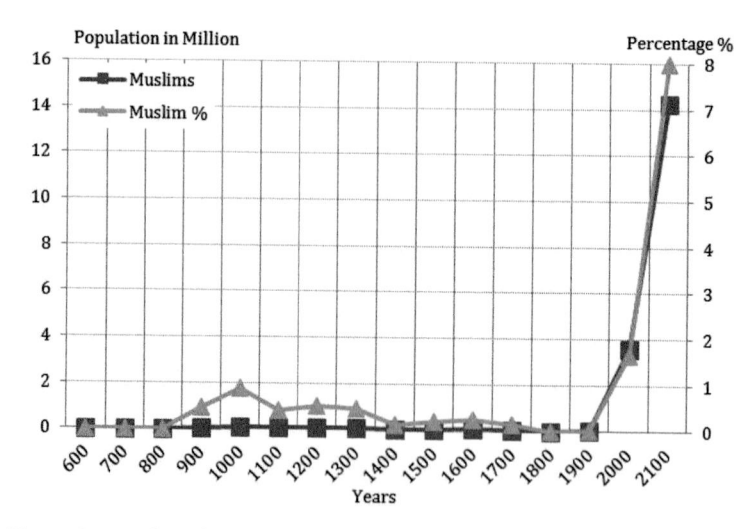

Figure 4.5a Plot of centennial estimates of the Muslim population and its percentage of the total population in Central Europe: 600–2100 AD (1–1500 H).

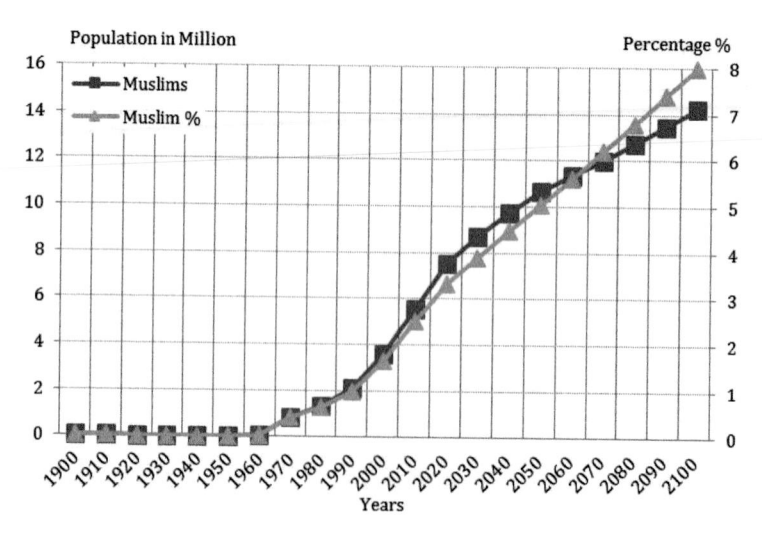

Figure 4.5b Plot of decennial estimates of the Muslim population and its percentage of the total population in Central Europe: 1900–2100 AD (1320–1520 H).

4.5.1 Austria

The Ottoman Empire besieged Vienna twice but was unable to conquer it. The first time was in 1528, after conquering Budapest, which caused the Austrian Empire to pay annual tribute (Jizya) to the Ottoman Empire. The second attempt was in 1683 resulted in the defeat of the Ottoman Empire and the subsequent shrinkage of their empire.

Currently, the Republic of Austria has an area of 83,871 sq. km and its map is presented in Fig. 4.5.1. According to census data as shown in Table 4.5.1c, the Muslim population was in double digits or 0.00% in 1869, 1880 and 1890, who lived in the capital Vienna. The 1900 census showed an increase of Muslims to 1,252 as follows: 889 in Vienna, 358 in Graz, two in Eisenerz and one in each of Baden, Modling and Leoben. The 1910 census recorded 1,044 Muslims in current Austria, mostly in Vienna and Graz, but distributed more than before.

Figure 4.5.1 Map of the Republic of Austria.

The number of Turks increased from none in 1934, to 112 in 1951, to 217 in 1961, to 16,423 in 1971, or almost three-quarters of all recorded Muslims in that year. The reason for this substantial increase is that Austria was heavily reliant on labor immigration for post-war reconstruction and economic expansion. Thus, Austria signed a bilateral agreement with Turkey in 1964 to recruit Turks as guest workers. The number of Turks increased to 59,900 in 1981, to 118,579 in 1991. The number of Muslims increased afterwards due to the settlement of Muslims from former Yugoslav republics, Iraqis, Afghanis and other from Muslims countries some as refugees and some for economic prosperity. By 2011, the number of first and second-generation Turks reached 280,441.

Statistics Austria resumed collection of the number of Muslims in the 1971 census. Accordingly, the number of Muslims increased to 22,267 or 0.3% in 1971 and all were not Austrian citizens, 66% were Turks, and 21% were from former Yugoslavia. The number increased again to 76,939 or 1.0% in 1981 of which 90% were foreigners, 69% were Turks, and 14% were from former Yugoslavia. The number then increased to 158,776 or 2.0% in 1991, 91% were foreigners, 68% were Turks and 15% from former Yugoslavia (mainly Bosnia and Kosovo). The number of Muslims continued to increase to 338,988 or 4.2% in 2001, of which 72% were foreigners, 36% were Turks and 29% were from former Yugoslavia. Table 4.5.1a summarizes the distribution of Muslims based on citizenship between 1971 and 2001. In addition, the fraction of those born abroad who gained Austrian citizenship as of 2001 (2011), are 25% (45%) of those born in Turkey, 14% (37%) of those born in Bosnia, 47% (54%) of those born in Iran, and 53% (64%) of those born in Egypt.

As summarized in Table 4.5.1b, the 2001 census showed that almost a third (28%) of Muslims were born in Austria, over a third (34.86%) in Turkey. The census also indicated that almost all those born in Turkey, half of those born in Bosnia and three-quarter of those born in Macedonia were Muslim.

Statistics Austria switched to register-based census in 2011, which did not contain information on religious affiliation. However, data is available about country of birth since 2002. We use this data to estimate the Muslims population in 2006 onward as shown in Table 4.5.1c. Accordingly, the Muslim population increased to 0.40 million or 4.8% in 2006, to 0.43 million or 5.2% in 2011, to 0.54 million or 6.2% in 2016 and 0.58 million or 6.5% in 2018.

Table 4.5.1a Distribution of the Muslim population in Austria based on country of citizenship, 1971–2001

	1971	1981	1991	2001
Turkish Citizens	16,423	59,900	118,579	127,226
TC Muslims	14,638	53,091	107,986	123,028
% Muslims	89.13	88.63	91.07	96.70
Former Yugoslavia Citizens	93,337	125,890	197,886	322,261
FYC Muslims	4,650	10,918	24,164	98,702
% Muslims	4.98	8.67	12.21	30.63
All Muslims	22,267	76,939	158,776	338,988
% TC Muslims	65.74	69.00	68.01	36.29
% FYC Muslims	20.88	14.19	15.22	29.12
% Non-Austrian Citizens Muslims	100.00	90.39	90.66	71.67
NAC Muslims	22,267	69,545	143,939	242,936

Table 4.5.1b Distribution of the Muslim population in Austria based on country of birth according to the 2001 Census

Country of Birth	Total	Muslims	% Birth	% All Muslims
Austria	7,029,527	95,252	1.36	28.10
Turkey	125,026	118,174	94.52	34.86
Bosnia	134,402	60,541	45.04	17.86
Serbia/Montenegro/Kosovo	143,077	21,453	14.99	6.33
Macedonia	13,948	10,402	74.58	3.07
Other	586,946	33,166	5.65	9.78
Total	8,032,926	338,988		

Table 4.5.1c Beginning of year estimate of the Muslim population in Austria based on country of birth

	2002	2006	2011	2016	2018
Bosnia	135,104	147,001	149,679	162,021	166,752
Turkey	126,828	152,499	158,535	160,184	160,313
Syria	1,784	2,341	3,046	33,569	46,963
Afghanistan	2,594	4,710	8,428	36,607	44,356
Kosovo	16,151	22,844	27,135	31,215	32,339
Macedonia	14,882	19,315	21,134	24,247	26,114
Iran	11,262	12,023	13,254	19,796	23,067
Iraq	3,183	3,740	4,870	15,319	15,986
Egypt	10,059	12,139	12,531	13,806	14,562
Pakistan	2,356	3,469	4,258	6,124	6,241
Somalia	142	289	1,204	4,482	5,566
Others	13,554	19,171	22,798	30,935	33,921
Total	337,899	399,541	426,872	538,305	576,180

Thus, as summarized in Table 4.5.1d, assuming that the percentage of Muslims will increase by one percentage point per decade, the Muslim population is expected to reach one million or 11% by 2050 and 1.4 million or 16% by 2100.

4.5.2 Czechia

The Czech Republic has an area of 78,867 sq. km and its map is presented in Fig. 4.5.2. It used to be united with Slovakia under the name of Czechoslovakia from 1918 until 1993, when they split peacefully with the fall of Communism. It adopted the shorter name of Czechia in 2016. The presence of Muslims here remained negligible until the start of the twenty-first century. The 1880 Austrian census showed three Muslims in current Czech lands, two in the capital Prague and one in Litomerice, just north of Prague. So, three Muslims in Bohemia region and none in Moravia and Silesia regions, the latter is mostly in current Poland. The 1900 census recorded three distributed among the cities of Teplice, Žižkov and Brno. In 1910, the number of Muslims increased to seventeen.

Table 4.5.1d Evolution of the Muslim population in Austria

Year	Population	Muslims	%	Source
1869	4,497,880	—	0.00	[AT890]c
1880	4,963,528	17	0.00	[AT880]c
1890	5,417,360	35	0.00	[AT890]c
1900	6,003,845	1,252	0.02	[AT900]c
1910	6,648,310	1,044	0.02	[AT910]c
1961	7,073,807	217	0.00	[AT01]e
1971	7,491,526	22,267	0.30	[AT01]c
1981	7,555,338	76,939	1.02	[AT01]c
1991	7,795,786	158,776	2.04	[AT01]c
2001	8,032,926	338,988	4.22	[AT01]c
2006	8,254,298	399,541	4.84	[AT11]e
2011	8,375,164	426,872	5.16	[AT11]e
2016	8,700,471	538,305	6.19	[AT11]e
2018	8,822,267	576,180	6.53	[AT11]e
2050	9,131,000	1,004,000	11.00	es
2100	8,677,000	1,388,000	16.00	es

The 1921 Czechoslovak census indicated the presence of 185 Muslims, who we assume lived in the capital Prague. As shown in Table 4.5.2 and according to census data, the number of Muslims increased from 495 or 0.01% in 1991, to 3,699 or 0.04% in 2001, to 3,358 or 0.06% in 2011. In the last census the religious affiliation was not collected from almost half of the population (4.66 million)! Thus, assuming that the percentage of Muslims will continue to increase by 0.02 of a percentage point per decade, the Muslim population is expected to increase to 15,000 or 0.2% by 2050 and 25,000 or 0.24% by 2100.

Figure 4.5.2 Map of the Czech Republic.

Table 4.5.2 Evolution of the Muslim population in the Czechia

Year	Population	Muslims	%	Source
1869	7,617,230	—	0.00	[AT890]c
1880	8,222,013	3	0.00	[AT880]c
1890	8,665,421	7	0.00	[AT890]c
1900	9,372,214	3	0.00	[AT900]c
1910	10,078,637	17	0.00	[AT910]c
1921	10,004,170	185	0.00	[ROD]c
1991	8,641,598	495	0.01	[CZ]c
2001	9,328,079	2,676	0.04	[CZ]c
2011	5,899,759	3,385	0.06	[CZ]c
2050	10,546,000	15,000	0.14	es
2100	10,274,000	25,000	0.24	es

4.5.3 Germany

Currently, the Federal Republic of Germany has an area of 357,022 sq. km and its map is presented in Fig. 4.5.3. The first Muslims were part of the Ottoman embassy in Berlin in the eighteenth century. Indeed, the 1843 Prussian census indicated the presence of 9 Muslims or less than 0.00% of the total population of 15,536,734. At the time, Prussia consisted of Germany and half Poland. The 1880 Prussian census indicated that there were 285 or less than 0.00% of the population of 27,279,111 belonging to religions other than Christianity or Judaism. Table 4.5.3a presents the number of Turks living in Berlin from 1878 to 1945, which shows that until the middle of the twentieth century, Muslims remained at 0.00% of the total population.

Figure 4.5.3 Map of the Federal Republic of Germany.

As German economy boomed, the country needed more manpower, many of whom came from Muslim countries, especially Turkey. The first agreement to supply Germany with "Guest Workers" was signed with Turkey in 1961, then Morocco in 1963, and Tunisia in 1965. In 1970s, unification of the guest workers with their families was allowed, which further increased the number of Muslims in Germany. In 1990s the number of naturalized people accelerated as immigration laws were relaxed, and by 2001, those born in Germany were given the German nationality. Table 4.5.3b presents the evolution of Turkish nationals living in Germany from 1961 to 1996, which shows multiplicative increase between 1961 and 1981 due to immigration. The decrease in the 1990s is due to the fact that more and more Turks were able to obtain the German citizenship.

Table 4.5.3a Evolution of the Turkish population in Berlin: 1878–1946

	1878	1893	1917	1925	1933	1938	1946
Turks	41	198	2,046	1,164	585	3,310	79

Table 4.5.3b Evolution of the resident Turkish population in Germany between 1961 and 2011 [DEB, EU]

	1961	1966	1971	1976	1981	1986
Turks	7,116	161,000	652,800	1,079,300	1,546,300	1,425,721

Table 4.5.3b (Continued)

	1991	1996	2001	2006	2011
Turks	1,779,586	2,049,060	1,998,534	1,764,041	1,731,688

According to census of West Germany, the number of Muslims increased from 788,643 (out of 60,651,000) in 1970 to 1,650,952 (out of 61,077,042) in 1987. Considering that the number of Muslims was negligible in East Germany during the same period (less than 2,000 according to [KET]), then the percentage of Muslims increased from 1.0% in 1970 to 2.2% in 1987 of the total population of the current Germany. According to the 2011, the Muslim population decreased to 1,524,174 or 1.9% of the total population. The latter data however, is deemed unreliable by the census organizers due to the format of the question and the fact that one-sixth of the population did not respond to the religion question [DE11]. The format of the question consisted of two parts, the first was compulsory, while the second was voluntary:

- What religious society do you belong to? Roman Catholic, Orthodox, Evangelical, Evangelical Free Churches, Jewish communities, other religious societies under public law. Or no religious society under public law, for which the respondent is referred to the second voluntary question:
- Which of the following religions, persuasions or beliefs do you adhere to? Christianity, Judaism, Islam, Buddhism, Hinduism, other, or no religion.

66.8% were Christians (3.8% Christian not under public law society), while 17.4% did not answer the second question. To adjust the under-estimate of Muslims, [DE70] subtracts the number of those who did not answer (17.4% or 13,958,227) from the total population 80,219,695, to obtain 66,261,468, then divide the number of Muslims who responded 1,524,174, by the latter to obtain the ration 2.30%. This approach does not take into account the two-step questionnaire. Here, we would like to determine how many of those who did not respond were Muslim. To do this, we subtract 66.8% and 17.4% of the total population 80,219,695 to obtain 12,674,712 as the number of those who answered the second question. We then divide the number of Muslims (1,524,174) by the latter to obtain the ratio 12.03%. We multiply it by the number of those who did not answer (17.4% or 13,958,227) and add it to the number of Muslims to obtain 3,202,695 which is 3.99% of the total population. We take this as the estimate of the number of Muslims according to the 2011 census. Thus, assuming that the percentage of Muslims will continue to increase by half of a percentage point per decade, the Muslim population is expected to increase to almost five million or 6% by 2050 and exceed six million or 8.5% by 2100. A summary of the data is provided in Table 4.5.3c.

4.5.4 Hungary

Islam entered Hungary during the eleventh century with the arrival of Muslim Tatars and Muslims from Morocco and Al-Andalus, reaching in number about 100,000 or 12.5% of the total population at the start of the thirteenth century. They remained until King Charles I (Charles Robert who reigned from 1310 to 1342) forced them to become Christians.

Table 4.5.3c Evolution of the Muslim population in Germany

Year	Population	Muslims	%	Source
1843	28,498,136	9	0.00	[DE43]c
1852	30,492,792	—	0.00	[EUH]c
1861	34,670,792	<2	0.00	[EUH]c
1871	39,127,976	<72	0.00	[EUH]c
1880	44,766,183	<366	0.00	[EUH]c
1890	49,428,470	198	0.00	[DEB]e
1919	59,189,678	2,046	0.00	[DEB]e
1925	62,410,619	1,164	0.00	[DEB]e
1933	65,362,115	585	0.00	[DEB]e
1939	79,375,281	3,310	0.00	[DEB]e
1946	65,137,274	79	0.00	[DEB]e
1961	73,668,454	7,116	0.01	[DET]e
1970	77,550,269	788,463	1.02	[DE70]c
1987	77,780,338	1,650,952	2.12	[ROD]c
2011	80,219,695	3,202,695	3.99	[DE11]c
2050	80,104,000	4,806,000	6.00	es
2100	74,741,000	6,353,000	8.50	es

After the Battle of Mohács in 1526 in which the King of Hungary Louis II died, the Ottomans emerges victorious and started conquering current Hungary. This was during the time of Sultan Sulaiman I al-Qanuni (Arabic for lawgiver) or the Magnificent ben Selim I, the tenth Sultan of the Ottoman Empire who reigned from 1520 to 1566. This battle marked the end of the Hungarian Empire. Eventually, the Muslim Ottomans occupied Budapest in 1529. The Muslims remained in control of current Hungary until they were forced out in 1699 after their unsuccessful siege of Vienna in 1683. Thus, the Muslim rule in Hungary lasted up to 173 years. Most Muslim settlers left Hungary after the defeat of the Ottomans. Currently, Hungary has an area of 93,028 sq. km and its map is presented in Fig. 4.5.4.

As summarized in Table 4.5.4., and based on the census data, there were no Muslims in the Kingdom of Hungary in 1850. The census of 1870 recorded 45 Muslims (41 in Budapest and four in Sármellék). It was reduced to twelve by 1880 (seven in Budapest and five in Tolna) then increased to 411 or 0.01% in 1910 (389 in Budapest, mostly Bosnian soldiers), then 468 or 0.01% in 1920, to 291 or 0.00% in 1930, to 3,201 or 0.04% in 2001, to 5,579 or 0.08% in 2011. The last census did not collect religious affiliation from almost a third of the population or 2.70 million people. Thus, assuming that the percentage of Muslims will continue to rise by 0.02 of a percentage point per decade, the Muslim population is expected to reach 14,000 or 0.2% by 2050 and 18,000 or 0.2% by 2100.

Figure 4.5.4 Map of Hungary.

Table 4.5.4 Evolution of the Muslim population in Hungary

Year	Population	Muslims	%	Source
1200	800,000	100,000	12.50	[KET76]es
1850	4,180,362	—	0.00	[EUH]c
1870	5,011,310	45	0.00	[HU870]c
1880	5,329,191	12	0.00	[HU880]c
1910	7,612,114	411	0.01	[HU910]c
1920	7,986,875	468	0.01	[HU920]c
1930	8,685,109	291	0.00	[HU930]c
2001	9,093,982	3,201	0.04	[HU11]c
2011	7,238,784	5,579	0.08	[HU11]c
2050	8,470,000	14,000	0.16	es
2100	6,857,000	18,000	0.26	es

4.5.5   Italy

The Italian Republic has an area of 301,340 sq. km and its map is presented in Fig. 4.5.5. It consists of the Italian Peninsula and the big islands of Sicily (25,708 sq. km) and Sardinia (24,090 sq. km). Muslims captured the Island of Pantelleria (81 sq. km) in 700, during the Umayyad Empire, after they conquered Tunisia. Muslims lost it in 1123, when Roger II of Sicily (son of Roger I) conquered the Island. The island is situated between Tunisia and Sicily, and its name is stemmed from the Arabic name given by its first Muslim conquerors: Bintul Riyah, or Daughter of the Wind. The Ottomans then conquered the Island in 1553 until they lost it to the Italians in the nineteenth century. Thus, the Island was under Muslim control for over 800 years.

Figure 4.5.5   Map of the Italian Republic.

As for Sardinia, it was captured by Muslims in 809 during the reign of Ibrahim bnul Aghlab of the Aghlabid Dynasty. He was the first ruler of the dynasty, seceded from the Abbasid Empire and ruled from 800 to 812. The Aghlabid Muslim and Arab Dynasty existed from 800 to 909, its capital was al-Qayrawan, in current Tunisia, and controlled the latter, plus North Algeria and East Libya. The Muslim control of Sardinia continued under the Fatimid Dynasty, which replaced the Aghlabid Dynasty. The Fatimid Dynasty lasted from 909 to 1171 and moved their capital to Cairo (current capital of Egypt) that they founded in 969. This Dynasty controlled North Africa, North Sudan, East Saudi Arabia, Palestine, Lebanon, Syria and Jordan. Muslims lost control over Sardinia in 1015. Thus, it was under Muslim control for 215 years.

Islam entered Sicily in 831 when its governor, Euphemius, plead for Aghlabid Amir Ziyadatul Allah I bnu Ibrahim (reined from 817 to 838, and was the third ruler after his father and brother Abdullah) to help him against the Byzantine Empire. Ziyadatul Allah I sent troops under the leadership of Asad bnul Furat. Consequently, the Muslims conquered Palermo (Northeast) in 831, followed by Messina (Northwest) in 843, then Enna (Middle) in 859, then Syracuse (Southeast) in 878, then Taormina (Northeast) in 902. The Muslim conquest of Sicily continued under the Fatimid Dynasty, and was culminated by capturing Rometta (Northeast) in 964, which was the last Byzantine toehold in the Island. This occurred during the reign of Muad al-Mu'iz li-Dinillah, who conquered Egypt, founded the city of Cairo, reigned from 953 to 975, and was the fourth caliph of the Dynasty. The majority of the Island's population by now was Muslim.

After successfully suppressing a revolt, the Fatimid Ruler appointed Hassan al-Kalbi as Emir of Sicily. He ruled from 948 to 964 and started the Kalbid Dynasty in the Island of Sicily which until 1053. Then divisions between Muslims in the Island caused the formation of several emirates in the Island.

In 1059, the Pope Nicholas II or Robert ben Tancred (Pope from 1059 to 1061), ordered the Island to be under Christian control. So, the conquest was led by his brother Roger (Roger I of Sicily), and the Island was lost to the Catholic Normans as follows: Messina in 1061, Palermo in 1072, Enna in 1087, and lastly Butera and Noto (Southern tip) in 1091. The percentage of Muslims continued to decrease as Muslims fled the Island or Christianized. Two centuries later, there were no Muslims in the Island. Thus, Sicily was under Muslim control for 96 to 264 years.

In the Italian Peninsula, Muslims occupied Brindisi (840 to 870), Taranto (840 to 880), Bari (841 to 871), Benevento (842–852), and arrived at Roma in 846, forcing the Pope to pay tribute (Jizya). Muslims did not stay long in the Peninsula (maximum of 40 years). There were other attacks and occupations by the Fatimids, Hafsids and the Ottoman later on, but the lands were under Muslim control for only about one year. This includes the northwestern parts of the Peninsula: the cities of Pisa and Genoa.

Table 4.5.5a Evolution of the Muslim population in Italy

Year	Population	Muslims	%	Source
1861	22,182,377	105	0.00	[SYB70]c
1881	28,953,480	4	0.00	[EUH, AT880]c
1901	32,965,504	222	0.00	[ROD]c
1911	35,845,048	<2,200	0.00	[SYB20]c
1931	40,309,621	<1,475	0.00	[SYB50]c
1981	56,479,285	15,373	0.03	[IT16]e
1991	56,744,119	108,759	0.19	[IT16]e
2001	56,960,692	451,471	0.79	[IT16]e
2006	58,064,214	963,013	1.66	[IT16]e
2011	59,364,690	1,482,530	2.50	[IT16]e
2016	60,665,551	1,780,585	2.94	[IT16]e
2050	54,382,000	3,807,000	7.00	es
2100	39,993,000	4,799,000	12.00	es

As shown in Table 4.5.5a, censuses before 1931 show negligible existence of Muslims in Italy as the population declared its religion other than Christian or Jewish made up 0.00% of the total population. The 1910 Austrian census in lands that belonged to the Austrian Empire, but now are in Italy recorded 207 Muslims distributed as follows: 191 in Triest, eleven in Merano, four in Bozen and one in Gorz.

Muslims started migrating to Italy in the second half of the twentieth century as Italy needed more workers, many of whom came from North Africa and the Balkans. The immigration intensified from Albania after 1991, with the collapse of Communism. A 2012 survey by the National Institute of Statistics (Istat) found that 29% of foreign nationals, 42% of Albanian foreign nationals as well as the majority of Bosnians and Macedonians were Muslim [IT12]. Table 4.5.5b presents the distribution of the estimated Muslim population per regions of Italy. The estimates are based on the assumptions that foreign nationals from Muslim majority countries are Muslim as well as those from Bosnia and Macedonia and half of those from Albania. The total was adjusted by those who obtained Italian citizenship. Over a quarter of Muslims live in the region of Lombardy where they make up almost 5% of the population. The next largest distributions are in Emilia-Romagna (13%), Veneto (11%), Piedmont (9%), Tuscany (7%) and Lazio (7%) where the capital Rome is located. Thus, the Muslim population in Italy increased from 15,000 or 0.03% in 1981, to 0.11 million or 0.2% in 1991, to 0.45 million or 0.8% in 2001, to 0.96 million or 1.7% in 2006, to 1.5 million or 2.5% in 2011, to 1.8 million or 2.9% in 2016.

Thus, assuming that the percentage of Muslims will continue to increase by one percentage point per decade, the Muslim population is expected to near four million or 7% by 2050 and near five million or 12% by 2100.

Table 4.5.5b Beginning of year evolution of regional Muslim population in Italy since 1981. P: Total population, M: Muslim population, M%: Percentage of Muslim population

		1981	1991	2001	2006	2011	2016
Piedmont	P	4,481,045	4,307,852	4,219,421	4,279,510	4,364,309	4,404,246
	M%	2,351	11,117	39,032	89,838	134,918	158,828
	M	0.05	0.26	0.93	2.10	3.09	3.61
Acosta Valley	P	112,349	115,275	119,273	123,969	126,761	127,329
	M%	54	390	1,140	2,622	4,027	4,445
	M	0.05	0.34	0.96	2.12	3.18	3.49
Liguria	P	1,814,312	1,684,514	1,578,998	1,574,148	1,574,132	1,571,053
	M%	316	3,206	11,332	22,882	37,325	46,119
	M	0.02	0.19	0.72	1.45	2.37	2.94
Lombardy	P	8,876,976	8,849,595	9,004,084	9,341,231	9,663,872	10,008,349
	M%	2,712	25,774	111,469	255,255	391,696	460,535
	M	0.03	0.29	1.24	2.73	4.05	4.60
South Tyrol (Bozen)	P	429,748	438,916	461,101	479,354	501,815	520,891
	M%	133	818	4,860	10,337	17,995	21,530
	M	0.03	0.19	1.05	2.16	3.59	4.13
Trentine	P	442,296	448,280	474,310	498,537	522,486	538,223
	M%	193	1,163	5,623	13,570	20,203	21,047
	M	0.04	0.26	1.19	2.72	3.87	3.91
Veneto	P	4,334,922	4,372,865	4,508,580	4,701,951	4,851,958	4,915,123
	M%	926	11,869	47,964	117,705	175,941	189,997
	M	0.02	0.27	1.06	2.50	3.63	3.87
Friuli-Venezia Giulia	P	1,234,244	1,198,178	1,181,238	1,201,522	1,220,849	1,221,218
	M%	110	931	6,872	18,405	31,510	35,567
	M	0.01	0.08	0.58	1.53	2.58	2.91
Emilia-Romagna	P	3,954,014	3,905,465	3,966,295	4,146,766	4,331,343	4,448,146
	M%	2,151	14,302	49,616	132,075	203,470	233,108
	M	0.05	0.37	1.25	3.19	4.70	5.24

Table 4.5.5b (Continued)

		1981	1991	2001	2006	2011	2016
Tuscany	P	3,577,778	3,531,357	3,494,857	3,570,259	3,668,948	3,744,398
	M%	516	7,034	31,911	70,040	110,617	131,551
	M	0.01	0.20	0.91	1.96	3.01	3.51
Umbria	P	805,503	810,623	824,187	853,259	884,010	891,181
	M%	176	1,816	8,861	21,464	31,649	33,321
	M	0.02	0.22	1.08	2.52	3.58	3.74
Marche	P	1,410,068	1,426,238	1,464,056	1,499,237	1,541,950	1,543,752
	M%	493	3,456	15,781	41,755	61,361	65,223
	M	0.03	0.24	1.08	2.79	3.98	4.22
Lazio	P	4,986,695	5,130,191	5,116,344	5,246,505	5,481,572	5,888,472
	M%	1,101	5,843	44,511	50,738	92,107	132,060
	M	0.02	0.11	0.87	0.97	1.68	2.24
Abruzzo	P	1,215,378	1,245,844	1,261,300	1,283,830	1,307,273	1,326,513
	M%	107	1,133	6,906	15,906	25,834	30,578
	M	0.01	0.09	0.55	1.24	1.98	2.31
Molise	P	328,253	330,557	321,468	318,297	314,342	312,027
	M%	15	180	724	1,488	2,249	3,786
	M	0.00	0.05	0.23	0.47	0.72	1.21
Campania	P	5,447,243	5,621,030	5,708,137	5,741,383	5,765,850	5,850,850
	M%	799	4,744	18,491	23,447	33,078	57,577
	M	0.01	0.08	0.32	0.41	0.57	0.98
Apulia	P	3,861,444	4,025,392	4,026,054	4,033,405	4,053,668	4,077,166
	M%	360	3,038	13,905	20,493	28,579	38,426
	M	0.01	0.08	0.35	0.51	0.71	0.94
Basilicata	P	611,183	610,970	599,404	589,480	579,358	573,694
	M%	48	373	1,449	2,472	3,380	4,832
	M	0.01	0.06	0.24	0.42	0.58	0.84
Calabria	P	2,061,008	2,075,886	2,018,722	1,978,390	1,962,856	1,970,521
	M%	114	2,220	7,007	12,619	18,574	27,864
	M	0.01	0.11	0.35	0.64	0.95	1.41
Sicily	P	4,905,866	4,970,805	4,978,068	4,969,155	5,005,657	5,074,261
	M%	2,614	7,729	19,302	32,788	46,293	68,140
	M	0.05	0.16	0.39	0.66	0.92	1.34
Sardinia	P	1,588,960	1,644,286	1,634,795	1,634,026	1,641,681	1,658,138
	M%	83	1,621	4,720	7,116	11,730	16,056
	M	0.01	0.10	0.29	0.44	0.71	0.97

4.5.6 Liechtenstein

The Principality of Liechtenstein has an area of 160 sq. km and its map is presented in Fig. 4.5.6. Muslims started immigrating here in the 1960s. As shown in Table 4.5.6, based on census data, the Muslim population increased from none before 1960 to eight or 0.04% in 1970, to 421 or 1.7% in 1980, to 689 or 2.4% in 1990, to 1,593 or 5.0% in 2000, to 1,960 or 5.6% in 2010 and 2,215 or 6.1% in 2015. Thus, assuming that the percentage of Muslims will continue to increase by one percentage point per decade, the Muslim population is expected to increase to 4,000 or 10% in 2050 and 7,000 or 15% by 2100.

Table 4.5.6 Evolution of the Muslim population in Liechtenstein

Year	Population	Muslims	%	Source
1861	7,866	—	0.00	[LIH]c
1868	8,001	—	0.00	[LIH]c
1874	8,089	—	0.00	[LIH]c
1880	8,410	—	0.00	[LIH]c
1891	8,537	—	0.00	[LIH]c
1901	8,766	—	0.00	[LIH]c
1911	9,670	—	0.00	[LIH]c
1921	10.439	—	0.00	[LIH]c
1930	12,073	—	0.00	[LIH]c
1970	21,350	8	0.04	[LI00]c
1980	25,135	421	1.67	[LI00]c
1990	28,766	689	2.40	[LI00]c
2000	31,932	1,593	4.99	[LI00]c
2010	35,224	1,960	5.56	[LI15]c
2015	36,369	2,215	6.09	[LI15]c
2050	40,000	4,000	10.00	es
2100	45,000	7,000	15.00	es

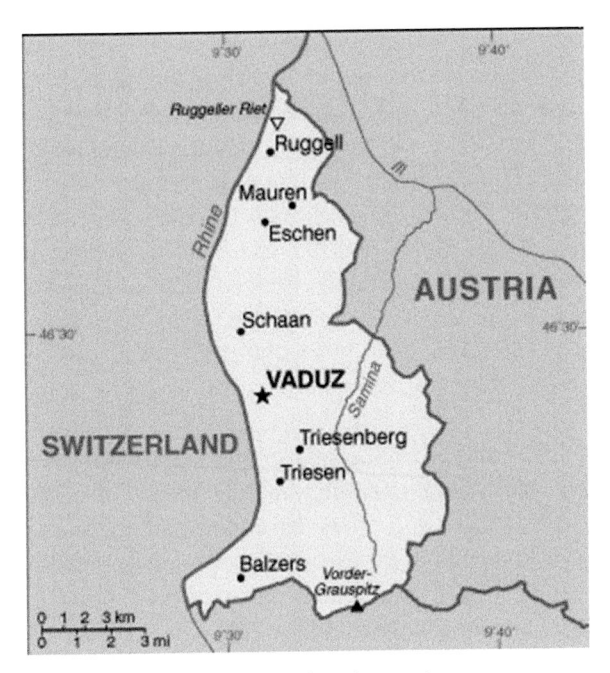

Figure 4.5.6 Map of the Principality of Liechtenstein.

4.5.7 Malta

The Republic of Malta is an island nation with a total area of 316 sq. km that consists mainly of three islands: Malta (246 sq. km), Gozo (67 sq. km), and Cumino (3 sq. km). A map of these islands is presented in Fig. 4.5.7. It was captured by Muslims of the Aghlabid Dynasty from the Byzantine Empire in 870. This occurred during the reign of Mohammed II bnu Ahmad who reigned from 863 to 875 and was the eighth ruler of the Aghlabid Dynasty. Eventually the majority of the population was Muslims.

After Sicily fell to the Christian Nomads, Roger I ben Tancred seized the island in 1091, after 221 years of Muslim control. But Muslims were forced to leave the Island or Islam, which reduced their number to zero, but the Maltese language remains an Arabic dialect. The Maltese Archipelago was eventually conquered by the British in 1802 until 1964, when it gained its independence.

The census of 1901 shows the presence of 12 Muslims in the Islands, or 0.01% of the total population. Until the 1990s, the presence of Muslims in the Maltese Islands remained negligible and consisted mostly of Libyan nationals. The 1957 recorded twenty Libyans living in the islands, making up 0.01% of the total population. In 1985, the number of those born in a Muslim majority country consisted of 216 in Libya and 13 in Egypt. In 1995, that number consisted only of 508 in Libya. In 2005, the number jumped to 2,640 including 755 in Libya, 446 in Egypt, 281 in Tunisia, 179 in Syria, 147 in Morocco, 139 in Turkey, 122 in Bosnia, 97 in Pakistan and 76 in Albania. However, according to EVS survey, the percentage of Muslims was 0.16% in 2008.

Hence, the number of Muslims in Malta increased from five in 1881 to twelve in 1901, to twenty in 1957, to 229 or 0.1% in 1985 and 508 or 0.1% in 1995. Thus, assuming that the percentage of Muslims will continue to increase by a tenth of a percentage point per decade, the Muslim population is expected to increase to 3,000 or 0.6% by 2050 and 4,000 or 1.1% by 2100. The data is summarized in Table 4.5.7.

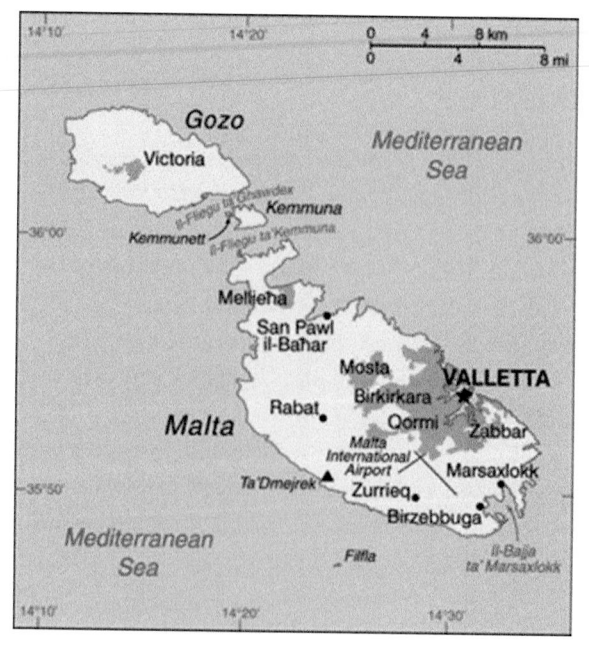

Figure 4.5.7 Map of the Republic of Malta.

Table 4.5.7 Evolution of the Muslim population in Malta

Year	Population	Muslims	%	Source
1881	149,782	5	0.00	[MT881]c
1901	207,890	12	0.01	[GB901]c
1957	319,620	20	0.01	[UNE]e
1985	345,418	229	0.07	[UNE]e
1995	378,132	508	0.13	[UNE]e
2008	420,796	670	0.16	[EVS]s
2050	427,000	3,000	0.60	es
2100	358,000	4,000	1.10	es

4.5.8 Poland

Muslim Tatars moved to Poland and its vicinity in the early thirteenth century as the Tatar Golden Horde Empire was converting to Islam. The Kingdom of Poland later used Muslim Tatars as part of its army power, and they were given formal permission to construct mosques and schools in 1569. Their number was estimated at 70,000 or 1.4% in 1591, which increased according to a census to 100,000 or 1.9% in 1631. However, their number was severely reduced later to 2,000 or 0.04% in 1867.

According to the 1897 Russian Empire census, the Muslim population Russian Poland (Privinslinsky province) was 4,903 or 0.05% of the total population. The province covered a third of current Poland and consisted of the following ten governorates: Kalish (Kalisz), Kielce (Keltsy or Kaletskaya), Lomza (Lomzha or Lomzhinskaya), Lublin (Lyublin), Petrokov (Petrokovskaya), Plotsk (Plock), Radom, Sedlets (Siedlce or Sedletskaya), Suvalki (Suwalki, one third in current Poland, while rest lays in Lithuania), and Warsaw. About half of current Poland (north and east) was under Prussia, which did not have much Muslims in the nineteenth century. The rest of Poland was under the Austro-Hungarian Empire, which in 1880 census recorded only one Muslim in the city of Neu-Sandec (Nowy Sacz) near the border with current Slovakia. Data of the 1897 census for the three governorates is summarized in Table 4.5.8a. For comparison, the table also contains an estimate based on the 1897 of the total population living in the current border of Latvia and the current area of the country.

Table 4.5.8a 1897 Census data for partial territory covering current Poland

Governorate	Population	Muslims	% Muslims	Area (km²)
Kalish	840,597	214	0.03	11,375
Kielce	761,995	96	0.01	10,093
Lomza	579,592	479	0.08	12,087
Lublin	1,160,662	464	0.04	16,838
Petrokov	1,403,901	311	0.02	12,248
Plotsk	553,633	266	0.05	10,878
Radom	814,947	65	0.01	12,352
Sedlet	772,146	669	0.09	14,336
Suvalki	582,913	786	0.13	12,551
Warsaw	1,931,867	1,553	0.08	14,564
Total	9,402,253	4,903	0.05	127,321
Poland	24,750,000			312,685

Currently, the Republic of Poland has an area of 312,685 sq. km and its map is presented in Fig. 4.5.8. The 2002 census provided ethnic data from which 3,537 belonged to Muslim majority ethnicities including 495 Tatar, 459 Arabs, 312 Syrians, 273 Algerians, 232 Turks, 229 Palestinians, 186 Iranians, 174 Lebanese, 132 Egyptians, 113 Moroccans, 111 Afghan and 102 Tunisians. The 2011 recorded 1,242 Muslims registered with the following organizations: Muslim Religious Union in Poland (1,132 members), Association of Muslim Unity (sixty members) and Islamic Assembly of Ahl-ul Bayt (fifty members). By 2013, the membership changed to 826, 63 and 51, respectively. According to the European Values Survey (EVS), the percentage of Muslims was 0.03% in 1999 and 0.09% in 2008. Hence, the Muslim population increased from 12,000 or 0.03% in 1999 to 35,000 or 0.09% in 2008. Thus, assuming that the percentage of Muslims will increase by 0.05 of a percentage point per decade, the Muslim population is expected to remain around 0.1 million throughout the second half of this century, comprising 0.3% of the total population by 2050 and 0.6% by 2100. A summary of the evolution of the Muslim population in Poland is presented in Table 4.5.8b.

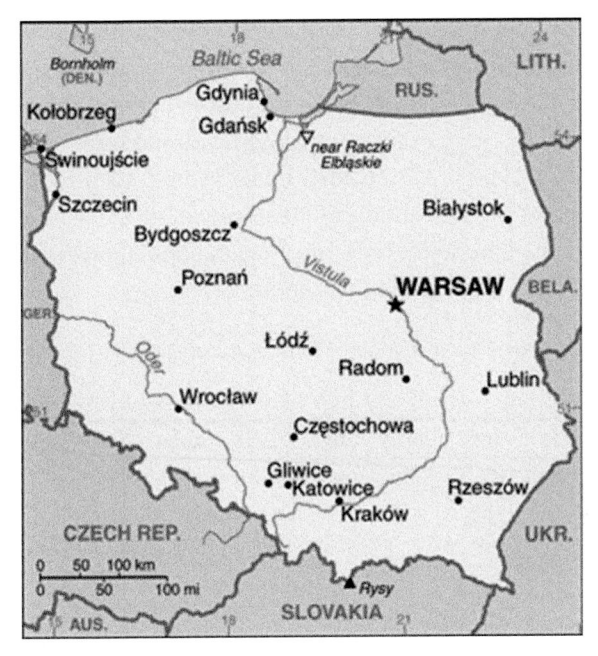

Figure 4.5.8 Map of the Republic of Poland.

Table 4.5.8b Evolution of the Muslim population in Poland

Year	Population	Muslims	%	Source
1591	4,900,000	70,000	1.43	[PLH]es
1631	5,300,000	100,000	1.89	[PLH]e
1867	5,097,000	2,000	0.04	[SYB70]c
1897	9,402,253	4,903	0.05	[SU]c
1999	38,515,000	11,500	0.03	[EVS]s
2008	38,526,000	35,000	0.09	[EVS]s
2050	32,295,000	100,000	0.30	es
2100	23,033,000	127,000	0.55	es

4.5.9 San Marino

The Republic of San Marino is an enclave in the Northeast of the Italian Peninsula with a total area of 61 sq. km and its map is presented in Fig. 4.5.9. As of 1976, there were no Muslims here. The largest ten nationalities of Muslim majority countries since 2012 is provided in Table 4.3.9a. Based on these numbers and as shown in Table 4.3.9b, the Muslim population increased from 104 or 0.3% in 2012, to 125 or 0.4% in 2015 and 146 or 0.4% in 2018. Thus, assuming that the percentage of Muslims will increase by a tenth of a percentage point per decade, the Muslim population is expected to stay below 300 throughout this century.

Table 4.5.9a Year-end evolution of Muslim nationalities in San Marino since 2012

	2012	2013	2014	2015	2016	2017	2018
Albania	49	51	58	61	67	72	77
Morocco	18	17	18	20	21	21	19
Tunisia	9	8	8	7	8	7	8
Syria	—	—	—	1	1	6	7
Senegal	1	3	4	5	5	6	6
Kyrgyzia	4	4	7	7	7	6	6
Turkey	5	4	5	5	5	5	5
Egypt	6	5	5	5	5	5	4
Gambia	—	—	2	2	3	4	4
Iran	8	6	6	6	4	3	3
Bosnia	2	2	2	2	3	2	3
Others	2	2	2	3	4	5	4
Total	104	102	117	124	133	142	146

Table 4.5.9b Evolution of the Muslim population in San Marino

Year	Population	Muslims	%	Source
1976	19,149	—	0.00	[KET76]es
2012	33,562	104	0.31	[SM]e
2015	34,006	125	0.37	[SM]e
2018	34,590	146	0.42	[SM]e
2050	35,000	250	0.70	es
2100	32,000	400	1.20	es

Figure 4.5.9 Map of the Republic of San Marino.

4.5.10 Slovakia

After conquering Budapest, Hungary in 1529, the Muslim Ottomans besieged Bratislava, the current capital of Slovakia, but were unable to conquer it. They captured Nitra (Southwest) in 1663, but lost it by 1685, after losing the Battle for Vienna in 1683. Currently, the Slovak Republic has an area of 49,035 sq. km and its map is presented in Fig. 4.5.10.

The Hungarian census recorded no Muslims in 1850 and 18780, two Muslims in 1870 (one in Zólyom and one in Bars) and five Muslims in 1910 (one in Gomor and four in Szepes). The 1921 Czechoslovak census indicated the presence of 185 Muslims, who we assume lived in the capital Prague, which is now the capital of the Czech Republic. Censuses inquiring on religious affiliation of the population were held every decade since 1991. Accordingly, the Muslim population increased from 180 or 0.00% in 1991, to 1,212 or 0.02% in 2001, to 1,935 or 0.04% in 2011. Thus, assuming that the percentage of Muslims will continue to increase by 0.02 of a percentage point per decade, the Muslim population is expected to reach 6,000 or 0.1% by 2050 and 8,000 or 0.2% by 2100. The data is summarized in Table 4.5.10 below.

Figure 4.5.10 Map of the Slovak Republic.

Table 4.5.10 Evolution of the Muslim population in Slovakia

Year	Population	Muslims	%	Source
1850	2,442,000	—	0.00	[EUH]c
1870	2,481,811	2	0.00	[HU870]c
1880	2,477,521	—	0.00	[HU880]c
1910	2,916,657	5	0.00	[HU910]c
1921	2,993,859	—	0.00	[SK]c
1991	4,356,500	180	0.00	[SK]c
2001	5,218,857	1,212	0.02	[SK]c
2011	4,825,599	1,935	0.04	[SK]c
2050	4,984,000	6,000	0.12	es
2100	3,829,000	8,000	0.22	es

4.5.11 Switzerland

The Swiss Confederation has an area of 41,277 sq. km and its map is presented in Fig. 4.5.11. There were no Muslims here according to the 1837 and 1850 censuses. The number of non-Christian religions combined increased from 387 or 0.01% in 1930 to 1,430 or 0.03% in 1950. As shown in Table 4.5.11 and based on census data, the Muslim population grew to 2,703 or 0.1% in 1960, to 16,353 or 0.3% in 1970, to 56,600 or 0.9% in 1980, to 152,200 or 2.3% in 1990, to 310,807 or 4.5% in 2000. The increase is due to Muslim immigrants coming mainly for economic reason.

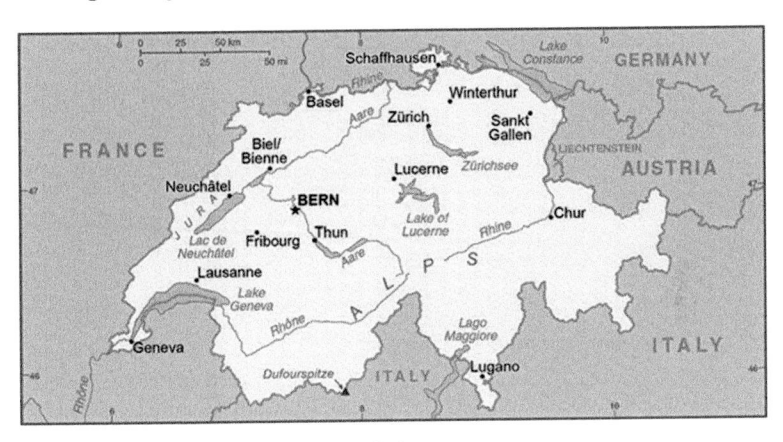

Figure 4.5.11 Map of the Swiss Confederation.

Starting in 2010, the census is conducted and evaluated on an annual basis in a new form by the Federal Statistical Office (FSO). In order to ease the burden on the population, the information is primarily drawn from population registers and supplemented by sample surveys. Only a small proportion of the population (about 5%) is now surveyed in writing or by telephone. The reference day is December 31st. In this new census system, religion data is based on permanent resident population aged fifteen and over living in private households. However, this criterion underestimates the Muslim population by almost a third. For example, according to this method the size of the Muslim population was 11,078 or 0.24% in 1970, 34,476 or 0.71% in 1980, 86,898 or 1.60% in 1990 and 210,580 or 3.72% in 2000. The Muslim population then increased to 0.30 million or 4.6% in 2010, to 0.35 million or 5.1% in 2015 and 0.38 million or 5.5% in 2017. Thus, assuming that the percentage of Muslims will increase by one percentage point per decade, the Muslim population is expected to reach 0.9 million or 9% by 2050 and 1.5 million or 14% by 2100.

Table 4.5.11 Evolution of the Muslim population in Switzerland

Year	Population	Muslims	%	Source
1837	2,192,098	—	0.00	[CHH]c
1850	2,391,741	—	0.00	[CHH]c
1930	4,077,099	<387	<0.01	[CH30]c
1950	4,714,992	<1,430	<0.03	[CH50]c
1960	5,417,670	2,703	0.05	[UN64]c
1970	6,245,496	16,353	0.26	[UN73]c
1980	6,296,863	56,625	0.90	[CH90]c
1990	6,771,788	152,217	2.25	[CH90]c
2000	6,972,244	310,807	4.46	[UN]c
2010	6,388,868	295,796	4.63	[CH]s
2011	6,515,897	320,958	4.93	[CH]s
2012	6,575,723	328,011	4.99	[CH]s
2013	6,657,112	341,572	5.13	[CH]s
2014	6,829,610	346,208	5.07	[CH]s
2015	6,907,818	351,961	5.10	[CH]s
2016	6,981,381	362,973	5.20	[CH]s
2017	6,937,561	379,748	5.47	[CH]s
2050	9,818,000	884,000	9.00	es
2100	11,041,000	1,546,000	14.00	es

4.5.12 Regional Summary and Conclusion

Central Europe has the least concentration of Muslims among the five regions spanning the European continent. Although Muslims captured parts of this region as early as the eighth century, the Muslim population in this region was almost nonexistent prior to WWII. The increase was at first through migration as countries in this region needed to compensate for their labor shortage in post WWII era. The Muslim population continued to increase significantly towards the end of last century and is expected to continue this trend. It is now

3% of the total population and is expected to reach 7% by the end of this century. The main cause of increase is due to migration of Muslims, then their natural increase and finally conversion to Islam by natives. The following tables present centennial data from 600 AD to 2100 AD (or approximately 1 H to 1500 H) in Table 4.5a and decennial data from 1790 AD to 2100 AD (or 1210 H to 1520 H) in Tables 4.5b and 4.5c for current countries in Central Europe. The data includes total population in thousands (P), the percentage of which is Muslim (M%), the corresponding Muslim population in thousands (M), and the annual population growth rate (APGR, or G%) of the total population in this region.

4.6 Europe's Summary and Conclusion

Europe was the third continent that Islam entered after Asia and Africa. Since the first century of Islam, this continent was the battleground of fierce fights between Muslims and Christians. Currently, only 3% of the world Muslim population lives in Europe, which is expected to decrease to 2% by the end of this century. Nevertheless, the number of Muslims is increasing rapidly in Europe, currently 6% of the total population, and expected to exceed 12% by the end of this century. The return of Islam to Europe since the second half of the last century is due to economic reasons: need by locals of workforce to stimulate local economy and need of immigrants for better life. The following tables present centennial data from 600 AD to 2100 AD (or approximately 1 H to 1500 H) in Table 4.6a and decennial data from 1790 AD to 2100 AD (or 1210 H to 1520 H) in Tables 4.6b and 4.6c for the five regions of Europe. The data includes total population in thousands (P), the percentage of which is Muslim (M%), the corresponding Muslim population in thousands (M), and the annual population growth rate (APGR, or G%) of the total population in Europe and each of its five regions. A choropleth map of Europe illustrating the presence of Muslims is presented in Fig. 4.6.0, where white is less than 1% of the district's population and the darker the region, the more the percentage of Muslims.

Table 4.5a Centennial estimates of the Muslim population (×1000) in Central Europe: 600–2100 AD (1–1500 H)

		600	700	800	900	1000	1100	1200	1300
Austria	P	500	550	600	650	700	800	900	1,100
	M%	—	—	—	—	—	—	—	—
	M	—	—	—	—	—	—	—	—
Czechia	P	780	810	840	870	900	1,150	1,400	1,650
	M%	—	—	—	—	—	—	—	—
	M	—	—	—	—	—	—	—	—
Germany	P	3,300	3,350	3,400	3,450	3,500	5,000	6,500	8,000
	M%	—	—	—	—	—	—	—	—
	M	—	—	—	—	—	—	—	—
Hungary	P	420	440	460	480	500	650	800	950
	M%	—	—	—	—	—	—	12.50	10.00
	M	—	—	—	—	—	—	100	95
Italy	P	6,200	5,900	5,600	5,300	5,000	6,000	7,000	8,000
	M%	—	0.01	0.01	1.00	2.00	1.00	0.00	0.00
	M	—	1	1	53	100	60	—	—
Liechtenstein	P	—	—	—	—	—	—	1	1
	M%	—	—	—	—	—	—	—	—
	M	—	—	—	—	—	—	—	—
Malta	P	10	15	18	20	21	20	19	18
	M%	—	—	1.00	20.00	50.00	40.00	—	—
	M	—	—	—	4	11	8	—	—
Poland	P	880	960	1,040	1,120	1,200	1,800	2,400	3,000
	M%	—	—	—	—	—	—	—	0.50
	M	—	—	—	—	—	—	—	15
SanMarino	P	1	1	1	1	1	1	1	1
	M%	—	—	—	—	—	—	—	—
	M	—	—	—	—	—	—	—	—
Slovakia	P	321	328	335	342	349	450	550	650
	M%	—	—	—	—	—	—	—	—
	M	—	—	—	—	—	—	—	—
Switzerland	P	300	300	300	300	300	370	440	510
	M%	—	—	—	—	—	—	—	—
	M	—	—	—	—	—	—	—	—
Total	P	12,712	12,654	12,594	12,533	12,471	16,241	20,011	23,880
	M%	0.00	0.00	0.01	0.45	0.89	0.42	0.50	0.46
	M	—	1	1	57	111	68	100	110
	G%		−0.005	−0.005	−0.005	−0.005	0.264	0.209	0.177

Table 4.5a (Continued)

		1400	1500	1600	1700	1800	1900	2000	2100
Austria	P	1,200	1,500	1,800	2,100	3,064	6,004	8,069	8,677
	M%	—	—	—	—	0.00	0.00	4.22	16.00
	M	—	—	—	—	—	—	341	1,388
Czechia	P	1,900	2,161	3,242	3,242	4,400	9,372	10,289	10,274
	M%	—	—	—	—	—	—	0.04	0.24
	M	—	—	—	—	—	—	4	25
Germany	P	9,500	12,000	16,000	15,000	21,715	54,388	81,401	74,741
	M%	—	—	—	—	—	0.00	3.00	8.50
	M	—	—	—	—	—	1	2,442	6,353
Hungary	P	1,050	1,150	1,250	1,500	3,200	6,854	10,221	6,857
	M%	—	—	—	—	—	0.01	0.04	0.26
	M	—	—	—	—	—	1	4	18
Italy	P	9,000	10,500	13,100	13,300	17,237	33,172	56,692	39,993
	M%	0.00	0.00	0.00	0.00	0.00	0.00	0.79	12.00
	M	—	—	—	—	—	—	448	4,799
Liechtenstein	P	2	3	4	5	6	10	33	45
	M%	—	—	—	—	—	—	4.99	15.00
	M	—	—	—	—	—	—	2	7
Malta	P	17	15	20	80	114	185	394	358
	M%	—	—	—	—	—	0.01	0.13	1.10
	M	—	—	—	—	—	—	1	4
Poland	P	3,500	4,000	5,000	6,000	9,040	24,750	38,557	23,033
	M%	1.00	1.50	2.00	1.00	0.04	0.05	0.03	0.55
	M	35	60	100	60	4	12	12	127
SanMarino	P	2	3	4	5	7	9	27	30
	M%	—	—	—	—	—	—	0.01	1.20
	M	—	—	—	—	—	—	—	—
Slovakia	P	750	839	1,258	1,258	2,200	2,783	5,399	3,829
	M%	—	—	—	—	—	—	0.02	0.22
	M	—	—	—	—	—	—	1	8
Switzerland	P	580	650	1,000	1,200	1,250	3,315	7,144	11,041
	M%	—	—	—	—	—	—	4.46	14.00
	M	—	—	—	—	—	—	319	1,546
Total	P	27,501	32,821	42,678	43,690	62,232	140,843	218,226	178,877
	M%	0.13	0.18	0.23	0.14	0.01	0.01	1.64	7.98
	M	35	60	100	60	4	14	3,572	14,275
	G%	0.141	0.177	0.263	0.023	0.354	0.817	0.438	−0.199

Table 4.5b Decennial estimates of the Muslim population (×1000) in Central Europe: 1790–1940 AD (1210–1360 H)

		1790	1800	1810	1820	1830	1840	1850	1860
Austria	P	3,046	3,064	3,054	3,202	3,477	3,650	3,880	4,076
	M%	0.00	0.00	0.00	0.00	0.00	0.00	0.00	0.00
	M	—	—	—	—	—	—	—	—
Czechia	P	4,200	4,400	4,600	5,000	5,300	5,816	6,135	6,437
	M%	—	—	—	—	—	—	—	—
	M	—	—	—	—	—	—	—	—
Germany	P	21,000	21,715	22,110	24,905	28,045	31,126	33,746	36,049
	M%	—	—	—	—	—	0.00	0.00	0.00
	M	—	—	—	—	—	—	—	—
Hungary	P	2,800	3,200	3,600	4,146	4,300	4,500	4,700	4,900
	M%	—	—	—	—	—	—	—	—
	M	—	—	—	—	—	—	—	—
Italy	P	16,257	17,237	17,943	18,967	21,975	24,351	24,964	26,328
	M%	0.00	0.00	0.00	0.00	0.00	0.00	0.00	0.00
	M	0.163	0.172	0.179	0.190	0.220	0.244	0.250	0.263
Liechtenstein	P	5	6	6	6	7	7	8	7
	M%	—	—	—	—	—	—	—	—
	M	—	—	—	—	—	—	—	—
Malta	P	113	114	93	112	120	114	123	134
	M%	—	—	—	—	—	—	—	—
	M	—	—	—	—	—	—	—	—
Poland	P	8,572	9,040	9,351	10,426	11,110	12,468	13,000	15,585
	M%	0.04	0.04	0.04	0.04	0.04	0.04	0.04	0.04
	M	3	4	4	4	4	5	5	6
SanMarino	P	6	7	7	7	7	7	8	7
	M%	—	—	—	—	—	—	—	—
	M	—	—	—	—	—	—	—	—
Slovakia	P	2,100	2,200	2,300	2,350	2,400	2,420	2,442	2,460
	M%	—	—	—	—	—	—	—	—
	M	—	—	—	—	—	—	—	—
Switzerland	P	1,000	1,250	1,558	1,836	2,188	2,251	2,393	2,511
	M%	—	—	—	—	—	—	—	—
	M	—	—	—	—	—	—	—	—
Total	P	59,099	62,232	64,622	70,957	78,928	86,711	91,398	98,494
	M%	0.01	0.01	0.01	0.01	0.01	0.01	0.01	0.01
	M	4	4	4	4	5	5	6	7
	G%		0.517	0.377	0.935	1.065	0.940	0.526	0.748

Table 4.5b (Continued)

		1870	1880	1890	1900	1910	1920	1930	1940
Austria	P	4,498	4,964	5,417	6,004	6,648	6,535	6,760	6,653
	M%	0.00	0.00	0.00	0.00	0.00	0.00	0.00	0.00
	M	—	—	—	—	—	—	—	—
Czechia	P	7,617	8,222	8,665	9,372	10,079	10,010	10,674	11,159
	M%	—	—	—	—	0.00	0.00	0.00	0.00
	M	—	—	—	—	—	—	—	—
Germany	P	39,231	43,500	47,607	54,388	62,884	60,894	65,084	69,835
	M%	0.00	0.00	0.00	0.00	0.00	0.00	0.00	0.00
	M	—	—	—	1	1	2	1	3
Hungary	P	5,011	5,329	6,009	6,854	7,612	7,987	8,685	9,316
	M%	0.00	0.00	0.00	0.01	0.01	0.01	0.00	0.00
	M	—	—	—	1	1	1	0.261	—
Italy	P	27,437	29,116	30,947	33,172	35,442	37,143	40,310	43,787
	M%	0.00	0.00	0.00	0.00	0.00	0.00	0.00	0.00
	M	0.274	0.291	0.309	0.332	0.354	0.371	0.403	0.438
Liechtenstein	P	8	9	9	10	11	12	10	13
	M%	—	—	—	—	—	—	—	—
	M	—	—	—	—	—	—	—	—
Malta	P	142	150	165	185	212	212	242	269
	M%	—	—	—	0.01	0.01	0.01	0.01	0.01
	M	—	—	—	—	—	—	—	—
Poland	P	16,865	17,145	22,854	24,750	26,644	23,968	28,204	30,021
	M%	0.04	0.04	0.05	0.05	0.05	0.01	0.01	0.01
	M	7	7	11	12	13	2	3	3
SanMarino	P	7	8	8	9	11	12	13	15
	M%	—	—	—	—	—	—	—	—
	M	—	—	—	—	—	—	—	—
Slovakia	P	2,482	2,478	2,595	2,783	2,917	2,994	3,324	3,554
	M%	—	—	—	—	—	—	—	—
	M	—	—	—	—	—	—	—	—
Switzerland	P	2,669	2,809	2,918	3,315	3,753	3,880	4,066	4,261
	M%	—	—	—	—	—	—	—	—
	M	—	—	—	—	—	—	—	—
Total	P	105,967	113,728	127,195	140,843	156,212	153,646	167,372	178,882
	M%	0.01	0.01	0.01	0.01	0.01	0.00	0.00	0.00
	M	8	8	12	14	15	6	4	7
	G%	0.731	0.707	1.119	1.019	1.036	−0.166	0.856	0.665

Table 4.5c Decennial estimates of the Muslim population (×1000) in Central Europe: 1950–2100 AD (1370–1520 H)

		1950	1960	1970	1980	1990	2000	2010	2020
Austria	P	6,936	7,071	7,516	7,610	7,724	8,069	8,410	9,006
	M%	0.00	0.00	0.30	1.02	2.04	4.22	5.16	8.00
	M	—	0.283	23	78	158	341	434	721
Czechia	P	8,903	9,590	9,818	10,349	10,341	10,289	10,537	10,709
	M%	0.00	0.00	0.00	0.00	0.01	0.04	0.06	0.08
	M	—	—	—	—	1	4	6	9
Germany	P	69,966	73,414	78,578	78,283	79,054	81,401	80,827	83,784
	M%	0.00	0.01	1.02	1.50	2.12	3.00	3.99	4.50
	M	—	7	801	1,174	1,676	2,442	3,225	3,770
Hungary	P	9,338	10,001	10,366	10,754	10,377	10,221	9,927	9,660
	M%	0.00	0.00	0.00	0.01	0.02	0.04	0.08	0.10
	M	—	—	—	1	2	4	8	10
Italy	P	46,599	49,700	53,519	56,349	57,048	56,692	59,325	60,462
	M%	0.00	0.00	0.00	0.03	0.19	0.79	2.50	4.00
	M	0.466	0.497	0.535	17	108	448	1,483	2,418
Liechtenstein	P	14	16	21	26	29	33	36	38
	M%	—	0.02	0.04	1.67	2.40	4.99	5.56	7.00
	M	—	—	—	—	1	2	2	3
Malta	P	312	313	321	331	362	394	414	442
	M%	0.01	0.01	0.01	0.07	0.07	0.13	0.16	0.30
	M	—	—	—	—	—	1	1	1
Poland	P	24,824	29,614	32,639	35,540	37,960	38,557	38,330	37,847
	M%	0.01	0.01	0.02	0.02	0.03	0.03	0.09	0.15
	M	2	3	7	7	11	12	34	57
SanMarino	P	13	15	19	21	24	27	31	34
	M%	—	—	—	—	—	0.01	0.31	0.41
	M	—	—	—	—	—	—	—	—
Slovakia	P	3,437	4,140	4,539	4,997	5,288	5,399	5,404	5,460
	M%	—	—	—	—	0.00	0.02	0.04	0.06
	M	—	—	—	—	—	1	2	3
Switzerland	P	4,668	5,281	6,151	6,284	6,653	7,144	7,809	8,655
	M%	—	0.05	0.26	0.90	2.25	4.46	4.63	6.00
	M	—	3	16	57	150	319	362	519
Total	P	175,009	189,156	203,487	210,545	214,861	218,226	221,050	226,096
	M%	0.00	0.01	0.42	0.63	0.98	1.64	2.51	3.32
	M	3	14	848	1,334	2,107	3,572	5,557	7,511
	G%	−0.219	0.777	0.730	0.341	0.203	0.155	0.129	0.226

Table 4.5c (*Continued*)

		2030	2040	2050	2060	2070	2080	2090	2100
Austria	P	9,176	9,212	9,131	8,976	8,845	8,729	8,653	8,677
	M%	9.00	10.00	11.00	12.00	13.00	14.00	15.00	16.00
	M	826	921	1,004	1,077	1,150	1,222	1,298	1,388
Czechia	P	10,745	10,626	10,546	10,430	10,217	10,109	10,168	10,274
	M%	0.10	0.12	0.14	0.16	0.18	0.20	0.22	0.24
	M	11	13	15	17	18	20	22	25
Germany	P	83,136	82,004	80,104	77,962	76,467	75,402	74,770	74,741
	M%	5.00	5.50	6.00	6.50	7.00	7.50	8.00	8.50
	M	4,157	4,510	4,806	5,068	5,353	5,655	5,982	6,353
Hungary	P	9,338	8,907	8,470	8,071	7,660	7,309	7,051	6,857
	M%	0.12	0.14	0.16	0.18	0.20	0.22	0.24	0.26
	M	11	12	14	15	15	16	17	18
Italy	P	59,031	57,160	54,382	50,557	46,815	44,094	41,947	39,993
	M%	5.00	6.00	7.00	8.00	9.00	10.00	11.00	12.00
	M	2,952	3,430	3,807	4,045	4,213	4,409	4,614	4,799
Liechtenstein	P	39	40	40	41	42	43	44	45
	M%	8.00	9.00	10.00	11.00	12.00	13.00	14.00	15.00
	M	3	4	4	4	5	6	6	7
Malta	P	449	440	427	415	400	382	367	358
	M%	0.40	0.50	0.60	0.70	0.80	0.90	1.00	1.10
	M	2	2	3	3	3	3	4	4
Poland	P	36,945	35,283	33,295	31,168	28,798	26,417	24,510	23,033
	M%	0.20	0.25	0.30	0.35	0.40	0.45	0.50	0.55
	M	74	88	100	109	115	119	123	127
SanMarino	P	34	34	34	32	31	30	30	30
	M%	0.50	0.60	0.70	0.80	0.90	1.00	1.10	1.20
	M	—	—	—	—	—	—	—	—
Slovakia	P	5,403	5,217	4,984	4,735	4,443	4,168	3,974	3,829
	M%	0.08	0.10	0.12	0.14	0.16	0.18	0.20	0.22
	M	4	5	6	7	7	8	8	8
Switzerland	P	9,185	9,551	9,818	10,012	10,205	10,428	10,699	11,041
	M%	7.00	8.00	9.00	10.00	11.00	12.00	13.00	14.00
	M	643	764	884	1,001	1,123	1,251	1,391	1,546
Total	P	223,482	218,473	211,229	202,399	193,922	187,110	182,213	178,877
	M%	3.89	4.46	5.04	5.61	6.19	6.79	7.39	7.98
	M	8,682	9,750	10,642	11,345	12,003	12,710	13,465	14,275
	G%	−0.116	−0.227	−0.337	−0.427	−0.428	−0.358	−0.265	−0.185

Table 4.6a Centennial estimates of the Muslim population (×1000) in Europe: 600–2100 AD (1–1500 H)

		600	700	800	900	1000	1100	1200	1300
Balkans	P	4,733	4,687	4,638	4,592	4,543	4,878	5,218	5,558
	M%	—	—	—	0.02	—	—	—	0.28
	M	—	—	—	1	—	—	—	16
Eastern Europe	P	3,463	3,658	3,853	4,048	4,246	5,613	7,434	8,755
	M%	—	0.65	1.30	1.95	2.59	3.12	3.83	4.67
	M	—	24	50	79	110	175	285	409
Western Europe	P	10,942	11,157	11,373	11,588	11,805	14,314	16,825	19,342
	M%	—	—	7.50	14.84	18.48	17.26	14.81	12.65
	M	—	—	853	1,719	2,181	2,470	2,492	2,446
Northern Europe	P	2,548	2,736	2,925	3,115	3,304	3,964	4,620	5,276
	M%	—	—	—	—	—	—	—	—
	M	—	—	—	—	—	—	—	—
Central Europe	P	12,712	12,654	12,594	12,533	12,471	16,241	20,011	23,880
	M%	—	0.00	0.01	0.45	0.89	0.42	0.50	0.46
	M	—	1	1	57	111	68	100	110
Total	P	34,398	34,892	35,383	35,876	36,369	45,011	54,108	62,811
	M%	—	0.07	2.55	5.17	6.60	6.03	5.32	4.75
	M	—	24	903	1,856	2,402	2,713	2,877	2,980
	G%		0.014	0.014	0.014	0.014	0.213	0.184	0.149

Table 4.6a (*Continued*)

		1400	1500	1600	1700	1800	1900	2000	2100
Balkans	P	5,898	6,240	7,688	8,288	14,945	32,511	64,397	38,177
	M%	1.77	5.82	11.89	17.33	13.68	11.38	12.70	12.04
	M	104	363	914	1,436	2,045	3,700	8,176	4,598
Eastern Europe	P	10,071	12,672	15,805	21,096	40,940	108,296	216,602	163,474
	M%	5.58	6.96	8.16	9.37	5.01	4.58	6.92	11.89
	M	562	882	1,290	1,977	2,053	4,959	14,989	19,435
Western Europe	P	22,258	25,198	31,000	36,212	46,454	75,078	136,910	135,084
	M%	8.91	5.60	2.69	0.24	0.00	0.01	3.17	13.69
	M	1,983	1,410	835	88	—	5	4,342	18,487
Northern Europe	P	5,932	6,561	9,468	13,459	20,843	51,298	87,169	117,568
	M%	—	—	—	—	—	0.00	2.57	14.42
	M	—	—	—	—	—	—	2,240	16,950
Central Europe	P	27,501	32,821	42,678	43,690	62,232	140,843	218,226	178,877
	M%	0.13	0.18	0.23	0.14	0.01	0.01	1.64	7.98
	M	35	60	100	60	4	14	3,572	14,275
Total	P	71,660	83,492	106,639	122,745	185,414	408,025	723,305	633,179
	M%	3.75	3.25	2.94	2.90	2.21	2.13	4.61	11.65
	M	2,685	2,716	3,139	3,561	4,102	8,679	33,318	73,744
	G%	0.132	0.153	0.245	0.141	0.412	0.789	0.573	−0.133

Table 4.6b Decennial estimates of the Muslim population (×1000) in Europe: 1790–1940 AD (1210–1360 H).

		1790	1800	1810	1820	1830	1840	1850	1860
Balkans	P	14,031	14,945	15,727	16,515	17,634	18,787	20,174	22,528
	M%	13.94	13.68	13.32	13.20	13.07	13.19	13.48	13.05
	M	1,956	2,045	2,094	2,180	2,306	2,478	2,719	2,939
Eastern Europe	P	38,240	40,940	43,633	47,106	52,500	57,470	62,261	69,171
	M%	4.94	5.01	5.11	5.18	5.19	5.17	5.17	5.14
	M	1,889	2,053	2,229	2,440	2,727	2,972	3,216	3,557
Western Europe	P	46,213	46,454	49,671	52,046	56,011	57,356	61,435	64,473
	M%	0.00	0.00	0.00	0.00	0.00	0.00	0.00	0.00
	M	—	—	—	—	—	—	1	1
Northern Europe	P	18,352	20,843	23,216	26,840	30,773	32,381	33,984	36,422
	M%	—	—	—	—	—	—	—	—
	M	—	—	—	—	—	—	—	—
Central Europe	P	59,099	62,232	64,622	70,957	78,928	86,711	91,398	98,494
	M%	0.01	0.01	0.01	0.01	0.01	0.01	0.01	0.01
	M	4	4	4	4	5	5	6	7
Total	P	175,936	185,414	196,869	213,463	235,847	252,705	269,253	291,087
	M%	2.19	2.21	2.20	2.17	2.14	2.16	2.21	2.23
	M	3,849	4,102	4,327	4,625	5,037	5,455	5,942	6,503
	G%		0.525	0.599	0.809	0.997	0.690	0.634	0.780

Table 4.6b (Continued)

		1870	1880	1890	1900	1910	1920	1930	1940
Balkans	P	24,398	26,489	29,594	32,511	36,002	36,572	42,018	46,172
	M%	12.85	13.05	12.54	11.38	11.17	11.45	7.79	7.77
	M	3,135	3,457	3,711	3,700	4,022	4,187	3,273	3,589
Eastern Europe	P	77,132	86,401	96,260	108,296	129,379	134,484	150,110	163,965
	M%	4.95	4.96	4.57	4.58	4.63	3.82	3.82	4.21
	M	3,820	4,289	4,401	4,959	5,997	5,135	5,740	6,907
Western Europe	P	66,034	69,021	72,020	75,078	79,151	81,163	88,273	92,112
	M%	0.00	0.00	0.01	0.01	0.02	0.05	0.12	0.14
	M	1	2	4	5	13	44	108	133
Northern Europe	P	39,723	44,181	48,402	51,298	58,065	61,712	64,100	68,528
	M%	—	0.00	0.00	0.00	0.00	0.01	0.04	0.07
	M	—	—	—	—	1	5	23	49
Central Europe	P	105,967	113,728	127,195	140,843	156,212	153,646	167,372	178,882
	M%	0.01	0.01	0.01	0.01	0.01	0.00	0.00	0.00
	M	8	8	12	14	15	6	4	7
Total	P	313,255	339,820	373,472	408,025	458,810	467,577	511,874	549,659
	M%	2.22	2.28	2.18	2.13	2.19	2.01	1.79	1.94
	M	6,963	7,756	8,129	8,679	10,047	9,376	9,148	10,685
	G%	0.734	0.814	0.944	0.885	1.173	0.189	0.905	0.712

Table 4.6c Decennial estimates of the Muslim population (×1000) in Europe: 1950–2100 AD (1370–1520 H)

		1950	1960	1970	1980	1990	2000	2010	2020
Balkans	P	48,785	54,244	58,928	63,986	66,853	64,397	61,813	59,040
	M%	8.09	8.32	9.82	10.40	12.42	12.70	12.67	13.02
	M	3,946	4,512	5,785	6,654	8,302	8,176	7,830	7,688
Eastern Europe	P	155,777	179,782	196,621	209,027	221,437	216,602	209,353	209,087
	M%	4.05	4.09	4.70	4.94	5.49	6.92	7.37	7.93
	M	6,313	7,357	9,245	10,321	12,159	14,989	15,422	16,573
Western Europe	P	97,344	105,910	116,349	125,791	131,232	136,910	148,690	151,726
	M%	0.25	0.39	0.71	1.71	1.92	3.17	4.93	6.29
	M	243	415	831	2,150	2,525	4,342	7,338	9,543
Northern Europe	P	72,412	75,611	80,377	82,372	84,111	87,169	93,824	100,326
	M%	0.14	0.14	0.22	0.32	0.76	2.57	4.10	5.56
	M	102	106	181	267	639	2,240	3,851	5,578
Central Europe	P	175,009	189,156	203,487	210,545	214,861	218,226	221,050	226,096
	M%	0.00	0.01	0.42	0.63	0.98	1.64	2.51	3.32
	M	3	14	848	1,334	2,107	3,572	5,557	7,511
Total	P	549,328	604,703	655,762	691,721	718,493	723,305	734,730	746,275
	M%	1.93	2.05	2.58	3.00	3.58	4.61	5.44	6.28
	M	10,607	12,405	16,889	20,727	25,732	33,318	39,998	46,893
	G%	−0.006	0.960	0.811	0.534	0.380	0.067	0.157	0.156

Table 4.6c (*Continued*)

		2030	2040	2050	2060	2070	2080	2090	2100
Balkans	P	56,602	53,958	51,091	47,996	44,931	42,291	40,116	38,177
	M%	13.44	13.63	13.65	13.53	13.37	13.03	12.53	12.04
	M	7,607	7,354	6,972	6,493	6,008	5,512	5,028	4,598
Eastern Europe	P	202,865	194,699	187,797	180,952	173,818	168,673	165,966	163,474
	M%	8.39	8.85	9.33	9.83	10.34	10.86	11.38	11.89
	M	17,018	17,222	17,517	17,783	17,969	18,317	18,884	19,435
Western Europe	P	153,037	152,836	150,641	146,642	142,363	139,208	137,061	135,084
	M%	7.20	8.12	9.04	9.98	10.92	11.85	12.77	13.69
	M	11,024	12,412	13,622	14,629	15,540	16,493	17,500	18,487
Northern Europe	P	104,505	107,662	110,201	111,932	113,360	114,767	116,084	117,568
	M%	6.65	7.73	8.83	9.93	11.05	12.16	13.28	14.42
	M	6,949	8,327	9,729	11,118	12,521	13,960	15,422	16,950
Central Europe	P	223,482	218,473	211,229	202,399	193,922	187,110	182,213	178,877
	M%	3.89	4.46	5.04	5.61	6.19	6.79	7.39	7.98
	M	8,682	9,750	10,642	11,345	12,003	12,710	13,465	14,275
Total	P	740,491	727,629	710,958	689,921	668,395	652,050	641,438	633,179
	M%	6.93	7.57	8.23	8.90	9.58	10.27	10.96	11.65
	M	51,279	55,065	58,482	61,369	64,041	66,992	70,299	73,744
	G%	−0.078	−0.175	−0.232	−0.300	−0.317	−0.248	−0.164	−0.130

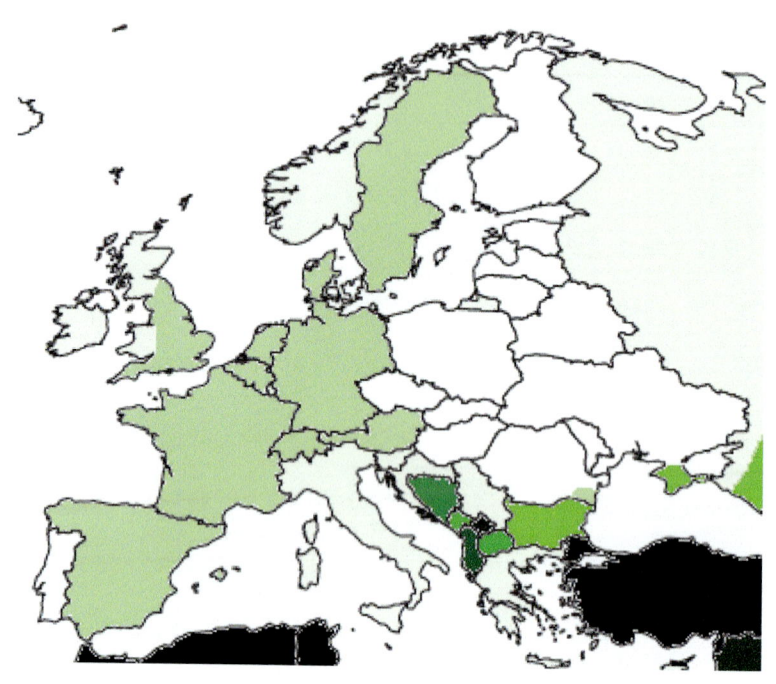

Figure 4.6.0 A choropleth map of Europe illustrating the presence of Muslims.

References

[AD] Department of Statistics of the Government of Andorra (2016). Demographics and Population - Population: Population by Nationality. Data Bank [Online]. Retrieved from http://www.estadistica.ad/

[AL11] National Institute of Statistics of Albania (INSTAT) (2012). *Population and Housing Census 2011* (Part 1: Main Results). Tirana: INSTAT.

[AT11] Statistics Austria (2015). *Population at the beginning of the year, country of birth by time section.* STATcube–Statistical Database of Statistics Austria [Online]. Retrieved at http://statcube.at/

[AT01] Statistics Austria (2007). *2001 Volkszählung Textband: Die demographische, soziale und wirtschaftliche, Struktur der österreichischen Bevölkerung* [2001 Census Textbook: The Demographic, Social and Economic Structure of the Austrian Population]. Vienna: Statistics Austria.

[AT910] Österreichische Statistik (1912). *Die Ergebnisse der Volkszählung vom 31. December 1910 in den im Reichsrathe vertretenen Königreiche* und Ländern (1. Heft. die Summarischen Ergebnisse der Volkszählung) [The Results of the Census of 31 December 1910 in the Kingdoms and Countries Represented in the Reichsrat (Vol. 1. Summary of the Results of the Census)]. Vienna: Aus der Kaiserlich-Koniglichen hof- und Staatsdruckerei.

[AT900] Österreichische Statistik (1902). *Die Ergebnisse der Volkszählung vom 31. December 1900 in den im Reichsrathe vertretenen Königreiche und Ländern* (1. Heft. die Summarischen Ergebnisse der Volkszählung) [The Results of the Census of 31 December 1900 in the Kingdoms and Countries Represented in the Reichsrat (Vol. 1. Summary of the Results of the Census)]. Vienna: Aus der Kaiserlich-Koniglichen hof- und Staatsdruckerei.

[AT890] Österreichische Statistik (1892). *Die Ergebnisse der Volkszählung vom 31. December 1890 in den im Reichsrathe vertretenen Königreiche und Ländern* (1. Heft. die Summarischen Ergebnisse der Volkszählung) [The Results of the Census of 31 December 1890 in the Kingdoms and Countries Represented in the Reichsrat (Vol. 1. Summary of the Results of the Census)]. Vienna: Aus der Kaiserlich-Koniglichen hof- und Staatsdruckerei.

[AT880] Österreichische Statistik (1880). *Die Bevölkerung der im Reichsrathe vertretenen Königreiche und Länder nach Religion, Bildungsgrad, Umgangssprache und nach ihren Gebrechen* (2. Heft der "Ergebnisse der Volkszählung und der mit derselben verbundenen Zählung der häuslichen Nutzthiere vom 31. December 1880") [The Population of the Kingdoms and Countries Represented in the Reichsrat According to Religion, Degree of Education, Spoken Language, and Ethnicity (Vol. 2. The Results of the Census and of the Count of the Domestic Population of 31 December 1880 within the Kingdoms and Countries Represented in the Reichsrat)]. Vienna: Aus der Kaiserlich-Koniglichen hof- und Staatsdruckerei.

[ATH] Walker, F. (1891). The Census of Austria. *Publications of the American Statistical Association*, 2(16), 444–449.

[BA13] Agency for Statistics of Bosnia and Herzegovina (BHAS) (2016). *Census of Population, Households and Dwellings in Bosnia and Herzegovina, 2013–Final Results*. Sarajevo: BHAS.

[BA91] Donia, R. J. & Fine, J. V. A. (1994). *Bosnia and Hercegovina: A Tradition Betrayed*. New York, NY: Columbia University Press.

[BA876] Tanovic, M. T., Pasalic, S. & Golijanin, J. (2014). Demographic development of Bosnia and Herzegovina from the Ottoman period till

1991 and the modern demographic problems. *Procedia - Social and Behavioral Sciences*, 120, 238–247.

[BEH] Eggerickx, T., Poulain, M. & Kesteloot, C. (2002). *La Population Allochtone en Belgique* [Foreign-Born Population in Belgium]. Monographie N° 3 de Recensement Général de la Population et des Logements au 1er Mars 1991. Bruxelles: Institut National de Statistique.

[BG11] National Statistical Institute (NSI) (2011). *2011 Population Census in the Republic of Bulgaria*. Sofia: NSI.

[BG] Eminov, A. (1997). *Turkish and other Muslim minorities in Bulgaria*. New York, NY: Routledge.

[BY] National Statistical Committee of the Republic of Belarus (NSCRB) (2011). *Population Census 2009* (Vol. III: Ethnic Composition of the Population of the Republic of Belarus). Minsk: NSCRB.

[CH] Office Fédéral de la Statistique (OFS) (2017, January 31). Population résidante permanente âgée de 15 ans ou plus selon l'appartenance la religieuse (je-f-01.08.02.01). *Relevé structurel du recensement fédéral de la population*. Neuchâtel: FSO. Online file retrieved from http://www.statistique.admin.ch

[CH90] Bovay, C. (2004). *Le Paysage Religieux en Suisse*. Recensement Fédéral de la Population 2000. Neuchâtel: Federal Statistical Office.

[CH50] Federal Statistical Office (FSO) (1956). *Recensement Fédéral de la Population 1950* (Vol. 24: Suisse, Tableaux Iere Partie). Berne: FSO.

[CH30] Federal Statistical Office (FSO) (1935). *Eidgenössische Volkszählung 1. Dezember 1930* (Band 21: Schweiz, Tabellenteil) [Federal Census, December 1, 1930 (Vol. 21: Switzerland, Table Part)]. Bern: FSO.

[CHH] Chaix, P. (1854). Summary of the Last Census of Switzerland. *Journal of the Royal Geographical Society of London*, 24, 313–318.

[CP14] Russian Federation Federal State Statistics Service (GKS) (2015). 4.1 Национальный состав населения [Ethnic composition of the population]. Перепись населения в Крымском федеральном округе [Census in the Crimean Federal District]. Online table retrieved from http://www.gks.ru/

[CPH] Illarionov, A. (2014, February 11). Этнический состав населения Крыма за три века [The ethnic composition of the population of Crimea for three centuries]. *LiveJournal*. Retrieved from http://aillarionov.livejournal.com/607335.html

[CP] Tulsky, M. (2003). The Results of the 2001 Census in Ukraine [in Russian]. *Demoscope Weekly*. No. 113–114.

[CZ] Czech Statistical Office (CSO) (2014). Table 1-18 Population by religious belief and sex by 1921, 1930, 1950, 1991, 2001 and 2011 censuses. *Czech Demographic Handbook 2013*. Prague: CSO.

[DE11] Statistisches Bundesamt (DESTATIS) (2013, May 31). Pressekonferenz: Zensus 2011–Fakten zur Bevölkerung in Deutschland. *Statement von Präsident Roderich Egeler*. [Online]. [Press Conference: Census 2011–Facts on Population in Germany. Statement by President Roderich Egeler]. Retrieved from http://www.destatis.de/

[DE70] Forschungsgruppe Weltanschauungen in Deutschland (fowid) (2012, March 29). *Religionszugehörigkeit, Deutschland, Bevölkerung, 1970–2011* [Online]. [Research Group Worldviews in Germany (fowid). *Religious affiliation, Germany, Population, 1970–2011*.] Retrieved from http://fowid.de/

[DET] Statistisches Bundesamt (DESTATIS) (2012). *GENESIS-Online Datenbank*. Retrieved from www-genesis.destatis.de

[DEB] Böer, I., Haerkötter, R., & Kappert, P. (eds.) (2002). *Türken in Berlin 1871–1945: eine Metropole in den Erinnerungen osmanischer und türkischer Zeitzeugen*. [Turks in Berlin 1871–1945: A Metropolis in the Memories of Ottoman and Turkish Eyewitnesses]. Berlin: Walter de Gruyter.

[DE43] Hebeler, B. (1847). Statistics of Prussia. *Journal of the Statistical Society of London*, 10(2), 154–186.

[DE80] JSSL (1885). Statistical Results of the Last Census of France, Prussia, Austria, Italy, Switzerland, and Belgium. *Journal of the Statistical Society of London*, 48(2), 288–297.

[DK] Statistics Denmark (2018). FOLK2: Population 1. January by Sex, Age, Ancestry, Country of Origin and Citizenship. Population and Elections: Population in Denmark. *StatBank Denmark* [Online Database]. Retrieved from www.statbank.dk

[DKH] Statistics Denmark (1960–1980) *Statistical Yearbook*. Copenhagen: Statistics Denmark.

[EE] Statistics Estonia (2013). Religious Affiliation, Demographic and Ethno-Cultural Characterisitcs of the Population, Population and Housing Census 2011. *Statistical Database* [Online Database]. Retrieved from pub.stat.ee

[EEH] Viikberg, J. K. (1999). *Eesti rahvaste raamat. Rahvusvähemused, -rühmad ja -killud* [The Book of Nationalities in Estonia: Minorities and Ethnic Groups]. Tallinn: Eesti Entsüklopeediakirjastus, [Tallinn: Estonian Encyclopedia Publishers]

[ES2] Observatorio Andalusí (2005–2018). *Estudio demográfico de la Población musulmana*. Madrid: Unión de Comunidades Islámicas de España

[ES1] Instituto Nacional de Estadistica (1950–2018). *Extranjeros residentes en España, Clasificados por comunidad autónoma, provincia de residencia y nacionalidad*. Anuarios Estadisticos. Retrieved from http://www.ine.es/

[ES50] Instituto Nacional de Estadística (1954). *Censo de la población de España y Territorios de su Soberania y Protectorado, segun el Empadronamiento Realizado el 31 de Diciembre de 1950* (Tomo III. Clasificaciones de la poblacion de hecho de la peninsula e islas adyacentes, obtenidas mediante una muestra del 10 por 100). Madrid: Sucesores de Rivadeneyra

[EU] European Commission (2016). Population by Citizenship and by Country of Birth. Population: International Migration and Asylum. *Eurostat Database* [Online]. Retrieved from http://epp.eurostat.ec.europa.eu/portal/page/portal/population/data/database

[EUH] Bertillon, J. (1899). *Statistique internationale résultant des recensements de la population exécutés dans les divers pays de l'Europe pendant le XIXe siècle et les époques précédentes, établie conformément au vœu de l'Institut international de statistique*. Paris: G. Masson

[EVS] European Values Study Foundation at Tilburg University [The Netherlands] (2015). *European Values Study, ZA4804: EVS 1981–2008 Longitudinal Data File*. Cologne: Leibniz Institute for the Social Sciences (GESIS). Retrieved from http://zacat.gesis.org/. More info on EVS is available at http://www.europeanvaluesstudy.eu/

[EW2] Office for National Statistics (ONS) (2012). *Religion in England and Wales 2011*. London: The Stationary Office (TSO).

[EW] Office for National Statistics (ONS) (2003). *Census 2001 - National Report for England and Wales*. London: The Stationary Office (TSO).

[FI] Statistics Finland (2018). 016 – Population by religious community, age and sex in 2000 to 2017. *Statistics Finland's PX-Web databases* [Online]. Retrieved from http://pxweb2.stat.fi/database/StatFin/vrm/vaerak/vaerak_en.asp

[FIH] Official Statistics of Finland (1996). Appendix Table 8: Religious Affiliation of the Population. *Population Structure, Annual Review*. Helsinki: Statistics Finland.

[FR] Institut National de la Statistique et des Études Économiques (INSEE) (2014). *Répartition des étrangers par nationalité*. [Online].

[FR21] Institut National de la Statistique et des Études Économiques (INSEE) (2013). *Recensements de 1851 à 1921 (Données de la Statistique Générale de la France (SGF))* [Online Database].

[FR99] Institut National de la Statistique et des Études Économiques (INSEE) (2011). R6: Étrangers Selon la Nationalité (17 Postes) de 1946 à 1999. Tableaux Rétrospectifs à Partir des Recensements Antérieurs à 1999. *INSEE RÉSULTATS*, 121 [Online].

[FO] Statistics Faroe Islands (2014) CS 10.1.2 Population by religious faith, educational attainment, occupation, country of birth, year of arrival in the country and place of usual residence. *Statbank* » Census 2011 » *Religion* » Religion, education and country of birth [Online Database]. Retrieved from http://www.hagstova.fo/en/statbank

[GB911] His Majesty's Stationary Office (HMSO) (1917). *Census of England and Wales. 1911 (*General Report with Appendices). London: Darling & Son.

[GB901] General Register Office (GRO) (1906). *Census of the British Empire. 1901. Report with Summary and Detailed Tables for the Several Colonies, &c., Area, Houses, and Population; also Population Classified by Ages, Condition as to Marriage, Occupation, Birthplaces, Religions, Degrees of Education, and Infirmities.* London: Darling & Son.

[GB851] Great Britain Census Office (1854). *Census of Great Britain, 1851: Religious Worship in England and Wales.* London: G. Routledge.

[GG] Guernsey Islamic Charitable Trust (GICT) (2010). *Islam in Guernsey.* [Online]. Retrieved from islamiguernsey.com

[GI12] HM Government of Gibraltar (2015). *Census of Gibraltar 2012.* Gibraltar: HM Government of Gibraltar.

[GI970] Government of Gibraltar (1971). *Report on the Census of Gibraltar 1970.* Gibraltar: Government of Gibraltar.

[GI961] Government of Gibraltar (1961). *Report on the Census of Gibraltar, Taken on 3rd October, 1961.* Gibraltar: The Garrison Library Printing Office

[GI951] Government of Gibraltar (1951). *Report on the Census of Gibraltar, Taken on 3rd July, 1951.* Gibraltar: The Garrison Library Printing Office

[GI931] Government of Gibraltar (1931). *The Census of Gibraltar, Taken on the 26th April, 1931.* Gibraltar: The Gibraltar Garrison Library Printing Works

[GI921] Government of Gibraltar (1921). *The Census of Gibraltar, Taken on the 19th June, 1921.* Gibraltar: Garrison Library Printing Establishment

[GI911] Government of Gibraltar (1911). *The Census of Gibraltar, Taken on the 2nd April, 1911*. Gibraltar: Garrison Library Printing Establishment

[GI901] Government of Gibraltar (1901). *The Census of Gibraltar, Taken on the 31st March, 1901*. Gibraltar: Garrison Library Printing Establishment

[GI881] Civil Secretary's Office (1881). *General Abstract of Gibraltar, 1881*. Gibraltar: Gibraltar National Archives. Retrieved from http://www.nationalarchives.gi/

[GI871] Civil Secretary's Office (1871). *General Abstract of Gibraltar, 1871*. Gibraltar: Gibraltar National Archives. Retrieved from http://www.nationalarchives.gi/

[GI868] Civil Secretary's Office (1869). *Census of the Population Ordinance Gibraltar, 1868*. Gibraltar: Gibraltar National Archives. Retrieved from http://www.nationalarchives.gi/

[GI834] Secretary's Office (1834). *Census. Gibraltar.* Gibraltar: Gibraltar National Archives. Retrieved from http://www.nationalarchives.gi/

[GI816] Secretary's Office (1816). *1816 Return of Inhabitants of Gibraltar*. Gibraltar: Gibraltar National Archives. Retrieved from http://www.nationalarchives.gi/

[GI814] Secretary's Office (1814). *1814 Return of Inhabitants of Gibraltar*. Gibraltar: Gibraltar National Archives. Retrieved from http://www.nationalarchives.gi/

[GI791] Secretary's Office (1792). *1791 Aplphabetic List of Inhabitants of Gibraltar*. Gibraltar: Gibraltar National Archives. Retrieved from http://www.nationalarchives.gi/

[GR51] Office National de Statistique de Gréce (ONSG) (1958). Résultats du Recensement de la Population Effectué le 7 Avril 1951. Athénes: Imprimerie Nationale.

[HR11] Croatian Bureau of Statistics (CBS) (2012). *Census of Population, Households and Dwellings 2011*. Zagreb: CBS.

[HR01] Central Bureau of Statistics (CBS) (2005). *Statistical Yearbook 2005*. Zagreb: CBS.

[HRH] Duvnjak, N. (1999). Muslim Community in the Republic of Croatia. *Religion in Eastern Europe*. XIX(3), 1–18.

[HR81] Crkvencic, I. (1998). Croatian Ethnic Territory and the Multiethnic Composition of Croatia as a Result of Population Migration. *Drustvena Istrazivanja: Journal for General Social Issues*. 7(1–2), 109–125.

[HU11] Központi Statisztikai Hivatal (KSH) [Hungarian Central Statistical Office] (2014). *2011. Évi Népszámlálás (10. Vallás, felekezet)* [2011

Population Census (Vol. 10. Religions and Denominations). Budapest: KSH.

[HU930] A Magyar Kir. Központi Statisztikai Hivatal [Hungarian Central Statistical Office] (1936). *Az 1930. Évi, Népszámlálás.* (IV. Rész. A *Népesség Foglalkozása A Főbb Demográfiái Adatokkal Egybevetve S A Népesség Ház- És Földbirtok Viszonyai. V. Rész. Részletes Demografia És A Lakóházak És Lakások Adatai*) [1930 Population Census (Vol. IV. Main Occupation of the Population. Vol. V. Detailed Demoraphic and Housing Characteristics)]. Budapest: *Stephaneum Nyomda Részvénytársaság*.

[HU920] A Magyar Kir. Központi Statisztikai Hivatal [Hungarian Central Statistical Office] (1928). *Az 1920. Évi, Népszámlálás. Ötödik Rész. Részletes Demografia.* [1920 population census, detailed demography]. Budapest: Pesti Könyvnyomda-Részvénytársaság.

[HU910] A Magyar Kir. Központi Statisztikai Hivatal [Hungarian Central Statistical Office] (1916). *A Magyar Szent Korona Országaiban 1910. Évi, Népszámlálása. Ötödik Rész. Részletes Demografia.* [1910 population census, detailed demography]. Budapest: Pesti Könyvnyomda Részvénytársaság.

[HU880] Az Országos Magyar Kir. Statisztikai Hivatal. [Hungarian National Statistical Office] (1882). *A Magyar Korona Országaiban Az 1881. Év Elején Végrehajtott Népszámlálás Eredményei Némely Hasznos Házi Állatok Kimutatásával Együtt.* [Results of the population census taken at the beginning of the year 1881 in the lands belonging to the Hungarian crown]. Budapest: Athenaeum E. Társulat.

[HU870] Az Országos Magyar Kir. Statisztikai Hivatal. [Hungarian National Statistical Office] (1871). *A Magyar Korona Országaiban Az 1870. Év Elején Végrehajtott Népszámlálás Eredményei A Hasznos Házi Állatok Kimutatásával Együtt.* [Results of the population census taken at the beginning of the year 1870 in the lands belonging to the Hungarian crown]. Pest: Druckei des Athenaeum.

[IE16] Central Statistics Office (2017). *Census 2016 Summary Results–Part 1.* Dublin: CSO.

[IE06] Central Statistics Office (2007). *Census 2006* (Vol. 13: Religion). Dublin: The Stationary Office.

[IE891] Her Majesty's Stationary Office (1892). *Census of Ireland, 1891* (Part II. General Report, with Illustrative Maps and Diagrams, Tables, and Appendix). Dublin: Alexander Thom & Co.

[IE881] Her Majesty's Stationary Office (1882). *Census of Ireland, 1881* (Part II. General Report, with Illustrative Maps and Diagrams, Tables, and Appendix). Dublin: Alexander Thom & Co.

[IE871] Her Majesty's Stationary Office (1872). *Census of Ireland, 1871* (Part I. Area, Houses, and Population, also the Ages, Civil, Condition, Occupations, Birthplaces, Religion, and Education of the People. Dublin: Alexander Thom & Co.

[IE861] Her Majesty's Stationary Office (1863). *Census of Ireland for the Year 1861* (Part I. Area, Population, and Number of Houses, by Townlands and Electoral Divisions. Dublin: Alexander Thom & Co.

[IM] IOMT (2009, May 5). Facebook Campaign for Manx Mosque Faces Opposition. *Isle of Man Today*.

[IS] Statistics Iceland (2018). *Populations by Religious Organizations 1998–2018*. Statistics–Population: Religious Organizations. [Online Database]. Retrieved from www.statice.is

[IT16] Italian National Institute of Statistics (Istat) (2018). *I.Stat: Data Warehouse of the Institute*. Onlien database retrieved from http://dati.istat.it/

[JE] States of Jersey Statistics Unit (SJSU) (2012). Report on the 2011 Jersey Census. St. Helier: SJSU.

[KET96] Kettani, A. (1996). Challenges to the organization of Muslim communities in Western Europe: The Political Dimension. In W. A. R. Shadid & P. S. Van Koningsveld (eds.). *Political Participation and Identities of Muslims in Non-Muslim States* (pp. 14–35). Kampen: Kok Pharos Publishing House.

[KET86] Kettani, A. (1986). *Muslim Minorities in the World Today*. London: Mansell Publishing.

[KET76] Kettani, A. (1976). *Muslims in Europe and the Americas*. [Arabic]. Beirut: Dar Idriss.

[KO] Kosovo Agency of Statistics (KAS) (2012). *Kosovo Population and Housing Census 2011 Final Results*. Priština: KAS.

[LI15] Liechtenstein Statistical Office (LSO) (2016). *Volkszählung 2015 - Erste Ergebnisse* [Census 2015 - Preliminary Results]. Vaduz: LSO.

[LI00] Liechtenstein Statistical Office (LSO) (2005). *Liechtensteinische Volkszählung 2000* (Band 2: Religion und Hauptsprache) [Liechtenstein 2000 Census (Vol. 2: Religion and Language)]. Vaduz: LSO.

[LIH] Amt für Statistik des Fürstentum Liechtenstein (1962). *Wohnbevölkerung Volkszählungen 1812–1930*. Vaduz: SFL

[LT] Statistics Lithuania (2013). Ethnicity, Mother Tongue and Religious Affiliation. *Lithuania 2011 Population Census*. Vilnius: Statistics Lithuania.

[LT23] Department of General Statistics (DGS) (1923). *Population de la Lithuanie, Données du Premier Recensement du 17 Septembre 1923.* Kaunas: DGS.

[LUH] Als, G. (1975). *La Population du Luxembourg.* 1974 Country Monographs, CICRED (Comite International de Coordination de Recherches Nationales de Demographie) Series. Luxembourg: STATEC (Institut National de la Statistique et des Études Économiques du Grand-Duché de Luxembourg). Retrieved from http://www.cicred.org/Eng/Publications/content/3MonographNational/Index.html

[LV] Statistics Latvia (2012). Table TSG11-06: Resident Population on March 1, 2011 by Ethnicity, Sex, and Age Group. *Statistical Database* [Online]. Retrieved from http://data.csb.gov.lv/

[LVH] Martuzāns, B. (2002). *ROOTS = SAKNES* (Vol. 2: Diverse Worlds a Human Lived in: Ethnicity, Religion, Estate, Occupation). [Online]. Retrieved from http://www.roots-saknes.lv/

[MAS23] Massignon, L. (1923). *Annuaire du Monde Musulman, Statistique, Historique, Social et Économique.* Paris: Ernest Leroux.

[MAS55] Massignon, L. (1955). *Annuaire du Monde Musulman, Statistique, Historique, Social et Économique* (4ème éd.). Paris: Presses Universitaires de France.

[MC16] Division des Statistiques des Etudes Economiques (2017). *Principauté de Monaco Recensement de la Population 2016.* Monte Carlo: MULTIPRINT.

[MC08] Division des Statistiques des Etudes Economiques (2009). *Recensement General de la Population 2008 Principauté de Monaco.* Monte Carlo: MULTIPRINT.

[MC61] Girardeau, E. (1962). La population de Monaco et les migrations. *Population*, 17(3), 491–504.

[MD14] National Bureau of Statistics of the Republic of Moldova (NBS) (2017). *Results of Population and Housing Census in the Republic of Moldova in 2014*, Tables: Characteristics–Population 2 (population by communes, religion, citizenship). Chişinău: NBS.

[MD04] National Bureau of Statistics of the Republic of Moldova (NBS) (2006). *Population Census 2004* (Vol. 1: Demographic, National, Linguistic, Cultural Characteristics Statistical Compilation). Chişinău: NBS.

[ME11] Montenegro Statistical Office (MSO) (2011). *Census of population, households and dwellings in Montenegro 2011: Population of*

Montenegro by Sex, Type of Settlement, Ethnicity, Religion and Mother Tongue, per Municipalities. Podgorica: MSO.

[ME03] Montenegro Statistical Office (MSO) (2009). *Statistical Yearbook 2009*. Podgorica: MSO.

[MK] State Statistical Office (SSO) (2012). *Statistical Yearbook of the Republic of Macedonia 2012*. Skopje: SSO.

[MT881] Cousin, G. (1882). *Census of the Islands of Malta, Gozo and Comino, Taken on the 3rd April 1881*. Malta: Government Printing Officie.

[MV] Ministry of Planning and National Development (MPND) (2008). *Analytical Report 2006: Population and Housing Census 2006*. Male: MPND, and email correspondence in February 2017 with Ms. Ihrisha Abdul Wahid from the National Bureau of Statistics of the Maldives.

[ND2] Northern Ireland Statistics and Research Agency (NISRA) (2013). Table QS218NI: Religion - Full Detail. Ethnicity, Identity, Language and Religion–Northern Ireland. *Census 2011 Statistics*. [Online]. Retrieved from http://www.nisra.gov.uk/Census/

[ND] Northern Ireland Statistics and Research Agency (NISRA) (2003). Table UV116: Religion (Full Detail). *Census 2001 Results Univariate Tables*. [Online]. Retrieved from http://www.nisra.gov.uk/Census/

[NLO] Statistics Netherlands (2018). Population; Sex, Origin and Generation, 1 January. *STATLINE* [Online Database]. Retrieved from http://statline.cbs.nl/

[NLE] Statistics Netherlands (2012). Population, Households and Population Dynamics; from 1899. Population by Nationality. Population on 1 January. *STATLINE* [Online Database]. Retrieved from http://statline.cbs.nl/

[NL] FORUM - Institute for Multicultural Affairs (2010). *Factbook 2010: The Position of Muslims in the Netherlands–Facts and Figures*. Utrecht: FORUM.

[NL971] Van Hemert, M. M. J. (1979). *Monografieën Volkstelling 1971* (Vol. 13. Kerkelijke gezindten: een analyse op basis van de Volkstelling 1971) [*1971 Census Monographs* (Vol. 13. Church Concepts: An Analysis Based on the 1971 Census). Amsterdam: Staatsuitgeverij.

[NO] Statistics Norway (1981–2018). Members of Religious and Life Stance Communities outside the Church of Norway, by Religion/Life Stance. Per 1 January. *Statistical Yearbook*. Oslo: Statistics Norway.

[NOC] Statistics Norway (2005). Population Structure - Population, by Country of Birth. *StatBank* [Online Database]. Retrieved from http://statbank.ssb.no

[PLH] Szajkowski, B. (1999). An Old Muslim Community of Poland: The Tatars. *International Institute for the Study of Islam in the Modern World (ISIM) Newsletter,* 4/99, p. 27.

[RS] Statistical Office of the Republic of Serbia (SORS) (2013). *2011 Census of Population, Households and Dwellings in the Republic of Serbia–Population* (Book 4: Religion, Mother Tongue and Ethnicity–Data by Municiplalities and Cities). Belgrade: SROS.

[RO11] Negruti, S. (2014). The evolution of the religious structure in Romania since 1859 to the present day. *Romanian Statistical Review Supplement,* 62(6), 39–47.

[RO77] Varga, Á. (1999). *Hungarians in Transylvania between 1870 and 1995.* Budapest: Teleki László Foundation.

[ROH] Grigore, G. (1999). Muslims in Romania. *International Institute for the Study of Islam in the Modern World (ISIM) Newsletter,* 3/99, p. 34.

[RO30] Manuila, S. (1940). *Studiu etnografic asupra populaţiei României.* Bucureşti: Institutul Central de Statistica.

[ROD] Radončić, E. (2013). *World Almanac of the demographic history of Muslims: Rich cartographic representation of the geographical distribution of Muslim populations* (M. Durmić, Trans.). Sarajevo: Edin Radončić.

[RU10] Federal State Statistics Service (Rosstat) (2012). *2010 All-Russia Population Census* (Vol. 4: National Composition, Knowledge of Languages and Citizenship). [Russian]. Moscow: Rosstat.

[RU] Information and Publishing Center "Statistics of Russia" (Rosstat) (2005). *2002 All-Russia Population Census* (Vol. 4: National Composition, Knowledge of Languages and Citizenship). Moscow: Rosstat.

[SE] Statistics Sweden (2016). Foreign-Born Persons in Sweden by Country of Birth, Age and Gender. Year 2000–2015. Population Statistics: Foreign-Born Persons. *Statistical Database* [Online]. Retrieved from http://www.ssd.scb.se/

[SE95] Statistics Sweden (1996). *Population Statistics 1995* (Part 3: Distribution by Sex, Age and Citizenship). [Swedish]. Stockholm: Statistics Sweden.

[SE930] Sveriges Officiella Statistik (1937). *Folkräkningen den 31 December 1930* (V, Trosbekännelse. Främmande stam. Främmande språk m.m.). Stockholm: Kungl. boktryckeriet, P.A. Norstedt & Söner. [Swedish Official Statistics (1937). The Census of 31 December 1930 (V. Religion, nationality, language, etc.)].

[SE920] Sveriges Officiella Statistik (1925). *Folkräkningen den 31 December 1920* (II. Befolkningsagglomerationer, rosbekännelse, stamskillnad, utrikes födelseort, främmande statsborgarskap, lyten m. m.). Stockholm: Kungl. boktryckeriet, P.A. Norstedt & Söner. [Swedish Official Statistics (1925). The Census of 31 December 1920 (II. Population distribution by household, creed, place of birth, citizenship, etc.)].

[SE910] Sveriges Officiella Statistik (1918). *Folkräkningen den 31 December 1910* (IV. Folkmängdens fördelning efter hushåll, trosbekännelse, födelseort m. m.). Stockholm: Kungl. boktryckeriet, P.A. Norstedt & Söner [Swedish Official Statistics (1918). The Census of 31 December 1910 (IV. Population distribution by household, creed, place of birth, etc.)].

[SI02] Dolenc, D., Žnidaršič, E., Ilić, M., Ložar, B., Novak, T., Flander, A. O., Šircelj, M., Šter, D., Švajncer, T. & Žnidaršič, T. (2003). *Census of Population, Households and Housing, Slovenia, 31. March 2002.* Rapid Reports 92(2): Population. Ljubljana: Statistical Office of the Republic of Slovenia.

[SI91] Vertot, N., Žnidaršič, E., Ilić, M., Šter, D., Povhe, J. & Garvas, T. (2001). *Censuses in Slovenia 1948–1991 and Census 2002.* Ljubljana: Statistical Office of the Republic of Slovenia.

[SI53] Federal Bureau of Statistics (FBS) (1959). *Final Results of the People's Federal Republic of Yugoslavia Population Census 1953* (Vol. 1: Vital and Ethnic Characteristics). [Serbian]. Belgrade: FBS.

[SK] Statistical Office of the Slovak Republic (SOSR) (2012). *The 2011 Population and Housing Census Basic data, Population by Religion.* Bratislava: SOSR. And e-mail correspondence with Mr. J. Šedivý from SOSR in April 2011 for 1991 and 2001 data, and with Ms. A. Kondášová from SOSR in March 2013 for 2011 data.

[SM] Ufficio Informatica, Tecnologia, dati e Statistica (UITDS) (2018). *Supplemento al Bollettino di Statistica.* Borgo Maggiore: UITDS.

[SQ2] National Records of Scotland (NRS) (2013). *2011 Census: Key Results on Population, Ethnicity, Identity, Language, Religion, Health, Housing and Accommodation in Scotland–Release 2A.* Edinburgh: NRS.

[SQ] Office of the Chief Statistician (2005). *Analysis of Religion in the 2001 Census - Summary Report.* Edinburgh: Scottish Executive.

[SU] Demoscope Analysis and Information System (2012). Census of the Russian Empire, the USSR, and the 15 New Independent States [Russian]. *Demoscope Weekly.* Retrieved from http://demoscope.ru/weekly/pril.php

[SYB70] Paxton, J. (ed.) (1970). *The Statesman's Year-Book.* London: MacMillan.

[SYB31] Epstein, M. (ed.) (1931). *The Statesman's Year-Book.* London: MacMillan.

[SYB10] Scott-Keltie, J. (ed.) (1910). *The Statesman's Year-Book.* London: MacMillan.

[SYB00] Scott-Keltie, J. (ed.) (1900). *The Statesman's Year-Book.* London: MacMillan.

[SYB885] Scott-Keltie, J. (ed.) (1885). *The Statesman's Year-Book.* London: MacMillan.

[UKH2] Le Chatelier, A. (1907). Les Musulmans en Angleterre et dans les Possessions Anglaises d'Europe. *Revue du Monde Musulman*, 3(9), 132–140.

[UKH1] Bouvat, L. (1906). Les Musulmans Anglais. *Revue du Monde Musulman*, 1(2), 270–272.

[UN] United Nations Statistics Division, UNSD Demographic Statistics (2018). Population by Religion, Sex and Urban/Rural Residence. *UNdata* [Online Database]. Retrieved from http://data.un.org

[UNE] United Nations Statistics Division, UNSD Demographic Statistics, (2018). Population by National and/or Ethnic Group, Sex and Urban/Rural Residence, *UNdata* [Online Database]. Retrieved from http://data.un.org

[UN04] United Nations, Department of International Economic and Social Affairs, Statistical Office (2006). Table 6 - Population by Religion, Sex and Urban/Rural Residence: Each Census, 1985–2004. *Demographic Yearbook Special Census Topics* (Vol. 2b: Ethnocultural Characteristics). New York, NY: United Nations.

[UN88] United Nations, Department of International Economic and Social Affairs, Statistical Office (1990). Table 29 - Population by Religion and Sex: Each Census, 1979–1988. *Demographic Yearbook 1988*. New York, NY: United Nations.

[UN79] United Nations, Department of International Economic and Social Affairs, Statistical Office (1980). Table 29 - Population by Religion, Sex and Urban/Rural Residence: Each Census, 1970–1979. *Demographic Yearbook 1979.* New York, NY: United Nations.

[UN73] United Nations, Department of International Economic and Social Affairs, Statistical Office (1974). Table 31 - Population by Religion, Sex and Urban/Rural Residence: Each Census, 1965–1973. *Demographic Yearbook 1973.* New York, NY: United Nations.

[UN64] United Nations, Department of International Economic and Social Affairs, Statistical Office (1965). Table 32 - Population by Religion and Sex: Each Census, 1955–1964. *Demographic Yearbook 1964*. New York, NY: United Nations.

[UN63] United Nations, Department of International Economic and Social Affairs, Statistical Office (1964). Table 11 - Population by Religion and Sex: Each Census, 1955–1963. *Demographic Yearbook 1963*. New York, NY: United Nations.

[UN56] United Nations, Department of International Economic and Social Affairs, Statistical Office (1957). Table 8 - Population by Religion and Sex: Each Census, 1945–1955. *Demographic Yearbook 1956*. New York, NY: United Nations.

[VV] Károly, M. (2000). Demográfiai Jellemzők, Társadalmi Mutatók [Demographic Characteristics, Social Indicators]. In G. M. Irén and M. Zsuzsa (eds.), *Vajdasági Marasztaló*, [Vojvodina Condemnation] (pp. 23–52). Sabotica: Magyarságkutató Tudományos Társaság [The Scientific Association for Hungarology Researches].

[YU] Kicošev, S. (2006). The Ethnic and Religious Structure of the Population of Serbia and Montenegro. In W. Lukan and L. Trgovcevic (eds.). *Serbia and Montenegro: Space and Population - History - Language and Literature - Culture - Politics - Society - Economy–Law* [German]. (pp. 55–72). Vienna: LIT-Verlag.

[YU31] Federal Statistical Office of Yugoslavia (FSOY) (1931). *Present Population by Religion and Mother Tongue, 1931 Census*. Belgrade: FSOY.

[YU21] Federal Statistical Office of Yugoslavia (FSOY) (1921). *Present Population (Civil and Military, Permanent and Fleeting Present) in Their Native Language and Religion, 1921 Census*. Belgrade: FSOY.

Chapter 5

Islam in the Americas

Muslims arrived at the Americas as early as the tenth century [MRO]; more than five centuries before Christopher Columbus "discovery" of "the new world" in 1492. Some Muslims also came with the Spanish conquistadors to the Americas. Like what happened in the Iberian Peninsula, all Muslims were forced to abandon their religion. In addition, between 1615 and 1700 Muslim Moriscos from Spain established a colony in the lands currently occupied by Colombia and Venezuela, but they were obliterated by the Spaniards. Other Muslims came with the waves of enslaved black Africans during the seventeenth and eighteenth centuries; some of whom were Muslims. They led revolts against the inhumane treatment in Haiti in 1757, Jamaica in 1831, and Brazil in 1835. However, all these revolts were crushed, and Islam was obliterated again due to various oppression tactics.

Slavery was abolished by the British in 1834, French in 1848, and Dutch in 1863. However, slavery was replaced by indentured labor mostly from British India by the Brits and Dutch and Java by the Dutch. Some of this labor was Muslim and their decedents remain in the continent, mostly in former British and Dutch colonies in the Caribbean. Most notably are Suriname, Guyana, and Trinidad; the three countries with the largest Muslim percentage in the continent. While 13% of Suriname's population is Muslim, the portion for Guyana and Trinidad is 7% each. The indentured labor system was

---

The World Muslim Population: Spatial and Temporal Analyses
Houssain Kettani
Copyright © 2020 Jenny Stanford Publishing Pte. Ltd.
ISBN 978-981-4800-31-0 (Hardcover), 978-0-429-42253-9 (eBook)
www.jennystanford.com

abolished by the British in 1917, and the other colonial powers followed suit afterwards.

During the second half of the nineteenth century, Muslims emigrated from Syria, Lebanon and the Balkans (North America), all of which were under the Ottoman Empire. Many were fleeing the wars and avoiding enlistment in the Turkish army. Then Muslim Tatars came from Poland to North America in the early twentieth century. This was followed by remarkable conversion rate of African descents to Islam throughout the continent. This conversion started in the United States in 1920s, where Black Muslims constitute a third of the Muslim population and where 80% of the continent's Muslim population lives. Another wave of Muslim immigrants started in mid-twentieth century from Palestine after the declaration of Israel.

In the 1960s, immigration laws in the US and Canada changed to welcome immigrants from all over the world and not just Whites. Since then the number of Muslims is fast growing due to immigration, but also higher birthrate and conversions. Immigration from the Middle East and Indian subcontinent towards all parts of the continent also continues. About half of Muslims in this continent live in the United States, a quarter live in Canada, and the remainder is spread sparsely throughout the rest of the continent. However, the countries with the largest Muslim minority starting with the largest are Suriname (13%), Guyana (7%), and Trinidad (5%).

The Muslim population in the Western Hemisphere, both in size and in percentage, remains the lowest in comparison with other continents. It has likely increased from none in 1800, to 0.06 million or 0.04% in 1900, to 2.6 million or 0.3% in 2000, to 6.4 million or 0.6% in 2020, and is projected to reach twelve million or 1.0% by 2050 and 23 million or 2% by 2100. A plot of decennial estimates of the Muslim population and its percentage with respect to the total population in the Americas from 1900 to 2100 is provided in Fig. 5.0a. This shows that the Muslim population in this region was barely increasing until 1940 and is increasing slowly but at a faster rate in both number and percentage afterwards and towards the end of this century.

We divided the Americas into six regions; the data for each is included in a separate section and are sorted in terms of the percentage of Muslims in descending order. These regions are

Southern Caribbean Islands (Section 5.1), North America (Section 5.2), Central Caribbean Islands (Section 5.3), South America (Section 5.4), Northern Caribbean Islands (Section 5.5), and Central America (Section 5.6). In Section 5.7, the total population in each of the five regions of the Americas and the corresponding percentage and number of Muslims is presented centennially in Table 5.7a from 600 to 2100 and decennially in Tables 5.7b and 5.7c from 1790 to 2100.

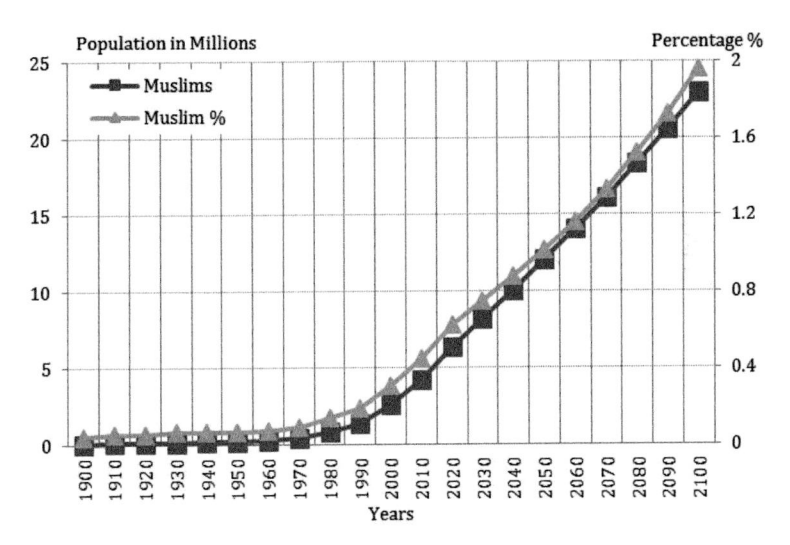

Figure 5.0a Plot of decennial estimates of the Muslim population and its percentage of the total population in the Americas: 1900–2100 AD (1320–1520 H).

5.1 Muslims in Southern Caribbean Islands

This region consists of ten island nations and territories south of Dominica and spread over a quarter of a million square kilometers of the western Atlantic Ocean and the eastern Caribbean Sea, with total land area of 9,684 sq. km, almost half (49.2%) of which is the Island of Trinidad. Thus, this region consists of Aruba, Barbados, Caribbean Netherlands, Curaçao, Dominica, Grenada, Martinique, Saint Lucia, Saint Vincent and the Grenadines, and Trinidad and

Tobago. The latter has the third largest percentage of Muslims in the Americas, after Suriname and Guyana. This region has the second highest density of territories in the world after the Central Caribbean Islands region. A map of this region is presented in Fig. 5.1a.

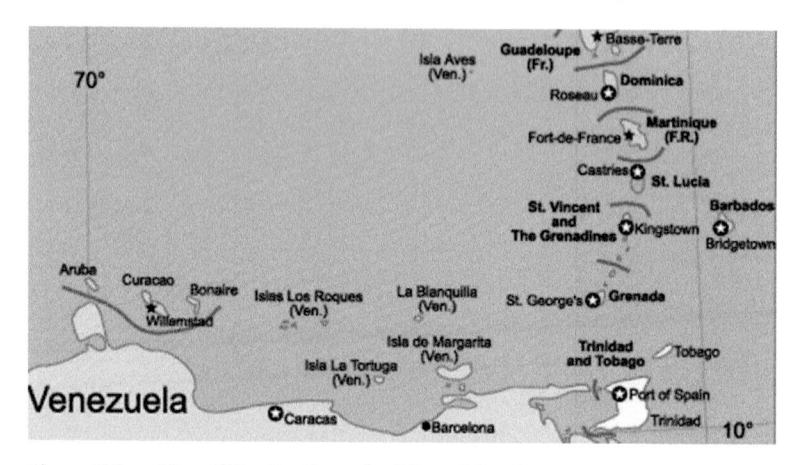

Figure 5.1a Map of the Southern Caribbean Islands.

Muslims arrived here as explorers, centuries before the arrival of the Europeans, then as escapees from Spanish Catholic oppression, then as black slaves brought by the Europeans from Africa. However, Islam went extinct from these migrations. Towards the end of the nineteenth century, the Brits and Dutch brought workers from India and Java, some of whom were Muslim, and Islam remains among their descendants till this day. The French also brought Muslims from their African colonies to their colonies in this region. Thus, the Muslim population has likely increased from none in 1800, to 11,000 or 1.3% in 1900, to 79,000 or 3.0% in 2000, to 83,000 or 2.9% in 2020, and is projected to decrease to 79,000 or 2.9% by 2050 and 56,000 or 2.7% by 2100.

A plot of decennial estimates of the Muslim population and its percentage with respect to the total population in this region from 1900 to 2100 is provided in Fig. 5.1b. The corresponding individual data for each country in this region is discussed below. In Section 5.1.11, the total population in each country in this region and the corresponding percentage and number of Muslims is presented centennially in Table 5.1a from 600 to 2100 and decennially in Tables 5.1b and 5.1c from 1790 to 2100.

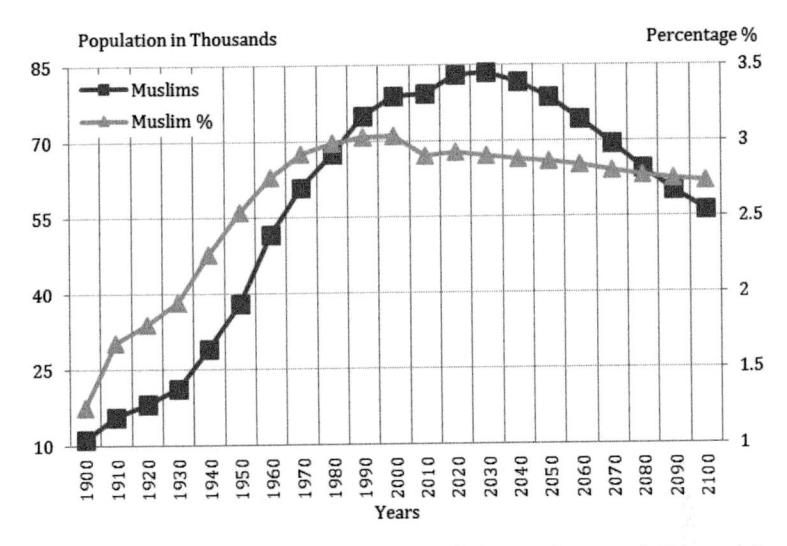

Figure 5.1b Plot of decennial estimates of the Muslim population and its percentage of the total population in Southern Caribbean Islands: 1900–2100 AD (1320–1520 H).

5.1.1 Aruba

It was conquered by the Dutch in 1636. It was part of the formation of the Netherland Antilles in 1952, but then seceded in 1986. However, Aruba remains a constituent country of the Kingdom of the Netherlands. It consists of one island with an area of 160 sq. km. And its map is presented in Fig. 5.1.1. According to census data as shown in Table 5.1.1, the Muslim population was nonexistent until the second half of the twentieth century. It increased from 17 or 0.03% in 1972, to 19 or 0.03% in 1981, to 218 or 0.3% in 1991, to 326 or 0.3% in 2010. The total population increased to 90,508 in 2000; however, the corresponding census did not include Islam in its religion questionnaire. Thus, assuming that the percentage of Muslims will increase by 0.05 of a percentage point each decade, the Muslim population is expected to remain in few hundreds throughout this century.

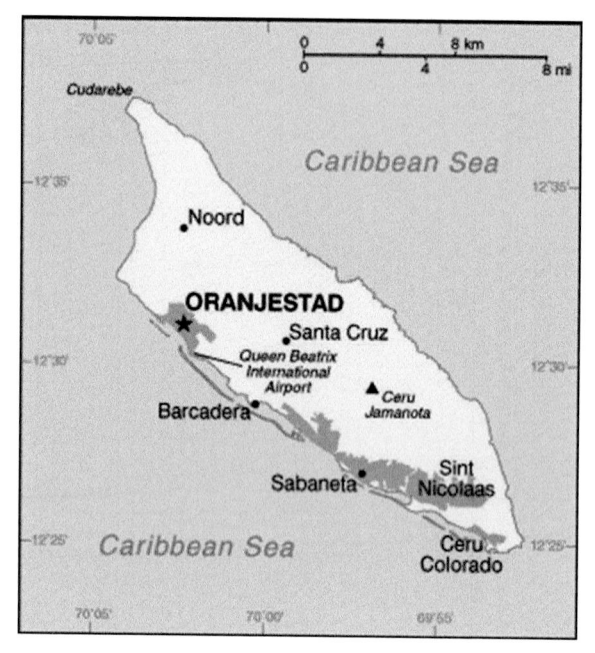

Figure 5.1.1 Map of Aruba.

Table 5.1.1 Evolution of the Muslim population in Aruba

Year	Population	Muslims	%	Source
1870	3,881	—	0.00	[SYB873]c
1891	7,886	—	0.00	[SYB894]c
1920	8,265	—	0.00	[SYB922]c
1972	57,905	17	0.03	[AW]c
1981	60,312	19	0.03	[AW]c
1991	66,609	218	0.33	[UN04]c
2010	100,969	326	0.32	[AW10]c
2050	109,000	500	0.50	es
2100	102,000	800	0.75	es

5.1.2 Barbados

This island was first visited by the Portuguese who founded uninhabited and named it Los Barbados, from the number of bearded fig trees which they found, and then they abandoned it. It was then claimed by the British in 1625 and gained its independence from the UK in 1966. It consists of one island with an area of 430 sq. km and its map is presented in Fig. 5.1.2. The first Muslims were indentured labor brought from India in the nineteenth century and were estimated to be fifteen Bengalis in 1921.

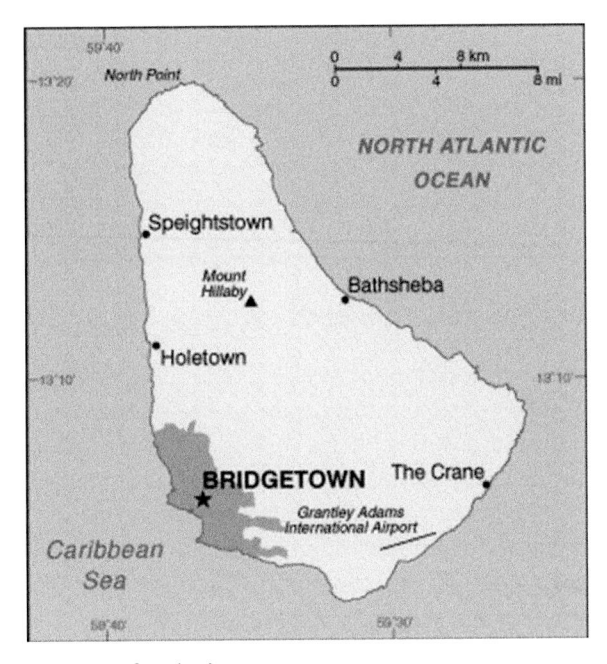

Figure 5.1.2 Map of Barbados.

According to census data as shown in Table 5.1.2, the Muslim population increased from none in 1871, to one in 1911, to 58 or 0.03% in 1946, to 336 or 0.1% in 1960, to 773 or 0.3% in 1980, to 1,047 or 0.4% in 1990, to 1,657 or 0.7% in 2000, to 1,605 or 0.7% in 2010. Thus, assuming that the percentage of Muslims will increase by a tenth of a percentage point each decade, the Muslim population is expected to remain less than 4,000 throughout this century.

Table 5.1.2 Evolution of the Muslim population in Barbados

Year	Population	Muslims	%	Source
1871	162,042	—	0.00	[GB871]c
1911	172,337	1	0.00	[GB911]c
1921	156,774	15	0.01	[KET01A]es
1946	192,563	58	0.03	[UN56]c
1960	232,333	336	0.14	[UN63]c
1980	237,609	773	0.33	[UN88]c
1990	240,636	1,047	0.44	[BB00]c
2000	241,822	1,657	0.69	[BB00]c
2010	223,417	1,605	0.72	[BB10]c
2050	277,000	3,000	1.10	es
2100	215,000	3,400	1.60	es

5.1.3 Caribbean Netherlands

These are three special municipalies of the Netherlands that were formed in 2010 after the disillusion of the Netherlands Antilles (1954–2010). It consists of three main islands: Bonaire, Sint Eustatius and Saba, also referred to by their acronym as BES Islands. The first is the largest and most inhabited, with 90% of the land and 80% of the population. It was conquered by the Dutch in 1634, has an area of 294 sq. km, comprising the main island of Bonaire and the uninhabited island of Little Bonaire (6 sq. km). The second, Sint Eustatius, also known as Statia for short, was conquered by the Dutch in 1636 and consists of one island with total area of 21 sq. km. The third island is Saba, was conquered by the Dutch in 1640 and consists of one island with total area of 13 sq. km. A map of the Caribbean Netherlands is presented in Fig. 5.1.3.

The first Muslims were recorded in the census of 1992 and all lived on Bonaire. Muslims arrived to Saba in affiliation with Saba University School of Medicine that was established in 1992. Based on census data as shown in Table 5.1.3, the Muslim population was nonexistent until towards the end of the twentieth century. It increased from increased from none in 1981 and before, to 26 or 0.2% in 1992, to 100 or 0.7% in 2001. A 2013 survey showed that 0.8% of Caribbean Netherlands is Muslim, their majority resided in Saba (5.9% of Saba's population). A 2018 survey showed that the fraction of Muslims was reduced to 0.5%, which is attributed to a cleanup of population register, where some people were registered as residents, but further investigation showed that they were no longer living on the islands. Thus, assuming that the percentage of Muslims will increase by a tenth of a percentage point per decade, then the Muslim population is expected to exceed 200 or 0.8% in 2050 and 400 or 1.3% by 2100.

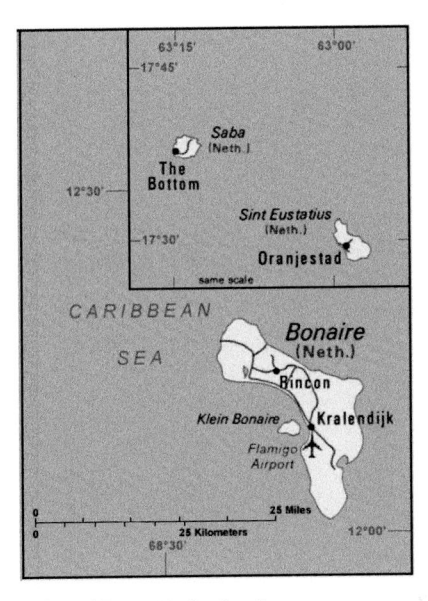

Figure 5.1.3 Map of Caribbean Netherlands.

Table 5.1.3a Evolution of the Muslim population in Caribbean Netherlands

Year	Population	Muslims	%	Source
1870	7,662	—	0.00	[SYB873]c
1891	7,502	—	0.00	[SYB894]c
1920	10,095	—	0.00	[SYB922]c
1971	10,300	—	0.00	[AN]c
1981	11,076	—	0.00	[AN]c
1992	13,156	26	0.26	[AN]c
2001	14,337	102	0.60	[AN]c
2013	23,296	187	0.80	[BQ]s
2018	25,052	125	0.50	[BQ]s
2050	29,706	240	0.80	es
2100	32,602	420	1.30	es

5.1.4 Curaçao

Land Curaçao is a constituent country within the Kingdom of the Netherlands. It was conquered by the Dutch in 1634 and seceded from the Netherlands Antilles in 2010 after its disillusion in the same year. It has an area of 444 sq. km, comprising the main island of Curaçao and the uninhabited island of Little Curaçao (2 sq. km). A map of Curaçao is presented in Fig. 5.1.4.

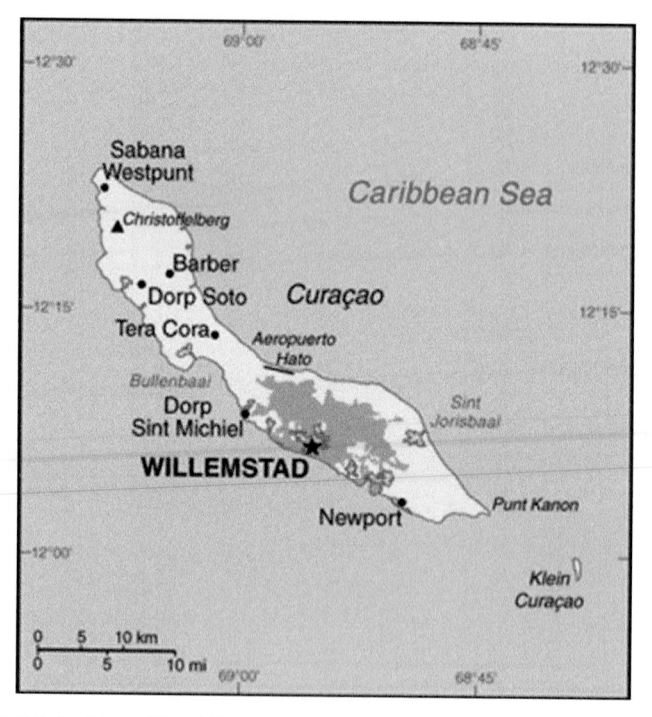

Figure 5.1.4 Map of Land Curaçao.

The first Muslims were Lebanese merchants and East Indians from Suriname. However, the Muslim population was nonexistent until the second half of the twentieth century. Based on census data as shown in Table 5.1.4; the Muslim population decreased from 530 or 0.4% in 1971, to 370 or 0.3% in 1981, to 352 or 0.2% in 1992, and then bounced back to 512 or 0.4% in 2001 and 751 or 0.5% in 2011. The decrease was due to social unrest that happened in the Island in 1969 which hurt the Muslim population financially. The Muslim population is expected to continue to increase by a tenth of a percentage point per decade but will remain less than 3,000 throughout this century.

Table 5.1.4 Evolution of the Muslim population in Curaçao

Year	Population	Muslims	%	Source
1870	21,089	—	0.00	[SYB873]c
1891	26,584	—	0.00	[SYB894]c
1920	32,709	—	0.00	[SYB922]c
1971	143,800	530	0.37	[AN]c
1981	147,388	370	0.25	[AN]c
1992	144,097	352	0.24	[AN]c
2001	130,304	512	0.39	[AN]c
2011	149,711	751	0.50	[CW]c
2050	176,000	1,600	0.90	es
2100	175,000	2,400	1.40	es

5.1.5 Dominica

The Commonwealth of Dominica consists of one island with an area of 751 sq. km and its map is presented in Fig. 5.1.5. It was conquered by the British in 1763 and gained its independence from the UK in 1978. According to census data as shown in Table 5.1.5, the Muslim population increased from none in 1911 and before, to two in 1946, to 54 or 0.07% in 1981 to 139 or 0.20% in 2001. Thus, assuming that the percentage of Muslims will increase by 0.05 of a percentage point each decade, the Muslim population is expected to remain in few hundreds throughout this century.

Table 5.1.5 Evolution of the Muslim population in Dominica

Year	Population	Muslims	%	Source
1871	27,178	—	0.00	[GB871]c
1911	33,863	—	0.00	[GB911]c
1946	47,624	2	0.00	[UN56]c
1981	72,947	54	0.07	[UN88]c
2001	69,055	142	0.21	[UN]c
2050	71,000	300	0.45	es
2100	51,000	400	0.70	es

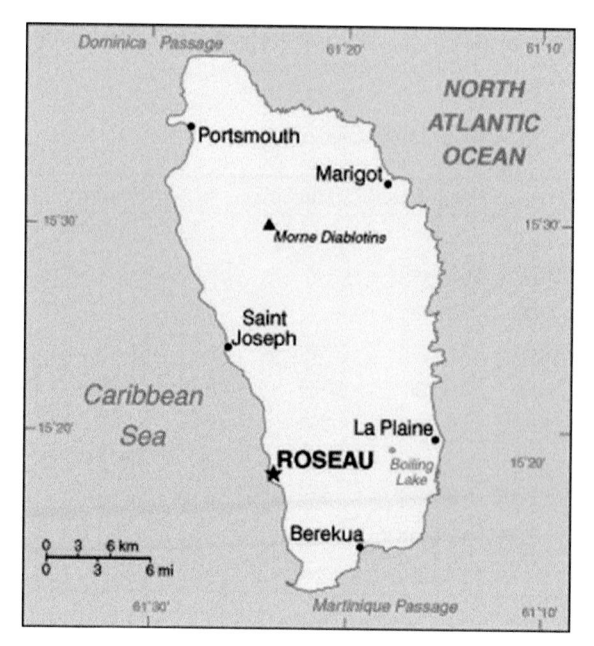

Figure 5.1.5 Map of the Commonwealth of Dominica.

5.1.6 Grenada

It has a total area of 344 sq. km and mainly consists of seven islands, four of which are small and uninhabited, and all are an extension to the south of the Grenadines Island Chain. The largest islands are Grenada (308 sq. km), Carriacou (27.5 sq. km), and Petite Martinique (3.7 sq. km). A map of these islands is presented in Fig. 5.1.6. These islands were conquered by the British in 1762 and gained its independence from the UK in 1974.

The first Muslims were indenture labor brought by the British from India in 1857. According to census data, the Muslim population increased from none in 1851 to 142 or 0.45% in 1861, to 152 or 0.40% in 1871, then decreased to 98 or 0.23% in 1881, to 78 or 0.15% in 1891, to 43 or 0.07% in 1901, to twenty or 0.03% in 1911, to nine or 0.02% in 1921, to one or 0.00% in 1946, it then increased as new Muslim immigrants started flowing in to 76 or 0.09% in 1981, to 237 or 0.23% in 2001, to 410 or 0.4% in 2011. In the Island of Carriacou, the Muslim population decreased from eight in 2001 to seven in 2011. Thus, assuming that the percentage of Muslims will increase by one tenth of a percentage point per decade, the Muslim population is expected to remain around one thousand throughout the second half of this century. A summary of the data is provided in Table 5.1.6.

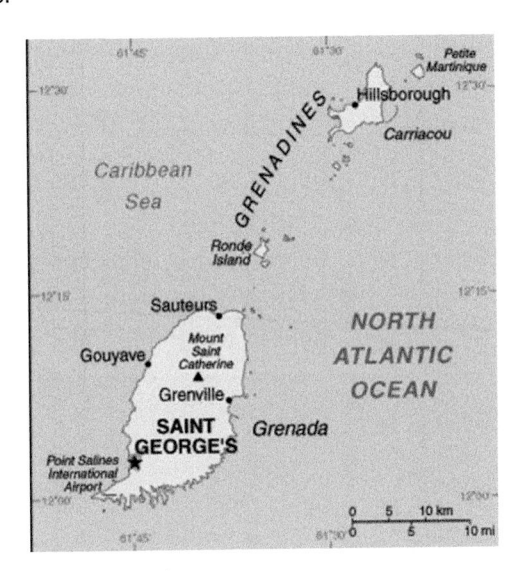

Figure 5.1.6   Map of Grenada.

Table 5.1.6   Evolution of the Muslim population in Grenada

Year	Population	Muslims	%	Source
1851	32,671	—	0.00	[GB861]e
1861	31,900	142	0.45	[GB861]c
1871	37,684	152	0.40	[BC]c
1881	42,403	98	0.23	[BC]c
1891	53,209	78	0.15	[BC]c
1901	63,438	43	0.07	[BC]c
1911	66,750	20	0.03	[BC]c
1921	66,302	9	0.02	[BC]c
1946	72,100	1	0.00	[UN56]c
1981	87,889	76	0.09	[UN88]c
2001	101,232	237	0.23	[GD11]c
2011	104,135	410	0.39	[GD11]c
2050	116,000	900	0.80	es
2100	86,000	1,100	1.30	es

5.1.7 Martinique

This is an Overseas Department of France which consists of one island with an area of 1,128 sq. km and its map is presented in Fig. 5.1.7. It was given its name by Columbus when he visited the Island in 1502. It was taken over by the French in 1635 as part of their "Compagnie des Iles d'Amerique" then became part of the French royal domain in 1674. The British attacked or conquered the Island in 1666, 1667, 1762–1763, 1793–1801 and 1809–1814. The French used the Island at first to cultivate cotton and tobacco, then sugarcane in 1650 and coffee in 1723. The original Arawaks and Caribs were exterminated and replaced by enslaved Africans whose numbers reached 60,000 in 1736. Some of those were Muslims but there is no record of the size and Islam was not preserved among their descendants.

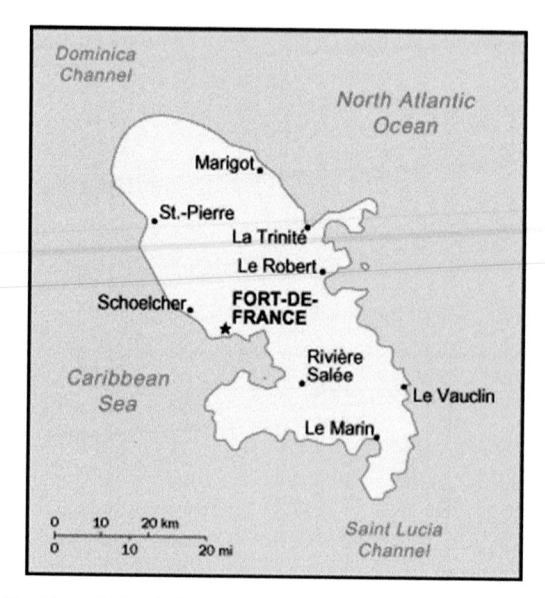

Figure 5.1.7 Map of Martinique.

Slavery was abolished in 1848 and in 1851 France started bringing indentured laborers here from its colonies in India. Then in 1861, France signed a convention with the British to bring more workers from the British Indian territories to its Caribbean possessions. This Indian immigration lasted 32 years until 1883, bringing a total of 25,509 Indians to Martinique. Of these 11,077

Indians returned home after the expiration of their contracts and 14,432 Indians remained in Martinique, which was further reduced to 4,665 in 1900 including their descendants. It is estimated that 14.5% of these numbers were Muslims. In other words, a total of 3,700 Muslims were brought from India over the 32-year period, of whom 660 remained by 1900, but by 1950 no Muslim Indians were present in Martinique, either because of their departure of the Island, or because Islam was not preserved among their children.

In October 1780, Martinique and its neighboring islands were hit by the Great Hurricane of 1780 which claimed more than 20,000 lives in total, 9,000 in Martinique and destroyed its former capital St. Pierre. In 1902, Mount Pelée volcano erupted in the north of the Island, destroying the former capital St Pierre and claiming more than 30,000 lives. The Capital was abandoned as a result and the Volcano erupted again in 1929.

The third wave of Muslim migration started in 1920s from Palestine and the fourth in 1970s from Senegal and Mali. By 1982, the number of Muslims was 500 or 0.15% in 1982. The number then increased to 1,000 or 0.25% in 2010. Thus, assuming that the percentage of Muslims will increase by 0.05 of a percentage point each decade, the Muslim population is expected to remain less than 2,000 throughout this century. A summary of the data is provided in Table 5.1.7.

Table 5.1.7 Evolution of the Muslim population in Martinique

Year	Population	Muslims	%	Source
1850	129,681	—	0.00	[KET01A]es
1900	203,781	660	0.32	[KET01A]es
1982	328,566	500	0.15	[KET86]es
2010	394,173	1,000	0.25	[PEW]es
2050	332,000	1,500	0.45	es
2100	234,000	1,600	0.70	es

5.1.8 Saint Lucia

It consists of one island with an area of 616 sq. km and its map is presented in Fig. 5.1.8. It was conquered by the British in 1814 and gained its independence from the UK in 1979. The first Muslims were indenture labors brought by the British from India in 1859. Between 1859 and 1860; 1,200 indenture labors were brought from India and the 1891 census shows that Muslims make up just over 10% of East Indians.

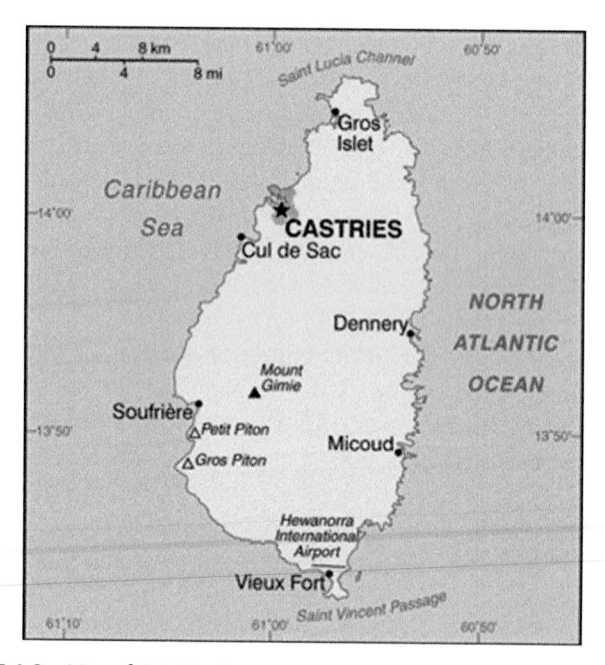

Figure 5.1.8 Map of Saint Lucia.

Thus, based on census data as illustrated in Table 5.1.8, the Muslim population increased from none in 1859, to 131 or 0.4% in 1871, to 199 or 0.5% in 1891, but then continued to decrease to 162 or 0.32% in 1901, to 67 or 0.14% in 1911, to 45 or 0.09% in 1921, to ten or 0.01% in 1946. It then gradually grew to 25 or 0.02% in 1980, then 81 or 0.06% in 1991, then 222 or 0.14% in 2001, but decreased to 174 or 0.11% in 2010. Thus, assuming that the percentage of Muslims will increase by 0.02 of a percentage point each decade, the Muslim population is expected to remain in few hundreds throughout this century.

Table 5.1.8 Evolution of the Muslim population in Saint Lucia

Year	Population	Muslims	%	Source
1851	24,290	—	0.00	[GB861]e
1871	31,610	131	0.41	[GB871]c
1891	42,220	199	0.47	[BC]c
1901	49,883	162	0.32	[GB901]c
1911	48,637	67	0.14	[BC]c
1921	51,505	45	0.09	[BC]c
1946	78,900	10	0.01	[UN56]c
1980	112,525	25	0.02	[UN88]c
1991	132,586	81	0.06	[UN04]c
2001	154,695	222	0.14	[UN]c
2010	163,010	174	0.11	[LC]c
2050	182,000	300	0.18	es
2100	122,000	350	0.28	es

5.1.9 Saint Vincent and the Grenadines

It has a total area of 389 sq. km and consists of the largest island of Saint Vincent (344 sq. km), and over a score of smaller islands referred to as the Grenadines. Some of these Cays belong to the country of Grenada to the south. The islands of Saint Vincent and the Grenadines were conquered by the British in 1783 and gained their independence from the UK in 1979. A map of this island chain is presented in Fig. 5.1.9.

The first Muslims were indenture labor brought by the British from India in 1861. The labor force consisted of 263 individuals, out of which fifteen were Muslim. The number of Muslims decreased to fourteen or 0.03% in 1911, then two in 1921, to one in 1931 and 1946. It then bounced back to sixteen or 0.02% in 1980, to 77 or 0.07% in 2001 and 111 or 0.1% in 2012. The corresponding censuses also show that all Muslims live in the island of St. Vincent and none in the Grenadines. A summary of the data is provided in Table 5.1.9. Thus, assuming that the percentage of Muslims will continue to increase by 0.03 of a percentage point per decade, the Muslim population is expected to remain in few hundreds throughout this century.

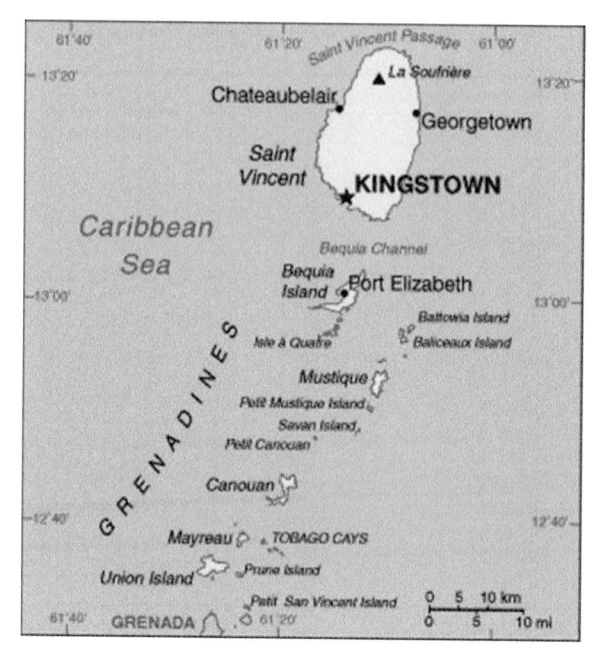

Figure 5.1.9 Map of Saint Vincent and the Grenadines.

Table 5.1.9 Evolution of the Muslim population in Saint Vincent and the Grenadines

Year	Population	Muslims	%	Source
1851	27,200	—	0.00	[GB861]e
1861	31,755	15	0.05	[GB861]c
1911	41,877	14	0.03	[GB911]c
1921	44,447	2	0.00	[BC]c
1931	47,961	1	0.00	[BC]c
1946	61,700	1	0.00	[UN56]c
1980	96,864	16	0.02	[UN88]c
2001	104,623	77	0.07	[VC01]c
2012	104,092	111	0.11	[VC12]c
2050	109,000	250	0.23	es
2100	75,000	300	0.38	es

5.1.10 Trinidad and Tobago

The Republic of Trinidad and Tobago has a total area of 5,128 sq. km and its map is presented in Fig. 5.1.10. It consists of two main islands: Trinidad (4,768 sq. km with 96% of the population) and Tobago (300 sq. km with 4% of the population) and over a score of smaller uninhabited islands. The British conquered Trinidad in 1797 and Tobago in 1802 and they both gained their independence from the UK in 1962. Currently, the islands have the third largest percentage of Muslims in the Americas, after Suriname and Guyana. The first Muslims were indenture labor brought by the British from India in 1844. Inquiry on religious adherence in the national census is available since 1891. Censuses since 1851 also include ethnic affiliation. In 1891, almost all Muslims arrived from India, and made little more than 12% of the total East Indian population. We use this percentage to get an estimate of the Muslim population prior to 1891.

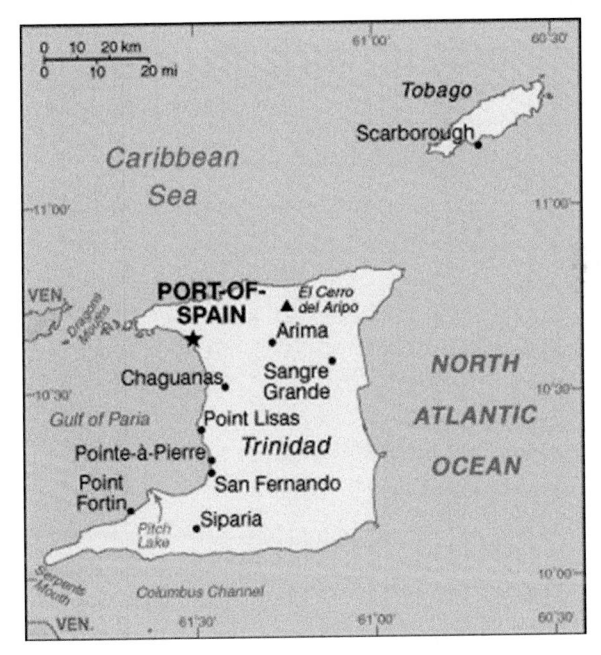

Figure 5.1.10 Map of Republic of Trinidad and Tobago.

Thus, there were no Muslims in the islands in 1844, when the first census was conducted. Soon after, thousands of indentured laborers were brought from India. Accordingly, the Muslim population increased from 479 or 0.6% in 1851, to 1,619 or 1.6% in 1861, to 3,291 or 2.6% in 1871, to 5,858 or 3.4% in 1881, to 8,638 or 4.0% in 1891, to 11,478 or 3.8% in 1901, to 14,957 or 4.5% in 1911, to 17,698 or 4.9% in 1921, to 20,992 or 5.1% in 1931, to 32,615 or 5.9% in 1946, to 49,736 or 6.0% in 1960, to 58,271 or 6.3% in 1970, to 63,333 or 6.1% in 1980, to 65,732 or 5.9% in 1990, then decreased slightly to 64,648 or 5.9% in 2000, then increased slightly to 65,705 or 5.6% in 2011.

The vast majority of the Muslim population lives in the Island of Trinidad. According to census data and as shown in Table 5.1.10b, the number of Muslims in Tobago increased from none in 1901, to 65 or 0.2% in 1960, to 98 or 0.3% in 1970, to 163 or 0.3% in 1990, to 233 or 0.5% in 2000, to 345 or 0.6% in 2011.

Table 5.1.10a Evolution of the Muslim population in Trinidad and Tobago

Year	Population	Muslims	%	Source
1844	74,193	—	0.00	[TTH]e
1851	82,978	479	0.58	[TTH]e
1861	99,848	1,619	1.62	[TTH]e
1871	126,692	3,291	2.60	[TTH]e
1881	171,179	5,858	3.42	[TTH]e
1891	218,381	8,638	3.95	[TTH]c
1901	273,899	11,478	3.83	[TTH]c
1911	333,552	14,957	4.48	[TTH]c
1921	365,913	17,698	4.84	[TTH]c
1931	412,783	20,992	5.09	[TTH]c
1946	557,288	32,615	5.85	[TTH]c
1960	827,957	49,736	6.01	[TTH]c
1970	931,071	58,271	6.26	[TTH]c
1980	1,045,042	63,333	6.06	[UN88]c
1990	1,114,395	65,732	5.90	[TT00]c
2000	1,099,602	64,648	5.88	[TT00]c
2011	1,175,749	65,705	5.59	[TT11]c
2050	1,344,000	70,000	5.20	es
2100	969,000	46,000	4.70	es

Thus, following census data and assuming that the percentage of Muslims will continue to decrease by 0.05 of a percentage point per decade, the Muslim population is expected to reach 70,000 or 5.2% by 2050 and 46,000 or 4.7% by 2100. The data is summarized in Table 5.1.10a.

Table 5.1.10b   Evolution of the Muslim population in the Island of Tobago

Year	Population	Muslims	%	Source
1901	18,751	—	0.00	[GB01]c
1960	33,333	65	0.20	[TT90]c
1970	38,754	98	0.25	[TT90]c
1990	48,600	163	0.34	[TT90]c
2000	43,257	233	0.54	[TT00]c
2011	54,764	345	0.63	[TT11]c

5.1.11   Regional Summary and Conclusion

Muslims arrived here as explorers, centuries before the arrival of the Europeans, then as escapees from Spanish Catholic oppression, then as black slaves brought by the Europeans from Africa. However, Islam went extinct from these migrations. Towards the end of the nineteenth century, the Brits and Dutch brought workers from India and Java, some of whom were Muslim, and Islam remains among their descendants till this day. The French also brought Muslims from their African colonies to their colonies in this region. This region has the highest concentration of Muslims in the Americas and is expected to remain just below 3% throughout this century. The following tables present centennial data from 600 AD to 2100 AD (or approximately 1 H to 1500 H) in Table 5.1a and decennial data from 1790 AD to 2100 AD (or 1210 H to 1520 H) in Tables 5.1b and 5.1c for current countries in Southern Caribbean Islands. The data includes total population in thousands (P), the percentage of which is Muslim (M%), the corresponding Muslim population in thousands (M), and the annual population growth rate (APGR, or G%) of the total population in this region.

Table 5.1a Centennial estimates of the Muslim population (×1000) in Southern Caribbean Islands: 600–2100 AD (1–1500 H)

		600	700	800	900	1000	1100	1200	1300
Aruba	P	—	—	—	—	—	—	—	—
	M%	—	—	—	—	—	—	—	—
	M	—	—	—	—	—	—	—	—
Barbados	P	—	—	—	—	—	—	—	—
	M%	—	—	—	—	—	—	—	—
	M	—	—	—	—	—	—	—	—
Caribbean Netherlands	P	—	—	—	—	—	—	—	—
	M%	—	—	—	—	—	—	—	—
	M	—	—	—	—	—	—	—	—
Curaçao	P	—	—	—	—	—	—	—	—
	M%	—	—	—	—	—	—	—	—
	M	—	—	—	—	—	—	—	—
Dominica	P	—	—	—	1	2	2	2	3
	M%	—	—	—	—	—	—	—	—
	M	—	—	—	—	—	—	—	—
Grenada	P	—	—	—	—	—	—	—	—
	M%	—	—	—	—	—	—	—	—
	M	—	—	—	—	—	—	—	—
Martinique	P	1	1	1	1	1	2	2	2
	M%	—	—	—	—	—	—	—	—
	M	—	—	—	—	—	—	—	—
St. Lucia	P	—	—	—	1	2	2	2	3
	M%	—	—	—	—	—	—	—	—
	M	—	—	—	—	—	—	—	—
SVG	P	—	—	—	1	2	2	2	3
	M%	—	—	—	—	—	—	—	—
	M	—	—	—	—	—	—	—	—
Trinidad &Tobago	P	5	6	6	7	7	8	8	9
	M%	—	—	—	—	—	—	—	—
	M	—	—	—	—	—	—	—	—
Total	P	7	8	8	11	14	17	17	21
	M%	—	—	—	—	—	—	—	—
	M	—	—	—	—	—	—	—	—
	G%		0.128	0.000	0.291	0.257	0.146	0.024	0.230

Table 5.1a (*Continued*)

		1400	1500	1600	1700	1800	1900	2000	2100
Aruba	P	1	1	1	1	2	10	91	102
	M%	—	—	—	—	—	—	0.33	0.75
	M	—	—	—	—	—	—	—	1
Barbados	P	—	—	—	150	88	196	272	215
	M%	—	—	—	—	—	—	0.69	1.60
	M	—	—	—	—	—	—	2	3
Caribbean Netherlands	P	—	1	1	1	3	8	14	33
	M%	—	—	—	—	—	—	0.60	1.30
	M	—	—	—	—	—	—	—	—
Curaçao	P	1	1	1	2	14	31	132	175
	M%	—	—	—	—	—	—	0.39	1.40
	M	—	—	—	—	—	—	1	2
Dominica	P	3	3	1	1	18	29	70	51
	M%	—	—	—	—	—	—	0.21	0.70
	M	—	—	—	—	—	—	—	—
Grenada	P	—	1	1	1	25	63	103	86
	M%	—	—	—	—	—	0.07	0.26	1.30
	M	—	—	—	—	—	—	—	1
Martinique	P	2	3	3	24	95	204	387	234
	M%	—	—	—	—	—	0.32	0.25	0.70
	M	—	—	—	—	—	1	1	2
St. Lucia	P	3	3	1	3	16	50	157	122
	M%	—	—	—	—	—	0.32	0.14	0.28
	M	—	—	—	—	—	—	—	—
SVG	P	3	3	1	3	15	42	108	75
	M%	—	—	—	—	—	0.05	0.07	0.38
	M	—	—	—	—	—	—	—	—
Trinidad &Tobago	P	9	10	4	10	30	274	1,267	969
	M%	—	—	—	—	—	3.83	5.88	4.70
	M	—	—	—	—	—	10	75	46
Total	P	22	24	13	196	306	906	2,600	2,062
	M%	—	—	—	—	—	1.26	3.04	2.73
	M	—	—	—	—	—	11	79	56
	G%	0.019	0.104	−0.653	2.743	0.440	1.088	1.057	−0.232

Table 5.1b Decennial estimates of the Muslim population (×1000) in Southern Caribbean Islands: 1790–1940 AD (1210–1360 H)

		1790	1800	1810	1820	1830	1840	1850	1860
Aruba	P	2	2	2	2	3	2	3	3
	M%	—	—	—	—	—	—	—	—
	M	—	—	—	—	—	—	—	—
Barbados	P	87	88	90	92	102	122	136	153
	M%	—	—	—	—	—	—	—	—
	M	—	—	—	—	—	—	—	—
Caribbean Netherlands	P	3	3	4	4	5	5	5	6
	M%	—	—	—	—	—	—	—	—
	M	—	—	—	—	—	—	—	—
Curaçao	P	9	11	11	12	15	12	16	19
	M%	—	—	—	—	—	—	—	—
	M	—	—	—	—	—	—	—	—
Dominica	P	16	18	26	20	19	19	22	25
	M%	—	—	—	—	—	—	—	—
	M	—	—	—	—	—	—	—	—
Grenada	P	28	25	22	28	29	30	33	32
	M%	—	—	—	—	—	—	—	0.45
	M	—	—	—	—	—	—	—	0
Martinique	P	97	95	92	90	110	120	130	136
	M%	—	—	—	—	—	—	—	0.01
	M	—	—	—	—	—	—	—	—
St. Lucia	P	15	16	17	16	13	21	25	27
	M%	—	—	—	—	—	—	—	0.41
	M	—	—	—	—	—	—	—	—
SVG	P	10	15	19	25	22	27	27	32
	M%	—	—	—	—	—	—	—	0.05
	M	—	—	—	—	—	—	—	—
Trinidad & Tobago	P	16	30	47	39	32	73	83	100
	M%	—	—	—	—	—	—	0.58	1.62
	M	—	—	—	—	—	—	—	2
Total	P	288	306	333	332	350	429	478	533
	M%	—	—	—	—	—	—	0.10	0.36
	M	—	—	—	—	—	—	—	2
	G%		0.683	0.854	−0.034	0.611	2.097	1.042	1.081

Table 5.1b (*Continued*)

		1870	1880	1890	1900	1910	1920	1930	1940
Aruba	P	4	6	7	10	9	8	13	31
	M%	—	—	—	—	—	—	—	—
	M	—	—	—	—	—	—	—	—
Barbados	P	162	172	183	196	172	157	159	179
	M%	—	—	—	—	—	0.01	0.01	0.01
	M	—	—	—	—	—	—	—	—
Caribbean Netherlands	P	8	10	8	8	10	10	8	7
	M%	—	—	—	—	—	—	—	—
	M	—	—	—	—	—	—	—	—
Curaçao	P	21	25	27	31	33	33	50	67
	M%	—	—	—	—	—	—	—	—
	M	—	—	—	—	—	—	—	—
Dominica	P	27	28	27	29	34	37	40	45
	M%	—	—	—	—	—	—	—	—
	M	—	—	—	—	—	—	—	—
Grenada	P	38	42	53	63	67	66	75	88
	M%	0.40	0.23	0.15	0.07	0.03	0.02	0.02	0.01
	M	—	—	—	—	—	—	—	—
Martinique	P	153	162	181	204	184	244	235	247
	M%	0.10	0.20	0.30	0.32	0.30	0.20	0.10	0.10
	M	—	—	1	1	1	—	—	—
St. Lucia	P	32	39	42	50	49	52	57	71
	M%	0.41	0.41	0.47	0.32	0.14	0.09	0.09	0.01
	M	0	0	0	0	0	0	0	0
SVG	P	36	41	41	42	42	44	48	57
	M%	0.05	0.05	0.05	0.05	0.03	0.00	0.00	0.00
	M	—	—	—	—	—	—	—	—
Trinidad & Tobago	P	127	171	218	274	334	366	413	492
	M%	2.60	3.42	3.95	3.83	4.48	4.84	5.09	5.84
	M	3	6	9	10	15	18	21	29
Total	P	607	696	787	906	933	1,018	1,098	1,284
	M%	0.62	0.93	1.20	1.26	1.67	1.80	1.94	2.26
	M	4	6	9	11	16	18	21	29
	G%	1.303	1.360	1.257	1.412	0.298	0.877	0.765	1.573

Table 5.1c Decennial estimates of the Muslim population (×1000) in Southern Caribbean Islands: 1950–2100 AD (1370–1520 H)

		1950	1960	1970	1980	1990	2000	2010	2020
Aruba	P	38	54	59	60	62	91	102	107
	M%	—	—	0.03	0.03	0.33	0.33	0.32	0.35
	M	—	—	—	—	—	—	—	—
Barbados	P	211	231	239	252	261	272	282	287
	M%	0.03	0.14	0.20	0.33	0.44	0.69	0.72	0.80
	M	—	—	—	1	1	2	2	2
Caribbean Netherlands	P	7	8	10	11	13	14	21	26
	M%	—	—	—	—	0.26	0.60	0.80	0.50
	M	—	—	—	—	—	—	—	—
Curaçao	P	100	127	144	148	147	132	149	164
	M%	—	0.10	0.37	0.25	0.24	0.39	0.50	0.60
	M	—	—	1	—	—	1	1	1
Dominica	P	51	60	71	75	70	70	71	72
	M%	0.01	0.01	0.07	0.07	0.07	0.21	0.25	0.30
	M	—	—	—	—	—	—	—	—
Grenada	P	77	90	94	89	96	103	106	113
	M%	0.01	0.01	0.09	0.09	0.09	0.23	0.39	0.50
	M	—	—	—	—	—	—	—	1
Martinique	P	222	282	325	325	358	387	395	375
	M%	0.10	0.10	0.15	0.15	0.25	0.25	0.25	0.30
	M	—	—	—	—	1	1	1	1
St. Lucia	P	83	90	104	118	138	157	174	184
	M%	0.01	0.01	0.02	0.02	0.06	0.14	0.11	0.12
	M	—	—	—	—	—	—	—	—
SVG	P	67	81	91	101	107	108	108	111
	M%	0.00	0.00	0.00	0.02	0.02	0.07	0.11	0.14
	M	—	—	—	—	—	—	—	0.155
Trinidad & Tobago	P	646	848	945	1,085	1,221	1,267	1,328	1,399
	M%	5.85	6.01	6.26	6.06	5.90	5.88	5.59	5.50
	M	38	51	59	66	72	75	74	77
Total	P	1,502	1,870	2,082	2,265	2,475	2,600	2,736	2,838
	M%	2.54	2.77	2.93	2.99	3.03	3.04	2.90	2.93
	M	38	52	61	68	75	79	79	83
	G%	1.568	2.200	1.074	0.839	0.886	0.492	0.504	0.377

Table 5.1c (*Continued*)

		2030	2040	2050	2060	2070	2080	2090	2100
Aruba	P	110	111	109	107	107	106	104	102
	M%	0.40	0.45	0.50	0.55	0.60	0.65	0.70	0.75
	M	—	—	1	1	1	1	1	1
Barbados	P	289	287	277	265	252	240	227	215
	M%	0.90	1.00	1.10	1.20	1.30	1.40	1.50	1.60
	M	3	3	3	3	3	3	3	3
Caribbean Netherlands	P	28	29	30	30	31	31	32	33
	M%	0.60	0.70	0.80	0.90	1.00	1.10	1.20	1.30
	M	—	—	—	—	—	—	—	—
Curaçao	P	171	175	176	176	177	177	176	175
	M%	0.70	0.80	0.90	1.00	1.10	1.20	1.30	1.40
	M	1	1	2	2	2	2	2	2
Dominica	P	73	73	71	68	65	60	55	51
	M%	0.35	0.40	0.45	0.50	0.55	0.60	0.65	0.70
	M	—	—	—	—	—	—	—	—
Grenada	P	116	116	116	112	107	100	93	86
	M%	0.60	0.70	0.80	0.90	1.00	1.10	1.20	1.30
	M	1	1	1	1	1	1	1	1
Martinique	P	368	356	332	306	286	269	253	234
	M%	0.35	0.40	0.45	0.50	0.55	0.60	0.65	0.70
	M	1	1	1	2	2	2	2	2
St. Lucia	P	189	188	182	173	161	147	133	122
	M%	0.14	0.16	0.18	0.20	0.22	0.24	0.26	0.28
	M	—	—	—	—	—	—	—	—
SVG	P	113	112	109	104	98	90	82	75
	M%	0.17	0.20	0.23	0.26	0.29	0.32	0.35	0.38
	M	—	—	—	—	—	—	—	—
Trinidad & Tobago	P	1,413	1,392	1,344	1,273	1,191	1,109	1,037	969
	M%	5.40	5.30	5.20	5.10	5.00	4.90	4.80	4.70
	M	76	74	70	65	60	54	50	46
Total	P	2,870	2,838	2,745	2,613	2,472	2,330	2,193	2,062
	M%	2.91	2.88	2.86	2.84	2.81	2.77	2.75	2.73
	M	83	82	78	74	69	64	60	56
	G%	0.112	−0.113	−0.336	−0.491	−0.556	−0.594	−0.606	−0.618

5.2 Muslims in North America

This region consists of five countries and territories: Bermuda, Canada, Greenland, Saint Pierre and Miquelon, and the United States of America. In the nineteenth century, the Muslim population consisted of small number of immigrants and converts. During the first half of the twentieth century, the Muslim population picked up momentum as more immigrants poured in and more African Americans started returning to Islam. In the 1960s, immigration laws in US and Canada changed to welcome immigrants from all over the world and not just Whites. Since then the number of Muslims is fast growing due to immigration, but also higher birthrate and conversions. Thus, the Muslim population has likely increased from none in 1800, to 23,000 or 0.03% in 1900, to 2.1 million or 0.7% in 2000, to 5.8 million or 1.6% in 2020, and is projected to reach eleven million or 2.6% by 2050 and 22 million or 4.5% by 2100.

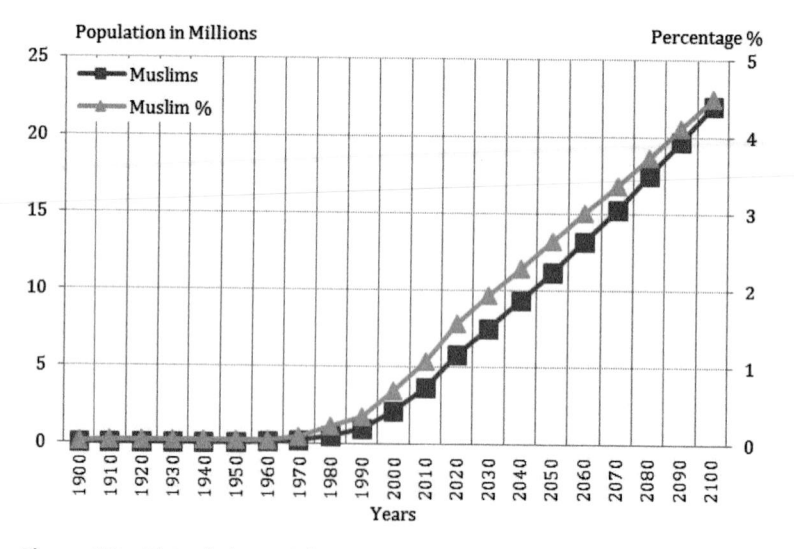

Figure 5.2 Plot of decennial estimates of the Muslim population and its percentage of the total population in North America: 1900–2100 AD (1320–1520 H).

A plot of decennial estimates of the Muslim population and its percentage with respect to the total population in this region from 1900 to 2100 is provided in Fig. 5.2. The corresponding individual data for each country in this region is discussed below. In Section 5.2.6, the total population in each country in this region and the corresponding percentage and number of Muslims is presented centennially in Table 5.2a from 600 to 2100 and decennially in Tables 5.2b and 5.2c from 1790 to 2100.

5.2.1 Bermuda

It has an area of 54 sq. km and consists of seven main islands surrounded by more than three hundred much smaller islands, only twenty are inhabited. Its map is presented in Fig. 5.2.1. It derives its name from J. Bermudez, a Spaniard who first sighted the islands in 1522, who found them uninhabited. The Islands were conquered by the British in 1609 and remain an overseas territory of the UK.

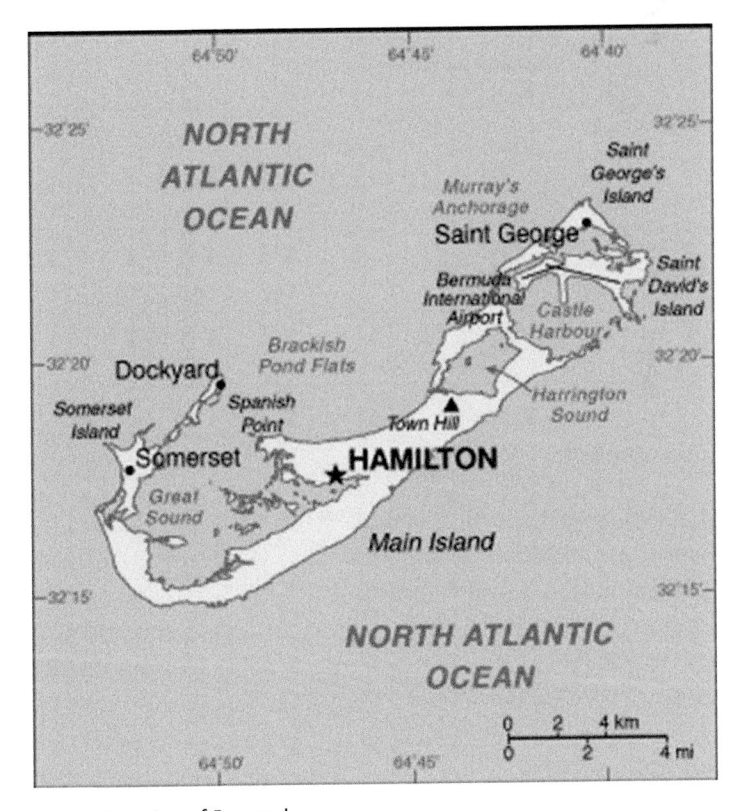

Figure 5.2.1 Map of Bermuda.

Islam started when many of African descent converted to Islam starting in the 1950s, influenced by the movement of Nation of Islam in the United States. According to census data as shown in Table 5.2.1, the Muslim population increased from none in 1911 and before, to less than fourteen or 0.03% in 1950, to 393 or 0.73% in 1980 to 489 or 0.8% in 1991, to 604 or 1.0% in 2000, to 616 or 1.0% in 2010. Thus, assuming that the percentage of Muslims will continue to increase by 0.05 of a percentage point per decade, the Muslim population is expected to remain around 700 throughout this century, reaching 1.5% by 2100.

Table 5.2.1   Evolution of the Muslim population in Bermuda

Year	Population	Muslims	%	Source
1871	12,121	—	0.00	[GB871]c
1901	20,961	—	0.00	[GB901]c
1911	23,839	—	0.00	[GB911]c
1950	36,263	<14	<0.03	[UN56]c
1980	53,547	393	0.73	[UN88]c
1991	58,094	489	0.84	[UN]c
2000	61,449	604	0.98	[UN]c
2010	60,279	616	1.02	[BM]c
2050	53,000	700	1.25	es
2100	37,000	600	1.50	es

5.2.2   Canada

It has an area of 9,984,670 sq. km, and is the second largest country in the world, after Russia. It consists of ten provinces: Alberta (661,848 sq. km or 6.6%), British Columbia (944,735 sq. km or 9.5%), Manitoba (647,797 sq. km or 6.5%), New Brunswick (72,908 sq. km or 0.7%), Newfoundland and Labrador (405,212 sq. km or 4.1%), Nova Scotia (55,284 sq. km or 0.6%), Nunavut, Quebec (1,542,056 sq. km or 15.4%), Saskatchewan (651,036 sq. km or 6.5%), Ontario (1,076,395 sq. km or 10.8%), and Prince Edward Island (5,660 sq. km or 0.1%), and three territories: Northwest Territories (1,346,106 sq. km or 13.5%), Nunavut (2,093,190 sq. km or 21.0%), and Yukon (482,443 sq. km or 4.8%). A map of Canada is presented in Fig. 5.2.2.

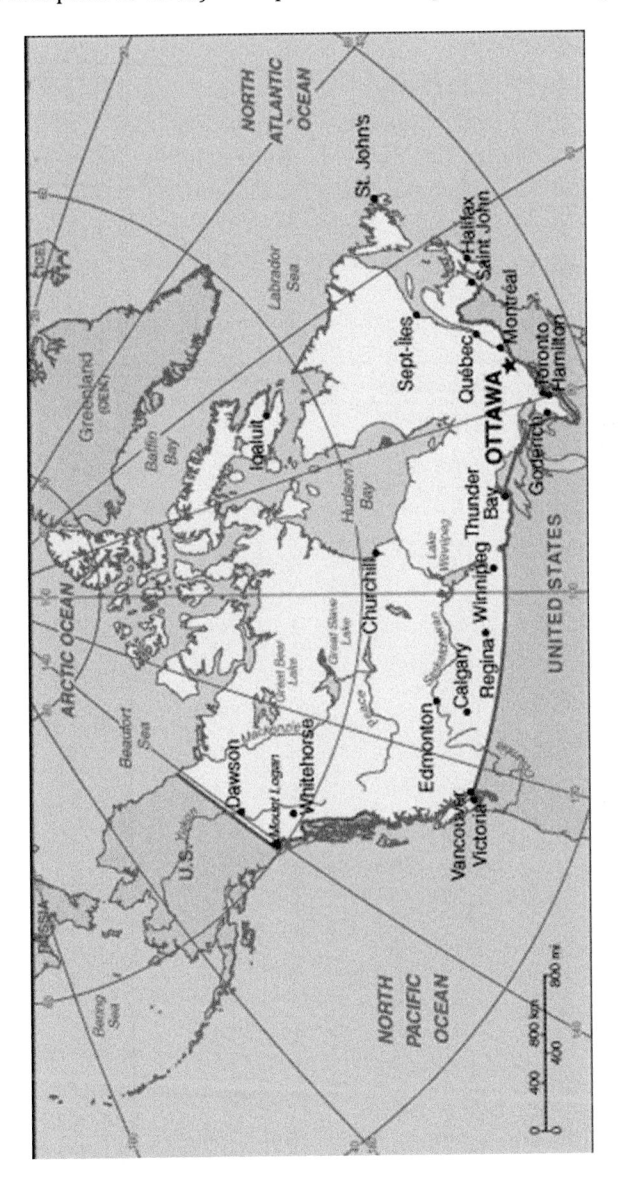

Figure 5.2.2 Map of Canada.

The number of Muslims increased from none in 1851 and 1861, to thirteen in Ontario in 1871, to 47 in 1901 (six in British Columbia, one in Manitoba, fifteen in Ontario and fifteen in Yukon), to 797 in 1911, to 478 in 1921 (63 in Alberta, 82 in British Columbia, 31 in Manitoba, ten in New Brunswick, 40 in Nova Scotia, 77 in Ontario, 31 in Quebec and 144 in Saskatchewan), to 645 in 1931 (126 in Alberta, 136 in British Columbia, 36 in Manitoba, 9 in New Brunswick, 37 in Nova Scotia, 55 in Ontario, 45 in Quebec, and 193 in Saskatchewan). The data for all Canada is summarized in Table 5.2.2a and data for each Canadian province and territory is summarized in Table 5.2.2b.

Table 5.2.2a Evolution of the Muslim population in Canada

Year	Population	Muslims	%	Source
1851	2,046,477	—	0.00	[CA871]c
1861	3,073,881	—	0.00	[CA871]c
1871	3,468,706	13	0.00	[CA871]c
1901	5,418,663	47	0.00	[CA]c
1911	7,221,662	797	0.00	[CA]c
1921	8,800,249	478	0.00	[CA]c
1931	10,376,379	645	0.00	[CA]c
1951	14,009,429	1,800	0.01	[CA]es
1961	18,238,247	5,800	0.03	[CA]es
1971	21,568,310	33,430	0.16	[CA]es
1976	22,992,605	69,300	0.30	[CA]es
1981	24,343,180	98,160	0.40	[CA]c
1991	27,296,859	253,260	0.93	[CA]c
2001	30,007,094	579,640	1.96	[CA01]c
2011	32,852,325	1,053,945	3.21	[CA11]c
2050	45,669,000	3,654,000	8.00	es
2100	56,953,000	7,973,000	14.00	es

Thus, the Muslim population increased from single digit in 1951, to double digits between 1871 and 1901, to triple digits between 1911 and 1931, but remained at 0.00% of the total population. After WWII, the estimated Muslim population increased to 1,800 or 0.01%

in 1951, and 5,800 or 0.03% in 1961. After the Immigration Act of 1967, which introduced the Points System, Canada became more welcoming to people from all over the world, and not only Europeans, as long as they will benefit the country. Thus, the estimated Muslim population jumped to 33,430 or 0.2% in 1971, to 69,300 or 0.3% in 1976, and based on census data to 98,160 or 0.4% in 1981, to 253,260 or 0.9% in 1991, to 579,640 or 2.0% in 2001, to 1.05 million or 3.2% in 2011. The 2001 was the last census to collect information on religious adherence, which started being collected in 2011 as national survey. Muslims are found in every Canadian province and territory, but over half of Canadian Muslims live in Ontario (55%), over a fifth live in Quebec (23%), over a tenth live in Alberta (11%) and almost a tenth live in British Columbia (8%).

Thus, assuming that the Muslim population will continue to increase by 1.2 of a percentage point per decade, then the Muslim population is expected to increase to close to four million or 8.0% by 2050 and eight million or 14.0% by 2100.

5.2.3   Greenland

Is a north Arctic island with area 2,166,086 sq. km with more than four-fifth of which is covered by ice sheet throughout the year and it is the largest island in the world (other than continents) and is a self-governing overseas administrative division of Denmark since 1979. Its map is presented in Fig. 5.2.3.

Yearly citizenship data of the population is provided by the Statistics Greenland since 1977. Table 5.2.3a shows the populations of citizens from Muslim majority countries since 1976. This data is constructed to estimate the Muslim population in Greenland. Accordingly, and as shown in Table 5.2.3b, the Muslim population in this island remains less than ten or 0.02% of the total population. These Muslims migrate temporarily from Denmark for work or educational training. Thus, assuming that the percentage of Muslims will increase by 0.01 of a percentage point per decade, the Muslim population is expected to remain less than fifty or 0.1% of the total population throughout this century.

Table 5.2.2b Evolution of the Muslim population in Canada per province and territory. P: Total population, M: Muslim population, M%: Percentage of Muslim population, MR%: Muslim Ratio

		1971	1981	1991	2001	2011
Alberta	P	1,627,875	2,237,724	2,545,553	2,974,807	3,567,980
	M%	0.14	0.75	1.22	1.65	3.18
	M	2,310	16,865	31,000	49,045	113,445
	MR%	6.91	17.18	12.24	8.46	10.76
British Columbia	P	2,184,620	2,744,467	3,282,061	3,907,738	4,324,455
	M%	0.06	0.46	0.76	1.44	1.83
	M	1,335	12,715	24,930	56,220	79,310
	MR%	3.99	12.95	9.84	9.70	7.53
Manitoba	P	988,245	1,026,241	1,091,942	1,119,583	1,174,345
	M%	0.06	0.19	0.32	0.46	1.06
	M	590	1,925	3,520	5,095	12,405
	MR%	1.76	1.96	1.39	0.88	1.18
New Brunswick	P	634,560	696,403	723,900	729,498	735,835
	M%	0.03	0.05	0.03	0.17	0.36
	M	220	315	250	1,275	2,640
	MR%	0.66	0.32	0.10	0.22	0.25
Newfoundland & Labrador	P	522,100	567,681	568,475	512,930	507,270
	M%	0.01	0.02	0.05	0.12	0.24
	M	70	100	300	630	1,200
	MR%	0.21	0.10	0.12	0.11	0.11
Northwest Territories	P	20,000	30,000	36,564	37,360	40,800
	M%	0.10	0.07	0.12	0.48	0.67
	M	20	20	45	180	275
	MR%	0.06	0.02	0.02	0.03	0.03
Nova Scotia	P	788,965	847,442	899,942	908,007	906,175
	M%	0.07	0.09	0.16	0.39	0.94
	M	570	790	1,435	3,550	8,500
	MR%	1.71	0.80	0.57	0.61	0.81
Nunavut	P	14,805	15,740	21,085	26,505	31,695
	M%	—	—	0.05	0.09	0.16
	M	—	—	10	25	50
	MR%	—	—	0.00	0.00	0.00

Table 5.2.2b (Continued)

		1971	1981	1991	2001	2011
Ontario	P	7,703,105	8,625,107	10,084,885	11,410,046	12,651,790
	M%	0.25	0.60	1.44	3.09	4.60
	M	19,375	52,110	145,560	352,530	581,950
	MR%	57.96	53.09	57.47	60.82	55.22
Prince Edward Island	P	111,635	122,506	129,765	135,294	137,375
	M%	0.05	0.06	0.05	0.14	0.48
	M	60	70	60	195	655
	MR%	0.18	0.07	0.02	0.03	0.06
Quebec	P	6,027,765	6,438,403	6,895,963	7,237,479	7,732,520
	M%	0.14	0.19	0.65	1.50	3.15
	M	8,380	12,115	44,930	108,620	243,430
	MR%	25.07	12.34	17.74	18.74	23.10
Saskatchewan	P	926,240	968,313	988,928	978,933	1,008,760
	M%	0.05	0.12	0.12	0.23	1.00
	M	490	1,120	1,185	2,230	10,040
	MR%	1.47	1.14	0.47	0.38	0.95
Yukon	P	18,390	23,150	27,797	28,674	33,320
	M%	0.05	0.04	0.11	0.21	0.12
	M	10	10	30	60	40
	MR%	0.03	0.01	0.01	0.01	0.00

Table 5.2.3a Year-end evolution of Muslim nationalities in Greenland since 1976

	1976	1981	1986	1991	1996	2001	2006	2011	2016
Morocco	—	—	—	—	—	—	—	2	3
Iran	—	—	—	1	1	1	1	1	3
Pakistan	—	—	—	—	—	1	2	1	2
Turkey	1	—	—	1	—	—	1	2	1
Total	1	—	1	2	1	2	4	6	9

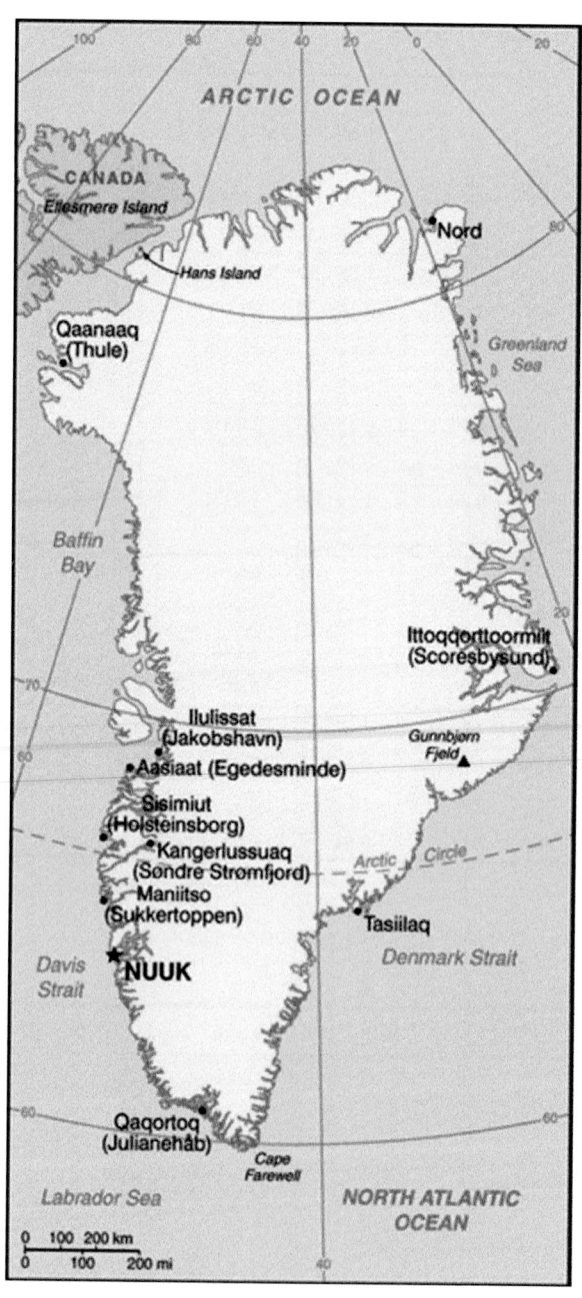

Figure 5.2.3 Map of Greenland.

Table 5.2.3b Evolution of the Muslim population in Greenland

Year	Population	Muslims	%	Source
1976	49,717	1	0.00	[GL]e
1981	51,435	—	0.00	[GL]e
1986	53,733	1	0.00	[GL]e
1991	55,381	2	0.00	[GL]e
1996	55,967	1	0.00	[GL]e
2001	56,512	2	0.00	[GL]e
2006	56,642	4	0.01	[GL]e
2011	56,746	6	0.01	[GL]e
2016	55,859	9	0.02	[GL]e
2050	53,000	30	0.05	es
2100	41,000	40	0.10	es

5.2.4 Saint Pierre and Miquelon

The Overseas Collectivity of Saint Pierre and Miquelon is a self-governing territorial overseas collectivity of France. It is off the coast of the Canadian province of Newfoundland and has a total area of 242 sq. km. Shared between two main islands: Saint Pierre (25 sq. km) which includes the capital with the same name, and the larger island of Saint Miquelon to the north. The map of these islands is presented in Fig. 5.2.4. The islands were first sighted by the Portuguese in 1520 and became French Possession in 1536. However, the islands were not permanently inhabited until the end of the seventeenth century. The permanent population was recorded as four in 1670 and 22 in 1691.

Table 5.2.4 Evolution of the Muslim population in Saint Pierre and Miquelon

Year	Population	Muslims	%	Source
1974	5,840	—	0.00	[KET76]es
2011	6,080	10	0.16	[PEW]es
2050	5,000	10	0.16	es
2100	4,000	10	0.16	es

As shown in Table 5.2.4, the Muslim population in this archipelago increased from none in 1974, to ten or 0.2% in 2011. Thus, assuming that the percentage of Muslims will remain fixed, the Muslim population is expected to remain around ten throughout this century.

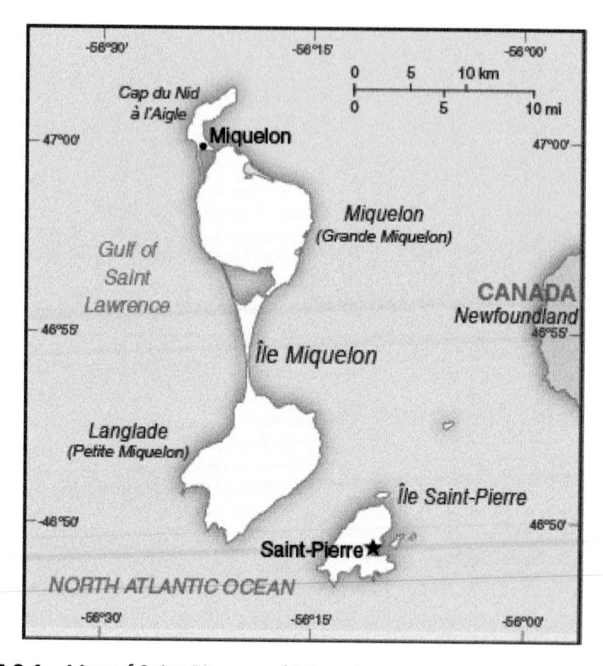

Figure 5.2.4 Map of Saint Pierre and Miquelon.

5.2.5 United States of America

It has an area of 9,629,091 sq. km which is divided into fifty states and the Federal District of Columbia, where the capital Washington is located. The area includes the states of Alaska (1,717,854 sq. km) and Hawaii (10,931 sq. km), however other territories under U.S. control such as Puerto Rico are not included in the area of the U.S. but are included in separate entries. Thus, the United States is the fourth largest country in the world, after Russia, Canada and China. USA however, has a comparable size to China. A map of the US is presented in Fig. 5.2.5.

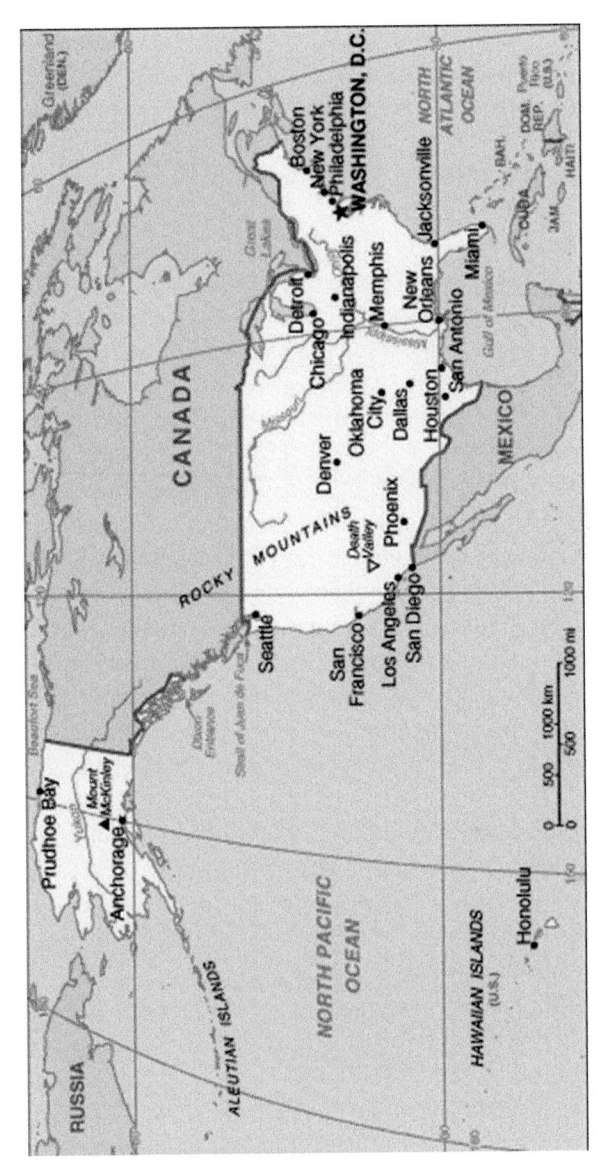

Figure 5.2.5 Map of the United States of America.

Unlike its neighbor to the north, the United States does not inquire about religious affiliation in its census, which occurs at the beginning of each decade. Thus, the estimated Muslim population varies widely. It was estimated by [MAS23] in 1920 at 10,000 or 0.01% of the total population. A. Kettani on the other hand, estimated that it increased from 200,000 or 0.1% in 1950, to one million or 0.5% in 1970, to three million or 1.3% in 1980 [KET76, KET86]. However, the method for coming up with such estimates were not revealed.

According to President Clinton in his 2000 Ramadan Greeting Speech to Muslims [US00], the number increased to six million or 2.1% in 2000. Furthermore, according to President Obama in his 2009 Cairo Speech to Muslims [US10], the number has increased to seven million or 2.3% in 2010. These rough estimates are generally considered an upper bound on the number of Muslim population living in the United States. The last two estimates are also in agreement with estimates by the Council on American Islamic Relation (CAIR), which is based on the number of Muslims affiliated with each mosque, multiplied by three to account for non-practicing Muslims. The study was published by I. Bagby in 2001 and 2011, and also found that almost a fifth (19%) of Muslims in America are converts and almost two-thirds (64%) are foreign-born [USS12]. The number of mosques according to the same study doubles almost every decade, increasing from 105 in 1960, to 232 in 1970, to 527 in 1980, to 962 in 1994, to 1,209 in 2000, to 2,106 in 2011. The number of Muslims who pray at least once a year (Eid) at a mosque were estimated to be half a million in 1994, two million in 2001 and 2.6 million in 2011. In addition, PEW 2011 Survey [USS11] found that the number of Muslims who go to a mosque at least once a year is 81% of the total number of Muslims in the US. Thus, the arbitrary three times factor assumed by Bagby is not valid.

In [US90], J. Weeks estimated the Muslim population based on census ancestry data, and added 2% of African American population, based on another survey. This led to the estimate 2.5 million or 1.0% in 1990 and 3.4 million or 1.2% in 2000. This study clearly overestimates the number of Muslims as children of immigrants may not preserve Islam, especially if a Muslim married a non-Muslim.

Other survey data based on random telephone number dialing deduced that the percentage of Muslims is between 0.3% in 1990 and 0.6% in 2008 (adjusted by excluding from the total those who refused to answer). For instance, the American Religious Identification Survey (ARIS) put the percentage of Muslims as 0.31%

in 1990, 0.56% in 2001, and 0.62% in 2008 [USS8]. Yearly Gallop Poll surveys from 2000 to 2011 consistently put the number at 0.5%. PEW surveys put the percentage at 0.76% in 2007 and 0.89% in 2011 [USS11]. PEW surveys also found that nearly two-thirds of the Muslim population in the United States is foreign-born, but this share is constantly decreasing (65% in 2007 and 63% in 2011 and 61% in 2014). These surveys clearly underestimate the number of Muslims since the Muslim population is not uniformly distributed throughout the country.

Here we follow a new criterion to monitor the increase of the number of Muslims in the United States since 1850. It is based on the number of foreign-born population provided by the US Census Bureau [US06]. We record the population born in available countries with Muslim majority as follows:

- For Asia these are: Afghanistan, Bangladesh, Iran, Pakistan, Indonesia, Malaysia, the Middle East, Turkey and one-tenth of the population born in India. The percentage born in the aforementioned countries with respect to the total born in Asia is 17.8% in 2000, 17.6% in 1990, 19.8% in 1980, 22.5% in 1970 and 27.6% in 1960. Thus, we assume that one-sixth (16.7%) of those born in Asia are Muslim.
- For Africa, the countries are Algeria, Egypt, Morocco, Libya, Sudan, Tunisia, and from 1980 to 2000: Senegal, Sierra Leone, Somalia, and half of those born in Nigeria, and from 1990 to 2000: Gambia and Guinea. The percentage born in the aforementioned countries with respect to the total born in Africa is 38.1% in 2000, 38.8% in 1990, 43.0% in 1980, 40.4% in 1970 and 45.0% in 1960. Thus, we assume that one-third (33.3%) of those born in Africa are Muslim.
- For Europe these are: Albania, Bosnia and Kosovo. The percentage born in the aforementioned countries with respect to the total born in Europe is 3.0% in 2000, 1.2% in 1990, 1.1% in 1980 and 1970, 0.9% in 1960, 0.7% in 1930 and 0.5% in 1920. Thus, we consider this category negligible and therefore not include it in the estimate of Muslims.

Thus, as detailed in Table 5.2.5a the Muslim population from 1850 to 1970 was estimated from the foreign-born data as one-sixth of those born in Asia and one-third of those born in Africa.

Table 5.2.5a Estimate of Muslims in USA based on foreign-born population: 1850 to 1970

1850	1860	1870	1880	1890	1900	1910	1920	1930	1960	1970
					Asia-born					
1,135	36,796	64,565	107,630	113,383	120,248	191,484	237,950	275,665	490,996	824,887
					Asia-born Muslims (1/6)					
189	6,133	10,761	17,938	18,897	20,041	31,914	39,658	45,944	81,833	137,481
					Africa-born					
551	526	2,657	2,204	2,207	2,538	3,992	16,126	18,326	35,355	80,143
					Africa-born Muslims (1/3)					
184	175	886	735	736	846	1,331	5,375	6,109	11,785	26,714
					Muslims					
373	6,308	11,647	18,673	19,633	20,887	33,245	45,034	52,053	93,618	164,196
					Muslim %					
0.00	0.02	0.03	0.04	0.03	0.03	0.04	0.04	0.04	0.05	0.08

The distribution of Muslims per state is provided in Table 5.2.5c. For 1980 and 1990, the data was estimated from ancestry data collected in the corresponding census year. In 1980, the ancestry that was considered Muslims were: Arab, Egypt, Iraq, Jordan, Morocco, Palestine, Saudi Arabia, Syria, Turkey, Middle East, Indonesia and Pakistan. In 1990, these regions were added: Algeria, Yemen, Sudan, Afghanistan, and Bangladesh. The data for 2000 onward was obtained by a religious census sponsored by the Association of Statisticians of American Religious Bodies (ASARB). Almost 90% of Muslims in USA are found in ten states as follows: 16% in Texas, 15% in New York, 14% in Illinois, 11% in California, 8% in Virginia, 6% in Florida, 6% in New Jersey, 5% in Michigan, 3% in Pennsylvania and 2% in Georgia.

Table 5.2.5b Evolution of the Muslim population in the United States of America

Year	Population	Muslims	%	Source
1850	23,191,876	373	0.00	[US06]e
1860	31,443,321	6,308	0.02	[US06]e
1870	39,818,449	11,647	0.03	[US06]e
1880	50,155,783	18,673	0.04	[US06]e
1890	62,947,714	19,633	0.03	[US06]e
1900	75,994,575	20,887	0.03	[US06]e
1910	91,972,266	33,245	0.04	[US06]e
1920	105,710,620	45,034	0.04	[US06]e
1930	122,775,046	52,053	0.04	[US06]e
1960	179,323,175	93,618	0.05	[US06]e
1970	203,211,926	164,196	0.08	[US06]e
1980	226,542,199	449,516	0.20	[US80]e
1990	253,148,054	740,917	0.29	[US90]e
2000	281,423,231	1,559,294	0.55	[US00]c
2010	308,745,538	2,600,082	0.84	[US10]c
2050	379,419,000	7,588,000	2.00	es
2100	433,854,000	14,100,000	3.25	es

Table 5.2.5c Evolution of the Muslim population in USA per state. *P*: Total population, *M*: Muslim population, *M%*: Percentage of Muslim population

		1980	1990	2000	2010
Alabama	P	3,894,025	4,040,587	4,447,100	4,779,736
	M%	0.04	0.08	0.17	0.21
	M	1,452	3,390	7,670	10,258
Alaska	P	401,851	550,043	626,931	710,231
	M%	0.06	0.07	0.22	0.13
	M	249	398	1,381	924
Arizona	P	2,716,546	3,665,228	5,130,632	6,392,017
	M%	0.17	0.21	0.23	0.13
	M	4,485	7,563	11,857	8,557
Arkansas	P	2,286,357	2,350,725	2,673,398	2,915,918
	M%	0.05	0.06	0.08	0.13
	M	1,101	1,527	2,044	3,746
California	P	23,667,764	29,760,021	33,871,653	37,253,956
	M%	0.38	0.54	0.77	0.73
	M	89,770	159,679	259,762	272,814
Colorado	P	2,889,735	3,294,394	4,301,997	5,029,196
	M%	0.18	0.20	0.35	0.33
	M	5,332	6,509	14,855	16,738
Connecticut	P	3,107,564	3,287,116	3,405,584	3,574,097
	M%	0.16	0.23	0.87	0.38
	M	5,078	7,437	29,647	13,418
Delaware	P	594,338	666,168	783,600	897,934
	M%	0.16	0.24	0.47	0.79
	M	966	1,573	3,691	7,124
District of Columbia	P	638,432	606,900	572,059	601,723
	M%	0.23	0.44	10.57	0.67
	M	1,457	2,654	60,479	4,032
Florida	P	9,746,961	12,937,926	15,982,820	18,801,310
	M%	0.21	0.28	0.20	0.88
	M	20,150	36,566	31,661	164,846
Georgia	P	5,462,982	6,478,216	8,186,517	9,687,653
	M%	0.07	0.13	0.47	0.54
	M	3,741	8,599	38,882	52,578
Hawaii	P	964,691	1,108,229	1,211,537	1,360,301
	M%	0.13	0.14	0.05	0.05
	M	1,213	1,600	609	616
Idaho	P	944,127	1,006,749	1,293,956	1,567,582
	M%	0.05	0.07	0.03	0.11
	M	506	753	363	1,732

		1980	1990	2000	2010
Illinois	P	11,427,409	11,430,602	12,419,570	12,830,632
	M%	0.22	0.37	1.01	2.80
	M	24,673	42,044	125,203	359,264
Indiana	P	5,490,210	5,544,159	6,080,506	6,483,802
	M%	0.11	0.13	0.18	0.22
	M	5,854	7,275	11,002	14,573
Iowa	P	2,913,808	2,776,755	2,926,382	3,046,355
	M%	0.08	0.10	0.16	0.21
	M	2,201	2,757	4,717	6,528
Kansas	P	2,364,236	2,477,574	2,688,814	2,853,118
	M%	0.10	0.13	0.13	0.27
	M	2,350	3,225	3,470	7,744
Kentucky	P	3,660,324	3,685,296	4,042,209	4,339,367
	M%	0.05	0.08	0.12	0.26
	M	1,662	2,812	4,696	11,123
Louisiana	P	4,206,116	4,219,973	4,468,958	4,533,372
	M%	0.10	0.13	0.29	0.22
	M	4,165	5,279	13,050	9,806
Maine	P	1,125,043	1,227,928	1,274,923	1,328,361
	M%	0.05	0.11	0.06	0.10
	M	568	1,303	809	1,332
Maryland	P	4,216,933	4,781,468	5,296,485	5,773,552
	M%	0.22	0.36	1.00	0.63
	M	9,454	17,355	52,867	36,484
Massachusetts	P	5,737,093	6,016,425	6,349,097	6,547,629
	M%	0.26	0.35	0.65	0.33
	M	15,001	20,846	41,497	21,768
Michigan	P	9,262,044	9,295,297	9,938,480	9,883,640
	M%	0.37	0.46	0.81	1.22
	M	34,383	42,812	80,515	120,351
Minnesota	P	4,075,970	4,375,099	4,919,485	5,303,925
	M%	0.09	0.11	0.25	0.32
	M	3,647	4,973	12,305	16,796
Mississippi	P	2,520,770	2,573,216	2,844,656	2,967,297
	M%	0.05	0.05	0.14	0.17
	M	1,151	1,242	3,919	5,012
Missouri	P	4,916,766	5,117,073	5,596,683	5,988,927
	M%	0.08	0.12	0.35	0.20
	M	4,033	6,065	19,359	11,708

Table 5.2.5c (Continued)

		1980	1990	2000	2010
Montana	P	786,690	799,065	902,195	989,415
	M%	0.06	0.06	0.07	0.03
	M	435	451	614	333
Nebraska	P	1,569,825	1,578,385	1,711,265	1,826,341
	M%	0.08	0.14	0.18	0.34
	M	1,303	2,164	3,115	6,156
Nevada	P	800,508	1,201,833	1,998,257	2,700,551
	M%	0.18	0.26	0.11	0.06
	M	1,466	3,118	2,291	1,700
New Hampshire	P	920,610	1,109,252	1,235,786	1,316,470
	M%	0.11	0.17	0.31	0.12
	M	1,023	1,871	3,782	1,616
New Jersey	P	7,365,011	7,730,188	8,414,347	8,791,894
	M%	0.38	0.67	1.43	1.83
	M	28,287	51,466	120,724	160,666
New Mexico	P	1,303,302	1,515,069	1,819,046	2,059,179
	M%	0.13	0.16	0.14	0.20
	M	1,645	2,476	2,604	4,116
New York	P	17,558,165	17,990,455	18,976,821	19,378,102
	M%	0.40	0.65	1.18	2.03
	M	69,947	117,463	223,968	392,953
North Carolina	P	5,880,095	6,628,637	8,046,451	9,535,483
	M%	0.06	0.11	0.25	0.27
	M	3,621	7,263	20,137	26,045
North Dakota	P	652,717	638,800	642,200	672,591
	M%	0.11	0.08	0.14	0.09
	M	711	534	902	636
Ohio	P	10,797,603	10,847,115	11,353,143	11,536,504
	M%	0.14	0.20	0.36	0.29
	M	15,004	22,160	41,281	33,408
Oklahoma	P	3,025,487	3,145,585	3,450,654	3,751,351
	M%	0.13	0.13	0.18	0.20
	M	3,900	4,014	6,145	7,392
Oregon	P	2,633,156	2,842,321	3,421,432	3,831,074
	M%	0.19	0.21	0.15	0.10
	M	4,908	5,913	5,224	3,988
Pennsylvania	P	11,864,720	11,881,643	12,281,054	12,702,379
	M%	0.17	0.23	0.58	0.63
	M	20,044	26,960	71,190	80,487

		1980	1990	2000	2010
Rhode Island	P	947,154	1,003,464	1,048,319	1,052,567
	M%	0.35	0.43	0.17	0.14
	M	3,319	4,279	1,827	1,444
South Carolina	P	3,120,729	3,486,703	4,011,848	4,625,364
	M%	0.06	0.09	0.14	0.13
	M	1,807	3,204	5,761	5,792
South Dakota	P	690,768	696,004	754,844	814,180
	M%	0.08	0.10	0.01	0.16
	M	523	704	50	1,332
Tennessee	P	4,591,023	4,877,185	5,689,262	6,346,105
	M%	0.05	0.09	0.32	0.24
	M	2,313	4,629	18,464	15,384
Texas	P	14,225,513	16,986,510	20,851,790	25,145,561
	M%	0.14	0.22	0.55	1.68
	M	20,422	37,759	114,999	421,972
Utah	P	1,461,037	1,722,850	2,233,198	2,763,885
	M%	0.08	0.11	0.16	0.18
	M	1,190	1,933	3,645	4,998
Vermont	P	511,456	562,758	608,827	625,741
	M%	0.08	0.10	0.02	0.05
	M	410	552	100	300
Virginia	P	5,346,797	6,187,358	7,078,483	8,001,024
	M%	0.19	0.46	0.72	2.66
	M	10,284	28,459	51,021	213,032
Washington	P	4,132,353	4,866,692	5,894,141	6,724,540
	M%	0.16	0.19	0.26	0.28
	M	6,547	9,135	15,553	19,092
West Virginia	P	1,950,186	1,793,477	1,808,350	1,852,994
	M%	0.09	0.12	0.08	0.10
	M	1,713	2,181	1,528	1,908
Wisconsin	P	4,705,642	4,891,769	5,363,704	5,686,986
	M%	0.08	0.12	0.15	0.26
	M	3,792	5,677	7,796	14,744
Wyoming	P	469,557	4,891,769	493,782	563,626
	M%	0.06	0.01	0.05	0.13
	M	260	316	263	716
USA (Total)	P	226,542,199	253,148,054	281,423,231	308,745,538
	M%	0.20	0.29	0.55	0.84
	M	449,516	740,917	1,559,294	2,600,082

Hence, as summarized in Table 5.2.5b, the Muslim population in USA remained negligible at 0.0% of the total population until the second half of the twentieth century, likely increasing in number from 400 in 1850, to 9,000 in 1860, to 12,000 in 1870, to around 20,000 in 1880 to 1900, to 33,000 in 1910, to 45,000 in 1920, to 52,000 in 1930. As a result of the United States opening its doors to immigration from parts of the world other than the white race after the Immigration Act of 1965, the Muslim population increased from 0.1 million or 0.1% in 1960 to 0.2 million or 0.1% in 1970, to 0.4 million or 0.2% in 1980, to 0.7 million or 0.3% in 1990, to 1.6 million or 0.6% in 2000, to 2.6 million or 0.8% in 2010.

Thus, assuming that the percentage of Muslims will continue to increase by a quarter of a percentage point per decade, the Muslim population is expected to increase to almost eight million or 2% by 2050 and exceed fourteen million or 3.3% by 2100.

5.2.6 Regional Summary and Conclusion

North America has the second highest concentration of Muslims among five regions of the Americas but has the highest number of Muslims. In the 1960s, immigration laws in US and Canada changed to welcome immigrants from all over the world and not just Whites. Since then the number of Muslims is fast growing due to immigration, but also higher birthrate and conversions. The percentage of Muslims has exceeded 1% of the total population and is expected to pass 4% by the end of this century. The following tables present centennial data from 600 AD to 2100 AD (or approximately 1 H to 1500 H) in Table 5.2a and decennial data from 1790 AD to 2100 AD (or 1210 H to 1520 H) in Tables 5.2b and 5.2c for current countries in North America. The data includes total population in thousands (P), the percentage of which is Muslim (M%), the corresponding Muslim population in thousands (M), and the annual population growth rate (APGR, or G%) of the total population in this region.

Muslims in North America | 673

Table 5.2a Centennial estimates of the Muslim population (×1000) in North America: 600–2100 AD (1–1500 H)

		600	700	800	900	1000	1100	1200	1300	1400	1500	1600	1700	1800	1900	2000	2100
Bermuda	P	—	—	—	—	—	—	—	—	—	—	—	8	8	18	65	37
	M%	—	—	—	—	—	—	—	—	—	—	—	—	—	—	0.98	1.50
	M	—	—	—	—	—	—	—	—	—	—	—	—	—	—	1	1
Canada	P	132	138	144	152	160	180	200	220	240	250	250	200	240	5,371	30,588	56,953
	M%	—	—	—	—	—	—	—	—	—	—	—	—	—	0.00	1.96	14.00
	M	—	—	—	—	—	—	—	—	—	—	—	—	—	0	600	7,973
Greenland	P	1	1	1	1	1	1	1	1	2	3	4	5	6	12	56	41
	M%	—	—	—	—	—	—	—	—	—	—	—	—	—	—	0.01	0.10
	M	—	—	—	—	—	—	—	—	—	—	—	—	—	—	—	—
SPM	P	—	—	—	—	—	—	—	—	—	—	—	—	1	7	6	4
	M%	—	—	—	—	—	—	—	—	—	—	—	—	—	—	—	0.16
	M	—	—	—	—	—	—	—	—	—	—	—	—	—	—	—	—
USA	P	1,040	1,120	1,180	1,240	1,300	1,450	1,600	1,750	1,875	2,000	1,500	1,000	5,309	76,212	281,711	433,854
	M%	—	—	—	—	—	—	—	—	—	—	—	—	—	0.03	0.55	3.25
	M	—	—	—	—	—	—	—	—	—	—	—	—	—	23	1,549	14,100
Total	P	1,173	1,259	1,325	1,393	1,461	1,631	1,801	1,971	2,117	2,253	1,754	1,213	5,562	81,620	312,427	490,889
	M%	—	—	—	—	—	—	—	—	—	—	—	—	—	0.03	0.69	4.50
	M	—	—	—	—	—	—	—	—	—	—	—	—	—	23	2,150	22,074
	G%	0.071	0.051	0.050	0.048	0.110	0.099	0.090	0.071	0.062	-0.250	-0.369	1.523	2.686	1.342	0.452	

Table 5.2b Decennial estimates of the Muslim population (×1000) in North America: 1790–1940 AD (1210–1360 H)

		1790	1800	1810	1820	1830	1840	1850	1860
Bermuda	P	8	8	8	8	8	9	11	11
	M%	—	—	—	—	—	—	—	—
	M	—	—	—	—	—	—	—	—
Canada	P	300	240	631	741	1,500	1,800	2,436	3,230
	M%	—	—	—	—	—	—	—	—
	M	—	—	—	—	—	—	—	—
Greenland	P	5	6	6	7	7	8	9	10
	M%	—	—	—	—	—	—	—	—
	M	—	—	—	—	—	—	—	—
SPM	P	1	1	1	1	1	1	2	3
	M%	—	—	—	—	—	—	—	—
	M	—	—	—	—	—	—	—	—
USA	P	3,930	5,309	7,240	9,638	12,866	17,070	23,192	31,443
	M%	—	—	—	—	—	0.00	0.00	0.02
	M	—	—	—	—	—	—	—	6
Total	P	4,243	5,562	7,886	10,395	14,382	18,888	25,650	34,697
	M%	—	—	—	—	—	0.00	0.00	0.02
	M	—	—	—	—	—	—	—	6
	G%		2.708	3.490	2.763	3.247	2.725	3.060	3.021

Table 5.2b (*Continued*)

		1870	1880	1890	1900	1910	1920	1930	1940
Bermuda	P	12	14	15	18	19	20	28	35
	M%	—	—	—	—	—	—	—	—
	M	—	—	—	—	—	—	—	—
Canada	P	3,689	4,325	4,833	5,371	7,207	8,788	10,377	11,507
	M%	0.00	0.00	0.00	0.00	0.01	0.01	0.01	0.01
	M	—	—	—	—	1	1	1	1
Greenland	P	10	10	11	12	13	14	17	18
	M%	—	—	—	—	—	—	—	—
	M	—	—	—	—	—	—	—	—
SPM	P	5	5	6	7	4	4	4	4
	M%	—	—	—	—	—	—	—	—
	M	—	—	—	—	—	—	—	—
USA	P	39,818	50,156	62,975	76,212	92,229	106,022	123,203	132,165
	M%	0.03	0.04	0.03	0.03	0.04	0.04	0.04	0.04
	M	12	20	19	23	37	42	49	53
Total	P	43,534	54,509	67,840	81,620	99,472	114,848	133,629	143,730
	M%	0.03	0.04	0.03	0.03	0.04	0.04	0.04	0.04
	M	12	20	19	23	38	43	50	54
	G%	2.269	2.248	2.188	1.849	1.978	1.437	1.515	0.729

Table 5.2c Decennial estimates of the Muslim population (×1000) in North America: 1950–2100 AD (1370–1520 H)

		1950	1960	1970	1980	1990	2000	2010	2020
Bermuda	P	37	45	53	58	62	65	65	62
	M%	0.01	0.01	0.73	0.73	0.84	0.98	1.02	1.10
	M	—	—	—	—	1	1	1	1
Canada	P	13,733	17,847	21,374	24,417	27,541	30,588	34,148	37,742
	M%	0.01	0.03	0.16	0.40	0.93	1.96	3.21	4.40
	M	1	5	34	98	256	600	1,096	1,661
Greenland	P	23	31	46	50	56	56	57	57
	M%	—	—	—	—	0.00	0.01	0.01	0.02
	M	—	—	—	—	—	—	—	—
SPM	P	5	5	6	6	6	6	6	6
	M%	—	—	—	—	—	—	0.16	0.16
	M	—	—	—	—	—	—	—	—
USA	P	158,804	186,721	209,513	229,476	252,120	281,711	309,011	331,003
	M%	0.04	0.05	0.08	0.20	0.29	0.55	0.84	1.25
	M	64	93	168	459	731	1,549	2,596	4,138
Total	P	172,603	204,649	230,992	254,007	279,785	312,427	343,287	368,870
	M%	0.04	0.05	0.09	0.22	0.35	0.69	1.08	1.57
	M	65	99	202	557	988	2,150	3,693	5,799
	G%	1.831	1.703	1.211	0.950	0.967	1.103	0.942	0.719

Table 5.2c (*Continued*)

		2030	2040	2050	2060	2070	2080	2090	2100
Bermuda	P	60	58	53	49	46	43	40	37
	M%	1.15	1.20	1.25	1.30	1.35	1.40	1.45	1.50
	M	1	1	1	1	1	1	1	1
Canada	P	40,834	43,486	45,669	47,779	50,096	52,348	54,540	56,953
	M%	5.60	6.80	8.00	9.20	10.40	11.60	12.80	14.00
	M	2,287	2,957	3,654	4,396	5,210	6,072	6,981	7,973
Greenland	P	57	56	53	51	49	46	43	41
	M%	0.03	0.04	0.05	0.06	0.07	0.08	0.09	0.10
	M	—	—	—	—	—	—	—	—
SPM	P	6	6	5	5	5	4	4	4
	M%	0.16	0.16	0.16	0.16	0.16	0.16	0.16	0.16
	M	—	—	—	—	—	—	—	—
USA	P	349,642	366,572	379,419	391,495	404,174	415,197	424,470	433,854
	M%	1.50	1.75	2.00	2.25	2.50	2.75	3.00	3.25
	M	5,245	6,415	7,588	8,809	10,104	11,418	12,734	14,100
Total	P	390,599	410,177	425,200	439,379	454,369	467,639	479,097	490,889
	M%	1.93	2.29	2.64	3.01	3.37	3.74	4.12	4.50
	M	7,532	9,373	11,243	13,205	15,315	17,491	19,716	22,074
	G%	0.572	0.489	0.360	0.328	0.335	0.288	0.242	0.243

5.3 Muslims in Central Caribbean Islands

This region consists of ten island nations and territories between Guadeloupe and the Virgin Islands spread over a tenth of a million square kilometers of the western Atlantic Ocean and the eastern Caribbean Sea, with total land area of 2,668 sq. km, over half (54%) of which is the twin main islands of Guadeloupe. Thus, this region consists of Anguilla, Antigua and Barbuda, Guadeloupe, Montserrat, Saint Barthélemy, Saint Kitts and Nevis, Saint Martin, Sint Maarten, the British Virgin Islands and the United States Virgin Islands. This region has the highest density of territories in the world. A map of this region is presented in Fig. 5.3a.

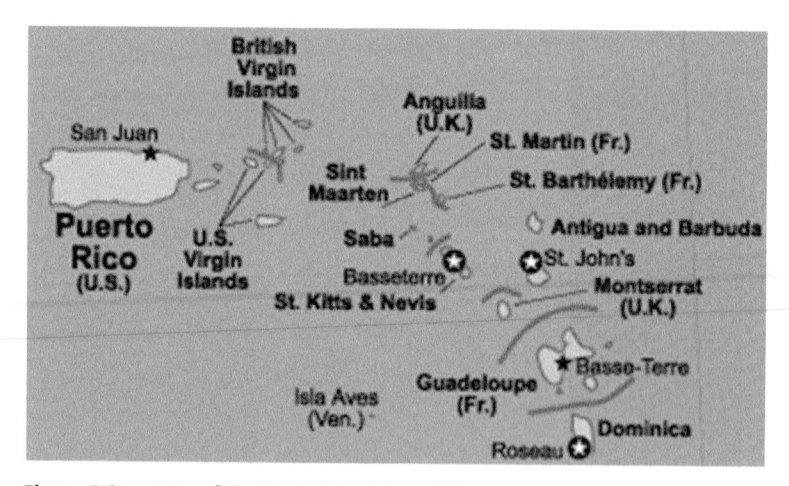

Figure 5.3a Map of the Central Caribbean Islands.

Muslims arrived here as explorers, centuries before the arrival of the Europeans, then as escapees from Spanish Catholic oppression, then as black slaves brought by the Europeans from Africa. However, Islam went extinct from these migrations. Towards the end of the nineteenth century, the Brits and Dutch brought workers from India and Java, some of whom were Muslim, and Islam remains among their descendants till this day. The French also brought Muslims from their African colonies to their colonies in this region.

Thus, the Muslim population has likely increased from none in 1900, to 2,000 or 0.7% in 2000, to 3,000 or 0.8% in 2020, and is projected to reach 4,000 or 1.0% by 2050 and 5,000 or 1.5% by 2100. A plot of decennial estimates of the Muslim population and its percentage with respect to the total population in this region from 1900 to 2100 is provided in Fig. 5.3b. The corresponding individual data for each country in this region is discussed below. In Section 5.1.13, the total population in each country in this region and the corresponding percentage and number of Muslims is presented centennially in Table 5.3a from 600 to 2100 and decennially in Tables 5.3b and 5.3c from 1790 to 2100.

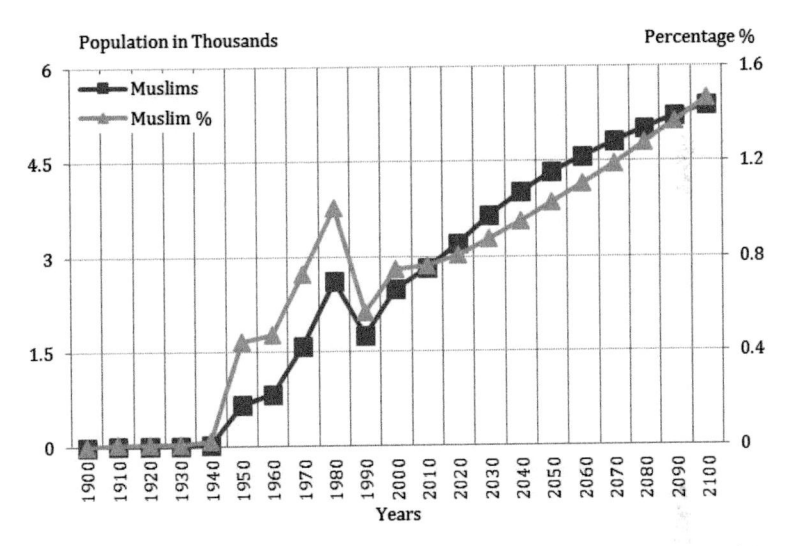

Figure 5.3b Plot of decennial estimates of the Muslim population and its percentage of the total population in Central Caribbean Islands: 1900–2100 AD (1320–1520 H).

5.3.1 Anguilla

It was conquered by the British in 1650 and remains an overseas territory of the UK. It has an area of 91 sq. km which comprises the main island of Anguilla and several mostly tiny and uninhabited nearby islands. The island's name means eel and it used to be named Snake Island due to its shape. A map of the island is presented in Figs. 5.3.1 and 5.3.5.

First Muslims arrived at this island in the 1990s and census data showed the presence of no Muslims in 1992 and before. The number of Muslims reached 36 or 0.3% in 2001 and 43 or 0.3% in 2011. Thus, assuming that the percentage of Muslims will continue to increase by 0.01 of a percentage point per decade, the Muslim population is expected to remain less than sixty throughout this century, reaching 0.4% by the end of this century. A summary of the data is provided in Table 5.3.1.

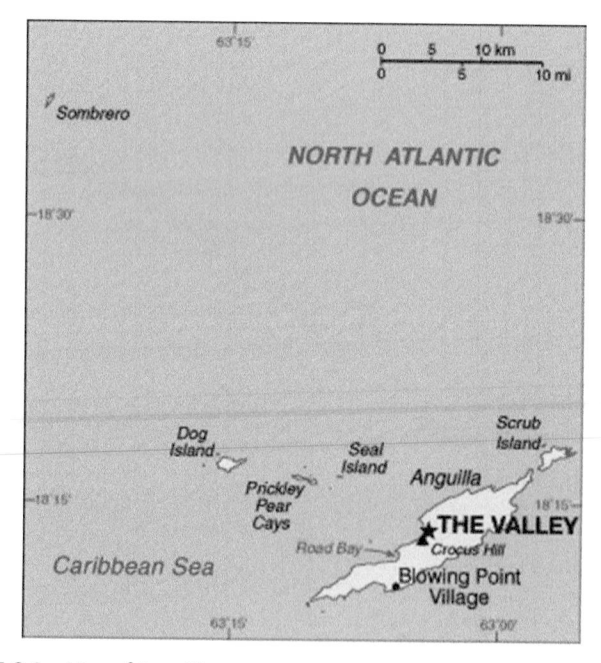

Figure 5.3.1 Map of Anguilla.

Table 5.3.1 Evolution of the Muslim population in Anguilla

Year	Population	Muslims	%	Source
1911	4,100	—	0.00	[GB911]c
1946	5,036	—	0.00	[UN56]c
1992	8,960	—	0.00	[AI01]c
2001	11,391	36	0.32	[AI01]c
2011	13,532	43	0.32	[AI11]c
2050	17,000	60	0.36	es
2100	14,000	60	0.41	es

5.3.2 Antigua and Barbuda

The Island of Antigua was sighted by Columbus in 1493, who named it after a church in Seville, Spain, called Santa Maria La Antigua. It was first inhabited by few English in 1632, and many colonists arrived to it in 1663. It was declared a British Possession in 1666. The islands gained their independence from the UK in 1981. They have an area of 443 sq. km which comprises the two main islands of Antigua (280 sq. km) and Barbuda (161 sq. km) and several much smaller uninhabited nearby islands. About 98% of the total population lives in Antigua while only 2% lives in Barbuda. A map of these islands is presented in Fig. 5.3.2.

The first Muslims arrived from Syria in early twentieth century. According to census data as shown in Table 5.3.2, the Muslim population increased from none in 1911 and before, to nineteen or 0.1% in 1946, to 228 or 0.3% in 2001, to 208 or 0.3% in 2011. Thus, assuming that the percentage of Muslims will increase by 0.05 of a percentage point each decade, the Muslim population is expected to remain less than one thousand throughout this century, reaching 0.7% of the total population by 2100.

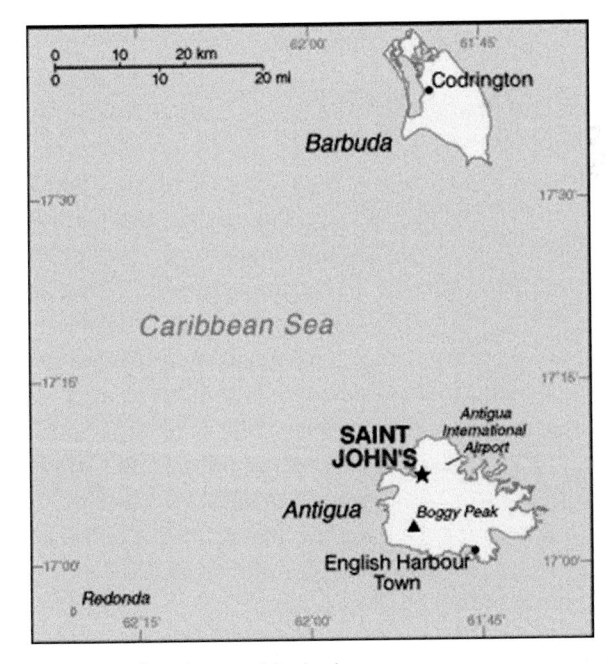

Figure 5.3.2 Map of Antigua and Barbuda.

Table 5.3.2 Evolution of the Muslim population in Antigua and Barbuda

Year	Population	Muslims	%	Source
1871	35,157	—	0.00	[GB871]c
1911	32,269	—	0.00	[GB911]c
1946	41,800	19	0.05	[UN56]c
2001	75,557	228	0.30	[AG01]c
2011	80,113	208	0.26	[AG11]c
2020	96,000	300	0.30	es
2050	111,000	500	0.45	es
2100	102,000	700	0.70	es

5.3.3 Guadeloupe

It is an overseas region of France which was captured in 1814. It has an area of 1,628 sq. km comprising six inhabited islands and dozens of much smaller islands around them. The inhabited islands are the main islands Basse-Terre (848 sq. km) on the west and Grande-Terre (587 sq. km) on the east, the nearby smaller islands of Marie-Galante (158 sq. km), La Desirade (21 sq. km), Terre-de-Bas (7 sq. km) and Terre-de-Haut (6 sq. km). A map of Guadeloupe is presented in Fig. 5.3.3.

Guadeloupe was given its name by Columbus when he visited the Island in 1493 on his second voyage after "Santa Maria de la Guadeloupe" in Extremadura, Spain. The Island was taken over by the French in 1635 as part of their "Compagnie des Iles d'Amerique" then became part of the French royal domain in 1674 as a dependency of Martinique, then became a separate colony in 1775. The British conquered the Island in 1759–1763 and in 1810–1816.

The original Caribs were exterminated and replaced by enslaved Africans whose numbers reached 93,000 in 1848 who were freed based on the abolition of slavery in the same year. Some of those were Muslims but there is no record of the size and Islam was not preserved among their descendants. In 1854, France started bringing indentured laborers here from its colonies in India. Then in 1861, France signed a convention with the British to bring more workers from the British Indian territories to its Caribbean possessions. This Indian immigration lasted 35 years until 1889, bringing a total of

42,326 Indians to Guadeloupe. Of these 9,460 Indians returned home after the expiration of their contracts and 32,866 Indians remained in Guadeloupe. It is estimated that 14.5% of these numbers were Muslims. In other words, a total of 6,140 Muslims were brought from India over the 35-year period, of whom 4,770. Few of their descendants are Muslim.

The third wave of Muslim migration was by Arabs from the Middle East before WWII and the fourth in 1970s from Senegal and Guadeloupian returnees from France who embraced Islam there. Thus, estimates of the Muslim population changed from 3,200 or 2.3% in 1889 to 3,000 or 1.5% in 1908. The Muslim population then changed to 1,000 or 0.3% in 1990, to 2,000 or 0.5% in 2008 and 2,500 or 0.6% in 2018. The data is summarized in Table 5.3.3. Thus, assuming that the percentage of Muslims will increase by a tenth of a percentage point per decade, the Muslim population in Guadeloupe is expected to reach 3,000 or 0.9% by 2050 and 5,000 or 1.4% by 2100.

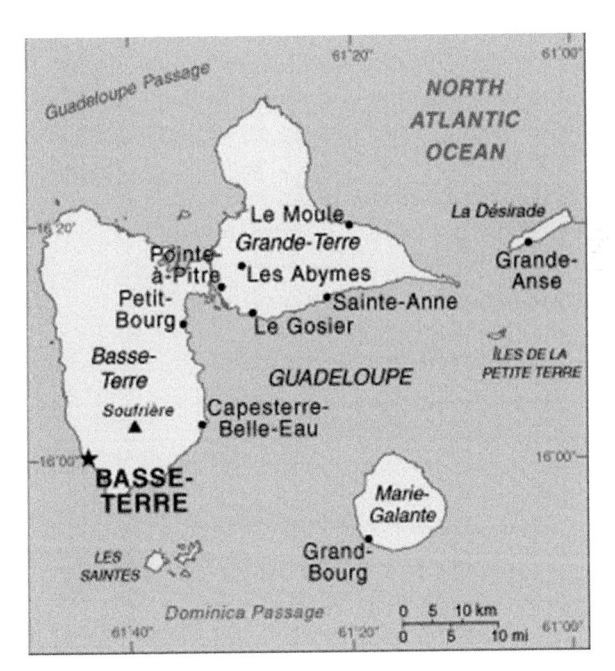

Figure 5.3.3   Map of Guadeloupe.

Table 5.3.3 Evolution of the Muslim population in Guadeloupe

Year	Population	Muslims	%	Source
1850	121,041	—	0.00	[KET01A]es
1889	142,294	3,200	2.25	[JAN]es
1908	200,000	3,000	1.50	[ARH]es
1990	353,431	1,000	0.28	[KET01A]es
2008	403,355	2,000	0.50	[PEW]es
2018	390,704	2,500	0.64	[GW]es
2050	387,000	3,000	0.90	es
2100	339,000	5,000	1.40	es

5.3.4 Montserrat

This island was sighted by Columbus in 1493, who named it after a mountain in Catalonia, to which it resembles. The island was conquered by the British in 1632 and remains an overseas territory of the UK. It consists of one island with an area of 102 sq. km and its map is presented in Fig. 5.3.4. About two thirds of the population left the Island after the 1995 Soufrière Hills volcano eruption.

According to census data as shown in Table 5.3.4, the Muslim population increased from none in 1946 and before, to fourteen or 0.1% in 1980, then decreased to none in 1991, and increased to seven or 0.2% in 2001 and then to eight or 0.2% in 2011. Thus, assuming that the percentage of Muslims will remain constant at 0.2% of the total population, the Muslim population is expected to remain less than ten throughout this century.

Table 5.3.4 Evolution of the Muslim population in Montserrat

Year	Population	Muslims	%	Source
1871	8,693	—	0.00	[GB871]c
1911	12,196	—	0.00	[GB911]c
1946	13,378	—	0.00	[UN56]c
1980	11,519	14	0.12	[UN88]c
1991	11,314	0	0.00	[MS01]c
2001	3,793	7	0.18	[MS01]c
2011	4,922	8	0.16	[MS11]c
2050	4,200	8	0.20	es
2100	2,900	6	0.20	es

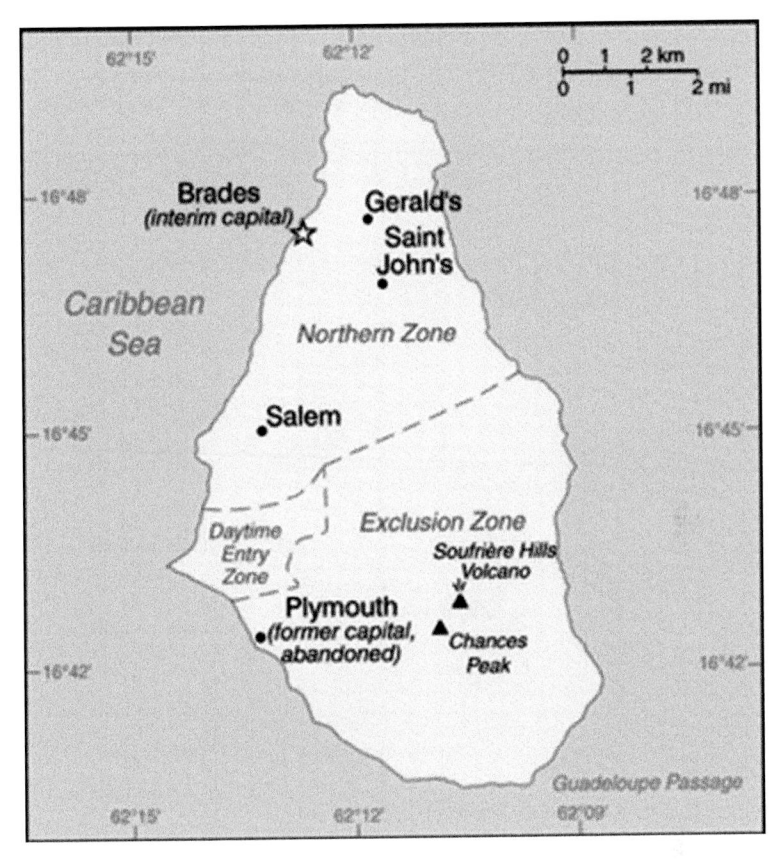

Figure 5.3.4 Map of Montserrat.

5.3.5 Saint Barthélémy

Also known as St. Bart for short, was conquered by France in 1648, then sold to Sweden in 1784, then purchased back in 1878. It seceded from Guadeloupe in 2007 and remains an overseas collectivity of France. It has an area of 21 sq. km comprises the main island surrounded by several much smaller islands. Its map is presented in Fig. 5.3.5. As shown in Table 5.3.5, estimates of the Muslim population increased from none in 1999 and before to six or 0.07% in 2008. Thus, assuming that their number will increase by one tenth of a percentage point per decade, then the Muslim population is expected to remain less than one hundred throughout this century, reaching 1% of the total population by 2100.

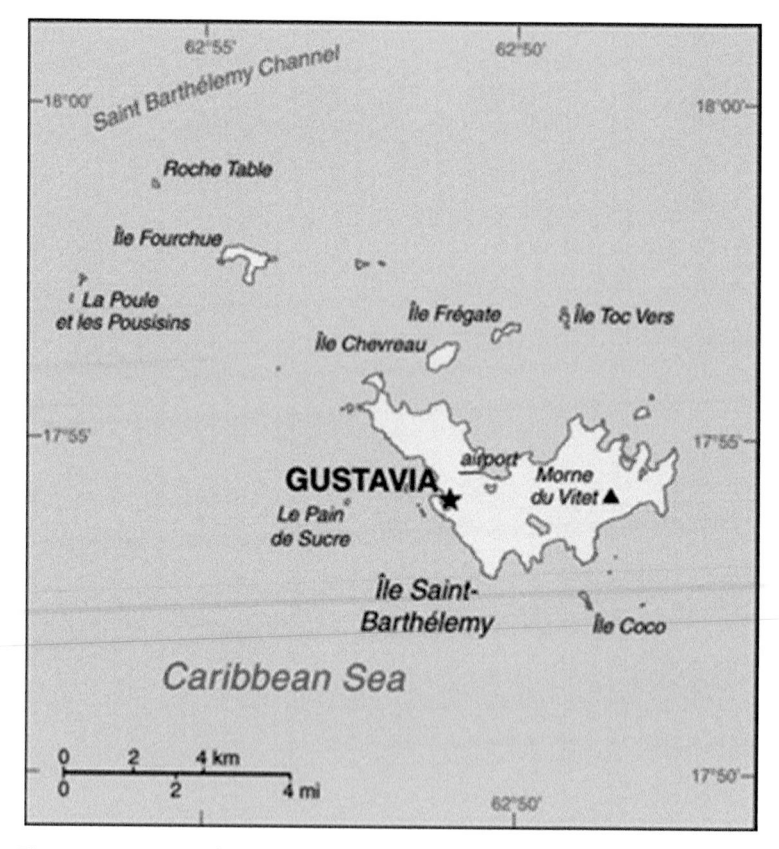

Figure 5.3.5 Map of the Overseas Collectivity of Saint Barthélémy.

Table 5.3.5 Evolution of the Muslim population in Saint Barthélémy

Year	Population	Muslims	%	Source
1999	6,852	—	0.00	[SX]es
2008	8,673	6	0.07	[SX]es
2020	9,600	20	0.20	es
2050	11,000	50	0.50	es
2100	10,000	100	1.00	es

5.3.6 Saint Kitts and Nevis

The Federation of Saint Kitts and Nevis was conquered by the British in 1623 and gained its independence from the UK in 1983. It has an area of 261 sq. km comprising two islands: Saint Kitts also known more formally as Saint Christopher (168 sq. km with 80% of the population) and Nevis (93 sq. km), in a shape of exclamation mark! A map of these islands is presented in Fig. 5.1.6. Muslims started migrating in 1970s, most of whom are medical students. According to census data as shown in Table 5.3.6, the Muslim population has increased from none in 1970 and before, to seven or 0.02% in 1980 to 45 or 0.11% in 1991, to 129 or 0.28% in 2001, to 244 or 0.5% in 2011. About a third of the Muslims population (82) lives in the Island of Nevis, where they constitute 0.7% of the population.

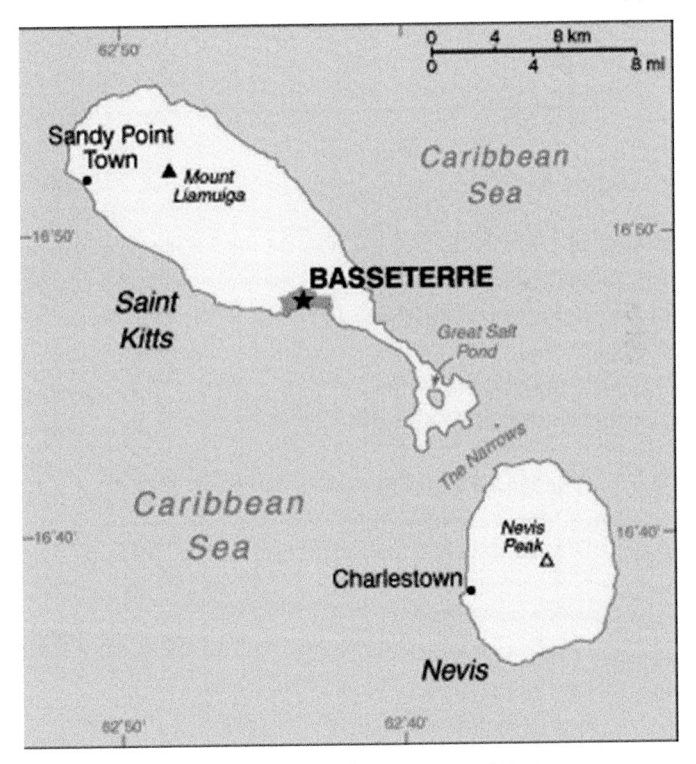

Figure 5.3.6 Map of the Federation of Saint Kitts and Nevis.

Thus, assuming that the percentage of Muslims will increase by a fifth of a percentage point each decade, the Muslim population is expected to remain less than one thousand throughout this century, reaching 1.3% of the total population by 2050 and 2.3% by 2100.

Table 5.3.6 Evolution of the Muslim population in Saint Kitts and Nevis

Year	Population	Muslims	%	Source
1911	42,900	—	0.00	[UN56]c
1946	41,200	—	0.00	[UN56]c
1960	47,530	—	0.00	[KN91]c
1970	42,279	—	0.00	[KN91]c
1980	42,738	7	0.02	[KN91]c
1991	39,585	45	0.11	[KN91]c
2001	44,839	129	0.29	[KN01]c
2011	47,169	244	0.52	[KN11]c
2050	56,000	700	1.30	es
2100	43,000	1,000	2.30	es

5.3.7 Saint Martin

The Island on which it is located was divided between France and the Netherlands in 1648. Whereby, the French north occupying 57% of the Island was named Saint Martin, while the southern 43% part was named Sint Maarten. It seceded from Guadeloupe in 2007 and remains an overseas collectivity of France and has an area of 54 sq. km. The map of Saint Martin is presented in Fig. 5.3.7.

Table 5.3.7 Evolution of the Muslim population in Saint Martin

Year	Population	Muslims	%	Source
1974	6,191	—	0.00	[SX]es
1982	8,072	—	0.00	[SX]es
1990	28,518	77	0.27	[SX]es
1999	29,078	294	1.01	[SX]es
2008	36,661	400	1.10	[SX]es
2050	51,000	800	1.50	es
2100	60,000	1,200	2.00	es

The French government does not keep record of religion in any of its censuses. Thus, as an estimate, we can apply the same percentage of Muslims in Sint Maarten, which has a similar population size and culture, to estimate the Muslim population in Saint Martin. Hence, Muslims arrived here in the 1980s. They increased from none in 1982 and before, to 77 or 0.3% in 1990, to 294 or 1.0% in 1999 and 400 or 1.1% in 2008. The data is summarized in Table 5.3.7. Thus, assuming that the number of Muslims will increase by one tenth of a percentage point per decade, then the Muslim population is expected to remain around one thousand throughout this century, reaching 2% of the total population by 2100.

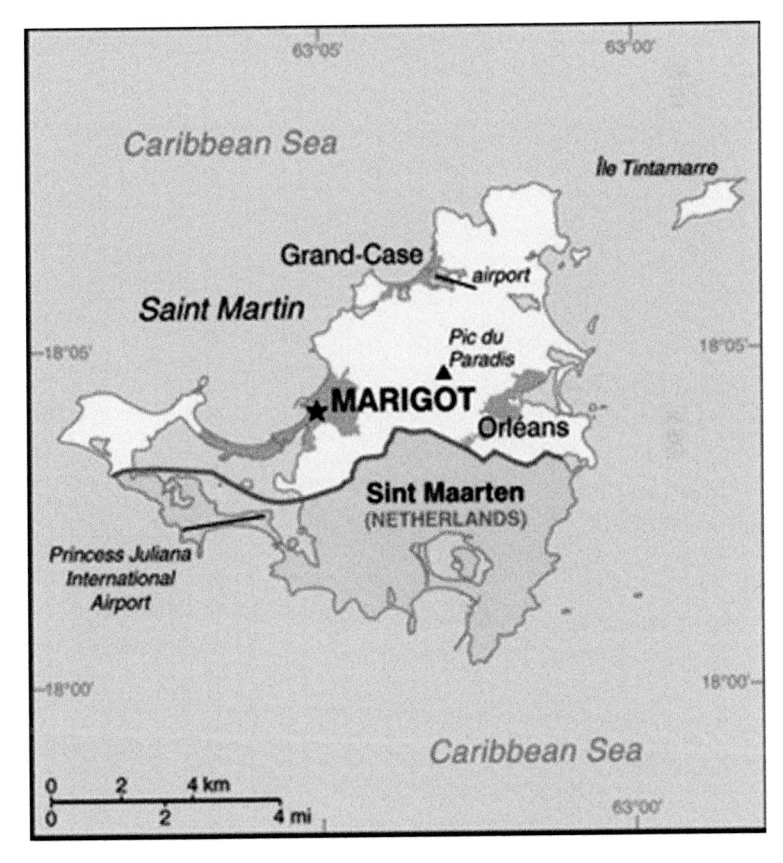

Figure 5.3.7 Map of the Overseas Collectivity of Saint Martin.

5.3.8 Sint Maarten

Land Sint Maarten is a constituent country within the Kingdom of the Netherlands. The island on which it is located was divided between France and the Netherlands in 1648. Whereby, the French north occupying 57% of the Island was named Saint Martin, while the southern 43% part was named Sint Maarten. It has an area of 34 sq. km, and it seceded in 2010 from the Netherlands Antilles (1954–2010) after its disillusion in the same year. The map of Sint Maarten is presented in Fig. 5.3.8. Muslims arrived here in the 1980s. Indeed, based on census data, the Muslim population increased from none in 1981 and before, to 88 or 0.3% in 1992, to 310 or 1.0% in 2001 and 377 or 1.1% in 2011. The data is summarized in Table 5.3.8.

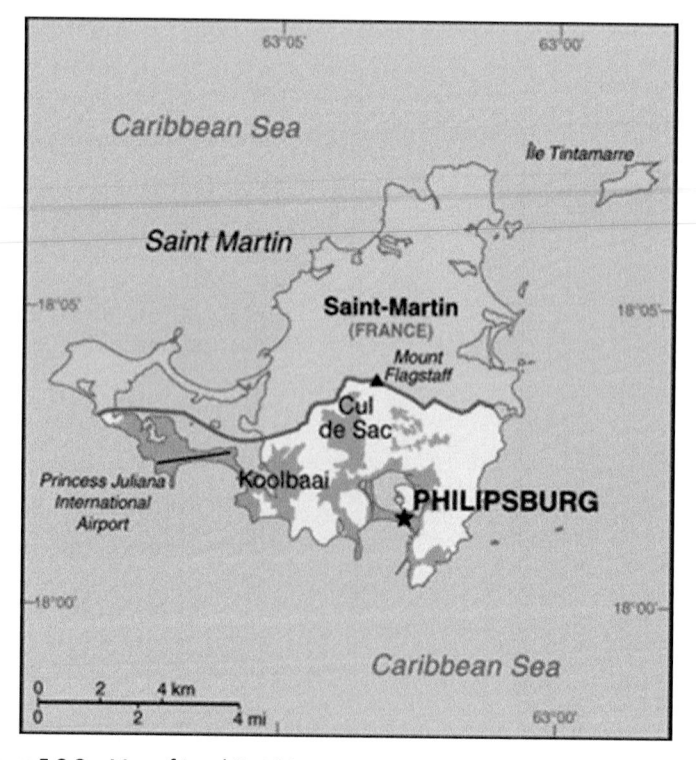

Figure 5.3.8 Map of Land Sint Maarten.

Table 5.3.8 Evolution of the Muslim population in Sint Maarten

Year	Population	Muslims	%	Source
1870	2,850	—	0.00	[SYB73]c
1891	3,881	—	0.00	[SYB94]c
1920	2,633	—	0.00	[SYB22]c
1971	6,385	—	0.00	[AN]c
1981	13,156	—	0.00	[AN]c
1992	32,221	88	0.27	[AN]c
2001	30,380	310	1.02	[AN]c
2011	32,813	377	1.15	[UN]c
2050	55,000	800	1.50	es
2100	69,000	1,400	2.00	es

Thus, assuming that the Muslim population will continue to increase by one tenth of a percentage point per decade, then the Muslim population is expected to increase to 800 or 1.5% by 2050 and 1,400 or 2.0% by 2100.

5.3.9 British Virgin Islands

The Virgin Islands were sighted by Christopher Columbus in 1493 who named them Santa Ursula y las Once Mil Virgenes, shortened to Las Virgenes, after Saint Ursula and her 11,000 virgins! Historically they included BVI, USVI which used to be Danish until 1917, and the Puerto Rican Islands of Vieques and Culebra, which were Spanish until 1898.

The BVI was conquered by the British from the Dutch in 1672 and remains an overseas territory of the UK but uses the US Dollar as its official currency. It has an area of 151 sq. km comprised of sixteen inhabited and more than twenty uninhabited islands. The main islands are Tortola (56 sq. km), Anegada (38 sq. km), Virgin Gorda (21 sq. km), and Jost van Dyke (5 sq. km). However, most of the population lives in Tortola (83%) and Virgin Gorda (13%). The map of these islands is presented in Fig. 5.3.9.

According to census data as shown in Table 5.3.9, the Muslim population increased from none in 1946 and before, to 34 or 0.3% in 1980, to 102 or 0.6% in 1991, to 196 or 0.9% in 2001 and 266 or 1.0% in 2010. In the last census, the Muslim Population was distributed as follows: 256 in Tortola (1.1% of its population), six in Anegada (2.3% of its population) and four in Virgin Gorda (0.1%). Thus, assuming that the percentage of Muslims will continue to increase by a tenth of a percentage point per decade, the Muslim population is expected to remain less than 500 throughout this century.

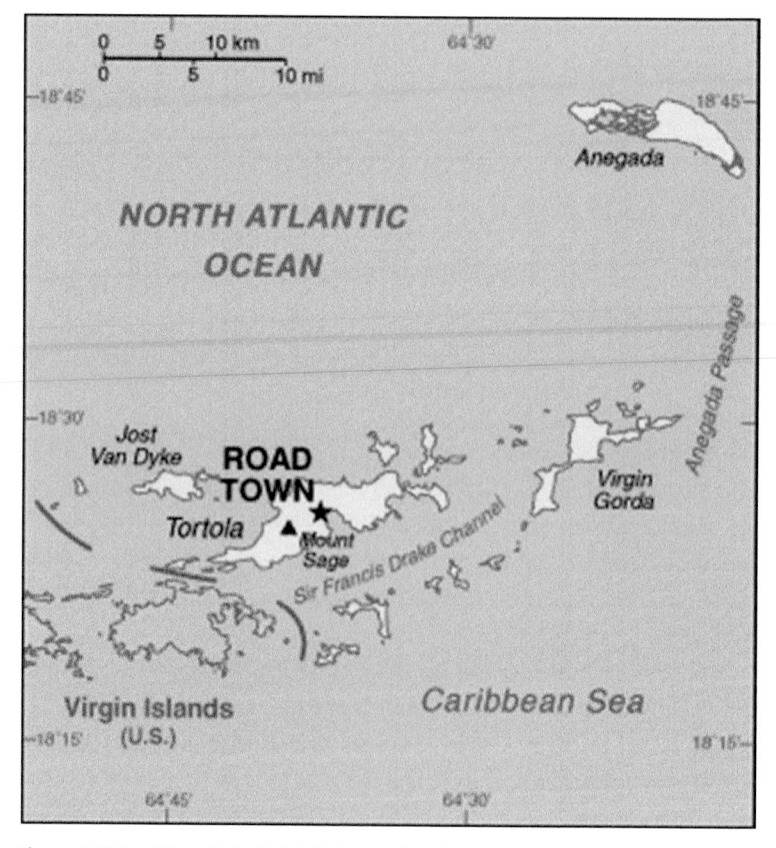

Figure 5.3.9 Map of the British Virgin Islands.

Table 5.3.9 Evolution of the Muslim population in the British Virgin Islands

Year	Population	Muslims	%	Source
1871	6,651	—	0.00	[GB871]c
1911	5,562	—	0.00	[GB911]c
1946	6,500	—	0.00	[UN56]c
1980	10,888	34	0.31	[UN88]c
1991	15,934	102	0.64	[VG01]c
2001	22,533	196	0.87	[VG01]c
2010	27,371	266	0.97	[VG10]c
2020	33,000	400	1.10	es
2050	32,000	400	1.50	es
2100	25,000	500	1.90	es

5.3.10 United States Virgin Islands

It was purchased by the United States from the Denmark in 1917 who controlled it since 1672 as Danish Wiest India Islands. The islands remain an organized and unincorporated territory of the United States. It has a total area of 346 sq. km, consisting of three main islands: Saint Croix (215 sq. km and 49% of the total population), Saint Thomas (81 sq. km and 47%), and Saint John (51 sq. km and 4%), and about fifty much smaller islands, most of which are uninhabited. A map of these islands is presented in Fig. 5.3.10.

Danish censuses prior to 1901 showed that there were no Muslims in the islands. The first Muslim was recorded in the census of 1901, his name was William Park, born in China, was 59 years old, arrived to the islands in 1888. He lived in St. Thomas and his wife was Luthern who was born in St. Croix. He was not recorded in the following census, but one Muslim was recorded again, his name was Nagib Hamdy, born in Syria and arrived the prior year when he was sixteen years old and lived in Charlotte Amalie. More Muslims arrived here from Palestine in the mid twentieth century and some African descent converted to Islam. Thus, estimates for the Muslim population increased from 1,500 or 2.4% in 1970, to 2,000 or 2.1% in 1980, then decreased to 1,300 or 1.2% in 2000 (300 in St. Thomas, and 1,000 in St. Croix), to 1,200 or 1.1% in 2010. Thus, assuming that the percentage of Muslims will remain fixed at 1.1% with respect to the total population, the Muslim population is expected to remain less than one thousand during the second half of this century. A summary of the data is provided in Table 5.3.10.

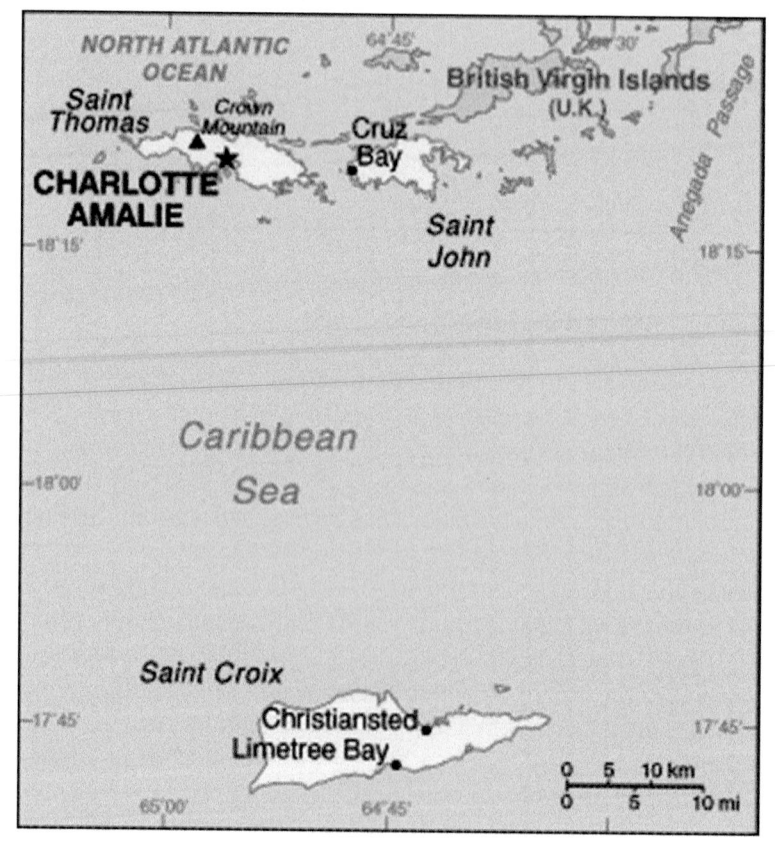

Figure 5.3.10 Map of the United States' Virgin Islands.

Table 5.3.10 Evolution of the Muslim population in the United States Virgin Islands

Year	Population	Muslims	%	Source
1835	43,178	—	0.00	[VIH]c
1850	39,614	—	0.00	[VIH]c
1860	38,231	—	0.00	[VIH]c
1870	37,821	—	0.00	[VIH]c
1880	33,763	—	0.00	[VIH]c
1890	32,786	—	0.00	[VIH]c
1901	30,527	1	0.00	[VIH]c
1911	27,086	1	0.00	[VIH]c
1970	62,468	1,500	2.39	[KET76]es
1980	96,569	2,000	2.07	[KET86]es
2000	108,612	1,300	1.19	[VI1,VI2]es
2010	106,405	1,200	1.13	[VI]es
2050	86,000	1,000	1.10	es
2100	40,000	500	1.10	es

5.3.11 Regional Summary and Conclusion

Muslims arrived here as explorers, centuries before the arrival of the Europeans, then as escapees from Spanish Catholic oppression, then as black slaves brought by the Europeans from Africa. However, Islam went extinct from these migrations. Towards the end of the nineteenth century, the Brits and Dutch brought workers from India and Java, some of whom were Muslim, and Islam remains among their descendants till this day. The French also brought Muslims from their African colonies to their colonies in this region. The percentage of Muslims is expected to increase to 1.5% by the end of this century. The following tables present centennial data from 600 AD to 2100 AD (or approximately 1 H to 1500 H) in Table 5.3a and decennial data from 1790 AD to 2100 AD (or 1210 H to 1520 H) in Tables 5.3b and 5.3c for current countries in Southern Caribbean Islands. The data includes total population in thousands (P), the percentage of which is Muslim (M%), the corresponding Muslim population in thousands (M), and the annual population growth rate (APGR, or G%) of the total population in this region.

Table 5.3a Centennial estimates of the Muslim population (×1000) in Central Caribbean Islands from 600 to 2100 (1H to 1500H)

		600	700	800	900	1000	1100	1200	1300
Anguilla	P	—	—	—	—	—	—	—	—
	M%	—	—	—	—	—	—	—	—
	M	—	—	—	—	—	—	—	—
Antigua Barbuda	P	—	—	—	—	—	—	—	—
	M%	—	—	—	—	—	—	—	—
	M	—	—	—	—	—	—	—	—
Guade- loupe	P	1	1	1	1	1	2	2	2
	M%	—	—	—	—	—	—	—	—
	M	—	—	—	—	—	—	—	—
Mont- serrat	P	—	—	—	—	—	—	—	—
	M%	—	—	—	—	—	—	—	—
	M	—	—	—	—	—	—	—	—
St. Bart	P	—	—	—	—	—	—	—	—
	M%	—	—	—	—	—	—	—	—
	M	—	—	—	—	—	—	—	—
SKN	P	1	2	3	4	5	5	5	6
	M%	—	—	—	—	—	—	—	—
	M	—	—	—	—	—	—	—	—
St. Martin	P	—	—	—	—	—	—	—	—
	M%	—	—	—	—	—	—	—	—
	M	—	—	—	—	—	—	—	—
Sint Maarten	P	—	—	—	—	—	—	—	—
	M%	—	—	—	—	—	—	—	—
	M	—	—	—	—	—	—	—	—
BVI	P	1	1	1	1	1	1	1	1
	M%	—	—	—	—	—	—	—	—
	M	—	—	—	—	—	—	—	—
USVI	P	1	1	1	1	1	1	1	1
	M%	—	—	—	—	—	—	—	—
	M	—	—	—	—	—	—	—	—
Total	P	4	5	6	7	8	8	8	9
	M%	—	—	—	—	—	—	—	—
	M	—	—	—	—	—	—	—	—
	G%		0.223	0.182	0.154	0.134	0.000	0.000	0.118

Table 5.3a (Continued)

		1400	1500	1600	1700	1800	1900	2000	2100
Anguilla	P	—	—	—	1	3	4	11	14
	M%	—	—	—	—	—	—	0.32	0.41
	M	—	—	—	—	—	—	—	—
Antigua Barbuda	P	—	—	—	16	38	35	76	102
	M%	—	—	—	—	—	—	0.30	0.70
	M	—	—	—	—	—	—	—	1
Guadeloupe	P	2	3	3	21	110	176	422	339
	M%	—	—	—	—	—	2	—	1
	M	—	—	—	—	—	3	2	5
Montserrat	P	—	—	—	5	25	12	5	3
	M%	—	—	—	—	—	—	0.18	0.20
	M	—	—	—	—	—	—	—	—
St. Bart	P	—	—	—	1	2	3	7	10
	M%	—	—	—	—	—	—	—	1.00
	M	—	—	—	—	—	—	—	—
SKN	P	6	6	3	9	35	43	44	43
	M%	—	—	—	—	—	—	—	2
	M	—	—	—	—	—	—	—	1
St. Martin	P	—	—	—	1	3	4	29	60
	M%	—	—	—	—	—	—	1	2
	M	—	—	—	—	—	—	—	1
Sint Maarten	P	—	—	—	1	2	4	33	69
	M%	—	—	—	—	—	—	1.02	2.00
	M	—	—	—	—	—	—	—	1
BVI	P	1	1	1	1	11	5	20	25
	M%	—	—	—	—	—	—	1	2
	M	—	—	—	—	—	—	—	—
USVI	P	1	1	1	3	33	31	109	40
	M%	—	—	—	—	—	—	1	1
	M	—	—	—	—	—	—	1	—
Total	P	9	10	7	37	151	139	334	367
	M%	—	—	—	—	—	—	0.75	1.47
	M	—	—	—	—	—	—	2	5
	G%	0.033	0.032	−0.330	1.686	1.402	−0.083	0.874	0.093

Table 5.3b Decennial estimates of the Muslim population (×1000) in Central Caribbean Islands from 1790 to 1940 (1210 H to 1360 H)

		1790	1800	1810	1820	1830	1840	1850	1860
Anguilla	P	2	3	3	3	3	3	3	4
	M%	—	—	—	—	—	—	—	—
	M	—	—	—	—	—	—	—	—
Antigua Barbuda	P	24	38	40	33	34	36	37	37
	M%	—	—	—	—	—	—	—	—
	M	—	—	—	—	—	—	—	—
Guadeloupe	P	107	110	113	116	120	120	121	138
	M%	—	—	—	—	—	—	0.01	0.10
	M	—	—	—	—	—	—	—	—
Montserrat	P	23	25	26	18	14	22	7	8
	M%	—	—	—	—	—	—	—	—
	M	—	—	—	—	—	—	—	—
St. Bart	P	1	2	5	2	2	2	2	3
	M%	—	—	—	—	—	—	—	—
	M	—	—	—	—	—	—	—	—
SKN	P	34	35	32	29	28	33	33	34
	M%	—	—	—	—	—	—	—	—
	M	—	—	—	—	—	—	—	—
St. Martin	P	3	3	3	3	3	3	3	3
	M%	—	—	—	—	—	—	—	—
	M	—	—	—	—	—	—	—	—
Sint Maarten	P	2	2	3	3	3	3	3	3
	M%	—	—	—	—	—	—	—	—
	M	—	—	—	—	—	—	—	—
BVI	P	10	11	7	6	5	8	9	6
	M%	—	—	—	—	—	—	—	—
	M	—	—	—	—	—	—	—	—
USVI	P	30	33	36	39	42	43	41	40
	M%	—	—	—	—	—	—	—	—
	M	—	—	—	—	—	—	—	—
Total	P	130	151	156	137	135	154	140	137
	M%	—	—	—	—	—	—	—	—
	M	—	—	—	—	—	—	—	—
	G%		1.527	0.271	−1.270	−0.119	1.284	−0.979	−0.201

Table 5.3b (*Continued*)

		1870	1880	1890	1900	1910	1920	1930	1940
Anguilla	P	4	4	4	4	4	4	5	6
	M%	—	—	—	—	—	—	—	—
	M	—	—	—	—	—	—	—	—
Antigua Barbuda	P	35	35	37	35	32	30	31	40
	M%	—	—	—	—	—	—	—	0.05
	M	—	—	—	—	—	—	—	—
Guadeloupe	P	142	149	160	176	206	223	260	299
	M%	1.00	2.00	2.25	1.50	1.50	1.00	0.50	0.28
	M	1	3	4	3	3	2	1	1
Montserrat	P	9	10	12	12	12	12	13	13
	M%	—	—	—	—	—	—	—	—
	M	—	—	—	—	—	—	—	—
St. Bart	P	2	3	3	3	3	3	2	3
	M%	—	—	—	—	—	—	—	—
	M	—	—	—	—	—	—	—	—
SKN	P	40	41	44	43	39	34	36	32
	M%	—	—	—	—	—	—	—	—
	M	—	—	—	—	—	—	—	—
St. Martin	P	3	4	3	4	4	4	5	7
	M%	—	—	—	—	—	—	—	—
	M	—	—	—	—	—	—	—	—
Sint Maarten	P	3	3	4	4	3	3	2	2
	M%	—	—	—	—	—	—	—	—
	M	—	—	—	—	—	—	—	—
BVI	P	7	5	5	5	6	5	4	5
	M%	—	—	—	—	—	—	—	—
	M	—	—	—	—	—	—	—	—
USVI	P	38	34	33	31	27	25	22	25
	M%	—	—	—	—	0.04	0.04	0.04	0.04
	M	—	—	—	—	—	—	—	—
Total	P	141	138	144	139	130	120	121	132
	M%	—	—	—	—	0.01	0.01	0.01	0.02
	M	—	—	—	—	—	—	—	—
	G%	0.270	−0.165	0.384	−0.300	−0.715	−0.818	0.090	0.882

Table 5.3c Decennial estimates of the Muslim population (×1000) in Central Caribbean Islands from 1950 to 2100 (1370 H to 1520 H)

		1950	1960	1970	1980	1990	2000	2010	2020
Anguilla	P	5	6	7	7	9	11	13	15
	M%	—	—	—	—	—	0.32	0.32	0.33
	M	—	—	—	—	—	—	—	—
Antigua Barbuda	P	46	54	64	62	63	76	88	98
	M%	0.05	0.05	0.05	0.30	0.30	0.30	0.26	0.30
	M	—	—	—	—	—	—	—	—
Guade-loupe	P	210	275	322	337	389	422	406	400
	M%	0.28	0.28	0.28	0.28	0.28	0.40	0.50	0.60
	M	1	1	1	1	1	2	2	2
Mont-serrat	P	14	12	12	12	11	5	5	5
	M%	—	—	—	0.12	—	0.18	0.16	0.20
	M	—	—	—	—	—	—	—	—
St. Bart	P	2	2	2	3	5	7	9	10
	M%	—	—	—	—	—	—	0.07	0.20
	M	—	—	—	—	—	—	—	—
SKN	P	46	51	45	43	40	44	49	53
	M%	—	—	—	0.02	0.11	0.29	0.52	0.70
	M	—	—	—	—	—	—	—	—
St. Martin	P	3	4	6	9	32	29	38	39
	M%	—	—	—	—	0.27	1.01	1.10	1.20
	M	—	—	—	—	—	—	—	—
Sint Maarten	P	1	3	7	13	29	33	34	43
	M%	—	—	—	—	0.27	1.02	1.15	1.20
	M	—	—	—	—	—	—	—	1
BVI	P	7	8	10	11	17	20	28	30
	M%	—	—	—	0.31	0.64	0.87	0.97	1.10
	P	—	—	—	—	—	—	—	—
USVI	P	27	33	65	99	104	109	106	104
	M%	2.39	2.39	2.39	2.39	1.19	1.19	1.13	1.10
	M	1	1	2	2	1	1	1	1
Total	P	151	174	217	259	309	334	370	397
	M%	0.44	0.47	0.73	1.01	0.56	0.75	0.76	0.81
	M	1	1	2	3	2	2	3	3
	G%	1.355	1.394	2.228	1.795	1.754	0.778	1.023	0.703

Table 5.3c *(Continued)*

		2030	2040	2050	2060	2070	2080	2090	2100
Anguilla	P	16	16	17	16	16	15	14	14
	M%	0.34	0.35	0.36	0.37	0.38	0.39	0.40	0.41
	M	—	—	—	—	—	—	—	—
Antigua Barbuda	P	105	109	111	111	109	107	105	102
	M%	0.35	0.40	0.45	0.50	0.55	0.60	0.65	0.70
	M	—	—	—	1	1	1	1	1
Guadeloupe	P	400	398	387	374	365	358	351	339
	M%	0.70	0.80	0.90	1.00	1.10	1.20	1.30	1.40
	M	3	3	3	4	4	4	5	5
Montserrat	P	5	4	4	4	4	3	3	3
	M%	0.20	0.20	0.20	0.20	0.20	0.20	0.20	0.20
	M	—	—	—	—	—	—	—	—
St. Bart	P	10	11	11	11	11	11	10	10
	M%	0.30	0.40	0.50	0.60	0.70	0.80	0.90	1.00
	M	—	—	—	—	—	—	—	—
SKN	P	56	57	56	54	52	49	46	43
	M%	0.90	1.10	1.30	1.50	1.70	1.90	2.10	2.30
	M	1	1	1	1	1	1	1	1
St. Martin	P	43	47	51	53	55	57	59	60
	M%	1.30	1.40	1.50	1.60	1.70	1.80	1.90	2.00
	M	1	1	1	1	1	1	1	1
Sint Maarten	P	48	52	55	58	60	63	66	69
	M%	1.30	1.40	1.50	1.60	1.70	1.80	1.90	2.00
	M	1	1	1	1	1	1	1	1
BVI	P	32	32	32	30	29	27	26	25
	M%	1.20	1.30	1.40	1.50	1.60	1.70	1.80	1.90
	M	—	—	—	—	—	—	—	—
USVI	P	101	94	86	77	69	60	51	40
	M%	1.10	1.10	1.10	1.10	1.10	1.10	1.10	1.10
	M	1	1	1	1	1	1	1	—
Total	P	415	423	422	414	404	394	381	367
	M%	0.88	0.95	1.03	1.11	1.19	1.28	1.37	1.47
	M	4	4	4	5	5	5	5	5
	G%	0.448	0.188	−0.027	−0.248	−0.248	−0.268	−0.323	−0.380

5.4 Muslims in South America

This region consists of fourteen countries and territories: Argentina, Bolivia, Brazil, Chile, Colombia, Ecuador, Falkland Islands, French Guiana, Guyana, Paraguay, Peru, Suriname, Uruguay, and Venezuela. There was a strong pre-Columbian Muslim presence in this region, then by the Moriscos in the seventeenth century, and later by African slaves in the nineteenth century, all were crushed by the Spaniards and Portuguese. Towards the end of the nineteenth century, waves of Arab Muslims migrated here from Greater Syria (Syria and Lebanon) due to economic hardship and avoiding enlistment in the Ottoman army, the British brought workers from India to Guyana, and the Dutch brought workers from India and Java to Suriname, while the French brought prisoners from Algeria and Morocco who fought their occupation there. Thus, the Muslim population has likely increased from none in 1800, to 20,000 or 0.05% in 1900, to 0.3 million or 0.1% in 2000, to 0.4 million or 0.1% in 2020, and is projected to reach 0.5 million or 0.1% by 2050 and 0.6 million or 0.1% by 2100.

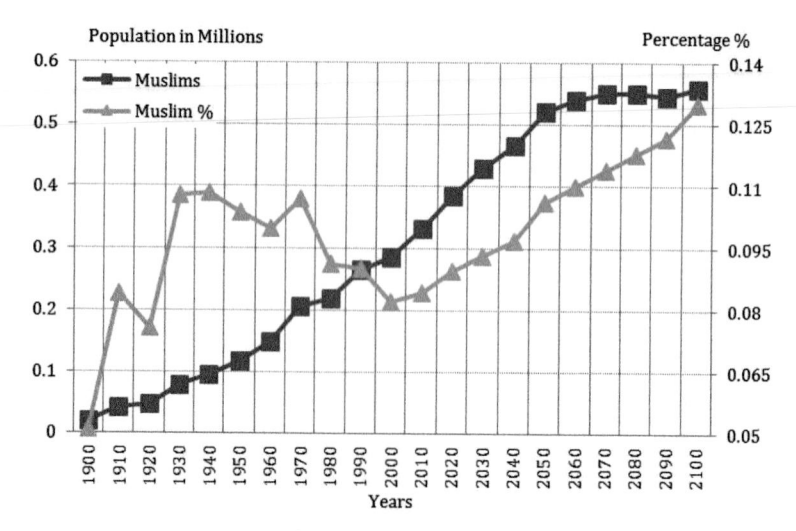

Figure 5.4 Plot of decennial estimates of the Muslim population and its percentage of the total population in South America: 1900–2100 AD (1320–1520 H).

A plot of decennial estimates of the Muslim population and its percentage with respect to the total population in this region from

1900 to 2100 is provided in Fig. 5.4. The corresponding individual data for each country in this region is discussed below. In Section 5.4.15, the total population in each country in this region and the corresponding percentage and number of Muslims is presented centennially in Table 5.4a from 600 to 2100 and decennially in Tables 5.4b to 5.4c from 1790 to 2100.

5.4.1 Argentina

The Argentine Republic has an area of 2,789,400 sq. km and its map is presented in Fig. 5.4.1. The Spanish arrived here in 1516 and established the capital Buenos Aires in 1580. Argentina gained its independence from Spain in 1816. The first Muslims arrived here from Greater Syria in 1880, due to economic hardship and avoiding enlistment in the Ottoman army. Thus, census-based estimates of the Muslim population increased from none in 1869, when the first census was held, to 962 or 0.02% in 1895, to 9,445 or 0.12% in 1914, to 18,764 or 0.12% in 1947, then dropped to 14,262 or 0.08% in 1960. Large scale migration from Greater Syria lasted until 1955 and included Muslim, Jewish and Christian Arabs, with more of the latter. Thus, the vast majority of Muslims are descendants of this region.

Argentinean national censuses after 1960 did not include information about religious affiliation. However, according to World Values Survey (WVS), the percentage of Muslims was 0.20% in 1991, 0.09% in 1995 and 0.19% in 2013. Thus, assuming that the percentage of Muslims will remain at 0.2% of the total population, the Muslim population is expected to remain around 0.1 million throughout this century. A summary of the data is provided in Table 5.4.1.

Table 5.4.1 Evolution of the Muslim population in Argentina

Year	Population	Muslims	%	Source
1869	1,830,214	—	0.00	[KET86]es
1895	4,044,911	962	0.02	[ROD]c
1914	7,903,662	9,445	0.12	[ROD]c
1947	15,779,238	18,764	0.12	[UN56]c
1960	17,879,969	14,262	0.08	[ROD]c
1991	33,193,000	65,000	0.20	[WVS]s
2013	42,538,000	80,000	0.19	[WVS]s
2050	54,867,000	110,000	0.20	es
2100	56,802,000	114,000	0.20	es

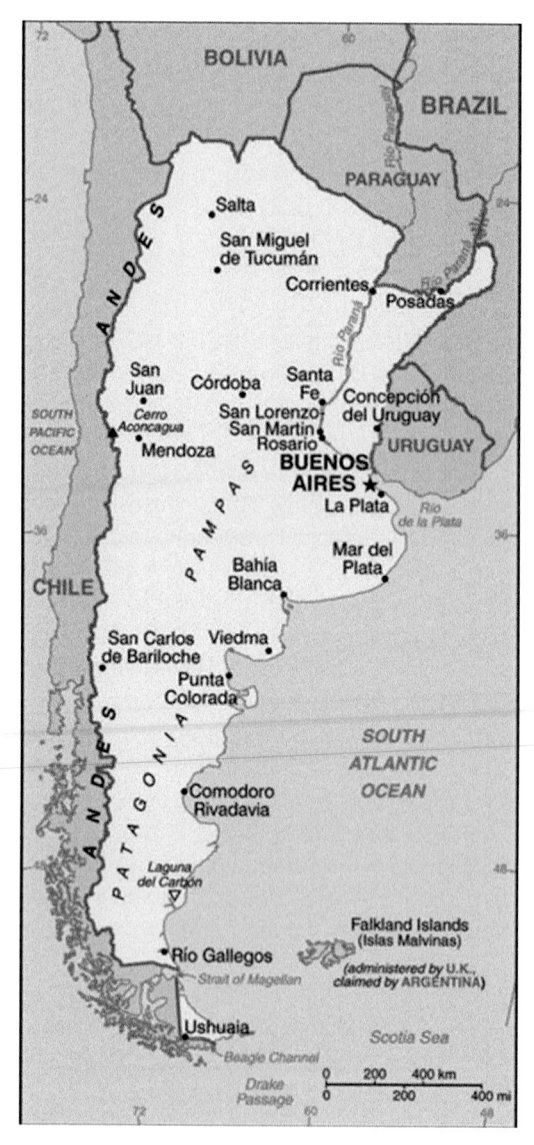

Figure 5.4.1 Map of the Argentine Republic.

5.4.2 Bolivia

The Plurinational State of Bolivia has an area of 1,098,581 sq. km and its map is presented in Fig. 5.4.2. It was conquered by Spain in 1533 and gained its independence in 1825. The first Muslims came in the 1920s, from Palestine. However, the number of Muslims remains negligible at 0.01% of the total population. Estimates for their number increased from 500 in 1976, to 1,000 in 1992 and 2011. Thus, assuming that the percentage of Muslims will remain constant, the Muslim population is expected to remain less than 2,000 throughout this century. The data is summarized in Table 5.4.2.

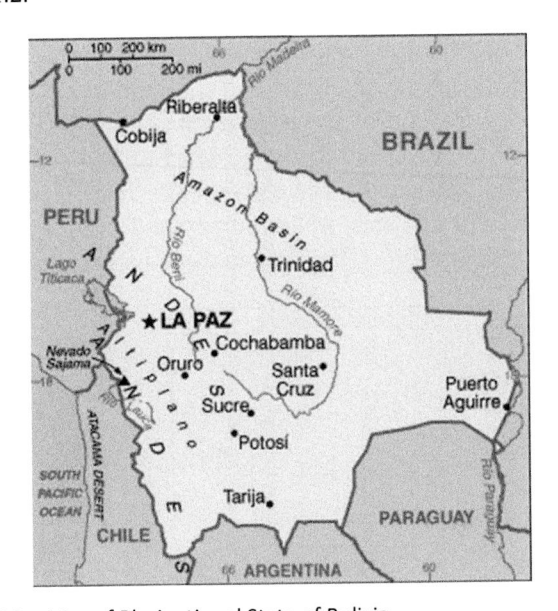

Figure 5.4.2 Map of Plurinational State of Bolivia.

Table 5.4.2 Evolution of the Muslim population in Bolivia

Year	Population	Muslims	%	Source
1976	4,613,486	500	0.01	[KET86]es
1992	6,420,792	1,000	0.02	[KET01C]es
2012	10,027,262	1,000	0.01	[DOS11]es
2020	11,548,000	1,200	0.01	es
2050	15,840,000	1,600	0.01	es
2100	17,391,000	1,800	0.01	es

5.4.3 Brazil

The Federative Republic of Brazil has an area of 8,514,877 sq. km and is the fifth largest country in the world. Its map is presented in Fig. 5.4.3. It was conquered by Portugal in 1500 and gained its independence in 1822. The first post-Columbian wave of Muslims came from Portugal in the sixteenth century. They were the remnants of Muslims in the Iberian Peninsula and were called Moriscos. But anyone who was discovered as Muslim was forced to be Christian or be burned alive. Thus, this wave was lost.

The second wave came as the Portuguese brought slaves from Africa towards the end of the eighteenth century. Many of these slaves were Muslims. They rebelled in 1835 to gain some freedom, but they were crushed by the Portuguese. Many Muslims who survived returned to Africa, where they have descendants in Benin and neighboring countries. Thus, Islam went extinct again in Brazil.

Figure 5.4.3 Map of the Federative Republic of Brazil.

The third wave came in 1860, as many Arabs, some of whom were Muslims, fled the Ottoman Empire from Syria and Lebanon. However, according to census data, the Muslim population remained less than 0.01% of the total population until 1950, changing in from 300 in 1890, to 123 in 1900, to 3,053 in 1940, to 3,454 in 1950, to 7,745 in 1960. The Muslim population then increased to 0.02%, numbering 22,449 in 1991, then 27,239 or 0.02% in 2000, and 35,167 or 0.02% in 2010. Nevertheless, the results of these censuses are widely challenged by Muslims in Brazil and independent researchers. They claim that the number of Muslims is ten to hundred times what these censuses report. Thus, from census data, and assuming that the percentage of Muslims will increase by 0.01 of a percentage point every half century, the Muslim population is expected to remain less than 0.1 million throughout this century, comprising 0.04% of the population by 2100. The census and projected data are summarized in Table 5.4.3.

Table 5.4.3 Evolution of the Muslim population in Brazil

Year	Population	Muslims	%	Source
1890	14,333,915	300	0.00	[ROD]c
1900	16,626,991	123	0.00	[BRH]c
1940	39,177,880	3,053	0.01	[BRH]c
1950	51,806,591	3,454	0.01	[BRH]c
1960	70,119,071	7,745	0.01	[ROD]c
1991	146,229,473	22,449	0.02	[ROD]c
2000	169,488,903	27,239	0.02	[BR00]c
2010	189,870,280	35,167	0.02	[BR10]c
2050	228,980,000	69,000	0.03	es
2100	180,683,000	72,000	0.04	es

5.4.4 Chile

The Republic of Chile has an area of 756,102 sq. km and its map is presented in Fig. 5.4.4. It was conquered by Spain in 1540 and gained its independence in 1818. The first Muslims came from Syria, Lebanon, and Palestine towards the end of the nineteenth century. They were fleeing economic hardship and enlistment in the Ottoman army. But still they were referred to as Turks!

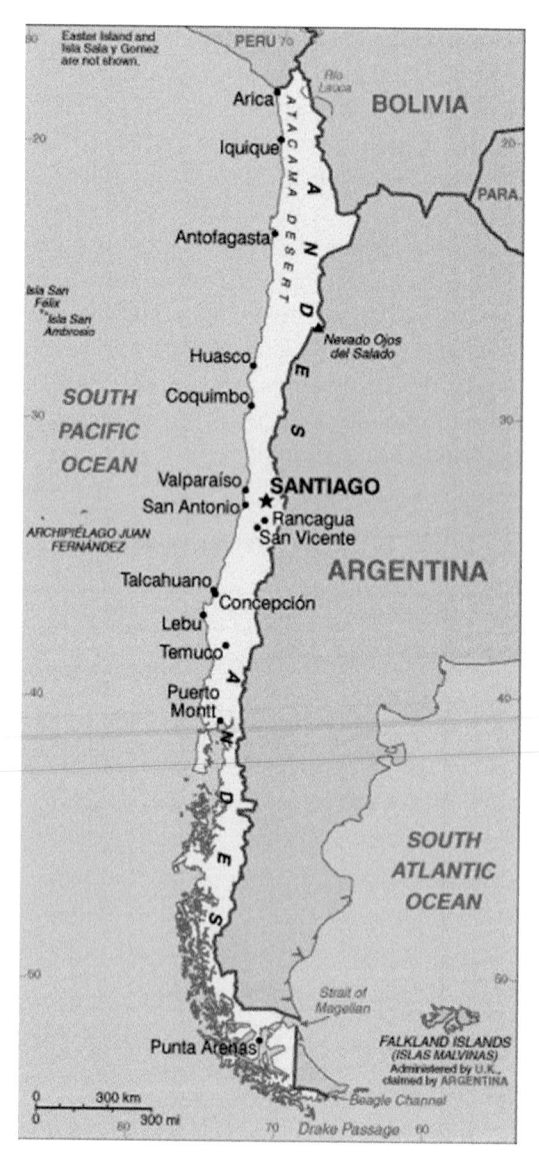

Figure 5.4.4 Map of the Republic of Chile.

According to census data, The Muslim population remained at 0.00% of the total population throughout the nineteenth century, increasing from two in 1865 and 1875, to 29 in 1885, to 58 in 1895. It then oscillated from 1,498 or 0.04% in 1907, to 402 or 0.01%

in 1920, to 956 or 0.02% in 1952, to 522 or 0.01% in 1960, and continued increasing since then to 1,431 or 0.02% in 1970 and to 2,894 or 0.03% in 2002, to 3,288 or 0.03% in 2012. The last two censuses refer to population aged 15 and older. Thus, from census data, and assuming that the percentage of Muslims will increase by 0.01 of a percentage point every half century, the Muslim population is expected to remain less than 10,000 or 0.05% throughout this century. A summary of the data is provided in Table 5.4.4.

Table 5.4.4 Evolution of the Muslim population in Chile

Year	Population	Muslims	%	Source
1865	1,819,223	2	0.00	[CLH]e
1875	2,075,971	2	0.00	[CLH]e
1885	2,507,005	29	0.00	[CLH]e
1895	2,695,625	58	0.00	[CLH]c
1907	3,231,022	1,498	0.04	[CLH]c
1920	3,720,235	402	0.01	[CLH]c
1952	5,760,848	956	0.02	[UN56]c
1960	7,374,115	522	0.01	[CLH]c
1970	7,953,242	1,431	0.02	[UN79]c
2002	11,226,309	2,894	0.03	[CL02]c
2012	12,907,543	3,288	0.03	[CL12]c
2050	20,319,000	8,000	0.04	es
2100	17,332,000	9,000	0.05	es

5.4.5 Colombia

The Republic of Colombia has an area of 1,109,104 sq. km and its map is presented in Fig. 5.4.5. It was conquered by Spain in 1499, gained its independence in 1819 as part of the former Gran Columbia, and then emerged from the collapse of the latter in 1830. Gran Columbia used to include current Ecuador, Columbia, Panama, and Venezuela.

Between 1615 and 1700 Muslim Moriscos from Spain established a colony here and in Venezuela. However, they were obliterated by the Spaniards. Towards the end of the nineteenth century Muslims started emigrating from Syria and Lebanon. Estimates for the Muslim population increased from 500 or 0.01% in 1921 to 10,000 or 0.04% in 1973. According to World Values Survey (WVS), the percentage of Muslims was 0.03% in 2005 and 0.07% in 2012. Thus, assuming that the Muslim population will increase by 0.02 of a percentage point every other decade, the Muslim population is expected to remain around 0.1 million during the second half of this century, comprising 0.3% of the total population by 2100. The data is summarized in Table 5.4.5.

Figure 5.4.5   Map of the Republic of Colombia.

Table 5.4.5   Evolution of the Muslim population in Colombia

Year	Population	Muslims	%	Source
1921	3,753,800	500	0.01	[VEH]es
1973	22,551,811	10,000	0.04	[KET86]es
2005	43,286,000	13,000	0.03	[WVS]s
2012	46,881,000	33,000	0.07	[WVS]s
2050	55,958,000	90,000	0.16	es
2100	45,221,000	118,000	0.26	es

5.4.6 Ecuador

The Republic of Ecuador has an area of 283,561 sq. km and its map is presented in Fig. 5.4.6. It was conquered by Spain in 1533, gained its independence in 1822 as part of the former Gran Columbia, and then emerged from the collapse of the latter in 1830. In 1908, the Muslim population consisted of twenty Chinese Muslims. This increased to 100 in 1974, with the arrival of Muslims from Syria Lebanon and Palestine. Muslims then increased to 500 or 0.01% in 1982, to 2,000 or 0.02% in 2005. Thus, assuming that the percentage of Muslims will increase by 0.01 of a percentage point every half century, the Muslim population is expected to remain less than 10,000 throughout this century. A summary of the data is provided in Table 5.4.6.

Figure 5.4.6 Map of the Republic of Ecuador.

Table 5.4.6 Evolution of the Muslim population in Ecuador

Year	Population	Muslims	%	Source
1908	1,396,000	20	0.00	[ARH]es
1974	6,521,710	100	0.00	[KET76]es
1982	8,060,712	500	0.01	[KET86]es
2001	12,156,608	500	0.00	[EC]es
2010	14,483,499	2,000	0.01	[PEW]es
2050	23,316,000	5,000	0.02	es
2100	24,483,000	7,000	0.03	es

5.4.7 Falkland Islands (Islas Malvinas)

The Islands were uninhabited when they were first sighted by the British in 1592, but there is evidence that the Islands were settled sometime in the past by Indians from South America. The first recorded landing on the Falkland Islands occurred in 1690 and was made at Bold Cove near Port Howard on West Falkland to replenish the water supplies of British ship "Welfare" commanded by John Strong, who named the stretch of water between West and East Falkland "Falkland Sound" after Lord Falkland, who was a financial supporter of Strong's voyage, Treasurer to the Navy and shortly to become First Lord of the Admiralty.

The Islands were first settled by the French in 1764 who established Fort St. Louis settlement with a population of thirty, increasing to 130 the following year. This settlement was later renamed Port Louis in 1830 and is located in Berkeley Sound on the East Island, north of the capital Stanley. The French named the Islands Isles Malouines after St. Malo, the port from which their expedition set out. The French however, ceded the Islands to Spain in 1767 who named them Islas Malvinas. But the British also claimed the Islands in 1765 who established a settlement in Port Egmont on Saunders Island, northwest of the West Island, about 200Km from the French settlement. The British left in 1774, and the Spaniards left in 1811, leaving the Islands deserted from permanent population. The Islands were claimed by Argentina in 1820 after its independence from Spain in 1816. By 1832, the population of the Islands was twenty.

The Islands were conquered by the British in 1833, who established the capital Port Stanley in 1845, named after the British Secretary of State for the Colonies, Edward Stanley. The Islands remain an overseas territory of the UK and have an area of 12,173 sq. km which comprises two main islands of East (6,605 sq. km) and West (4,532 sq. km) Falkland and almost a thousand of much smaller islands. However, almost all population lives in East Falkland. A map of these islands is presented in Fig. 5.4.7.

According to census data as shown in Table 5.4.7, the Muslim population changed from none before 1980s to nine or 0.3% in 2006, to two or 0.1% in 2012, to seven or 0.2% in 2016, all live in East Falkland. Based on country of birth, those born in a Muslim majority country changed from none prior to 1980, to two (Iran and UAE) or 0.1% in 1986, to three or 0.1% in 1991 (one in Senegal and two in UAE), 1996 (two in Indonesia and one in UAE) and 2001 (Iran, Malaysia and UAE). The number increased to eighteen in 2006, born in thirteen majority Muslim countries. Thus, assuming that the percentage of Muslims will increase by a tenth of a percentage point per decade, then the Muslim population is expected to remain less than thirty throughout this century.

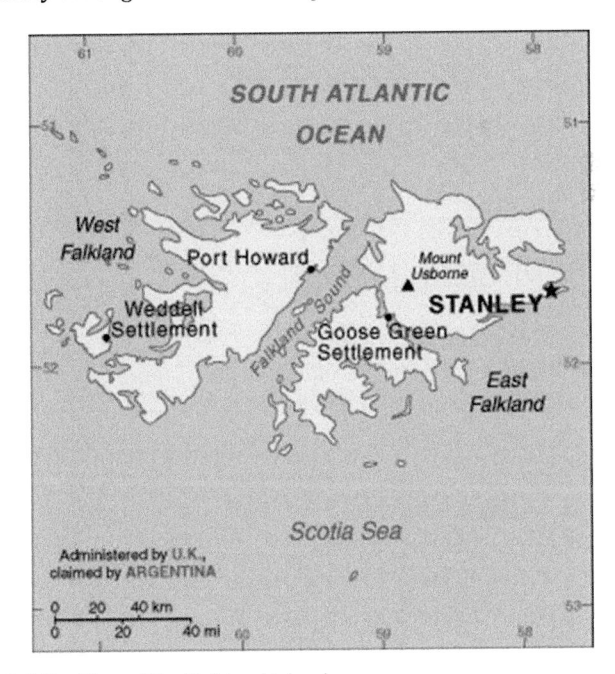

Figure 5.4.7 Map of the Falkland Islands.

Table 5.4.7 Evolution of the Muslim population in the Falkland Islands

Year	Population	Muslims	%	Source
1833	22	—	0.00	[FKE]e
1842	52	—	0.00	[FK842]c
1843	111	—	0.00	[FK843]c
1851	383	—	0.00	[FKE]e
1901	2,253	—	0.00	[GB901]c
1911	3,275	—	0.00	[GB911]c
1921	2,094	—	0.00	[FKE]e
1931	2,392	—	0.00	[FK931]c
1953	2,230	—	0.00	[FKE]e
1962	2,172	—	0.00	[FKE]e
1972	1,957	—	0.00	[FKE]e
1980	1,813	—	0.00	[FK80]c
1986	1,916	2	0.10	[FKE]e
1991	2,091	3	0.10	[FKE]e
1996	2,564	3	0.10	[FKE]e
2001	2,913	3	0.10	[FK06]e
2006	2,955	9	0.30	[FK06]c
2012	2,621	2	0.08	[FK16]c
2016	3,004	7	0.23	[FK16]c
2050	3,200	15	0.50	es
2100	2,700	30	1.00	es

5.4.8 French Guiana (Guyane)

It is an overseas department of France that it acquired in 1667 and has a total area of 83,534 sq. km and its map is presented in Fig. 5.4.8. The slavery was abolished in 1848, thus the French tried to imitate the British indentured labor system. Hence, in 1850, Indians, Chinese, Malay, and Africans were brought to work on Sugarcane plantations, among whom were many Muslims. France also used the territory as a prison between 1852 and 1951, including its notorious Devil's Island prison. Many Muslim prisoners of war who fought the French occupation in Morocco and Algeria were sent to prisons here. Thus, by 1907, the Muslim population reached 1,570 or 6.5% of the total population, all but 482 were prisoners. The Muslim population increased to 2,500 or 9.0% in 1951, to 4,000 in 1974 and 1982, or 7.3% and 5.5%, respectively. The Muslim population however, continued to decrease since then to 1,000 or 0.6% in 1999, to 750 or 0.3% in 2010. Thus, assuming that the percentage of Muslims will remain constant, the Muslim population is expected to remain less than 3,000 throughout this century. The data is summarized in Table 5.4.8.

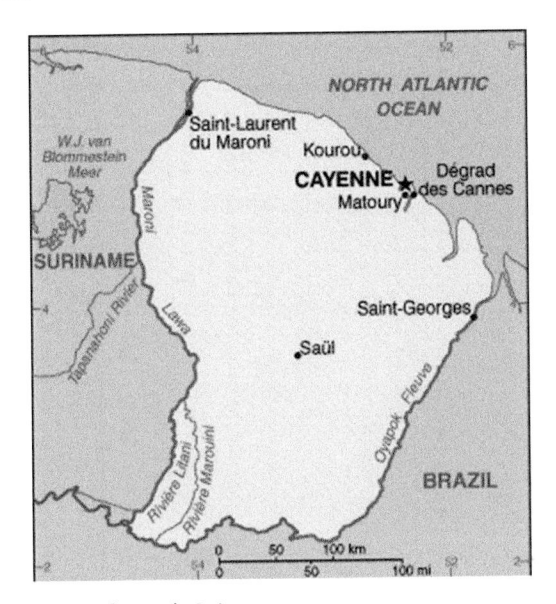

Figure 5.4.8 Map of French Guiana.

Table 5.4.8 Evolution of the Muslim population in French Guiana

Year	Population	Muslims	%	Source
1850	16,817	—	0.00	[KET76]es
1907	24,000	1,570	6.54	[ARH]es
1954	27,863	2,500	8.97	[KET76]es
1974	55,125	4,000	7.26	[KET76]es
1982	73,022	4,000	5.48	[KET86]es
1999	156,790	1,000	0.64	[KET01A]es
2010	229,040	750	0.33	[GF]es
2050	545,000	2,000	0.33	es
2100	927,000	3,000	0.33	es

5.4.9 Guyana

The Cooperative Republic of Guyana has an area of 214,696 sq. km and its map is presented in Fig. 5.4.9. It was conquered by the British in 1815 and gained its independence from the UK in 1966, when it changed its name from British Guiana. It has the second largest percentage of Muslims in the Americas after Suriname. Muslims entered the country in 1838 under the indenture system that was introduced by the British Empire in 1835. Accordingly, the British brought slave like labor force from India to exploit the British dependencies. Accordingly, 406 East Indian labor was brought in 1838, but 235 were sent back in 1843.

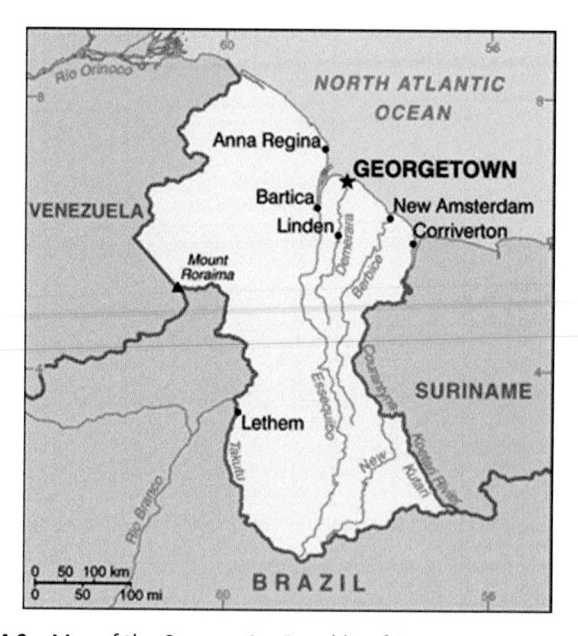

Figure 5.4.9 Map of the Cooperative Republic of Guyana.

Ethnic censuses were carried out since 1841. Later religious censuses carried since 1891. The 1891 census showed that 11% of the East Indian population was Muslim. This ratio was used to estimate the Muslim population from the censuses prior to 1891. Thus, based on census data, the Muslim population increased from none in 1831, to 35 or 0.04% in 1841, to 845 or 0.6% in 1851, to 2,552 or 1.6% in 1861, to 5,000 or 2.6% in 1871, to 9,000 or 3.6%

in 1881, to 12,000 or 4.2% in 1891, to 18,000 or 6.2% in 1911, to 18,000 or 6.2% in 1921, to 22,000 or 7.0% in 1931, to 29,000 or 7.9% in 1946, to 49,000 or 8.8% in 1960, and then peaked to 64,000 or 9.1% in 1970. It continued decreasing since then however due to migration to more prosperous countries. Thus, the Muslim population decreased to 66,000 or 8.9% in 1980, to 58,000 or 8.0% in 1991, to 55,000 or 7.3% in 2002 and 51,000 or 6.8% in 2012. According to DHS survey data, the percentage of Muslims was 7.07% in 2009. Thus, assuming that the percentage of Muslims will continue to decrease by a quarter of a percentage point per decade, the Muslim population is expected to decrease to 50,000 or 5.8% by 2050 and 25,000 or 4.5% by 2100. The data is summarized in Table 5.4.9.

Table 5.4.9 Evolution of the Muslim population in Guyana

Year	Population	Muslims	%	Source
1831	98,000	—	0.00	[GYH]e
1841	98,154	35	0.04	[GYH]e
1851	135,994	845	0.62	[GYH]e
1861	155,907	2,552	1.64	[GYH]e
1871	193,491	4,943	2.55	[GYH]e
1881	252,186	9,192	3.64	[GYH]e
1891	278,328	11,691	4.20	[GYH]c
1911	296,041	18,217	6.15	[GYH]c
1921	297,691	18,410	6.18	[GYH]c
1931	310,933	21,792	7.01	[GYH]c
1946	368,881	29,281	7.94	[GYH]c
1960	560,406	49,297	8.80	[GYH]c
1970	579,658	63,687	10.99	[GYH]c
1980	758,619	66,122	8.72	[UN88]c
1991	723,673	57,669	7.97	[GY02]c
2002	751,223	54,554	7.26	[GY12]c
2012	746,955	50,572	6.77	[GY12]c
2050	825,000	47,000	5.75	es
2100	531,000	24,000	4.50	es

5.4.10 Paraguay

The Republic of Paraguay has an area of 406,752 sq. km and its map is presented in Fig. 5.4.10. It was conquered by Spain in 1537, when its capital Asunción was founded, and gained its independence in 1811. In 1908, the Muslim population consisted of 300 individuals or 0.04% who arrived from the Ottoman Empire. The number of Muslims decreased to 100 or 0.01% in 1920, and then increased to 1,000 or 0.04% in 1972, to 1,000 or 0.03% in 1982. According to census data, the Muslim population ten years and over decreased from 1,200 or 0.04% in 1992 to 872 or 0.02% in 2002. Thus, assuming that the percentage of Muslims will remain fixed, the Muslim population is expected to remain less than 2,000 throughout this century. A summary of the data is provided in Table 5.4.10.

Figure 5.4.10 Map of the Republic of Paraguay.

Table 5.4.10 Evolution of the Muslim population in Paraguay

Year	Population	Muslims	%	Source
1908	715,800	300	0.04	[ARH]es
1920	699,000	100	0.01	[VEH]es
1972	2,357,955	1,000	0.04	[KET76]es
1982	3,029,830	1,000	0.03	[KET86]es
1992	2,939,309	1,200	0.04	[UN04]c
2002	3,855,397	872	0.02	[UN]c
2050	9,102,000	1,800	0.02	es
2100	8,734,000	1,700	0.02	es

5.4.11 Peru

The Republic of Peru has an area of 1,285,216 sq. km and its map is presented in Fig. 5.4.11. It was conquered by Spain in 1533 and gained its independence in 1821. The number of Muslims remained negligible afterwards at 0.00% of the total population, and per census data increased from 109 in 1940 to 196 in 1961. Later estimates increased to 500 in 1972, to 1,000 in 1981, to 3,000 or 0.01% in 1993. According to the World Values Survey (WVS), the percentage of Muslims was 0.08% in 2012. Thus, assuming that the percentage of Muslims will increase by 0.01 of a percentage point per decade, the Muslim population is expected to increase to 50,000 or 0.1% by 2050 and 70,000 or 0.2% by 2100. The data is summarized in Table 5.4.11.

Table 5.4.11 Evolution of the Muslim population in Peru

Year	Population	Muslims	%	Source
1940	7,023,111	109	0.00	[ROD]c
1961	10,420,357	196	0.00	[KET01C]c
1972	14,121,564	500	0.00	[KET76]es
1981	17,762,231	1,000	0.00	[KET86]es
1993	22,639,443	3,000	0.01	[KET01C]es
2012	29,987,800	24,000	0.08	[PEW]es
2050	40,374,000	48,000	0.12	es
2100	39,158,000	67,000	0.17	es

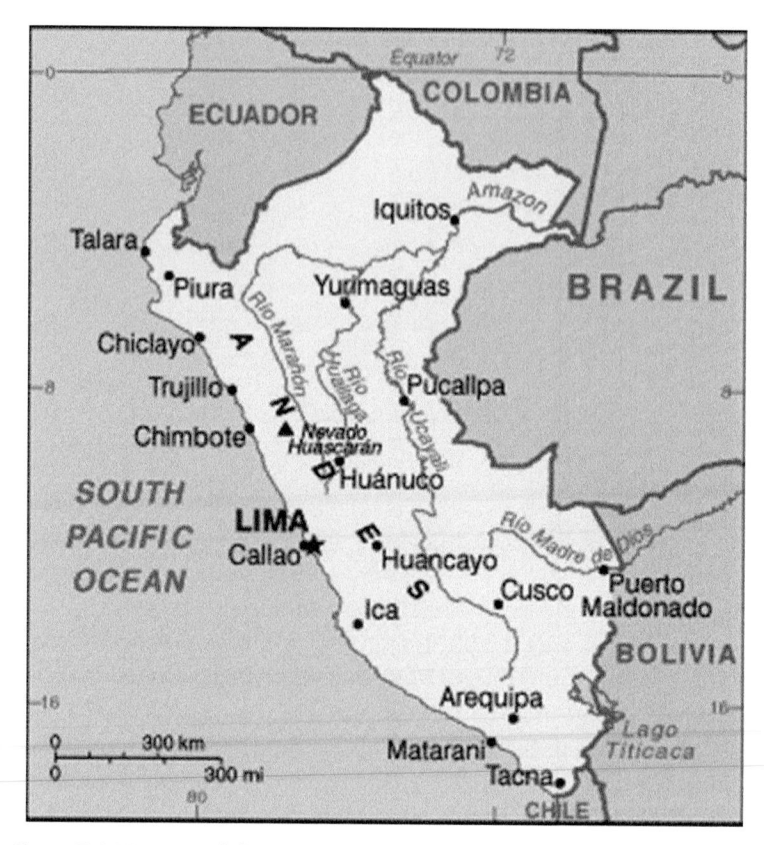

Figure 5.4.11   Map of the Republic of Peru.

5.4.12   Suriname

The Republic of Suriname has an area of 163,820 sq. km and its map is presented in Fig. 5.4.12. It was conquered by the Netherlands in 1667 and gained its independence in 1975, when it changed its name from Dutch Guiana. It has the largest percentage of Muslims in the Americas. The Dutch abolished slavery in 1863, but replaced it with indentured labor, first from India between 1873, and second between 1893 and 1940 from the island of Java, in current Indonesia, which was Dutch occupied. Most of the Javanese were Muslims, while a substantial portion of Indians were Muslim.

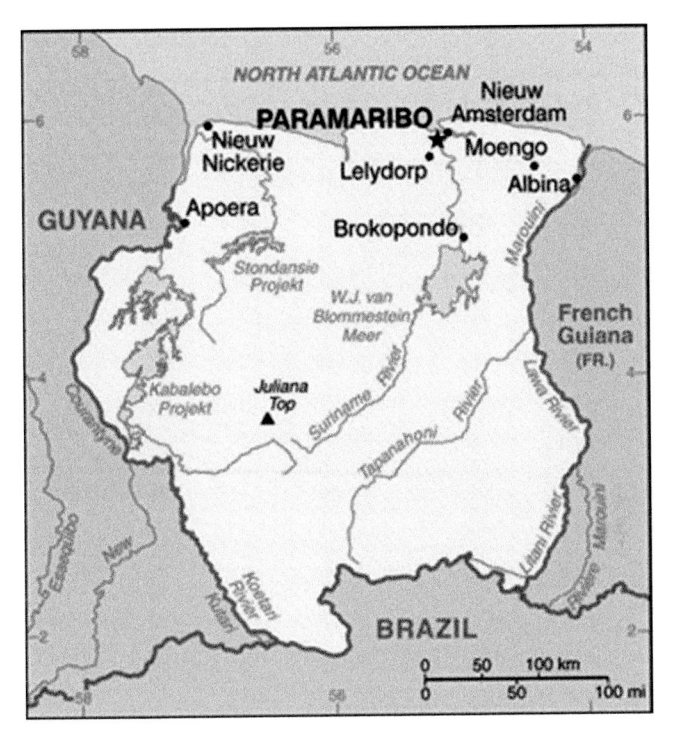

Figure 5.4.12   Map of the Republic of Suriname.

Based on official estimates, the Muslim population increased from none in 1873, to 1,629 or 2.9% in 1887, to 2,594 or 4.1% in 1895, to 3,918 or 5.7% in 1900, to 6,071 or 8.3% in 1904, to 10,584 or 12.47% in 1911, to 15,431 or 13.6% in 1920, to 35,675 or 23.6% in 1929, to 52,940 or 25.0% in 1949. Post-independence censuses that inquired about religious adherence showed that the Muslim population changed to 65,000 or 20.6% in 1964, to 74,000 or 19.5% in 1972, to 70,000 or 19.6% in 1980, to 66,307 or 16.0% in 2004, to 75,053 or 14.3% in 2012 (including 14,161 Ahmadiyya). The decrease happened as hundreds of thousands of the population, about ten percent of which was Muslim, left to the Netherlands. Thus, assuming that the percentage of Muslims will continue to decrease by one tenth of a percentage point per decade, the Muslim population is expected to remain less than 0.1 million throughout this century. The data is summarized in Table 5.4.12.

Table 5.4.12 Evolution of the Muslim population in Suriname

Year	Population	Muslims	%	Source
1873	50,000	—	0.00	[KET76]es
1887	57,141	1,629	2.85	[SYB890]es
1895	63,000	2,594	4.12	[SYB900]es
1900	68,968	3,918	5.68	[SYB903]es
1904	73,542	6,071	8.26	[SYB910]es
1911	86,233	10,584	12.27	[SYB915]es
1920	113,181	15,431	13.63	[SYB920]es
1929	151,350	35,675	23.57	[SYB931]es
1949	211,804	52,940	24.99	[SYB950]es
1964	315,279	64,842	20.57	[SR80]c
1972	379,607	74,170	19.54	[SR80]c
1980	354,860	69,638	19.62	[SR80]c
2004	415,625	66,307	15.95	[SR]c
2012	524,556	75,053	14.31	[SR]c
2050	680,000	95,000	13.90	es
2100	613,000	82,000	13.40	es

5.4.13 Uruguay

The Oriental Republic of Uruguay has an area of 176,215 sq. km and its map is presented in Fig. 5.4.13. It was conquered by Spain in 1624 and gained its independence in 1811. In 1908, the Muslim population consisted of 500 individuals or 0.05% who arrived from the Ottoman Empire. This number decreased to 100 or 0.01% in 1908, then increased to 1,000 or 0.04% in 1975, to 1,000 or 0.03% in 1985, to 2,000 or 0.06% in 1996, and then decreased to 400 or 0.01% in 2004 and 2011. Thus, assuming that the percentage of Muslims will increase by 0.01 of a percentage point every four decades, the Muslim population is expected to remain around one thousand throughout this century. A summary of the data is provided in Table 5.4.13.

Table 5.4.13 Evolution of the Muslim population in Uruguay

Year	Population	Muslims	%	Source
1908	1,042,686	500	0.05	[ARH]es
1924	1,553,000	100	0.01	[VEH]es
1975	2,788,429	1,000	0.04	[KET76]es
1985	2,955,241	1,000	0.03	[KET86]es
1996	3,163,763	2,000	0.06	[KET01D]es
2004	3,241,003	400	0.01	[DOS07]es
2011	3,286,314	400	0.01	[DOS12]es
2050	3,639,000	1,100	0.03	es
2100	3,182,000	1,300	0.04	es

Figure 5.4.13 Map of the Oriental Republic of Uruguay.

5.4.14 Venezuela

The Bolivarian Republic of Venezuela has an area of 912,050 sq. km and its map is presented in Fig. 5.4.14. It was conquered by Spain in 1522, gained its independence in 1811 as part of the former Gran Colombia, and then emerged from the collapse of the latter in 1830. Gran Columbia used to include current Bolivia, Colombia, Panama, and Venezuela.

Figure 5.4.14 Map of the Bolivarian Republic of Venezuela.

Between 1615 and 1700 Muslim Moriscos from Spain established a colony here and in Colombia. However, they were obliterated by the Spaniards. Towards the end of the nineteenth century Muslims started emigrating from Syria and Lebanon. A 1920 estimate put the number of Muslims at 500 or 0.02% of the total population. According to ISS survey, the percentage of Muslims increased to 0.07% in 2003. Thus, assuming that the percentage of Muslims will continue to increase by 0.01 of a percentage point per decade, the Muslim population is expected to exceed 40,000 or 0.1% by 2050 and 60,000 or 0.2% by 2100. The data is summarized in Table 5.4.14.

Table 5.4.14 Evolution of the Muslim population in Venezuela

Year	Population	Muslims	%	Source
1920	2,479,525	500	0.02	[VEH]es
2003	25,858,000	19,000	0.07	[ISS]s
2050	37,023,000	44,000	0.12	es
2100	34,241,000	58,000	0.17	es

5.4.15 Regional Summary and Conclusions

Muslims started to move here at the end of the nineteenth century from regions that were under the Ottoman Empire. The Muslim population has been steadily increasing but remains negligible. Many Muslims lost their religion to assimilate while others married Christians and their descendants were not Muslim. The Muslim population is expected to continue to increase but remain well below one percent of the total population throughout this century. The following tables present centennial data from 600 AD to 2100 AD (or approximately 1 H to 1500 H) in Table 5.4a and decennial data from 1790 AD to 2100 AD (or 1210 H to 1520 H) in Tables 5.4b and 5.4c for current countries in South America. The data includes total population in thousands (P), the percentage of which is Muslim (M%), the corresponding Muslim population in thousands (M), and the annual population growth rate (APGR, or G%) of the total population in this region.

Table 5.4a Centennial estimates of the Muslim population (×1000) in South America: 600–2100 AD (1–1500 H)

		600	700	800	900	1000	1100	1200	1300
Argentina	P	340	360	380	400	414	450	490	530
	M%	—	—	—	—	—	—	—	—
	M	—	—	—	—	—	—	—	—
Bolivia	P	675	720	765	810	852	930	1,000	1,080
	M%	—	—	—	—	—	—	—	—
	M	—	—	—	—	—	—	—	—
Brazil	P	540	580	620	660	700	760	820	880
	M%	—	—	—	—	—	—	—	—
	M	—	—	—	—	—	—	—	—
Chile	P	470	500	530	560	591	650	700	750
	M%	—	—	—	—	—	—	—	—
	M	—	—	—	—	—	—	—	—
Colombia	P	750	800	840	880	934	1,000	1,100	1,200
	M%	—	—	—	—	—	—	—	—
	M	—	—	—	—	—	—	—	—
Ecuador	P	310	330	350	370	387	420	455	490
	M%	—	—	—	—	—	—	—	—
	M	—	—	—	—	—	—	—	—
Falklands	P	—	—	—	—	—	—	—	—
	M%	—	—	—	—	—	—	—	—
	M	—	—	—	—	—	—	—	—
Guyane	P	5	6	6	7	7	8	8	9
	M%	—	—	—	—	—	—	—	—
	M	—	—	—	—	—	—	—	—
Guyana	P	5	6	7	8	9	10	11	12
	M%	—	—	—	—	—	—	—	—
	M	—	—	—	—	—	—	—	—
Paraguay	P	87	93	99	105	111	120	130	140
	M%	—	—	—	—	—	—	—	—
	M	—	—	—	—	—	—	—	—
Peru	P	2,000	2,250	2,500	2,750	3,000	3,200	3,400	3,600
	M%	—	—	—	—	—	—	—	—
	M	—	—	—	—	—	—	—	—
Suriname	P	5	6	6	7	7	8	8	9
	M%	—	—	—	—	—	—	—	—
	M	—	—	—	—	—	—	—	—
Uruguay	P	35	37	39	41	43	46	50	54
	M%	—	—	—	—	—	—	—	—
	M	—	—	—	—	—	—	—	—
Venezuela	P	444	472	500	528	556	600	650	700
	M%	—	—	—	—	—	—	—	—
	M	—	—	—	—	—	—	—	—
Total	P	5,666	6,160	6,642	7,126	7,611	8,202	8,822	9,454
	M%	—	—	—	—	—	—	—	—
	M	—	—	—	—	—	—	—	—
	G%		0.084	0.075	0.070	0.066	0.075	0.073	0.069

Table 5.4a (Continued)

		1400	1500	1600	1700	1800	1900	2000	2100
Argentina	P	570	600	360	409	552	4,693	36,871	56,802
	M%	—	—	—	—	—	0.02	0.20	0.20
	M	—	—	—	—	—	1	74	114
Bolivia	P	1,160	1,235	741	843	900	1,696	8,418	17,391
	M%	—	—	—	—	—	—	0.01	0.01
	M	—	—	—	—	—	—	1	2
Brazil	P	940	1,000	800	1,250	3,754	17,438	174,790	180,683
	M%	—	—	—	—	—	0.00	0.02	0.04
	M	—	—	—	—	—	—	35	72
Chile	P	800	857	519	591	750	2,959	15,342	17,332
	M%	—	—	—	—	—	0.00	0.03	0.05
	M	—	—	—	—	—	—	5	9
Colombia	P	1,300	1,354	812	924	1,160	3,998	39,630	45,221
	M%	—	—	—	—	—	0.00	0.03	0.26
	M	—	—	—	—	—	0	12	118
Ecuador	P	525	561	337	383	480	1,400	12,681	24,483
	M%	—	—	—	—	—	—	0.00	0.03
	M	—	—	—	—	—	—	1	7
Falklands	P	—	—	—	0	0	2	3	3
	M%	—	—	—	—	—	—	0.30	1.00
	M	—	—	—	—	—	—	—	—
Guyane	P	9	10	11	12	17	20	163	927
	M%	—	—	—	—	—	6.54	0.64	0.33
	M	—	—	—	—	—	1	1	3
Guyana	P	13	14	15	30	60	301	747	531
	M%	—	—	—	—	—	4.20	7.26	4.50
	M	—	—	—	—	—	13	54	24
Paraguay	P	150	161	96	110	137	440	5,323	8,734
	M%	—	—	—	—	—	0.04	0.02	0.02
	M	—	—	—	—	—	0	1	2
Peru	P	3,800	4,000	1,300	1,300	1,315	3,648	26,460	39,158
	M%	—	—	—	—	—	—	0.04	0.17
	M	—	—	—	—	—	—	11	67
Suriname	P	9	10	10	20	55	82	471	613
	M%	—	—	—	—	—	5.68	15.95	13.40
	M	—	—	—	—	—	5	75	82
Uruguay	P	58	62	37	42	53	915	3,320	3,182
	M%	—	—	—	—	—	0.05	0.01	0.04
	M	—	—	—	—	—	—	—	1
Venezuela	P	750	806	484	550	680	2,542	24,192	34,241
	M%	—	—	—	—	—	0.00	0.07	0.17
	M	—	—	—	—	—	—	17	58
Total	P	10,084	10,670	5,522	6,464	9,913	40,135	348,412	429,301
	M%	—	—	—	—	—	0.05	0.08	0.13
	M	—	—	—	—	—	20	286	558
	G%	0.065	0.056	−0.659	0.158	0.428	1.398	2.161	0.209

Table 5.4b Decennial estimates of the Muslim population (×1000) in South America: 1790–1940 AD (1210–1360 H)

		1790	1800	1810	1820	1830	1840	1850	1860
Argentina	P	421	552	550	534	700	926	1,100	1,304
	M%	—	—	—	—	—	—	—	—
	M	—	—	—	—	—	—	—	—
Bolivia	P	800	900	1,000	1,100	1,200	1,300	1,374	1,461
	M%	—	—	—	—	—	—	—	—
	M	—	—	—	—	—	—	—	—
Brazil	P	3,307	3,754	4,262	4,507	5,519	6,425	7,234	8,703
	M%	—	—	—	—	—	—	—	—
	M	—	—	—	—	—	—	—	—
Chile	P	740	750	760	771	965	1,181	1,410	1,700
	M%	—	—	—	—	—	—	—	—
	M	—	—	—	—	—	—	—	—
Colombia	P	1,140	1,160	1,180	1,206	1,500	1,686	2,065	2,295
	M%	—	—	—	—	—	—	—	—
	M	—	—	—	—	—	—	—	—
Ecuador	P	470	480	490	500	600	700	816	910
	M%	—	—	—	—	—	—	—	—
	M	—	—	—	—	—	—	—	—
Falklands	P	—	—	—	—	—	—	—	1
	M%	—	—	—	—	—	—	—	—
	M	—	—	—	—	—	—	—	—
Guyane	P	15	17	19	21	24	20	17	20
	M%	—	—	—	—	—	—	—	0.01
	M	—	—	—	—	—	—	—	—
Guyana	P	35	60	77	77	66	98	128	148
	M%	—	—	—	—	—	0.04	0.62	1.64
	M	—	—	—	—	—	—	1	2
Paraguay	P	134	137	140	143	200	300	350	370
	M%	—	—	—	—	—	—	—	—
	M	—	—	—	—	—	—	—	—
Peru	P	1,314	1,315	1,316	1,317	1,737	1,890	2,001	2,482
	M%	—	—	—	—	—	—	—	—
	M	—	—	—	—	—	—	—	—
Suriname	P	50	55	60	65	60	57	64	54
	M%	—	—	—	—	—	—	—	—
	M	—	—	—	—	—	—	—	—
Uruguay	P	52	53	54	55	70	100	132	230
	M%	—	—	—	—	—	—	—	—
	M	—	—	—	—	—	—	—	—
Venezuela	P	660	680	700	718	880	945	1,324	1,564
	M%	—	—	—	—	—	—	—	—
	M	—	—	—	—	—	—	—	—
Total	P	9,137	9,913	10,608	11,014	13,520	15,629	18,015	21,240
	M%	—	—	—	—	—	0.00	0.00	0.01
	M	—	—	—	—	—	—	1	2
	G%		0.815	0.678	0.375	2.050	1.449	1.421	1.647

Table 5.4b (Continued)

		1870	1880	1890	1900	1910	1920	1930	1940
Argentina	P	1,796	2,462	3,376	4,693	6,836	8,861	11,896	14,169
	M%	—	0.01	0.02	0.02	0.12	0.12	0.12	0.12
	M	—	—	1	1	8	11	14	17
Bolivia	P	1,495	1,559	1,626	1,696	1,837	2,136	2,397	2,690
	M%	—	—	—	—	—	—	0.01	0.01
	M	—	—	—	—	—	—	—	—
Brazil	P	9,930	11,794	14,334	17,438	22,216	30,636	33,568	41,236
	M%	—	—	0.00	0.00	0.00	0.00	0.01	0.01
	M	—	—	—	—	—	—	3	4
Chile	P	1,945	2,264	2,608	2,959	3,317	3,754	4,287	5,024
	M%	0.00	0.00	0.00	0.00	0.04	0.01	0.01	0.02
	M	—	—	—	—	1	—	—	1
Colombia	P	2,392	2,879	3,369	3,998	4,890	6,213	7,914	9,174
	M%	—	—	0.00	0.00	0.01	0.01	0.01	0.01
	M	—	—	—	—	—	1	1	1
Ecuador	P	1,013	1,128	1,257	1,400	1,837	2,136	1,944	2,466
	M%	—	—	—	—	0.00	0.00	0.00	0.00
	M	—	—	—	—	—	—	—	—
Falklands	P	1	2	2	2	2	2	2	2
	M%	—	—	—	—	—	—	—	—
	M	—	—	—	—	—	—	—	—
Guyane	P	19	18	19	20	21	22	23	24
	M%	1.00	2.00	4.00	6.54	6.54	6.54	6.54	6.54
	M	—	—	1	1	1	1	2	2
Guyana	P	193	252	278	301	296	298	311	344
	M%	2.55	3.64	4.20	4.20	6.15	6.18	7.01	7.94
	M	5	9	12	13	18	18	22	27
Paraguay	P	384	402	420	440	554	699	880	1,111
	M%	—	—	—	0.04	0.04	0.01	0.01	0.01
	M	—	—	—	—	—	—	—	—
Peru	P	2,606	2,953	3,346	3,648	4,137	4,690	5,480	6,440
	M%	—	—	—	—	0.00	0.00	0.00	0.00
	M	—	—	—	—	—	—	—	—
Suriname	P	51	70	71	82	88	113	152	167
	M%	—	2.85	4.12	5.68	12.27	13.63	23.57	24.99
	M	—	2	3	5	11	15	36	42
Uruguay	P	343	464	686	915	1,081	1,371	1,713	1,965
	M%	—	—	—	0.05	0.05	0.01	0.01	0.01
	M	—	—	—	—	1	—	—	—
Venezuela	P	1,732	2,005	2,222	2,542	2,596	2,480	3,300	3,851
	M%	—	—	—	0.00	0.01	0.02	0.03	0.04
	M	—	—	—	—	—	—	1	2
Total	P	23,901	28,252	33,614	40,135	49,708	63,410	73,868	88,663
	M%	0.02	0.04	0.05	0.05	0.08	0.08	0.11	0.11
	M	5	12	16	20	42	48	80	96
	G%	1.180	1.672	1.738	1.773	2.139	2.434	1.527	1.826

Table 5.4c Decennial estimates of the Muslim population (×1000) in South America: 1950–2100 AD (1370–1520 H)

		1950	1960	1970	1980	1990	2000	2010	2020
Argentina	P	17,038	20,482	23,881	27,897	32,619	36,871	40,896	45,196
	M%	0.12	0.08	0.10	0.15	0.20	0.20	0.19	0.20
	M	20	16	24	42	65	74	78	90
Bolivia	P	3,082	3,657	4,484	5,580	6,865	8,418	10,049	11,673
	M%	0.01	0.01	0.01	0.01	0.02	0.01	0.01	0.01
	M	—	—	—	1	1	1	1	1
Brazil	P	53,975	72,179	95,113	120,694	149,003	174,790	195,714	212,559
	M%	0.01	0.01	0.01	0.01	0.02	0.02	0.02	0.02
	M	5	7	10	12	30	35	39	43
Chile	P	6,599	8,133	9,783	11,419	13,275	15,342	17,063	19,116
	M%	0.02	0.01	0.02	0.02	0.03	0.03	0.03	0.03
	M	1	1	2	2	4	5	5	6
Colombia	P	11,982	16,058	21,480	26,901	33,103	39,630	45,223	50,883
	M%	0.02	0.03	0.04	0.03	0.03	0.03	0.07	0.10
	M	2	5	9	8	10	12	32	51
Ecuador	P	3,470	4,544	6,069	7,989	10,231	12,681	15,011	17,643
	M%	0.00	0.00	0.00	0.01	0.01	0.00	0.01	0.01
	M	—	—	—	—	1	1	2	2
Falklands	P	2	2	2	2	2	3	3	3
	M%	—	—	—	—	0.10	0.30	0.08	0.23
	M	—	—	—	—	—	—	—	—
Guyane	P	25	32	48	67	116	163	233	299
	M%	8.97	8.97	7.26	5.48	5.48	0.64	0.33	0.33
	M	2	3	3	4	6	1	1	1
Guyana	P	407	572	705	780	743	747	749	787
	M%	7.94	8.80	10.99	8.72	7.97	7.26	6.77	6.50
	M	32	50	77	68	59	54	51	51
Paraguay	P	1,473	1,904	2,475	3,182	4,223	5,323	6,248	7,133
	M%	0.04	0.04	0.04	0.04	0.04	0.02	0.02	0.02
	M	1	1	1	1	2	1	1	1
Peru	P	7,777	10,155	13,460	17,548	22,071	26,460	29,028	32,972
	M%	0.00	0.00	0.00	0.00	0.01	0.04	0.08	0.09
	M	—	—	1	1	2	11	23	30
Suriname	P	215	288	368	360	405	471	529	587
	M%	23.00	20.57	19.54	19.62	18.00	15.95	14.31	14.20
	M	49	59	72	71	73	75	76	83
Uruguay	P	2,239	2,539	2,810	2,915	3,110	3,320	3,359	3,474
	M%	0.04	0.04	0.04	0.03	0.03	0.01	0.01	0.02
	M	1	1	1	1	1	—	—	1
Venezuela	P	5,482	8,142	11,396	15,183	19,633	24,192	28,440	28,436
	M%	0.04	0.05	0.05	0.06	0.06	0.07	0.08	0.09
	M	2	4	6	9	12	17	23	26
Total	P	113,765	148,686	192,074	240,515	295,398	348,412	392,544	430,760
	M%	0.10	0.10	0.11	0.09	0.09	0.08	0.08	0.09
	M	118	148	206	220	266	286	331	385
	G%	2.493	2.677	2.560	2.249	2.055	1.651	1.193	0.929

Table 5.4c (Continued)

		2030	2040	2050	2060	2070	2080	2090	2100
Argentina	P	49,056	52,297	54,867	56,709	57,704	57,934	57,596	56,802
	M%	0.20	0.20	0.20	0.20	0.20	0.20	0.20	0.20
	M	98	105	110	113	115	116	115	114
Bolivia	P	13,240	14,653	15,840	16,756	17,364	17,644	17,636	17,391
	M%	0.01	0.01	0.01	0.01	0.01	0.01	0.01	0.01
	M	1	1	2	2	2	2	2	2
Brazil	P	223,852	229,059	228,980	224,412	215,858	204,531	192,347	180,683
	M%	0.02	0.02	0.03	0.03	0.03	0.03	0.03	0.04
	M	45	46	69	67	65	61	58	72
Chile	P	19,458	20,157	20,319	20,075	19,550	18,854	18,076	17,332
	M%	0.03	0.03	0.04	0.04	0.04	0.04	0.04	0.05
	M	6	6	8	8	8	8	7	9
Colombia	P	53,417	55,336	55,958	55,408	53,813	51,344	48,311	45,221
	M%	0.12	0.14	0.16	0.18	0.20	0.22	0.24	0.26
	M	64	77	90	100	108	113	116	118
Ecuador	P	19,819	21,762	23,316	24,416	25,037	25,203	24,981	24,483
	M%	0.01	0.01	0.02	0.02	0.02	0.02	0.02	0.03
	M	2	2	5	5	5	5	5	7
Falklands	P	3	3	3	3	3	3	3	3
	M%	0.30	0.40	0.50	0.60	0.70	0.80	0.90	1.00
	M	—	—	—	—	—	—	—	—
Guyane	P	378	461	545	629	711	790	863	927
	M%	0.33	0.33	0.33	0.33	0.33	0.33	0.33	0.33
	M	1	2	2	2	2	3	3	3
Guyana	P	822	835	825	795	749	688	613	531
	M%	6.25	6.00	5.75	5.50	5.25	5.00	4.75	4.50
	M	51	50	47	44	39	34	29	24
Paraguay	P	7,950	8,614	9,102	9,400	9,478	9,348	9,081	8,734
	M%	0.02	0.02	0.02	0.02	0.02	0.02	0.02	0.02
	M	2	2	2	2	2	2	2	2
Peru	P	36,031	38,552	40,374	41,410	41,591	41,097	40,237	39,158
	M%	0.10	0.11	0.12	0.13	0.14	0.15	0.16	0.17
	M	36	42	48	54	58	62	64	67
Suriname	P	632	664	680	684	679	664	641	613
	M%	14.10	14.00	13.90	13.80	13.70	13.60	13.50	13.40
	M	89	93	95	94	93	90	87	82
Uruguay	P	3,569	3,624	3,639	3,613	3,544	3,441	3,316	3,182
	M%	0.02	0.02	0.03	0.03	0.03	0.03	0.04	0.04
	M	1	1	1	1	1	1	1	1
Venezuela	P	33,626	35,819	37,023	37,538	37,412	36,722	35,624	34,241
	M%	0.10	0.11	0.12	0.13	0.14	0.15	0.16	0.17
	M	34	39	44	49	52	55	57	58
Total	P	461,854	481,837	491,472	491,847	483,493	468,263	449,327	429,301
	M%	0.09	0.10	0.11	0.11	0.11	0.12	0.12	0.13
	M	430	466	522	541	551	551	546	558
	G%	0.697	0.424	0.198	0.008	−0.171	−0.320	−0.413	−0.456

5.5 Muslims in Northern Caribbean Islands

This region consists of eight island nations and territories in the Caribbean basin from Puerto Rico to its north and west. These are the Bahamas, Cayman Islands, Cuba, Dominican Republic, Haiti, Jamaica, Puerto Rico and Turks and Caicos. A map of this region is presented in Fig. 5.5a. Some Muslims came to this region in the Sixteenth

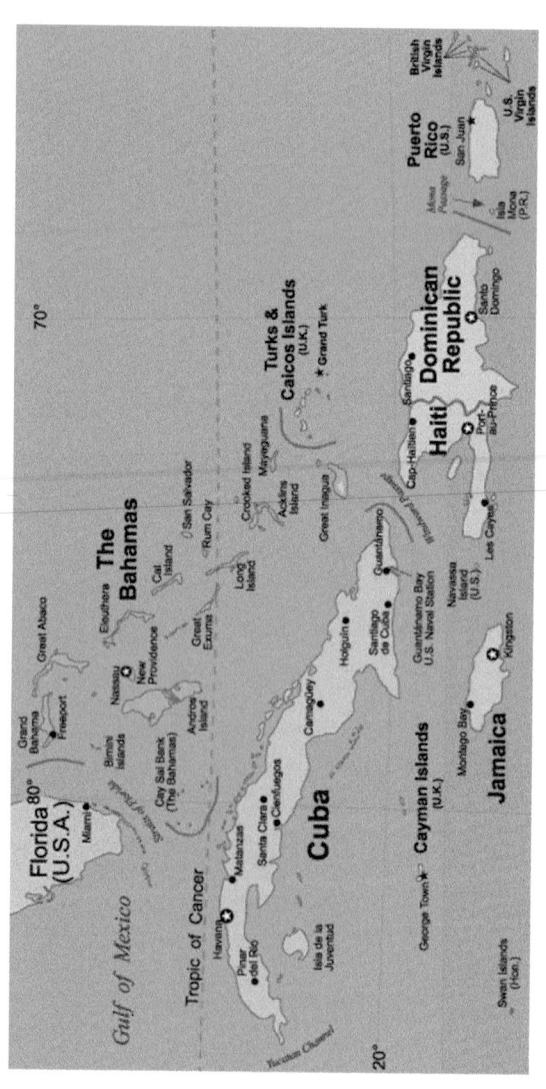

Figure 5.5a Map of the Northern Caribbean Islands.

century from Spain with the Spanish Conquistadors. However, they were exterminated when the Inquisition discovered them. The other wave of Muslims came towards the end of the nineteenth century from Syria and Lebanon, then towards the middle of the twentieth century from Palestine after the declaration of Israel. Thus, the Muslim population has likely increased from none in 1800, to 1,000 or 0.02% in 1900, to 13,000 or 0.04% in 2000, to 25,000 or 0.1% in 2020, and is projected to reach 40,000 or 0.1% by 2050 and 51,000 or 0.1% by 2100.

A plot of decennial estimates of the Muslim population and its percentage with respect to the total population in this region from 1900 to 2100 is provided in Fig. 5.4. The corresponding individual data for each country in this region is discussed below. In Section 5.4.9, the total population in each country in this region and the corresponding percentage and number of Muslims is presented centennially in Table 5.4a from 600 to 2100 and decennially in Tables 5.4b and 5.4c from 1790 to 2100.

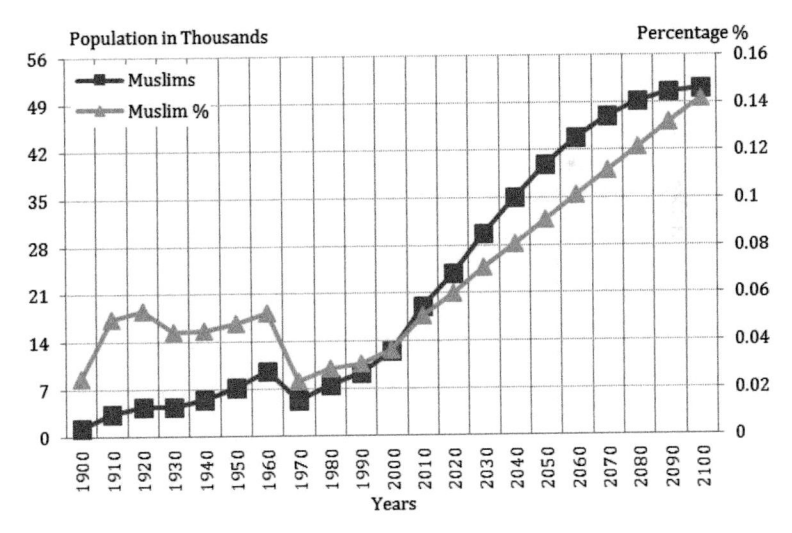

Figure 5.5b Plot of decennial estimates of the Muslim population and its percentage of the total population in Northern Caribbean Islands: 1900–2100 AD (1320–1520 H).

5.5.1 The Bahamas

The Commonwealth of the Bahamas became a British colony in 1783 and attained its independence from the UK in 1973. It consists of over three thousand islands, islets and cays (about 10% of which are inhabited) with total area of 13,943 sq. km. Most of the population lives on three islands: New Providence (207 sq. km with 70% of the total population), Grand Bahama (1,372 sq. km with 15% of the total population), and Abaco (1,681 sq. km with 5% of the total population). The largest island though is Andros with area 5,959 sq. km but 2% of the total population. A map of the islands of the Bahamas is presented in Fig. 5.5.1.

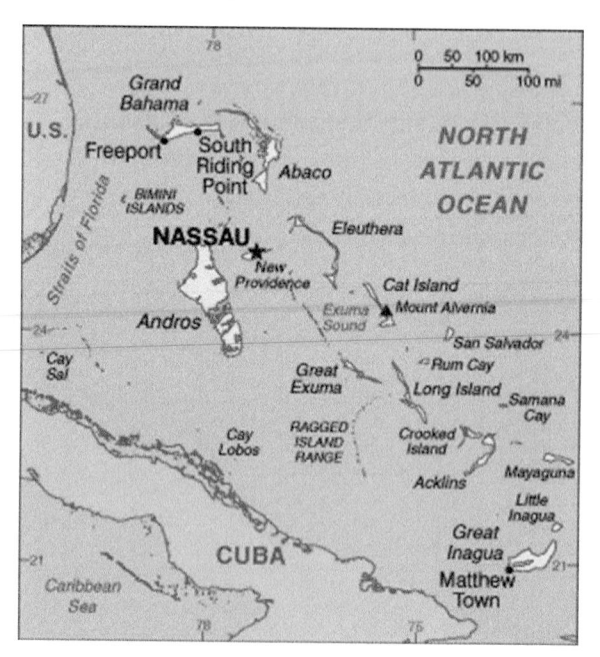

Figure 5.5.1 Map of the Commonwealth of the Bahamas.

We used country of birth to estimate the number of Muslims in these islands from 1970 to 1990. Accordingly, in 1970 there were 74 Muslims or 0.04% of the total population, consisting of 28 born in Egypt, 23 born in each of Malaysia and Pakistan. In 1980, the number decreased to 56 or 0.03%, including thirteen in Egypt, seven in each of Sierra Leone and Pakistan, five in each of Iran and Malaysia. In

1990, the number increased to 91 or 0.04%, including 18 in Malaysia, 16 in Pakistan, ten in Turkey, and nine in Egypt. The first census to inquire on religious data was held in 2000, recording 292 Muslims or 0.10% of the total population. The Muslim population increased in 2010 to 306 but decreased in percentage to 0.09%. Thus, assuming that the percentage of Muslims will remain constant at 0.1% of the total population, the Muslim population is expected to remain round 500 throughout this century. The data is summarized in Table 5.5.1.

Table 5.5.1 Evolution of the Muslim population in the Bahamas

Year	Population	Muslims	%	Source
1970	168,812	74	0.04	[UNE]e
1980	209,505	56	0.03	[UNE]e
1990	255,049	91	0.04	[UNE]e
2000	303,611	292	0.10	[BS00]c
2010	342,411	306	0.09	[BS10]c
2050	463,000	500	0.10	es
2100	460,000	500	0.10	es

5.5.2 Cayman Islands

It was largely uninhabited when it was conquered by the British in 1655 and remains an overseas territory of the UK. It has an area of 264 sq. km which comprises three islands: Grand Cayman (197 sq. km), Cayman Brac (39 sq. km), and Little Cayman (29 sq. km). However, almost all the population (96%) lives in Grand Cayman. A map of these islands is presented in Fig. 5.5.2.

The 1911 census showed that there were no Muslims in these islands. The number of Muslims was estimated at 200 or 0.5% in 1999. According to 2007 pre-census enumeration data, the number of Muslims was 411 or 0.8% of the total population. However, the number of Muslims dropped to 212 or 0.4% in the 2010 census. The reason may be due to the 2008 global economic crises. Thus, assuming that the percentage of Muslims will continue to increase by 0.05 of a percentage point per decade, the Muslim population is expected to remain less than one thousand, comprising 0.9% of the total population by 2100. The data is summarized in Table 5.5.2.

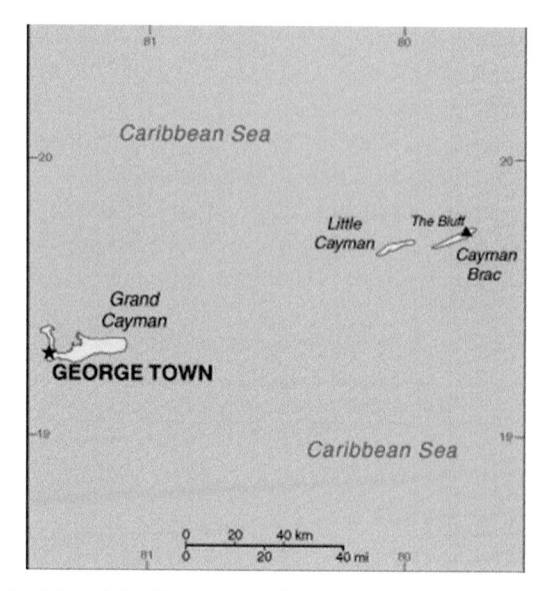

Figure 5.5.2 Map of the Cayman Islands.

Table 5.5.2 Evolution of the Muslim population in the Cayman Islands

Year	Population	Muslims	%	Source
1911	5,564	—	0.00	[GB911]c
1999	39,020	200	0.51	[KET01A]es
2007	51,598	411	0.80	[KY07]c
2010	53,475	212	0.40	[KY10]c
2050	85,000	500	0.60	es
2100	103,000	900	0.85	es

5.5.3 Cuba

The Republic of Cuba changed hands from Spain to the United States in 1898 and gained its independence from the latter in 1902. It has an area of 110,860 sq. km and consists mostly of the island of Cuba, but also includes Isla de la Juventud (2,149 sq. km), and over three thousands of much smaller islands surrounding the main island, almost all are uninhabited. A map of Cuba is presented in Fig. 5.5.3.

In 1907 there were 2,500 Muslims or 0.1% of the total population. They were mostly workers from India and China. Estimates for the Muslim population doubled to 5,000 or 0.1% in 1953 and then decreased due to the communist revolution to 1,000 or 0.01% in 1970, 1981 and 2002. It then increased to 3,000 or 0.07% in 2012. The first mosque was opened in old Havana in 2015.

Thus, assuming that the percentage of Muslims will increase by 0.01 of a percentage point per decade, the Muslim population is expected to remain less than 8,000 or 0.1% throughout of this century. The data is summarized in Table 5.5.3.

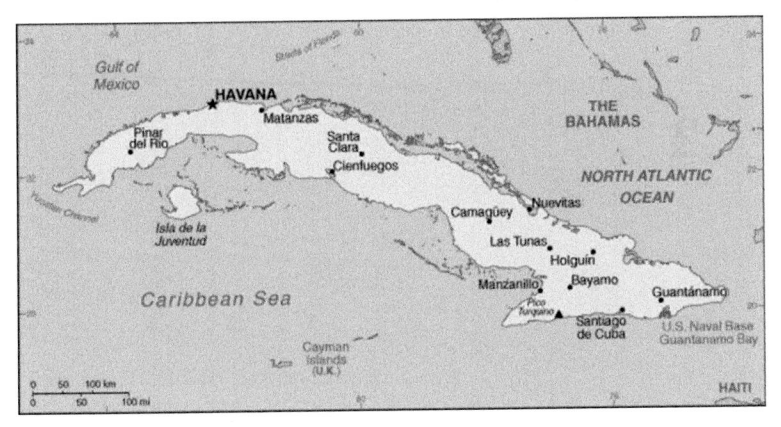

Figure 5.5.3 Map of the Republic of Cuba.

Table 5.5.3 Evolution of the Muslim population in Cuba

Year	Population	Muslims	%	Source
1907	2,048,980	2,500	0.12	[ARH]es
1953	5,829,029	5,000	0.09	[KET76]es
1970	8,569,121	1,000	0.01	[KET76]es
1981	9,723,605	1,000	0.01	[KET86]es
2002	11,177,743	1,000	0.06	[KET01A]es
2012	11,167,325	3,000	0.03	[DOS13]es
2050	10,162,000	7,000	0.07	es
2100	6,671,000	8,000	0.12	es

5.5.4 Dominican Republic

It gained its independence from Spain in 1865 and has an area of 48,670 sq. km. It consists of two-thirds of the island of Hispaniola which it shares with Haiti. Its map is presented in Fig. 5.5.4. Estimates for the Muslim population increased from 100 or 0.00% in 1970 and 500 or 0.01% in 1993. According to ISS survey, the percentage of Muslims was 0.10% in 2005 and 0.05% in 2008. Thus, assuming that the percentage of Muslims will continue to increase by 0.01 of a percentage point per decade, the number of Muslims is expected to remain less than 15,000 or 0.1% throughout of this century. The data is summarized in Table 5.5.4.

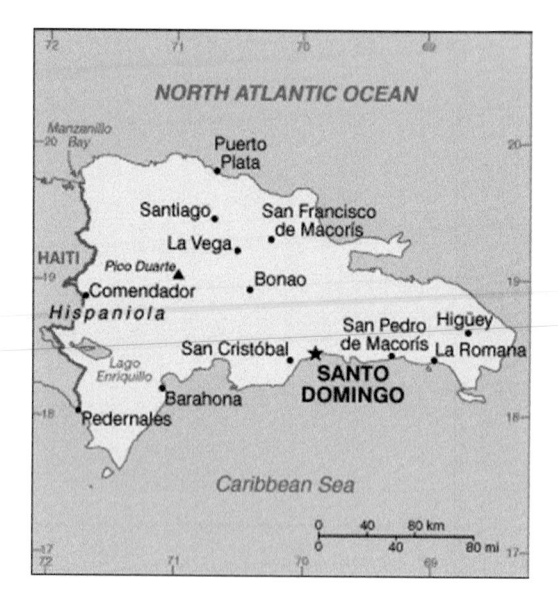

Figure 5.5.4 Map of the Dominican Republic.

Table 5.5.4 Evolution of the Muslim population in the Dominican Republic

Year	Population	Muslims	%	Source
1970	4,009,458	100	0.00	[KET76]es
1993	7,293,390	500	0.01	[KET01A]es
2008	9,636,000	5,000	0.05	[ISS]s
2050	12,796,000	12,000	0.10	es
2100	11,013,000	15,000	0.14	es

5.5.5 Haiti

The Republic of Haiti gained its independence from France in 1804 and has an area of 27,750 sq. km. It consists of a third of the island of Hispaniola which it shares with the Dominican Republic, together with small islands close to the shore. A map of Haiti is presented in Fig. 5.5.5. In the 1970s there were few Muslims who came from Morocco and Syria. According to the 2003 Census, the number of Muslims increased to 2,013 or 0.02% of the total population. The recent increase is due to the return of Haitians to Islam. Thus, assuming that the percentage of Muslims will continue to increase by 0.01 of a percentage point per decade, the Muslim population is expected to remain less than 19,000 or 0.1% throughout of this century. The data is summarized in Table 5.5.5.

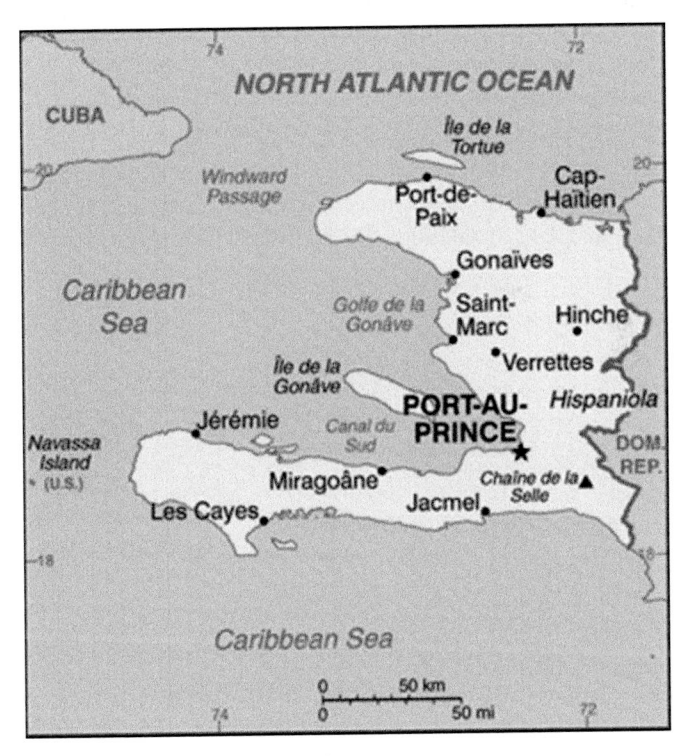

Figure 5.5.5 Map of the Republic of Haiti.

Table 5.5.5 Evolution of the Muslim population in Haiti

Year	Population	Muslims	%	Source
1971	4,329,991	100	0.00	[KET76]es
2003	8,373,750	2,013	0.02	[HT03]c
2050	14,878,000	12,000	0.08	es
2100	14,760,000	19,000	0.13	es

5.5.6 Jamaica

It was seized by the British in 1655 and gained its independence from the UK in 1962. It consists of one main island with area of 10,991 sq. km and its map is presented in Fig. 5.5.6. The first Muslims arrived here from India under the British indentured labor system in the second half of the nineteenth century. In 1845, the first ship arrived with 261 people from India. Census data indicate that the number of Muslims increased from 672 or 0.1% in 1911, to 2,238 or 0.1% in 1982, but decreased to 1,024 or 0.04% in 2001 and then increased to 1,513 or 0.06% in 2011. Thus, assuming that the percentage of Muslims will continue to increase by 0.02 of a percentage point per decade, the Muslim population is expected to remain less than five thousand or 0.2% throughout this century. The data is summarized in Table 5.5.6.

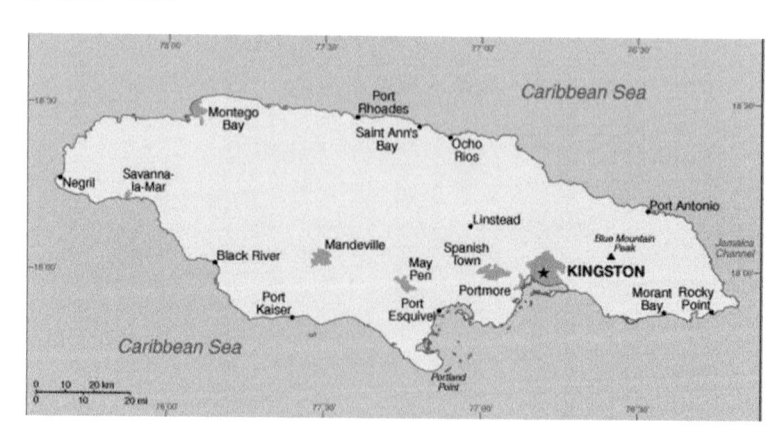

Figure 5.5.6 Map of Jamaica.

Table 5.5.6 Evolution of the Muslim population in Jamaica

Year	Population	Muslims	%	Source
1844	377,433	—	0.00	[GB861]e
1911	831,383	672	0.08	[GB911]c
1982	1,929,265	2,238	0.12	[UN88]c
2001	2,523,811	1,024	0.04	[UN]c
2011	2,622,779	1,513	0.06	[JM]c
2050	2,960,000	4,100	0.14	es
2100	1,793,000	4,300	0.24	es

5.5.7 Puerto Rico

The Commonwealth of Puerto Rico is unincorporated, organized territory of the United States with area 13,790 sq. km. Its map is presented in Fig. 5.5.7. It has been under US control since 1898, when it was captured from Spain. It is an archipelago that includes the main island of Puerto Rico and over a hundred smaller islands mostly to the east of the main island and almost all are uninhabited, the largest of which are Vieques (348 sq. km), Culebra (30 sq. km), and Mona (57 sq. km but uninhabited).

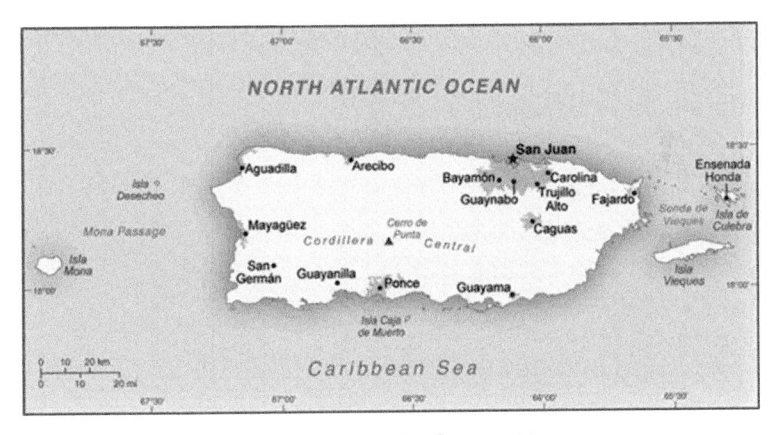

Figure 5.5.7 Map of the Commonwealth of Puerto Rico.

The first Muslims arrived around the mid-twentieth century from Palestine as a result of the declaration of Israel. Thus, estimates of the Muslim population increased from none in 1940, to 2,000 or 0.07% in 1970, to 3,000 or 0.09% in 1980, to 4,500 or 0.13% in 1990, to 5,000 or 0.13% in 2010. Thus, assuming that the percentage of Muslims will continue to increase by 0.01 of a percentage point per decade, the Muslim population is expected to decrease to 4,000 in 2050 and 3,000 in 2100, comprising 0.2% of the total population. The data is summarized in Table 5.5.7.

Table 5.5.7 Evolution of the Muslim population in Puerto Rico

Year	Population	Muslims	%	Source
1940	1,869,255	—	0.00	[KET76]es
1970	2,712,033	2,000	0.07	[KET76]es
1980	3,196,520	3,000	0.09	[KET86]es
1990	3,522,037	4,500	0.13	[KET01A]es
2010	3,725,789	5,000	0.13	[PR]es
2050	2,445,000	4,000	0.17	es
2100	1,217,000	3,000	0.22	es

5.5.8 Turks and Caicos

It was conquered by the British in 1783 and remains an overseas territory of the UK. It was part of the Bahamas until it separated in 1848. It has an area of 948 sq. km, consisting of seven inhabited islands and over three hundred much smaller islands. Over three-quarter of the population lives in the Island of Providenciales with an area of 122 sq. km and locally referred to as Provo. Other inhabited islands are Grand Turk, 17 sq. km with a sixth of the population, North Caicos (116 sq. km) and South Caicos (21 sq. km), with 4% of the population each, Middle Caicos (144 sq. km), Salt Cay (7 sq. km) and Parrot Cay (6 sq. km). A map of these islands is presented in Fig. 5.5.8.

Table 5.5.8 Evolution of the Muslim population in Turks and Caicos

Year	Population	Muslims	%	Source
1871	4,723	—	0.00	[GB871]c
1911	5,615	—	0.00	[GB911]c
1980	7,404	—	0.00	[TC12]c
1990	11,465	—	0.00	[TC12]c
2001	19,886	—	0.00	[TC12]c
2012	31,458	40	0.13	[TC12]c
2050	50,000	250	0.50	es
2100	54,000	550	1.00	es

According to census data, there were no Muslims in these islands before the start of this century. The 2012 census recorded forty Muslims or 0.1% of the total population. Their distribution was as follows: 26 in Provo, nine in North Caicos, five Muslims in Grand Turk and none in the other islands. Thus, assuming that the percentage of Muslims will increase by a tenth of a percentage point per decade, then the Muslim population is expected to exceed 200 or 0.5% by 2050 and 500 or 1.0% by 2100. The data is summarized in Table 5.5.8.

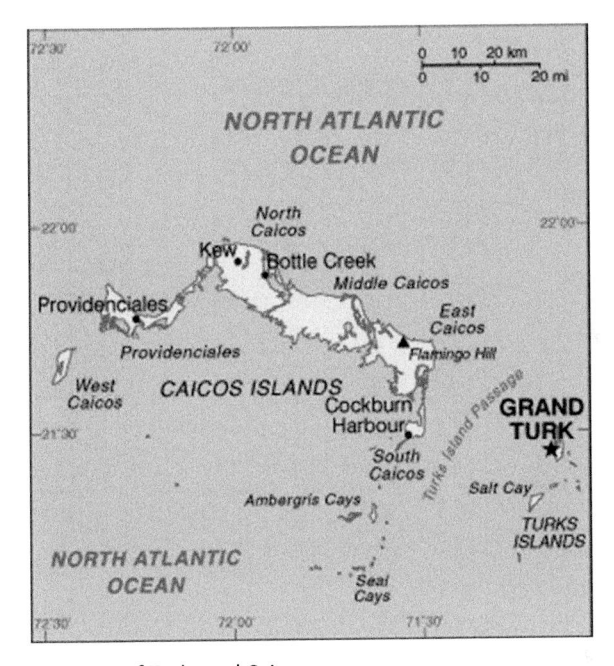

Figure 5.5.8 Map of Turks and Caicos.

5.5.9 Regional Summary and Conclusion

Northern Caribbean Islands has the second least concentration of Muslims among the five regions spanning the Americas. The Muslim population from Europe and Africa became almost nonexistent prior to the twentieth century. However, it has been steadily increasing but remains negligible. It is expected to continue to increase but remain well below one percent of the total population throughout this century. The following tables present centennial data from 600 AD to 2100 AD (or approximately 1 H to 1500 H) in Table 5.5a and decennial data from 1790 AD to 2100 AD (or 1210 H to 1520 H) in Tables 5.5b and 5.5c for current countries in Northern Caribbean Islands. The data includes total population in thousands (P), the percentage of which is Muslim (M%), the corresponding Muslim population in thousands (M), and the annual population growth rate (APGR, or G%) of the total population in this region.

Table 5.5a Centennial estimates of the Muslim population (×1000) in Northern Caribbean Islands: 600–2100 AD (1–1500 H)

		600	700	800	900	1000	1100	1200	1300
Bahamas	P	—	—	—	—	—	1	1	1
	M%	—	—	—	—	—	—	—	—
	M	—	—	—	—	—	—	—	—
Caymans	P	—	—	—	—	—	—	—	—
	M%	—	—	—	—	—	—	—	—
	M	—	—	—	—	—	—	—	—
Cuba	P	55	59	63	67	71	77	83	89
	M%	—	—	—	—	—	—	—	—
	M	—	—	—	—	—	—	—	—
Dominican Republic	P	8	9	9	10	10	11	12	13
	M%	—	—	—	—	—	—	—	—
	M	—	—	—	—	—	—	—	—
Haiti	P	69	73	77	81	85	93	101	109
	M%	—	—	—	—	—	—	—	—
	M	—	—	—	—	—	—	—	—
Jamaica	P	2	3	4	5	6	7	8	9
	M%	—	—	—	—	—	—	—	—
	M	—	—	—	—	—	—	—	—
Puerto Rico	P	25	26	27	28	29	32	35	38
	M%	—	—	—	—	—	—	—	—
	M	—	—	—	—	—	—	—	—
Turks & Caicos	P	—	—	—	—	—	—	—	—
	M%	—	—	—	—	—	—	—	—
	M	—	—	—	—	—	—	—	—
Total	P	160	171	181	192	202	221	240	259
	M%	—	—	—	—	—	—	—	—
	M	—	—	—	—	—	—	—	—
	G%		0.067	0.058	0.060	0.051	0.090	0.083	0.077

Table 5.5a (Continued)

		1400	1500	1600	1700	1800	1900	2000	2100
Bahamas	P	1	1	1	1	14	54	298	460
	M%	—	—	—	—	—	—	0.10	0.10
	M	—	—	—	—	—	—	—	—
Caymans	P	—	—	—	—	1	5	42	103
	M%	—	—	—	—	—	—	0.51	0.85
	M	—	—	—	—	—	—	—	1
Cuba	P	95	103	41	103	520	1,658	11,126	6,671
	M%	—	—	—	—	—	0.05	0.01	0.12
	M	—	—	—	—	—	1	1	8
Dominican Republic	P	14	15	6	15	80	515	8,471	11,013
	M%	—	—	—	—	—	—	0.03	0.14
	M	—	—	—	—	—	—	3	15
Haiti	P	117	124	49	124	600	1,560	8,464	14,760
	M%	—	—	—	—	—	—	0.03	0.13
	M	—	—	—	—	—	—	3	19
Jamaica	P	11	13	15	20	295	756	2,655	1,793
	M%	—	—	—	—	—	0.07	0.04	0.24
	M	—	—	—	—	—	1	1	4
Puerto Rico	P	40	42	17	42	210	959	3,669	1,217
	M%	—	—	—	—	—	—	0.13	0.22
	M	—	—	—	—	—	—	5	3
Turks & Caicos	P	—	—	—	1	1	5	20	54
	M%	—	—	—	—	—	—	—	1.00
	M	—	—	—	—	—	—	—	1
Total	P	278	298	129	307	1,721	5,512	34,746	36,071
	M%	—	—	—	—	—	0.02	0.04	0.14
	M	—	—	—	—	—	1	13	51
	G%	0.071	0.070	−0.835	0.862	1.725	1.164	1.841	0.037

Table 5.5b Decennial estimates of the Muslim population (×1000) in Northern Caribbean Islands: 1790–1940 AD (1210–1360 H)

		1790	1800	1810	1820	1830	1840	1850	1860
Bahamas	P	13	14	15	16	17	19	28	35
	M%	—	—	—	—	—	—	—	—
	M	—	—	—	—	—	—	—	—
Caymans	P	1	1	1	1	2	2	2	2
	M%	—	—	—	—	—	—	—	—
	M	—	—	—	—	—	—	—	—
Cuba	P	480	520	560	605	705	1,008	1,186	1,397
	M%	—	—	—	—	—	—	—	—
	M	—	—	—	—	—	—	—	—
Dominican Republic	P	75	80	85	89	125	130	146	200
	M%	—	—	—	—	—	—	—	—
	M	—	—	—	—	—	—	—	—
Haiti	P	550	600	650	723	800	900	938	1,000
	M%	—	—	—	—	—	—	—	—
	M	—	—	—	—	—	—	—	—
Jamaica	P	291	295	300	305	311	377	399	441
	M%	—	—	—	—	—	—	0.02	0.03
	M	—	—	—	—	—	—	—	—
Puerto Rico	P	190	210	230	248	288	400	495	583
	M%	—	—	—	—	—	—	—	—
	M	—	—	—	—	—	—	—	—
Turks & Caicos	P	1	1	2	2	3	3	3	4
	M%	—	—	—	—	—	—	—	—
	M	—	—	—	—	—	—	—	—
Total	P	1,601	1,721	1,843	1,990	2,249	2,839	3,197	3,663
	M%	—	—	—	—	—	—	0.00	0.00
	M	—	—	—	—	—	—	—	—
	G%		0.723	0.682	0.766	1.228	2.328	1.187	1.361

Table 5.5b (Continued)

		1870	1880	1890	1900	1910	1920	1930	1940
Bahamas	P	39	44	48	54	56	53	60	66
	M%	—	—	—	—	—	—	—	—
	M	—	—	—	—	—	—	—	—
Caymans	P	2	4	4	5	5	5	6	5
	M%	—	—	—	—	—	—	—	—
	M	—	—	—	—	—	—	—	—
Cuba	P	1,331	1,432	1,541	1,658	2,219	2,997	3,837	4,566
	M%	—	—	0.01	0.05	0.12	0.12	0.09	0.09
	M	—	—	—	1	3	4	3	4
Dominican Republic	P	242	311	400	515	688	879	1,256	1,674
	M%	—	—	—	—	—	—	—	—
	M	—	—	—	—	—	—	—	—
Haiti	P	1,150	1,273	1,409	1,560	1,809	2,124	2,422	2,751
	M%	—	—	—	—	—	—	—	—
	M	—	—	—	—	—	—	—	—
Jamaica	P	510	581	639	756	831	858	1,009	1,237
	M%	0.04	0.05	0.06	0.07	0.08	0.09	0.10	0.11
	M	—	—	—	1	1	1	1	1
Puerto Rico	P	645	736	840	959	1,126	1,312	1,552	1,880
	M%	—	—	—	—	—	—	—	—
	M	—	—	—	—	—	—	—	—
Turks & Caicos	P	5	5	5	5	6	6	5	6
	M%	—	—	—	—	—	—	—	—
	M	—	—	—	—	—	—	—	—
Total	P	3,924	4,386	4,886	5,512	6,740	8,234	10,147	12,185
	M%	0.01	0.01	0.01	0.02	0.05	0.05	0.04	0.04
	M	—	—	1	1	3	4	4	5
	G%	0.690	1.111	1.081	1.205	2.012	2.002	2.089	1.830

Table 5.5c Decennial estimates of the Muslim population (×1000) in Northern Caribbean Islands: 1950–2100 AD (1370–1520 H)

		1950	1960	1970	1980	1990	2000	2010	2020
Bahamas	P	79	110	169	211	256	298	355	393
	M%	—	0.01	0.04	0.03	0.04	0.10	0.09	0.10
	M	—	—	—	—	—	—	—	—
Caymans	P	6	8	9	16	25	42	57	66
	M%	—	—	—	0.10	0.30	0.51	0.40	0.45
	M	—	—	—	—	—	—	—	—
Cuba	P	5,920	7,141	8,713	9,849	10,597	11,126	11,226	11,327
	M%	0.09	0.09	0.01	0.01	0.01	0.01	0.03	0.04
	M	5	6	1	1	1	1	3	5
Dominican Republic	P	2,365	3,294	4,500	5,804	7,133	8,471	9,695	10,848
	M%	—	—	0.00	0.01	0.01	0.03	0.05	0.06
	M	—	—	—	1	1	3	5	7
Haiti	P	3,221	3,866	4,676	5,643	7,038	8,464	9,949	11,403
	M%	—	—	0.00	0.01	0.02	0.03	0.04	0.05
	M	—	—	—	1	1	3	4	6
Jamaica	P	1,403	1,629	1,876	2,163	2,420	2,655	2,810	2,961
	M%	0.12	0.12	0.12	0.12	0.12	0.04	0.06	0.08
	M	2	2	2	3	3	1	2	2
Puerto Rico	P	2,218	2,295	2,632	3,091	3,403	3,669	3,580	2,861
	M%	0.01	0.05	0.07	0.09	0.09	0.13	0.13	0.14
	M	0	1	2	3	3	5	5	4
Turks & Caicos	P	5	6	6	8	12	20	33	39
	M%	—	—	—	—	—	0.10	0.13	0.20
	M	—	—	—	—	—	—	—	—
Total	P	15,217	18,348	22,580	26,785	30,885	34,746	37,705	39,897
	M%	0.05	0.05	0.02	0.03	0.03	0.04	0.05	0.06
	M	7	10	5	8	9	13	19	24
	G%	2.222	1.871	2.075	1.708	1.424	1.178	0.817	0.565

Table 5.5c (*Continued*)

		2030	2040	2050	2060	2070	2080	2090	2100
Bahamas	P	427	450	463	468	470	469	464	460
	M%	0.10	0.10	0.10	0.10	0.10	0.10	0.10	0.10
	M	—	—	—	—	—	—	—	—
Caymans	P	73	80	85	88	91	95	99	103
	M%	0.50	0.55	0.60	0.65	0.70	0.75	0.80	0.85
	M	0	0	1	1	1	1	1	1
Cuba	P	11,142	10,765	10,162	9,382	8,611	7,929	7,265	6,671
	M%	0.05	0.06	0.07	0.08	0.09	0.10	0.11	0.12
	M	6	6	7	8	8	8	8	8
Dominican Republic	P	11,770	12,416	12,796	12,885	12,716	12,319	11,733	11,013
	M%	0.07	0.08	0.09	0.10	0.11	0.12	0.13	0.14
	M	8	10	12	13	14	15	15	15
Haiti	P	12,733	13,916	14,878	15,491	15,719	15,621	15,282	14,760
	M%	0.06	0.07	0.08	0.09	0.10	0.11	0.12	0.13
	M	8	10	12	14	16	17	18	19
Jamaica	P	3,048	3,045	2,960	2,808	2,597	2,341	2,066	1,793
	M%	0.10	0.12	0.14	0.16	0.18	0.20	0.22	0.24
	M	3	4	4	4	5	5	5	4
Puerto Rico	P	2,905	2,689	2,445	2,176	1,921	1,682	1,450	1,217
	M%	0.15	0.16	0.17	0.18	0.19	0.20	0.21	0.22
	M	4	4	4	4	4	3	3	3
Turks & Caicos	P	44	48	50	52	52	53	53	54
	M%	0.30	0.40	0.50	0.60	0.70	0.80	—	1.00
	M	—	—	—	—	—	—	—	1
Total	P	42,143	43,410	43,840	43,351	42,177	40,509	38,413	36,071
	M%	0.07	0.08	0.09	0.10	0.11	0.12	0.13	0.14
	M	30	35	40	44	47	49	51	51
	G%	0.548	0.296	0.099	−0.112	−0.274	−0.404	−0.531	−0.629

5.6 Muslims in Central America

This region consists of eight countries: Belize, Costa Rica, El Salvador, Guatemala, Honduras, Mexico, Nicaragua, and Panama. Some Muslims came to this region in the Sixteenth century from Spain with the Spanish Conquistadors. However, they were exterminated when they were discovered by the Spanish Inquisition. The other wave of Muslims came towards the end of the nineteenth century from Syria and Lebanon, then towards the middle of the twentieth century from Palestine. But the Muslim population remains almost negligible with respect to the total population. Thus, the Muslim population has likely increased from none in 1900, to 39,000 or 0.03% in 2000, to 81,000 or 0.04% in 2020, and is projected to reach 0.14 million or 0.1% by 2050 and 0.22 million or 0.1% by 2100.

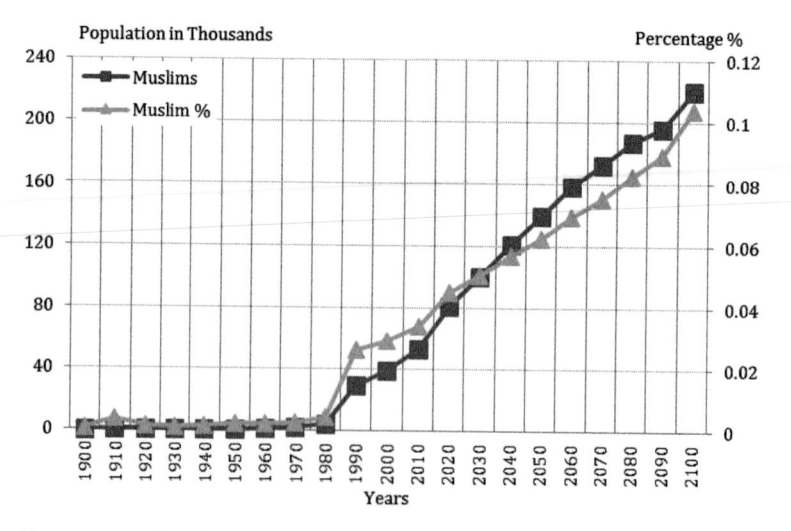

Figure 5.6 Plot of decennial estimates of the Muslim population and its percentage of the total population in Central America: 1900–2100 AD (1320–1520 H).

A plot of decennial estimates of the Muslim population and its percentage with respect to the total population in this region from 1900 to 2100 is provided in Fig. 5.6. The corresponding individual data for each country in this region is discussed below. In Section

5.6.9, the total population in each country in this region and the corresponding percentage and number of Muslims is presented centennially in Table 5.6a from 600 to 2100 and decennially in Tables 5.6b and 5.6c from 1790 to 2100.

5.6.1 Belize

It gained its independence from the UK in 1981 when it also changed its name from British Honduras. It has an area of 22,966 sq. km and its map is presented in Fig. 5.6.1. According to census data as shown in Table 5.6.1, the Muslim population increased from none in 1871, to two in 1911, to twelve or 0.02% in 1946, to 110 or 0.08% in 1980, to 159 or 0.09% in 1991, to 266 or 0.1% in 2000, to 622 or 0.2% in 2010. Thus, assuming that the percentage of Muslims will increase by 0.05 of a percentage points per decade, the Muslim population is expected to remain less than four thousands throughout this century, comprising 0.4% of the total population by 2050 and 0.7% by 2100.

Figure 5.6.1 Map of Belize.

Table 5.6.1 Evolution of the Muslim population in Belize

Year	Population	Muslims	%	Source
1870	24,710	—	0.00	[GB871]c
1911	40,458	2	0.00	[GB911]c
1946	59,163	12	0.02	[UN56]c
1980	138,709	110	0.08	[UN88]c
1991	178,403	159	0.09	[UN04]c
2000	247,427	266	0.11	[BZ]c
2010	320,425	622	0.19	[BZ]c
2050	571,000	2,000	0.40	es
2100	620,000	4,000	0.65	es

5.6.2 Costa Rica

The Republic of Costa Rica gained its independence from Spain in 1821. It has an area of 51,100 sq. km and its map is presented in Fig. 5.6.2. First Muslims came from Palestine in early twentieth century. As shown in Table 5.6.2, censuses in the nineteenth century showed no presence of Muslims in Costa Rica. Later estimates showed an increase from none in 1908, to 100 or 0.01% in 1973, to 500 or 0.01% in 2000 or 2011, but remain at 0.01% of the total population since 1970s. Thus, assuming that the percentage of Muslims will increase by 0.01 of a percentage points per decade, then the Muslim population is expected to reach few thousands or 0.1% throughout the second half of this century.

Table 5.6.2 Evolution of the Muslim population in Costa Rica

Year	Population	Muslims	%	Source
1864	120,499	—	0.00	[CR864]c
1883	182,073	—	0.00	[CR883]c
1892	243,205	—	0.00	[CR892]c
1908	351,200	—	0.00	[ARH]es
1973	1,871,780	100	0.01	[KET76]es
2000	3,810,179	500	0.01	[KET01B]es
2011	4,301,712	500	0.01	[CR]es
2050	5,773,000	3,000	0.05	es
2100	4,798,000	5,000	0.10	es

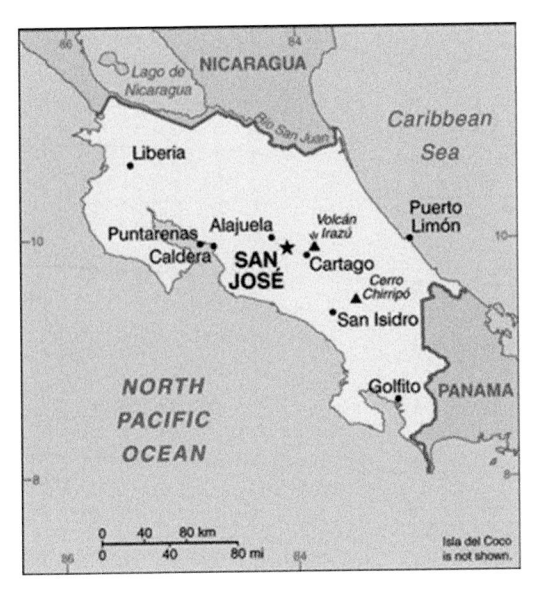

Figure 5.6.2 Map of the Republic of Costa Rica.

5.6.3 El Salvador

The Republic of El Salvador gained its independence from Spain in 1821. It has an area of 21,041 sq. km and its map is presented in Fig. 5.6.3. The first Muslims came from Palestine. As shown in Table 5.6.3, estimates for the Muslim population increased from none in 1908, to 100 in 1971, to 500 or 0.01% in 1992, to 1,300 or 0.02% in 2007. Thus, assuming that the percentage of Muslims will increase by 0.01 of a percentage point per decade, the Muslim population is expected to remain at few thousands or 0.1% of the total population throughout the second half of this century.

Table 5.6.3 Evolution of the Muslim population in El Salvador

Year	Population	Muslims	%	Source
1908	948,000	10	0.00	[ARH]es
1971	3,554,648	100	0.00	[KET76]es
1992	5,120,411	500	0.01	[KET01B]es
2007	5,744,113	1,300	0.02	[SV]es
2050	6,937,000	4,000	0.06	es
2100	4,766,000	5,000	0.11	es

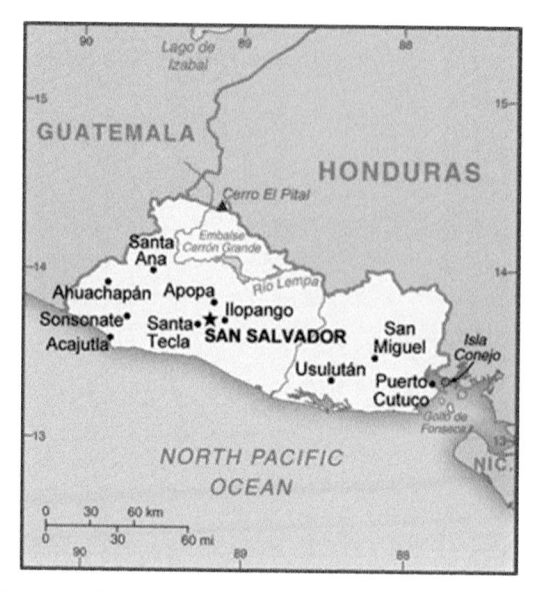

Figure 5.6.3 Map of the Republic of El Salvador.

5.6.4 Guatemala

The Republic of Guatemala gained its independence from Spain in 1821. It has an area of 108,889 sq. km and its map is presented in Fig. 5.6.4. The first Muslims came from Palestine. As shown in Table 5.6.4, estimates for the Muslim population increased from twenty in 1914, to 200 in 1973, to 1,000 or 0.01% in 1994, to 1,200 or 0.01% in 2002. Thus, assuming that the percentage of Muslims will increase by 0.01 of a percentage point every other decade, the Muslim population is expected to reach 8,000 or 0.03% by 2050 and 19,000 or 0.06% by 2100.

Table 5.6.4 Evolution of the Muslim population in Guatemala

Year	Population	Muslims	%	Source
1914	1,180,000	20	0.00	[ARH]es
1973	5,160,221	200	0.00	[KET76]es
1994	8,331,874	1,000	0.01	[KET01B]es
2002	11,237,196	1,200	0.01	[GT]es
2050	26,921,000	8,000	0.03	es
2100	31,270,000	19,000	0.06	es

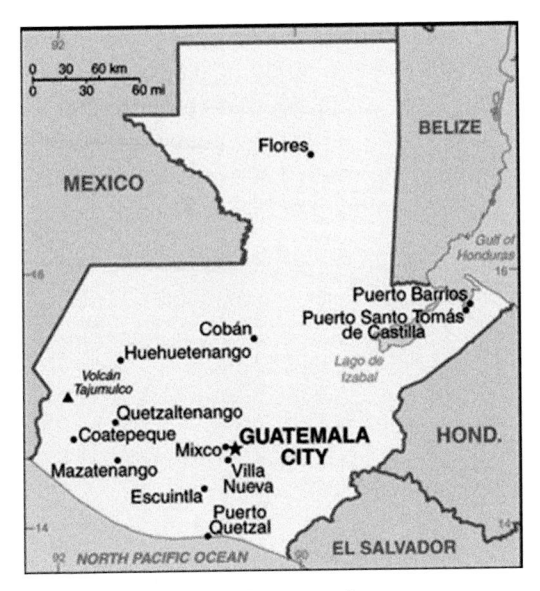

Figure 5.6.4 Map of the Republic of Guatemala.

5.6.5 Honduras

The Republic of Honduras gained its independence from Spain in 1821. It has an area of 112,090 sq. km and its map is presented in Fig. 5.6.5. The first Muslims came from Syria, Lebanon and Palestine. By 1910, there were only ten Muslims in the country. The 1945 census indicated that the Muslim population consisted of 28 individuals or

Table 5.6.5 Evolution of the Muslim population in Honduras

Year	Population	Muslims	%	Source
1910	553,446	10	0.00	[ARH]es
1945	1,200,542	28	0.00	[UN56]c
1974	2,656,900	100	0.00	[KET76]es
2001	6,076,885	1,000	0.02	[KET01B]es
2011	8,143,564	2,000	0.02	[DOS12]es
2050	13,831,000	8,000	0.06	es
2100	14,325,000	16,000	0.11	es

0.00% of the total population. Later estimates increased the number to 100 or 0.00% in 1974, to 1,000 or 0.02% in 2001, to 2,000 or 0.02% in 2011. Thus, assuming that the percentage of Muslims will increase by 0.01 of a percentage point per decade, the Muslim population is expected to reach 8,000 or 0.06% by 2050 and 16,000 or 0.11% by 2100. A summary of the data is provided in Table 5.6.5.

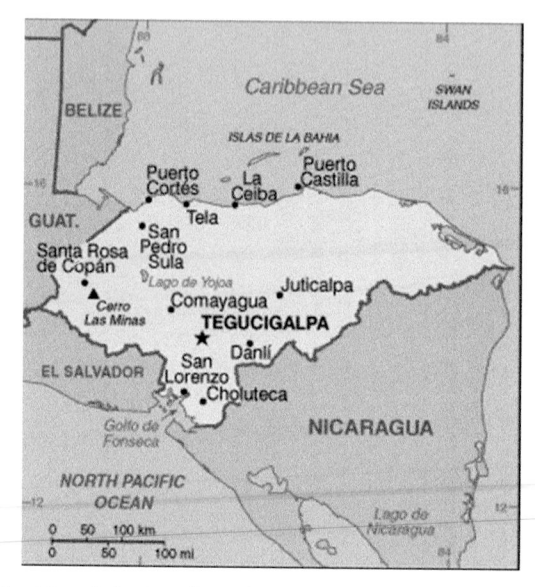

Figure 5.6.5 Map of the Republic of Honduras.

5.6.6 Mexico

The United Mexican States gained its independence from Spain in 1821. It has an area of 1,964,375 sq. km and its map is presented in Fig. 5.6.6. Towards the end of the nineteenth century Muslims emigrated from Syria. Based on census data as shown in Table 5.6.6, the Muslim population increased from 162 in 1895, to 602 in 1910, to 1,421 in 2000, to 3,760 in 2010, but still constitutes 0.00% of the Mexican population. Thus, assuming that the percentage of Muslims will increase by 0.01 of a percentage point every half century, the Muslim population is expected to reach 16,000 or 0.01% by 2050 and 28,000 or 0.02% by 2100.

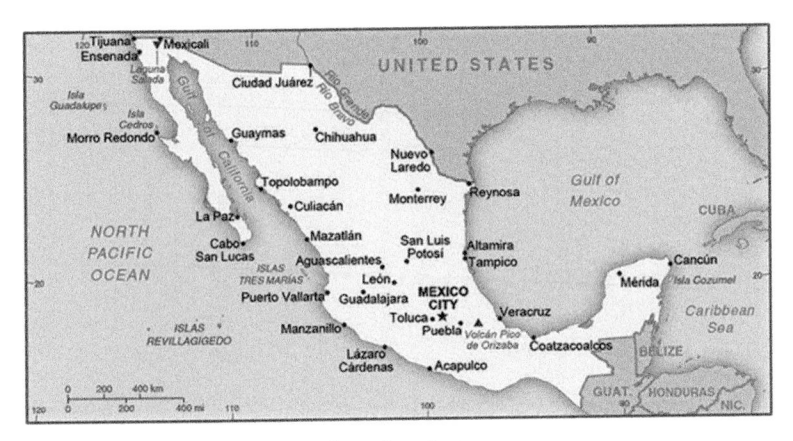

Figure 5.6.6 Map of the United Mexican States.

Table 5.6.6 Evolution of the Muslim population in Mexico

Year	Population	Muslims	%	Source
1895	12,632,425	162	0.00	[MX]c
1910	15,160,377	602	0.00	[MX]c
2000	84,061,824	1,421	0.00	[MX]c
2010	109,284,029	3,760	0.00	[MX10]c
2020	134,837,000	13,000	0.01	es
2050	155,151,000	16,000	0.01	es
2100	141,510,000	28,000	0.02	es

5.6.7 Nicaragua

The Republic of Nicaragua gained its independence from Spain in 1821. It has an area of 130,370 sq. km and its map is presented in Fig. 5.6.7. The first Muslims came from Palestine. Estimates of Muslims increased from ten in 1906, to eighteen in 1920 (census), to 150 or 0.01% in 1971, to 500 or 0.01% in 1995. According to the 2005 census, the Muslim population increased to 321 or 0.01%. Thus, assuming that the percentage of Muslims will increase by 0.01 of a percentage point per decade, the Muslim population is expected to reach 4,000 or 0.05% by 2050 and 8,000 or 0.10% by 2100. The data is summarized in Table 5.6.7.

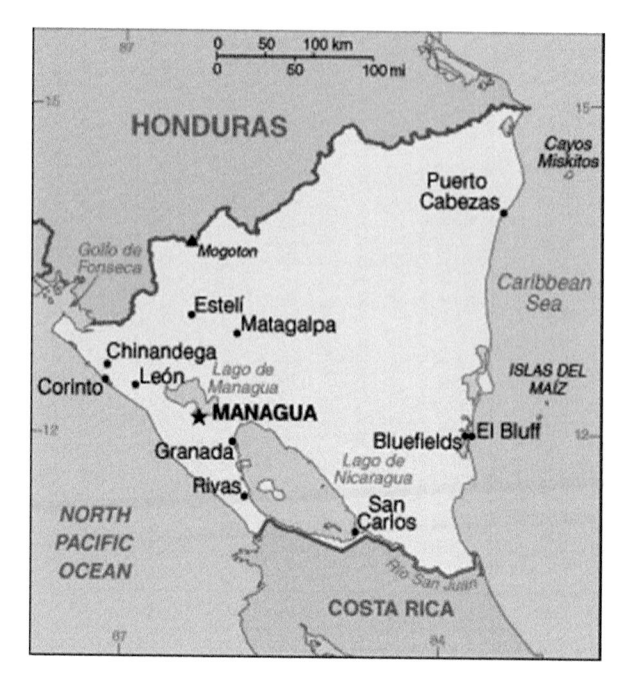

Figure 5.6.7 Map of the Republic of Nicaragua.

Table 5.6.7 Evolution of the Muslim population in Nicaragua

Year	Population	Muslims	%	Source
1906	505,000	10	0.00	[ARH]es
1920	638,119	18	0.00	[NI920]c
1971	1,877,952	150	0.01	[KET76]es
1995	4,357,099	500	0.01	[KET01B]es
2005	5,142,098	321	0.01	[NI]c
2020	6,418,000	1,000	0.02	es
2050	8,531,000	4,000	0.05	es
2100	8,124,000	8,000	0.10	es

5.6.8 Panama

The Republic of Panama gained its independence from Columbia in 1903. It has an area of 75,420 sq. km and its map is presented in Fig. 5.6.8. The first Muslims in Panama were Chinese workers who were brought for building the Panama Canal between 1904 and 1914. They were then followed by Palestinians after the middle of the twentieth century. Thus, estimates for the Muslim population increased none in 1904, to twenty of 0.01% in 1911, to 500 or 0.04% in 1970 and 1,000 or 0.05% in 1980. According to LAPOP survey, the percentage of Muslims was 1.02% in 1991. Thus, assuming that the percentage of Muslims will increase by a tenth of a percentage point per decade, the Muslim population is expected to be around 0.1 million throughout the second half of this century, comprising 1.6% of the total population by 2050 and 2.1% by 2100. The data is summarized in Table 5.6.8.

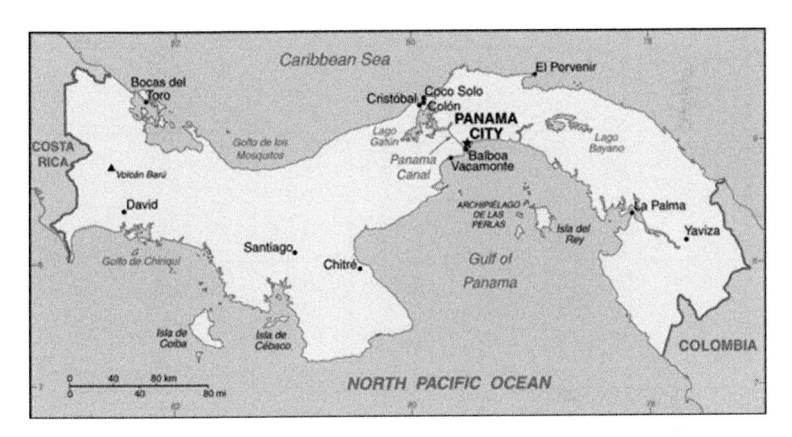

Figure 5.6.8 Map of the Republic of Panama.

Table 5.6.8 Evolution of the Muslim population in Panama

Year	Population	Muslims	%	Source
1904	290,000	—	—	[KET86]es
1911	336,742	20	0.01	[ARH]es
1970	1,428,082	500	0.04	[KET76]es
1980	1,805,287	1,000	0.06	[KET86]es
1991	2,539,140	26,000	1.02	[LAPOP]s
2020	4,231,000	55,000	1.30	es
2050	5,853,000	94,000	1.60	es
2100	6,440,000	135,000	2.10	es

Table 5.6a Centennial estimates of the Muslim population (×1000) in Central America: 600–2100 AD (1–1500 H)

		600	700	800	900	1000	1100	1200	1300
Belize	P	3	3	3	3	3	3	3	3
	M%	—	—	—	—	—	—	—	—
	M	—	—	—	—	—	—	—	—
Costa Rica	P	9	11	13	15	17	19	21	23
	M%	—	—	—	—	—	—	—	—
	M	—	—	—	—	—	—	—	—
El Salvador	P	150	160	170	180	192	210	230	250
	M%	—	—	—	—	—	—	—	—
	M	—	—	—	—	—	—	—	—
Guatemala	P	380	400	420	440	461	500	540	580
	M%	—	—	—	—	—	—	—	—
	M	—	—	—	—	—	—	—	—
Honduras	P	85	90	95	100	105	115	125	135
	M%	—	—	—	—	—	—	—	—
	M	—	—	—	—	—	—	—	—
Mexico	P	3,500	3,750	4,000	4,250	4,500	5,100	5,700	6,300
	M%	—	—	—	—	—	—	—	—
	M	—	—	—	—	—	—	—	—
Nicaragua	P	116	123	130	137	144	157	170	183
	M%	—	—	—	—	—	—	—	—
	M	—	—	—	—	—	—	—	—
Panama	P	57	61	65	69	73	80	87	94
	M%	—	—	—	—	—	—	—	—
	M	—	—	—	—	—	—	—	—
Total	P	4,300	4,598	4,896	5,194	5,495	6,184	6,876	7,568
	M%	—	—	—	—	—	—	—	—
	M	—	—	—	—	—	—	—	—
	G%		0.067	0.063	0.059	0.056	0.118	0.106	0.096

Table 5.6a (Continued)

		1400	1500	1600	1700	1800	1900	2000	2100
Belize	P	3	3	3	3	3	37	247	620
	M%	—	—	0.01	0.01	—	0.00	0.11	0.65
	M	—	—	—	—	—	—	—	4
Costa Rica	P	25	27	16	48	61	297	3,962	4,798
	M%	—	—	0.01	0.01	—	—	0.01	0.10
	M	—	—	—	—	—	—	—	5
El Salvador	P	265	278	167	190	235	766	5,888	4,766
	M%	—	—	0.01	0.01	—	—	0.02	0.11
	M	—	—	—	—	—	—	1	5
Guatemala	P	620	668	401	456	500	1,300	11,651	31,270
	M%	—	—	0.01	0.01	—	—	0.01	0.06
	M	—	—	—	—	—	—	1	19
Honduras	P	145	152	91	103	128	544	6,575	14,325
	M%	—	—	0.01	0.01	—	—	0.02	0.11
	M	—	—	—	—	—	—	1	16
Mexico	P	6,900	7,500	2,500	4,500	6,250	13,607	98,900	141,510
	M%	—	—	0.01	0.01	—	0.00	0.00	0.02
	M	—	—	—	—	—	—	1	28
Nicaragua	P	196	209	125	142	178	478	5,069	8,124
	M%	—	—	0.01	0.01	—	—	0.01	0.10
	M	—	—	—	—	—	—	1	8
Panama	P	100	106	64	72	90	263	3,030	6,440
	M%	—	—	0.01	0.01	—	—	1.10	2.10
	M	—	—	—	—	—	—	33	135
Total	P	8,254	8,943	3,367	5,514	7,446	17,292	135,322	211,854
	M%	—	—	0.01	0.01	—	0.00	0.03	0.10
	M	—	—	—	1	—	—	39	220
	G%	0.087	0.080	−0.977	0.493	0.300	0.843	2.057	0.448

Table 5.6b Decennial estimates of the Muslim population (×1000) in Central America: 1790–1940 AD (1210–1360 H)

		1790	1800	1810	1820	1830	1840	1850	1860
Belize	P	3	3	4	4	7	10	15	26
	M%	—	—	—	—	—	—	—	—
	M	—	—	—	—	—	—	—	—
Costa Rica	P	60	61	62	63	70	78	115	120
	M%	—	—	—	—	—	—	—	—
	M	—	—	—	—	—	—	—	—
El Salvador	P	230	235	240	248	280	330	366	400
	M%	—	—	—	—	—	—	—	—
	M	—	—	—	—	—	—	—	—
Guatemala	P	450	500	550	600	650	700	850	950
	M%	—	—	—	—	—	—	—	—
	M	—	—	—	—	—	—	—	—
Honduras	P	96	128	150	175	200	225	250	275
	M%	—	—	—	—	—	—	—	—
	M	—	—	—	—	—	—	—	—
Mexico	P	6,100	6,250	6,400	6,587	6,782	6,944	7,662	8,296
	M%	—	—	—	—	—	—	—	—
	M	—	—	—	—	—	—	—	—
Nicaragua	P	174	178	182	186	230	260	300	320
	M%	—	—	—	—	—	—	—	—
	M	—	—	—	—	—	—	—	—
Panama	P	88	90	92	94	110	125	135	174
	M%	—	—	—	—	—	—	—	—
	M	—	—	—	—	—	—	—	—
Total	P	7,202	7,446	7,680	7,957	8,329	8,672	9,693	10,560
	M%	—	—	—	—	—	—	—	—
	M	—	—	—	—	—	—	—	—
	G%		0.334	0.309	0.355	0.457	0.404	1.113	0.857

Table 5.6b (*Continued*)

		1870	1880	1890	1900	1910	1920	1930	1940
Belize	P	25	27	31	37	40	45	51	57
	M%	—	—	—	0.00	0.01	0.01	0.02	0.02
	M	—	—	—	—	—	—	—	—
Costa Rica	P	137	182	243	297	363	420	504	633
	M%	—	—	—	—	—	—	—	—
	M	—	—	—	—	—	—	—	—
El Salvador	P	492	570	700	766	946	1,168	1,460	1,633
	M%	—	—	—	—	0.00	0.00	0.00	0.00
	M	—	—	—	—	—	—	—	—
Guatemala	P	1,080	1,225	1,365	1,300	1,441	2,005	2,300	2,600
	M%	—	—	—	—	0.00	0.00	0.00	0.00
	M	—	—	—	—	—	—	—	—
Honduras	P	300	307	332	544	553	740	854	1,108
	M%	—	—	—	—	0.00	0.00	0.00	0.00
	M	—	—	—	—	—	—	—	—
Mexico	P	9,219	10,399	11,643	13,607	15,160	14,409	16,553	19,815
	M%	—	—	—	0.00	0.00	0.00	0.00	0.00
	M	—	—	—	—	1	—	—	—
Nicaragua	P	337	379	425	478	553	634	683	830
	M%	—	—	—	—	0.00	0.00	0.00	0.00
	M	—	—	—	—	—	—	—	—
Panama	P	176	201	230	263	337	487	467	623
	M%	—	—	—	—	0.01	0.01	0.01	0.01
	M	—	—	—	—	—	—	—	—
Total	P	11,766	13,290	14,969	17,292	19,393	19,908	22,872	27,298
	M%	—	—	—	0.00	0.00	0.00	0.00	0.00
	M	—	—	—	—	1	—	—	—
	G%	1.081	1.218	1.189	1.443	1.147	0.262	1.388	1.769

Table 5.6c Decennial estimates of the Muslim population (×1000) in Central America: 1950–2100 AD (1370–1520 H)

		1950	1960	1970	1980	1990	2000	2010	2020
Belize	P	69	92	122	144	188	247	322	398
	M%	0.08	0.08	0.08	0.08	0.09	0.11	0.19	0.25
	M	—	—	—	—	—	—	1	1
Costa Rica	P	946	1,331	1,847	2,390	3,119	3,962	4,577	5,094
	M%	—	—	0.01	0.01	0.01	0.01	0.01	0.02
	M	—	—	—	—	—	—	—	1
El Salvador	P	2,200	2,766	3,673	4,591	5,270	5,888	6,184	6,486
	M%	0.00	0.00	0.00	0.01	0.02	0.02	0.02	0.03
	M	—	—	—	—	1	1	1	2
Guatemala	P	3,115	4,211	5,622	7,283	9,264	11,651	14,630	17,916
	M%	0.00	0.00	0.00	0.01	0.01	0.01	0.01	0.02
	M	—	—	—	1	1	1	1	4
Honduras	P	1,547	2,039	2,717	3,678	4,955	6,575	8,317	9,905
	M%	0.00	0.00	0.00	0.01	0.01	0.02	0.02	0.03
	M	—	—	—	—	—	1	2	3
Mexico	P	27,945	37,772	51,494	67,761	83,943	98,900	114,093	128,933
	M%	0.00	0.00	0.00	0.00	0.00	0.00	0.00	0.01
	M	—	—	1	1	1	1	3	13
Nicaragua	P	1,295	1,773	2,407	3,266	4,173	5,069	5,824	6,625
	M%	0.00	0.00	0.01	0.01	0.01	0.01	0.01	0.02
	M	—	—	—	—	—	1	1	1
Panama	P	860	1,133	1,519	1,978	2,471	3,030	3,643	4,315
	M%	0.04	0.04	0.04	0.06	1.02	1.10	1.20	1.30
	M	—	—	1	1	25	33	44	56
Total	P	37,976	51,117	69,400	91,092	113,384	135,322	157,591	179,670
	M%	0.00	0.00	0.00	0.00	0.03	0.03	0.03	0.04
	M	1	1	2	4	29	39	53	81
	G%	3.301	2.972	3.058	2.720	2.189	1.769	1.523	1.311

Table 5.6c (*Continued*)

		2030	2040	2050	2060	2070	2080	2090	2100
Belize	P	468	526	571	603	624	631	629	620
	M%	0.30	0.35	0.40	0.45	0.50	0.55	0.60	0.65
	M	1	2	2	3	3	3	4	4
Costa Rica	P	5,468	5,692	5,773	5,724	5,574	5,340	5,062	4,798
	M%	0.03	0.04	0.05	0.06	0.07	0.08	0.09	0.10
	M	2	2	3	3	4	4	5	5
El Salvador	P	6,779	6,931	6,937	6,786	6,474	6,006	5,409	4,766
	M%	0.04	0.05	0.06	0.07	0.08	0.09	0.10	0.11
	M	3	3	4	5	5	5	5	5
Guatemala	P	21,213	24,261	26,921	29,093	30,632	31,462	31,625	31,270
	M%	0.02	0.03	0.03	0.04	0.04	0.05	0.05	0.06
	M	4	7	8	12	12	16	16	19
Honduras	P	11,449	12,780	13,831	14,575	14,977	15,027	14,776	14,325
	M%	0.04	0.05	0.06	0.07	0.08	0.09	0.10	0.11
	M	5	6	8	10	12	14	15	16
Mexico	P	140,876	149,759	155,151	157,156	156,314	153,104	147,930	141,510
	M%	0.01	0.01	0.01	0.01	0.01	0.01	0.01	0.02
	M	14	15	16	16	16	15	15	28
Nicaragua	P	7,392	8,049	8,531	8,807	8,875	8,750	8,485	8,124
	M%	0.03	0.04	0.05	0.06	0.07	0.08	0.09	0.10
	M	2	3	4	5	6	7	8	8
Panama	P	4,928	5,442	5,853	6,151	6,343	6,447	6,475	6,440
	M%	1.40	1.50	1.60	1.70	1.80	1.90	2.00	2.10
	M	69	82	94	105	114	122	130	135
Total	P	198,571	213,441	223,567	228,895	229,812	226,766	220,391	211,854
	M%	0.05	0.06	0.06	0.07	0.08	0.08	0.09	0.10
	M	100	121	139	158	172	187	196	220
	G%	1.000	0.722	0.463	0.236	0.040	−0.133	−0.285	−0.395

Table 5.7a Centennial estimates of the Muslim population (×1000) in the Americas: 600–2100 AD (1–1500 H)

		600	700	800	900	1000	1100	1200	1300	1400	1500	1600	1700	1800	1900	2000	2100
Southern Caribbean	P	7	8	8	11	14	17	17	21	22	24	13	196	306	906	2,600	2,062
	M%	—	—	—	—	—	—	—	—	—	—	—	—	—	1.26	3.04	2.73
	M	—	—	—	—	—	—	—	—	—	—	—	—	—	11	79	56
North America	P	1,173	1,259	1,325	1,393	1,461	1,631	1,801	1,971	2,117	2,253	1,754	1,213	5,562	81,620	312,427	490,889
	M%	—	—	—	—	—	—	—	—	—	—	—	—	—	0.03	0.69	4.50
	M	—	—	—	—	—	—	—	—	—	—	—	—	—	23	2,150	22,074
Central Caribbean	P	4	5	6	7	8	8	8	9	9	10	7	37	151	139	334	367
	M%	—	—	—	—	—	—	—	—	—	—	—	—	—	—	0.75	1.47
	M	—	—	—	—	—	—	—	—	—	—	—	—	—	—	2	5
South America	P	5,666	6,160	6,642	7,126	7,611	8,202	8,822	9,454	10,084	10,670	5,522	6,464	9,913	40,135	348,412	429,301
	M%	—	—	—	—	—	—	—	—	—	—	—	—	—	0.05	0.08	0.13
	M	—	—	—	—	—	—	—	—	—	—	—	—	—	20	286	558
Northern Caribbean	P	160	171	181	192	202	221	240	259	278	298	129	307	1,721	5,512	34,746	36,071
	M%	—	—	—	—	—	—	—	—	—	—	—	—	—	0.02	0.04	0.14
	M	—	—	—	—	—	—	—	—	—	—	—	—	—	1	13	51
Central America	P	4,300	4,598	4,896	5,194	5,495	6,184	6,876	7,568	8,254	8,943	3,367	5,514	7,446	17,292	135,322	211,854
	M%	—	—	—	—	—	—	—	—	—	—	0.01	0.01	—	0.00	0.03	0.10
	M	—	—	—	—	—	—	—	—	—	—	—	1	—	—	39	220
Total	P	11,310	12,201	13,058	13,923	14,791	16,263	17,764	19,283	20,764	22,198	10,792	13,731	25,099	145,604	833,841	1,170,543
	M%	—	—	—	—	—	—	—	—	—	—	0.00	0.00	—	0.04	0.31	1.96
	M	—	—	—	—	—	—	—	—	—	—	—	1	—	56	2,569	22,965
	G%		0.076	0.068	0.064	0.061	0.095	0.088	0.082	0.074	0.067	-0.721	0.241	0.603	1.758	1.745	0.339

Table 5.7b Decennial estimates of the Muslim population (×1000) in the Americas: 1790–1940 AD (1210–1360 H)

		1790	1800	1810	1820	1830	1840	1850	1860	1870	1880	1890	1900	1910	1920	1930	1940
Southern Caribbean	P	288	306	333	332	350	429	478	533	607	696	787	906	933	1,018	1,098	1,284
	M%	—	—	—	—	—	—	0.10	0.36	0.62	0.93	1.20	1.26	1.67	1.80	1.94	2.26
	M	—	—	—	—	—	—	0	2	4	6	9	11	16	18	21	29
North America	P	4,243	5,562	7,886	10,395	14,382	18,888	25,650	34,697	43,534	54,509	67,840	81,620	99,472	114,848	133,629	143,730
	M%	—	—	—	—	—	0.00	0.00	0.02	0.03	0.04	0.03	0.03	0.04	0.04	0.04	0.04
	M	—	—	—	—	—	—	—	6	12	20	19	23	38	43	50	54
Central Caribbean	P	130	151	156	137	135	154	140	137	141	138	144	139	130	120	121	132
	M%	—	—	—	—	—	—	—	—	—	—	—	—	0.01	0.01	0.01	0.02
	M	—	—	—	—	—	—	—	—	—	—	—	—	—	—	—	—
South America	P	9,137	9,913	10,608	11,014	13,520	15,629	18,015	21,240	23,901	28,252	33,614	40,135	49,708	63,410	73,868	88,663
	M%	—	—	—	—	—	0.00	0.00	0.01	0.02	0.04	0.05	0.05	0.08	0.08	0.11	0.11
	M	—	—	—	—	—	—	1	2	5	12	16	20	42	48	80	96
Northern Caribbean	P	1,601	1,721	1,843	1,990	2,249	2,839	3,197	3,663	3,924	4,386	4,886	5,512	6,740	8,234	10,147	12,185
	M%	—	—	—	—	—	—	0.00	0.00	0.01	0.01	0.01	0.02	0.05	0.05	0.04	0.04
	M	—	—	—	—	—	—	—	—	—	—	1	1	3	4	4	5
Central America	P	7,202	7,446	7,680	7,957	8,329	8,672	9,693	10,560	11,766	13,290	14,969	17,292	19,393	19,908	22,872	27,298
	M%	—	—	—	—	—	—	—	—	—	—	—	0.00	0.00	0.00	0.00	0.00
	M	—	—	—	—	—	—	—	—	—	—	—	—	1	—	—	—
Total	P	22,600	25,099	28,504	31,824	38,966	46,611	57,173	70,830	83,873	101,270	122,238	145,604	176,376	207,537	241,734	273,292
	M%	—	—	—	—	—	0.00	0.00	0.02	0.03	0.04	0.04	0.04	0.06	0.06	0.06	0.07
	M	—	—	—	—	—	—	2	11	21	39	45	56	99	114	156	185
	G%	1.049	1.272	1.102	2.025	1.791	2.043	2.142	1.690	1.885	1.882	1.749	1.917	1.627	1.525	1.227	

Table 5.7c Decennial estimates of the Muslim population (×1000) in the Americas: 1950–2100 AD (1370–1520 H)

		1950	1960	1970	1980	1990	2000	2010	2020
Southern Caribbean	P	1,502	1,870	2,082	2,265	2,475	2,600	2,736	2,838
	M%	2.54	2.77	2.92	2.99	3.03	3.04	2.90	2.93
	M	38	52	61	68	75	79	79	83
North America	P	172,603	204,649	230,992	254,007	279,785	312,427	343,287	368,870
	M%	0.04	0.05	0.09	0.22	0.35	0.69	1.08	1.57
	M	65	99	202	557	988	2,150	3,693	5,799
Central Caribbean	P	151	174	217	259	309	334	370	397
	M%	0.44	0.47	0.73	1.01	0.56	0.75	0.76	0.81
	M	1	1	2	3	2	2	3	3
South America	P	113,765	148,686	192,074	240,515	295,398	348,412	392,544	430,760
	M%	0.10	0.10	0.11	0.09	0.09	0.08	0.08	0.09
	M	118	148	206	220	266	286	331	385
Northern Caribbean	P	15,217	18,348	22,580	26,785	30,885	34,746	37,705	39,897
	M%	0.05	0.05	0.02	0.03	0.03	0.04	0.05	0.06
	M	7	10	5	8	9	13	19	24
Central America	P	37,976	51,117	69,400	91,092	113,384	135,322	157,591	179,670
	M%	0.00	0.00	0.00	0.00	0.03	0.03	0.03	0.04
	M	1	1	2	4	29	39	53	81
Total	P	341,213	424,844	517,346	614,924	722,236	833,841	934,234	1,022,432
	M%	0.07	0.07	0.09	0.14	0.19	0.31	0.45	0.62
	M	230	310	477	858	1,369	2,569	4,178	6,375
	G%	2.220	2.192	1.970	1.728	1.609	1.437	1.137	0.902

Muslims in Central America | 769

Table 5.7c (Continued)

		2030	2040	2050	2060	2070	2080	2090	2100
Southern Caribbean	P	2,870	2,838	2,745	2,613	2,472	2,330	2,193	2,062
	M%	2.91	2.88	2.86	2.84	2.81	2.77	2.75	2.73
	M	83	82	79	74	69	65	60	56
North America	P	390,599	410,177	425,200	439,379	454,369	467,639	479,097	490,889
	M%	1.93	2.29	2.64	3.01	3.37	3.74	4.12	4.50
	M	7,532	9,373	11,243	13,205	15,315	17,491	19,716	22,074
Central Caribbean	P	415	423	422	414	404	394	381	367
	M%	0.88	0.95	1.03	1.11	1.19	1.28	1.37	1.47
	M	4	4	4	5	5	5	5	5
South America	P	461,854	481,837	491,472	491,847	483,493	468,263	449,327	429,301
	M%	0.09	0.10	0.11	0.11	0.11	0.12	0.12	0.13
	M	430	466	522	541	551	551	546	558
Northern Caribbean	P	42,143	43,410	43,840	43,351	42,177	40,509	38,413	36,071
	M%	0.07	0.08	0.09	0.10	0.11	0.12	0.13	0.14
	M	30	35	40	44	47	49	51	51
Central America	P	198,571	213,441	223,567	228,895	229,812	226,766	220,391	211,854
	M%	0.05	0.06	0.06	0.07	0.08	0.08	0.09	0.10
	M	100	121	139	158	172	187	196	220
Total	P	1,096,453	1,152,127	1,187,246	1,206,500	1,212,728	1,205,900	1,189,802	1,170,543
	M%	0.75	0.88	1.01	1.16	1.33	1.52	1.73	1.96
	M	8,179	10,081	12,027	14,027	16,159	18,348	20,574	22,965
	G%	0.699	0.495	0.300	0.161	0.051	−0.056	−0.134	−0.163

5.6.9 Regional Summary and Conclusion

Central America has the least concentration of Muslims among the five regions spanning the Americas. The Muslim population in this region was almost nonexistent prior to the twentieth century. However, it has been steadily increasing but remains negligible at less than 0.1% throughout this century. The following tables present centennial data from 600 AD to 2100 AD (or approximately 1 H to 1500 H) in Table 5.6a and decennial data from 1790 AD to 2100 AD (or 1210 H to 1520 H) in Tables 5.6b and 5.6c for current countries in Central America. The data includes total population in thousands (P), the percentage of which is Muslim (M%), the corresponding Muslim population in thousands (M), and the annual population growth rate (APGR, or G%) of the total population in this region.

5.7 The Americas' Summary and Conclusion

Muslims came to the Americas as early as the tenth century. Many waves followed, but all went extinct due to European persecution, either by the Spaniards under the name of Catholicism, or by others while fighting slaves' uprising. The number of Muslims is increasing steadily since the second half of the last century and is expected to remain so for the next three centuries. It is currently below 1% of the total population of this continent and is expected to reach 2% by the end of this century. The following tables present centennial data from 600 AD to 2100 AD (or approximately 1 H to 1500 H) in Table 5.7a and decennial data from 1790 AD to 2100 AD (or 1210 H to 1520 H) in Tables 5.7b and 5.7c for the five regions of the Americas. The data includes total population in thousands (P), the percentage of which is Muslim (M%), the corresponding Muslim population in thousands (M), and the annual population growth rate (APGR, or G%) of the total population in the Americas and each of its five regions. A choropleth map of the Americas illustrating the presence of Muslims is presented in Fig. 5.7.0, where white is less than 1% of the district's population and the darker the region, the more is the percentage of Muslims.

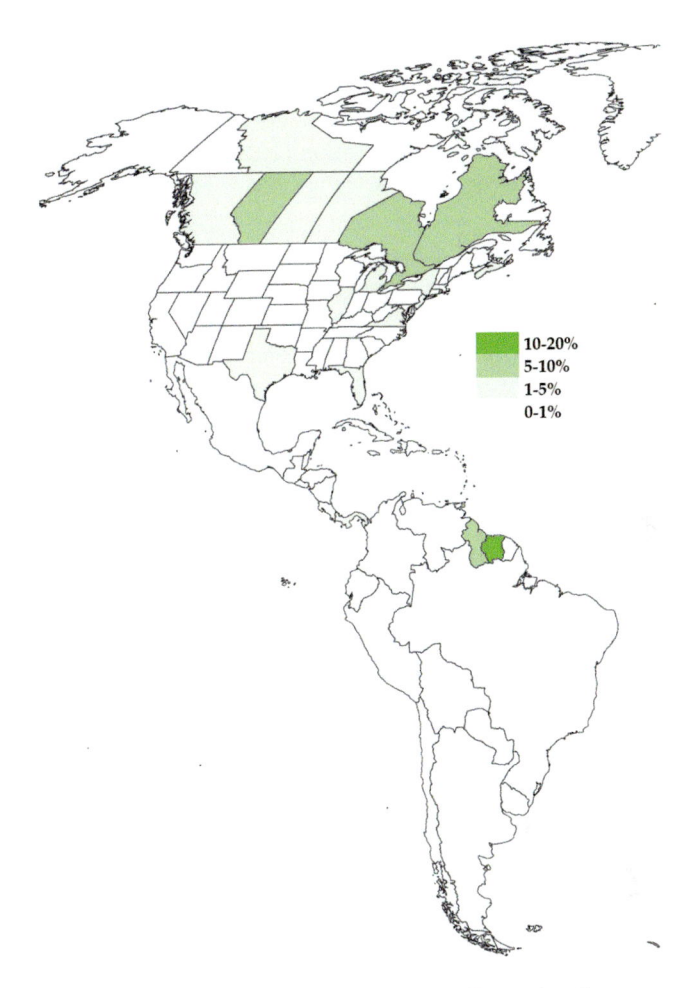

Figure 5.7.0 A choropleth map of the Americas illustrating the presence of Muslims.

References

[AG11] Statistics Division of the Ministry of Finance and the Economy (SD/MFE) (2014). *Antigua and Barbuda 2011 Population and Housing Census* (Book of Statistical Tables I). St. John's: SD/MFE.

[AG01] Statistics Division of the Ministry of Finance and the Economy (SD/MFE) (2004). *2001 Census of Population and Housing, Summary Social,*

Economic, Demographic, and Housing Characteristics (Vol. I: Antigua & Barbuda Summary). St. John's: SD/MFE.

[AI11] Anguilla's preliminary census findings #5: "Who are we?–Ethnic composition and religious affiliation (2014, November 20). *Anguilla News*. Retrieved from http://www.anguillanews.com/

[AI01] CARICOM (Caribbean Community) Capacity Development Programme (CCDP) (2009). *2000 Round of Population and Housing Census of the Caribbean Community: National Census Report, Anguilla*. Georgetown: The CARICOM Secretariat.

[AN] Central Bureau of Statistics (CBS) (2002). *Fourth Population and Housing Census Netherlands Antilles 2001*. Willemstad: CBS. And e-mail correspondence with Mr. O. Girigori from CBS in January 2011.

[ARH] Djinguiz, M. (1908). L'Islam dans l'Amérique Centrale et dans l'Amérique du Sud. *Revue du Monde Musulman*, 6(10), 314–318.

[AW10] Central Bureau of Statistics–Aruba (CBS) (2011). *Fifth Population and Housing Census Aruba*. Oranjestad: CBS. And e-mail communication with Mr. F.C. Eelens from CBS in October 2011.

[AW] Central Bureau of Statistics–Aruba (CBS) (2002). *The People of Aruba, Continuity and Change: Census 2000 Special Reports*. Oranjestad: CBS.

[BB10] Barbados Statistical Service (BSS) (2013). *2010 Population and Housing Census*, Volume 1. Warrens: BSS.

[BB00] CARICOM (Caribbean Community) Capacity Development Programme (CCDP) (2009). *2000 Round of Population and Housing Census of the Caribbean Community: National Census Report, Barbados*. Georgetown: The CARICOM Secretariat.

[BC] Central Statistical Office of Trinidad (CSO) (1964). *British Caribbean Census Reports*. Port of Spain: CSO.

[BM10] Government of Bermuda, Department of Statistics (GBDS) (2012). *2010 Census of Population and Housing Report*. Hamilton: GBDS.

[BM60] The Bermuda Government (1961). Census of Bermuda 23rd October 1960: Report of census & statistical tables compiled in accordance with the Census Act, 1950. Hamilton: The Bermuda Government

[BQ] Statistics Netherlands (2019). *Caribbean Netherlands; religious denomination, personal characteristics*. StatLine Open Data. Retrieved from https://opendata.cbs.nl/statline/

[BR10] Instituto Brasileiro de Geografia e Estatística (IBGE) (2012). *Censo Demográfico 2010: Características Gerais da População, Religião e Pessoas com Deficiência*. Rio de Janeiro: IBGE.

[BR00] Instituto Brasileiro de Geografia e Estatística (IBGE) (2003). *Demographic Census 2000: Population and Household Characteristics–Universe Results.* Rio de Janeiro: IBGE.

[BRH] Lamounier, B., Camargo, C. P. F., Berquo, E. S., Madeira, F. R., Santos, J. L. F., Duarte, J. C., Yunes, J., Lopes, J. R. B., Marcilio, M. L., Levy, M. S. F., Goncalves, M. A. I., Patarra, N. L., & Singer, P. I. (1975). *La Population du Brésil.* 1974 Country Monographs, CICRED (Comite International de Coordination de Recherches Nationales de Demographie) Series. Paris: CICRED. Retrieved from http://www.cicred.org/Eng/Publications/content/3MonographNational/Index.html

[BS10] Department of Statistics of the Bahamas (DSB) (2012). *The 2010 Census of Population and Housing Report - All Bahamas.* Nassau: DSB.

[BS00] Department of Statistics of the Bahamas (DSB) (2008). *The 2000 Census of Population and Housing Report - All Bahamas.* Nassau: DSB.

[BZ] Statistical Institute of Belize (SIB) (2013). *Belize Population and Housing Census 2010, Country Report.* Belmopan: SIB.

[CA11] Statistics Canada (2013). *2011 National Household Survey: Data Tables.* Retrieved online from http://www12.statcan.gc.ca/nhs-enm/2011/dp-pd/dt-td/Index-eng.cfm with catalogue number 99-010-X2011032.

[CA01] Statistics Canada (2003). *Religion in Canada.* 2001 Census: Analysis Series. Ottawa: Minister of Industry.

[CA] Hamadani, D. H. (2001). Islam in Canada. In A. M. M'Bow & A. Kettani (eds.), *Islam and Muslims in the American Continent.* (pp. 61–103). Beirut: Center of Historical, Economical and Social Studies.

[CA871] Statistics of Canada (1878). *Censuses of Canada 1608 to 1876* (Volume V). Ottawa: Maclean, Roger & Co.

[CL12] Instituto Nacional de Estadística de Chile (INE) (2013). *Resultados XVIII Censo de Población 2012.* Santiago de Chile: Empresa Periodística La Nación S.A.

[CL02] Instituto Nacional de Estadística de Chile (INE) (2003). *Censo 2002–Síntesis de Resultados.* Santiago de Chile: Empresa Periodística La Nación S.A.

[CLH] Elhamalawy, S (2010, March 4). The Muslim Community of Chile. *The Muslim Observer.* Retrieved from http://muslimmedianetwork.com/mmn/?p=5893

[CR] Leff, A. (2008, September 9). Ramadan Off to Quiet Start in Costa Rica. *Tico Times.*

[CR864] Seccion de Estadistica (1868). *Censo General de la Republica de Costa-Rica (27 de Noviembre de 1864.)* (N. IX.X.XI.XII. Cuadros que Comprenden la Poblacion de la Republica por Confesiones, Enfermedades, Nacionalidad y Residencia). San Jose: Imprenta Nacional.

[CR883] Seccion de Estadistica (1885). *Censo de la Republica de Costa-Rica Levantado el 30 de Noviembre del Ano de 1883* (Part II: Nacionalidad, Religión, Profesiones, Grado de Instruccion). San Jose: Tipografia Nacional.

[CR892] Direccion General de Estadistica (1893). *Censo General de de la Republica de Costa Rica Levantado Bajo la Administracion del Licenciado Don Jose J. Rodriguez el 18 de Febrero de 1892.* San Jose: Tipografia Nacional.

[CW] Central Bureau of Statistics Curaçao (CBS) (2013). *Table D-5. Population by Religion, Age Group and Sex, Census 2011.* Willemstad: CBS.

[DOS13] United States Department of State (DOS), Bureau of Democracy, Human Rights, and Labor (2013). *2013 Annual Report on International Religious Freedom.* Washington, DC: DOS. Retrieved from http://www.state.gov/j/drl/rls/irf/

[DOS12] United States Department of State (DOS), Bureau of Democracy, Human Rights, and Labor (2012). *2012 Annual Report on International Religious Freedom.* Washington, DC: DOS. Retrieved from http://www.state.gov/j/drl/rls/irf/

[DOS11] United States Department of State (DOS), Bureau of Democracy, Human Rights, and Labor (2011). *2011 Annual Report on International Religious Freedom.* Washington, DC: DOS. Retrieved from http://www.state.gov/j/drl/rls/irf/

[DOS07] United States Department of State (DOS), Bureau of Democracy, Human Rights, and Labor (2007). *2007 Annual Report on International Religious Freedom.* Washington, DC: DOS. Retrieved from http://www.state.gov/j/drl/rls/irf/

[EC] Suquillo, Y. J. (2002). *Islam in Ecuador.* IslamAwareness.net.

[FK16] Falkland Islands Government. Policy and Economic Development Unit (2017). *2016 Census Report.* Stanley: Policy Unit. And e-mail correspondence with Ms. M. Daly, Statistics and Performance Officer with the Policy Unit, FIG in April 2017.

[FK06] Falkland Islands Government (2006). *Falkland Islands Census Statistics 2006.* Stanley: Policy Unit.

[FKE] Falkland Islands Government (1996). *Census Reports–Population by birth place 1833–1996.* Jane Cameron National Archives. Electronic file retrieved from http://www.fig.gov.fk/archives/index.php/online-collections/people/census-information-early-settlers

[FK80] Falkland Islands Government (1980). *Report of Census 1980.* Stanley: Policy Unit. And e-mail correspondence with Ms. M. Daly, Statistics and Performance Officer with the Policy Unit, FIG in April 2017.

[FK931] Falkland Islands Government (1932). *Report of Census, 1931.* Stanley: Government Printing Office.

[FK843] Falkland Islands Government (1843). *Statistical table showing the names, occupations, etc, of the inhabitants of the Falkland Islands on the 31st of March 1843.* Jane Cameron National Archives [G1; 1–4]. Handwritten report retrieved from http://www.fig.gov.fk/archives/index.php/online-collections/people/census-information-early-settlers

[FK842] Falkland Islands Government (1842). *Statistical table showing the names, occupations, etc, of the inhabitants of the Falkland Islands in January 1842.* Jane Cameron National Archives [B1; 150–152]. Handwritten report retrieved from http://www.fig.gov.fk/archives/index.php/online-collections/people/census-information-early-settlers

[GB911] His Majesty's Stationary Office (HMSO) (1917). *Census of England and Wales. 1911.* (General Report with Appendices). London: Darling & Son.

[GB901] General Register Office (GRO) (1906). *Census of the British Empire. 1901. Report with Summary and Detailed Tables for the Several Colonies, &c., Area, Houses, and Population; also Population Classified by Ages, Condition as to Marriage, Occupation, Birthplaces, Religions, Degrees of Education, and Infirmities.* London: Darling & Son.

[GB871] Her Majesty's Stationary Office (HMSO) (1873). *Census of England and Wales for the Year 1871* (General Report. Vol. IV). London: George Edward Eyre and William Spottiswoode.

[GB861] Coke, C. A. (1864). *Census of the British Empire: Compiled from Official Returns for the Year 1861, with Its Colonies and Foreign Possessions.* London: Harrison.

[GD] CARICOM (Caribbean Community) Capacity Development Programme (CCDP) (2009). *2000 Round of Population and Housing Census of the Caribbean Community: National Census Report, Grenada.* Georgetown: The CARICOM Secretariat.

[GD11] Central Statistics Office (2019). *Census Population Density – Census Submission Final*. Online file retrieved from https://www.finance.gd/

[GF] Calmont, A. (2008). Les Africains en Guyane. *EchoGéo* [Online], 6. doi:10.4000/echogeo.6333. Retrieved from http://echogeo.revues.org/6333

[GL] Statistics Greenland (2013). Table BEEST6: Population by Year, Citizenship, Age and Gender. Population: Population in Greenland. *Database for Greenland* [Online]. Retrieved from bank.stat.gl

[GT] Sandoval, J. (2005, June 5). Musulmanes en Guatemala. *Prensa Libre*, No. 48.

[GY12] Bureau of Statistics (BOS) [Guyana] (2016). *2012 Population & Housing Census, Compendium 2: Population Composition*. Georgetown: BOS.

[GY02] Bureau of Statistics (BOS) [Guyana] (2007). *Population & Housing Census 2002 - Guyana National Report*. Georgetown: BOS.

[GYH] Hamid, A. & Baksh, W. (2001). Islam in Guyana. In A. M. M'Bow & A. Kettani (eds.), *Islam and Muslims in the American Continent*. (pp. 353–421). Beirut: Center of Historical, Economical and Social Studies.

[GW] Clara (2017, May 19). Musulmans et Juifs de Guadeloupe. 100% Antilles. A blog retrieved from http://100-pour-cent-antilles.com/musulmans-et-juifs-de-guadeloupe

[HT03] Institut Haïtien de Statistique et d'Informatique (IHSI) (2009). Grandes Leçons Socio-Demographiques Tirees du 4e RGPH. Port-au-Prince: IHSI.

[ISS] International Social Survey Programme (2015). *International Social Survey*. Cologne: Leibniz Institute for the Social Sciences (GESIS). Retrieved from http://zacat.gesis.org/. More info on ISS is available at http://www.issp.org/

[JAN] Jansen, H. (1897). *Verbreitung des Islâms: mit angabe der verschiedenen riten, sekten und religiösen bruderschaften in den verschiedenen ländern der erde 1890 bis 1897. Mit benutzung der neuesten angaben (zählungen, berechnungen, schätzungen und vermutungen). [German for Dissemination of Islam: with number of various rites, sects and religious brotherhoods in the various countries of the earth 1890 to 1897. With use of information on the latest disclosures (censuses, calculations, estimates and guesses)]*. Berlin: Selbstverlag des Verfassers.

[JM] Statistical Institute of Jamaica (SIJ) (2012). *Population and Housing Census 2011 Jamaica* (Vol. I: General Report). Kingston: SIJ.

[KET01D] Kettani, A. (2001). Islam in Argentina, Paraguay and Uruguay. In A. M. M'Bow & A. Kettani (eds.), *Islam and Muslims in the American*

Continent. (pp. 565–607). Beirut: Center of Historical, Economical and Social Studies.

[KET01C] Kettani, A. (2001). Islam in the Andean States. In A. M. M'Bow & A. Kettani (eds.), *Islam and Muslims in the American Continent*. (pp. 509–563). Beirut: Center of Historical, Economical and Social Studies.

[KET01B] Kettani, A. (2001). Islam in Central America. In A. M. M'Bow & A. Kettani (eds.), *Islam and Muslims in the American Continent*. (pp. 461–508). Beirut: Center of Historical, Economical and Social Studies.

[KET01A] Kettani, A. (2001). Islam in the Caribbean. In A. M. M'Bow & A. Kettani (eds.), *Islam and Muslims in the American Continent*. (pp. 231–291). Beirut: Center of Historical, Economical and Social Studies.

[KET86] Kettani, A. (1986). *Muslim Minorities in the World Today.* London: Mansell Publishing.

[KET76] Kettani, A. (1976). *Muslims in Europe and the Americas*. [Arabic]. Beirut: Dar Idriss.

[KN11] Nevis Department of Statistics (2017). *The Nevis Statistical Digest 2017*. Charlestown: Nevis Island Administration. And e-mail correspondence with Mr. C. Phipps from St. Kitts Department of Statistics in May 2018.

[KN01] CARICOM (Caribbean Community) Capacity Development Programme (CCDP) (2009). *2000 Round of Population and Housing Census of the Caribbean Community: National Census Report, St. Kitts and Nevis*. Georgetown: The CARICOM Secretariat.

[KN91] Statistical Division of the Planning Unit (SDPU) (2003). *St. Kitts and Nevis Demography Digest: 2002*. Basseterre: SDPU.

[KY10] Ministry of Finance, Tourism and Development (MFTD) (2011). *The Cayman Islands' 2010 Census of Population and Housing Report*. George Town: MFTD.

[KY07] Economics and Statistics Office (ESO) (2009). *Statistical Compendium 2008* (Vol. 11: Vital Population Statistics). George Town: ESO.

[LAPOP] Vanderbuilt University (2015). *The Latin American Public Opinion Project (LAPOP) System for Online Data Analysis (SODA)*. Retrieved from http://lapop.ccp.ucr.ac.cr/

[LC] Central Statistics Office (CSO) (2011). 2010 Population and Housing Census, Preliminary Report. Castries: CSO.

[MRO] Mroueh, Y. (1996). *Pre-Columbian Muslims in the Americas*. Report of the Preparatory Committee for International Festivals to Celebrate the Millennium of the Muslims Arrival to the Americas (996–1996 CE). Burton, MI: As-Sunnah Foundation of America.

[MS11] Statistics Department of Montserrat (SDM) (2012). *Census 2011: Montserrat at a Glance*. Brades: SDM. And e-mail correspondence with Ms. U. Haynes from SDM in June 2012.

[MS01] CARICOM (Caribbean Community) Capacity Development Programme (CCDP) (2009). *2000 Round of Population and Housing Census of the Caribbean Community: National Census Report, Montserrat*. Georgetown: The CARICOM Secretariat.

[MX10] Instituto Nacional de Estadística, Geografía e Informática (INEGI) (2011). *Censo de Población y Vivienda 2010: Tabukados del Cuestionario Básico*. Mexico City: INEGI.

[MX] Instituto Nacional de Estadística, Geografía e Informática (INEGI) (2005). *La Diversidad Religiosa en México*. Censo General de Población y Vivienda 2000. Mexico City: INEGI.

[NI] Instituto Nacional de Estadísticas y Censos (INEC) (2006). *Resumen Censal–Censos Nacionales, 2005: Poblacion, Vivienda, Hogar*. VIII Censo de Población y IV de Vivienda. Managua: INEC.

[NI920] La Oficina Central (1920). *Censo de 1920*. Managua: La Oficina Central.

[PEW] Pew Research Center (PRC) (2009). *Mapping the Global Muslim Population, A Report on the Size and Distribution of the World's Muslim Population*. Washington, DC: PRC.

[PR] James, A. (2010, September 16). Eidul Fitr in Puerto Rico. *The Muslim Observer*.

[ROD] Radončić, E. (2013). *World Almanac of the demographic history of Muslims: Rich cartographic representation of the geographical distribution of Muslim populations* (M. Durmić, Trans.). Sarajevo: Edin Radončić.

[SR] General Bureau of Statistics–Suriname (GBS) (2013). *Suriname Census 2012* (Vol. I: Demographic and Social Characteristics). Paramaribo: GBS.

[SR80] Jamaludin, I. & Kettani A. (2001). Islam in Surinam. In A. M. M'Bow & A. Kettani (eds.), *Islam and Muslims in the American Continent*. (pp. 423–460). Beirut: Center of Historical, Economical and Social Studies.

[SV] Lemus, E. (2008, May 8). Ser Musulmán en El Salvador. *BBC Mundo*.

[SX] Clifford, J. (2011). E-mail correspondence with Mr. J. Clifford from the Islamic Center of Sint Maarten in January 2011.

[SYB950] Steinberg, S. H. (ed.) (1946–1968). *The Statesman's Year-Book.* London: MacMillan.

[SYB931] Epstein, M. (ed.) (1927–1945). *The Statesman's Year-Book.* London: MacMillan.

[SYB922] Scott-Keltie, J. & Epstein M. (eds.) (1922). *The Statesman's Year-Book.* London: MacMillan.

[SYB920] Scott-Keltie, J. & Epstein M. (eds.) (1920). *The Statesman's Year-Book.* London: MacMillan.

[SYB915] Scott-Keltie, J. (ed.) (1915). *The Statesman's Year-Book.* London: MacMillan.

[SYB910] Scott-Keltie, J. (ed.) (1910). *The Statesman's Year-Book.* London: MacMillan.

[SYB903] Scott-Keltie, J. (ed.) (1903). *The Statesman's Year-Book.* London: MacMillan.

[SYB900] Scott-Keltie, J. (ed.) (1900). *The Statesman's Year-Book.* London: MacMillan.

[SYB894] Scott-Keltie, J. (ed.) (1894). *The Statesman's Year-Book.* London: MacMillan.

[SYB890] Scott-Keltie, J. (ed.) (1890). *The Statesman's Year-Book.* London: MacMillan.

[SYB873] Martin, F. (ed.) (1873). *The Statesman's Year-Book.* London: MacMillan.

[TC12] Strategic Policy and Planning Department (2013). *The Turks and Caicos Islands 2012 Census of Population and Housing Tables (Grand Turk to Pine Cay).* Grand Turk: Statistical Office. And Email communication with Ms. S.V. Williams from Turks & Caicos Statistics Department in August 2019.

[TT11] Central Statistical Office (CSO) (2012). *Trinidad and Tobago 2011 Population and Housing Census Demographic Report.* Port of Spain: CSO.

[TT00] CARICOM (Caribbean Community) Capacity Development Programme (CCDP) (2009). *2000 Round of Population and Housing Census of the Caribbean Community: National Census Report, Trinidad and Tobago.* Georgetown: The CARICOM Secretariat.

[TT90] Ibrahim, M. (2001). Islam in Trinidad and Tobago. In A. M. M'Bow & A. Kettani (eds.), *Islam and Muslims in the American Continent.* (pp. 293–351). Beirut: Center of Historical, Economical and Social Studies.

[TTH] Harewood, J. (1975). The Population of Trinidad and Tobago. 1974 Country Monographs, CICRED (Comite International de Coordination de Recherches Nationales de Demographie) Series. Paris: CICRED. Retrieved from http://www.cicred.org/Eng/Publications/content/3MonographNational/Index.html

[UN] United Nations Statistics Division, UNSD Demographic Statistics (2018). Population by Religion, Sex and Urban/Rural Residence. *UNdata* [Online Database]. Retrieved from http://data.un.org

[UNE] United Nations Statistics Division, UNSD Demographic Statistics (2017). Population by National and/or Ethnic Group, Sex and Urban/Rural Residence, *UNdata* [Online Database]. Retrieved from http://data.un.org

[UN04] United Nations, Department of International Economic and Social Affairs, Statistical Office (2006). Table 6 - Population by Religion, Sex and Urban/Rural Residence: Each Census, 1985–2004. *Demographic Yearbook Special Census Topics* (Vol. 2b: Ethnocultural Characteristics). New York, NY: United Nations.

[UN88] United Nations, Department of International Economic and Social Affairs, Statistical Office (1990). Table 29 - Population by Religion and Sex: Each Census, 1979–1988. *Demographic Yearbook 1988*. New York, NY: United Nations.

[UN79] United Nations, Department of International Economic and Social Affairs, Statistical Office (1980). Table 29 - Population by Religion, Sex and Urban/Rural Residence: Each Census, 1970–1979. *Demographic Yearbook 1979*. New York, NY: United Nations.

[UN63] United Nations, Department of International Economic and Social Affairs, Statistical Office (1964). Table 11 - Population by Religion and Sex: Each Census, 1955–1963. *Demographic Yearbook 1963*. New York, NY: United Nations.

[UN56] United Nations, Department of International Economic and Social Affairs, Statistical Office (1957). Table 8 - Population by Religion and Sex: Each Census, 1945–1955. *Demographic Yearbook 1956*. New York, NY: United Nations.

[USS12] Bagby, I. (2012). *The American Mosque 2011: Basic Characteristics of the American Mosque Attitudes of Mosque Leaders.* Report Number 1 from the US Mosque Study 2011. Washington, DC: Council on American-Islamic Relations (CAIR).

[USS11] Pew Research Center (PRC) (2011). *Muslim Americans: No Signs of Growth in Alienation or Support for Extremism.* Washington, DC: PRC.

[USS] U. S. Census Bureau (2015). *American Community Survey* [online]. Retrieved from http://www.census.gov/acs/

[USS8] U.S. Census Bureau (2011). *The 2012 Statistical Abstract: The National Data Book*. Washington, DC: U.S. Census Bureau.

[USS6] Gibson, C. & Jung, K. (2006). *Historical Census Statistics on the Foreign-Born Population of the United States: 1850–2000*. Population Division Working Paper No. 81. Washington, DC: U.S. Census Bureau.

[US10] Grammich, C., Hadaway, K., Houseal, R., Jones, D. E., Krindatch, A., Stanley, R. & Taylor, R. H. (2012). *2010 U.S. Religion Census: Religious Congregations & Membership Study*. Lenexa, KS: Association of Statisticians of American Religious Bodies. Data retrieved from http://www.rcms2010.org

[US00] Jones, D. E., Doty, S., Grammich, C., Horsch, J. E., Houseal, R., Lynn, M., Marcum, J. P., Sanchagrin, K. M. & Taylor, R. H. (2002). *Religious Congregations and Membership in the United States 2000: An Enumeration by Region, State and County Based on Data Reported by 149 Religious Bodies*. Nashville, TN: Glenmary Research Center. Data retrieved from http://www.rcms2010.org

[US90] US Bureau of the Census (1992). *1990 Census of Population: Detailed Ancestry Groups for States*. Washington, DC: U.S. Government Priniting Office.

[US80] US Bureau of the Census (1983). *1980 Census of Population: Ancestry of the population by State: 1980-Supplementary Report*. Washington, DC: U.S. Government Priniting Office.

[US09] CQ Transcriptwire (2009, June 4). President Obama Addresses Muslim World in Cairo. *The Washington Post*.

[US90] Weeks, J. R. (2003). Estimating the Muslim Population in the United States Using Census 2000 Data. *Espace, Populations, Scocietes*, 1, 89–101.

[US01] Nasser, M. K. (2000, November 28). Clinton Sends World Muslims a Message of Respect as Ramadan Begins. *Al-Bawaba*.

[VC12] SVG Statistical Office (2015). *St. Vincent & the Grenadines Population & Housing Census Report 2012*. Kingstown: SVG Statistical Office.

[VC01] CARICOM (Caribbean Community) Capacity Development Programme (CCDP) (2009). *2000 Round of Population and Housing Census of the Caribbean Community:* Volume of Basic Tables. Georgetown: The CARICOM Secretariat.

[VEH] Paul, R. (1924). Les Musulmans en Amérique. *Journal de la Société des Américanistes,* 16(1): 435–436.

[VG10] Central Statistics Office (CSO) (2016). *Virgin Islands 2010 Population and Housing Census Report.* Road Town: CSO.

[VG01] CARICOM (Caribbean Community) Capacity Development Programme (CCDP) (2009). *2000 Round of Population and Housing Census of the Caribbean Community: National Census Report, The British Virgin Islands.* Georgetown: The CARICOM Secretariat.

[VI] Leedy, J. (2011, March 6). Faith Matters: Islam in the USVI. *St. Croix Source.*

[VI2] Muhammad, T. (2007, June 13). Islam on US Virgin Islands. *Black Star News.*

[VI1] Caribbean Muslim Forum (CMF) (2005). St. Croix Muslim Community Profile. *Caribbean Muslim Directory–Community Profiles.* Valsayn: CMF.

[VIH] Statistics Denmark (1850–1911). *West Indian Census.* Copenhagen: Rigsarkivet (Danish National Archives). Retrieved from https://www.sa.dk/en/

[WVS] World Values Survey Association at the University of Aberdeen [The United Kingdom] (2015). *World Values Survey Online Data Analysis.* Retrieved from http://www.worldvaluessurvey.org/WVSOnline.jsp

Chapter 6

Islam in Oceania

Oceania consists of tens of thousands of islands with total land area of 8.54 million square kilometers, over 90% of which is mainland Australia; spread over eighty million square kilometers of the Pacific Ocean. The total population of this continent increased from 2.7 million in 1870, to 12.7 million in 1950, to 42 million in 2020, and is projected to reach 57 million by 2050, and 72 million by 2100.

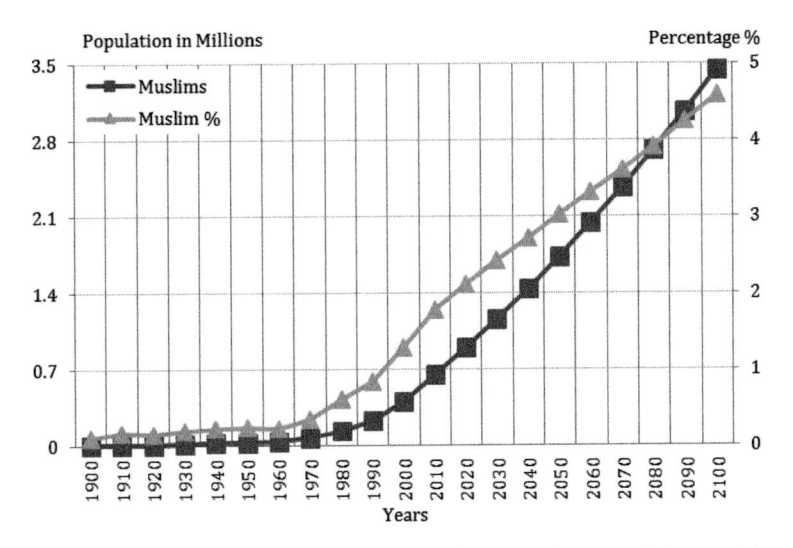

Figure 6.0a Plot of decennial estimates of the Muslim population and its percentage of the total population in Oceania: 1900–2100 AD (1320–1520 H).

The World Muslim Population: Spatial and Temporal Analyses
Houssain Kettani
Copyright © 2020 Jenny Stanford Publishing Pte. Ltd.
ISBN 978-981-4800-31-0 (Hardcover), 978-0-429-42253-9 (eBook)
www.jennystanford.com

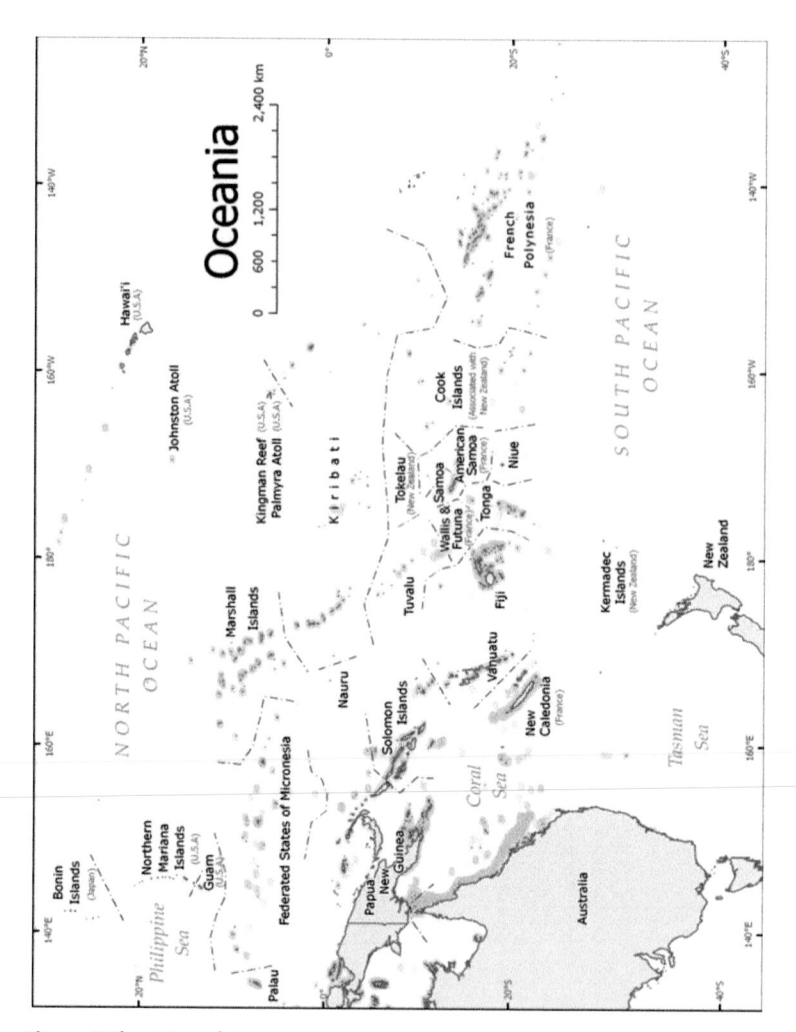

Figure 6.0b Map of Oceania.

Therefore, it is the least populated continent in the world. A plot of decennial estimates of the Muslim population and its percentage with respect to the total population in Oceania from 1900 to 2100 is provided in Fig. 6.0a. Muslims entered Oceania as early as the sixteenth century as Islam spread through current Indonesia. However, until 1950s the number of Muslims remained few. Remarkable presence of Muslims started in 1960s with migration of Muslims for economic means mainly to Australia. Noticeable increase of Muslims in other parts of this continent started in 1990s, such as New Zealand, Northern Mariana Islands, and Palau. A map of Oceania is presented in Fig. 6.0b. A substantial increase in the number and percentage of Muslims is expected throughout this century as depicted in Fig. 6.0a.

Thus, the Muslim population in this continent has likely increased from none in 1800, to 6,000 or 0.1% in 1900, to 0.41 million or 1.3% in 2000, to 0.9 million or 2.1% in 2020, and is projected to reach 1.7 million or 3.0% by 2050 and 3.4 million or 4.6% by 2100.

We divided Oceania into four regions; the data for each is included in a separate section and are sorted in terms of the percentage of Muslims in descending order. These regions are Australasia (Section 6.1), Melanesia (Section 6.2), Micronesia (Section 6.3), and Polynesia (Section 6.4). In Section 6.5, the total population in each of the four regions of Oceania and the corresponding percentage and number of Muslims is presented centennially in Table 6.5a from 600 to 2100 and decennially in Tables 6.5b to 6.5c from 1790 to 2100.

6.1 Muslims in Australasia

This region consists of the two island nations of Australia and New Zealand. Muslims arrived here as early as the sixteenth century from current Indonesia; however, their numbers remained few. After WWII, there was noticeable increase in the number of Muslims as Australia needed more labor force. Thus, the Muslim population has likely increased from none in 1800, to 4,000 or 0.08% in 1900, to 0.34 million or 1.5% in 2000, to 0.83 million or 2.8% in 2020, and is projected to reach 1.6 million or 4.3% by 2050 and 3.3 million or 6.8% by 2100.

Australasia has the largest concentration of Muslims among four regions covering Oceania: Australasia, Melanesia, Micronesia and Polynesia. The fraction of the population living in this region out of the total population of Oceania increased from a quarter in the beginning of the eighteenth century, to a third towards its middle, to 75% in 1880, then peaked at 83% in 1930, then started decreasing afterwards, reaching 73% in 2010, and is expected to reach 69% towards the end of this century. The fraction of Muslims living in Australasia out of the total Muslim population of Oceania increased steadily from 11% in 1950 to almost 90% by 2010 and is expected to reach 97% by the end of this century.

A plot of decennial estimates of the Muslim population and its percentage with respect to the total population in this region from 1900 to 2100 is provided in Fig. 6.1. This shows that the Muslim population was negligible until 1970; then started its steady and rapid increase afterwards, which is expected to continue through this century. The corresponding individual data for each country in this region is discussed below. In Section 6.1.3, the total population in each country in this region and the corresponding percentage and number of Muslims is presented centennially in Table 6.1a from 600 to 2100 and decennially in Tables 6.1b and 6.1c from 1790 to 2100.

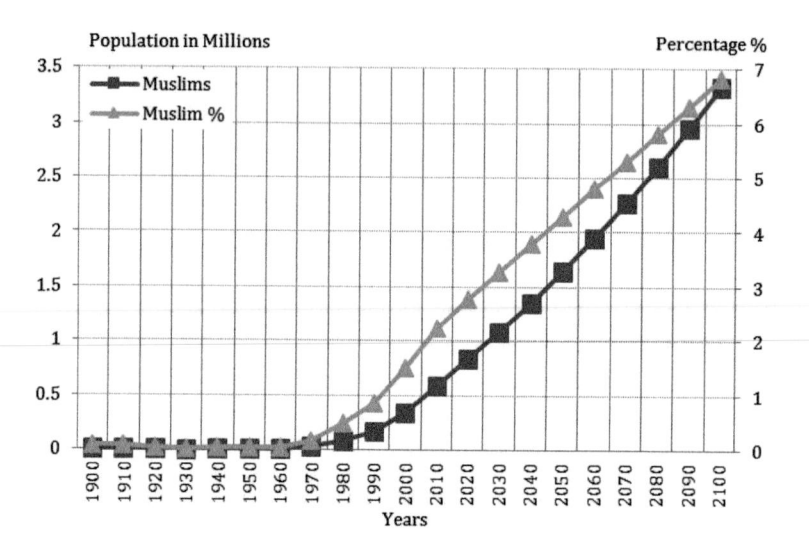

Figure 6.1 Plot of decennial estimates of the Muslim population and its percentage of the total population in Australasia: 1900–2100 AD (1320–1520 H).

6.1.1 Australia

The Commonwealth of Australia has an area of 7,687,809 sq. km and is the sixth largest country in the world. It was conquered by the British in 1770 and gained its independence from the UK in 1901. Geographically, it consists of the Mainland Australia surrounded by over 8,200 islands, including Tasmania (68,018 sq. km), Melville (5,786 sq. km), Kangaroo (4,416 sq. km), Groote Eylandt (2,285 sq. km), Bathurst (1,693 sq. km), Fraser (1,653 sq. km), Flinders (1,359 sq. km), King (1,091 sq. km), and Mornington (1,002 sq. km). A map of Australia is presented in Fig. 6.1.1a.

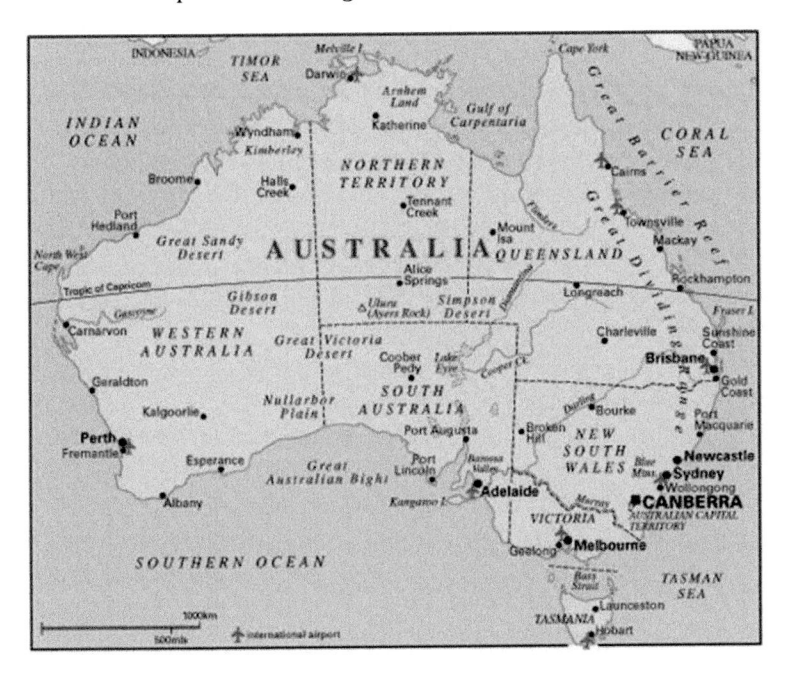

Figure 6.1.1a Map of the Commonwealth of Australia.

Politically, the Commonwealth of Australia consists of six states, three territories, and seven external territories. The states are: New South Wales (800,809 sq. km), Queensland (1,729,958 sq. km), South Australia (984,179 sq. km), Tasmania (68,018 sq. km), Victoria (227,496 sq. km), and Western Australia (2,526,574 sq. km). The territories are: Australian Capital (2,358 sq. km), Jervis Bay (70 sq. km) and Northern (1,348,199 sq. km). The external territories: Australian Antarctic (5,896,500 sq. km), and the islands of Christmas (136 sq. km), Cocos or Keeling (14 sq. km) and Norfolk (35 sq. km), and the uninhabited islands of Ashmore & Cartier (199 sq. km), Coral Sea (10 sq. km) and Heard & McDonald (372 sq. km). Almost a third of the Australian population lives in New South Wales, which includes the largest Australian city of Sydney. A fourth of the

Australian population lives in Victoria, which includes the second largest Australian city of Melbourne. Finally, a tenth of the Australian Population lives in Western Australia.

Muslims moved to Australia from current Indonesia starting in the sixteenth century. But their numbers remained small. The British then brought Malays between 1850 and 1930 to work as pearl divers and in sugar cane plantations. Then Afghans, between 1862 and 1930, to help breeding camels and steering them through the Australian desert for economic purpose. Then Indians between 1879 and 1916, as indentured labor on sugar cane plantations.

Albanian refugees settled here between 1920 s and 1930s. After WWII, then Turkish Cypriots in 1950s and 1960s, followed by Turks between 1968 and 1972, and then Lebanese Muslims in 1970s. After the relaxation of immigration laws in 1970s, more Muslims emigrated here from many countries.

Based on census data as shown in Table 6.1.1a, the Muslim population decreased from 561 or 0.05% in 1861, to 382 or 0.02% in 1871, then increased to 616 or 0.03% in 1881, to 1,847 or 0.06% in 1891, to 3,641 or 0.10% in 1901, to 3,908 or 0.09% in 1911, to 2,868 or 0.05% in 1921, to 1,877 or 0.03% in 1933, then to 2,704 or 0.04% in 1947. A substantial increase occurred between the censuses of 1947 and 1971 happened towards the end of the 1960s as can be seen from the censuses of 1954, 1961, and 1966. In these censuses, Muslims were included in non-Christian religions, excluding Judaism. Thus, the number of Muslims continued to increase to 22,311 or 0.2% in 1971, to 45,200 or 0.4% in 1976, to 0.08 million or 0.6% in 1981, to 0.11 million or 0.9% in 1986, to 0.15 million or 1.0% in 1991, to 0.20 million or 1.2% in 1996, to 0.28 million or 1.7% in 2001, to 0.34 million or 1.9% in 2006, to 0.48 million or 2.4% in 2011 and 0.60 million or 2.8% in 2016. Thus, assuming that the percentage of Muslims will continue to increase by half of a percentage point per decade, the Muslim population is expected to reach 1.5 million or 4.5% by 2050 and three million or 7% by 2100.

Almost half of the Muslim population lives in New South Wales, and they increased steadily to 4% of the total population. Almost one-third of the Muslim population lives in Victoria, and they increased to 4% of the total population as well. Thus, 80% of Australian Muslims live in the southeast of Australia, between Sydney and Melbourne. This percentage used to be 90% in 1981, but it is decreasing slowly as Muslims are spreading around to other states. The distribution of Muslims per state and territory since 1861 is summarized in Table 6.1.1b. Prior to 1911, data of ACT and Jervis Bay Territory were included in NSW. Prior to 1997 Jervis Bay territory was included with the ACT, while Christmas and Cocos (Keeling) Islands were excluded from population estimates for Australia. Population of Norfolk Island is still excluded from the total population and does not have any Muslim residents.

Table 6.1.1a Evolution of the Muslim population in Australia including Christmas, Cocos and Norfolk Islands

Year	Population	Muslims	%	Source
1861	1,153,641	561	0.05	[AUH1]c
1871	1,663,882	382	0.02	[AUH1]c
1881	2,252,617	616	0.03	[AUH1]c
1891	3,121,111	1,847	0.06	[AUH1]c
1901	3,717,610	3,641	0.10	[AUH1]c
1911	4,335,888	3,908	0.09	[AUH2]c
1921	5,343,476	2,868	0.05	[AUH2]c
1933	5,780,891	1,877	0.03	[AUH2]c
1947	6,754,534	2,704	0.04	[AUH2]c
1971	11,944,978	22,311	0.19	[AU]c
1976	11,955,442	45,200	0.38	[AU]c
1981	12,981,135	76,792	0.59	[AU]c
1986	11,698,101	109,523	0.94	[AU]c
1991	15,088,070	147,487	0.98	[AU]c
1996	16,148,080	200,885	1.24	[AU]c
2001	16,933,660	281,579	1.66	[AU]c
2006	17,500,119	340,400	1.95	[AU]c
2011	19,497,668	476,291	2.44	[AU]c
2016	21,269,716	604,242	2.84	[AU]c
2050	32,814,000	1,477,000	4.50	es
2100	42,877,000	3,001,000	7.00	es

Table 6.1.1b Evolution of the Muslim population in Australia per state and territory since 1861. Until 1901, ACT was included in NSW

		1861	1871	1881	1891	1901	1911	1921	1933	1947	1971
Australian Capital	P						1672	2,546	8,060	15,488	134,914
	M%						0.06	—	—	—	0.13
	M						1	—	—	—	177
	MR%						0.03	—	—	—	0.79
New South Wales	P	350,860	498,035	737,771	1,109,926	1,337,812	1,614,265	2,067,213	2,300,612	2,646,282	4,326,705
	M%	0.10	0.04	0.05	0.05	0.08	0.05	0.03	0.02	0.02	0.23
	M	350	200	400	528	1,072	819	692	523	584	9,808
	MR%	62.39	51.81	64.94	28.39	29.06	20.96	24.13	27.86	21.60	43.96
Northern Territory	P	—	201	3,451	4,898	4,096	2,820	3,448	3,333	8,882	75,293
	M%	—	—	—	0.27	1.17	1.35	0.75	1.38	0.29	0.30
	M	—	—	—	13	48	38	26	46	26	224
	MR%	—	—	—	0.70	1.30	0.97	0.91	2.45	0.96	1.00
Queensland	P	29,746	117,403	212,179	391,872	503,266	586,017	737,969	816,488	981,346	1,696,506
	M%	0.04	0.02	0.02	0.05	0.09	0.12	0.10	0.07	0.07	0.08
	M	12	27	35	210	435	692	729	547	641	1,351
	MR%	2.14	6.99	5.68	11.29	11.79	17.71	25.42	29.14	23.71	6.06

South Australia	P	126,830	185,425	265,623	305,445	355,111	390,117	495,160	580,949	646,073	1,100,933
	M%	—	—	—	0.10	0.13	0.11	0.06	0.05	0.04	0.06
	M	—	—	—	299	449	440	274	267	263	628
	MR%	—	—	—	16.08	12.17	11.26	9.55	14.22	9.73	2.81
Tasmania	P	89,977	99,328	115,705	141,493	169,095	184,812	209,494	195,551	228,894	365,177
	M%	—	0.00	—	0.10	0.02	0.01	0.01	0.00	0.01	0.04
	M	—	4	—	142	27	10	17	2	33	133
	MR%	—	—	—	7.63	0.73	0.26	0.59	0.11	1.22	0.60
Victoria	P	522,392	716,003	847,291	1,117,528	1,184,957	1,283,760	1,505,222	1,582,154	1,836,935	3,279,347
	M%	0.04	0.02	0.01	0.02	0.04	0.03	0.02	0.01	0.03	0.27
	M	189	125	111	241	467	391	304	159	632	8963
	MR%	33.69	32.38	18.02	12.96	12.66	10.01	10.60	8.47	23.37	40.17
Western Australia	P	15,593	23,315	29,708	49,186	180,045	272,425	323,985	381,152	452,229	966,103
	M%	0.06	0.13	0.24	0.87	0.66	0.56	0.25	0.09	0.12	0.11
	M	10	30	70	427	1,191	1,517	826	333	525	1,027
	MR%	1.78	7.77	11.36	22.96	32.29	38.82	28.80	17.74	19.42	4.60

Table 6.1.1b (Continued)

		1976	1981	1986	1991	1996	2001	2006	2011	2016
Australian Capital	P	171,011	197,790	221,386	252,274	269,078	272,991	290,916	330,197	363,398
	M%	0.24	0.37	0.57	0.74	0.92	1.28	1.50	2.25	2.72
	M	408	730	1,255	1,868	2,466	3,488	4,372	7,432	9,882
	MR%	0.90	0.95	1.15	1.27	1.23	1.24	1.29	1.56	1.64
New South Wales	P	4,234,085	4,614,812	4,839,939	5,196,743	5,510,803	5,650,936	5,890,497	6,383,190	6,823,466
	M%	0.52	0.83	1.19	1.50	1.86	2.49	2.87	3.44	3.92
	M	22,206	38,527	57,551	77,825	102,288	140,907	168,785	219,377	267,658
	MR%	49.12	50.17	52.55	52.77	51.10	50.17	49.69	46.13	44.35
Northern Territory	P	77,867	96,939	125,875	150,995	166,747	169,753	159,883	181,103	193,500
	M%	0.25	0.33	0.38	0.42	0.46	0.56	0.68	0.88	1.21
	M	193	317	479	630	768	945	1,090	1,589	2,338
	MR%	0.43	0.41	0.44	0.43	0.38	0.34	0.32	0.33	0.39
Queensland	P	1,786,448	2,013,575	2,273,258	2,668,816	3,009,311	3,158,541	3,445,831	3,937,135	4,257,888
	M%	0.10	0.12	0.16	0.21	0.31	0.47	0.59	0.86	1.05
	M	1,717	2,457	3,731	5,605	9,421	14,990	20,322	34,047	44,880
	MR%	3.80	3.20	3.41	3.80	4.71	5.34	5.98	7.16	7.44

	P	1,119,296	1,142,534	1,178,275	1,246,279	1,279,083	1,274,639	1,332,780	1,453,815	1,534,609
South Australia	M%	0.09	0.13	0.21	0.25	0.38	0.59	0.79	1.34	1.86
	M	1,031	1,456	2,486	3,092	4,798	7,478	10,517	19,511	28,547
	MR%	2.28	1.90	2.27	2.10	2.40	2.66	3.10	4.10	4.73
	P	352,644	354,899	374,698	403,761	412,988	399,177	418,341	450,753	462,812
Tasmania	M%	0.04	0.10	0.15	0.15	0.20	0.22	0.25	0.38	0.54
	M	135	369	569	623	807	865	1,050	1,708	2,497
	MR%	0.30	0.48	0.52	0.42	0.40	0.31	0.31	0.36	0.41
	P	3,160,063	3,366,220	3,430,346	3,758,114	3,956,175	4,061,038	4,382,108	4,907,673	5,399,867
Victoria	M%	0.56	0.87	1.11	1.32	1.69	2.28	2.50	3.11	3.65
	M	17,622	29,355	37,965	49,617	67,047	92,742	109,370	152,775	197,032
	MR%	38.98	38.23	34.66	33.64	33.49	33.02	32.20	32.13	32.65
	P	1,002,805	1,120,815	1,236,697	1,411,088	1,540,835	1,593,074	1,708,878	2,022,085	2,230,383
Western Australia	M%	0.19	0.32	0.44	0.58	0.82	1.22	1.42	1.93	2.27
	M	1,894	3,581	5,487	8,227	12,583	19,456	24,185	39,117	50,647
	MR%	4.19	4.66	5.01	5.58	6.29	6.93	7.12	8.23	8.39

6.1.1.1 Antarctica

Antarctica is the world's southernmost continent resting in the Antarctic Circle and surrounded by the Southern Ocean. With fourteen million sq. km of area, it's the fifth largest continent. Almost 98% of the continent is covered by ice averaging two kilometers in thickness. Almost six million sq km or 42% of the area is claimed by Australia, while New Zealand claims almost half million sq km or 3% of the total area of Antarctica. A map of the Antarctica is presented in Fig. 6.1.1.1. The Continent is governed by the Antarctic Treaty System which was first signed in 1959.

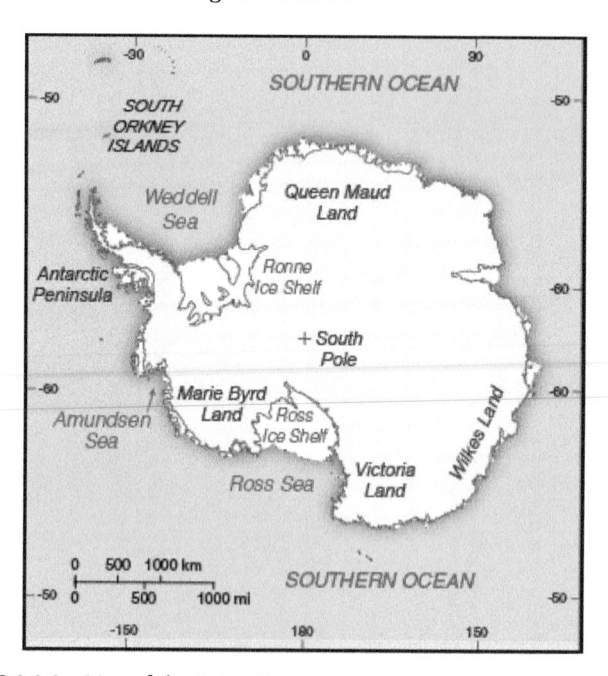

Figure 6.1.1.1 Map of the Antarctica.

Antarctica has no indigenous inhabitants, and its population is comprised mainly of scientific research staff from around thirty countries scattered in around forty permanent research bases. The number of residents varies, from around a thousand in the harsh Antarctic winter to almost five thousand in the milder summer months of October to February. Sightings of Antarctica by European expeditions were recorded as early as the fifteenth century, but permanent bases were only established after WWII. Pakistan established a base in 1991, but only operates during the summer. No Muslims live in Antarctica based on national data for the year-round population and as summarized in Table 6.1.1.1.

Table 6.1.1.1 Evolution of the Muslim population in Antarctica

Year	Population	Muslims	%	Source
1939	0	—	0.00	[CIA]e
1999	964	—	0.00	[CIA]e
2009	1,106	—	0.00	[CIA]e
2017	1,036	—	0.00	[CIA]e

6.1.1.2 Christmas Island

The Territory of the Christmas Island is a territory of Australia, located in the Indian Ocean. It is located over 1,500 Km northwest Australia and over 300 Km southwest of Java. The island has an area of 136 sq. km, about two-third of which is a national park. A map of the Island is presented in Fig. 6.1.1.2.

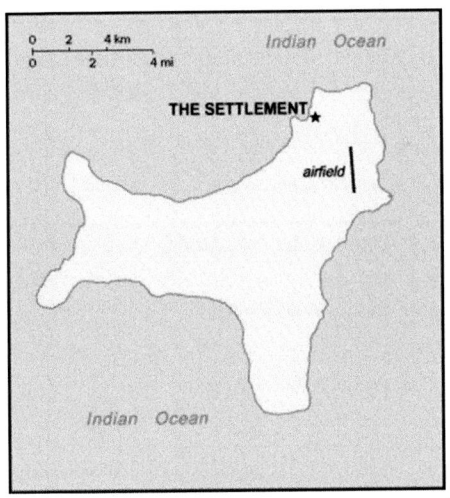

Figure 6.1.1.2 Map of the Australian Territory of Christmas Island.

The first mention of Christmas Island appears in a map published in Holland in 1666, in which it is called Moni Island, although it is believed that the British Captain William Mynors of the East India Company had sighted the island on Christmas Day in

1643 and had named it accordingly. In 1888 it was annexed by the British and placed under the supervision of the Straits Settlements Government for administrative purposes. Following upon this, a small settlement was established at Flying Fish Cove by Mr. G. Clunies-Ross of Cocos (Keeling) Islands. In 1891 Sir John Murray and Mr. Clunies-Ross were granted a 99-year lease of the Island. This lease was transferred to the Christmas Island Phosphate Co. Ltd. in 1897, following the discovery of large deposits of phosphate of lime on the Island. In 1900, Christmas Island was incorporated for administrative purposes with the Settlement of Singapore and the laws of Singapore were generally applied to the Island. Eventually, the Island was transferred by the British to the Commonwealth of Australia in 1958.

As shown in Table 6.1.1.2, census data show that the Muslim population decreased sharply from 730 or 27% in 1981, to 139 or 8% in 1986. The total population decreased by more than a third in the same period due to migration to Australia. The total population peaked at 3,391 in 1966, but continued to decrease, reaching 1,275 in 1991, but started increasing and fluctuating due to asylum seekers and related activities to deal with the issue of immigration by the Australian government. The Muslim population changed to 298 or 23% in 2001, then 266 or 22% in 2006, to 305 or 25% in 2011 and 357 or 27% in 2016.

Table 6.1.1.2 Evolution of the Muslim population in Christmas Island

Year	Population	Muslims	%	Source
1891	100	10	10.00	[JAN]es
1981	2,738	730	26.66	[AU]c
1986	1,672	139	8.31	[AU]c
2001	1,286	298	23.17	[AU]c
2006	1,203	266	22.11	[AU]c
2011	1,210	305	25.21	[AU]c
2016	1,328	357	26.88	[AU]c

6.1.1.3 Cocos (Keeling) Islands

The Territory of the Cocos (Keeling) Islands is a territory of Australia, located in the Indian Ocean, southwest of Christmas

Island and approximately midway between Australia and Sri Lanka: about 2,100 Km northwest of Australia, 2,700 Km southeast of Sri Lanka, 1,100 Km southwest of Sumatra and Java, and 1,000 Km southwest of Christmas Island. The territory consists of two separate atolls comprising 27 coral islands with a total area of 14 sq. km, of which two are inhabited. The largest of the islands is West Island. It comprises almost half of the total area (6 sq. km) but has one sixth of the total population and mostly ethnic European. The most populated island is Home Island, has an area of 1 sq. km. It has the rest of the population which is mostly ethnic Malay and Muslim. A map of the Islands is presented in Fig. 6.1.1.3.

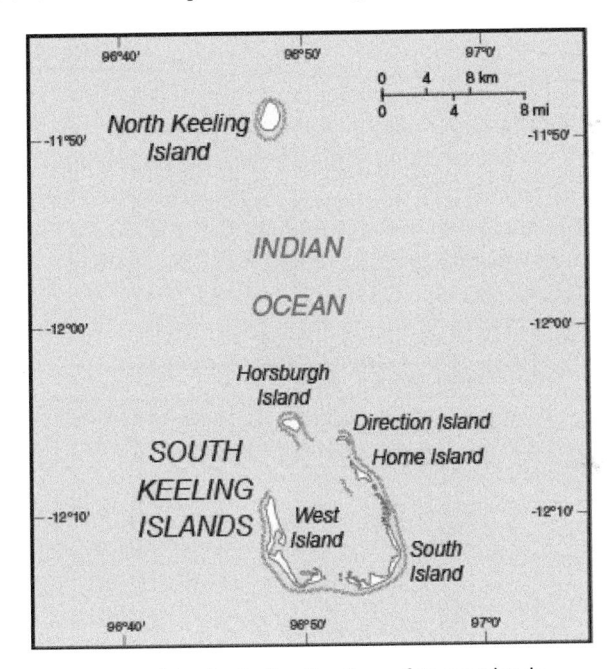

Figure 6.1.1.3 Map of the Australian Territory of Cocos Islands.

The islands were discovered in 1609 by the British Captain William Keeling of the East India Company. They were uninhabited and remained so until 1826 when the first settlement was established on the main atoll by an Englishman, Alexander Hare, with forty Malay women, but he quitted the islands in about 1831. In the meantime, a second settlement was formed on the main atoll by John Clunies-Ross, a Scottish seaman and adventurer, who landed with several boat-loads of Malay seamen. In 1857, the islands were annexed to the British Empire, and in 1878 responsibility for their supervision was transferred from the Colonial Office to the Government of Ceylon, and subsequently, in 1882, to the Government of the Straits Settlements. By indenture in 1886, Queen Victoria granted the land comprised in the islands to John Clunies-Ross in perpetuity, and until 1946, the patriarchal rule of the head of the Clunies-Ross family was recognized. The head of the family had semi-official status as resident magistrate and representative of the Government. However, in 1946, when the islands became a dependency of the Colony of Singapore, a Resident Administrator, responsible to the Governor of Singapore, was appointed. The islands were transferred to the Commonwealth of Australia in 1955 and remain since then under its authority.

The majority of the population is Muslim Malay. As shown in Table 6.1.1.3, census data show that the Muslim population fluctuated from 418 or 73% in 1961, to 472 or 71% in 1966, to 314 or 67% in 1981, to 401 or 64% in 1986, to 409 or 70% in 2001, to 432 or 76% in 2006, to 419 or 80% in 2011 and 403 or 79% in 2016.

Table 6.1.1.3 Evolution of the Muslim population in Cocos Islands

Year	Population	Muslims	%	Source
1891	554	55	10.00	[JAN]es
1961	570	418	73.33	[AU]c
1966	664	472	71.08	[AU]c
1981	470	314	66.81	[AU]c
1986	627	401	63.96	[AU]c
2001	581	409	70.40	[AU]c
2006	566	432	76.33	[AU]c
2011	526	419	79.66	[AU]c
2016	509	403	79.17	[AU]c

6.1.1.4 Norfolk Island

Norfolk Island is an external territory of Australia, located in the Pacific Ocean, 1,400 Km east of Australia, over 700 Km northwest of New Zealand and over 700 Km south of New Caledonia. The area of Norfolk Island is 35 sq. km, including Phillip and Nepean Islands, which are uninhabited. A map of the Islands is presented in

Fig. 6.1.1.4. The Islands were claimed by the British in 1788, as part of their settlement in Australia. The island then served as a convict penal settlement until 1794, when it was abandoned until 1856, when permanent residence on the island for civilians began. In 1901, the island became a part of the Commonwealth of Australia which it has remained until this day. As shown in Table 6.1.1.4., census data show that there were no Muslims living in these islands, until 2016, when three Muslim males were recorded, constituting 0.2% of the total population.

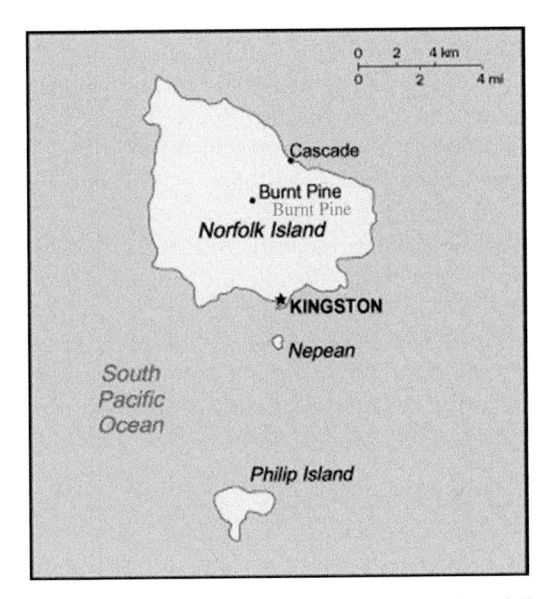

Figure 6.1.1.4 Map of the Australian External Territory of Norfolk Island.

Table 6.1.1.4 Evolution of the Muslim population in Norfolk Island

Year	Population	Muslims	%	Source
1954	942	—	0.00	[AUH2]c
1961	844	—	0.00	[AUH2]c
1971	1,683	—	0.00	[AU]c
1981	1,849	—	0.00	[NF]c
1991	2,285	—	0.00	[NF]c
2001	2,601	—	0.00	[NF]c
2011	2,302	—	0.00	[NF]c
2016	1,582	3	0.19	[AU]c

6.1.2   New Zealand

It was conquered by the British in 1840 and gained its independence from the UK in 1907. It has a total area of 267,710 sq. km and consists of two main islands and over a thousand much smaller mostly uninhabited islands, covering over four million sq. km of the South Pacific Ocean. The two main islands are South Island (151,215 sq. km) with almost a fourth of the population and a tenth of the Muslim population, and North Island (113,729 sq. km) with three-quarters of the population and nine-tenth of the Muslim population. The third populous island is Waiheke (92 sq. km), with few thousand residents. Other populated islands with few hundred inhabitants are Stewart (1,746 sq. km), Chatham (900 sq. km), and Great Barrier (285 sq. km). The largest of the remaining islands are the uninhabited Auckland (510 sq. km), and Resolution (208 sq. km). A map of New Zealand and its regional councils are presented in Fig. 6.1.2a and 6.1.2b.

The first Muslims were Gujarati traders from India. By 1950s, Muslims settled here from Albania and Bosnia. Then after 1987 military coup d'état in Fiji, many Indian Muslims migrated to New Zealand, as the Indian community felt that it is targeted by the coup. After 1993, a couple thousand Somali refugees joined the Muslim community. According to census data as shown in Table 6.1.2a, the Muslim population remained at or below 0.01% of the total population until the 1960s, changing from four in 1861, to seventeen in 1874 (one in Auckland and sixteen in Otago regions), to 39 in 1878 (three in Auckland, one in Wellington, three in Otago and 32 in Canterbury regions), to seven in 1881 (one in Otago and six in Canterbury regions), to 43 in 1896, to 41 in 1901, to 17 in 1906, to 12 in 1911, to 47 in 1916 (one in Taranaki, seven in Hawke's Bay, seventeen in Auckland, thirteen in Wellington, two in Otago and seven in Canterbury regions), to 65 in 1921, to 76 in 1926, to 51 in 1936, to 67 in 1945, to 205 in 1951, to 200 in 1956, to 260 in 1961. The Muslim population then continued its steady increase to 551or 0.02% in 1966, to 779 or 0.03% in 1971, to 1,415 or 0.05% in 1976, to 2,004 or 0.08% in 1981, to 2,544 or 0.09% in 1986, to 5,772 or 0.20% in 1991 to 13,548 or 0.43% in 1996, to 23,631 or 0.73% in 2001, to 36,072 or 1.0% in 2006 and to 46,146 or 1.2% in 2013 and 61,455 or 1.4% in 2018. Thus, assuming that the percentage of Muslims will continue to increase by half of a percentage point per decade, then the Muslim population is expected to reach 0.2 million or 3.0% by 2050 and 0.3 million or 5.5% by 2100.

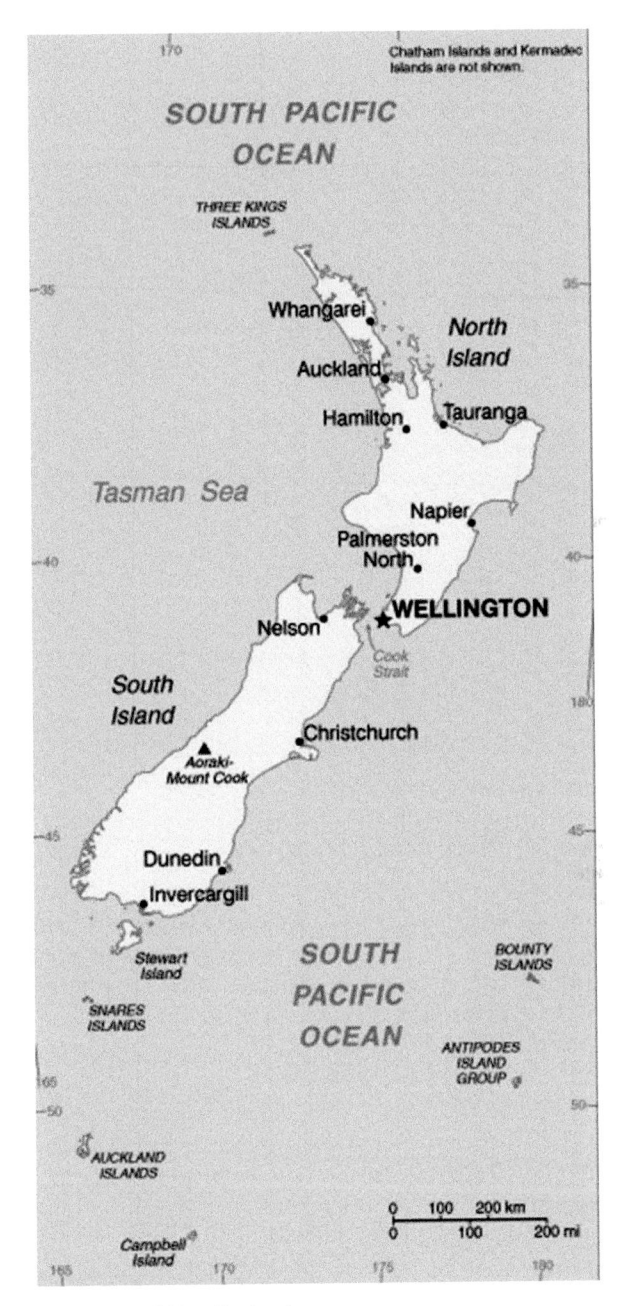

Figure 6.1.2a Map of New Zealand.

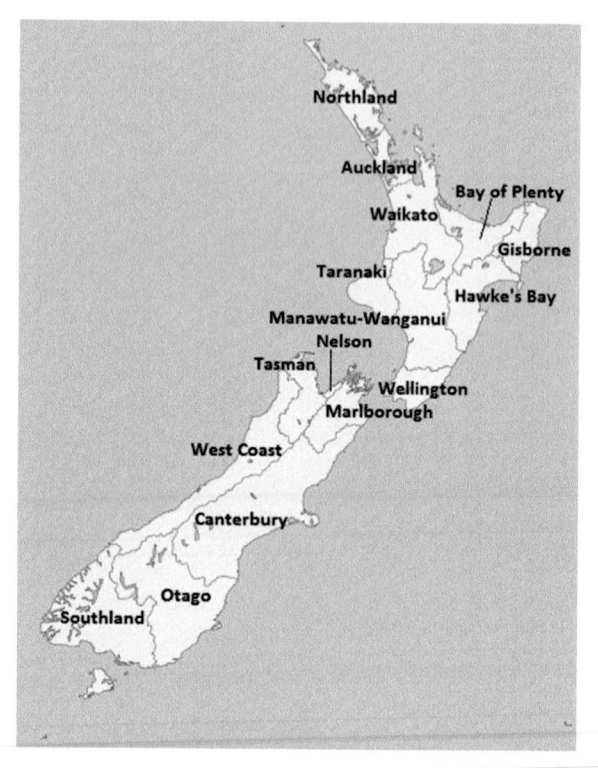

Figure 6.1.2b Map of New Zealand's Regional Councils.

The North Island includes over three-quarters of the New Zealand population and over nine-tenth of the Muslim population. This island includes the capital Wellington and the largest city Auckland. The island consists of nine Regional Councils based on the 1989 Local Government Act that changed all the Boundaries in New Zealand. These regions are from north to south: Northland (13,941 sq. km), Auckland (5,600 sq. km), where three-quarter of Muslims on these islands live, Waikato (25,598 sq. km), Bay of Plenty (12,447 sq. km), Gisborne (8,351 sq. km), Hawke's Bay (14,164 sq. km), Taranaki (7,273 sq. km), Manawatu-Whanganui (22,215 sq. km) and Wellington (8,124 sq. km), where a tenth of Muslims live. The distribution of Muslims per region is summarized in Table 6.1.2b. As shown in Table 6.1.2d, the Muslim population in North Island increased from none in 1881, to 38 or 0.01% in 1916, to 1,150 or 0.1% in 1976, to 1,443 or 0.1% in 1981, to 5,556 or 0.2% in 1991, to 11,961 or 0.5% in 1996, to 21,051 or 0.9% in 2001, to 32,097 or 1.2% in 2006, and 41,562 or 1.5% in 2013.

Table 6.1.2a Evolution of the Muslim population in New Zealand

Year	Population	Muslims	%	Source
1861	99,021	4	0.00	[NZ861]c
1874	345,000	17	0.00	[NZH]c
1878	458,000	39	0.01	[NZH]c
1881	534,000	7	0.00	[NZH]c
1896	776,100	43	0.01	[NZH]c
1901	815,900	41	0.01	[NZH]c
1906	936,300	17	0.00	[NZH]c
1911	1,058,300	12	0.00	[NZH]c
1916	1,190,441	47	0.00	[NZH]c
1921	1,120,124	65	0.01	[NZH]c
1926	1,274,911	76	0.01	[NZH]c
1936	1,405,477	51	0.00	[NZH]c
1945	1,469,038	67	0.00	[NZH]c
1951	1,791,424	205	0.01	[NZH]c
1956	1,984,241	200	0.01	[NZH]c
1961	2,196,730	260	0.01	[NZH]c
1966	2,446,768	551	0.02	[NZH]c
1971	2,512,079	779	0.03	[NZH]c
1976	2,632,905	1,415	0.05	[NZH]c
1981	2,570,424	2,004	0.08	[NZH]c
1986	2,959,737	2,544	0.09	[NZH]c
1991	3,064,299	6,096	0.20	[NZH]c
1996	3,148,710	13,545	0.43	[NZH]c
2001	3,229,572	23,637	0.73	[NZH]c
2006	3,501,048	36,072	1.03	[NZ13]c
2013	3,728,130	46,146	1.24	[NZ13]c
2018	4,386,960	61,455	1.40	[NZ18]c
2050	5,608,000	168,000	3.00	es
2100	6,008,000	330,000	5.50	es

Table 6.1.2b Evolution of the Muslim population in North Island of New Zealand per Regional Council since 1976. P: Total population, M: Muslim population, M%: Percentage of Muslim population, MR%: Muslim Ratio in percentage

Region		1976	1981	1991	1996	2001	2006	2013
Northland	P	107,013	113,994	112,830	114,627	117,720	121,821	125,031
	M%	0.01	0.02	0.05	0.07	0.20	0.16	0.22
	M	7	24	54	75	234	201	270
	MR%	0.61	1.66	0.97	0.63	1.11	0.63	0.65
Auckland	P	797,406	827,408	849,504	920,364	991,533	1,131,417	1,244,649
	M%	0.08	0.10	0.45	0.93	1.54	2.09	2.50
	M	644	822	3,825	8,526	15,318	23,688	31,158
	MR%	56.00	56.96	68.84	71.32	72.76	73.80	74.97
Waikato	P		261,939	300,525	303,831	307,614	325,161	352,110
	M%		0.03	0.09	0.22	0.49	0.67	0.83
	M		84	273	678	1,518	2,166	2,937
	MR%		5.82	4.91	5.67	7.21	6.75	7.07
Bay of Plenty	P	472,083	226,823	184,845	193,584	204,687	219,009	230,037
	M%	0.02	0.01	0.05	0.07	0.18	0.31	0.28
	M	92	33	90	141	375	669	642
	MR%	8.00	2.29	1.62	1.18	1.78	2.08	1.54
Gisborne	P	48,147	53,295	39,684	38,481	38,022	37,338	36,219
	M%	0.01	0.02	0.02	0.04	0.13	0.18	0.27
	M	4	9	9	15	51	69	96
	MR%	0.35	0.62	0.16	0.13	0.24	0.21	0.23
Hawke's Bay	P	145,061	137,840	126,612	123,453	123,606	127,593	130,869
	M%	0.01	0.02	0.07	0.24	0.25	0.35	0.58
	M	11	21	93	297	309	441	753
	MR%	0.96	1.46	1.67	2.48	1.47	1.37	1.81
Taranaki	P	107,071	103,798	98,193	93,219	88,659	88,281	95,340
	M%	0.02	0.01	0.05	0.09	0.20	0.26	0.41
	M	24	12	45	81	174	231	393
	MR%	2.09	0.83	0.81	0.68	0.83	0.72	0.95
Manawatu Wanganui	P		181,940	205,584	200,340	190,890	192,795	194,457
	M%		0.04	0.13	0.28	0.32	0.48	0.67
	M		81	258	567	615	924	1,305
	MR%		5.61	4.64	4.74	2.92	2.88	3.14
Wellington	P	591,612	412,147	363,765	363,771	370,530	396,192	418,926
	M%	0.06	0.09	0.25	0.43	0.66	0.94	0.96
	M	368	357	909	1,575	2,460	3,708	4,008
	MR%	32.00	24.74	16.36	13.17	11.68	11.55	9.64

Table 6.1.2c Evolution of the Muslim population in South Island of New Zealand per Regional Council since 1976. *P*: Total population, *M*: Muslim population, *M%*: Percentage of Muslim population, *MR%*: Muslim Ratio in percentage

Region		1976	1981	1991	1996	2001	2006	2013
Tasman	P			30,096	32,691	34,827	38,256	41,664
	M%			0.02	0.05	0.14	0.15	0.11
	M			6	15	48	57	45
	MR%			1.12	0.94	1.86	1.44	0.98
Nelson	P	75,562	65,934	32,889	35,115	35,757	37,299	40,962
	M%	0.01	0.00	0.05	0.12	0.14	0.21	0.22
	M	11	3	18	42	51	78	90
	MR%	4.15	0.56	3.35	2.64	1.97	1.97	1.96
Marlborough	P	35,030	37,557	32,043	33,762	34,584	36,702	38,316
	M%	0.02	0.46	0.03	0.04	0.13	0.20	0.15
	M	6	171	9	15	45	72	57
	MR%	2.26	32.02	1.68	0.94	1.74	1.81	1.24
West Coast	P	24,049	34,178	28,416	27,687	26,106	26,709	27,741
	M%	0.01	0.01	0.04	0.05	0.16	0.19	0.24
	M	3	3	12	15	42	51	66
	MR%	1.13	0.56	2.23	0.94	1.62	1.28	1.44
Canterbury	P	428,586	421,618	402,249	415,518	423,321	462,462	481,989
	M%	0.04	0.07	0.09	0.24	0.42	0.61	0.64
	M	181	279	366	999	1,758	2,838	3,075
	MR%	68.30	52.25	68.16	62.71	67.98	71.50	67.08
Otago	P	188,903	183,566	163,938	164,625	160,134	170,451	179,607
	M%	0.03	0.03	0.06	0.28	0.34	0.43	0.59
	M	55	51	96	453	549	729	1,065
	MR%	20.75	9.55	17.88	28.44	21.23	18.37	23.23
Southland	P	108,860	107,905	92,484	87,039	80,985	80,364	83,313
	M%	0.01	0.03	0.03	0.06	0.11	0.18	0.22
	M	9	27	30	54	93	144	186
	MR%	3.40	5.06	5.59	3.39	3.60	3.63	4.06

The South Island includes almost one-fourth of the New Zealand population and almost one-tenth of the Muslim population. The island consists of seven regional councils based on the 1989 Local Government Act that changed all the Boundaries in New Zealand. These regions are from north to south: Tasman (9,786 sq. km), Nelson (445 sq. km), Marlborough (12,484 sq. km), West Coast (23,336 sq. km), Canterbury (45,346 sq. km), where over two-third of Muslims on this island live, Otago (31,990 sq. km), where almost a quarter of Muslims live, and Southland (34,347 sq. km). The distribution of Muslims per region is summarized in Table 6.1.2c. As shown in Table 6.1.4e, the Muslim population in South Island increased from seven in 1881, to nine or 0.01% in 1916, to 265 or 0.03% in 1976, to 534 or 0.1% in 1981, to 537 or 0.1% in 1991, to 1,587 or 0.2% in 1996, to 2,586 or 0.3% in 2001, to 3,969 or 0.5% in 2006, and 4,584 or 0.5% in 2013. The census data is summarized in Table 6.1.2e.

Table 6.1.2d   Evolution of the Muslim population in the North Island

Year	Population	Muslims	%	M.Ratio%
1874	111,934	1	0.00	5.88
1878	158,208	4	0.00	10.26
1881	193,047	0	0.00	0.00
1916	642,348	38	0.01	80.85
1976	2,268,393	1,150	0.05	81.27
1981	2,319,184	1,443	0.06	72.99
1991	2,281,542	5,556	0.24	91.19
1996	2,351,670	11,955	0.51	88.24
2001	2,433,261	21,054	0.87	89.06
2006	2,639,607	32,097	1.22	89.00
2013	2,827,638	41,562	1.47	90.07

Table 6.1.2e   Evolution of the Muslim population in the South Island

Year	Population	Muslims	%	M.Ratio%
1874	187,451	16	0.01	94.12
1878	256,008	35	0.01	89.74
1881	296,644	7	0.00	100.00
1916	448,377	9	0.00	19.15
1976	860,990	265	0.03	18.73
1981	850,758	534	0.06	27.01
1991	782,115	537	0.07	8.81
1996	796,437	1,593	0.20	11.76
2001	795,714	2,586	0.32	10.94
2006	852,243	3,969	0.47	11.00
2013	893,592	4,584	0.51	9.93

6.1.3 Regional Summary and Conclusion

Australasia has the largest number and concentration of Muslims among the four regions spanning Oceania. The Muslim population in this region was few thousands by the start of the twentieth century and only picked up momentum in its second half through migration for economic and political reasons. The Muslim population continued to increase significantly towards the end of last century and is expected to continue this trend, currently more than 2% of the total population and is expected to be near 7% by the end of this century. The main cause of increase is due to migration of Muslims, then their natural increase and finally conversion to Islam by natives. The following tables present centennial data from 600 AD to 2100 AD (or approximately 1 H to 1500 H) in Table 6.1a and decennial data from 1790 AD to 2100 AD (or 1210 H to 1520 H) in Tables 6.1b and 6.1c for current countries in Australasia. The data includes total population in thousands (P), the percentage of which is Muslim (M%), the corresponding Muslim population in thousands (M), and the annual population growth rate (APGR, or G%) of the total population in this region.

6.2 Muslims in Melanesia

This region consists of five island nations covering over five million square kilometers of the South Pacific Ocean: Fiji, New Caledonia, Papua New Guinea, Solomon Islands, and Vanuatu. The first Muslims arrived here in the nineteenth century by the British and the French. The former brought Indians to Fiji as indenture workers; while the latter brought prisoners of war from Algeria to New Caledonia. Thus, the Muslim population in this region has likely increased to 2,000 or 0.2% in 1900, to 60,000 or 0.8% in 2000, to 58,000 or 0.5% in 2020, and is projected to reach 77,000 or 0.4% by 2050 and 103,000 or 0.4% by 2100.

A plot of decennial estimates of the Muslim population and its percentage with respect to the total population in this region from 1900 to 2100 is provided in Fig. 6.2. The corresponding individual data for each country in this region is discussed below. In Section 6.2.6, the total population in each country in this region and the corresponding percentage and number of Muslims is presented centennially in Table 6.2a from 600 to 2100 and decennially in Tables 6.2b and 6.2c from 1790 to 2100.

Table 6.1a Centennial estimates of the Muslim population (×1000) in Australasia: 600–2100 AD (1–1500 H)

		600	700	800	900	1000	1100	1200	1300
Australia	P	380	385	390	395	400	410	420	430
	M%	—	—	—	—	—	—	—	—
	M	—	—	—	—	—	—	—	—
New Zealand	P	6	7	8	9	10	20	40	60
	M%	—	—	—	—	—	—	—	—
	M	—	—	—	—	—	—	—	—
Total	P	386	392	398	404	410	430	460	490
	M%	—	—	—	—	—	—	—	—
	M	—	—	—	—	—	—	—	—
	G%		0.015	0.015	0.015	0.015	0.048	0.067	0.063

Table 6.1a (Continued)

		1400	1500	1600	1700	1800	1900	2000	2100
Australia	P	440	450	450	450	305	3,774	18,991	42,877
	M%	—	—	—	—	—	0.10	1.66	7.00
	M	—	—	—	—	—	4	315	3,001
New Zealand	P	80	100	100	100	105	816	3,859	6,008
	M%	—	—	—	—	—	0.01	0.73	5.50
	M	—	—	—	—	—	—	28	330
Total	P	520	550	550	550	410	4,590	22,850	48,885
	M%	—	—	—	—	—	0.08	1.50	6.82
	M	—	—	—	—	—	4	343	3,332
	G%	0.059	0.056	0.000	0.000	−0.293	2.415	1.605	0.761

Table 6.1b Decennial estimates of the Muslim population (×1000) in Australasia: 1790–1940 AD (1210–1360 H)

		1790	1800	1810	1820	1830	1840	1850	1860
Australia	P	317	305	292	294	310	410	605	1,154
	M%	—	—	0.00	0.01	0.02	0.03	0.04	0.05
	M	—	—	—	—	—	—	—	1
New Zealand	P	110	105	100	95	90	82	88	154
	M%	—	—	—	—	—	—	—	—
	M	—	—	—	—	—	—	—	—
Total	P	427	410	392	389	400	492	693	1,308
	M%	—	—	0.00	0.01	0.02	0.03	0.03	0.04
	M	—	—	—	—	—	—	—	1
	G%		−0.391	−0.465	−0.078	0.292	2.078	3.418	6.350

Table 6.1b (Continued)

		1870	1880	1890	1900	1910	1920	1930	1940
Australia	P	1,664	2,253	3,183	3,774	4,455	5,436	6,575	7,165
	M%	0.02	0.03	0.06	0.10	0.09	0.05	0.03	0.04
	M	—	1	2	4	4	3	2	3
New Zealand	P	256	534	669	816	1,058	1,272	1,493	1,637
	M%	—	0.00	0.00	0.01	0.01	0.01	0.01	0.01
	M	—	—	—	—	—	—	—	—
Total	P	1,920	2,787	3,852	4,590	5,513	6,707	8,068	8,802
	M%	0.02	0.02	0.05	0.08	0.07	0.04	0.03	0.03
	M	0	1	2	4	4	3	2	3
	G%	3.840	3.724	3.237	1.753	1.834	1.960	1.846	0.871

Table 6.1c Decennial estimates of the Muslim population (×1000) in Australasia: 1950–2100 AD (1370–1520 H)

		1950	1960	1970	1980	1990	2000	2010	2020
Australia	P	8,177	10,242	12,793	14,588	16,961	18,991	22,155	25,500
	M%	0.04	0.04	0.19	0.59	0.98	1.66	2.44	3.00
	M	3	4	24	86	166	315	541	765
New Zealand	P	1,908	2,373	2,818	3,147	3,398	3,859	4,370	4,822
	M%	0.01	0.01	0.03	0.08	0.20	0.73	1.24	1.50
	M	—	—	1	3	7	28	54	72
Total	P	10,085	12,615	15,611	17,735	20,359	22,850	26,525	30,322
	M%	0.03	0.03	0.16	0.50	0.85	1.50	2.24	2.76
	M	3	4	25	89	173	343	595	837
	G%	1.361	2.238	2.131	1.275	1.380	1.155	1.491	1.338

Table 6.1c (Continued)

		2030	2040	2050	2060	2070	2080	2090	2100
Australia	P	28,177	30,572	32,814	34,950	36,961	38,900	40,855	42,877
	M%	3.50	4.00	4.50	5.00	5.50	6.00	6.50	7.00
	M	986	1,223	1,477	1,748	2,033	2,334	2,656	3,001
New Zealand	P	5,173	5,433	5,608	5,727	5,837	5,928	5,971	6,008
	M%	2.00	2.50	3.00	3.50	4.00	4.50	5.00	5.50
	M	103	136	168	200	233	267	299	330
Total	P	33,350	36,005	38,422	40,677	42,798	44,828	46,826	48,885
	M%	3.27	3.77	4.28	4.79	5.30	5.80	6.31	6.82
	M	1,090	1,359	1,645	1,948	2,266	2,601	2,954	3,332
	G%	0.952	0.766	0.650	0.570	0.508	0.463	0.436	0.430

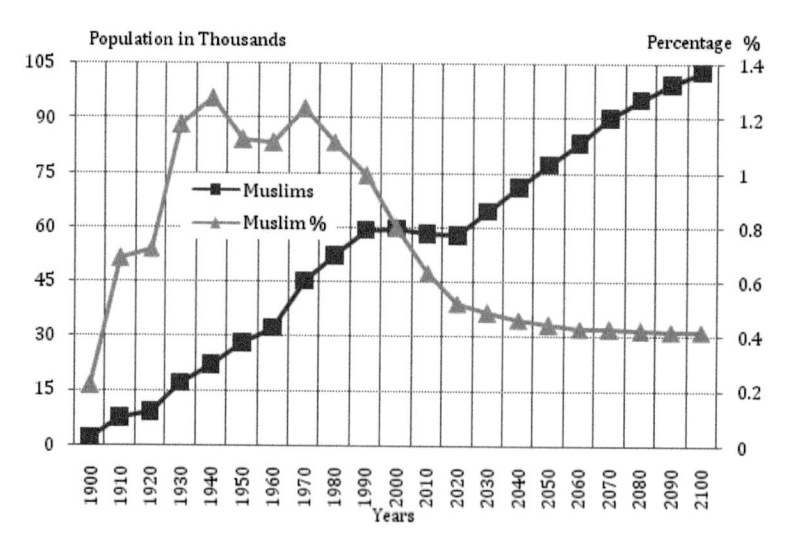

Figure 6.2 Plot of decennial estimates of the Muslim population and its percentage of the total population in Melanesia: 1900–2100 AD (1320–1520 H).

6.2.1 Fiji

The Republic of Fiji has a total area of 18,333 sq. km and consists of 844 islands covering over 1.3 million sq. km of the South Pacific Ocean. About 106 of these islands are inhabited, the largest of which are Viti Levu (10,429 sq. km) where the capital Suva that includes all the population but fifth lives, Vanua Levu (5,556 sq. km), Taveuni (470 sq. km) which includes one-sixth of the total population, Kadavu (411 sq. km), Gau (140 sq. km), and Koro (140 sq. km). A map of these islands is presented in Fig. 6.2.1. The Fiji Islands were conquered by the British in 1874 and gained their independence from the UK in 1970.

Muslims entered the Islands in 1879 under the indenture system that was introduced by the British Empire. Accordingly, the British brought slave like labor force from the Indian sub-continent to exploit the British dependencies. Ethnic censuses were carried out since 1881. Later religious censuses carried since 1946, and they show that about a tenth of the Indian population was Muslim in 1921. Using this data, we can estimate the Muslim population between 1881 and 1901.

Based on census data, the Muslim population increased from none in 1879, to 59 or 0.05% in 1881, to 747 or 0.6% in 1891, to 1,711 or 1.4% in 1901, to 5,302 or 3.8% in 1911, to 6,435 or 4.1% in 1921, to 11,290 or 5.7% in 1936, to 16,932 or 6.5% in 1946, to 25,394 or 7.3% in 1956, to 37,107 or 7.8% in 1966, to 44,321 or 7.7% in 1976, to 56,001 or 7.8% in 1986, then decreased to 54,323 or 7.0% in 1996, then 52,505 or 6.3% in 2007 and 50,925 or 5.8% in 2017. The decrease is due to 1987 military coup to curtail perceived Indian dominance over natives. This caused an exodus of Indians from the Islands towards Australia and New Zealand, some of whom were Muslims. The 2007 census also showed that 16.3% (almost one sixth) of Indians are Muslim, and 97.2% of all Muslims are Indians. Thus, assuming that the percentage of Muslims will remain constant, the Muslim population is expected to remain less than 70,000 throughout this century. The data is summarized in Table 6.2.1.

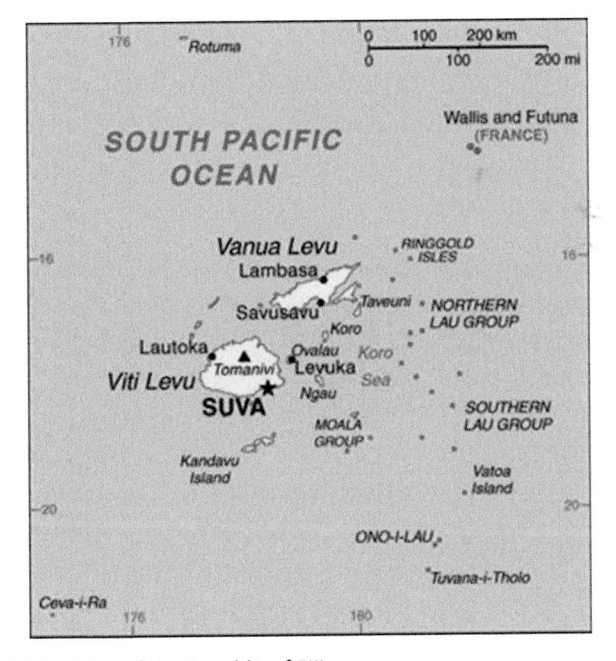

Figure 6.2.1 Map of the Republic of Fiji.

Table 6.2.1 Evolution of the Muslim population in Fiji

Year	Population	Muslims	%	Source
1879	125,000	—	0.00	[KET86]es
1881	127,486	59	0.05	[FJ]e
1891	121,180	747	0.62	[FJ]e
1901	120,124	1,711	1.42	[FJ]e
1911	139,541	5,302	3.80	[GB911]c
1921	157,266	6,435	4.09	[KET86]c
1936	198,379	11,290	5.69	[KET86]c
1946	259,638	16,932	6.52	[UN56]c
1956	345,737	25,394	7.34	[UN63]c
1966	476,479	37,107	7.79	[FJ17]c
1976	574,299	44,321	7.72	[FJ17]c
1986	715,333	56,001	7.83	[FJ17]c
1996	775,077	54,323	7.01	[FJ17]c
2007	837,271	52,505	6.27	[FJ17]c
2017	884,887	50,925	5.75	[FJ17]c
2050	1,071,000	62,000	5.75	es
2100	1,067,000	61,000	5.75	es

6.2.2 New Caledonia

The Territory of New Caledonia and Dependencies is a territorial collectivity of France, which conquered it in 1853. It has an area of 18,576 sq. km, comprising the main island or Grand Terre (16,372 sq. km), Lifou (1,197 sq. km), Maré (642 sq. km), l'Île des Pins or Isle of Pines (152 sq. km), Ouvéa (132 sq. km), and hundreds of much smaller islands, covering a million sq. km of the South Pacific Ocean. A map of these islands is presented in Fig. 6.2.2.

The first Muslims were 116 Algerians who arrived in 1872 as prisoners of war by the French occupying forces in Algeria. They were sent to the penal settlement in Isle of Pines. They were transferred to Ducos in the main island in 1881. By the end of the nineteenth century, their number was reduced to 24, as some were able to return to Algeria.

At the turn of the twentieth century, several Muslims from Java, Indonesia settled here. Their number increased to 1,483 in 1911, to 2,098 in 1921, to 6,000 in 1931, to 8,641 in 1946, but was reduced to 4,070 in 1969 and increased to 7,000 in 2009. In the 1950s, more Muslims settled here from former French colonies, especially Djibouti, due to the growth of the nickel industry. However, over 80% of the Muslim population are of Malay decent (mainly Java).

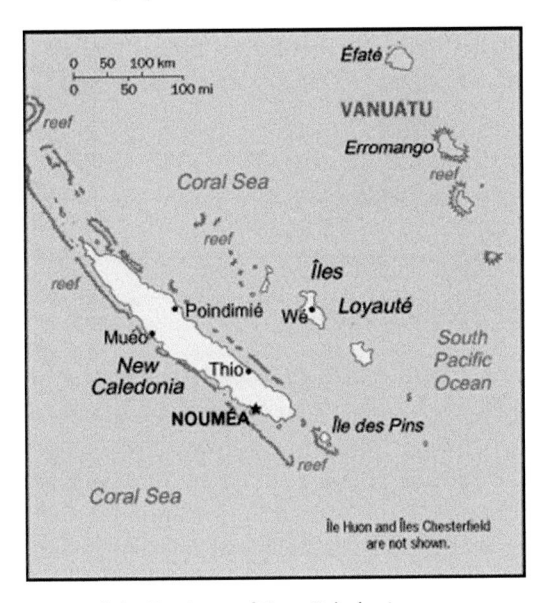

Figure 6.2.2 Map of the Territory of New Caledonia.

Thus, with the absence of official statistics on religious adherence and as shown in Table 6.2.2, estimates for the Muslim population fluctuated from none before 1872 to 116 or 0.2% in 1880, to 24 or 0.04% in 1901, to 1,483 or 2.9% in 1911, to 2,098 or 4.4% in 1921, to 6,000 or 10.5% in 1031, to 8,641 or 13.8% in 1946, to 4,070 or 4.1% in 1969. By 2014, the Association of Muslims of New Caledonia put the estimate at 3,000 or 1% of the total population, of which only 200 to 300 are practicing Muslims. Thus, assuming that the percentage of Muslims will remain constant at 1% of the total population, the Muslim population is expected to remain around 3,500 throughout the second half of this century.

Table 6.2.2 Evolution of the Muslim population in New Caledonia

Year	Population	Muslims	%	Source
1872	60,000	—	0.00	[NCH]es
1880	68,600	116	0.17	[NCH]es
1901	54,400	24	0.04	[NCH]es
1911	50,608	1,483	2.93	[SYB17]es
1921	47,505	2,098	4.42	[SYB22]es
1931	57,165	6,000	10.50	[SYB30]es
1946	62,700	8,641	13.79	[SYB50]es
1969	100,579	4,070	4.05	[SYB70]es
2014	268,767	3,000	1.12	[NC]es
2050	347,000	3,500	1.00	es
2100	356,000	3,500	1.00	es

6.2.3 Papua New Guinea

The Independent State of Papua New Guinea was occupied by the British in 1885, then transferred to Australia in 1902, and gained independence from Australia in 1975. Most of the country is situated in the eastern third of the island of New Guinea, which it shares with Indonesia and is the second largest Island in the world after Greenland. It has an area of 462,840 sq. km, comprising the eastern half of the Island of New Guinea which is 85% of the country's area, in addition to hundreds of islands surrounding this part, the largest of which are New Britain (35,144 sq. km), Bougainville (9,318 sq. km), New Ireland (7,404 sq. km), Manus (2,100 sq. km), Fergusson (1,437 sq. km), New Hanover (1,186 sq. km) and Normanby (1,000 sq. km). A map of this country is presented in Fig. 6.2.3.

Islam entered here through Indonesia, but the number of Muslims remains small. In 1890, there were no Muslims in PNG. In 1907, the Muslim population consisted of 525 individuals; all are Malay except twelve Chinese, two Tagales, and one Indian. This constituted 0.08% of the total population. It decreased to 500 or 0.06% in 1921, all Malays, but dropped to 120 or 0.00% by 1971. The 1990 census was the first to include Islam in the list of religions inquired. Hence, according to census data, the Muslim population increased from 440 or 0.01% in 1990 to 756 or 0.01% in 2000, to 1,352 or 0.02% in

2011. The last increase is due large number of conversion by locals. Thus, assuming that the percentage of Muslims will increase by 0.01 of a percentage point per decade, the Muslim population is expected to increase to 9,000 or 0.06% by 2050 and 24,000 or 0.12% by 2100. The data is summarized in Table 6.2.3.

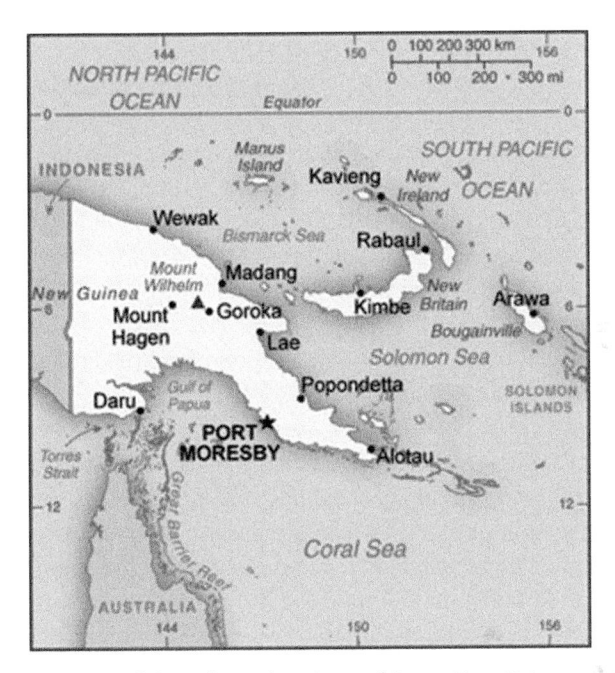

Figure 6.2.3 Map of the Independent State of Papua New Guinea.

Table 6.2.3 Evolution of the Muslim population in Papua New Guinea

Year	Population	Muslims	%	Source
1890	460,000	—	0.00	[JAN]es
1907	677,000	525	0.08	[PGH]es
1921	900,000	500	0.06	[MAS23]es
1971	2,489,936	120	0.00	[PG11]es
1990	3,582,333	440	0.01	[PG00]c
2000	5,171,548	756	0.01	[PG00]c
2011	7,229,880	1,352	0.02	[PG11]c
2050	14,204,000	9,000	0.06	es
2100	19,783,000	24,000	0.12	es

6.2.4 Solomon Islands

It was conquered by the British in 1893 and gained its independence from the UK in 1978. It has an area of 28,230 sq. km, comprising over 900 islands that covers over two million sq. km of the South Pacific Ocean, about a hundred of which are inhabited. The largest islands are Guadalcanal (5,302 sq. km) where the capital Honiara is, Malaita (4,307 sq. km) the most populated, Makira (3,190 sq. km), Santa Isabel (2,999 sq. km), Choiseul (2,971 sq. km), New Georgia (2,145 sq. km), Maramasike (700 sq. km), Kolombangara (685 sq. km), Vella Lavella (670 sq. km), Rennell (630 sq. km), Santa Cruz (519 sq. km), Vangunu (520 sq. km), Shortland (420 sq. km), Rendova (400 sq. km), Vanikoro (190 sq. km), San Jorge (184 sq. km), Vonavona (180 sq. km), Ranongga (145 sq. km), Tetepare (120 sq. km), Pavuvu (120 sq. km), Nggatokae (110 sq. km), Wagina (86 sq. km), Ulawa (66 sq. km), Ulawa (65 sq. km), Ugi (42 sq. km), and Savo (31 sq. km). A map of these islands is presented in Fig. 6.2.4.

As shown in Table 6.2.4, the Muslim population was estimated to be 70 or 0.02% in 1999. The 2009 census was the first to include information about Muslims, according to which it was 316 or 0.06% of the total population. They are scattered in small numbers in the islands of Guadalcanal, Malaita, and Pavuvu. Thus, assuming that the percentage of Muslims will increase by 0.05 of a percentage point per decade, the Muslim population is expected to increase to 3,000 or 0.3% by 2050 and 12,000 or 0.5% by 2100.

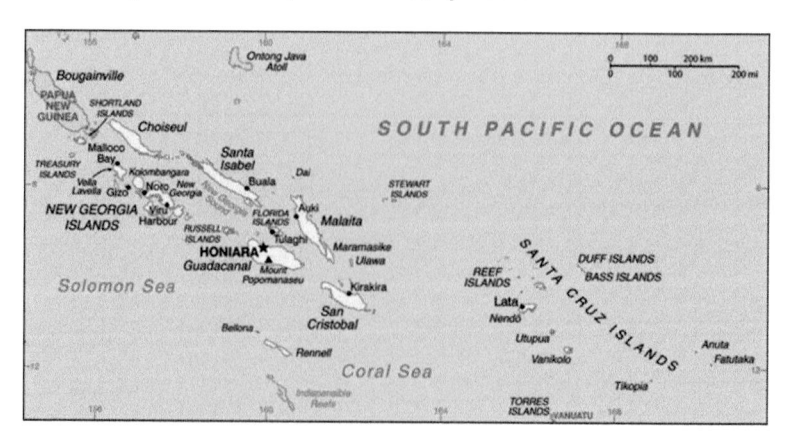

Figure 6.2.4 Map of the Solomon Islands.

Table 6.2.4 Evolution of the Muslim population in Solomon Islands

Year	Population	Muslims	%	Source
1999	409,042	70	0.02	[RIS]es
2009	501,885	316	0.06	[SB]c
2050	1,290,000	3,000	0.25	es
2100	2,410,000	12,000	0.50	es

6.2.5 Vanuatu

The Republic of Vanuatu gained its independence in 1980 from France and the UK, when it changed its name from New Hebrides. It was under joint British and France administration since 1906. It has a total area of 12,281 sq. km, comprising over eighty islands, covering about half a million sq. km of the South Pacific Ocean. The largest islands are Santo (3,956 sq. km), with a sixth of the population, Malekula (2,041 sq. km), with a tenth of the population, Efate (900 sq. km) with almost a third of the population and the capital Port Vila, Erromango (888 sq. km), Ambrym (678 sq. km), Tanna (550 sq. km), with one-eighth of the population, Pentecost (490 sq. km), Epi (444 sq. km), Gaua (342 sq. km), Vanualava (314 sq. km), Aneityum (159 sq. km), Hiu (51 sq. km) and Ureparapara (39 sq. km). A map of these islands is presented in Fig. 6.2.5.

Islam was introduced to Vanuatu by Hussein Nabanga who converted to Islam while training to be a Christian missionary in India. Islam is made up of Ni-Vanuatu converts and is spreading very fast. It started with the return of Hussein Nabanga in 1978, reaching twenty or 0.01% in 1989, to forty or 0.02% in 1999. According to the 2009 census, the Muslim population was 103 or 0.04% of the total population. They are located in two southern islands as follows:

- Efate: 36 or 0.05% of the total population of 65,734.
- Tanna: 67 or 0.23% of the total population of 28,799.

Figure 6.2.5 Map of the Republic of Vanuatu.

Thus, assuming that the percentage of Muslims will increase by 0.02 of a percentage point per decade, the Muslim population is expected to increase to less than a thousand or 0.1% by 2050 and exceed two thousands or 0.2% by 2100. A summary of the estimates is provided in Table 6.2.5.

Table 6.2.5 Evolution of the Muslim population in Vanuatu

Year	Population	Muslims	%	Source
1967	78,088	—	0.00	[VU]es
1979	111,251	1	0.00	[VU]es
1989	142,944	20	0.01	[VU]es
1999	186,678	40	0.02	[RIS]es
2009	233,539	103	0.04	[VU09]c
2050	557,000	700	0.12	es
2100	968,000	2,100	0.22	es

6.2.6 Regional Summary and Conclusion

Muslims started coming to Melanesia at the end of the nineteenth century and were brought by the British as indenture workers and by the French as prisoners. The Muslim population is increasing slowly but is expected to remain well less than 1% of the total population throughout this century. The following tables present centennial data from 600 AD to 2100 AD (or approximately 1 H to 1500 H) in Table 6.2a and decennial data from 1790 AD to 2100 AD (or 1210 H to 1520 H) in Tables 6.2b and 6.2c for current countries in Melanesia. The data includes total population in thousands (P), the percentage of which is Muslim (M%), the corresponding Muslim population in thousands (M), and the annual population growth rate (APGR, or G%) of the total population in this region.

Table 6.2a Centennial estimates of the Muslim population (×1000) in Melanesia: 600–2100 AD (1–1500 H)

		600	700	800	900	1000	1100	1200	1300	1400	1500	1600	1700	1800	1900	2000	2100
Fiji	P	87	90	94	99	105	112	120	130	140	150	160	170	190	120	811	1,067
	M%	—	—	—	—	—	—	—	—	—	—	—	—	—	1.42	7.01	5.75
	M	—	—	—	—	—	—	—	—	—	—	—	—	—	2	57	61
New Caledonia	P	54	56	58	60	62	64	66	68	70	72	75	80	90	54	217	356
	M%	—	—	—	—	—	—	—	—	—	—	—	—	—	0.04	1.00	1.00
	M	—	—	—	—	—	—	—	—	—	—	—	—	—	—	2	4
PNG	P	220	235	250	265	280	295	310	325	340	355	370	385	418	587	5,848	19,783
	M%	—	—	—	—	—	—	—	—	—	—	—	—	—	0.04	0.01	0.12
	M	—	—	—	—	—	—	—	—	—	—	—	—	—	—	1	24
Solomons	P	78	80	82	84	86	88	90	92	95	100	105	110	120	140	413	2,410
	M%	—	—	—	—	—	—	—	—	—	—	—	—	—	—	0.01	0.50
	M	—	—	—	—	—	—	—	—	—	—	—	—	—	—	—	12
Vanuatu	P	23	25	27	29	31	33	35	37	39	41	43	45	50	8	185	968
	M%	—	—	—	—	—	—	—	—	—	—	—	—	—	—	0.02	0.22
	M	—	—	—	—	—	—	—	—	—	—	—	—	—	—	—	2
Total	P	462	486	511	537	564	592	621	652	684	718	753	790	868	910	7,473	24,583
	M%	—	—	—	—	—	—	—	—	—	—	—	—	—	0.22	0.80	0.42
	M	—	—	—	—	—	—	—	—	—	—	—	—	—	2	60	103
	G%	0.051	0.050	0.050	0.050	0.049	0.048	0.048	0.049	0.048	0.049	0.048	0.048	0.094	0.047	2.106	1.191

Muslims in Melanesia | 821

Table 6.2b Decennial estimates of the Muslim population (×1000) in Melanesia: 1790–1940 AD (1210–1360 H)

		1790	1800	1810	1820	1830	1840	1850	1860	1870	1880	1890	1900	1910	1920	1930	1940
Fiji	P	200	190	180	170	165	160	155	150	148	127	121	120	140	157	180	211
	M%	—	—	—	—	—	—	—	—	—	0.05	0.62	1.42	3.80	4.09	5.69	6.52
	M	—	—	—	—	—	—	—	—	—	—	1	2	4	6	10	14
New Caledonia	P	100	90	80	70	60	50	40	29	40	42	50	54	51	48	57	53
	M%	—	—	—	—	—	—	—	—	—	0.17	0.17	0.04	2.93	4.42	10.50	13.79
	M	—	—	—	—	—	—	—	—	—	—	—	—	1	2	6	7
PNG	P	400	418	440	460	500	540	530	510	500	490	480	587	719	880	1,077	1,318
	M%	—	—	—	—	—	—	—	—	—	—	—	0.04	0.08	0.06	0.06	0.06
	M	—	—	—	—	—	—	—	—	—	—	—	0	1	1	1	1
Solomons	P	125	120	115	110	105	100	95	90	85	80	89	140	150	151	94	95
	M%	—	—	—	—	—	—	—	—	—	—	—	—	—	—	—	—
	M	—	—	—	—	—	—	—	—	—	—	—	—	—	—	—	—
Vanuatu	P	55	50	45	40	35	30	25	20	15	12	10	8	15	22	30	40
	M%	—	—	—	—	—	—	—	—	—	—	—	—	—	—	—	—
	M	—	—	—	—	—	—	—	—	—	—	—	—	—	—	—	—
Total	P	880	868	860	850	865	880	845	799	788	751	750	910	1,074	1,257	1,438	1,717
	M%	—	—	—	—	—	—	—	—	—	0.02	0.11	0.22	0.69	0.72	1.17	1.27
	M	—	—	—	—	—	—	—	—	—	—	1	2	7	9	17	22
	G%	-0.137	-0.093	-0.117	0.175	0.172	-0.406	-0.560	-0.137	-0.480	-0.013	1.930	1.659	1.575	1.344	1.770	

Table 6.2c Decennial estimates of the Muslim population (×1000) in Melanesia: 1950–2100 AD (1370–1520 H)

		1950	1960	1970	1980	1990	2000	2010	2020	2030	2040	2050	2060	2070	2080	2090	2100
Fiji	P	289	393	521	635	729	811	860	896	966	1,027	1,071	1,099	1,111	1,110	1,095	1,067
	M%	6.52	7.34	7.79	7.72	7.83	7.01	6.27	5.75	5.75	5.75	5.75	5.75	5.75	5.75	5.75	5.75
	M	19	29	41	49	57	57	54	52	56	59	62	63	64	64	63	61
New Caledonia	P	65	78	105	145	170	217	254	285	311	332	347	354	359	360	359	356
	M%	13.79	4.05	4.05	2.00	1.00	1.00	1.00	1.00	1.00	1.00	1.00	1.00	1.00	1.00	1.00	1.00
	M	9	3	4	3	2	2	3	3	3	3	3	4	4	4	4	4
PNG	P	2,002	2,256	2,783	3,571	4,616	5,848	7,311	8,947	10,709	12,493	14,204	15,785	17,180	18,321	19,183	19,783
	M%	0.01	0.01	0.01	0.01	0.01	0.01	0.02	0.03	0.04	0.05	0.06	0.07	0.09	0.10	0.11	0.12
	M	—	—	—	—	—	1	1	3	4	6	9	11	15	18	21	24
Solomons	P	90	118	160	231	312	413	528	687	865	1,068	1,290	1,523	1,761	1,997	2,217	2,410
	M%	—	—	—	—	—	0.01	0.06	0.10	0.15	0.20	0.25	0.30	0.35	0.40	0.45	0.50
	M	—	—	—	—	—	—	—	1	1	2	3	5	6	8	10	12
Vanuatu	P	48	64	85	116	147	185	236	307	383	468	557	647	737	822	900	968
	M%	—	—	—	0.00	0.01	0.02	0.04	0.06	0.08	0.10	0.12	0.14	0.16	0.18	0.20	0.22
	M	—	—	—	—	—	—	—	—	—	—	1	1	1	1	2	2
Total	P	2,493	2,909	3,655	4,698	5,973	7,473	9,188	11,123	13,234	15,387	17,469	19,410	21,148	22,610	23,755	24,583
	M%	1.12	1.11	1.23	1.11	0.99	0.80	0.63	0.52	0.49	0.46	0.44	0.43	0.43	0.42	0.42	0.42
	M	28	32	45	52	59	60	58	58	65	71	77	83	90	95	99	103
	G%	3.732	1.541	2.282	2.511	2.402	2.241	2.065	1.911	1.738	1.507	1.269	1.054	0.858	0.669	0.494	0.343

6.3 Muslims in Micronesia

This region consists of seven island nations and territories covering over eleven million square kilometers of the southern and western Pacific Ocean. These are the Federated States of Micronesia, Guam, Kiribati, Marshall Islands, Nauru, Northern Mariana Islands, and Palau. By the start of the twentieth century, there were tens of Muslims in these islands brought by the occupying forces. However, a noticeable increase of the Muslim population happened in the 1990s, with immigrants coming for economic needs, and locals embracing Islam. Thus, the Muslim population in this region was negligible until the end of the twentieth century and has likely reached 1,000 or 0.2% in 2000 and is projected to reach 3,000 or 0.4% by 2100.

A plot of decennial estimates of the Muslim population and its percentage with respect to the total population in this region from 1900 to 2100 is provided in Fig. 6.3. The corresponding individual data for each country in this region is discussed below. In Section 6.3.8, the total population in each country in this region and the corresponding percentage and number of Muslims is presented centennially in Tables 6.3a from 600 to 2100 and decennially in Tables 6.3b and 6.3c from 1790 to 2100.

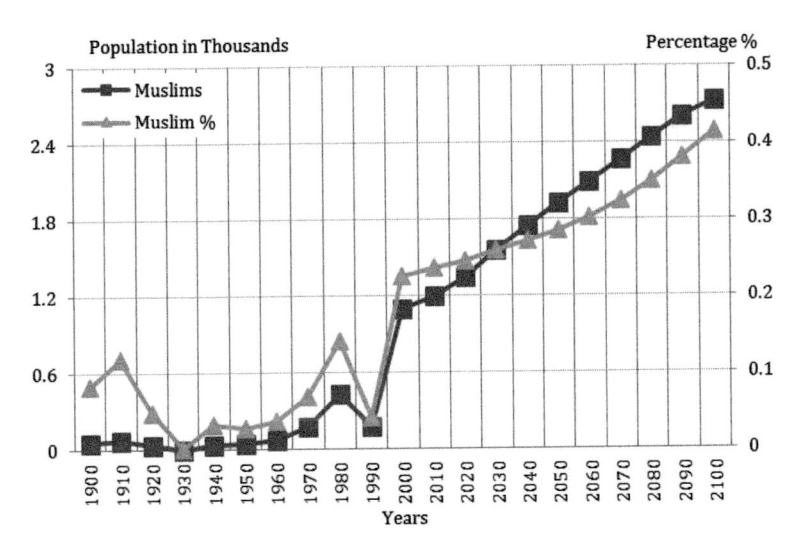

Figure 6.3 Plot of decennial estimates of the Muslim population and its percentage of the total population in Micronesia: 1900–2100 AD (1320–1520 H).

6.3.1 Federated States of Micronesia

It gained its independence in 1986 from the United States of America's administrated United Nations' Trust, which lasted since 1947, but remains in Compact of free Association with the United States. The islands were under Spanish control from 1886 to 1899, then purchased by Germany and remained under their control until 1914, when Japan took them over and remained there until 1945.

The states have a total area of 701 sq. km, comprising 607 islands spread through 2.6 million sq. km of the Western Pacific Ocean, and divided among four states: Chuuk (127 sq. km and half of the population), Kosrae (110 sq. km and a tenth of the population), Pohnpei (345 sq. km and a third of the population), and Yap (118 sq. km and a sixth of the population). A map of these islands is presented in Fig. 6.3.1.

In 1907, the Muslim population consisted of eleven Malay policemen working with the German forces in Pohnpei. This constituted 0.07% of the total population. The Muslim population increased to 26 or 0.02% in 1994 but decreased to two or 0.0% in 2010. Thus, assuming that the percentage of Muslims will increase by 0.01 of a percentage point per decade, the Muslim population is expected to be around one hundred or 0.1% of the total population throughout the second half of this century. The data is summarized in Table 6.3.1.

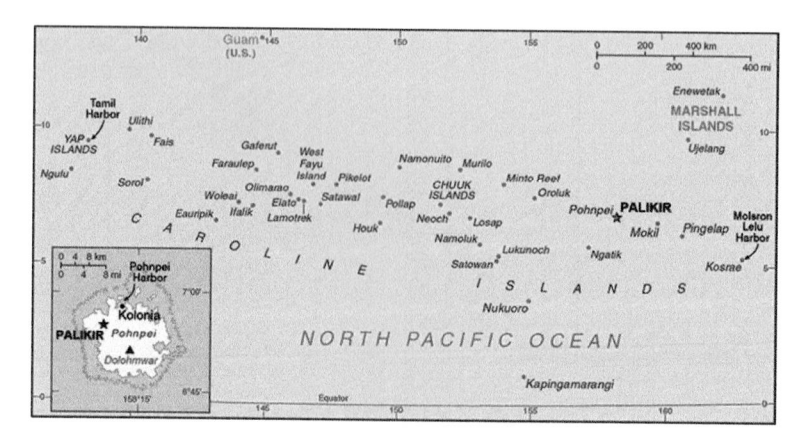

Figure 6.3.1 Map of the Federated States of Micronesia.

Table 6.3.1 Evolution of the Muslim population in the Federated States of Micronesia

Year	Population	Muslims	%	Source
1907	15,000	11	0.07	[PGH]es
1994	105,506	26	0.02	[UN04]c
2010	102,843	2	0.00	[FM]c
2050	139,000	70	0.05	es
2100	115,000	100	0.10	es

6.3.2 Guam

The Territory of Guam is an organized, unincorporated territory of the United States of America. It was ceded to the US by Spain in 1899 and occupied by Japan between 1941 and 1944. The Island Territory has an area of 541 sq. km and its map is presented in Fig. 6.3.2. As shown in Table 6.3.2, estimates of the Muslim population in this U.S. Territory decreased from 100 or 0.08% in 1990 to 50 or 0.03% in 2010. Thus, assuming that the percentage of Muslims will increase by 0.01 of a percentage point per decade, the Muslim population is expected to remain less than 300 or 0.1% throughout this century.

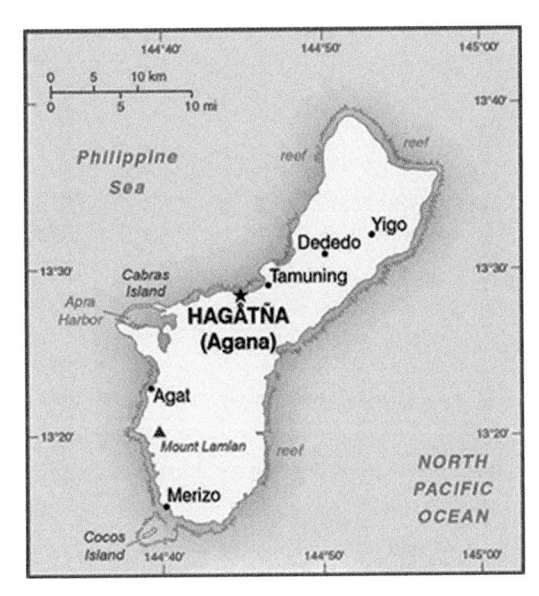

Figure 6.3.2 Map of the Territory of Guam.

Table 6.3.2 Evolution of the Muslim population in Guam

Year	Population	Muslims	%	Source
1990	133,152	100	0.08	[GU]es
2010	159,358	50	0.03	[GU]es
2050	193,000	150	0.07	es
2100	168,000	200	0.12	es

6.3.3 Kiribati

The Republic of Kiribati was conquered by the British in 1892 and gained its independence from the UK in 1979, when it changed its name from Gilbert Islands. It has a total area of 811 sq. km and consists of 33 coral atolls covering over five million squared kilometers of the Central Pacific Ocean. By far the largest island is Kirimati with area of 388 sq. km, but only 6% of the population. Half of the population lives in Tarawa Atoll (31 sq. km), where the capital Bairiki is located. There is 84 sq. km of uninhabited islands. A map of the Kiribati Islands is presented in Fig. 6.3.3.

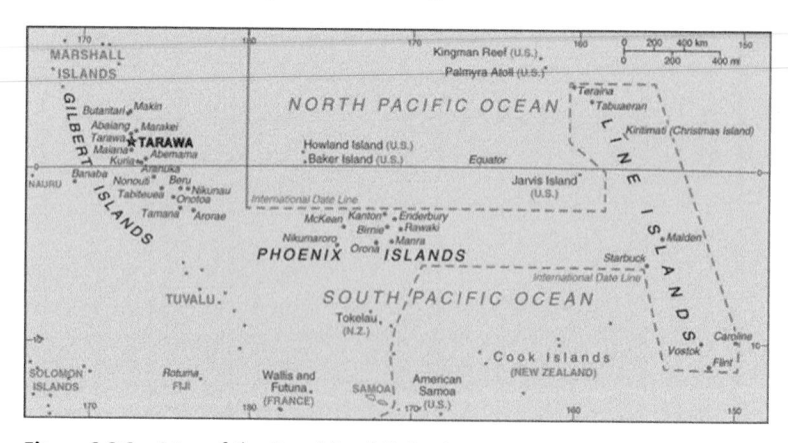

Figure 6.3.3 Map of the Republic of Kiribati.

Census is held here every five years, but Islam was added to the list of religions only in 2010. Accordingly, the Muslim population increased from 119 or 0.1% in 2010 to 139 or 0.1% in 2015. They were found on ten atolls, but more than half lives on South Tarawa and fifth lives on Betio. Thus, assuming that the percentage of

Muslims will increase by 0.05 of a percentage point per decade, the Muslim population is expected to increase to 500 or 0.3% by 2050 and 1,300 or 0.6% by 2100. The data is summarized in Tables 6.3.3a. and 6.3.3b.

Table 6.3.3a Evolution of the Muslim population in Kiribati

Year	Population	Muslims	%	Source
2010	102,846	119	0.12	[KI10]c
2015	110,136	139	0.13	[KI15]c
2050	177,000	500	0.30	es
2100	239,000	1,300	0.55	es

Table 6.3.3b Evolution of the Muslim population in Kiribati per island

Island		2010	2015	Island		2010	2015
Banaba	P	295	268	Nonout	P	2,683	2,743
	M	—	—		M	6	7
Makin	P	1,798	1,990	N. Tabiteuea	P	3,689	3,955
	M	1	—		M	4	—
Butaritari	P	4,346	3,224	S. Tabiteuea	P	1,290	1,306
	M	1	—		M	—	—
Marakei	P	2,872	2,799	Beru	P	2,099	2,051
	M	1	2		M	—	—
Abaiang	P	5,502	5,568	Nikunau	P	1,907	1,789
	M	5	—		M	—	5
N. Tarawa	P	6,102	6,629	Onotoa	P	1,519	1,393
	M	7	2		M	—	—
S. Tarawa	P	34,427	39,058	Tamana	P	951	1,104
	M	51	81		M	—	—
Betio	P	15,755	17,330	Arorae	P	1,279	1,011
	M	25.00	31		M	—	—
Maiana	P	2,027	1,982	Teeraina	P	1,690	1,712
	M	—	—		M	1	—
Abemama	P	3,213	3,262	Tabuaeran	P	1,960	2,315
	M	11	2		M	—	—
Kuria	P	980	1,046	Kiritimati	P	5,586	6,456
	M	—	—		M	6	9
Aranuka	P	1,057	1,125	Kanton	P	31	20
	M	—	—		M	—	—

P: Total population, *M*: Muslim population.

6.3.4 Marshall Islands

The Republic of the Marshall Islands gained its independence in 1986 from the United Sates of America's administrated United Nations Trust Territories of the Pacific Islands but remains in Compact of Free Association with the United States of America. The islands were under German control from 1885 to 1914, then Japanese control until 1944, then the Americans under the UN Trust.

The Marshall Islands have a total area of 181 sq. km, comprising 29 atolls and five isolated islands, covering over two million squared kilometers of the South Pacific Ocean. The largest atolls are Bokak (129 sq. km), Toke (94 sq. km), Kwajalein (16 sq. km), Ailinglaplap and Mili (15 sq. km each), Jaluit (11 sq. km), Likiep, Majuro, Maloelap (10 sq. km each). Majuro Atoll is the capital and has almost half of the total population. A map of these islands is presented in Fig. 6.3.4.

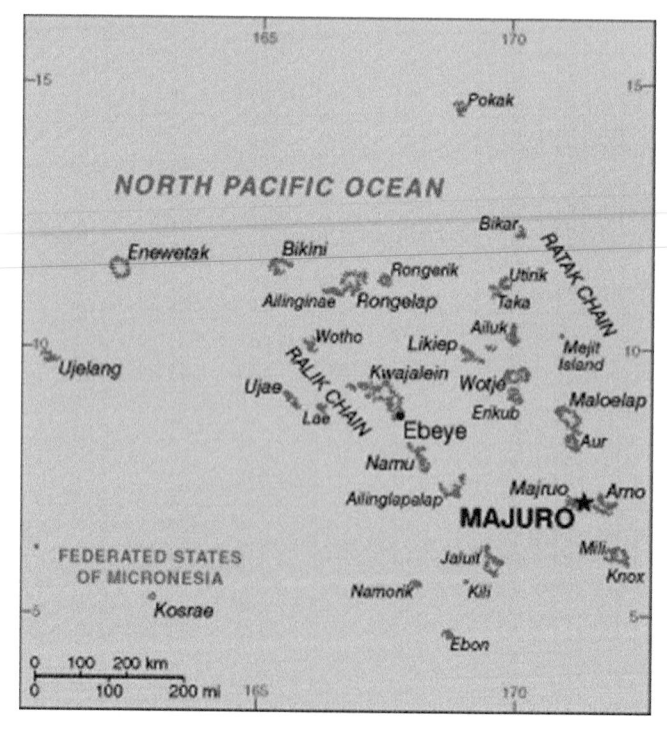

Figure 6.3.4 Map of the Republic of the Marshall Islands.

In 1907, the Muslim population consisted of fifty Malay or 0.4% of the total population. But the 2011 census indicated the presence of no Muslims (excluding Ahmadiyya Muslims). The Ahmadi Muslims established their first "mosque" on the Islands in 2012, which is located in Uliga at the east of Majuro Atoll. Thus, assuming that the percentage of Muslims will increase by 0.01 of a percentage point per decade, the Muslim population is expected to remain less than one hundred or 0.1% of the total population throughout this century. The data are summarized in Table 6.3.4.

Table 6.3.4 Evolution of the Muslim population in Marshall Islands

Year	Population	Muslims	%	Source
1907	13,000	50	0.38	[PGH]es
2011	53,158	—	0.00	[MH]c
2050	75,000	30	0.04	es
2100	63,000	60	0.09	es

6.3.5 Nauru

The Republic of Nauru consists of one island with surface area of 21 sq. km and its map is presented in Fig. 6.3.5. It was annexed by Germany in 1888. It was then controlled by Australia in 1923 under League of Nations Trust until it was occupied by Japan from 1942 to 1945. It then was part of Australian, New Zealand and UK administrated United Nations Trust until its independence in 1968. The Muslim population was reduced from 34 or 1.2% in 1947, to none since 1961, and the situation is expected to remain the same throughout this century. The census and estimate data are summarized in Table 6.3.5.

Table 6.3.5 Evolution of the Muslim population in Nauru

Year	Population	Muslims	%	Source
1947	2,855	34	1.19	[UN56]c
1961	4,613	—	0.00	[UN63]c
1983	7,674	—	0.00	[NR]c
1992	9,919	—	0.00	[NR]c
2002	10,065	—	0.00	[NR]c
2006	9,233	—	0.00	[NR]c
2011	9,945	—	0.00	[NR]c
2050	11,000	—	0.00	es
2100	6,000	—	0.00	es

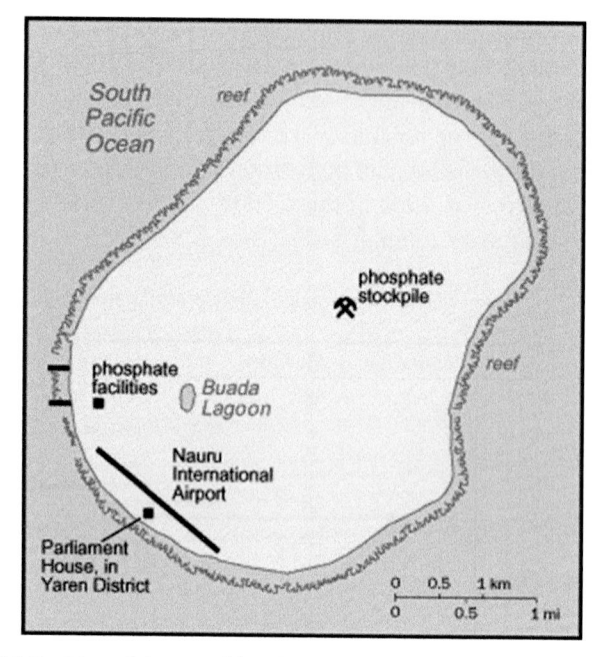

Figure 6.3.5 Map of the Republic of Nauru.

6.3.6 Northern Mariana Islands

The Commonwealth of the Northern Mariana Islands was established in 1976 in union with the United States of America. They were captured by the latter in 1944 from the Japanese and have been under US control since then. The islands have a total area of 457 sq. km; comprising fifteen islands only three are inhabited: Saipan, Tinian, and Rota. The largest of the islands are Saipan (115 sq. km), where the capital Saipan is located and about 90% of the population live, Tinian (101 sq. km), Rota (85 sq. km), Pagan (47 sq. km), Agrihan (44 sq. km), and Anatahan (31 sq. km). A map of these islands is presented in Fig. 6.3.6.

In 1907, the Muslim population consisted of ten Malays, or 0.02% of the total population. The censuses since 1980 included ethnic data on Bangladeshis which is taken here as an estimate of the Muslim population in the Islands. Accordingly, the Muslim population in the Islands increased from three or 0.02% in 1980, to

28 or 0.06% in 1990, to 459 or 0.78% in 1995, to 897 or 1.30% in 2000, but dropped to 501 or 0.9% in 2010. Thus, assuming that the percentage of Muslims will remain constant, the Muslim population is expected to remain less than six hundred during the second half of this century. A summary of the data is provided in Table 6.3.6.

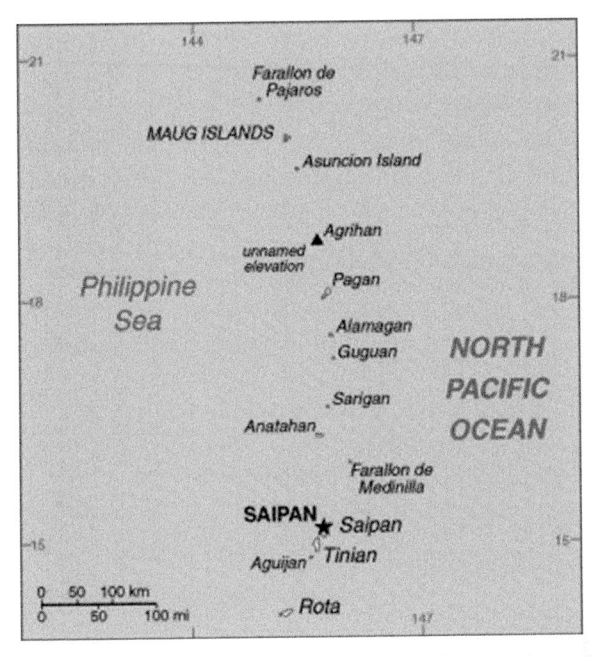

Figure 6.3.6 Map of the Commonwealth of the Northern Mariana Islands.

Table 6.3.6 Evolution of the Muslim population in Northern Mariana Islands

Year	Population	Muslims	%	Source
1907	3,000	10	0.33	[PGH]es
1980	16,780	3	0.02	[MP00]e
1990	43,345	28	0.06	[MP00]e
1995	58,846	459	0.78	[MP00]e
2000	69,221	897	1.30	[MP00]e
2010	53,883	501	0.93	[MP10]e
2050	62,000	600	0.93	es
2100	53,000	500	0.93	es

6.3.7 Palau

The Republic of Palau gained independence in 1994 from the US administered UN Trust of Pacific Islands, which it was under since 1947. It has an area of 416 sq. km spread over 0.6 million sq. km of the east Pacific. Palau consists of more than 340 islands, of which only eight are inhabited. These are Babeldaob (331 sq. km) with almost a third of the total population and where the capital Ngerulmud is located, Koror (18 sq. km) where the former capital Koror is located until 2006 and almost two-thirds of the total population, Peliliu (12 sq. km), Angaur (9 sq. km), Hatohobei, Helen Reef Kayangel, Pulo Anna and Sonsorol. A map of Palau is presented in Fig. 6.3.7.

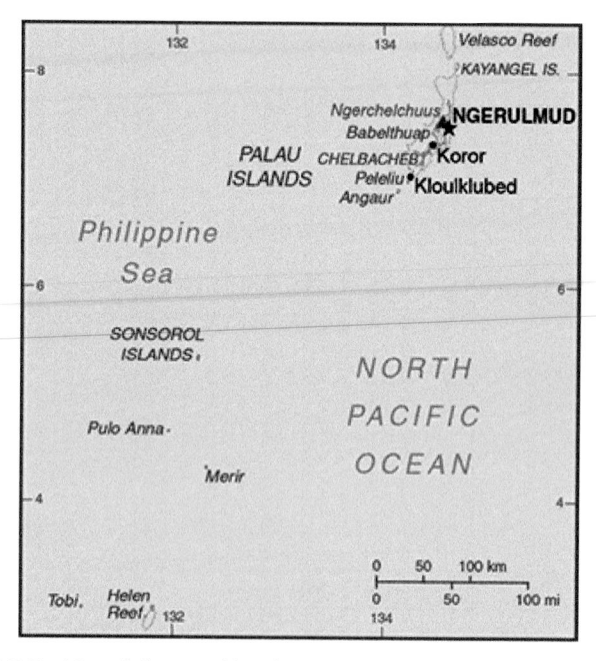

Figure 6.3.7 Map of the Republic of Palau.

In the 1990s, the number of Muslims increased due to the import of foreign workers. Most of Muslims are from Bangladesh. According to official statistics, the number of Muslims (Bangladeshis) changed from none in 1990 and before, to 47 or 0.3% in 1995, to 200 or 1.0% in 2000, to 425 or 2.1% in 2005, to 600 or 2.9% in 2010. The 2015 census recorded 524 Muslims or 3.0% of the total population, all but twelve are male. They are located in all major islands but three;

Kayangel, Sonsorol and Hatohobei, but more than half of the Muslim population (303) live in Kror. Thus, assuming that the percentage of Muslims will increase by a tenth of a percentage point per decade, the Muslim population is expected to remain less than six hundred throughout the second half of this century, comprising 3.3% of the total population by 2050 and 3.8% by 2100. A summary of the data is provided in Table 6.3.7.

Table 6.3.7 Evolution of the Muslim population in Palau

Year	Population	Muslims	%	Source
1990	15,122	—	0.00	[PW00]e
1995	17,225	47	0.27	[PW00]e
2000	19,129	200	1.05	[PW04]e
2005	19,907	425	2.13	[PW07]e
2010	20,879	600	2.87	[PW10]e
2015	17,661	524	2.97	[PW15]c
2050	18,000	600	3.30	es
2100	14,000	500	3.80	es

6.3.8 Regional Summary and Conclusion

A noticeable increase of the Muslim population in Micronesia happened towards the end of the twentieth century with immigrants coming for economic needs and locals embracing Islam. The Muslim population is increasing slowly but is expected to remain well less than 1% of the total population throughout this century. The following tables present centennial data from 600 AD to 2100 AD (or approximately 1 H to 1500 H) in Table 6.3a and decennial data from 1790 AD to 2100 AD (or 1210 H to 1520 H) in Tables 6.3b and 6.3c for current countries in Micronesia. The data includes total population in thousands (P), the percentage of which is Muslim (M%), the corresponding Muslim population in thousands (M), and the annual population growth rate (APGR, or G%) of the total population in this region. The total population estimate in each country since 1950 is based on the United Nations' World Population Prospects [UNP] while pre-1950 data is based on [PSH, MAD, OCE]. Other estimates and census data is used to fill in missing data from the aforementioned sources.

Table 6.3a Centennial estimates of the Muslim population (×1000) in Micronesia: 600–2100 AD (1–1500 H)

		600	700	800	900	1000	1100	1200	1300
FSM	P	25	26	27	34	37	40	45	50
	M%	—	—	—	—	—	—	—	—
	M	—	—	—	—	—	—	—	—
Guam	P	1	2	3	4	5	6	7	8
	M%	—	—	—	—	—	—	—	—
	M	—	—	—	—	—	—	—	—
Kiribati	P	12	12	13	16	17	18	19	20
	M%	—	—	—	—	—	—	—	—
	M	—	—	—	—	—	—	—	—
Marshalls	P	2	3	4	5	6	7	8	9
	M%	—	—	—	—	—	—	—	—
	M	—	—	—	—	—	—	—	—
Nauru	P	—	—	—	1	1	1	1	1
	M%	—	—	—	—	—	—	—	—
	M	—	—	—	—	—	—	—	—
CNMI	P	12	12	13	16	17	18	19	20
	M%	—	—	—	—	—	—	—	—
	M	—	—	—	—	—	—	—	—
Palau	P	3	3	4	4	5	5	6	6
	M%	—	—	—	—	—	—	—	—
	M	—	—	—	—	—	—	—	—
Total	P	55	58	64	80	87	95	104	114
	M%	—	—	—	—	—	—	—	—
	M	—	—	—	—	—	—	—	—
	G%		0.064	0.092	0.218	0.091	0.084	0.097	0.088

Table 6.3a (Continued)

		1400	1500	1600	1700	1800	1900	2000	2100
FSM	P	54	60	65	70	77	28	107	115
	M%	—	—	—	—	—	0.00	0.02	0.10
	M	—	—	—	—	—	—	—	—
Guam	P	9	10	11	12	4	10	155	168
	M%	—	—	—	—	—	—	0.05	0.12
	M	—	—	—	—	—	—	0	0
Kiribati	P	22	24	27	31	2	3	84	239
	M%	—	—	—	—	—	—	0.06	0.55
	M	—	—	—	—	—	—	—	1
Marshalls	P	10	11	12	13	14	13	51	63
	M%	—	—	—	—	—	0.38	—	0.09
	M	—	—	—	—	—	—	—	—
Nauru	P	1	1	1	1	1	1	10	6
	M%	—	—	—	—	—	—	—	—
	M	—	—	—	—	—	—	—	—
CNMI	P	22	24	27	30	—	3	57	53
	M%	—	—	—	—	—	—	1.30	0.93
	M	—	—	—	—	—	—	1	0
Palau	P	7	7	8	9	10	4	19	14
	M%	—	—	—	—	—	—	1.05	3.80
	M	—	—	—	—	—	—	—	1
Total	P	125	137	151	166	108	61	485	659
	M%	—	—	—	—	—	0.08	0.23	0.41
	M	—	—	—	—	—	—	1	3
	G%	0.089	0.096	0.095	0.096	−0.428	−0.570	2.070	0.307

Table 6.3a Centennial estimates of the Muslim population (×1000) in Micronesia: 1790–1940 AD (1210–1360 H)

		1790	1800	1810	1820	1830	1840	1850	1860
FSM	P	75	77	78	79	80	80	76	70
	M%	—	—	—	—	—	—	—	—
	M	—	—	—	—	—	—	—	—
Guam	P	4	4	4	5	5	5	6	7
	M%	—	—	—	—	—	—	—	—
	M	—	—	—	—	—	—	—	—
Kiribati	P	2	2	2	2	2	2	2	3
	M%	—	—	—	—	—	—	—	—
	M	—	—	—	—	—	—	—	—
Marshalls	P	14	14	14	15	15	15	14	13
	M%	—	—	—	—	—	—	—	—
	M	—	—	—	—	—	—	—	—
Nauru	P	1	1	1	1	1	1	1	1
	M%	—	—	—	—	—	—	—	—
	M	—	—	—	—	—	—	—	—
CNMI	P	—	—	—	1	1	1	1	1
	M%	—	—	—	—	—	—	—	—
	M	—	—	—	—	—	—	—	—
Palau	P	10	10	10	10	10	10	9	8
	M%	—	—	—	—	—	—	—	—
	M	—	—	—	—	—	—	—	—
Total	P	105	108	110	112	114	115	110	103
	M%	—	—	—	—	—	—	—	—
	M	—	—	—	—	—	—	—	—
	G%		0.264	0.176	0.173	0.170	0.063	−0.437	−0.610

Table 6.3b (*Continued*)

		1870	1880	1890	1900	1910	1920	1930	1940
FSM	P	60	40	30	28	29	30	30	30
	M%	—	—	—	0.00	0.07	0.07	0.00	0.00
	M	—	—	—	—	—	—	—	—
Guam	P	8	9	9	10	12	13	19	22
	M%	—	—	—	—	—	—	—	—
	M	—	—	—	—	—	—	—	—
Kiribati	P	3	3	3	3	3	3	30	32
	M%	—	—	—	—	—	—	—	—
	M	—	—	—	—	—	—	—	—
Marshalls	P	12	11	10	13	12	10	10	11
	M%	—	—	—	0.38	0.38	—	—	—
	M	—	—	—	—	—	—	—	—
Nauru	P	1	1	1	1	2	2	3	3
	M%	—	—	—	—	—	—	—	1.19
	M	—	—	—	—	—	—	—	0
CNMI	P	2	2	2	3	3	3	4	4
	M%	—	—	—	—	0.33	0.33	—	—
	M	—	—	—	—	—	—	—	—
Palau	P	7	6	5	4	4	6	6	6
	M%	—	—	—	—	—	—	—	—
	M	—	—	—	—	—	—	—	—
Total	P	93	71	60	61	65	67	101	108
	M%	—	—	—	0.08	0.12	0.05	0.00	0.03
	M	—	—	—	—	—	—	—	—
	G%	−1.093	−2.600	−1.740	0.198	0.545	0.372	4.071	0.660

Table 6.3c Decennial estimates of the Muslim population (×1000) in Micronesia: 1950–2100 AD (1370–1520 H)

		1950	1960	1970	1980	1990	2000	2010	2020
FSM	P	32	45	61	73	96	107	103	115
	M%	0.00	0.00	0.00	0.00	0.02	0.02	0.00	0.02
	M	—	—	—	—	—	—	—	—
Guam	P	60	67	84	104	130	155	159	169
	M%	—	0.10	0.20	0.40	0.08	0.05	0.03	0.04
	M	—	—	—	—	0.104	0.078	0.048	—
Kiribati	P	33	41	51	59	72	84	103	119
	M%	—	—	—	0.01	0.03	0.06	0.12	0.15
	M	—	—	—	—	—	—	—	—
Marshalls	P	13	15	20	31	47	51	56	59
	M%	—	—	—	—	—	—	—	0.01
	M	—	—	—	—	—	—	—	—
Nauru	P	3	4	7	8	10	10	10	11
	M%	1.19	—	—	—	—	—	—	—
	M	—	—	—	—	—	—	—	—
CNMI	P	7	10	13	17	46	57	54	58
	M%	—	—	0.01	0.02	0.06	1.30	0.93	0.93
	M	—	—	—	—	—	1	1	1
Palau	P	7	10	12	12	15	19	18	18
	M%	—	—	—	—	—	1.05	2.87	2.97
	M	—	—	—	—	—	—	1	1
Total	P	156	191	248	303	417	485	504	549
	M%	0.03	0.04	0.07	0.14	0.04	0.23	0.24	0.25
	M	—	—	—	—	—	1	1	1
	G%	3.682	2.065	2.607	2.011	3.172	1.512	0.381	0.862

Table 6.3c (*Continued*)

		2030	2040	2050	2060	2070	2080	2090	2100
FSM	P	127	135	139	141	139	132	124	115
	M%	0.03	0.04	0.05	0.06	0.07	0.08	0.09	0.10
	M	—	—	—	—	—	—	—	—
Guam	P	181	189	193	193	191	186	178	168
	M%	0.05	0.06	0.07	0.08	0.09	0.10	0.11	0.12
	M	—	—	—	—	—	—	—	—
Kiribati	P	139	157	177	194	209	222	233	239
	M%	0.20	0.25	0.30	0.35	0.40	0.45	0.50	0.55
	M	—	—	1	1	1	1	1	1
Marshalls	P	65	71	75	78	78	75	70	63
	M%	0.02	0.03	0.04	0.05	0.06	0.07	0.08	0.09
	M	—	—	—	—	—	—	—	—
Nauru	P	11	11	11	10	9	8	7	6
	M%	—	—	—	—	—	—	—	—
	M	—	—	—	—	—	—	—	—
CNMI	P	61	63	62	61	59	58	56	53
	M%	0.93	0.93	0.93	0.93	0.93	0.93	0.93	0.93
	M	1	1	1	1	1	1	1	—
Palau	P	18	18	18	17	16	16	15	14
	M%	3.10	3.20	3.30	3.40	3.50	3.60	3.70	3.80
	M	1	1	1	1	1	1	1	1
Total	P	603	645	674	694	702	698	684	659
	M%	0.26	0.27	0.29	0.30	0.32	0.35	0.38	0.41
	M	2	2	2	2	2	2	3	3
	G%	0.932	0.680	0.445	0.282	0.114	−0.047	−0.208	−0.375

6.4 Muslims in Polynesia

This region consists of nine island nations and territories covering over seven million square kilometers of the South Pacific Ocean. These are Cook Islands, French Polynesia, Niue, Samoa, American Samoa, Tokelau, Tonga, Tuvalu, and Wallis and Futuna. The presence of Muslims in this region remains low and is expected to near 4,000 or 0.5% by the end of this century.

A plot of decennial estimates of the Muslim population and its percentage with respect to the total population in this region from 1900 to 2100 is provided in Fig. 6.4. The corresponding individual data for each country in this region is discussed below. In Section 6.4.10, the total population in each country in this region and the corresponding percentage and number of Muslims is presented centennially in Table 6.4a from 600 to 2100 and decennially in Tables 6.4b and 6.4c from 1790 to 2100.

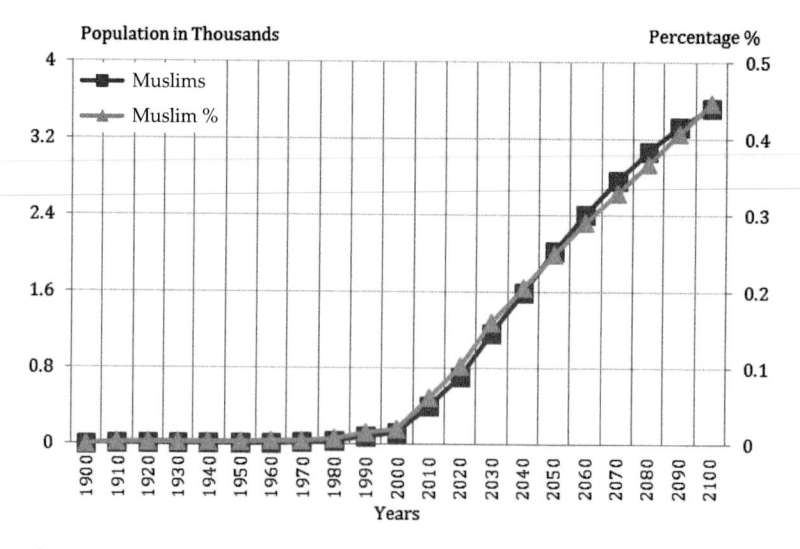

Figure 6.4 Plot of decennial estimates of the Muslim population and its percentage of the total population in Polynesia: 1900–2100 AD (1320–1520 H).

6.4.1 Cook Islands

These are self-governing entity in free association with New Zealand. It has an area of 237 sq. km comprising fifteen major islands spread over 2.2 million sq. km of South Pacific Ocean. Twelve of the islands are inhabited and the largest are Rarotonga (67 sq. km) where the capital Avarua is located and almost three-quarter of the total population, Mangaia (52 sq. km), Atiu (27 sq. km), Mitiaro (22 sq. km), Aitutaki (18 sq. km) and Penrhyn (10 sq. km). A map of these islands is presented in Fig. 6.4.1.

Figure 6.4.1 Map of the Cook Islands.

Based on census data, the first Muslims arrived here in late 1990s. Accordingly, as shown in Table 6.4.1, the Muslim population increased from none in 1996 and before, to three or 0.02% in 2001 to eight or 0.05% in 2006, to 26 or 0.17% in 2011. Thus, assuming that the percentage of Muslims will continue to increase by 0.05 of a percentage point per decade, the Muslim population is expected to remain less than one hundred throughout this century, comprising 0.4% of the total population by 2050 and 0.7% by 2100.

Table 6.4.1 Evolution of the Muslim population in Cook Islands

Year	Population	Muslims	%	Source
1911	12,598	—	0.00	[GB11]c
1991	17,518	—	0.00	[CK6]c
1996	18,071	—	0.00	[CK6]c
2001	15,017	3	0.02	[CK6]c
2006	15,324	8	0.05	[CK6]c
2011	14,974	26	0.17	[CK11]c
2050	17,000	70	0.40	es
2100	14,000	90	0.65	es

6.4.2 French Polynesia

The Overseas Lands of French Polynesia was established as a French protectorate in 1889. The conquest of the Islands by France started with Tahiti and Tahuata in 1842. French Polynesia has an area of 3,521 sq. km, comprising over 130 islands scattered over 2.5 million sq. km of the South Pacific Ocean. The largest of these islands are Tahiti (1,045 sq. km) where over two thirds of the population live and the capital Papeete is located, Nuku-Hiva (388 sq. km), Hiva-Oa (330 sq. km), Raiatea (167 sq. km), Moorea (134 sq. km), Tahaa (90 sq. km), Huahine (75 sq. km), Tubuai (45 sq. km), Rapa (40 sq. km), Rurutu (29 sq. km), and Bora Bora (29 sq. km). A map of these islands is presented in Fig. 6.4.2. As shown in Table 6.4.2, the presence of Muslims is negligible, which increased from none in 1907 and 1967, to about 500 or 0.2% in 2017. Almost all Muslims live in Tahiti, where the first mosque in the archipelago was inaugurated in 2013.

Table 6.4.2 Evolution of the Muslim population in French Polynesia

Year	Population	Muslims	%	Source
1907	30,600	—	0.00	[PGH]es
1967	61,519	—	0.00	[SYB70]es
2017	281,674	500	0.18	[PF]es
2050	311,000	1,600	0.50	es
2100	262,000	2.600	1.00	es

Thus, assuming that the percentage of Muslims will continue to increase by a tenth of a percentage point per decade, the Muslim population is expected to increase to over 1,500 or 0.5% in 2050 and almost 3,000 or 1.0% by 2100.

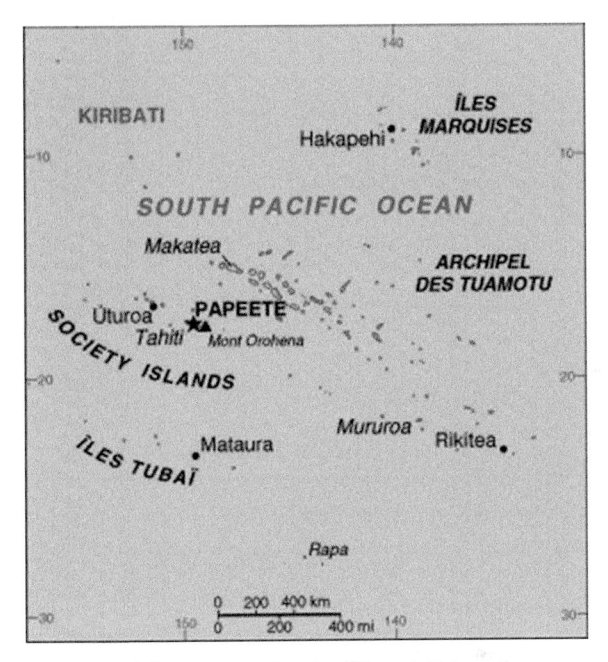

Figure 6.4.2   Map of the Overseas Lands of French Polynesia.

6.4.3   Niue

Is a self-governing Island in free association with New Zealand with total area 259 sq. km and its map is presented in Fig. 6.4.3. With extensive migration to New Zealand, the Island's population is in constant decline from a peak of 5,194 in 1966. The Island is expected to be under sea level by the end of this century. As shown in Table 6.4.3, census data indicate the presence of no Muslims in this island, and the situation is expected to remain the same throughout this century.

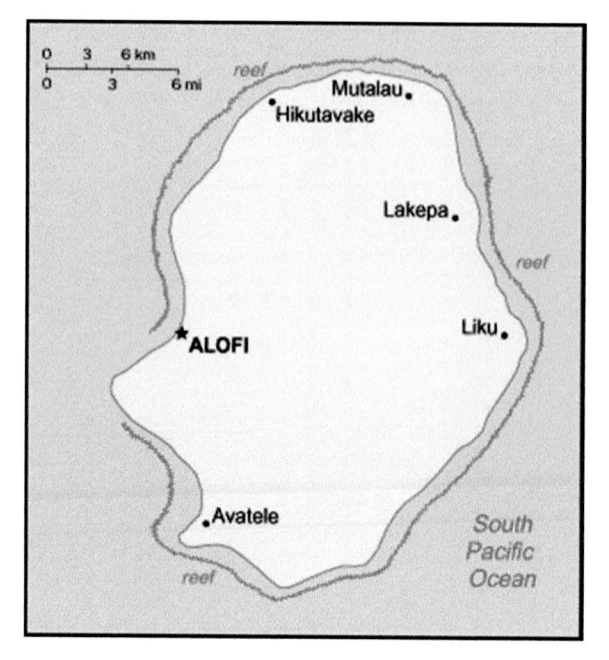

Figure 6.4.3 Map of Niue.

Table 6.4.3 Evolution of the Muslim population in Niue

Year	Population	Muslims	%	Source
1986	2,531	—	0.00	[NU]c
1991	2,239	—	0.00	[NU]c
1997	2,088	—	0.00	[NU]c
2001	1,736	—	0.00	[NU]c
2006	1,538	—	0.00	[NU]c
2011	1,460	—	0.00	[NU]c
2050	1,800	—	0.00	es
2100	1,700	—	0.00	es

6.4.4 Samoa

The Independent State of Samoa changed its name from Western Samoa in 1997. It gained its independence in 1962 from New Zealand, which took it from the Germans in 1914. The total area is 2,934 sq. km comprising two main islands where almost all the population lives and eight much smaller islands. The main islands are Savai'i (1,694 sq. km) and Upolu (1,091 sq. km) where the capital Apia is located and over three quarter of the population lives. The other inhabited islands are Manono (3 sq. km) and Apolima (1 sq. km), located between the two main islands, in Apolima Strait. A map of these islands is presented in Fig. 6.4.4.

In 1907, the Muslim population comprised of five Chinese Muslims, or 0.01% of the total population. According to Census data, the Muslim population increased from less than thirteen in 1945 to less than fifteen in 1951, remaining less than 0.02% of the total population. The number increased to 48 or 0.03% in 2001, then 61 or 0.03% in 2006, but decreased to 38 or 0.02% in 2011 then increased to 87 or 0.04% in 2016. Thus, assuming that the percentage of Muslims will increase by 0.02 of a percentage point per decade, the Muslim population is expected to reach 300 or 0.1% by 2050 and 600 or 0.2% by 2100. The data is summarized in Table 6.4.4.

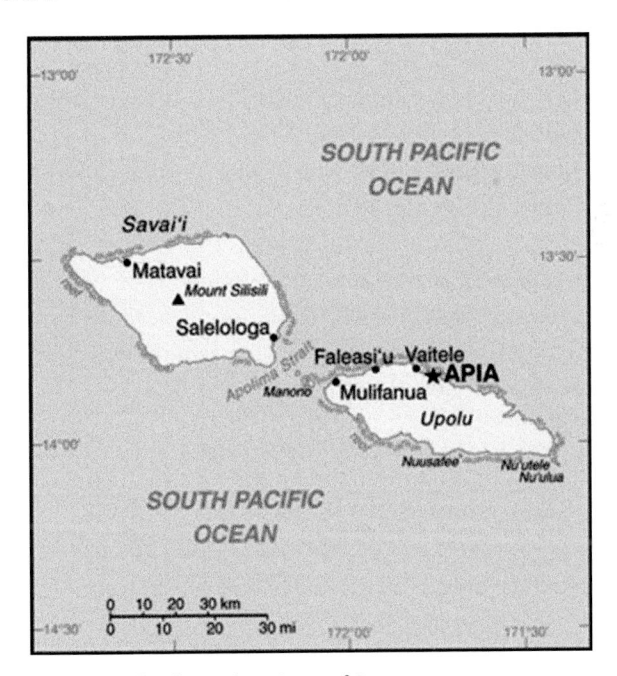

Figure 6.4.4 Map of Independent State of Samoa.

Table 6.4.4 Evolution of the Muslim population in Samoa

Year	Population	Muslims	%	Source
1907	37,000	5	0.01	[PGH]es
1945	67,821	<13	<0.02	[UN56]c
1951	84,909	<15	<0.02	[UN56]c
2001	176,710	48	0.03	[WS01]c
2006	180,741	61	0.03	[WS06]c
2011	186,340	38	0.02	[WS11]c
2016	195,749	87	0.04	[WS16]c
2050	267,000	300	0.10	es
2100	310,000	600	0.20	es

6.4.5 American Samoa

The Territory of American Samoa is an unincorporated and unorganized territory of the United States of America. It has a total area of 199 sq. km, consisting of seven islands, all but one is inhabited. These are Tutuila (142 sq. km), Ta'u (44 sq. km), Ofu (7 sq. km), Olosega (5 sq. km), Aunu'u and Swains and Rose (2 sq. km each) and Rose, which is small and uninhabited. Almost all the population lives in Tutuila, where the capital Pago Pago is located. A map of American Samoa is presented in Fig. 6.4.5.

As shown in Table 6.4.5, since 1990 the Muslim population is estimated at 0.03% of the total population. Thus, assuming that the percentage of Muslims will continue to increase by 0.01 of a percentage point per decade, the Muslim population is expected to remain less than fifty throughout this century.

Table 6.4.5 Evolution of the Muslim population in American Samoa

Year	Population	Muslims	%	Source
1990	46,773	12	0.03	[RIS]es
2010	55,519	15	0.03	[PEW]es
2050	54,000	40	0.08	es
2100	36,000	50	0.13	es

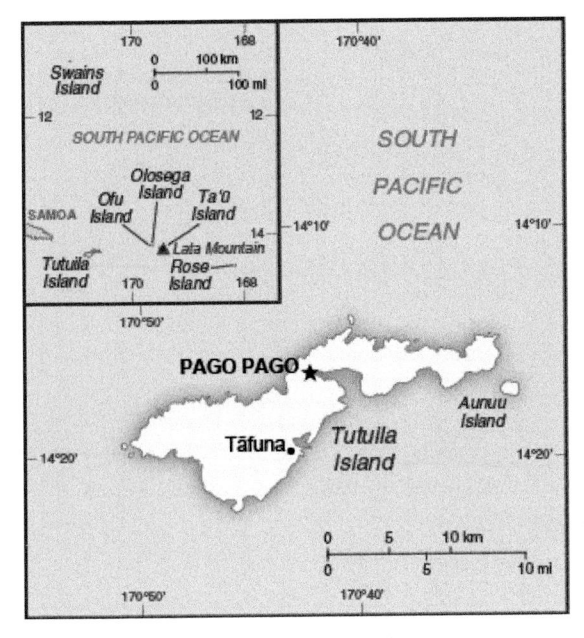

Figure 6.4.5 Map of the Territory of American Samoa.

6.4.6 Tokelau

Is a self-administrating territory of New Zealand with total area 12 sq. km consisting of three atolls: Nukunonu (6 sq. km), Atafu and Fakaofo (3 sq. km each). Each atoll has about a third of the total population. A map of these atolls is presented in Fig. 6.4.6. Its name was changed in 1946 from Union Islands to Tokelau Islands, then in 1976 to Tokelau. As shown in Table 6.4.6, the 2011 census and before show no presence of Muslims in these atolls, and the situation is expected to remain the same throughout this century.

Table 6.4.6 Evolution of the Muslim population in Tokelau

Year	Population	Muslims	%	Source
1945	1,388	—	0.00	[UN56]c
1961	1,870	—	0.00	[UN63]c
2006	1,466	—	0.00	[TK06]c
2011	1,411	—	0.00	[TK16]c
2016	1,499	—	0.00	[TK16]c
2050	1,600	—	0.00	es
2100	1,500	—	0.00	es

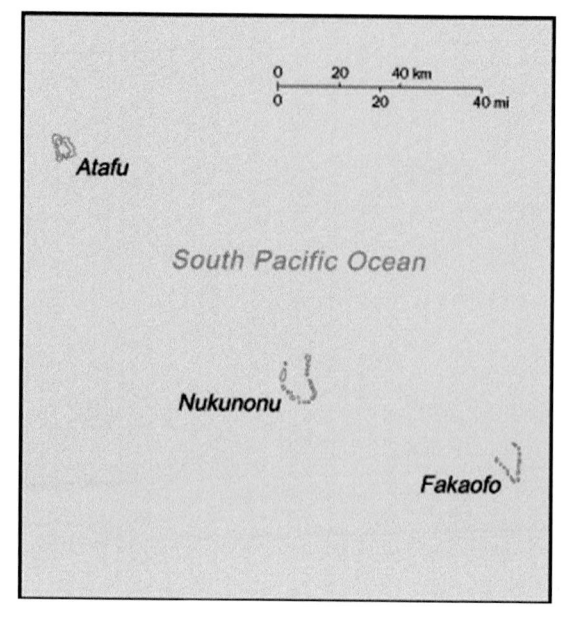

Figure 6.4.6 Map of Tokelau.

6.4.7 Tonga

The Kingdom of Tonga was occupied by the British in 1900 and gained its independence from the UK in 1970. It has a total area of 749 sq. km, comprising 176 islands scattered over 0.7 million sq. km of the South Pacific Ocean. Only 52 of these islands are inhabited, the largest of which is Tongatapu (259 sq. km) with over two-thirds of the total population and the capital Nuku'alofa and where all the Muslim population lives. A map of this island nation is presented in Fig. 6.4.7.

According to census data, the Muslim population increased from none in 1911, to 35 or 0.04% in 1996, to 47 or 0.05% in 2006, but decreased to 24 or 0.02% in 2011. Thus, assuming that the percentage of Muslims will remain fixed, the Muslim population is expected to remain less than fifty throughout this century. The data is summarized in Table 6.4.7.

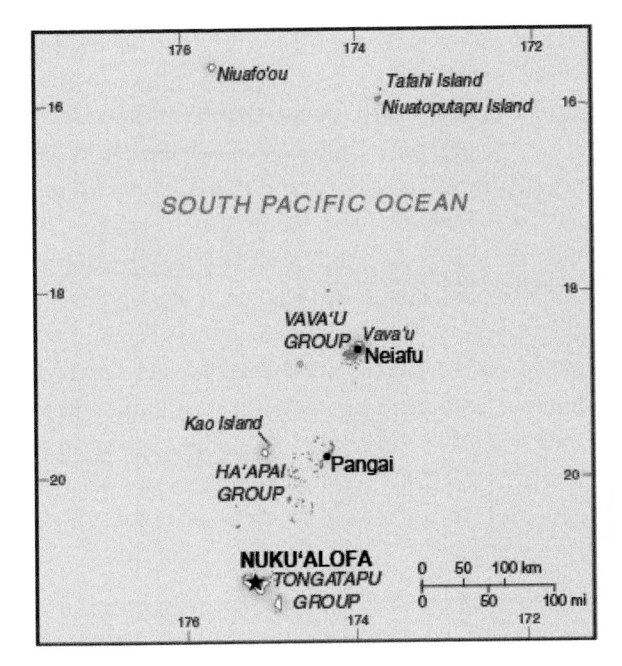

Figure 6.4.7 Map of the Kingdom of Tonga.

Table 6.4.7 Evolution of the Muslim population in Tonga

Year	Population	Muslims	%	Source
1911	23,017	—	0.00	[GB11]c
1996	96,020	35	0.04	[UN]c
2006	101,991	47	0.05	[TO6]c
2011	103,036	24	0.02	[TO11]c
2020	111,000	20	0.02	es
2050	134,000	30	0.02	es
2100	138,000	30	0.02	es

6.4.8 Tuvalu

It was occupied by the British in 1892, changed its name from Ellice Islands in 1974, and gained its independence from the UK in 1978. It has a total area of 26 sq. km, consisting of 124 islands spread over half a million square kilometers of South Pacific Ocean, and nine of which are inhabited. These are from south to north: Niulakita (0.4 sq. km), Nukulaelae (1.8 sq. km), Funafuti (2.8 sq. km) where the capital Funafuti is located and over half of the population lives, Nukufetau (3.0 sq. km), Vaitupu (5.6 sq. km), Nui (2.8 sq. km), Niutao (2.5 sq. km), Nanumaga (2.8 sq. km) and Nanumea (3.9 sq. km). The name "Tuvalu" means "group of eight" referring to the country's eight largest and traditionally inhabited islands. A map of these islands is located in Fig. 6.4.8.

As shown in Table 6.4.8, the presence of Muslims in these islands remains minimal, estimated at ten or 0.1% in 2002 and twenty or 0.2% in 2012. Thus, assuming that the percentage of Muslims will continue to increase by 0.05 of a percentage point per decade, the Muslim population is expected to reach sixty or 0.4% by 2050 and 130 or 0.7% by 2100.

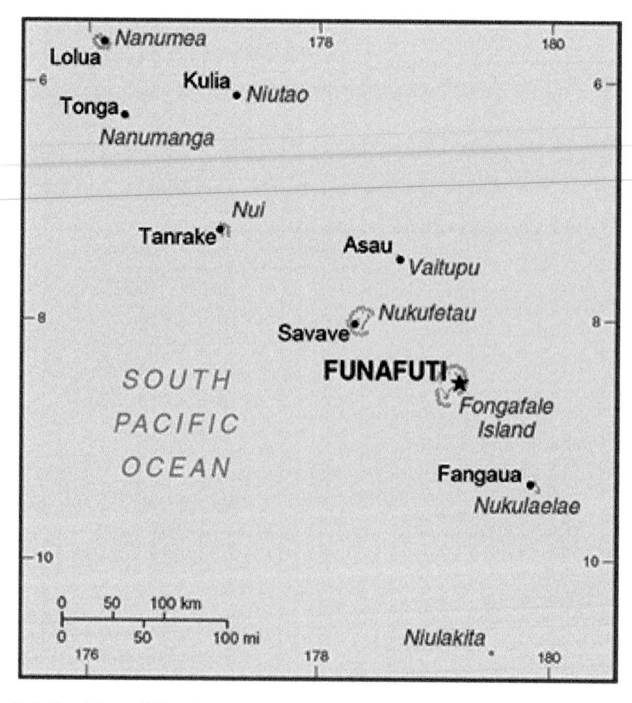

Figure 6.4.8 Map of Tuvalu.

Table 6.4.8 Evolution of the Muslim population in Tuvalu

Year	Population	Muslims	%	Source
2002	9,561	10	0.10	[PEW]es
2012	10,782	20	0.19	[TV]es
2050	16,000	60	0.40	es
2100	20,000	130	0.65	es

6.4.9 Wallis and Futuna

The Territory of the Wallis and Futuna was conquered by France in 1842 and remains an Overseas Territory of France since 1959. It has a total area of 142 sq. km, consisting of three relatively large islands and twenty islets. Only the two main islands of Wallis and Futuna are inhabited. The main islands are Futuna (83 sq. km), where almost a third of the population live, Wallis (78 sq. km), where over two-third of the population lives, and the capital Mata-Utu is located, and Alofi (32 sq. km). A map of these islands is presented in Fig. 6.4.9. As shown in Table 6.4.9, no Muslims live in this French Territory, and the situation is expected to remain the same throughout this century.

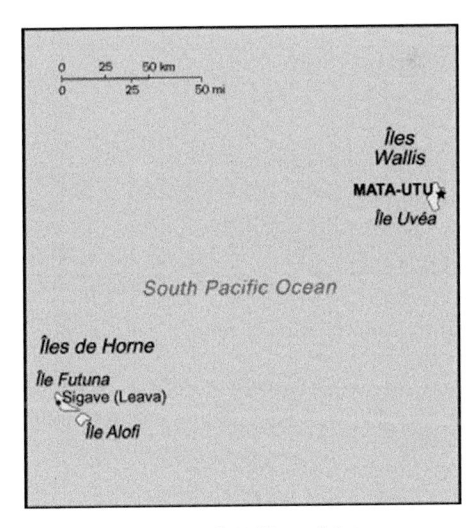

Figure 6.4.9 Map of the Territory of Wallis and Futuna.

Table 6.4a Centennial estimates of the Muslim population (×1000) in Polynesia: 600–2100 AD (1–1500 H)

		600	700	800	900	1000	1100	1200	1300
Cook Islands	P	1	2	3	4	5	6	7	7
	M%	—	—	—	—	—	—	—	—
	M	—	—	—	—	—	—	—	—
French Polynesia	P	27	29	31	33	35	37	40	44
	M%	—	—	—	—	—	—	—	—
	M	—	—	—	—	—	—	—	—
Niue	P	—	—	1	1	1	2	2	3
	M%	—	—	—	—	—	—	—	—
	M	—	—	—	—	—	—	—	—
Samoa	P	12	13	15	17	20	23	26	29
	M%	—	—	—	—	—	—	—	—
	M	—	—	—	—	—	—	—	—
American Samoa	P	—	—	1	1	1	1	1	2
	M%	—	—	—	—	—	—	—	—
	M	—	—	—	—	—	—	—	—
Tokelau	P	—	—	—	—	1	1	1	1
	M%	—	—	—	—	—	—	—	—
	M	—	—	—	—	—	—	—	—
Tonga	P	9	10	11	12	13	14	15	16
	M%	—	—	—	—	—	—	—	—
	M	—	—	—	—	—	—	—	—
Tuvalu	P	—	—	1	1	1	1	1	2
	M%	—	—	—	—	—	—	—	—
	M	—	—	—	—	—	—	—	—
Wallis & Futuna	P	—	—	1	1	1	1	1	2
	M%	—	—	—	—	—	—	—	—
	M	—	—	—	—	—	—	—	—
Total	P	50	56	63	70	78	86	95	104
	M%	—	—	—	—	—	—	—	—
	M	—	—	—	—	—	—	—	—
	G%		0.112	0.117	0.104	0.108	0.101	0.102	0.093

Table 6.4a (Continued)

		1400	1500	1600	1700	1800	1900	2000	2100
Cook Islands	P	8	8	9	9	11	8	18	14
	M%	—	—	—	—	—	—	0.02	0.65
	M	—	—	—	—	—	—	—	—
French Polynesia	P	48	50	55	60	70	31	241	262
	M%	—	—	—	—	—	—	0.00	1.00
	M	—	—	—	—	—	—	—	3
Niue	P	3	4	4	5	5	4	2	2
	M%	—	—	—	—	—	—	—	—
	M	—	—	—	—	—	—	—	—
Samoa	P	32	34	41	46	54	33	174	310
	M%	—	—	—	—	—	—	0.03	0.20
	M	—	—	—	—	—	—	—	1
American Samoa	P	2	2	3	3	4	6	58	36
	M%	—	—	—	—	—	—	0.03	0.13
	M	—	—	—	—	—	—	0	0
Tokelau	P	1	1	1	1	1	1	2	1
	M%	—	—	—	—	—	—	—	—
	M	—	—	—	—	—	—	—	—
Tonga	P	17	18	19	20	21	21	98	138
	M%	—	—	—	—	—	—	0.04	0.02
	M	—	—	—	—	—	—	—	—
Tuvalu	P	2	2	3	3	4	3	9	20
	M%	—	—	—	—	—	—	0.10	0.65
	M	—	—	—	—	—	—	—	—
Wallis & Futuna	P	2	2	3	3	3	6	15	6
	M%	—	—	—	—	—	—	—	—
	M	—	—	—	—	—	—	—	—
Total	P	114	121	137	150	174	111	616	789
	M%	—	—	—	—	—	—	0.02	0.45
	M	—	—	—	—	—	—	—	4
	G%	0.093	0.053	0.125	0.092	0.148	−0.446	1.714	0.246

Table 6.4b Decennial estimates of the Muslim population (×1000) in Polynesia: 1790–1940 AD (1210–1360 H)

		1790	1800	1810	1820	1830	1840	1850	1860
Cook Islands	P	10	11	12	13	14	15	16	15
	M%	—	—	—	—	—	—	—	—
	M	—	—	—	—	—	—	—	—
French Polynesia	P	71	70	69	68	67	66	62	40
	M%	—	—	—	—	—	—	—	—
	M	—	—	—	—	—	—	—	—
Niue	P	5	5	5	5	5	5	4	4
	M%	—	—	—	—	—	—	—	—
	M	—	—	—	—	—	—	—	—
Samoa	P	55	54	53	52	51	50	45	40
	M%	—	—	—	—	—	—	—	—
	M	—	—	—	—	—	—	—	—
American Samoa	P	4	4	4	5	5	5	5	4
	M%	—	—	—	—	—	—	—	—
	M	—	—	—	—	—	—	—	—
Tokelau	P	2	1	1	1	1	1	1	1
	M%	—	—	—	—	—	—	—	—
	M	—	—	—	—	—	—	—	—
Tonga	P	22	21	21	20	20	19	19	18
	M%	—	—	—	—	—	—	—	—
	M	—	—	—	—	—	—	—	—
Tuvalu	P	4	4	3	3	3	3	3	3
	M%	—	—	—	—	—	—	—	—
	M	—	—	—	—	—	—	—	—
Wallis & Futuna	P	3	3	4	4	4	4	4	4
	M%	—	—	—	—	—	—	—	—
	M	—	—	—	—	—	—	—	—
Total	P	175	174	172	170	169	167	158	129
	M%	—	—	—	—	—	—	—	—
	M	—	—	—	—	—	—	—	—
	G%		−0.092	−0.093	−0.094	−0.094	−0.125	−0.505	−2.069

Table 6.4b (Continued)

		1870	1880	1890	1900	1910	1920	1930	1940
Cook Islands	P	14	13	10	8	9	9	10	12
	M%	—	—	—	—	—	—	—	—
	M	—	—	—	—	—	—	—	—
French Polynesia	P	23	25	28	31	32	32	36	44
	M%	—	—	—	—	—	—	—	—
	M	—	—	—	—	—	—	—	—
Niue	P	4	3	2	4	4	4	4	4
	M%	—	—	—	—	—	—	—	—
	M	—	—	—	—	—	—	—	—
Samoa	P	34	35	37	33	38	36	40	56
	M%	—	—	—	—	0.01	0.01	0.01	0.01
	M	—	—	—	—	—	—	—	—
American Samoa	P	3	4	5	6	7	8	10	13
	M%	—	—	—	—	—	—	—	—
	M	—	—	—	—	—	—	—	—
Tokelau	P	1	1	1	1	1	1	1	1
	M%	—	—	—	—	—	—	—	—
	M	—	—	—	—	—	—	—	—
Tonga	P	18	18	19	21	23	25	29	34
	M%	—	—	—	—	—	—	—	—
	M	—	—	—	—	—	—	—	—
Tuvalu	P	3	2	2	3	3	4	4	5
	M%	—	—	—	—	—	—	—	—
	M	—	—	—	—	—	—	—	—
Wallis & Futuna	P	5	5	6	6	6	6	6	6
	M%	—	—	—	—	—	—	—	—
	M	—	—	—	—	—	—	—	—
Total	P	103	106	109	111	123	125	140	175
	M%	—	—	—	—	0.00	0.00	0.00	0.00
	M	—	—	—	—	—	—	—	—
	G%	−2.226	0.259	0.316	0.171	0.995	0.146	1.151	2.262

Table 6.4c Decennial estimates of the Muslim population (×1000) in Polynesia: 1950–2100 AD (1370–1520 H)

		1950	1960	1970	1980	1990	2000	2010	2020
Cook Islands	P	15	18	21	18	18	18	18	18
	M%	—	—	—	—	—	0.02	0.17	0.25
	M	—	—	—	—	—	—	—	—
French Polynesia	P	60	78	110	153	200	241	266	281
	M%	—	—	—	—	—	0.00	0.10	0.18
	M	—	—	—	—	—	—	—	1
Niue	P	5	5	5	3	2	2	2	2
	M%	—	—	—	—	—	—	—	—
	M	—	—	—	—	—	—	—	—
Samoa	P	82	109	143	156	163	174	186	198
	M%	0.01	0.01	0.01	0.01	0.03	0.03	0.02	0.04
	M	—	—	—	—	—	—	—	—
American Samoa	P	19	20	27	33	47	58	56	55
	M%	—	—	0.01	0.02	0.03	0.03	0.04	0.05
	M	—	—	—	—	—	—	—	—
Tokelau	P	2	2	2	2	2	2	1	1
	M%	—	—	—	—	—	—	—	—
	M	—	—	—	—	—	—	—	—
Tonga	P	47	62	84	93	95	98	104	106
	M%	—	—	—	0.01	0.02	0.04	0.02	0.02
	M	—	—	—	—	—	—	—	—
Tuvalu	P	5	5	6	8	9	9	11	12
	M%	—	—	—	—	0.01	0.10	0.19	0.25
	M	—	—	—	—	—	—	—	—
Wallis & Futuna	P	7	8	9	11	14	15	13	11
	M%	—	—	—	—	—	—	—	—
	M	—	—	—	—	—	—	—	—
Total	P	242	307	408	476	550	616	657	684
	M%	0.00	0.00	0.00	0.01	0.02	0.02	0.06	0.10
	M	—	—	—	—	—	—	—	1
	G%	3.212	2.393	2.851	1.537	1.449	1.140	0.635	0.402

Table 6.4c (*Continued*)

		2030	2040	2050	2060	2070	2080	2090	2100
Cook Islands	P	18	17	17	17	16	15	15	14
	M%	0.30	0.35	0.40	0.45	0.50	0.55	0.60	0.65
	M	—	—	—	—	—	—	—	—
French Polynesia	P	297	308	311	307	299	288	276	262
	M%	0.30	0.40	0.50	0.60	0.70	0.80	0.90	1.00
	M	1	1	2	2	2	2	2	3
Niue	P	2	2	2	2	2	2	2	2
	M%	—	—	—	—	—	—	—	—
	M	—	—	—	—	—	—	—	—
Samoa	P	220	244	267	285	300	312	315	310
	M%	0.06	0.08	0.10	0.12	0.14	0.16	0.18	0.20
	M	—	—	—	—	—	—	1	1
American Samoa	P	55	54	54	52	50	46	42	36
	M%	0.06	0.07	0.08	0.09	0.10	0.11	0.12	0.13
	M	—	—	—	—	—	—	—	—
Tokelau	P	1	2	2	2	2	2	2	1
	M%	—	—	—	—	—	—	—	—
	M	—	—	—	—	—	—	—	—
Tonga	P	116	126	134	140	145	146	144	138
	M%	0.02	0.02	0.02	0.02	0.02	0.02	0.02	0.02
	M	—	—	—	—	—	—	—	—
Tuvalu	P	13	15	16	17	18	19	20	20
	M%	0.30	0.35	0.40	0.45	0.50	0.55	0.60	0.65
	M	—	—	—	—	—	—	—	—
Wallis & Futuna	P	10	10	9	8	8	7	6	6
	M%	—	—	—	—	—	—	—	—
	M	—	—	—	—	—	—	—	—
Total	P	732	776	811	831	840	837	819	789
	M%	0.16	0.21	0.25	0.29	0.33	0.37	0.41	0.45
	M	1	2	2	2	3	3	3	4
	G%	0.676	0.595	0.439	0.236	0.106	−0.028	−0.216	−0.382

Table 6.4.9 Evolution of the Muslim population in Wallis and Futuna

Year	Population	Muslims	%	Source
1907	6,000	—	0.00	[PGH]es
2013	12,197	—	0.00	[PEW]es
2050	9,000	—	0.00	es
2100	6,000	—	0.00	es

6.4.10 Regional Summary and Conclusion

Polynesia has the least concentration of Muslims among the four regions spanning Oceania and has the least number of Muslims in the world. The number of Muslims is increasing slowly but is expected to remain less than 0.1% throughout this century. The following tables present centennial data from 600 AD to 2100 AD (or approximately 1 H to 1500 H) in Table 6.4a and decennial data from 1790 AD to 2100 AD (or 1210 H to 1520 H) in Tables 6.4b and 6.4c for current countries in Polynesia. The data includes total population in thousands (P), the percentage of which is Muslim (M%), the corresponding Muslim population in thousands (M), and the annual population growth rate (APGR, or G%) of the total population in this region.

6.5 Oceania's Summary and Conclusion

Oceania was that last continent to which Muslims arrived, and it remains the least populated by Muslim in number and percentage. The number of Muslims started increasing in the 1960s, passed 1% of the total population at the turn of this century and is expected to near 5% by the end of this century. The following tables present centennial data from 600 AD to 2100 AD (or approximately 1 H to 1500 H) in Table 6.5a and decennial data from 1790 AD to 2100 AD (or 1210 H to 1520 H) in Tables 6.5b and 6.5c for the four regions of Oceania. The data includes total population in thousands (P), the percentage of which is Muslim (M%), the corresponding Muslim population in thousands (M), and the annual population growth rate (APGR, or G%) of the total population in Oceania and each of its five regions.

Table 6.5a Centennial estimates of the Muslim population (×1000) in Oceania: 600–2100 AD (1–1500 H)

		600	700	800	900	1000	1100	1200	1300	1400	1500	1600	1700	1800	1900	2000	2100
Australasia	P	386	392	398	404	410	430	460	490	520	550	550	550	410	4,590	22,850	48,885
	M%	—	—	—	—	—	—	—	—	—	—	—	—	—	0.08	1.50	6.82
	M	—	—	—	—	—	—	—	—	—	—	—	—	—	4	343	3,332
Melanesia	P	462	486	511	537	564	592	621	652	684	718	753	790	868	910	7,473	24,583
	M%	—	—	—	—	—	—	—	—	—	—	—	—	—	0.22	0.80	0.42
	M	—	—	—	—	—	—	—	—	—	—	—	—	—	2	60	103
Micronesia	P	55	58	64	70	78	86	95	104	114	125	137	151	166	108	485	659
	M%	—	—	—	—	—	—	—	—	—	—	—	—	—	0.08	0.23	0.41
	M	—	—	—	—	—	—	—	—	—	—	—	—	—	1	1	3
Polynesia	P	50	56	63	70	78	86	95	104	114	121	137	150	174	111	616	789
	M%	—	—	—	—	—	—	—	—	—	—	—	—	—	—	0.02	0.45
	M	—	—	—	—	—	—	—	—	—	—	—	—	—	—	—	4
Total	P	953	992	1,036	1,090	1,139	1,202	1,280	1,360	1,443	1,526	1,590	1,656	1,560	5,672	31,425	74,916
	M%	—	—	—	—	—	—	—	—	—	—	—	—	—	0.10	1.29	4.59
	M	—	—	—	—	—	—	—	—	—	—	—	—	—	6	404	3,441
	G%	0.041	0.043	0.051	0.044	0.055	0.063	0.060	0.059	0.056	0.042	0.040	−0.060	1.291	1.712	0.869	

Table 6.5b Decennial estimates of the Muslim population (×1000) in Oceania: 1790–1940 AD (1210–1360 H)

		1790	1800	1810	1820	1830	1840	1850	1860	1870	1880	1890	1900	1910	1920	1930	1940
Australasia	P	427	410	392	389	400	492	693	1,308	1,920	2,787	3,852	4,590	5,513	6,707	8,068	8,802
	M%	—	—	0.00	0.01	0.02	0.03	0.03	0.04	0.02	0.02	0.05	0.08	0.07	0.04	0.03	0.03
	M	—	—	—	—	—	—	—	1	—	1	2	4	4	3	2	3
Melanesia	P	880	868	860	850	865	880	845	799	788	751	750	910	1,074	1,257	1,438	1,717
	M%	—	—	—	—	—	—	—	—	—	0.02	0.11	0.22	0.69	0.72	1.17	1.27
	M	—	—	—	—	—	—	—	—	—	—	1	2	7	9	17	22
Micronesia	P	105	108	110	112	114	115	110	103	93	71	60	61	65	67	101	108
	M%	—	—	—	—	—	—	—	—	—	—	—	0.08	0.12	0.05	0.00	0.03
	M	—	—	—	—	—	—	—	—	—	—	—	—	—	—	—	—
Polynesia	P	175	174	172	170	169	167	158	129	103	106	109	111	123	125	140	175
	M%	—	—	—	—	—	—	—	—	—	—	—	—	0.00	0.00	0.00	0.00
	M	—	—	—	—	—	—	—	—	—	—	—	—	—	—	—	—
Total	P	1,587	1,560	1,534	1,521	1,548	1,654	1,806	2,339	2,904	3,715	4,771	5,672	6,775	8,156	9,746	10,801
	M%	—	—	0.00	0.00	0.00	0.01	0.01	0.02	0.01	0.02	0.06	0.10	0.17	0.15	0.20	0.23
	M	—	—	—	—	—	—	—	1	—	1	3	6	12	12	19	25
	G%	0.000	−0.172	−0.170	−0.083	0.175	0.663	0.882	2.585	2.163	2.463	2.502	1.729	1.777	1.856	1.781	1.028

Oceania's Summary and Conclusion | 861

Table 6.5c Decennial estimates of the Muslim population (×1000) in Oceania: 1950–2100 AD (1370–1520 H)

		1950	1960	1970	1980	1990	2000	2010	2020	2030	2040	2050	2060	2070	2080	2090	2100
Australasia	P	10,085	12,615	15,611	17,735	20,359	22,850	26,525	30,322	33,350	36,005	38,422	40,677	42,798	44,828	46,826	48,885
	M%	0.03	0.03	0.16	0.50	0.85	1.50	2.24	2.76	3.27	3.77	4.28	4.79	5.30	5.80	6.31	6.82
	M	3	4	25	89	173	343	595	837	1,090	1,359	1,645	1,948	2,266	2,601	2,954	3,332
Melanesia	P	2,493	2,909	3,655	4,698	5,973	7,473	9,188	11,123	13,234	15,387	17,469	19,410	21,148	22,610	23,755	24,583
	M%	1.12	1.11	1.23	1.11	0.99	0.80	0.63	0.52	0.49	0.46	0.44	0.43	0.43	0.42	0.42	0.42
	M	28	32	45	52	59	60	58	58	65	71	77	83	90	95	99	103
Micronesia	P	156	191	248	303	417	485	504	549	603	645	674	694	702	698	684	659
	M%	0.03	0.04	0.07	0.14	0.04	0.23	0.24	0.25	0.26	0.27	0.29	0.30	0.32	0.35	0.38	0.41
	M	—	—	—	—	—	1	1	1	2	2	2	2	2	2	3	3
Polynesia	P	242	307	408	476	550	616	657	684	732	776	811	831	840	837	819	789
	M%	0.00	0.00	0.00	0.01	0.02	0.02	0.06	0.10	0.16	0.21	0.25	0.29	0.33	0.37	0.41	0.45
	M	—	—	—	—	—	—	—	1	1	2	2	2	3	3	3	4
Total	P	12,976	16,022	19,922	23,212	27,299	31,425	36,873	42,678	47,919	52,814	57,376	61,612	65,486	68,974	72,084	74,916
	M%	0.24	0.23	0.35	0.61	0.85	1.29	1.78	2.10	2.41	2.71	3.01	3.30	3.61	3.92	4.24	4.59
	M	31	37	70	141	232	404	655	897	1,157	1,433	1,726	2,036	2,362	2,701	3,060	3,441
	G%	1.834	2.108	2.179	1.528	1.622	1.408	1.599	1.462	1.158	0.973	0.829	0.712	0.610	0.519	0.441	0.385

References

[AU] Australian Bureau of Statistics (ABS) (1966–2017). *Census of Population and Housing*. Canberra: Commonwealth Government Printer.

[AUH1] Australian Data Archive (ADA) (2013). *Historical Census and Colonial Data Archive (HCCDA)* [Online Collection of Censuses Taken in Australia Up to 1901]. Retrieved from http://hccda.anu.edu.au/

[AUH2] Commonwealth Bureau of Census and Statistics (CBCS) (1911–1961). *Census of the Commonwealth of Australia*. Melbourne: Government Printer.

[CK11] Cook Islands Statistics Office (CISO) (2012). *Cook Islands 2011 Census of Population and Dwellings–Main Report*. Avarua: CISO. And e-mail correspondence with Ms. Tangimetua from CISO in June 2013.

[CK6] Cook Islands Statistics Office (CISO) (2008). *2006 Census of Population and Dwellings*. Avarua: CISO. And e-mail correspondence with Mr. K. Hosking from CISO in October/November 2009.

[FJ] Fiji Bureau of Statistics (FBS) (2012). Table 1.2A: Census Population of Fiji by Ethnicity & Table 1.10: Population by religion and Province of Enumeration, Fiji: 2007 Census. *Key Statistics*. Suva: FBS.

[FJ17] Fiji Bureau of Statistics (FBS) (2019). *Fiji 2017 Population and Housing Census*. Suva: FBS. Electronic file obtained through email correspondence with Mr. M. Qaloewai from FBS in July 2019.

[FM] FSM Office of Statistics, Budget, Overseas Development Assistance and Compact Management (SBOC) (2012). *Summary Analysis of Key Indicators from the FSM 2010 Census of Population and Housing*. Palikir: SBOC. And Version 1.0 of the public use dataset (April 2001), provided by the National Data Archive.

[GB911] His Majesty's Stationary Office (HMSO) (1917). *Census of England and Wales. 1911.* (General Report with Appendices). London: Darling & Son.

[GB871] Her Majesty's Stationary Office (HMSO) (1873). *Census of England and Wales for the Year 1871* (General Report. Vol. IV). London: George Edward Eyre and William Spottiswoode.

[GU] Babauta, L. (2009). Muslim Association of Guam, Creating Understanding and Harmony. *Guampedia*. [Online] Retrieved from http://guampedia.com/muslim-association-of-guam/

[JAN] Jansen, H. (1897). *Verbreitung des Islâms: mit angabe der verschiedenen riten, sekten und religiösen bruderschaften in den verschiedenen* ländern der erde 1890 bis 1897. Mit benutzung der neuesten angaben (zählungen, berechnungen, schätzungen und vermutungen). [German for Dissemination of Islam: with number of various rites, sects and religious brotherhoods in the various countries of the earth 1890 to 1897. With use of information on the latest disclosures (censuses, calculations, estimates and guesses)]. Berlin: Selbstverlag des Verfassers.

[KET86] Kettani, A. (1986). *Muslim Minorities in the World Today.* London: Mansell Publishing.

[KI15] National Statistics Office (NSO) (2016). *2015 Population and Housing Census* (Vol.1: Management Report and Basic Tables). Bairiki: NSO.

[KI10] National Statistics Office (NSO) (2012). *Report on the Kiribati 2010 Census of Population and Housing* (Vol. 1: Basic Information and Tables). Bairiki: NSO.

[MAS23] Massignon, L. (1923). *Annuaire du Monde Musulman, Statistique, Historique, Social et Économique.* Paris: Ernest Leroux.

[MH] Economic Policy, Planning, and Statistics Office (EPPSO) (2012). The RMI 2011 Census of Poplation and Housing Summary and Highlights Only. Majuro: EPPSO. And Version 1.0 of the public use dataset (April 2001), provided by the National Data Archive.

[MP10] Central Statistics Division (CSD) (2012). *2010 Census Demographic Profile Summary by District.* Saipan: CSD.

[MP00] Central Statistics Division (CSD) (2003). *The U.S. Commonwealth of the Northern Mariana Islands (CNMI) Statistical Yearbook 2002.* Saipan: CSD.

[NC] Association des Musulmans de Nouvelle-Calédonie (AMNC) (2017). *La communauté musulmane de Nouvelle-Calédonie.* AMNC Website, retrieved from http://www.amnc.org/

[NCH] Lonely Planet Publications (LPP) (2009). *Vanuatu & New Caledonia* (6th ed.). London: LPP.

[NF] Census Statistics of Norfolk Island (CSNI) (2011). *Norfolk Island Report on the 2011 Census on Population and Housing: Census Description, Analysis and Basic Tables.* Kingston: Photopress International. And e-mail correspondence with Mr. A. McNeal from CSNI in January 2012.

[NR] The Secretariat of the Pacific Community (SPC) Statistics and Demographic Programme & Nauru Bureau of Statistics (NBS) (2006). *2002 Nauru Census Main Report & Demographic Profile of the Republic of Nauru, 1992–2002.* Noumea: SPC. And e-mail correspondence with Mrs. I. Gadabu and Ms. L. Thoma from NBS in October 2009 and February 2013, respectively.

[NU] Niue Economics, Planning, Development and Statistics Unit & The Secretariat of the Pacific Community (SPC) Statistics and Demographic Programme (2008). *Niue Population Profile Based on 2006 Census of Population and Housing: A Guide for Planners and Policy-Makers*. Auckland: Ultimo Group. And e-mail correspondence with Mr. K. Vaha from Statistics Niue in June 2013.

[NZ] Statistics New Zealand (SNZ) (2014). *QuickStats about Culture and Identity–2013 Census*. Willington: SNZ.

[NZ861] Registrar General's Office (RGO) (1862). Census results and general statistics of New Zealand for 1861. Auckland: RGO.

[NZ18] Statistics New Zealand (SNZ) (2019). 2018 *Census totals by topic-national highlights*. Willington: SNZ.

[NZ13] Statistics New Zealand (SNZ) (2014). *QuickStats about Culture and Identity–2013 Census*. Willington: SNZ.

[NZH] Shepard, W. (2006). New Zealand Muslims and Their Organizations. *New Zealand Journal of Asian Studies*, 8(2), 8–44.

[OCE] Caldwell, J., Missingham, B., & Marck, J. (2001). The Population of Oceania in the Second Millennium. *Proceedings of the International Union for the Scientific Study of Population's Seminar on the History of World Population in the Second Millennium*, Florence, Italy, pp. 1–34.

[PEW] Pew Research Center (PRC) (2009). *Mapping the Global Muslim Population, A Report on the Size and Distribution of the World's Muslim Population*. Washington, DC: PRC.

[PF] Minority Rights Group International (2014, July 3). *State of the World's Minorities and Indigenous Peoples 2014 - Case study: Tahiti: Islamophobia in French Polynesia*. Available at: http://www.refworld.org/docid/53ba8dbb5.html

[PG11] Zocca, F. (2014, November 19). Islam in Papua New Guinea. *Catholic Reporter PNG*. Retrieved from https://catholicreporterpng.wordpress.com/2014/11/19/islam-in-png/

[PG00] Gibbs, P. (2006). Papua New Guinea. In M. Ernst (ed.), *Globalization and the Re-Shaping of Christianity in the Pacific Islands*. (pp. 81–158). Suva: Pacific Theological College.

[PGH] Djinguiz, M. (1908). L'Islam en Australie et en Polynésie. *Revue du Monde Musulman*. 4(1), 75–85.

[PSH] Lahmeyer, J. (2006). Population Statistics: Historical Demography of All Countries, Their Divisions and Towns [Online Database]. Retrieved from www.populstat.info

[PW15] Palau Statistics Office (PSO) (2016). *2015 Census of Population. Housing and Agriculture Tables*. Koror: PSO

[PW10] United States Department of State (DOS) (2013). *2012 Country Reports on Human Rights Practices–Palau*. Washington, DC: DOS.

[PW07] Freedom House (2007). Palau. *Freedom in the World Annual Report.* Washington DC: Freedom House.

[PW04] Bureau of Budget and Planning (BBP) (2004). *2002–2003 Statistical Yearbook.* Koror: BBP.

[PW00] Office of Planning and Statistics (OPS) (2000). *1999 Yearbook Tables: The Republic of Palau.* Koror: OPS.

[RIS] Regional Islamic Dawah Council of Southeast Asia and the Pacific (RISEAP) (1996). *Muslim Almanac: Asia Pacific*. Kuala Lumpur: Berita Publishing.

[SB] Solomon Islands National Statistics Office (SINSO) (2011). Report on 2009 Population & Housing Census (Vol. I: Basic Tables and Census Description). Honiara: SINSO. And Version 1.0 of the public use dataset (April 2001), provided by the National Data Archive.

[SYB70] Paxton, J. (ed.) (1970). *The Statesman's Year-Book.* London: MacMillan.

[SYB50] Steinberg, S. H. (ed.) (1950). *The Statesman's Year-Book.* London: MacMillan.

[SYB30] Epstein, M. (ed.) (1930). *The Statesman's Year-Book.* London: MacMillan.

[SYB22] Scott-Keltie, J. & Epstein M. (eds.) (1922). *The Statesman's Year-Book.* London: MacMillan.

[SYB17] Scott-Keltie, J. (ed.) (1917). *The Statesman's Year-Book.* London: MacMillan.

[TK16] Statistics New Zealand, & Tokelau National Statistics Office (TNSO) (2016). *Final Population Counts: 2016 Tokelau Census.* Apia: TNSO. And e-mail correspondence with Dr. J. Jasperse from TNSO in January 2017.

[TK06] Statistics New Zealand, & the Office for the Council of the Ongoing Government of Tokelau (OCOGT) (2006). *Tokelau 2006 Census of Population and Dwellings–Tabular Report.* Atafu: OCOGT. And e-mail correspondence with Mr. K. Kelekolio from Tokelau Statistics Unit of OCOGT in October 2009.

[TO11] Statistics Department Tonga (2012). *Tonga National Population and Housing Census 2011–Priliminary Result.* Nuku'alofa: SDT. And e-mail correspondence with Mr. S. Lolohea from SDT in June 2013.

[TO06] Statistics Department Tonga (2008). *Tonga 2006 Census of Population and Housing* (Vol. 1: Administrative Report and Basic Tables). Nuku'alofa: SDT.

[TV] E-mail correspondence with Ms. G. Alapati from Tuvalu Statistics in June 2013.

[UN] United Nations Statistics Division, UNSD Demographic Statistics (2018). Population by Religion, Sex and Urban/Rural Residence. *UNdata* [Online Database]. Retrieved from http://data.un.org

[UN04] United Nations, Department of International Economic and Social Affairs, Statistical Office (2006). Table 6 - Population by Religion, Sex and Urban/Rural Residence: Each Census, 1985–2004. *Demographic Yearbook Special Census Topics* (Vol. 2b: Ethnocultural Characteristics). New York, NY: United Nations.

[UN83] United Nations, Department of International Economic and Social Affairs, Statistical Office (1985). Table 29 - Population by Religion and Sex: Each Census, 1974–1983. *Demographic Yearbook 1983*. New York, NY: United Nations.

[UN71] United Nations, Department of International Economic and Social Affairs, Statistical Office (1972). Table 17 - Population by Religion, Sex and Urban/Rural Residence: Each Census, 1962–1971. *Demographic Yearbook 1971*. New York, NY: United Nations.

[UN63] United Nations, Department of International Economic and Social Affairs, Statistical Office (1964). Table 11 - Population by Religion and Sex: Each Census, 1955–1963. *Demographic Yearbook 1963*. New York, NY: United Nations.

[UN56] United Nations, Department of International Economic and Social Affairs, Statistical Office (1957). Table 8 - Population by Religion and Sex: Each Census, 1945–1955. *Demographic Yearbook 1956*. New York, NY: United Nations.

[VU09] Vanuatu National Statistics Office (VNSO) (2011). *2009 National Population and Housing Census* (Vol. 1: Basic Tables Report). Port Vila: VNSO. And e-mail correspondence with VNSO in September 2011.

[VU] KCNA (2006, December 14). Muslims in Vanuatu (Interview with M. Kaloas). *Kavkaz Center News Agency*.

[WS16] Samoa Bureau of Statistics (SBS) (2017). *2016 Census Brief No. 1: Population Snapshot and Housing Highlights*. Apia: SBS.

[WS11] Samoa Bureau of Statistics (SBS) (2012). *Population and Housing Census 2011 Analytical Report*. Apia: SBS.

[WS06] Samoa Bureau of Statistics (SBS) (2008). *Samoa Population and Housing Report 2006*. Apia: SBS.

[WS01] Samoa Statistics Department (SSD) (2002). *2001 Census of Population and Housing–Main Report*. Apia: SSD.

Chapter 7

World Summary

A summary for each continent for the purpose of comparison is presented in Tables 7.0a to 7.0c. These tables present data from 600 to 2100, showing the total population in each of the world's five continents, the corresponding percentage and number of Muslims, and the ratio of Muslims in each continent. Accordingly, the global data show that the world Muslim population has likely increased from 8.9 million or 3.5% of the total world population in 700 AD, to 19 million or 7.2% in 800 AD, to 28 million or 10.5% in 900 AD, to 36 million or 13.1% in 1000 AD, to 41 million or 13.4% in 1100 AD, to 46 million or 13.6% in 1200 AD, to 52 million or 14.2% in 1300 AD, to 58 million or 14.4% in 1400 AD, to 66 million or 15.1% in 1500 AD, to 84 million or 15.2% in 1600 AD, to 97 million or 16.0% in 1700 AD, to 122 million or 13.0% in 1800 AD, to 217 million or 13.7% in 1900, to 1.32 billion or 21.5% in 2000, to 2.0 billion or 25.2% in 2020, and is projected to reach 2.9 billion or 30.1% by 2050 and 3.9 billion or 36% by 2100.

Thus, the percentage of Muslims with respect to the world population has likely increased from 3% in 700 to 7% in 800, to 11% in 900, to 13% in 1000 to 1100, to 14% in 1200 to 1400, then 15% in 1500 and 1600 and 16% in 1700. However, the percentage of Muslims dropped to 13% in 1800 and increased to 14% in

---

The World Muslim Population: Spatial and Temporal Analyses
Houssain Kettani
Copyright © 2020 Jenny Stanford Publishing Pte. Ltd.
ISBN 978-981-4800-31-0 (Hardcover), 978-0-429-42253-9 (eBook)
www.jennystanford.com

1900. Nevertheless, a remarkable trend happened after World War II, by which the rate of increase in each decade became over one percentage point. This caused the percentage of Muslims with respect to the world population to reach 25% in 2020. This rate of increase is expected to be reduced to less than 0.5% towards the end of this century. Accordingly, Muslims increased from one out of eight of the world population in 1840, to one out of seven in 1950, to one out of six in 1970, to one out of five in 1990, to one out of four in 2020, to one out of three by 2075. A plot of centennial estimates of the world population, the world Muslim population and its percentage of the world population from is provided in Fig. 7.0a. A zoom in of this plot, providing a plot of decennial estimates of the world population, the world Muslim population and its percentage of the world population from 1900 to 2100 is provided in Fig. 7.0b. The latter shows a fast increase in the world population, and faster rate of the world Muslim population between 1950 and 2050, but the rate of increase is expected to decrease afterwards.

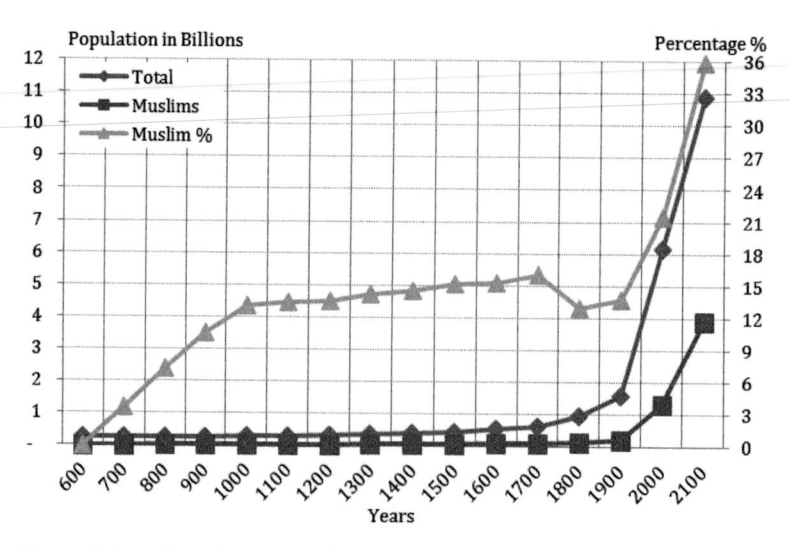

Figure 7.0a Plot of centennial estimates of the world population, world Muslim population and its percentage of the world population: 600–2100 AD (1–1500 H).

Most Muslims live in Asia, but this fraction of Muslims decreased from 75% throughout the nineteenth century and the first half of the twentieth century, to 70% towards the end of the twentieth century, to 67% in 2020, and is projected to reach 60% by 2050, and 48% by 2100. Africa comes second; with the fraction of Muslims living in it increased from 21% throughout the nineteenth century, to 27% towards the end of the twentieth century, to 30% in 2020, and is projected to reach 37% by 2050, and 49% by 2100. The fraction of the world Muslim population living in Europe was between 3% and 4% for the last two centuries but decreases to 2% throughout this century. As for the Americas, since 1950 less than 0.4% of the world Muslim population lives there and is expected to increase to 0.6% by the end of this century. Finally, less than 0.1% of the world Muslim population lives in Oceania and is expected to remain so throughout this century.

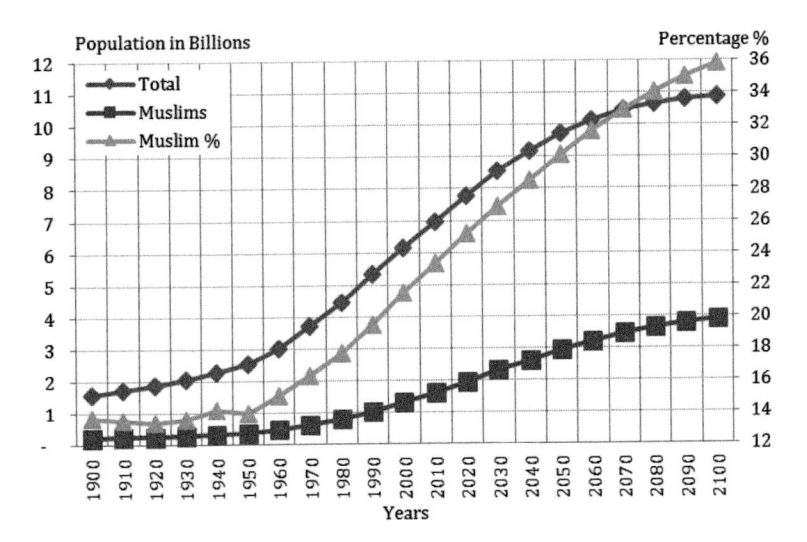

Figure 7.0b Plot of decennial estimates of the world population, world Muslim population and its percentage of the world population: 1900–2100 AD (1320–1520 H).

A choropleth map of the world illustrating the presence of Muslims in each country is presented in Fig. 7.0c. There are fifty Muslim majority countries in which 79% of the world Muslim population lives. So, a fifth of the world Muslim population lives as a minority.

Since the 1970, the world population annual growth (APGR) has been decreasing constantly, from 2.0% to 1.2% in the 2010. The corresponding APGR for the world Muslim population has also been decreasing as well, from 2.8% in 1970 to 1.2% in 2010. However, Muslim's growth is about 50% higher than the world's growth. This explains why the representation of Muslims with respect to the total world population is increasing at over one percentage point each decade. Both the world's and Muslim's APGR are expected to decrease substantially towards the end of this century, reaching 0.06% for the former, and 0.29% for the latter, which is about a triple. Thus, the growth of the fraction of Muslims to the world population is expected to remain around one percentage point per decade throughout this century.

Every attempt is sought to present reliable data, however, the statistics presented in this book, in the words of the French demographer Jean-Baptiste Moheau (1745–1794): "These estimates based on likelihoods, constitute a first step to the truth, and the only proper way to criticize them, is to displace them by more accurate ones."

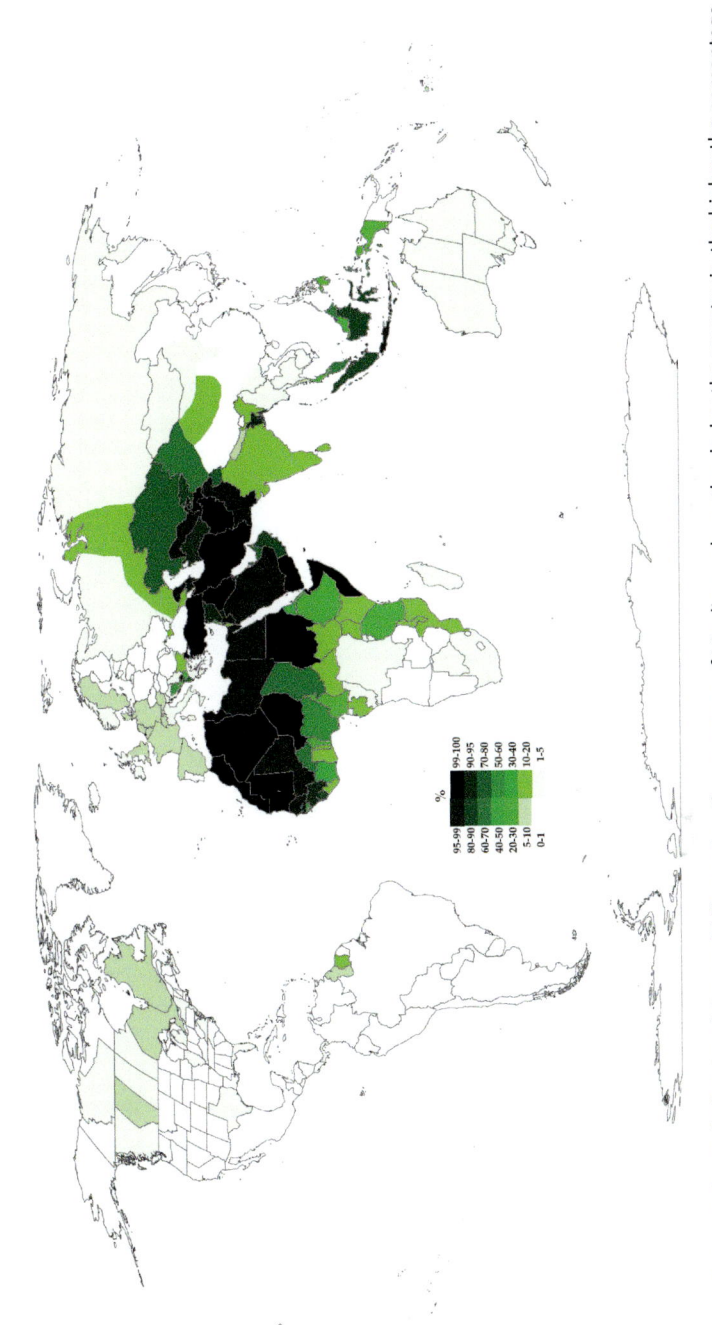

Figure 7.0c A choropleth map of the world illustrating the presence of Muslims, where the darker the country is, the higher the percentage of Muslims within it, whereas countries in white have less than 1% of their population as Muslims.

Table 7.0a Centennial estimates of the world Muslim population (×1000): 600–2100 AD (1–1500 H). *P*: Total population, *M*: Muslim population, *M%*: Percentage of Muslim population, *CG%*: Continental APGR, *MG%*: Muslim APGR, *MR%*: Muslim Ratio

		600	700	800	900	1000	1100	1200	1300
Asia	*P*	175,995	178,891	181,785	184,678	187,543	207,029	226,601	246,199
	M%	—	4.25	7.00	9.21	11.26	12.03	12.84	14.22
	M	—	7,599	12,727	17,015	21,111	24,897	29,085	35,020
	CG%		0.016	0.016	0.016	0.015	0.099	0.090	0.083
	MG%			0.516	0.290	0.216	0.165	0.155	0.186
	MR%		85.69	68.10	60.74	59.27	61.13	63.70	66.87
Africa	*P*	26,465	27,908	29,371	30,833	32,312	34,889	37,643	40,508
	M%	—	4.46	17.22	29.65	37.47	37.60	36.38	35.48
	M	—	1,244	5,058	9,142	12,108	13,118	13,695	14,371
	CG%		0.053	0.051	0.049	0.047	0.077	0.076	0.073
	MG%			1.402	0.592	0.281	0.080	0.043	0.048
	MR%		14.03	27.07	32.63	33.99	32.21	29.99	27.44
Europe	*P*	34,398	34,892	35,383	35,876	36,369	45,011	54,108	62,811
	M%	—	0.07	2.55	5.17	6.60	6.03	5.32	4.75
	M	—	24	903	1,856	2,402	2,713	2,877	2,980
	CG%		0.014	0.014	0.014	0.014	0.213	0.184	0.149
	MG%			3.614	0.720	0.258	0.122	0.059	0.035
	MR%		0.27	4.83	6.63	6.74	6.66	6.30	5.69

Region		1	2	3	4	5	6	7	8
Americas	P	11,310	12,201	13,058	13,923	14,791	16,263	17,764	19,283
	M%	—	—	—	—	—	—	—	—
	M	—	—	—	—	—	—	—	—
	CG%		0.076	0.068	0.064	0.061	0.095	0.088	0.082
	MG%		—	—	—	—	—	—	—
	MR%		—	—	—	—	—	—	—
Oceania	P	953	992	1,036	1,090	1,139	1,202	1,280	1,360
	M%	—	—	—	—	—	—	—	—
	M	—	—	—	—	—	—	—	—
	CG%		0.041	0.043	0.051	0.044	0.055	0.063	0.060
	MG%		—	—	—	—	—	—	—
	MR%		—	—	—	—	—	—	—
Total	P	249,120	254,884	260,632	266,400	272,154	304,394	337,397	370,161
	M%	—	3.48	7.17	10.52	13.09	13.38	13.53	14.15
	M	—	8,868	18,689	28,013	35,621	40,728	45,657	52,371
	CG%		0.023	0.022	0.022	0.021	0.112	0.103	0.093
	MG%		9.090	0.745	0.405	0.240	0.134	0.114	0.137

Table 7.0a (Continued)

		1400	1500	1600	1700	1800	1900	2000	2100
Asia	P	265,753	284,542	378,770	403,338	660,438	908,620	3,741,263	4,719,416
	M%	15.13	16.66	15.87	17.29	13.94	17.87	24.78	39.70
	M	40,214	47,413	60,111	69,746	92,073	162,362	926,986	1,873,690
	CG%	0.076	0.068	0.286	0.063	0.493	0.319	1.415	0.232
	MG%	0.138	0.165	0.237	0.149	0.278	0.567	1.742	0.704
	MR%	69.13	71.43	71.53	72.20	75.46	74.82	70.19	48.05
Africa	P	43,498	46,637	55,201	60,832	69,560	113,612	810,984	4,280,127
	M%	35.11	34.83	37.66	38.29	37.14	40.41	44.07	44.98
	M	15,271	16,244	20,788	23,293	25,838	45,909	357,401	1,925,333
	CG%	0.071	0.070	0.169	0.097	0.134	0.491	1.965	1.663
	MG%	0.061	0.062	0.247	0.114	0.104	0.575	2.052	1.684
	MR%	26.25	24.47	24.74	24.11	21.18	21.16	27.06	49.38
Europe	P	71,660	83,492	106,639	122,745	185,414	408,025	723,305	633,179
	M%	3.75	3.25	2.94	2.90	2.21	2.13	4.61	11.65
	M	2,685	2,716	3,139	3,561	4,102	8,679	33,318	73,744
	CG%	0.132	0.153	0.245	0.141	0.412	0.789	0.573	−0.133
	MG%	−0.104	0.011	0.145	0.126	0.141	0.749	1.345	0.794
	MR%	4.62	4.09	3.74	3.69	3.36	4.00	2.52	1.89

World Summary

Americas	P	20,764	22,198	10,792	13,731	25,099	145,604	833,841	1,170,543
	M%	—	—	0.00	0.00	—	0.04	0.31	1.96
	M	—	—	—	1	—	56	2,569	22,965
	CG%	0.074	0.067	−0.721	0.241	0.603	1.758	1.745	0.339
	MG%	—	—	—	—	—	—	3.821	2.191
	MR%	—	—	0.00	0.00	—	0.03	0.19	0.59
Oceania	P	1,443	1,526	1,590	1,656	1,560	5,672	31,425	74,916
	M%	—	—	—	—	—	0.10	1.29	4.60
	M	—	—	—	—	—	6	404	3,441
	CG%	0.059	0.056	0.042	0.040	−0.060	1.291	1.712	0.869
	MG%	—	—	—	—	—	—	4.237	2.138
	MR%	—	—	—	—	—	0.00	0.03	0.09
Total	P	403,119	438,395	552,993	602,302	942,071	1,581,533	6,140,818	10,878,181
	M%	14.43	15.14	15.20	16.04	12.95	13.72	21.51	35.84
	M	58,170	66,373	84,038	96,601	122,013	217,013	1,320,678	3,899,173
	CG%	0.085	0.084	0.232	0.085	0.447	0.518	1.357	0.572
	MG%	0.105	0.132	0.236	0.139	0.234	0.576	1.806	1.083

Table 7.0b Decennial estimates of the world Muslim population (×1000): 1790–1940 AD (1210–1360 H). *P*: Total population, *M*: Muslim population, *M%*: Percentage of Muslim population, *CG%*: Continental APGR, *MG%*: Muslim APGR, *MR%*: Muslim Ratio

		1790	1800	1810	1820	1830	1840	1850	1860
Asia	P	632,855	660,438	688,085	716,182	758,286	776,483	792,519	774,635
	M%	14.16	13.94	13.73	13.54	13.44	13.74	14.08	15.04
	M	89,619	92,073	94,456	96,981	101,885	106,653	111,550	116,477
	CG%		0.427	0.410	0.400	0.571	0.237	0.204	−0.228
	MG%		0.270	0.255	0.264	0.493	0.457	0.449	0.432
	MR%	75.64	75.46	75.31	75.26	75.17	75.43	75.23	75.06
Africa	P	67,998	69,560	71,114	72,524	75,575	77,921	81,192	84,472
	M%	36.79	37.14	37.47	37.59	37.87	37.59	37.93	38.10
	M	25,015	25,838	26,644	27,259	28,620	29,294	30,794	32,184
	CG%		0.227	0.221	0.196	0.412	0.306	0.411	0.396
	MG%		0.324	0.307	0.228	0.487	0.233	0.499	0.441
	MR%	21.11	21.18	21.24	21.15	21.12	20.72	20.77	20.74
Europe	P	175,936	185,414	196,869	213,463	235,847	252,705	269,253	291,087
	M%	2.19	2.21	2.20	2.17	2.14	2.16	2.21	2.23
	M	3,849	4,102	4,327	4,625	5,037	5,455	5,942	6,503
	CG%		0.525	0.599	0.809	0.997	0.690	0.634	0.780
	MG%		0.636	0.534	0.666	0.854	0.796	0.855	0.902
	MR%	3.25	3.36	3.45	3.59	3.72	3.86	4.01	4.19

World Summary

Americas	P	22,600	25,099	28,504	31,824	38,966	46,611	57,173	70,830
	M%	—	—	—	—	—	0.00	0.00	0.02
	M	—	—	—	—	—	—	2	11
	CG%		1.049	1.272	1.102	2.025	1.791	2.043	2.142
	MG%							21.579	17.781
	MR%	—	—	—	—	—	0.00	0.00	0.01
Oceania	P	1,587	1,560	1,534	1,521	1,548	1,654	1,806	2,339
	M%	—	—	—	—	—	—	—	—
	M	—	—	—	—	—	—	—	1
	CG%		−0.172	−0.170	−0.083	0.175	0.663	0.882	2.585
	MG%				23.093	7.478	6.859	6.763	8.680
	MR%				0.00	0.00	0.00	0.00	0.00
Total	P	900,977	942,071	986,106	1,035,514	1,110,221	1,155,373	1,201,944	1,223,363
	M%	13.15	12.95	12.72	12.44	12.21	12.24	12.34	12.68
	M	118,483	122,013	125,427	128,865	135,542	141,402	148,288	155,175
	CG%		0.446	0.457	0.489	0.697	0.399	0.395	0.177
	MG%		0.294	0.276	0.270	0.505	0.423	0.476	0.454

Table 7.0b *(Continued)*

		1870	1880	1890	1900	1910	1920	1930	1940
Asia	P	778,363	807,571	858,171	908,620	975,453	1,049,497	1,136,230	1,268,159
	M%	16.23	16.65	17.39	17.87	18.06	17.83	18.39	19.15
	M	126,361	134,426	149,262	162,362	176,187	187,085	208,899	242,856
	CG%	0.048	0.368	0.608	0.571	0.710	0.732	0.794	1.099
	MG%	0.815	0.619	1.047	0.841	0.817	0.600	1.103	1.506
	MR%	75.53	74.84	75.03	75.04	74.59	74.63	74.48	75.13
Africa	P	88,410	96,185	105,332	113,612	125,847	139,339	159,436	179,214
	M%	38.42	38.88	39.38	39.83	39.61	38.81	39.05	38.78
	M	33,964	37,400	41,484	45,256	49,853	54,081	62,265	69,498
	CG%	0.456	0.843	0.908	0.757	1.023	1.018	1.347	1.169
	MG%	0.538	0.964	1.037	0.870	0.967	0.814	1.409	1.099
	MR%	20.30	20.82	20.85	20.92	21.11	21.57	22.20	21.50
Europe	P	313,255	339,820	373,471	408,025	458,810	467,584	511,874	549,659
	M%	2.22	2.28	2.18	2.13	2.19	2.01	1.79	1.94
	M	6,963	7,756	8,129	8,679	10,047	9,376	9,148	10,685
	CG%	0.734	0.814	0.944	0.885	1.173	0.189	0.905	0.712
	MG%	0.684	1.078	0.470	0.654	1.464	−0.691	−0.246	1.553
	MR%	4.16	4.32	4.09	4.01	4.25	3.74	3.26	3.31

Americas	*P*	83,873	101,270	122,238	145,604	176,376	207,537	241,734	273,292
	M%	0.03	0.04	0.04	0.04	0.06	0.06	0.06	0.07
	M	21	39	45	56	99	114	156	185
	CG%	1.690	1.885	1.882	1.749	1.917	1.627	1.525	1.227
	MG%	6.722	6.069	1.569	2.186	5.646	1.436	3.121	1.699
	MR%	0.01	0.02	0.02	0.03	0.04	0.05	0.06	0.06
Oceania	*P*	2,904	3,715	4,771	5,672	6,775	8,156	9,746	10,801
	M%	—	—	—	0.08	0.12	0.05	0.00	0.03
	M	—	1	3	6	12	12	19	25
	CG%	2.163	2.463	2.502	1.729	1.777	1.856	1.781	1.028
	MG%	-5.501	8.966	12.163	7.569	6.776	0.328	4.654	2.707
	MR%	0.00	0.00	0.00	0.00	0.00	0.00	0.01	0.01
Total	*P*	1,266,806	1,348,560	1,463,984	1,581,533	1,743,261	1,872,113	2,059,021	2,281,125
	M%	13.21	13.32	13.59	13.68	13.55	13.39	13.62	14.17
	M	167,310	179,621	198,924	216,360	236,197	250,669	280,487	323,249
	CG%	0.349	0.625	0.821	0.772	0.974	0.713	0.952	1.024
	MG%	0.753	0.710	1.021	0.840	0.877	0.595	1.124	1.419

Table 7.0c Decennial estimates of the world Muslim population (×1000): 1950–2100 AD (1370–1520 H). *P*: Total population, *M*: Muslim population, *M%*: Percentage of Muslim population, *CG%*: Continental APGR, *MG%*: Muslim APGR, *MR%*: Muslim Ratio

		1950	1960	1970	1980	1990	2000	2010	2020
Asia	*P*	1,404,909	1,705,041	2,142,480	2,649,578	3,226,099	3,741,263	4,209,594	4,641,055
	M%	17.97	19.07	20.04	21.10	22.74	24.78	26.57	28.39
	M	252,456	325,124	429,418	559,073	733,714	926,986	1,118,280	1,317,568
	CG%	1.024	1.936	2.284	2.124	1.969	1.481	1.179	0.976
	MG%	0.388	2.530	2.782	2.639	2.718	2.338	1.876	1.640
	MR%	71.20	71.07	71.17	70.77	70.49	70.10	68.92	67.16
Africa	*P*	227,794	283,361	363,448	476,386	630,350	810,984	1,039,304	1,340,598
	M%	40.05	42.21	43.06	43.90	44.39	44.28	44.20	44.01
	M	91,236	119,610	156,497	209,155	279,791	359,076	459,378	589,954
	CG%	2.399	2.183	2.489	2.706	2.800	2.520	2.481	2.546
	MG%	2.722	2.708	2.688	2.900	2.910	2.495	2.463	2.502
	MR%	25.73	26.15	25.94	26.48	26.88	27.15	28.31	30.07
Europe	*P*	549,328	604,703	655,762	691,721	718,493	723,305	734,730	746,275
	M%	1.93	2.05	2.58	3.00	3.58	4.61	5.44	6.28
	M	10,607	12,405	16,889	20,727	25,732	33,318	39,998	46,893
	CG%	−0.006	0.960	0.811	0.534	0.380	0.067	0.157	0.156
	MG%	−0.073	1.566	3.086	2.048	2.163	2.584	1.827	1.590
	MR%	2.99	2.71	2.80	2.62	2.47	2.52	2.47	2.39

Americas	P	341,213	424,844	517,346	614,924	722,236	833,841	934,234	1,022,432
	M%	0.07	0.07	0.09	0.14	0.19	0.31	0.45	0.62
	M	230	310	477	858	1,369	2,569	4,178	6,375
	CG%	2.220	2.192	1.970	1.728	1.609	1.437	1.137	0.902
	MG%	2.161	3.009	4.311	5.871	4.668	6.292	4.865	4.226
	MR%	0.06	0.07	0.08	0.11	0.13	0.19	0.26	0.32
Oceania	P	12,976	16,022	19,922	23,212	27,299	31,425	36,873	42,678
	M%	0.03	0.04	0.07	0.14	0.04	0.23	0.24	0.25
	M	31	37	70	141	232	404	655	897
	CG%	1.834	2.108	2.179	1.528	1.622	1.408	1.599	1.462
	MG%	2.338	1.525	6.524	6.970	4.896	5.653	4.775	3.205
	MR%	0.01	0.01	0.01	0.02	0.02	0.03	0.04	0.05
Total	P	2,536,220	3,033,970	3,698,958	4,455,821	5,324,477	6,140,818	6,954,734	7,793,037
	M%	13.98	15.08	16.31	17.73	19.55	21.53	23.33	25.17
	M	354,561	457,486	603,352	789,955	1,040,838	1,322,354	1,622,488	1,961,687
	CG%	1.060	1.792	1.982	1.862	1.781	1.426	1.245	1.138
	MG%	0.925	2.549	2.768	2.695	2.758	2.394	2.045	1.898

Table 7.0c (Continued)

		2030	2040	2050	2060	2070	2080	2090	2100
Asia	P	4,974,092	5,188,949	5,290,263	5,289,195	5,206,455	5,068,328	4,900,833	4,719,416
	M%	30.11	31.78	33.47	35.05	36.47	37.73	38.82	39.70
	M	1,497,619	1,649,302	1,770,483	1,853,757	1,898,545	1,912,239	1,902,696	1,873,690
	CG%	0.693	0.423	0.193	-0.002	-0.158	-0.269	-0.336	-0.377
	MG%	1.281	0.965	0.709	0.460	0.239	0.072	-0.050	-0.154
	MR%	65.16	62.83	60.34	57.76	55.15	52.61	50.23	48.05
Africa	P	1,688,321	2,076,750	2,489,275	2,904,977	3,307,528	3,680,571	4,008,067	4,280,127
	M%	43.85	43.78	43.86	44.00	44.18	44.42	44.69	44.98
	M	740,295	909,208	1,091,695	1,278,072	1,461,385	1,634,799	1,791,356	1,925,333
	CG%	2.306	2.071	1.812	1.544	1.298	1.069	0.852	0.657
	MG%	2.270	2.055	1.829	1.576	1.340	1.121	0.915	0.721
	MR%	32.21	34.64	37.20	39.82	42.45	44.97	47.29	49.38
Europe	P	740,491	727,629	710,958	689,921	668,395	652,050	641,438	633,179
	M%	6.93	7.57	8.23	8.90	9.58	10.27	10.96	11.65
	M	51,279	55,065	58,482	61,369	64,041	66,992	70,299	73,744
	CG%	-0.078	-0.175	-0.232	-0.300	-0.317	-0.248	-0.164	-0.130
	MG%	0.894	0.712	0.602	0.482	0.426	0.450	0.482	0.478
	MR%	2.23	2.10	1.99	1.91	1.86	1.84	1.86	1.89

World Summary

Americas	P	1,096,453	1,152,127	1,187,246	1,206,500	1,212,728	1,205,900	1,189,802	1,170,543
	M%	0.75	0.88	1.01	1.16	1.33	1.52	1.73	1.96
	M	8,179	10,081	12,027	14,027	16,159	18,348	20,574	22,965
	CG%	0.699	0.495	0.300	0.161	0.051	-0.056	-0.134	-0.163
	MG%	2.491	2.092	1.764	1.539	1.415	1.271	1.145	1.100
	MR%	0.36	0.38	0.41	0.44	0.47	0.50	0.54	0.59
Oceania	P	47,919	52,814	57,376	61,612	65,486	68,974	72,084	74,916
	M%	0.26	0.27	0.29	0.30	0.32	0.35	0.38	0.41
	M	1,157	1,433	1,726	2,036	2,362	2,701	3,060	3,441
	CG%	1.158	0.973	0.829	0.712	0.610	0.519	0.441	0.385
	MG%	2.533	2.136	1.855	1.645	1.481	1.341	1.242	1.172
	MR%	0.05	0.05	0.06	0.06	0.07	0.07	0.08	0.09
Total	P	8,547,276	9,198,267	9,735,119	10,152,205	10,460,593	10,675,823	10,812,225	10,878,181
	M%	26.89	28.54	30.14	31.61	32.91	34.05	35.03	35.84
	M	2,298,528	2,625,089	2,934,413	3,209,261	3,442,492	3,635,080	3,787,984	3,899,173
	CG%	0.924	0.734	0.567	0.420	0.299	0.204	0.127	0.061
	MG%	1.585	1.328	1.114	0.895	0.702	0.544	0.412	0.289

Index

Abkhaz 29–30, 459, 480, 482
Abkhazia 29–30
ABS, *see* Afro Barometer Survey 5, 260, 263, 265, 326, 364, 366, 371
ABS survey data 261–262, 365, 371
Afghan 58, 61–62, 459, 473, 476, 480, 482, 508, 590, 788
Afghanistan 3, 33, 43–44, 46–47, 68–73, 206, 535, 555, 572, 665, 667
Africa 241–244, 246, 248, 250, 252, 254, 256, 264–266, 282, 346, 388–390, 392, 394, 665, 706
Africa-born Muslims 666
African American population 664
African population 497
Africans 497–499, 714
Afro Barometer Survey (ABS) 5, 260, 263, 265, 326, 364, 366, 371
Albania 206, 407, 410–411, 413–414, 425, 444–449, 583, 588, 592, 665, 800
Albanians 413, 422, 426, 428–430, 437, 439–440, 443, 465–466, 480, 482, 583
Algeria 241, 244, 246–247, 258, 266–271, 493, 495, 498, 505, 508, 665, 667, 702, 807, 812
Americans 93, 209, 289, 828, 852–857
American Samoa 840, 846–847
Andorra 490, 492–493, 526–531
Angola 360–362, 382–387
Anguilla 678–680, 696–701

answer 4, 80, 408, 577, 664
Antigua 678, 681–682, 696–701
AO, *see* Autonomous Oblas
Arab 58, 65–66, 129, 453–454, 465–466, 473, 476, 480, 482, 590, 667, 683, 706
Arabian Asia 12, 74–75, 77, 79, 81, 83, 85, 87, 89, 91, 93, 95, 97–104, 222–227
Argentina 702–703, 712, 726–731
Armenia 3, 14–19, 38–43
Artsakh 18–22
Aruba 627, 629–630, 646–651
ASARB, *see* Association of Statisticians of American Religious Bodies
Asia-born Muslims 666
Association of Statisticians of American Religious Bodies (ASARB) 667
ASSR, *see* Autonomous Soviet Socialist Republic
Australasia 785–787, 789, 791, 793, 795, 797, 799, 801, 803, 805, 807–809, 859–861
Australia 25, 784–791, 793–794, 796–799, 808–809, 814, 829
Austria 420, 432, 561, 569–573, 598–603
Autonomous Oblast (AO) 48, 51, 64, 362, 471
Autonomous Soviet Socialist Republic (ASSR) 48, 51, 56, 465, 478
Avar 20–21, 29–30, 59, 453–454, 456, 459, 465–466, 472, 476, 480, 482

Azerbaijan 3, 14, 16–22, 38–43
Azerbaijanis 16, 20, 29, 52, 453, 456, 459, 462, 465
Azeri 18, 20, 48, 50, 59, 61–62, 65–66, 454, 456, 459, 462–463, 466, 472, 480, 482

Bahamas 732, 734–735, 742, 744–749
Bahrain 3, 74–77, 98–103, 375
Balearic Islands 516–518
Balkan Peninsula 407, 409–413, 415, 417, 419, 421, 423, 425, 427, 429, 431, 433, 435, 437, 443–449
Balkans 407, 410, 443, 583, 604–609, 626
Balkar 52, 473, 476, 480, 482
Bangladesh 124, 152, 154, 156, 159, 180–185, 207–208, 665, 667, 832
Barbados 627, 630–632, 646–651
Barbuda 678, 681–682, 696–701
Bashkir 48, 50, 52–53, 59, 61–62, 65–66, 453–454, 456, 459, 462–463, 465–466, 472, 476, 480, 482
Belarus 450, 452–454, 458, 461, 484–489
Belgium 336, 342, 347, 490, 493–495, 526–531
Belize 750–752, 760–765
Benin 5, 272, 274–276, 298–303, 706
Berlin 575–576
Bermuda 652–654, 673–677
Bhutan 152, 156–157, 180–185
BiH, see Bosnia and Herzegovina
Bissau 287, 298–303
Bolivia 702, 704–705, 726–731
Bonan 190, 192–203
BOS, see Bureau of Statistics

Bosnia 407, 410, 416, 429, 444–449, 502, 535, 551, 555, 557, 570–572, 583, 588, 592, 665
Bosnia and Herzegovina (BiH) 410–411, 414–417
Bosniac 426, 429–430, 439–440, 443
Bosnians 414, 416, 428, 430, 583
Botswana 360, 363–364, 382–387
Brazil 625, 702, 706–707, 726–731
Britain 24–25, 323, 374–375
British India 124, 158–159, 625
British Virgin Islands 678, 691–693
Brunei 105, 107–108, 110, 120, 146–151
Bulgaria 47, 410–411, 417–419, 429, 431–433, 444–449, 467
Burkina 298–303
Burundi 334, 336–337, 347, 354–359

Caicos 732, 742–749
CAIR, see Council on American Islamic Relation
Caliph Omar 16, 19, 28, 43, 46, 74, 78, 81, 83, 86, 171, 241, 244, 251, 253
Cambodia 105, 108–110, 118, 146–151
Cameroon 334, 337–338, 343, 349, 354–359
Canada 188, 626, 652, 654, 656–658, 662, 672–677
Canary Islands 510, 512, 517–519, 525
Caribbean Netherlands 627, 632–633
Cayman Islands 732, 735–736
Caymans 744–749

census 24, 47–49, 55–56, 65, 124, 204–206, 415–416, 422–423, 428, 455–456, 461–462, 465, 569–573, 576–579, 588–590
  ethnic 5, 31
  first 34, 55, 77, 82, 85, 90, 95, 97, 120–121, 168, 255, 340, 644, 703, 735
  first post-independence 81
  national 643
  register-based 442, 570
  religious 667
Central Africa 242, 334–335, 337, 339, 341, 343, 345, 347, 349, 351, 353–359, 391–393
Central America 627, 750–751, 753, 755, 757, 759–770
Central Asia 11–12, 43, 45, 47, 49, 51, 53, 55, 57, 59, 61, 63, 65, 67–73, 222–227
Central Caribbean Islands 627–628, 678–679, 681, 683, 685, 687, 689, 691, 693, 695–701
Central Europe 561, 563, 565, 567–569, 571, 573, 575, 577, 579, 581, 583, 585, 587, 595–598, 604–609
Chad 244, 248–251, 266–271
Channel Islands 541, 547
Chechen 29–30, 48, 50, 52–53, 59, 453–454, 456, 459, 462–463, 465–466, 472, 476, 480, 482
Chile 702, 707–709, 726–731
China 11, 44, 158, 186–192, 204, 211, 214, 216–221, 662, 693, 737
Chinese Muslims 215, 711
choropleth map 124, 126, 155–156, 170, 275–277, 280–281, 284–290, 292–297, 309–312, 327, 337–340, 367–368, 370, 377–380, 870–871

Christians 142, 260, 309, 408, 410, 422, 462, 500, 510, 516–517, 577, 583, 597, 706
Christmas Island 795–797
CNMI, see Commonwealth of the Northern Mariana Islands
Cocos Islands 797–798
Colombia 625, 702, 709–710, 724, 726–731
Commonwealth of Australia 786–787, 796, 798–799
Commonwealth of the Northern Mariana Islands (CNMI) 830–831, 834–839
Comoros 304, 306–307, 319, 328–333
Comoros Islands 319
Congo 242, 340–343, 354–359
Cook Islands 840–842
Costa Rica 750, 752–753, 760–765
Côte 272, 280–281, 298–303
Council on American Islamic Relation (CAIR) 664
Crimea 407, 468, 478–483
Crimean Peninsula 468–469, 478–480
Crimean Tatars 478, 481
Croatia 410–411, 414, 419–421, 444–449
Cuba 732, 736–737, 744–749
Curaçao 627, 634–635, 646–651
Cyprus 14, 23–25, 27, 38–43, 408, 557
Czechia 561, 572, 574, 598–603

Darghin 59, 453–454, 480, 482
Dargin 52, 61, 65–66, 472, 476
Demographic and Health Surveys (DHS) 4, 63, 67, 81, 255–256, 306–307, 309–310, 314, 316, 326, 341–345, 347, 349–350, 381, 481

Denmark 532–537, 542, 562–567, 657, 693
DHS, see Demographic and Health Surveys
Djibouti 304, 307–308, 328–333, 813
Dominica 627, 635–636, 646–651
Dominican 744–749
Dongxiang 190, 192–203

East Africa 242, 304–307, 309, 311, 313, 315, 317, 319, 321, 323, 325, 327–333, 391–393, 507
Eastern Europe 450–451, 453, 455, 457, 459, 461, 463, 465, 467, 469, 471, 473, 475, 483–489, 604–609
Eastern Europe and Central Asia 411
East Thrace 35–36, 407, 410–411
East Timor 105, 110, 113, 142–143
Ecuador 702, 711–712, 726–731
Egypt 206–208, 244, 251–252, 262, 266–271, 505, 508, 570, 572, 581, 588, 592, 665, 667, 734–735
El Salvador 750, 753–754, 760–765
England 375, 557, 559–560
Eritrea 2, 241, 304, 309–310, 328–333
ESS, see European Social Survey
Estonia 450, 455–457, 484–489
Eswatini 360, 364–365, 382–387
Ethiopia 2, 304, 309–312, 328–333
European colonizers 248, 260, 265
Europeans 408, 628, 645, 657, 678, 695
European Social Survey (ESS) 4

European Values Study (EVS) 4, 37, 423–424, 494–495, 499–500, 502–503, 520, 545–546, 559, 589–591
European Values Survey 37, 423, 494, 500, 502, 559, 590
EVS, see European Values Study

Falkland Islands 702, 712–714
Faroe Islands 532, 537–538
Faroes 562–567
FBiH, see Federation of Bosnia and Herzegovina
Federal Statistical Office (FSO) 595
Federation of Bosnia and Herzegovina (FBiH) 414, 416–417
Fiji 800, 807, 810–812, 820–822
Filipino Muslims 127, 136
Finland 532, 538–541, 554, 562–567
France 246, 248, 254, 256–257, 263, 275–276, 280, 285, 291, 339–340, 490, 496, 498–500, 526–531, 682–683
Frontier Regions 172
FSO, Federal Statistical Office
Futuna 840, 851–858

Gabon 334, 344–345, 354–359
Gambia 272, 282, 298–303, 592, 665
Georgia 14, 28–31, 38–43, 667–668
Germany 337, 561, 575–576, 578, 598–603, 824, 829
Ghana 272, 283–285, 298–303
Gibraltar 241, 490, 500–502, 526–531
Greater Syria 702–703
Greece 15, 410–411, 413, 422–424, 429, 444–449

Greenland 533, 652, 657, 659–661, 673–677, 814
Grenada 407, 517, 627, 636–637, 641, 646–651
Guadeloupe 678, 682–685, 688, 696–701
Guam 823, 825–826, 834–839
Guatemala 750, 754–755, 760–765
Guernsey 532, 541–542, 547, 557, 562–567
Guinea 272, 285–287, 298–303, 509, 665
Guyana 625–626, 628, 643, 702, 716–717, 726–731
Guyane 714, 726–731

Haiti 625, 732, 738–740, 744–749
Herzegovina 411, 414–417
Honduras 750, 755–756, 760–765
Hong Kong 186, 188, 204–206, 212, 216–221
Hungary 420–421, 432, 436, 561, 577–579, 593, 598–603

Iberian Peninsula 407, 490, 492, 496, 500, 510, 516–517, 519, 625, 706
Iceland 532, 542–544, 562–567
Independent State of Samoa 844–845
India 152, 155–161, 173–174, 180–185, 316, 321–322, 374, 376, 628, 636, 638–641, 643–645, 682–683, 702, 740
  current 124, 157
Indian Ocean 304, 795–796
Indians 122, 136–138, 205, 222–227, 316, 638–639, 683, 712, 714, 720, 788, 807, 811, 814
Indian Subcontinent 11–12, 152–153, 155, 157, 159, 161, 163, 165, 167, 169, 171, 173, 175, 177, 179–185

Indonesia 105, 110–112, 114–116, 142, 146–151, 205–208, 212, 506–507, 509, 665, 667, 713, 813–814
Ingush 48, 50, 453–454, 472, 476, 480, 482
International Social Survey (ISS) 4, 725, 738
Iran 3, 14, 33–35, 38–43, 206, 505, 509, 535, 541, 555, 570, 572, 659, 665, 713
Iraq 33, 74, 77–79, 98–103, 188, 206, 495, 505, 508, 535, 551, 555, 557, 572, 667
Ireland 532, 544–546, 560, 562–567
Islam in Oceania 783–784, 786, 788, 790, 792, 794, 796, 798, 800, 802, 804, 806, 808, 810, 812
Island of Cyprus 14, 23–25, 27
Island of Mauritius 316–317, 319
Island of Rodrigues 316–317, 319
Island of Sumatra 112
Isle 532, 546–547, 557, 562–567
Israel 86–88, 97–103, 626, 733, 741
ISS, see International Social Survey
Italian Peninsula 407, 561, 580, 582, 591
Italy 253, 309, 423, 496, 561, 580, 582–584, 598–603
Ivoire 272, 280–281, 298–303

Jamaica 625, 732, 740, 744–749
Japan 5, 11, 186, 206–208, 214, 216–221, 824–825, 829
Java 110, 113, 116, 174, 360, 625, 628, 645, 678, 695, 702, 795, 797, 813
  island of 113, 720
Jersey 532, 541, 547–548, 557, 562–567

Jewish population 88, 248, 260, 265
Jordan 3, 74, 79–81, 86, 98–103, 581, 667

Kabardian 59, 453–454, 472, 480, 482
Kanem-Bornu Empire 244, 248, 261, 334, 337, 346
Karakalpak 18, 59, 61–62, 65–66, 453–454
Kazakh 48, 50, 52, 59, 61–62, 65–66, 190–203, 453–454, 456, 459, 462–463, 465–466, 468, 480, 482
Kazakhstan 43, 47–51, 55, 60, 68–73, 450, 467
  current 47–48, 52, 468
Kenya 304, 312–314, 328–333, 375
Kilwa Muslim Sultanate 306, 319
Kingdom 291, 506, 629, 634, 690
Kirghiz 191–203
Kiribati 823, 826–827, 834–839
Korea 209–210, 216–221
Kosovo 407, 410–411, 413, 425–426, 429, 444–449, 502, 570–572, 665
Kurd 16–17, 20–21, 29–30, 48, 50, 52, 61–62, 473, 476, 480, 482
Kuwait 3, 74, 82–83, 98–103
Kyrgyz 48, 50, 52–53, 58, 65–66, 190, 453–454, 456, 459, 462–463, 465–466, 473, 476, 480, 482
Kyrgyzstan 43, 48, 51–54, 56, 65, 68–73

Labuan Island 119–120
Laos 105, 118–119, 146–151
LAPOP, *see* Latin American Public Opinion Project

Latin American Public Opinion Project (LAPOP) 759
Latvia 450, 452, 455, 457–461, 484–489
Lebanon 3, 74, 83–84, 98–103, 505, 509, 535, 555, 702, 706–707, 709, 725, 733, 750, 755
Lesotho 360, 365–366, 382–387
Lezghin 48, 50, 59, 456, 459, 462–463, 465–466, 480, 482
Lezgin 20–21, 29–30, 52–53, 61–62, 65–66, 453–454, 472, 476
Liberia 272, 289–290, 298–303
Libya 3, 244, 253–254, 266–271, 588, 665
Liechtenstein 561, 586–587, 598–603
Lithuania 450, 452, 460–463, 484–489, 589
London 557
Luxembourg 490, 501, 503, 526–531

Macao 186, 188, 211–212, 216–221
Macedonia 410–411, 413, 423, 429, 444–449, 570–572, 583
Madagascar 304, 313, 315–316, 328–333
Madeira 510–511, 514–515
Madeira Islands 512
Madrid 517, 520, 524
Malawi 242, 360, 366–368, 380, 382–387
Malays 109, 121–122, 136–138, 140, 714, 788, 814, 830
Malaysia 105, 107, 110, 119, 122–123, 128, 146–151, 207–208, 212, 509, 665, 713, 734–735
Maldives 152, 161, 168–169, 180–185

Mali 244, 254–256, 266–271, 639
Malta 408, 561, 587–589, 598–603
Marshalls 834–839
Martinique 627, 638–639, 646–651, 682
Mauritania 244, 256–257, 266–271, 278
Mauritius 304, 316–319, 328–333
Mayotte 304, 319–320, 328–333
Medina 2–3, 11, 80, 92, 241
Mediterranean Islands 407, 561
Melanesia 785, 807, 809–811, 813, 815, 817, 819–822, 859–861
Mexico 750, 756–757, 760–765
Micronesia 785, 823–825, 827, 829, 831, 833–839, 859–861
Mindanao Island 129, 136
Moldova 47, 433, 450, 463–467, 480, 484–489
Monaco 490, 504–505, 526–531
Mongolia 43, 47, 54–56, 68–73, 450, 467
Montenegro 410, 413, 427–429, 444–449
Moroccans 492, 494, 499–501, 507, 519, 534, 542, 590
Morocco 244, 246, 257–260, 266–271, 493, 495, 498, 500, 505–510, 518, 555, 575, 577, 665, 667
Mozambique 242, 360, 369–371, 382–387
Muslim Arabs 123, 262
Muslim Armies 44, 152, 188
Muslim ethnicities 16, 56, 190, 428, 465
  largest 20, 29, 426, 429–430, 436, 439–440, 443, 498
Muslim immigrants 360, 595, 626, 636
Muslim Khanates 450, 467–468
Muslim migration 214, 639, 683

Muslim Moriscos 625, 709, 724
Muslim Ottomans 578, 593
Muslim population pre-WWI 557
Muslim refugees 210, 429
Muslim Sultanates 105, 112, 120, 128, 304, 336, 342, 347
Muslim Tatars 206, 452, 460–461, 561, 577, 589, 626
Muslim traders 248, 272, 306, 314, 319, 325, 334, 340, 351
Muslim troops 16, 19, 28, 33, 35, 78, 80–81, 83, 241, 244, 247, 251, 253, 258–259, 263
Myanmar 105, 123, 125–127, 146–151

Nagorno Karabakh Republic (NKR) 18, 20, 23
Namibia 360, 371–372, 382–387
National Institute of Statistics (NIS) 583
Nauru 823, 829–830, 834–839
Nepal 152, 169–171, 180–185
Netherlands 4, 110, 490, 506–509, 526–531, 629, 632, 634, 646–651, 688, 690, 720–721
Netherlands Antilles 629, 632, 634, 690
New Caledonia 798, 807, 812–814, 820–822
New South Wales 787–788
New York 667, 670
New Zealand 785, 798, 800, 802–806, 811, 840, 843–844, 847
Nicaragua 750, 757–758, 760–765
Niger 244, 261–262, 266–271
Nigeria 278, 343, 346–347, 349, 354–359, 665
NIS, *see* National Institute of Statistics
Niue 840, 843–844, 852–857

NKR, see Nagorno Karabakh Republic
Norfolk Island 789, 798–799
North Africa 241–242, 244–245, 247, 249, 251, 253, 255, 257, 259, 261, 263, 265–271, 391–393, 518, 583
North America 626–627, 652–653, 655, 657, 659, 661, 663, 665, 667, 669, 671–677, 766–769
Northern Caribbean 766–769
Northern Caribbean Islands 627, 732–733, 735, 737, 739, 741, 743–749
Northern Europe 532–533, 535, 537, 539, 541, 543, 545, 547, 549, 551, 553, 555, 557, 559–562, 604–609
Northern Ireland 557
Northern Mariana Islands 785, 823, 830–831
Northern Morocco 260, 519, 521
Northern Province 175
North Island 800, 802, 806
North Macedonia 429–431
North Morocco 500, 510, 517–518
Norway 532, 537, 548–549, 551, 553, 562–567

Oceania 783–786, 788, 790, 792, 794, 796, 798, 800, 802, 804, 806, 808, 810, 812, 858–862
Oman 3, 74, 85–86, 98–103, 304, 325, 375
Ottoman Empire 251, 253, 411, 413–414, 417, 419, 422, 425, 427, 429, 431, 436, 569, 722, 725
Ottomans 16, 23, 28, 36, 77, 79, 83, 86, 93, 411, 417, 425, 578, 580, 582

Pacific Islands 828, 832
Pakistan 152, 154–155, 159–160, 171–173, 180–185, 207–208, 495, 508, 535, 551, 555, 659, 665, 667, 734–735
Palau 785, 823, 832–839
Palestine 3, 81, 90, 97–103, 581, 626, 639, 705, 707, 711, 733, 741, 750, 752–755, 757
Palestinian Central Bureau of Statistics, see PCBS
Panama 709, 724, 750, 758–765
Papua New Guinea 110, 807, 814–815
Paraguay 702, 718–719, 726–731
Paris 321, 407, 496
Persian 16, 18, 29–30, 59, 61–62, 65–66, 78, 96, 453–454, 459, 480, 482
Peru 702, 719–720, 726–731
Pew Research Center (PRC) 1
Philippines 105, 127–130, 136, 146–151
Poland 452, 461, 561, 589–591, 598–603, 626
Polynesia 785, 840–841, 843, 845, 847, 849, 851–861
population exchange 26, 160
  forced 155, 159, 173
population growth 6, 175
Portugal 211, 258, 278, 287, 349, 362, 369, 490, 510, 513–514, 517, 526–531, 706
Portuguese 119, 142, 174, 278, 287, 316, 321, 325, 349, 361, 369, 374, 630, 702, 706
PRC, see Pew Research Center
Puerto Rico 662, 732, 741–742

Qatar 3, 74, 76, 90–91, 98–103
Queensland 787, 790, 792

Regional Council 800, 802, 804–806

religion question 419, 428, 430, 576
religious adherence 4–5, 55, 63, 136, 434, 456, 462, 465, 537, 643, 657, 721, 813
religious affiliation 4–5, 30, 82, 97, 145, 255, 262, 341, 423, 428, 434, 506, 570, 573, 579
religious census data 136
Republika Srpska (RS) 414, 416–417, 438–440
Réunion 304, 321, 328–333
Romania 47, 410–411, 431–436, 444–449, 465, 467
RS, *see* Republika Srpska
Russia 23, 44, 47, 51, 55, 60, 64, 450, 452, 467–469, 472, 474, 476–477, 480, 484–489
  country of 12, 410
  current 47, 450, 467, 471
Russian Empire 16, 19, 48, 51, 56, 452, 455, 457–458, 461, 463, 465, 468, 471, 477, 480
Russian Federation 468, 470–471
Russians 16, 18, 24, 28, 36, 49, 56, 60, 65, 67, 209, 450, 457, 462, 478
Rwanda 334, 347–348, 354–359

SADR, *see* Sahrawi Arab Democratic Republic
Sahrawi Arab Democratic Republic (SADR) 258
Salar 190, 192–203
Samoa 840, 844–846, 852–857
San Marino 598–603
Sardinia 407, 496, 580–581, 585
Saudi Arabia 3, 74, 76, 92–93, 667
Sayfawa Muslim Dynasty 248, 261, 337, 346
Senegal 272, 278, 291–292, 298–303, 495, 505, 592, 639, 665, 683, 713

Serbia 410–411, 413–414, 420, 426, 429, 436–438, 440, 444–449
Seychelles 304, 322–324, 328–333, 375
SFSR, *see* Soviet Federative Socialist Republic
Sicily 407, 580–582, 585, 587
  islands of 580–581
Sierra 298–303
Sierra Leone 272, 293–294, 509, 665
Singapore 105, 136–137, 139, 146–151, 796, 798
Sint Maarten 678, 689–691
Slovakia 561, 572, 593–594, 598–603
Slovenia 410, 441–449
Somalia 304, 323–325, 328–333, 508, 535, 551, 555, 557, 572, 665
South Africa 5, 360, 365, 371, 374, 376–378
South America 627, 702–703, 705, 707, 709, 711–713, 715, 717, 719, 721, 723, 725–731, 766–769
Southeast Asia 105–107, 109, 111, 113, 115, 117, 119, 121, 123, 125, 127, 129, 131, 145–151, 222–227
Southern Africa 242, 360–361, 363, 365, 367, 369, 371, 373, 375, 377, 379, 381–387, 389, 391–394
Southern Caribbean 766–769
Southern Region 140
South Island 800, 806
South Ossetia 28–31
South Pacific Ocean 800, 807, 810, 812, 816–817, 828, 840–842, 848, 850
South Sudan 262, 334, 350–351

Soviet Federative Socialist Republic (SFSR) 48, 51
Soviet Socialist Republic (SSR) 48, 51, 56, 60, 64, 465, 478
Soviet Union 16, 18, 45, 47, 51, 56, 60, 63–64, 67, 450, 452–453, 455–457, 460, 462–464, 478–479
Spain 257–258, 490, 492, 516–520, 522, 526–531, 681–682, 703–704, 707, 709, 711–712, 718–719, 722, 724, 752–757
Special Administrative Regions 188, 204, 211
Sri Lanka 152, 174–176, 178–185, 797
SSR, *see* Soviet Socialist Republic
St. Bart 685, 696–701
St. Helena 372, 374–375, 382–387
St. Lucia 646–651
Sudan 244, 262–263, 266–271, 350, 354–359, 508, 665, 667
Sumatra 105, 110, 112, 116, 120, 797
Suriname 507, 625–626, 628, 634, 643, 702, 716, 720–722, 726–731
Sweden 532, 554–556, 562–567, 685
Switzerland 561, 594, 596, 598–603
Syria 74, 79, 93, 95, 98–103, 505, 508, 555, 557, 588, 592, 702, 706–707, 709, 755–756
Syrian Arab Republic 93–94

Tabasaran 453–454, 473, 476, 480, 482
Taiwan 5, 186, 214–221
Tajik 48, 50, 52–53, 56, 58, 61–62, 65–66, 190–203, 453–454, 456, 459, 462–463, 465–466, 480, 482
Tajikistan 43–44, 52, 55–58, 60, 64–65, 68–73
Tanzania 304, 325–333
Tatar 18, 20–21, 48, 50, 52–53, 61–62, 65–66, 190–203, 434, 436, 453–457, 459, 462–463, 465–467, 476–477
Thailand 105, 140–142, 146–151
Timor-Leste 105, 142–143, 146–151
Tobago 628, 643–645, 648–651
Togo 272, 295–296, 298–303
Tokelau 840, 847–848, 852–857
Tonga 840, 848–849, 852–857
Transnistria 463–467
Trinidad 625–627, 643–644, 646–651
TRNC, *see* Turkish Republic of Northern Cyprus
Tunisia 244, 263–271, 493, 495, 498, 505, 508, 576, 580, 588, 592, 665
Turkey 12, 14–15, 35–43, 407, 410–411, 415, 423, 493, 495, 505–509, 555, 570–572, 575, 665, 667
 current 35–36, 94
Turkish Republic of Northern Cyprus (TRNC) 24
Turkmen 52–53, 58, 61–62, 65–66, 453–454, 456, 459, 462–463, 465–466, 473, 476, 480, 482
Turkmenistan 43–44, 60–63, 68–73
Turks 16, 29–30, 52–53, 65–66, 418–419, 426, 434, 456, 459, 465–466, 497–499, 501, 570, 575–576, 742–749
Tuvalu 840, 849–857

UAE 3, 85, 98–103, 713
Uganda 334, 351–359, 375
Uighur 52–53, 59, 61, 65–66, 190, 480

Ukraine 47, 432–433, 450, 452, 465, 467, 469, 477–489
United Kingdom 4, 372, 500, 532, 544, 557–560
United Nations 210, 253
United States 626, 654, 662–665, 672, 693, 736, 741, 824
United States Virgin Islands 678, 693, 695
Uruguay 702, 722–723, 726–731
Uyghur 191–203
Uzbek 48, 50, 52–53, 58, 61–62, 65–66, 190–203, 453–454, 456, 459, 462–463, 465–466, 472, 480, 482
Uzbekistan 43–44, 48, 52, 56, 60, 64–73, 478, 481

Vanuatu 807, 817–822
Venezuela 625, 702, 709, 724–731
Victoria 787–788, 791, 793
Vienna 569, 578, 593
Vietnam 105, 143–144, 146–151
Virgin Islands 678, 691
Vitebsk 452–453, 457–458, 469

Wallis 840, 851–858
Washington 671
west 11, 33, 77, 112–113, 119, 244, 360, 388–390, 423, 510, 518, 671, 682, 712–713, 732

West Africa 272–273, 275, 277, 279, 281, 283, 285, 287, 289, 291, 293, 295, 297–303, 344–345, 391–393
Western Europe 408, 410, 490–491, 493, 495, 497, 499, 501, 503, 505, 507, 509, 521, 525–531, 604–609
West Island 712, 797
West Mediterranean Islands 490, 496
World Values Survey (WVS) 4, 25, 28, 79, 81, 83–84, 93, 97, 205–206, 211, 247–248, 254, 703, 710, 719
WVS, *see* World Values Survey

Xinjiang Uyghur Autonomous Region 188, 191

Yemen 3, 74, 96–104, 206, 667
Yugoslavia 420, 428–430, 437, 441, 502, 570

Zambia 360, 379–380, 382–387
Zanzibar 325, 336, 342, 347, 351, 367, 374
Zimbabwe 360, 369, 380–387